Vascular Diagnosis with Ultrasound

Vascular Diagnosis with Ultrasound

Clinical References
with Case Studies

Michael Hennerici, M.D.
Professor and Chairman
Department of Neurology
Ruprecht-Karls-Universität Heidelberg
Klinikum Mannheim, Germany

Doris Neuerburg-Heusler, M.D.
Former Director
Department of Noninvasive Diagnostics
Aggertalklinik, Engelskirchen
Cologne, Germany

With contributions by
Thomas Karasch, M.D.
Wolfgang Rautenberg, M.D.

Foreword by A. Fronek, M.D., Ph.D.
and J. P. Mohr, M.D.

754 illustrations
57 tables

Thieme
Stuttgart · New York 1998

Translated by Christopher Kronen

This book is an authorized, updated, and revised translation of the 2nd German edition published and copyrighted 1995 by Georg Thieme Verlag, Stuttgart, Germany. Title of the German edition: Gefäßdiagnostik mit Ultraschall: Doppler- und farbkodierte Duplexsonographie der großen Körperarterien und -venen.

Any reference to or mention of manufacturers of specific brand names should not be interpreted as an endorsement or advertisement for any company or product.

Library of Congress Cataloging-in-Publication Data
Hennerici, M. (Michael)
    [Gefäßdiagnostik mit Ultraschall. English]
Vascular diagnosis with ultrasound : clinical references with case studies/Michael Hennerici, Doris Neuerburg-Heusler : with contributions by Thomas Karasch, Wolfgang Rautenberg : foreword by J. P. Mohr and Arnost Fronek : [translated by Christopher Kronen].
        p. cm.
    Updated and rev. translation of: Gefässdiagnostik mit Ultraschall.
    2. Aufl. 1995.
        Includes bibliographical references and index.
        ISBN 3-13-103831-4. — ISBN 0-86577-603-2
        1. Blood-vessels—Ultrasonic imaging. 2. Blood-vessels–
–Ultrasound imaging—Case studies. I. Neuerburg-Heusler, Doris. II. Title.
        [DNLM: 1. Vascular Diseases—ultrasonography. WG 500 H5 15 g 1997 a]
    RC691.6.U47H4613   1998
    616.1'307543—dc21
    DNLM/DLC   97-34733
    for Library of Congress   CIP

© 1998 Georg Thieme Verlag, Rüdigerstraße 14, D-70469 Stuttgart, Germany
Thieme Medical Publishers, Inc., 333 Seventh Avenue, New York, N.Y. 10001

Typesetting by Druckhaus Götz GmbH, D-71636 Ludwigsburg
typset on CCS-Textline [Linotronic 630]
Printed in Germany by Staudigl-Druck, D-86609 Donauwörth

ISBN 3-13-103831-4 (GTV, Stuttgart)
ISBN 0-86577-603-2 (TMP, New York)   3 4 5 6

Important Note: Medicine is an ever-changing science undergoing continual development. Research and clinical experience are continually expanding our knowledge, in particular our knowledge of proper treatment and drug therapy. Insofar as this book mentions any dosage or application, readers may rest assured that the authors, editors and publishers have made every effort to ensure that such references are in accordance **with the state of knowledge at the time of production of the book.**

Nevertheless this does not involve, imply, or express any guarantee or responsibility on the part of the publishers in respect of any dosage instructions and forms of application stated in the book. **Every user is requested to examine** carefully the manufacturers' leaflets accompanying each drug and to check, if necessary in consultation with a physician or specialist, whether the dosage schedules mentioned therein or the contraindications stated by the manufacturers differ from the statements made in the present book. Such examination is particularly important with drugs that are either rarely used or have been newly released on the market. **Every dosage schedule or every form of application used is entirely at the user's own risk and responsibility.** The authors and publishers request every user to report to the publishers any discrepancies or inaccuracies noticed.

Foreword

This splendid, thorough-beyond-thorough treatise sets a standard few competitors would attempt to match. Issued initially in German, now revised and thoroughly updated and finally available in English, the book covers the field of clinical color duplex ultrasound in a masterful fashion.

The authors of this volume are in the forefront of ultrasound diagnostics, specifically in the areas of the brain and extremities. They have pooled their specific expertise in different areas, and in addition they have utilized their long experience in presenting their results to different groups of medical professionals. A common denominator is the unusually high degree of didactic presentation that makes this text not only resourceful but also enjoyable to read and study.

The emphasis is on the application of ultrasound in the diagnosis of vascular diseases of the cerebral circulation as well as of the lower and upper extremities. There are several chapters that are not usually found in similar texts, for example, on cerebral veins, abdominal veins, tumor vascularization, and vascularization of the genitalia. The basic principles of hemodynamics as they relate to clinical symptomatology are described in excellent detail.

The carefully written chapters repay frequent rereading, layer by layer of information emerging from repeated review, with each chapter proceeding in an orderly fashion, building on the information before. The book has also been written to be used as a reference manual, its painstaking organization allowing those with neither the time nor the patience for a complete reading to skip quickly to the place of interest.

Not to be overlooked is the wealth of references (some even from 1997), including sources from both sides of the Atlantic, a rarely found feature.

For all these reasons, this text will be a valuable source for the beginner as well as a reference book for the experienced neurologist, internist, vascular surgeon, and angiologist. Those who practice ultrasonography will find it a reference book without parallel. In many centers already (and soon in even more) it bodes well to be the standard text on the subject.

In summary, it can be expected that this volume will become a standard text for years to come.

J. P. Mohr, M.D., *Arnost Fronek,* M.D., Ph.D.

Preface

Since the introduction of noninvasive ultrasound techniques in clinical practice in the early 1970s, the diagnosis and treatment of vascular diseases have made considerable progress: not only is the initial diagnosis from early stages of atherosclerosis as the major source of arterial obstructive lesions facilitated, but also its follow-up—whether spontaneous or during treatment in randomized prospective clinical trials, as well as in clinical practice—is state of the art in most modern industrialized societies with a high prevalence and incidence of myocardial infarction, stroke, and peripheral vascular disease. In addition, refined technologies have now helped make it possible to image noninvasively structural and functional abnormalities in both large and small vessels nearly everywhere in the human body with high accuracy. Thus ultrasound has become a standard technology for screening patients at risk for atherosclerosis in the absence of clinical symptoms, for a detailed diagnosis in symptomatic subjects and in many patients supposed to undergo percutaneous angioplasty or major vascular surgery such as carotid endarterectomy, coronary artery bypass, aortic aneurysm and peripheral artery surgery, and is currently in use for more sophisticated protocols including hyperacute stroke management.

The first German edition of this book (1988) was welcomed by beginners, as well as experienced sonographers for its strictly organized and illustrative composition. After a short introduction of the basic ultrasound principles and instruments used, the various arterial and venous systems (cerebral, peripheral, abdominal, etc.) and the ultrasound approach for their investigation are explained and separately presented for normal, variant, and pathological findings. Because of the rapidly developing technologies, in particular color-coded Doppler duplex sonography, a major revision was necessary in 1994, but the careful organization of the chapters remained unchanged.

Unique and particulary important is the collection of individual case histories and vascular findings, which now forms about one third of the book. This atlas illustrates the combined use of ultrasound technologies with other methods applied in clinical practice and may be useful for both the specialized investigator in the vascular laboratory, as well as clinicians unfamiliar with specific ultrasound tests. Therefore, rare findings and specific problems are only occasionally included, and repetitions are intentional. These exemplary case findings have been selected on the basis of extensive clinical and ultrasound experience with patients admitted to the Departments of Neurology and Angiology in Düsseldorf, Mannheim, and Engelskirchen.

This first English edition is a complete revision of the second German monograph and includes recent technological advances, data from large study trials, and current literature on research and addresses still experimental processes, such as the application of echo-contrast media, harmonic imaging, 3D and 4D imaging, flow volume measurements, intra-arterial and interventional applications. In addition, functional studies, in particular for the assessment of brain perfusion and neuronal activity, and studies of various venous systems have been added. There is a completely new chapter illustrating the utility of color flow duplex sonography in oncology, addressing the most important organ manifestations as well as tumor classifications.

This book would not have been possible without the help of Dr. Thomas Karasch, Department of Internal Medicine of the University of Cologne, and PD Dr. Wolfgang Rautenberg, formerly Department of Neurology, Mannheim, who is now in private practice in Düsseldorf. Their cooperation and expertise formed an essential basis of the expanded chapters on cerebral and abdominal studies, and they have contributed many images. Colleagues from the Aggertal Clinic of Vascular Disease and both university departments (Cologne and Mannheim) supplied additional material, in particular Prof. Dr. Roth (Engelskirchen) and Prof. Dr. Schwartz, PD Dr. Daffertshofer, PD Dr. Steinke, Dr. Meairs, and Dr. Ries (Mannheim). We are also most grateful to our vascular technicians Ms. Kuprion and Ms. Pflästerer (Mannheim) and our secretaries Ms. Eudenbach (Cologne) and Ms. Lorenz and Ms. Garcia-Knapp (Mannheim), who helped us with the enormous work on the manuscript and all the necessary changes and additional comments for this first English edition. We extend special thanks to the staff of Thieme for their generous support and for their meticulous editorial work. Last, but not least, we are most grateful to our families, to Marion Hennerici and Helmut Neuerburg especially – without their love and support this book would not have been possible.

Mannheim and Cologne, autumn 1997,
Michael Hennerici and Doris Neuerburg-Heusler

Addresses

Arnost Fronek, M.D., Ph.D.
Professor, Department of Surgery
and Bioengineering
University of California, San Diego
9500 Gilman Drive
La Jolla, CA 92093–0643, USA

Michael Hennerici, M.D.
Professor and Chairman
Department of Neurology
University of Heidelberg
Klinikum Mannheim
Theodor-Kutzer-Ufer
68135 Mannheim, Germany

Thomas Karasch, M.D.
Department of Internal Medicine III
University of Cologne
Joseph-Stelzmann-Str. 9
50924 Cologne, Germany

J. P. Mohr, M.D.
Sciarra Professor of Clinical Neurology
Director, Stroke Unit and Neurovascular-
Doppler Laboratory
Neurological Institute
Columbia Presbyterian Medical Center
New York, NY 10032, USA

Doris Neuerburg-Heusler, M.D.
Former Director of the Department
of Noninvasive Diagnostics
Aggertalklinik, Engelskirchen
Goethestraße 68
50968 Cologne, Germany

Wolfgang Rautenberg, M.D.
Ass. Professor of Neurology
Neurological Institute
Hohenzollernstr. 5
40211 Düsseldorf, Germany

Table of Contents

3 Intracranial Cerebral Arteries   89

4 Cerebral Veins
125

6 Veins

9 Vasculature of the Male Genitalia   307

10 Tumor Vascularization   323

Peripheral Veins   416

Abdominal Arteries   430

12 Glossary   449

References   459

Index   489

Ultrasound Procedures

Principles and Technical Basis

Sound waves above a frequency of 20 kHz are termed "ultrasound." Like all sound waves, ultrasound propagates through various media in the form of a pulsating pressure wave. In their simplest form, ultrasound waves are characterized by a cyclical alternation between higher and lower particle densities, the deviation of which from the resting state is defined as their amplitude, or sound intensity. Amplitude is a direct measure of the sound energy. The wavelength λ is a measure of the distance between two adjacent maximum or minimum values of a sine curve. The number of these waves per unit of time defines the frequency (Hz). Sound waves can either have a specific directional orientation, or can be bundled. They are subject to the laws of geometric optics (reflection, transmission, and refraction).

Two principles are used in ultrasound diagnosis: the *echo-impulse technique* and the *Doppler technique.*

The Echo-Impulse Principle

When the speed of ultrasound propagation through tissue is known, the depth of the echo's reflection can be measured (Fig. 1.**1**) by:

– *Amplitude modulation: A-mode;* or by
– *Brightness modulation: B-mode.*

When the ultrasound axes are shifted sequentially and laterally—which is technically possible in various ways with different arrangements and settings in the transducer components—it is possible to produce a two-dimensional ultrasound image in real time (Fig. 1.**1**).

With appropriate selection of different planes, blood vessel segments of interest may be displayed, both in longitudinal and cross-sections.

The quality of the image produced depends on the *image resolution:*

– Axial resolution
– Lateral resolution

As in the field of optics, "resolution" is defined as the smallest distance between two points at which they can still be depicted as separate. The *axial resolution* depends exclusively on the length of the ultrasound impulse. Since for B-mode ultrasound the ultrasound pulses consist of only 1–2 sinus wavelengths, the axial resolution lies in the range of the ultrasound wavelength λ (0.2–1 mm). Since this value is reciprocal to the ultrasound frequency (λ = c/f), the *axial resolution* improves with increasing ultrasound frequency.

The *lateral resolution* refers to the ability to depict as separate entities two points that lie at right angles to the direction of ultrasound propagation. This is primarily dependent on the width of the ultrasound beam. To be able to resolve points that lie close together, the width of the ultrasound beam has to be kept reasonably small. This is possible when the wavelength is kept small and the diameter of the transducer is kept as large as possible (ie., small phased-array transducers have a worse lateral resolution than large linear or curved-array transducers). For physical reasons, secondary wave fronts also form

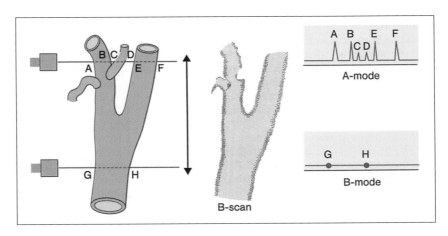

Fig. 1.**1** Schematic drawing illustrating examination conditions using the echo-impulse technique: A-mode and B-mode (echotomogram)

in addition to the main wave front and may cause interference in sub-optimal conditions.

The intensity of the reflected echo is assigned to a specific brightness level along a gray scale, providing a finely shaded image.

Transmission frequencies used range between 1 and 10 MHz, depending upon the specific area of application (for arteries located close to the skin > 7.5 MHz, for arteries located deeper within the body between 3 and 5 MHz, and for transcranial application < 2 MHz).

When selecting a transmission frequency, one should also note the following:

- The *axial resolution* is directly proportional to the ultrasound frequency.
- The *signal intensity,* however, depends on the attenuation of the ultrasound as it passes through tissue: the higher the ultrasound frequency is, the stronger the attenuation of the signal intensity will be.

The selection of a specific frequency for various diagnostic procedures always involves a compromise between the optimal ultrasound penetration depth and the ability to identify relevant tissue changes.

Ultrasound reflections occur when there is an interface between two media with differing levels of resistance to ultrasound. This *impedance,* also known as acoustic resistance, depends on the ultrasound propagation speed and the tissue density: the greater the difference in impedance, the stronger the reflection. Ultrasound is only optimally reflected to the receiver if it is perpendicularly incident to a medium. Tissue particles that are relatively small in relation to the wavelength (e.g., blood cells), and particles with differing impedance that lie very close to one another, cause *scattering* or *speckling.* Reflecting media that lie at an angle to the ultrasound propagation axis can only be recognized due to these scattering phenomena, which are accompanied by an attenuation in the echo intensity. *Speckling effects* result from extinction of reflexes from adjacent or closely located structures behind one another at a $\lambda/4$ distance. The stronger the reflections are at the interfaces, the less ultrasound energy is available to reach deeper tissue. If a total reflection occurs at an interface to air, bone, or calcium-containing tissue, then a so-called *ultrasound shadow* results. Most of the ultrasound energy is converted into kinetic energy within the tissue. This applies particularly to higher ultrasound frequencies, at which the proportion of energy absorbed increases. On the other hand, due to its shorter wavelength, high-frequency ultrasound is subject to less scattering at interfaces that lie very close to one another. This produces better local resolution.

The *interface structure* in vessel walls is a much better reflector of ultrasound than the pillar-like mixture seen in the blood cells, which for the most part produces a diffuse scattering of ultrasound.

The Doppler Principle

The speed of flow of the corpuscular elements in blood is defined by a principle named after Christian Andreas Doppler (1803–1852) (Eden 1986). The *Doppler shift,* $\Delta f(Hz)$, is proportional to both the flow velocity, v (cm/s), and the transmission frequency of the ultrasound, f (MHz). The physical context of the Doppler effect may well be familiar from everyday experience: when a truck blowing its horn approaches a stationary observer on a highway, the observer will hear the sound of the horn rising in pitch until the truck is directly in front of him. When the truck has passed and drives away from the observer with its horn still blowing, the pitch of the horn starts to fall. To analyze this phenomenon quantitatively, one has to know the value of the angle α between the ultrasound beam and the vascular axis. However, this parameter is not known in many of the diagnostic procedures used in clinical medicine.

$$\Delta f = 2 \, f \cdot v \cdot \cos \alpha / c$$

(c = velocity of flow through tissue ≈ 1540 m/s)

Ultrasound waves are produced by a vibrating crystal using a *piezoelectric effect.* This effect involves the characteristic that certain materials have of changing their form when subjected to an electric current—or vice versa, of producing an electric current due to changes in their form (depending on whether they are used as the transmitter or receiver). These piezoelectric crystals are placed in a specific arrangement *(array)* within the ultrasound probe.

In order to improve the *acoustic coupling* (impedance match), a gel has to be placed between the crystal and the surface of the skin.

When the vasculature is being examined, the reflection caused by the blood's corpuscular elements, mainly erythrocytes, plays a major role. Since the flow velocity through the different areas of any given cross-section of a blood vessel varies, the Doppler signal itself does not correspond to a single frequency, but contains a broad *frequency spectrum* (in the normal internal carotid artery, this ranges from 0.5 kHz to 3.5 kHz and < 120 cm/s respectively if a transmission frequency of 4 MHz is used).

Reflections at the blood vessel wall, appear as low-frequency artifacts with an intensity up to 30 times higher than that from moving blood elements. They may be eliminated using *high-pass filters* (100–400 Hz). However, it should be noted that slow velocities that are naturally part of the flow phenomena in a particular blood vessel will also be filtered out. It is therefore important for the investigator to apply the wall filter selectively, adapting its use according to the specific circumstances of each investigation (e.g., in the case of a pseudo-occlusion of the internal carotid artery). This diagnostic limitation can be compensated for using modern *echo-contrast media* (Fig. 1.**2**).

The larger the number of moving blood cells, the higher the amplitude of the Doppler signal will be.

Fig. 1.**2** The principle of conventional imaging **a** and harmonic imaging **b**. Only the echo enhancement agent, which contains microbubbles that oscillate nonlinearly in the acoustic field, emits echoes at double the transmitted frequency (blue). In this way, harmonic imaging offers the opportunity to suppress echoes detected from non-echo enhancement agent-bearing solid tissue (e.g., vessel wall). (Modified according to Burns)

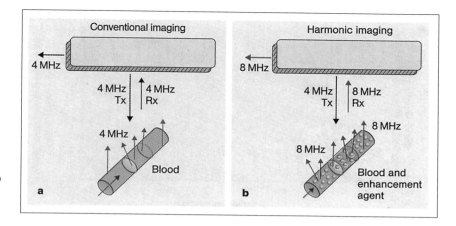

Holding the ultrasound probe at an increasingly steep angle to the vascular axis causes increasing echo reflections with minimal scattering: attenuation of signal intensity is minimal. On the other hand, the Doppler equation clearly shows that there is no Doppler shift at an angle of 90°, and that the Doppler shift is largest for a more or less flat position of the probe (cos 0° = 1, cos 180° = −1) (Fig. 1.**3**). Thus the ultrasound probe should be held so as to create an incident angle of the ultrasound beam of approximately 30°. This yields the optimal relationship between the Doppler effect (Hz) and signal intensity (dB).

Since the Doppler shift Δf only describes the difference between the emitted and reflected frequencies, it is not possible to deduce from this value alone the direction of the blood flow being examined. Ultrasound equipment therefore carries out a parallel analysis of the reflected signals using what is termed a *phase-quadrature detector*.

In all directional Doppler instruments, the direction of blood flow is indicated as a positive or negative signal, depending on the probe position and the vascular flow direction.

The Doppler shift lies in the audible range: usually, signals measured in the large blood vessels lie between 20 Hz and 18,000 Hz, and correspond to blood flow velocities of between 0.1 m/s and approximately 8 m/s. The absolute value of the Doppler shift at a constant blood flow velocity depends on the ultrasound frequency emitted (Fig. 1.**4b**). Whenever possible, it is better to investigate high Doppler shifts using a low-frequency transducer, while low Doppler shifts are better examined with a high-frequency transducer. The capacity for penetration, the axial resolution, and near-field and far-field investigations, sometimes stand in the way of this ideal.

If the angle α is known, or a sufficiently accurate approximation to its value can be deduced, one can calculate a simple transformation between the Doppler shift (kHz) and the flow velocity (m/s), as follows:

$$\frac{0.78 \cdot \Delta f \ (\text{kHz})}{f \ (\text{MHz}) \cdot \cos \alpha} = v (\text{m/s})$$

As an example, using a 4-MHz probe and an angle of 60° (cos 60° = 0.5), 3 kHz would yield a velocity of:

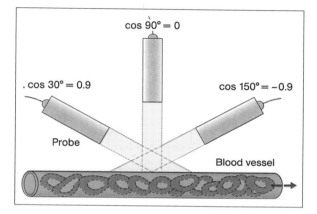

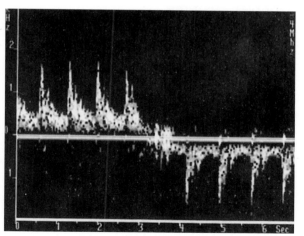

Fig. 1.**3**   The effect of various ultrasound angles on the measurement of the Doppler shift in a blood vessel

$$\frac{0.78 \cdot 3 \ (\text{kHz})}{4 \ (\text{MHz}) \cdot 0.5} = 1.17 \ \text{m/s}$$

To determine the frequency, the formula yields:

$$\Delta f \ (\text{kHz}) = 1.28 \ f \ (\text{MHz}) \cdot v (\text{m/s}) \cdot \cos \alpha$$

One can also use simple conversion graphics to calculate these values (Fig. 1.**4a**).

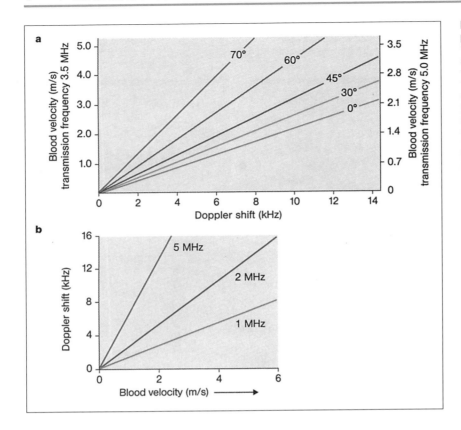

Fig. 1.**4**   The mutual interdependence of Doppler shift, flow velocity, angle of examination, and the emitted ultrasound frequency, **a** A conversion table for two transmission frequencies (3.5 and 5 MHz) to compare the Doppler frequencies (kHz) and the flow velocities (m/s), depending on the position of the angle formed by the Doppler axis and the flow axis. **b** The relationship between Doppler shift and flow velocity at different ultrasound emission frequencies. As expected from the Doppler equation, identical flow velocities produce different frequency shifts

Doppler Systems

Three types of Doppler systems, which differ both in technical terms and in the equipment, are used in medical diagnostics:

– Continuous wave Doppler systems
– Single-channel pulsed wave Doppler systems
– Multi-channel pulsed wave Doppler systems

Continuous Wave Doppler Sonography

Continuous wave (CW) Doppler equipment uses two *piezoelectric elements* with overlapping ultrasound fields that serve as transmitter and receiver (Fig. 1.**5** **a**). Due to the continuous and simultaneous emission and reflection of ultrasound waves, *no* exact *information about depth* can be obtained with this equipment. However, specific areas of interest can be selected using appropriate focusing.

Pulsed Wave Doppler Sonography

Pulsed wave Doppler systems are used to detect blood flow at a specifically defined depth. Using a single piezoelectric element, these systems function alternatingly as transmitter and receiver (Fig. 1.**5b**). By electronically selecting the length of the time interval between the transmitting and receiving pulses, it is possible to localize signals at different depths using the transit time of ultrasound through the tissue. This region is called the *information volume* or *sample volume,* and it can be shifted along the ultrasound axis. Its size is variable, and is determined both by the characteristics of the ultrasound beam and by the duration of the pulse, which is controlled by the oscillator. Shorter ultrasound packets during the transmission phase (duration approximately 0.5–2.0 μs), result in a smaller sample volume, with a corresponding increase in the axial resolution.

At the receiving end, changing the time delay causes displacement of the information volume along the ultrasound axis (sample volume localization).

To ensure unequivocal depth measurements, a new pulse to determine the velocity can only be sent after the ultrasound packet has returned. *The pulse repetition frequency* (PRF) is therefore defined as the transit time required for the pulse to travel to the investigation area and back again at a given ultrasound propagation velocity (c), with E referring to the penetration depth:

$$PRF_{max} = \frac{c}{2} \cdot E$$

The frequency at which individual ultrasound packets are emitted restricts the maximum Doppler shift that can be measured. The pulse repetition frequency has to be at least twice as large as the Doppler shift frequency to be recorded. For example, a PRF of 10 kHz allows detection of a Doppler shift of a maximum of 5 kHz.

Referred to as the *Nyquist theorem,* this rule applies to all events that cannot be observed continuously. For example, in a movie the rotation of a spoked wheel can be perceived correctly when the wheel begins to move. With further acceleration, the movement of the wheel increases up to a maximum velocity, beyond which it incorrectly appears that the wheels start to rotate backward, against the direction in which the vehicle is moving. This phenomenon is due to the pulse effect deriving from the camera technology: the higher a camera's pulse repetition speed, the higher the maximum velocity that can still be clearly depicted. Beyond this limit, errors occur. Different ultrasound transmission frequencies lead to different upper limits for the pulse repetition frequencies, and as a result, different threshold values for the flow velocities that can still be detected in different investigation areas. This is illustrated in Fig. 1.**6**. It should be noted that:

– With the pulsed wave process at a given depth, the application of low emission frequencies is preferable if high flow velocities are to be recorded.

While these statements with regard to the maximum Doppler frequency are correct, it should also be noted that the standardized Doppler frequencies have to be much lower than PRF/2 to ensure that they can still be analyzed accurately enough. The PRF depends upon the ultrasound propagation velocity in tissue and the distance between the transducer and the measuring point. It has to be selected at a frequency low enough to allow the ultrasound packet to return before the next one is sent. Applying the Nyquist theorem, measurement difficulties can arise particularly in relation to stenoses involving the deeper-lying blood vessels. (For example, if a PRF upper limit value of 15 kHz is given at a depth of 5 cm, this restricts the maximum ascertainable Doppler shift to 7.5 kHz.)

Doppler shifts can be registered even above the Nyquist limit. This measurement situation is termed *aliasing,* and it describes a degree of uncertainty in determining the velocity (Fig. 1.**6b**). In the above example, the impression was that the wheel was turning backward at almost maximum speed just above the Nyquist limit. The faster the wheel turns, the further the spokes tend to drift back to their original position, eventually coming to a standstill at twice the Nyquist speed. In fast Fourier transform (FFT) spectral analysis, the velocity is plotted in the opposite direction. Using a *baseline shift* and deleting the counterdirectional indication, the recordable velocity can be increased to a maximum value of twice the Nyquist limit. Multiple aliasing, however, can produce ambiguous diagrams. Alternatively, the pulse repetition rate can be raised *(high PRF),* so that instead of wait-

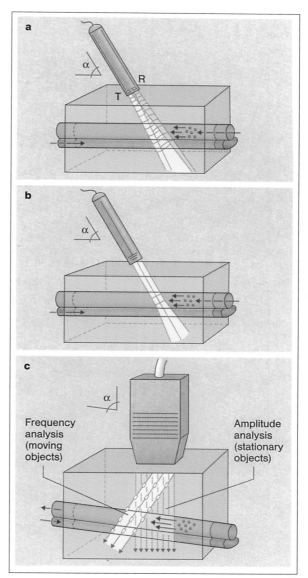

Fig. 1.**5a** The principle of continuous wave Doppler sonography, and **b** of pulsed wave Doppler sonography. The arrow shows the direction of flow, and angle α is formed by the ultrasound and vascular axes. T = transmission crystal, R = receiving crystal, **c** Color flow duplex sonography is a new procedure that allows nearly simultaneous display of flow signals and tissue structures. In contrast to conventional duplex sonography, two ultrasound processes are used to construct the B-mode and color-mode images: frequency-domain and amplitude-domain analyses (for details, see text)

ing for a transmitted pulse to return, a second pulse can be sent within a defined interval. When this procedure is extended, a transition into the continuous wave mode takes place.

In principle, blood flow velocities can be analyzed at different points within the cross-section of a blood vessel, provided a suitable pulsed wave Doppler system is available. Using a *single-channel system,*

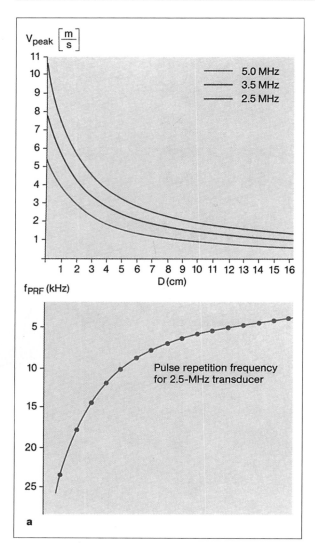

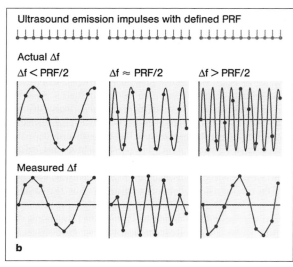

b

Fig. 1.**6a** The relationship between the positioning of the sample volume D (depth), the pulse repetition frequency (f_{PRF}), and the peak flow velocity (V peak) for different ultrasound frequencies (2.5–5 MHz), **b** Reconstruction of Doppler oscillations of various frequencies from individual high-frequency ultrasound emission impulses (•) with a defined pulse repetition frequency (PRF). The benchmark points for the reconstruction were the amplitude values of the sinus oscillation at the moment of each emitted ultrasound impulse. If the Doppler frequency shift Δf exceeds half of the PRF *(right)*, then the measured Doppler frequency will be interpreted incorrectly ("aliasing" effect) (adapted from Widder)

these recordings would have to be carried out sequentially during different heart cycles. Due to the variability of the blood flow velocity from one heartbeat to the next, however, this method involves potential errors. In addition, it is often very difficult with a single-channel system to determine the exact location of the sample volume—which is particularly important when trying to demonstrate altered blood flow velocities near the vessel wall while excluding artifacts. Although the equipment involved is much more expensive (Keller et al. 1976, Casty 1982, Hennerici and

Freund 1984, Reneman et al. 1986), *multi-channel pulsed wave Doppler systems,* using a parallel arrangement of several measuring channels, can simultaneously measure the velocity in various neighboring areas of a blood vessel and display the velocity distribution as a function of time for each heart cycle (Fig. 1.**7**). Provided the angle between the ultrasound beam and the vascular axis is known, an algorithm can then be used to calculate the flow volume quantitatively (Casty and Anlicker 1978, Hennerici and Freund 1984).

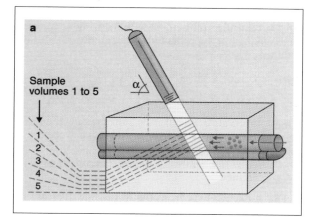

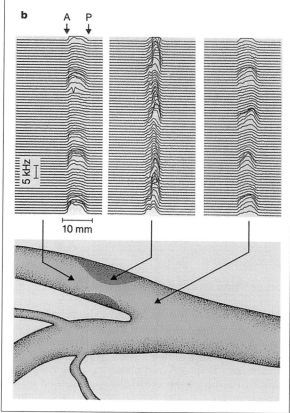

Fig. 1.**7 a** The principle of a multi-channel pulsed wave Doppler system, and **b** the display from a medium-grade stenosis of the internal carotid artery at the bifurcation

Anterior wall A   Posterior wall P

Duplex Systems

Conventional Duplex Sonography

In duplex systems, a continuous wave or pulsed wave Doppler system (single-channel or multi-channel) is permanently combined with an echo-impulse system (Barber et al. 1974, Blackshear et al. 1979, Hennerici 1983, Arbeille et al. 1984, Hennerici and Freund 1984). This type of equipment uses linear-array and sector transducers, mechanical rotation or oscillating systems, as well as annular phased-array techniques (Hatle and Angelsen 1985, Sold 1986) (pp. 17–18). The advantages and disadvantages of each system decide the field of application for each of the various diagnostic procedures and the choice of the specific Doppler method used. Usually, pulsed wave systems are used, although continuous wave systems are needed to study deep-lying areas relevant to cardiology and to analyze the pelvic and abdominal vasculature, where very high flow velocities can occur. In most systems, the angle between the Doppler ultrasound beam and the vascular axis can be modified continuously or incrementally. This allows one to locate the Doppler sample volume within the specific area of interest in a B-scan. Simultaneous use of echo-

impulse and Doppler sonography techniques may still cause quality reductions in both procedures. However, the technique is useful as an initial examination for orientation purposes.

Due to the split between the image and the Doppler pulses, a significant reduction in the pulse repetition rate has to be accepted even with pulsed wave Doppler sonography. This results in a reduction of the maximally detectable flow velocity. Under visual control in two dimensions, the sample volume is usually first positioned on an electronic beam; with the image in freeze-frame, a Doppler investigation is then carried out. If the Doppler transducer is not integrated with the B-mode transducer—and this is usually the case with linear and phased-array transducers, as well as with continuous wave Doppler transducers—one should be aware that differences between the two transducer transmission frequencies can lead to artifacts in positioning the sample volume. Unnoticed changes in the transducer position or in the area investigated between measurements are additional error sources.

More recent developments use *missing single estimators* (MSEs) to attempt real-time duplex imaging

(Angelsen and Kristofferson 1984). First, a sector image is constructed and saved in freeze-frame. Until the next image is created, the original spectrum from a selected Doppler flow analysis is displayed, and then, while the next image is being built up, a "synthesized" Doppler signal is interpolated into the first.

In the conventional duplex procedure, Doppler analysis is carried out using a sample volume selected on the basis of its position in the B-mode scan. Attempts to use what were termed *multigate procedures,* however, did not prove to be practicable. Although this method made it possible to evaluate Doppler sample volumes at different depths, producing a wealth of flow velocity graphics, the information proved to be confusing and ambiguous for diagnostic purposes (Keller et al. 1976, Hennerici and Freund 1984, Reneman et al. 1985, 1986). However, because their measurements are so precise, these systems are still of significance for scientific purposes.

Color Flow Duplex Systems

Color flow duplex systems are now the principal type of equipment used in clinical diagnosis (Burns 1993) (Fig. 1.**5 c**). Two technically different methods are commercially distributed. The *"frequency-domain"* method uses Doppler- and phase shifts of ultrasound echoes to display flow velocity superimposed on the B-mode image; i.e., this is a Doppler-related method. The *"time-domain"* method analyzes alteration of high-frequency echo signals for the display of movement patterns; i.e., this method is independent of Doppler-shifted signals. Due to the larger computational demands, this method is not widely distributed, and advantages, which are said to improve the quality of low flow velocity interpretation, have not yet been demonstrated (Herment and Dumee 1994; Herment and Guglielmi 1994).

Using the frequency-domain method, real-time spectral analysis of the individual Doppler signals cannot be performed, because of the extremely high demands on data-processing facilities. Instead, color-coded, two-dimensional flow velocity information is registered separately from the image information, although both types of data are displayed quasi-simultaneously on the screen. Black-and-white encoding is usually used to depict structures, while the flow velocity is presented in color-coded form. Technically, linear and phased-array systems can register a combination of both Doppler and image beams in a single pass, while annular array systems combine a quick succession of images in one direction with a slower Doppler component.

Instead of spectral analysis, the *autocorrelation principle* is used to determine three Doppler measurement parameters from the individual signal pattern, and the mean of these is displayed. In this process, it is not the frequency change in the individual ultrasound pulses that is displayed, but the phase change between two successive signals. If the velocity (v) of a blood component examined by ultrasound during analysis

time (t) remains constant between two ultrasound pulses, then the following relationship describes the phase change in the reflected ultrasound wave and the distance (d) it has covered:

$$v = \frac{d}{t}$$

If the interval between two ultrasound pulses is known, then simply measuring the phase change is sufficient to determine the velocity.

Usually, the following measurement parameters are determined:

- The mean velocity from the frequency distribution of all the frequencies, independent of their signal intensities *(velocity-mode);*
- The scatter range of frequencies around the mean *(variance-mode);*
- The total amplitude of the Doppler reflection within a sample volume *(amplitude-* or *power-energy mode}.*

Only a single value is calculated for each pixel with a color-coded Doppler signal (Mitchell 1990, Cape et al. 1991). This represents a significant limitation in the amount of available information when compared to what is possible with FFT signal intensity–weighted frequency analysis. It is possible to underestimate the true maximum velocity significantly, and the variability in flow velocities can also be significantly underestimated in turbulent flow conditions, when the velocity values within a sample volume are averaged out. In addition, artifacts due to wall movements are included in the calculated signal, and this can lead to serious displacement of the autocorrelation relationship (Fig. 1.**8),** since there is no objective method of noise reduction for individual frequencies.

The data are analyzed by a phase detector (Wells 1992). The phase change is given a digital value, and the value calculated for a specific sample volume over time is then displayed in color. The boundaries of the color scale used can be variably defined by the investigator, based on the expected flow velocities in relation to the Nyquist theorem. Equipment manufacturers all use the basic colors red and blue, but the representation of the direction of flow varies. In our illustrations, arterial blood flow is coded red, and venous blood flow is coded blue. The color saturation level increases with increasing flow velocities, and the boundaries between red and blue meet during aliasing. The intensity of the Doppler signal is represented by the color saturation level, and variable scales are available for high-amplitude and low-amplitude Doppler signals, so that better differentiation is possible in depicting amplitude fluctuations often seen in higher-intensity signals. A simpler grid is sufficient for lower-amplitude signals.

Application of the variance-mode uses data from the frequency-distribution per time interval analyzed subsequently. If with an increasing degree of stenosis the Doppler spectrum broadening increases at low intensities, the variance of the color-coded digital values

Fig. 1.**8** Comparing the parameters from an amplitude-weighted spectral analysis (fast Fourier transform [FFT] analysis) and a signal distribution calculated using autocorrelation. In the FFT spectrum, documentation of amplitude-weighted low and medium Doppler frequencies is distinct (e.g., wall movements and laminar flow within a blood vessel). In contrast, in color-coded duplex instruments using the frequency-domain analysis the individual frequency intensities are disregarded due to the overall signal statistics (amplitude, bandwidth, and average frequency). This should be borne in mind when interpreting color-coded flow signals (adapted from Fehske 1988)

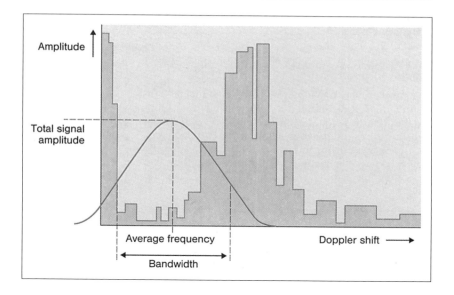

increases too. This condition is often displayed by green signals scattered throughout the color-coded signals.

In addition, an amplitude or power–energy mode has become available more recently for the display of flow-velocity signals irrespective of the angle of ultra-sound axis and without evidence of flow direction. This facilitates the separation of areas with flow from those without even slow flow, which favors the delineation of the extent and structure of an arterial lesion but often exaggerates its margins due to the poor spatial resolution (Fig. 1.**9**).

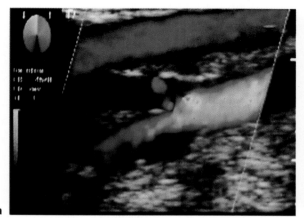

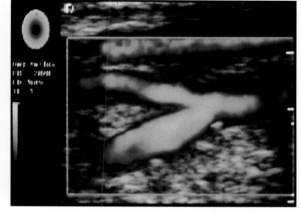

a

b

Fig. 1.**9** Velocity-mode **a** and power-energy mode **b** display of the carotid bifurcation. Note the significant improvement in the illustration of all areas of blood-flow velocity in the power-energy mode (despite a somewhat poorer spatial resolution) when compared to the velocity-mode image, which displays a selected interval of the pulsatile blood flow velocity within the diastole of the cardiac cycle with no flow/ very slow flow in the external carotid artery and the carotid bulb, moderate laminar flow in the internal and common carotid arteries, and secondary vortex flow separation at the branching point (coded in blue). Flow direction is only visible in **a** (note the large blue signal from the jugular vein), and not in **b**

Signal Analysis

Audio Signal Analysis

The simplest way to evaluate a Doppler signal is to interpret the audio impression reproduced by a loudspeaker (Barnes et al. 1981). The direction of flow can also be represented acoustically using two speakers, with flow away from the probe and flow toward the probe being distributed separately to the two speakers. Even with these simple methods, an experienced investigator can identify factors that are significant in evaluating a blood vessel's condition. Audio analysis is always an indispensable precondition for selecting and optimizing the signal that is subsequently to be investigated.

Zero-Crossing Counter

Using the directional Doppler technique, directions of flow "toward the probe" and "away from the probe" are registered on separate channels. These Doppler signals are depicted graphically as a two-dimensional diagram, with the value of the Doppler shift on the y-axis plotted against time on the x-axis. When the incident angle α is known and a constant baseline is maintained, both the Doppler shift itself and its corresponding velocity can be read off.

In fact, however, using what are termed zero-crossing counters (zero-crossers), which display less information than the envelope curve, only an average frequency for the Doppler spectrum is documented. Many studies have shown that this simple procedure does not even enable one to determine the average frequency of the Doppler spectrum corresponding to the average flow velocity (Peronneau et al. 1970, Reneman et al. 1973). Instead, a systematic error develops, which depends upon the form of the received signal and increases proportionally to the true average frequency (Fig. 1.**10a**). This is especially detrimental when stenoses are present. In the low-frequency range, high-amplitude signals—vascular wall movements, for example—are noticeable sources of error, and using suitable high-pass filters also excludes important information provided by the Doppler spectrum (e.g., areas of turbulence). For a long time, these disadvantages were outweighed by the advantages of having a simple and cost-effective technology. The envelope curve recording the analogue signal of the Doppler spectrum (known as the *analogue procedure*) provides an adequate overview of changes in the average flow velocity. This procedure is still widely used to provide general orientation.

Time-Interval Histograms

The time-interval histogram, which replaced amplitude-weighted spectral analysis up to the 1980 s (Sandmann et al. 1975, Baker et al. 1978) is also based on the zero-crossing counter principle. This method analyzes not only the time at which the Doppler signal crosses the baseline in a particular direction, but also the time interval to the previous crossing (Fig. 1.**10b**). This produces an image that contains more information than the envelope curve.

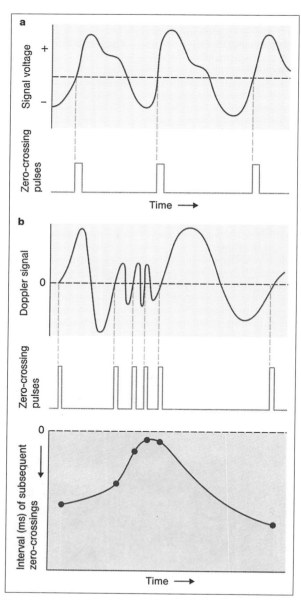

Fig. 1.**10a** Zero-crossing and **b** time interval histograms. An increase in the Doppler frequency leads to shorter time intervals between the zero-crossings (adapted from Zagzebski and Madsen 1982).

Spectral Analysis

The frequencies contained within a Doppler signal, the time at which they occur, and their intensity can be calculated and depicted using a *fast Fourier transform (FFT)* (Barnes et al. 1976, Felix et al. 1976, Lewis et al. 1978, Spencer and Reid 1979) (Fig. 1.**10 c).**

The basis of the Fourier transform is the so-called Fourier theorem, according to which a periodically repeating event can be almost exactly defined by superimposing various sinus waveforms. This technique is used today both for continuous wave and pulsed wave Doppler investigations—with the exception of color flow duplex methods, since they produce so much data and require so much computing power that three-dimensional signal analysis cannot be depicted.

The frequency of the FFT analysis is plotted as the ordinate, the time as the abscissa, and the amplitude of the signal is documented as a gray scale, or, alternatively, in color-coded format. The flow direction is indicated as a positive (above the baseline) or a negative signal (below the baseline).

Ideally, a computer would process the FFT analysis in real time and describe the Doppler spectrum precisely. For technical reasons, however, even approximate "real-time analysis" is complex and quite costly. Much of the equipment currently available tries to apply simpler signal-processing methods, using electronic or analog filters that analyze entire frequency bands in combined form, instead of just single frequencies. As a result, Doppler spectrum distortions can occur when a signal passes through several filters and is repeatedly calculated or displayed.

To analyze complex flow phenomena precisely, various parameters are calculated and displayed digitally, including:

- The maximum Doppler frequency *(peak frequency)*
- The average Doppler frequency *(mean frequency)*
- The Doppler frequency with the maximum amplitude *(mode frequency)*
- Bandwidth *(spectral broadening)*
- Spectral *window*

As can be seen from the individual *power spectra* (see the illustration under "Spectrum" in the Glossary, p. 456), all of these parameters vary from one moment of analysis to the next. Each power spectrum depicts the frequency distribution (on the abscissa) at a particular time, plotted against its amplitude (on the ordinate). The amplitudes are raised to powers of ten to provide clearer differentiation between the individual components.

It should be noted that the spectrum is particularly influenced by the area measured, as well as a variety of other factors, so that it is subject to significant variations, especially in pulsed wave Doppler systems. Due to laminar flow, a selected sample volume that is too small can incorrectly enlarge the bandwidth or reduce the spectral window, especially when different velocities are being analyzed (this also applies to vascular branching points). In addition, a clear distinc-

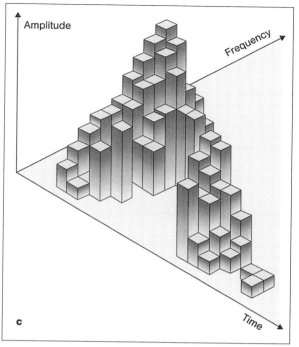

Fig. 1.**10 c**   A three-dimensional diagram showing the different components in a fast Fourier transform spectrum (adapted from Seitz and Kubale 1988)

tion between background noise and signal should be ensured by selecting an appropriate noise reduction filter—otherwise, low-amplitude but high-frequency signals reflecting significant stenosis may be missed. If the signal is weak, it can still be successfully differentiated in small steps by changing the amplitude scale *(compression)*. In contrast, the global amplification *(gain)* works like a loudspeaker's volume control, without any amplitude weighting of the Doppler spectra. Since the signal-to-noise ratio deteriorates with increasing signal amplification, care should be taken to make the settings as precise as possible—taking full advantage of global amplification to detect any remaining flow signals only makes sense for particular circumstances like differentiations between subtotal stenosis and complete blockage.

In the same way that certain growth processes are best documented using time-delay photography and rapid movements are best represented by slow-motion photography, various analysis times are also required for individual spectral components. With increasing analysis time (T), a Doppler shift (Δf) can therefore be precisely determined:

$$\Delta f = 1/T \text{ or } T = 1/\Delta f$$

To define a Doppler shift of 200 Hz (Δf) precisely, a 5-ms signal observation time would therefore be sufficient:

$$T = 1:200 \text{ Hz} = 5 \text{ ms}$$

When calculating the spectrum, care should be taken to select adequate time intervals for the analysis (between 4 ms and 40 ms). It is important to note that the analysis time is inversely proportional to the frequency resolution. This means that with a 4-ms analysis time, the discrimination capability is 280 Hz, and with 16 ms it is 70 Hz.

It should be noted that the precision of the measurement, which depends on the analysis time, is further affected by the ultrasound frequency and the transit distance (defined by the sample volume length). Imprecision in measurement that occurs during the observation time therefore has to be

balanced out by the *transit time effect*—i.e., in regions of increasing velocity, the observation time also has to increase. Nevertheless, the analysis time cannot be extended indefinitely, since flowing blood can only be considered constant for a limited time (approximately 10–15 ms). An excessive analysis time would reduce the precision of the velocity measurement.

Statistical analyses of Doppler signals are complex, and with the exception of specific investigations, have not been of any diagnostic significance (e.g., Laplace transforms).

Echo Contrast Media

Contrast media have been used experimentally in ultrasound diagnostics over the last 20 years (Burns 1994, Goldberg et al. 1994); a mixture of 5–10 ml 0.9% NaCl solution and 0–2 ml air well shaken was used as well as a similar foam of soluble gelatin (e.g., gelifundol) without complications (Görtler et al. 1995). It is only recently, however, that pharmaceutical manufacturers have expressed interest in developing such substances. For example, the use of galactose-based carriers of air microbubbles—e.g., Levovist—, sonicated albumin and polymercoated bubbles has been introduced into vascular sonography and echo-cardiography. Currently, additional substances are being developed for use in cardiovascular and gastrointestinal diagnostics (e.g., perfluorocarbon-exposed sonicated dextrose albumin) (Porter et al. 1996). It is likely that a whole range of contrast media, comparable to the assortment of medical isotopes with various characteristics, will soon become available.

Microencapsulated air bubbles (with a diameter of 8–20 µm) were first used to heighten the contrast of reflections at interfaces and to provide better documentation of wall movements in the left ventricle and in transcranial Doppler sonography at an early stage to improve the reduced signal amplitude of small bone fenestrations. In patients aged over 70, who often have hyperostosis, this reduced signal amplitude was technically responsible for inadequate registration in up to 25% of cases. This technology is now increasingly used for a significantly improved visualization of basal cerebral arteries and veins by means of 2-D and 3-D transcranial duplex sonography (Baumgartner et al. 1994, Otis et al. 1995, Delcker et al. 1997, Ries et al. 1997) (Fig. 1.**11**).

A diagnostic application that is currently under investigation is the imaging of small blood vessels in parenchymatous organs (pancreas, kidney, liver, heart, and brain). The goal is to achieve better detection of small tumors, and to distinguish between ischemia and infarction at an early stage. *Harmonic imaging* is of particular interest here (Burns 1996). Using a high signal amplitude in this procedure, the aim is to solve the problem of superimposed slow Doppler shifts resulting from the movements of

neighboring interface structures (see Fig. 1.**2**). Contrast media are used to reflect harmonic echo reflections (at exactly twice the original frequency). After appropriate adjustments to the system software in the duplex equipment, the harmonics alone can be analyzed, e.g., only 8 MHz is registered in the upper frequency range after an original 4 MHz emission. With appropriate signal analysis, only signals that are harmonics of the underlying frequency are separated out. Since these higher-frequency echoes are solely caused by small contrast-medium bubbles, and not by the surrounding tissue, they are necessarily located intravasally. Reflections from the surrounding tissue and the vessel wall itself remain suppressed. With this method, even very slow blood flow velocities in small blood vessels can be distinguished from artifacts.

One of the first contrast media with particles smaller than red blood cells that was also stable enough to pass through the lungs was Echovist. This preparation has been approved for use in several different countries, and is also used in transcranial monitoring and in detecting sources of embolism in cerebral sonography (see Fig. 3.**24**). It also appears to simplify the detection of the remaining blood flow in small distal blood vessels in cases of proximal occlusion of the leg arteries (e.g., in the plantar arteries and the anterior and posterior tibial arteries). This is especially helpful in patients with diabetes mellitus, since 10–15% of distal blood flow signals are not detectable using conventional Doppler sonography. Additional research aims to produce even smaller air bubbles (diameter 1 µm) that would circulate within the central blood flow for 3–5 minutes (Levovist) or even longer (20–30 minutes), and would be eliminated by Kupffer cells in the liver. These phagocytic cells are missing in areas that contain small tumors, which means that dark regions in an ultrasound image could serve as an early sign of neoplasm. Due to their long half-life, these contrast media are also potentially suitable for conducting investigations in very small vascular regions that have extensive proximal occlusions.

Other substances are available that have an adhesive effect on thrombotic material, strengthening its echogenic qualities. This could be especially significant when examining venous blood flow; no increase

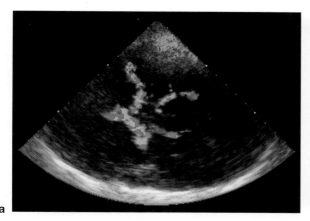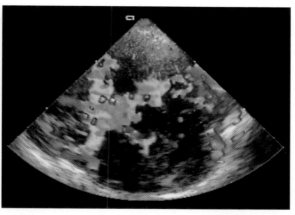

a   b

Fig. 1.**11**   Transcranial color-coded duplex sonogram of the anterior circle of Willis before **a** and 1 minute after administration of echo-contrast **b**. Note the display of small arterial and arteriolar (including capillary) branches, and parts of the transverse sinus (arrow) as well

in the echo reflection takes place in veins with a normal blood flow. Further possible benefits include being able to take better-informed decisions in the postoperative treatment of kidney transplant patients. Finally, there is some hope that quantitative analysis of the concentration of these contrast media might provide better ways of estimating the Doppler signal intensity—the goal being to advance the method from its current semiquantitative status toward a standard that at least shows relative changes with good quantitative resolution (Schwarz et al. 1993, Ranke et al. 1992, Sitzer et al. 1994).

New and Future Developments

Although definite clinical studies have not been performed and advantages and limitations have not yet been established, one of the most promising developments in diagnostic ultrasound for the evaluation of vascular disease, is the application of computer vision and optical/flow techniques to exploit the full potential of real time imaging and Doppler studies. Among these technologies, four-dimensional (4-D) analysis of wall motion and plaque movement for the study of atherogenesis needs to be mentioned. Rapid advances in this field, as well as in magnetic resonance technology suggest that the potential clinical utility of new developments should be assessed in relation to one another (Meairs et al. 1995).

Motion Analysis of Arterial Wall and Atherosclerotic Plaques

Experimental studies have identified areas of low-shear and recirculation which predispose to early atherogenesis: in the carotid system, the carotid bulb opposite the flow divider is the predominant location of plaques (Ku et al. 1985, Zarins et al. 1983, Steinke et al. 1990 a/b). Previous work on 3-dimensional ultrasound of the carotid arteries' geometry (Picot et al. 1993) has been extended to include a temporal 4-D evaluation of pulsatile flow-dynamics, plaque geometry and alterations of vessel wall segments (Meairs et al. 1995, Hennerici et al. 1995). Image processing of the 4-D ultrasound data sets has involved reconstruction of time-dependent geometrical features of carotid arteries and atherosclerotic plaques, as well as automated motion analysis using hierarchical bloc-matching algorithms. This technique provides excellent visualization of wall movement over time (Fig. 1.**12 a, b**) and plaque motion (Fig. 1.**12 c, d**). In patients with *symptomatic* carotid artery disease, it has enabled detection of complex rotation of wall plaque movement as evidence for intraplaque instability (Meairs et al. 1995). In contrast in patients with *asymptomatic* carotid plaques, a uniform plaque and wall motility could be demonstrated with low amplitude, unidirectional vectors. Other prospective studies, however, are necessary to confirm these promising initial findings and their clinical relevance.

Evaluation of Small Artery Disease by Computerized Doppler Analysis

While the evaluation of large artery and vein diseases is the major domain of duplex system analysis, the arteriolar and capillary areas cannot yet be adequately investigated. This is, however, a major disadvantage of the technology because both peripheral and cerebral circulation are often affected by small vessel disease. About one-third of ischemic strokes, for instance, are due to small vessel disease. In these patients the underlying vascular processes cannot be

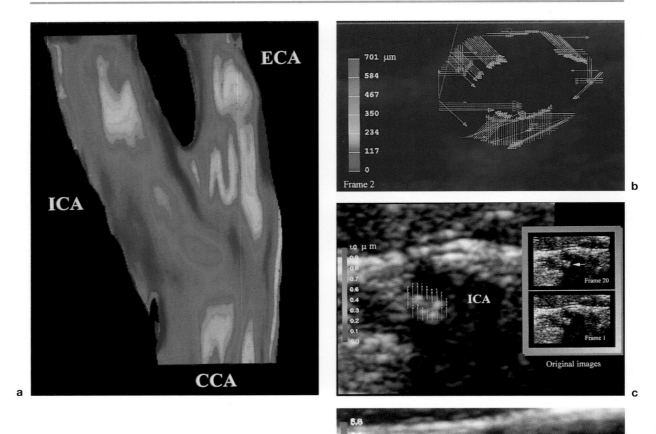

Fig. 1.**12 a-d**   3- and 4-dimensional imaging of motion analysis of arterial wall and atherosclerotic plaque, **a** and **b** show movement of the common carotid artery walls in a quantitative vector analysis on a spatial (2-D) with temporal (3-D) distribution (green and blue—small movement, yellow and red—large movement), **c** and **d** show plaque motion with vector-indicated direction and amplitude distribution in asymptomatic vs. symptomatic patients

directly identified in the absence of proximal large vessel atherogenesis and thrombus formation or any sources of embolism. Recent studies have succeeded in evaluating the distal vascular territory from hemodynamic parameters recorded from the proximal vessels. Bäzner et al. (1995) identified a significant alteration of low-frequency spontaneous oscillations (LSO) originating in the capillary arteriolar network in patients with small vessel disease. Low-frequency SO are characteristic features of a functional disturbance in cerebral autoregulation similar to that commonly seen in patients with proximal occlusion of the cerebral or peripheral arteries (Bollinger et al. 1991) with compensating maximum dilatation of the capillary vasculature, and in animal experiments (Dirnagl et al. 1993, Hundley et al. 1988, Schechner and Braverman 1992).

Equipment

Instruments and Applications

Unidirectional Doppler Instruments

A unidirectional or nondirectional ultrasound Doppler (the so-called "pocket Doppler") reproduces the Doppler signal acoustically, and is primarily used to measure the arterial systolic blood pressure in the extremities. Although this type of equipment does not show the direction of the blood flow, it can still provide a general orientation in venous diagnosis.

Continuous Wave and Pulsed Wave Doppler Instruments

Bidirectional Doppler ultrasound instruments with continuous ultrasound emission and an internal or externally connectable plotter, including an oscilloscope, are necessary to register flow velocity curves in the following blood vessels: *extracranial cerebral arteries, arteries in the extremities, penile arteries, extracranial veins, peripheral veins,* and *the scrotal vasculature.* The equipment displays and documents the direction of the blood flow and flow velocity graphs. Frequency shift analysis is carried out using analog procedures (zero-crossing counter), which is sufficient for a general, overall assessment.

Spectral analysis is increasingly used for quantitative measurements and the calculation of indices as a supplement in cerebral and peripheral artery studies when quantitative measurements are desired in addition to qualitative assessments.

Only pulsed wave Doppler instruments that provide spectral analysis from defined sample volume depths are useful for documentation of *intracranial and abdominal* flow signals.

Echo-Impulse (B-mode) Instruments

The use of real-time B-mode equipment *without* a prior functional examination using Doppler sonography or an accompanying Doppler-sonographic assessment of the flow velocity (duplex system) is an exception in vascular diagnosis.

It may be used for the evaluation of deep venous thrombosis, since it has been shown that an absence of venous compressibility by the transducer, widening of the lumen, and absent responses to breathing maneuvers are reliable parameters to establish the diagnosis. In contrast, the sole use of B-mode sonography for the examination of arterial disease is unsuitable and potentially misleading. However, in combination with functional Doppler sonography, the B-mode component of duplex systems provides valuable information.

Conventional and Color Flow Duplex Systems

Conventional duplex-system equipment consists of a B-mode scanner that incorporates a pulsed wave Doppler mode and often an additional continuous wave Doppler mode. In *color flow duplex sonography,* it is also possible to depict the velocity and the direction of flow in blood vessels with almost simultaneous superimposition on the B-mode image.

B-mode

The B-mode procedure serves to identify arterial and venous blood vessels and to position the sample volume for the Doppler mode accurately. Although it is possible to use CW Doppler sonography alone to evaluate blood vessels that lie close to the body's surface, such as the extracranial cerebral arteries and the accessible areas of the extremities, continuous tracing of the vasculature of the extremities is only feasible with the B-mode. Because of the deep location of the abdominal and pelvic blood vessels, they can only be examined using duplex sonography.

The advantage of the B-mode system is the information provided about morphological structures within the vascular lumen and the surrounding tissue such as:

- Anatomical topography
- Spatial relationship to neighboring structures
- Vascular caliber
- Vascular wall morphology
- Structures inside the lumen (plaque, thrombi)
- Compressibility (of the veins)

Doppler mode

Duplex-system transducers usually have a pulsed wave Doppler mode, and often also provide for switchable, integrated CW Doppler systems or a separate CW Doppler probe.

Using a spatially well-defined *sample volume,* the pulsed wave Doppler mode is able to detect local flow velocity phenomena that are mainly free of any superimposed artifacts.

To examine small-caliber arteries and veins (foot, lower arm, penis and scrotum), the pulsed wave Doppler ultrasound beam should be narrowly adjusted and precisely focused.

By analyzing the Doppler frequency spectrum, one can determine the hemodynamic effect of any wall deposits that are shown in B-mode images, calibrate arterial stenoses, and confirm any proximal and distal changes shown by the Doppler flow curve.

Duplex-system studies are useful for all types of blood vessel diseases affecting both arteries and veins.

Measurements made using the Doppler mode to produce qualitative and quantitative assessments of vascular stenoses and to localize and measure occlusions require very careful arterial scanning, which is time-consuming and—especially in small-caliber blood vessels—often unsuccessful. Usually, a screening test using continuous wave Doppler sonography is carried out to provide a general orientation, especially when dealing with the cerebral arteries and the arteries of the extremities, in order to identify the blood vessel segment for closer investigation.

Color flow Doppler mode

The advent of the color flow Doppler mode, with its visual and dynamic representation of the blood flow, made it very much easier to detect blood vessels, and in some regions made this type of analysis possible for the first time. In particular, it was not possible to identify the deep-lying vasculature or small-caliber arteries and veins within a reasonable amount of time before color-coded depiction of blood flow became available.

Technically, color flow duplex sonography exploits all the potential of conventional duplex systems, and in addition makes it possible to represent ultrasound frequency shifts in different colors when moving corpuscular blood components are encountered (p. 8). This information is superimposed on the B-mode image in a selectable display section. When the equipment is given a known setting that is appropriate to the area being examined and the inquiry being pursued, the various colors displayed can provide relevant information about the direction of the blood flow and the relative changes in its average velocity within the arteries.

It is best to select the color setting of the duplex equipment to display the physiological flow within a blood vessel as a constant and strong color, i.e., without any aliasing (pp. 5–6). In straight blood vessels, i.e., those that maintain a constant angle to the Doppler axis, the color representation throughout the blood vessel's course remains constant. In optimally accessible arterial segments, one can even detect in color-coded form the different average blood flow velocities, which present in a longitudinal profile with a fast component at the center of the blood vessel and slower components on its periphery.

The color coding parameters have to be adjusted properly, and equipment-dependent factors have to be appropriately preselected for the requirements of a specific analysis and the specific inquiry. In order to obtain optimal arterial color representation in segments of interest, the following color adjustments need to be selected:

– Color
– Gain or power
– Pulse repetition frequency (PRF)
– Filter
– Angle
– Baseline
– Scale
– Variance

Zones of pathologically elevated blood flow velocities can be quickly detected due to the distinctly lighter shade of the basic color that has been selected, or due to a color change produced by aliasing.

However, it should be noted that a change in the vascular axis relative to the transducer, as in the naturally curving course of an artery, will also lead to a change in the color flow Doppler information (Fig. 1.**13**). This means that not every color change is equivalent to a pathological condition.

Quantitative assessment of flow velocity still inevitably requires the information derived from PW or CW Doppler mode studies in the form of flow velocity curves, and integrated spectral analysis. Thus the ability to depict the frequency spectra accurately represents an important characteristic of all color-coded duplex systems, independent of proper functioning in the color flow Doppler mode. This includes precise resolution in space and time during the recording, visual display of frequency shifts, and accurate acoustic frequency representation.

Transducers, Probes, and Transmission Frequencies

Unidirectional Doppler Sonography

When measuring systolic blood pressure, a transmission frequency of 8–10 MHz should be used for the pen-shaped probes to record the Doppler signal on the foot arteries (posterior tibial artery or dorsal artery of

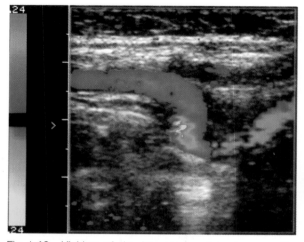

Fig. 1.**13** Kinking of the internal carotid artery as displayed in a color flow duplex system. Note the change of color at the top of the arterial loop due to the change of flow direction with view to the transducer despite physiological flow conditions

the foot, or both) and the arm arteries (radial artery or ulnar artery, or both). An overall orientation with regard to the corresponding veins is obtained at the same frequencies. For deeper-lying veins (common femoral vein, popliteal vein, subclavian vein), a transmission frequency of 4–5 MHz may prove more effective.

Continuous Wave and Pulsed Wave Doppler Sonography

The probe transmission frequencies are also selected according to the desired penetration depth. To examine the extracranial cerebral arteries, a transmission frequency of 4–5 MHz is optimal; the fronto-orbital terminal branches of the ophthalmic artery are best examined using 8–10 MHz transmission frequencies. In addition to the transmission frequency produced by continuous wave Doppler equipment, focusing also influences the optimal depth and width of the applied ultrasound.

The transmission frequency used by pulsed wave Doppler equipment to assess the intracranial arteries is almost 2 MHz, with a varying, selectable measurement depth. Equipment is available with varying sample volumes, depending on its generation and type. Due to ultrasound reflection near the bones, the sample volume there is larger than that for extracranial pulsed wave equipment (e.g., 4 × 10 × 10 mm). The ultrasound energy has to be kept consistently high because of the significant intensity loss when the signal passes through the skull. An adjustment between 10 and 100 mW/cm² is necessary, while extracranial equipment usually uses consistently lower energies than 5 mW/cm² (Hennerici et al. 1987).

To examine *peripheral arteries and veins* (pelvic blood vessels, subclavian artery, common femoral artery, popliteal artery, and corresponding veins), probes should be selected with a frequency of around 4–5 MHz, depending on the location and size of the blood vessels. Due to their location near the surface, the blood vessels of *the foot, lower arm,* and *penis* require piezoelectric crystals that emit a frequency of 8–10 MHz in the pen-shaped probes, in order to ensure correct ultrasound beam focusing within a small distance from the skin.

In the *abdominal region,* the normal pen-shaped probes are hardly ever used, due to their minimal penetration depth and inadequate orientation.

Conventional and Color Flow Duplex Sonography

Transducers

There is a basic distinction between rotating or oscillating mechanical transducers and electronic *phased-array* or *linear-array* transducers. The characteristic common to all of these transducers is the B-mode image construction used, which consists of individual scan lines that are adjacent to one another in either a parallel (linear) or sectoral (radial) arrangement. The

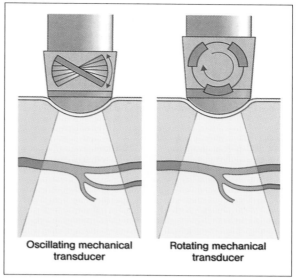

Oscillating mechanical transducer — Rotating mechanical transducer

Fig. 1.**14** The two most commonly used mechanical sector transducers. In the transducer on the left, the transmitting crystal oscillates, generating a new image with each oscillation. The transducer on the right is equipped with three or four rotating transformers, allowing images to be generated slightly faster

image quality increases with the density (number of scan lines) and also with the *frame rate* (number of image repetitions). Using the correct frequency for the image sequence is important for a flicker-free image. However, both parameters cannot be increased indefinitely, since their product yields the pulse repetition frequency (PRF), which in turn is determined by the required penetration depth.

$$PRF\ (Hz) = frame\ rate \times image\ line\ density$$

In *mechanical* transducers (Fig. 1.**14**), one or more piezoelectric crystals are moved along a circular path while *rotating* or *oscillating.* In this case, the ultrasound beam formed by the piezoelectric elements is rapidly passed through the field being examined in a circular sector of 45–90°. The individual scan lines are rapidly processed and assembled one after the other. *Electronic transducers (linear-array, curved-array),* by contrast, incorporate multiple small converters that can be excited either sequentially or segmentally, creating an ultrasound image of a rectangular or sector-shaped composition (Fig. 1.**15**).

Annular-array transducers use circular crystals that are mechanically stimulated from inside to outside. This process depicts information in a three-dimensional form, and due to the phase-displaced stimulation it allows focusing (Fig. 1.**16**) (Bushong and Archer 1991).

In *phased-array* transducers, an electronic time delay of approximately 10^{-9} seconds in stimulating each crystal facilitates electronic focusing (Fig. 1.**17**).

Adjustable *time-gain compensation* (TGC) provides selective amplification of attenuated reflections from tissue cross-sections at various depths. Also, the

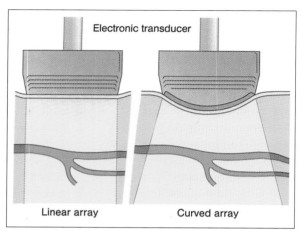

Fig. 1.**15** The linear-array (linear) and curved-array (concave) transducers, and their ultrasound fields. The contact area of the linear transducer is relatively long, while the concave transducer matches the body surface better. Despite its sector-shaped ultrasound field, the resolution of the curved-array transducer is nearly as good as that of linear scanners. This type of concave transducer is becoming increasingly popular

total amplification *(gain)* can usually be continuously regulated. The selected *image* should be enlargeable, and is displayed on the monitor in *real time*. A *scale*, consisting of electronically produced lines that indicate distance, allows measurement of the echo structures depicted.

The advantage of transducers that produce a sector image (using the *phased-array* principle) is a small contact area; however, they have the disadvantage that they can only assess the vascular structure orthogonally on a limited and selective basis. Transducers with a larger surface contact area *(linear array)* allow examination of a larger vascular segment under the optimal 90° angle. The "curved-array" transducer (Fig. 1.**15**) has a geometric structure in between that of a linear and a sector-shaped ultrasound field. The near field is broader and the depth resolution is also better than that of linear scanners.

In general, electronic ultrasound equipment is lighter, easier to use, and less susceptible to disturbances, but more expensive, and it often has a smaller ultrasound field than mechanical transducers. On the other hand, the mechanical transducer can produce a larger cross-section, is more sturdily constructed, and costs less, although it is usually heavier and larger.

The choice between the various transducers depends on the vascular segment to be examined. Blood vessels that are parallel to the body surface are preferably examined with linear scanners. On the other hand, if there is only a small ultrasound window available and a larger and deeper vascular segment needs to be examined, then sector transducers are necessary—e.g., to examine the abdomen, the pelvic vasculature, or the bifurcation of the carotid artery when it lies in a far cranial position.

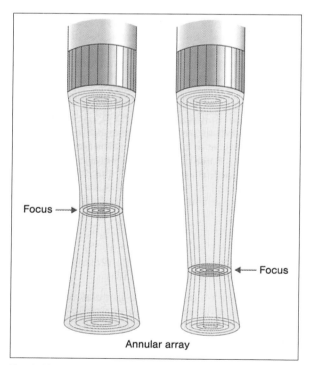

Fig. 1.**16** An annular-array transducer consists of several ring-shaped converters arranged concentrically. When an ultrasound pulse is transmitted, the external rings are mechanically activated shortly before the more centrally located ones. This process produces a three-dimensional ultrasound beam, the focus of which can be altered depending on the activation method used

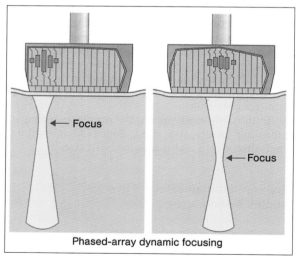

Fig. 1.**17** A phased-array transducer with electronic focusing. Delayed activation of the individual elements allows electronic focusing without using lenses. The near field can therefore be shortened, while at the same time the beam is concentrated on the desired investigation area

Fig. 1.**18** Intravasal sonography, showing (**a**) a normal common carotid artery, and (**b**) a heterogeneous plaque deposit

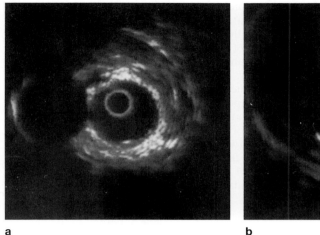
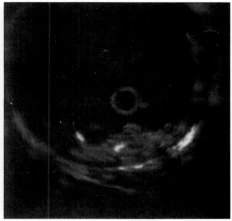

a   b

As a supplementary information source, intravasal ultrasound probes are also used in vascular diagnostics. These provide a better display of plaque structures using an additional image emanating from the center of the blood vessel (Fig. 1.**18**) (Caraye and White 1993).

Transmission Frequencies

B-mode

Transducers with transmission frequencies of 10.0–7.5 (–5) MHz are used to examine blood vessels that are close to the surface (arteries supplying the brain, penile and scrotal blood vessels, arteries and veins of the groin and the hollow of the knee, vasculature supplying tumors, e.g., in the thyroid gland or the breast). At the extremities, especially in their proximal parts, transmission frequencies of 5 MHz are best. Intracranial investigations require frequencies of between 1.5 MHz and 3.0 MHz.

To obtain a satisfactory B-mode image in the *abdominal region* (aorta, vena cava, renal artery and vein, mesenteric blood vessels, pelvic vasculature), frequencies between 3 and 5 MHz are normally used. Lower or higher frequencies of 2 MHz or 7 MHz, respectively, are only necessary in extremely adipose or thin patients (children).

The minimal penetration depth needed to image the blood vessels of the *male genitalia* that are situated close to the surface requires an initial transmission frequency of between 7–10 MHz to produce the B-mode image, allowing a high-resolution B-mode image of structures close to the transducer.

Doppler mode

The continuous wave or pulsed wave Doppler mode transmission frequency integrated into the transducer is independent of the transmission frequency that is used to form the B-mode image. The pulsed wave Doppler component should be narrowly adjusted and

precisely focused to assess small-caliber blood vessels correctly.

Although the continuous wave Doppler component has the advantage of recording blood flow velocities in hemodynamically effective high-grade stenoses, usually the pulsed wave Doppler mode in duplex systems detects high frequencies or velocities, without any physical restrictions (aliasing, pulse repetition frequency).

Care should be taken to place the sample volume in the center of the blood flow and at an angle of less than 60°.

The size of the sample volume can be varied. As a rule, an attempt is made to place a small sample volume (around 1.5 mm) in the central flow. Occasionally, a larger sample volume needs to be selected in order to gather all the available information about the specific circumstances of a flow phenomenon (as in CW Doppler sonography).

Color flow Doppler mode

To produce color, frequencies between 2 MHz and 5 MHz are used, which have to correspond neither to the frequency used by the transducers for B-mode image formation nor to the frequency used by the Doppler component. The detailed recommendations of the equipment manufacturer with regard to the transducer should be followed. Unfortunately, precise information is usually not available.

Possible Biological Effects of the Transmission Energy

Particularly when applying ultrasound *transorbitally* and for *monitoring* purposes, the examiner has to keep the energy level of ultrasound as low as possible. Ultrasound has been used for over 40 years in medical diagnostics without any signs of tissue damage being reported in human beings. Recently, however, intracavitary transducer placement and pulsed wave ultrasound procedures have led to several experimental investigations being carried out to assess further potential biological side-effects, and the mechanisms that might underlie them. Thermal effects and especially the biological consequences are currently a focus of attention.

Thermal effects are particularly relevant in transcranial Doppler and duplex sonography. Applying high ultrasound energies at the bone–tissue interface can lead to a significant elevation in temperature in this context. Animal experiments show that temperatures above 38.5 °C may be a cause for concern. Temperatures above 41 °C in situ can prove problematic when examining embryological or fetal tissue for more than 15 minutes (Bosward et al. 1993, WFUMB Symposium on Safety and Standardization of Ultrasound in Medicine 1992).

Calibration

Unidirectional Doppler Sonography

Since quantitative measurements are not made, it is not necessary to calibrate the equipment.

Continuous Wave and Pulsed Wave Doppler Sonography

To ensure the linearity and the symmetry of the Doppler system, internal or external calibration mechanisms check the linearity and symmetry using a four-position frequency switch. The equipment's specific calibration signal must be recorded at regular intervals.

Conventional and Color Flow Duplex Sonography

Unfortunately the system itself does not usually provide a method to test the transmission efficiency with the equipment. Apart from any necessary adjustments to the equipment, checking the apparatus by external means can be very cumbersome, and is not possible in the setting of a medical practice. The provision of appropriate features by manufacturers would be welcome.

Documentation

Unidirectional Doppler Sonography

When *measuring systolic blood pressure,* it is obligatory to indicate the numerical values obtained with reference to the area measured (cuff position) and the arteries examined (probe position).

Venous signals must be examined in relation to their sound phenomena and their reaction to any breathing maneuvers and distal or proximal compression. A description of the results obtained is sufficient. This method is sometimes used in bedside investigation.

Continuous Wave and Pulsed Wave Doppler Sonography

Doppler flow signals can be documented as an analog pulse curve or as a Doppler frequency spectrum over time. For *arterial* flow signals, the physiological flow should be recorded as a positive deviation from the baseline. It may be necessary to change the polarity to do this. Polarity of flow direction must be documented on all recordings. Baseline adjustment should adequately indicate any signal components that represent backflow. On two-channel recorders, the forward and the backward flows can be recorded on separate channels.

Interpretation of the spectrum requires adequate records of the Doppler frequencies measured. Using a recording speed of 25 or 50 mm/s, two to three heartbeats or more should be documented.

Abnormal Doppler signals are to be described in qualitative terms and the spectrum analyzed quantitatively (e.g., systolic and diastolic peak frequency or velocity). If functional tests are conducted, these have to be documented—e.g., using a bar above the flow curve to indicate when a compression test was done.

Documentation includes flow graphs for each blood vessel examined. Clear, separate labeling of each individual vascular segment is obligatory. Pathological changes such as stenoses must be indicated according to the sites of recordings, e.g. in prestenotic and poststenotic segments. Data obtained from within the stenosis must be appropriately marked. Segments to be examined in the different vascular areas are listed in Tables 1.1 to 1.3. A distinction is made between the normal results obtained in obligatory examinations and the documentation required if the findings show any pathology.

When using directional Doppler equipment for *venous diagnosis,* it is obligatory to record the flow tracings during breathing and during Valsalva maneuvers. If the results are pathological, then additional recordings and documentation of distal and proximal compression maneuvers should be carried out (see Table 1.3).

The recording width for each channel should be at least 4 cm. Adequate documentation should imply which of the individual recordings corresponds to which of the examined veins. The polarity should be such that the physiological flow direction is recorded above the baseline.

When using two-channel recording equipment, it may be useful to depict both flow directions simultaneously. A slow chart speed (2.5–5.0 mm/s) is recommended for documenting venous flow.

Table 1.**1**   Registration points for continuous wave Doppler-sonographic examination of the cerebral arteries

Registration points	Standard documentation adequate for quality assurance		Recommended extension of the investigation and documentation	
	Normal result	**Pathological result**	**Normal result**	**Pathological result**
Supratrochlear artery	+	+	+	
Supraorbital artery	Ø	Ø	Ø	+
These arteries, with compression of the external carotid artery	Ø	+	+	+
Common carotid artery	+	+	+	+
Continuous display of the transition zone from the common carotid artery to the internal carotid artery	Ø	Ø	+	+
Internal carotid artery				
Internal carotid artery, origin	Ø	+	+	+
Internal carotid artery, distal (optimal signal)	+	+	+	+
Internal carotid artery, proximal stenosis	2 registration points 1 Peak acceleration segments 2 Distal poststenotic segment		Complete depiction of the prestenotic and poststenotic segments	
External carotid artery				
Main branch from the common carotid artery (with compression test if differentiation is difficult)	+	+	+	+
Branches	Ø	Ø	Ø	e.g., – occipital artery as a collateral blood flow source – Distal to a stenosis or occlusion – Arteriovenous shunt
Vertebral artery Test for steal syndrome using upper arm compression and/or making a fist	1 registration point, at the origin **or** the mastoid region Ø	Continuous recording of the compression/ decompression phase	Origin at the subclavian artery **and** mastoid region Ø	Continuous recording of the compression/ decompression phase plus overflow type documentation
Cervical collaterals	Ø	Ø	Ø	When diagnosing proximal stenoses or vertebral artery occlusions
Subclavian artery	Proximally **or** distally	Proximally **and** distally	Proximally **or** distally	Proximally **and** distally

Table 1.**2**   Registration points for examining the arteries of the upper and lower extremities using continuous wave Doppler sonography

Registration points	Obligatory		Optional	
	Normal result	Pathological result	Normal result	Pathological result
Subclavian artery (proximal)	+	+	+	+
Subclavian artery (distal)	−	+	+	+
Axillary artery	−	−	−	+
Brachial artery	−	+	−	+
Radial artery	+	+	+	+
Ulnar artery	−	+	+	+
Digital arteries	−	−	−	+
Common femoral artery	+	+	+	+
Superficial femoral artery	−	−	−	+
Popliteal artery	+	+	+	+
Posterior tibial artery	+	+	+	+
Dorsal artery of the foot	−	+	+	+
Fibular (peroneal) artery	−	+	−	+
Digital arteries of the foot	−	+	−	+

Table 1.**3**   Registration points for examining the veins of the upper and lower extremities using continuous wave Doppler sonography

Registration points	Obligatory		Optional	
	Normal result	Pathological result	Normal result	Pathological result
1. External iliac vein	−	−	−	+
Common femoral vein	+	+	+	+
Superficial femoral vein	−	−	−	−
Great saphenous vein (near the orifice)	+	+	+	+
Popliteal vein	+	+	+	+
Posterior tibial vein	−	+	+	+
Small saphenous vein (near the orifice)	−	+	+	+
Communicating veins	−	+	−	+
2. Subclavian vein	−	+	+	+
Axillary vein	−	−	−	+
Brachial vein	−	+	−	+
3. Suprapubic collateral veins	−	+	−	+
Other collateral veins	−	+	−	+
Maneuvers during the examination				
Deep breathing	+	+	+	+
Valsalva*	+	+	+	+
Compression				
Proximal to the probe	−	+	+	+
Distal to the probe	−	+	+	−
Over the collateral veins	−	+	−	+
Tourniquet	−	−	−	+

* When a deep venous thrombosis is suspected, this maneuver should be used with great caution.

Conventional and Color Flow Duplex Sonography

Image documentation should be straightforward and self-explanatory. This is particularly important when the detection or exclusion of pathological vascular wall processes or intraluminal structures forms a significant portion of the diagnosis. In this case, anatomical structures and individual Doppler signals must be clearly labeled and named (Bönhof 1987).

Various kinds of data media can be directly connected to the ultrasound equipment and used to register image and spectral data. Color printers, video systems, or foil printers that make images directly available during the investigation are used. Systems that require additional image data processing are also employed—e.g., electronic and optic disks, slide films, and radiographs.

It is very advantageous to record representative examination segments on videotape along with the acoustic information regarding the Doppler shift. Documentation on videotape must be capable of being checked, and patient identification must be guaranteed. The cross-section used must be evident from the documentation.

Special documentation requirements for individual vascular segments are presented below.

Cerebral Arteries and Peripheral Arteries

Changes in the vascular wall should be characterized in terms of their location, extent, structure, and surface. In addition, blood vessel width, course, or observed pulsations are to be described. Abnormalities in the Doppler flow spectrum are recorded both quantitatively (peak frequency, angle-corrected peak velocity) and qualitatively (flow disturbance) (Tables 1.**4**, 1.**5**).

For normal conditions, the region examined should be documented in a longitudinal sonographic B-mode image, including the Doppler flow velocity spectrum. The normal flow direction is marked above the baseline in the frequency spectrum analysis.

Pathological findings are documented in two planes. The Doppler frequency spectrum must be given, and the location of the sample volume must be identified. In the case of stenoses, the Doppler frequency spectrum must be obtained from the area of maximum obstruction. When the extent of stenoses and occlusions can be accurately assessed, additional documentation of a frequency spectrum from the adjacent proximal and/or distal vascular segments is recommended.

Veins

In B-mode images, veins should be characterized by their location, extent, structure, and compressibility. Any deviations in the vascular width examined in profile, increases in the lumen during functional tests, and variations in the course of the vessel, should also be described.

In the Doppler waveform, the spontaneous flow, reactions during functional tests such as Valsalva maneuvers and other breathing maneuvers, and reactions to manual compression, should be assessed.

Abdominal Blood Vessels

The frequency spectrum of an intra-abdominal blood vessel with blood flowing in a physiological direction should be recorded above the baseline in the area of the abdominal arteries as well, in order to ensure consistency with the documentation format recommended above for peripheral and cerebral arteries (Widder et al. 1990). Arteries with retrograde blood flow are plotted with their frequency spectra below the baseline. Regular coding of arterial blood flow in red and venous portions in blue is not possible in the abdominal region because of the curving course of the abdominal vessels, resulting in different flow directions in relation to the transducer.

Vasculature of the Male Genitalia

In examinations using duplex sonography, representative Doppler frequency spectra should be provided for all the arteries that can be depicted, and particularly for both deep penile arteries.

Recordings of flow velocity after any provocation tests that are carried out, e.g., injections into the spongy body of the penis, should be specially noted.

Examination Reports

The written report of the examination and its results should summarize the clinical data, the problem to be investigated, the ultrasound findings obtained, their interpretation, and, if desired, also recommendations for further diagnosis and/or therapy.

A description of the findings can either be in written form, or can be presented graphically in a diagram. Table 1.**4** lists the *minimal* requirements for documenting findings in a clinical investigation using duplex sonography. An anatomical illustration of the vascular regions examined can incorporate pathological findings (plaques, stenoses, occlusions) and is especially useful for follow-up procedures. The minimum requirements may be expanded as appropriate in relation to the specific indication that led to the ultrasound examination and any relevant pathological findings (Table 1.**5**).

A comprehensive report of the findings obtained with duplex sonography describes the sonographically detected structures and also the frequency spectra or blood flow velocities derived from individual vascular segments. This initially purely descriptive account, making no attempt at histological assessment and following the criteria outlined in Table 1.5, is followed by a summary and evaluation of the findings, and the report closes with a diagnosis based on studies in either duplex or color duplex sonography, and the examining physician's signature.

Examination findings that include pathological changes should be recorded in more detail. In particular, the relevant parameters leading to the particular diagnosis should be documented.

To structure the examination protocols and ensure that they are kept uniform within the clinic and its laboratories, a tabular or schematic arrangement is often useful. Table 1.5 indicates the *contents* of such an arrangement, listing each individual component and providing a comprehensive description of the findings.

The acceptability of diagnoses based on duplex sonography is critically dependent on having good documentation of the examination findings, and not only from the forensic point of view. Although a sonographic examination can only be comprehensively assessed and evaluated by the investigator himself, the objective must nevertheless be to produce optimal documentation, giving a clear, pictorial reproduction of the steps involved in the investigation and obtaining the findings, on the basis of which the pathological findings, in particular, become comprehensible.

Table 1.**4** Formal requirements for proper image documentation in duplex sonography

Patient identification
Examination date
Indication leading to the examination
Formulation of the question
Blood vessel identification

B-mode image parameters
Section
– sagittal
– transverse
– frontal
– longitudinal

Doppler parameters
– Transmission frequency
– Pulse repetition frequency
– Filter adjustment
– Sample volume size
– Flow direction
– Baseline position

Table 1.**5** Content guidelines to consult when describing findings from a duplex sonography study

B-mode image component	**Doppler component**

Depiction
  Good
  Average
  Poor
	Flow detection

Anatomy
  Regular
  Variation/anomaly
  Kinking/coiling
  Aneurysm, dilatation
	Flow direction
	  Physiological
	  Retrograde
	  Alternating forward and backward (pendular)

Vascular caliber
  Interior wall diameter
  Exterior wall diameter (remaining lumen)
  Intima-media thickness
	Flow velocity
	  Flow curve amplitude
	  – Decreased
	  – Elevated
	  Frequency
	  – High
	  – Low
	  Spectral broadening

Pulsations
  Transverse
  Axial
  Systolic/diastolic
	Flow pattern
	  Parallel, laminar
	  Disturbed
	  Secondary vortex
	  Absence
	  Limited acceleration
	Disturbances in heart rhythm

Vascular wall structures
  Localization
  Form
  Size
  Surface contour
  – Continuous, smooth
  – Interrupted, rough
  Echo pattern
  – Strength (strong/weak)
  – Size (progressive/regressive)
  – Echogenicity (anechoic/echogenic)
  – Texture (homogeneous/heterogeneous)
  – Ultrasound shadow (present/absent)
  Plaque motion (uniform/discrepant)
  Plaque volume
  Embolic activity
	Additionally in veins:
	Flow modulation
	  Normal deep breathing
	  Valsalva maneuver
	Proximal/distal compression

Additionally in veins: **Compressibility**	

2 Extracranial Cerebral Arteries

Examination

Special Equipment and Documentation

Orbital Arteries and Neck Arteries

■ Continuous Wave and Pulsed Wave Doppler Sonography

Transmission frequencies of 4–5 MHz are used to examine the extracranial arteries supplying the brain, while frequencies between 8 and 10 MHz produce better results when investigating the orbital arteries (the indirect examination method). The use of pulsed wave Doppler sonography alone to examine the extracranial cerebral arteries has not gained widespread acceptance.

■ Conventional and Color Flow Duplex Sonography

For optimal image quality, transducers with transmission frequencies between 7.5 and 10 MHz should be used when examining the extracranial cerebral arteries and veins.

Examination Conditions

Patient and Examiner

The *patient* should sit relaxed on a comfortable seat or a tilted examination chair. It is best to place the patient in an examination chair with a backrest that can be lowered diagonally. The patient's head should be resting on an easily adjustable support, and should be firmly secured during compression maneuvers. Ultrasound examination can also be carried out with the patient lying in bed, but this has disadvantages for the examiner, since it does not provide adequate fixation of the patient's head (Neuerburg-Heusler 1986).

When an examination is being carried out using a duplex system, it is particularly important to be able to recline the patient's head adequately by providing additional support underneath the shoulders. The patient's *head position* must be adapted to the requirements of the region to be examined, and the position has to be corrected appropriately when any variations in the normal courses of the blood vessels are encountered.

The *examiner's* position is not fixed. In addition to the seated position behind the patient, seated or standing access to the patient from the side or from the front is also common. Steady movement of the probe and fixation of the patient's head during compression maneuvers should always be ensured.

Conducting the Examination

■ Continuous Wave and Pulsed Wave Doppler Sonography

The probe is guided by hand. The angle between the ultrasound beam and the blood vessel varies in a wide range and needs to be optimized by monitoring the audiosignal and the registered Doppler waveform or spectrum. Only the best signal should be recorded.

■ Conventional and Color Flow Duplex Sonography

When carrying out ultrasound imaging procedures, one should ensure that the blood vessels being examined are depicted in at least three planes (Fig. 2.1). A standardized examination sequence is recommended, e.g.:

– Transverse (cross-section)
– Longitudinal (longitudinal section)
– – Anterior
– – Posterolateral

These planes can only be obtained with consistency in the extracranial system (Fig. 2.2).

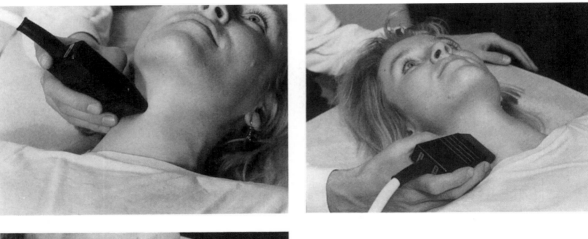

a

b

c

Fig. 2.**1**   Handling the transducer in duplex sonography of the neck arteries
a   Transverse (or cross-)sectional plane
b   Longitudinal sectional plane from anterior
c   Longitudinal sectional plane from posterolateral

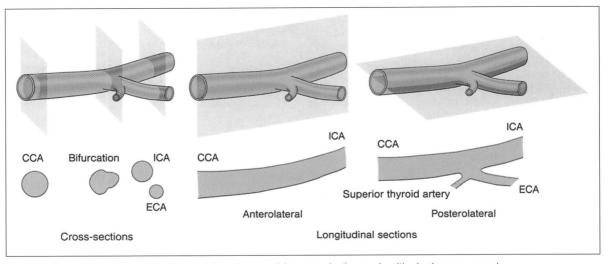

Fig. 2.**2**   Sectional planes used in examining the carotid system in the neck with duplex sonography

Examination Sequence

Using side-to-side comparisons, information should be gathered from the individual examination areas, following a set laboratory sequence. A logical sequence of examination steps is best, e.g., first recordings that yield only overview results (indirect Doppler sonography), or proceeding anatomically along the neck from caudal to cranial.

Orbital Arteries

■ Continuous Wave and Pulsed Wave Doppler Sonography

Examination of the orbital arteries is carried out with the Doppler probe and using the continuous wave Doppler is preferable. This is traditionally the oldest procedure used to detect obstructions in the carotid system.

The terminal branches of the ophthalmic artery are the supratrochlear artery and the supraorbital artery. Ultrasound evaluation of the supratrochlear artery takes place at the nasal canthus, while the supraorbital artery is assessed at the upper orbital margin.

The probe is placed on the nasal canthus, slightly above the eyeball, without applying any pressure, and using only a little contact gel. The examiner may choose to support the ball of his or her thumb on the patient's forehead, or to "rein in" the probe by using the cable to hold it gently in place (Figs. 2.**3**, 2.**4a**). Proceeding in a mediocranial direction, and only using slight variations in the angle of the probe, a search is made for the optimal acoustic signal, or the one with the highest amplitude on the oscilloscope. Often, a

slight change in position or a circular probe movement is necessary to "thread" the ultrasound beam through the thin blood vessels. A chart speed of 5 mm/s is sufficient for the data documentation, since spectrum analysis is not necessary.

Compression tests. Usually, the temporal artery or facial artery are simultaneously compressed ipsilaterally. Figure 2.4 shows the typical compression points.

A little practice is needed to master the technique, first palpating the pulsating arteries with the thumb and middle finger without applying any pressure, and then exerting firm pressure, moving *neither* the head *nor* the probe held in the other hand above the orbit. It may sometimes be necessary to apply pressure to the contralateral arteries, or with an additional examiner on both sides simultaneously.

■ Color Flow Duplex Sonography

Recently color flow duplex sonography has been suggested to be useful for visualization of the ophthalmic, central retinal, and posterior ciliary arteries. A variety of conditions affecting the orbital blood vessels have been studied, and embolism in particular was suggested to represent a challenge for future investigation (Hedges 1995).

Neck Arteries

Common Carotid Artery

■ Continuous Wave and Pulsed Wave Doppler Sonography

The examination of the extracranial carotid system usually starts with the proximal common carotid artery. The probe is placed flat above the clavicle, in a medio-

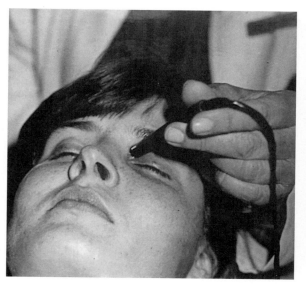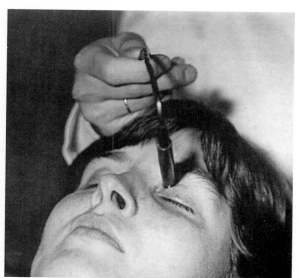

a   b

Fig. 2.**3**   Insonation of the orbital arteries: **a** by supporting the hand, and **b** "reining in" the probe

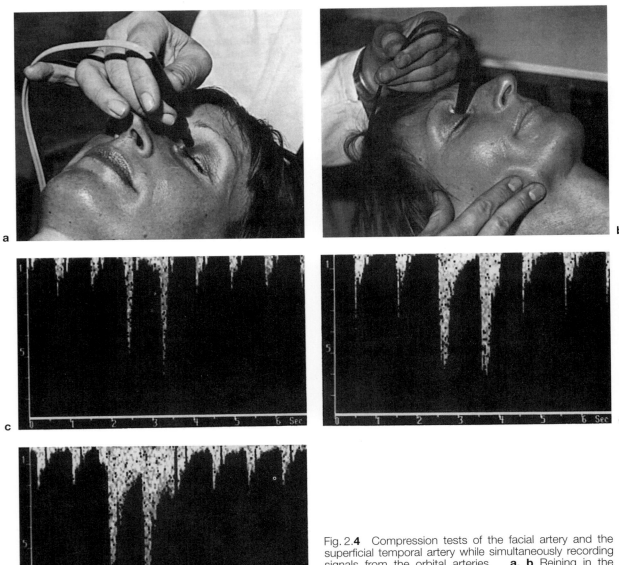

Fig. 2.**4** Compression tests of the facial artery and the superficial temporal artery while simultaneously recording signals from the orbital arteries. **a, b** Reining in the probe. **c–e** Doppler spectra of the supratrochlear artery while compressing the superficial temporal artery (**c**), the facial artery (**d**), and simultaneously compressing both blood vessels (**e**) for two heart circles respectively

cranial direction, lateral to the proximal part of the sternocleidomastoid muscle. Usually, the artery can be identified immediately at this location and can be differentiated from the branches of the external carotid artery and the branches of the thyrocervical trunk by the deep, slightly hollow sound it produces (Fig. 2.**5a**).

Very occasionally, the artery has to be located further cranially and medially to the sternocleidomastoid muscle. It can sometimes be extremely laterally displaced. If the signals received are asymmetric (different incident angles for applying the ultrasound on each side, due to differences in the course of the vessels), side-to-side comparison should be repeated more distally. To assess the proximal intrathoracic part of the blood vessel, the probe has to be turned up (flow toward the probe). An additional examination using pulsed wave Doppler (Fig. 11.**2**) or duplex sonography can provide clearer results (Fig. 11.**3**).

■ Conventional and Color Flow Duplex Sonography

To display a longitudinal section through the common carotid artery, the probe is placed above the clavicle, usually in a mediocranial direction, lateral to the proximal part of the sternocleidomastoid muscle (Fig. 2.**1b**). From there, a search is made for the bifurcation, which can be used as the topmost border. One can then examine the wall structure of the entire vascular segment that can be directly imaged above the clavicle. It is important to examine different longitudinal and cross-sectional images, in order to obtain a three-dimensional overview of the blood vessel's course that is as complete as possible (Fig. 2.2). Finally, the sample volume of the pulsed wave Doppler is placed within the lumen, and after the angle has been optimized, a Doppler spectrum should be documented. Particularly in unfavorable examination conditions (e.g., short, stocky neck, somnolent patient), color flow duplex sonography facilitates the procedure and easily displays the CCA in the neck up to the carotid bifurcation.

Internal and External Carotid Arteries

■ Continuous Wave and Pulsed Wave Doppler Sonography

Usually, it is possible to glide continuously with the probe from the common carotid artery in a cranial direction to the bifurcation, where a transition can be heard to the whipping, systolically weighted signal of the external carotid artery, or the soft low sound of the internal carotid artery. In a wide sinus, the sound of the internal carotid artery is not yet typical, but the velocity of the blood flow is usually noticeably slower than that of the common carotid artery. In addition, the sinus can be recognized as the center of a triangle formed by the common carotid artery (caudally), the internal carotid artery (laterally), and the external carotid artery (medially). In doubtful cases, it may be much simpler to begin by locating the submandibular component of the internal carotid artery and then proceed from there, following it caudally to the carotid sinus, which is clearly recognized by its "bubbling," low frequency sound and its low flow velocity (Widder 1985, Neuerburg-Heusler 1986).

The carotid bifurcation has to be located without fail, since obstructive lesions are most frequently found at the origin of the internal and external carotid arteries and at the bifurcation itself (Fig. 2.5b).

The positional relationship between the internal and external carotid arteries varies considerably. Most commonly (47%), the internal carotid artery lies directly dorsolateral to the external carotid artery. However, the internal and external carotid arteries are often superposed significantly at the bifurcation,

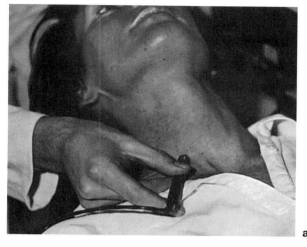

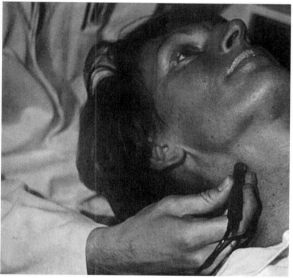

Fig. 2.**5** Doppler sonography **a** of the common carotid artery, **b** of the carotid bifurcation

so that a mixed signal is heard. If the internal carotid artery lies dorsally (38%), a more lateral approach is recommended. Only occasionally (6%), the internal carotid artery is found medial to the external carotid artery (Tismer and Böhlke 1986) (Fig. 2.**6**).

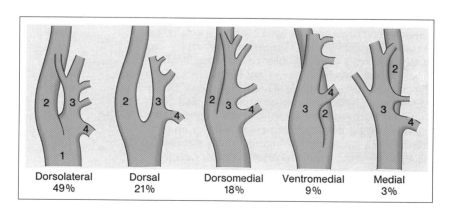

Fig. 2.**6** Variations in location of the internal carotid artery in relation to the external carotid artery (adapted from Faller)
1 Common carotid artery
2 Internal carotid artery
3 External carotid artery
4 Superior thyroid artery

| Dorsolateral 49% | Dorsal 21% | Dorsomedial 18% | Ventromedial 9% | Medial 3% |

The level of the carotid bifurcation is variable. When the location is atypical, it may help to look for the blood vessel at the same location on the contralateral side. However, the course of the arteries in the neck is not necessarily symmetrical bilaterally.

When the patient has a short, stocky neck, the examination becomes difficult, since the course of the vessel cranially can only be followed for a short distance. A complicating factor is often significant overlay by venous flow, which can be reduced by applying light pressure over a broad supraclavicular area (e.g., using a pencil held sideways), or by conducting the examination with the patient seated.

Identifying the branches of the external carotid artery can be facilitated by repeated manual compression or tapping of the superficial temporal artery, occipital artery, or facial artery. This produces retrograde pulse waves that are transmitted acoustically, and can be registered as small oscillations. Increased peripheral resistance may also be registered, as well as the hyperemic phase after releasing the compression of the blood vessel (von Reutern et al. 1976 a).

Identifying the individual branches of the external carotid artery only makes sense when there is any pathology. Using compression tests (von Reutern et al. 1976b), it is possible to identify:

– Superior thyroid artery: manual compression of the thyroid gland.
– Lingual artery: pressing the tongue against the palate.
– Facial artery: manual compression against the mandible.
– Ascendent pharyngeal artery: this is difficult to identify but acoustic modification of the Doppler signal may occur during swallowing.
– Occipital artery: manual compression at the mastoid process. It is important to differentiate the vertebral artery.
– Superficial temporal artery: preauricular manual compression.
– Maxillary artery: clenching the teeth is only occasionally effective, due to the superposition of the superficial temporal artery.

■ Conventional and Color Flow Duplex Sonography

The internal carotid artery is more difficult to display than the common carotid artery, and especially individual branches of the external carotid artery can hardly be imaged even if a color-coded image is made. Whereas both dividing arteries are only depicted in a single longitudinal plane in 40–60% of cases, this is usually straightforward in cross-section and with color. Separation of the internal carotid artery from the external carotid artery is only possible with conventional duplex sonography by incrementally identifying the Doppler signals from adjacent vascular segments. In a color flow duplex scan, a blood vessel branching off from the external carotid artery is visible occasionally (Fig. 2.**38**). Depending on the cir-

cumstances, assigning the arterial branches to the carotid bifurcation can be carried out either sequentially or simultaneously, depending on the anatomic variations. Again, one should attempt to obtain a good overview of the anatomic and flow relationships using longitudinal and cross-sections. Amplitude-mode studies may facilitate this issue even in difficult anatomical and pathological conditions (Steinke et al. 1996, Griewing et al. 1996). Individual compression tests in the external carotid artery are helpful for identification purposes in duplex sonography, and should be used when necessary, especially under pathological conditions. The characteristic flow signals described in the previous section, and anatomic structural classification using the B-mode, allow reliable identification of the entire extracranial carotid system.

▓ Vertebral Artery

■ Continuous Wave and Pulsed Wave Doppler Sonography

Examination of the atlas loop at the mastoid. With the patient's head turned to the side, the probe is placed below the mastoid process, directly lateral to the sternocleidomastoid muscle and in the direction of the contralateral orbit or ear (Fig. 2.**7a**). The sound characteristics of the signal are the same as in the internal carotid artery, but the volume is lower due to the smaller vascular caliber, the unfavorable angle for the investigation, and the greater distance from the transducer. An experienced investigator can differentiate the vertebral artery from the occipital artery purely acoustically, due to the differences in the sound qualities produced by blood vessels supplying the brain and those supplying skin and muscle. In addition, the occipital artery can be compressed by applying pressure broadly in the occipital region.

The vertebral artery can sometimes be mistaken for the internal carotid artery, especially in patients with thin necks or with an occlusion of the internal carotid artery. In contrast to the internal carotid artery, the vertebral artery cannot be continuously followed along its course, but can be detected only on a selective basis in the intervertebral spaces and in the proximal part of the neck.

Examination at the origin. Examining the vertebral artery at its origin is no easy task. It can be mistaken for branches of the thyrocervical trunk or of the common carotid artery. The direction of flow is cranial.

The soft sound of the artery, resembling the sound of the internal carotid artery, can be recognized acoustically. In addition, identification of the proper blood vessel should be ensured by repeatedly tapping the vertebral artery where it loops round the atlas (2.**7b**). Locating the artery at its origin is easiest when one glides in a cranial direction along the subclavian artery, holding the probe in a cranial-to-caudal direction.

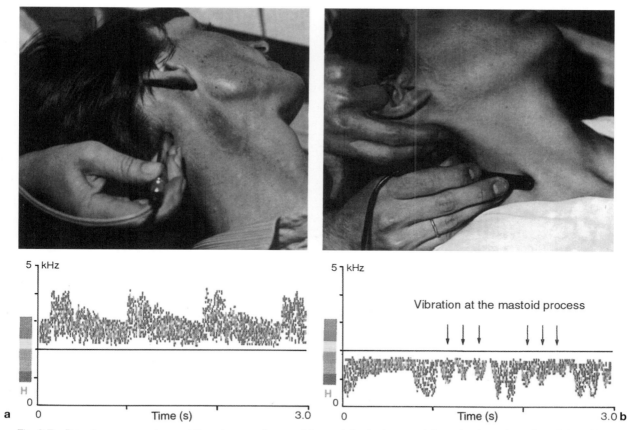

Fig. 2.**7** Doppler sonography and Doppler waveforms of the vertebral artery, **a** at the atlas loop, **b** at the origin, with a compression maneuver at the mastoid process

■ Conventional and Color Flow Duplex Sonography

Using duplex sonography in the same position as in Doppler sonography, the vertebral artery can be recorded at its origin and at the atlas loop (Fig. 2.**8**). In addition, its course in the neck and between the transverse processes of the cervical vertebral bodies can be examined particularly well using color flow duplex sonography (Fig. 2.**18**). By tipping the transducer slightly from anteromedial to lateral, the main bodies of the cervical vertebrae can be displayed in longitudinal section with the corresponding ultrasound shadow. The vertebral artery is seen in the intervertebral spaces. Depending on the specific anatomic configuration, the proximal segment can be depicted all the way to the origin. At the origin, as in the intertransversal segment, Doppler spectra are documented after positioning the sample volume (Ackerstaff et al. 1984, Trattnig et al. 1990, Touboul et al. 1987, Bartels et al. 1993, Ries et al. 1996).

Compression tests. To determine the direction of flow within the vertebral artery loop round the atlas, upper arm compression and/or clenching of the patient's fist can be used. This is important to determine indirectly whether or not the vertebral artery drains into the sub-clavian artery (Fig. 2.**9**), a characteristic finding in the *subclavian steal phenomenon*. The procedure is as follows:

1. Place the blood pressure cuff around the upper arm.
2. Locate the vertebral artery at the atlas loop, or at its origin from the subclavian artery, and register the flow waveform (the characteristic sound made by the vertebral artery may be different when its function has been altered to supplying the extremities instead of the brain, i.e., when it has a low diastolic flow velocity).
3. Interrupt the blood flow to the arm by exerting suprasystolic pressure by means of a cuff (if necessary, with help from an assistant) or, alternatively, the patient can make a tight clenched fist, which reduces the diastolic flow.
4. After the blood pressure cuff is released, register the hyperemic reaction, with an increase in the diastolic blood flow component in the vertebral artery. This phenomenon produces a marked acoustic impression lasting for several cardiac cycles.

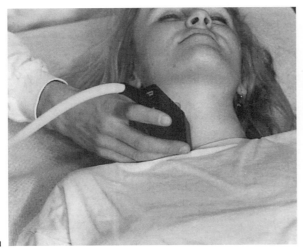

a
b

Fig. 2.**8**   Handling the transducer in duplex sonography of the vertebral arteries. Insonation **a** in the neck section and **b** at the atlas loop

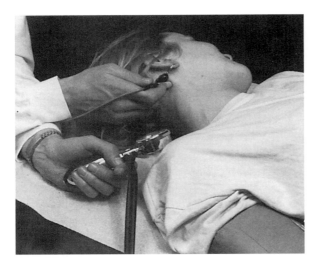

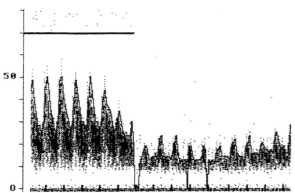

Fig. 2.**9**   Testing for the subclavian steal phenomenon: suprasystolic pressure of the upper arm through the use of a blood-pressure cuff causes an increase of flow velocity signals (–) from the vertebral artery at the mastoid process. After release, a sudden short retrograde flow signal is followed by gradual restoration of the pretest situation

Subclavian Artery

■ Continuous Wave and Pulsed Wave Doppler Sonography

A search is made for the subclavian artery in the supra-clavicular fossa. Flow direction may be toward the probe (proximal segment), or away from the probe (distal segment).

Usually, the triphasic forward and backward components of the pulse can be displayed (see Fig. 11.**12**, p. 360). The analogue curve resembles that of an artery supplying the extremities, with high peripheral resistance: systolic forward flow, early diastolic back-flow, and late diastolic forward flow. In doubtful cases, one can identify the artery by interrupting the blood flow in the upper arm, or using repeated compressions proximal to the elbow. Using pulsed wave Doppler equipment and a low transmission frequency, the evaluation of the subclavian artery is often easier and more extensive.

■ Conventional and Color Flow Duplex Sonography

The position used to examine the subclavian artery is the same in both Doppler sonography and duplex sonography. It is not possible to examine the vessel in cross-sections or in longitudinal sections rather than in diagonal sections due to the limited overview because of anatomic relationships and the large size of the transducer (Fig. 2.**10**).

Innominate Artery
(Brachiocephalic Trunk)

■ Continuous Wave and Pulsed Wave Doppler Sonography

The innominate artery can also be located, with the probe being applied in a mediocaudal direction within the right supraclavicular fossa. Since the penetration depth of ordinary continuous wave Doppler systems is not sufficient, pulsed wave Doppler equipment can also be used. The Doppler signal in this large, wide blood vessel that branches off into the common carotid artery (supplying the brain) and the subclavian artery (supplying the extremities) often sounds rather hollow and turbulent.

■ Conventional and Color Flow Duplex Sonography

In addition to the location previously described, one can also attempt to depict the innominate artery in longitudinal section directly above the jugular fossa, using color flow Doppler sonography to record the flow signal (2.**10**).

Thoracic Aorta

■ Color Flow Duplex Sonography and Transthoracic or Transesophageal Echo Cardiography

The aorta cannot be examined using continuous wave Doppler sonography. The duplex procedure is usually used for this vascular segment, especially for the

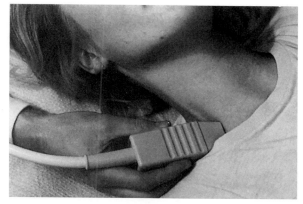
Fig. 2.**10** Handling the transducer in duplex sonography of the subclavian artery in the supraclavicular fossa

branching, proximal arteries supplying the brain, either transthoracically in echocardiographic sonography or—more recently—transesophageally, to obtain better access (Seward et al. 1990, Rauh et al. 1996) (Fig. 2.**47**).

Normal Findings

Orbital Arteries

Principle

The direction of flow in the fronto-orbital terminal branches of the ophthalmic artery is determined using bidirectional continuous wave or pulsed wave Doppler equipment. This so-called *indirect method* is used to obtain information about the patency of the internal carotid artery by analyzing the flow relationships within the *circulatory system of the ophthalmic artery* (Fig. 2.**11**).

Physiologically, the pressure in the two branches of the anastomosis of the ophthalmic artery—the internal carotid artery and the external carotid artery—is the same. Both branches form this important collateral, via the supratrochlear artery and the supraorbital artery, with the watershed above the orbits. If the

Doppler probe is placed on the medial canthus, it is usually possible to record blood flow signals from the carotid siphon directed toward the probe (Fig. 2.**4**).

Even in a physiological case, it is sometimes found that the watershed has been displaced to a location within the orbital cavities, so that it is not possible to document a definite signal, or only a very weak one when compared to the contralateral side (Fig. 2.**22**).

Anatomy and Findings

Anatomy

The ophthalmic artery originates from the frontal convexity of the carotid siphon (C_2), and then proceeds through the dura and the optic canal, entering the orbit. It follows a course around the optic nerve until it divides into different and very variable branches (Hayreh and Dass 1962 a, b):

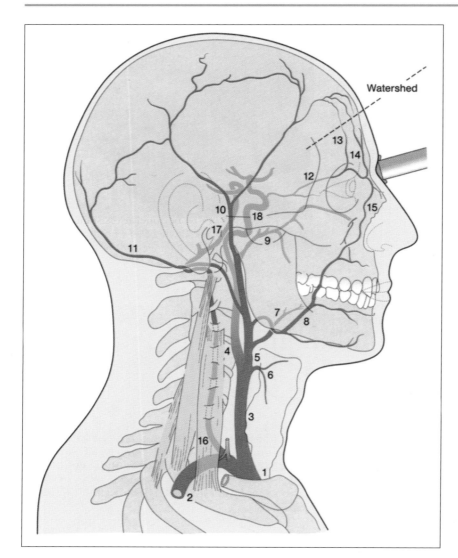

Fig. 2.**11** Anatomy of the large arteries supplying the brain
1 Innominate artery
2 Subclavian artery
3 Common carotid artery
4 Internal carotid artery
5 External carotid artery
6 Superior thyroid artery
7 Lingual artery
8 Facial artery
9 Maxillary artery
10 Superficial temporal artery
11 Occipital artery
12 Ophthalmic artery
13 Supraorbital artery
14 Supratrochlear artery
15 Angular artery
16 Vertebral artery
17 Basilar artery
18 Carotid siphon

– Laterally, through the lacrimal artery to an anastomosis with the middle meningeal artery
– Medially, through the anterior and posterior ethmoidal arteries to an anastomosis with the maxillary artery
– Rostrally, via the supratrochlear and supraorbital arteries to an anastomosis with the superficial temporal artery

The lumen of the ophthalmic artery is approximately 1 mm.

Findings

It is usually sufficient to record the supratrochlear artery. When the results are unclear, or the Doppler signal is poor, it is recommended to obtain additional data from the supraorbital artery. The supraorbital artery is often hypoplastic, but rarely (5%), when the supratrochlear artery is hypoplastic, it can take on a role as the terminal vascular system. The blood flow is directed toward the probe, and is usually registered as a negative deflection in relation to the baseline. Although physiological signals are usually recorded as positive deflections, this convention has become common, since it avoids the necessity to change the polarity during a given series of measurements. The diastolic component of the blood flow is relatively low, since the artery supplies both the skin and musculature. During *compression tests* of the branches of the external carotid artery (the superficial temporal and facial artery, or both), the velocity of the flow normally increases, since the counterpressure in the external carotid artery is reduced. This confirms *the physiological* (orthograde) *flow direction*.

Evaluation

Only the direction of flow is relevant; formally analyzing the Doppler spectrum is of secondary importance. The diastolic flow component, in particular, has different characteristics in each individual, and depends on the peripheral resistance in the skin and musculature.

The significance of any differences in the flow velocity spectrum between the two sides of the body should not be exaggerated. Such differences depend on various factors (e.g., the position of the probe in relation to the blood vessel, variability in the vascular caliber, pressure relationships in the circulatory system of the external and internal carotid arteries), and they cannot be properly evaluated as isolated phenomena. Compression tests are therefore indispensable in clearly determining the direction of flow in the ophthalmic artery.

Sources of Error

When examining a vascular loop (similar to evaluating the vertebral artery at the atlas loop), it is possible to misinterpret the blood flow direction if compression tests are not performed. They are necessary here to determine whether or not the blood is in fact flowing through the ophthalmic artery into the cranium—i.e., whether the blood flow originates in the branches of the external carotid artery. If this is the case, the blood flow should either decrease, cease, or change its direction when branches of the external carotid artery are compressed (i.e., retrograde perfusion). If the flow increases under compression, this is evidence of physiological or orthograde flow. If compression has no effect on the blood flow, no certain statement concerning the flow direction can be made. The latter is almost always the case when an anastomosis has formed from the ophthalmic artery through the maxillary artery.

Neck Arteries

Principle

■ Continuous Wave and Pulsed Wave Doppler Sonography

All flow velocity spectra from the neck arteries have to be differentiated acoustically and formally. Their shape is mainly affected by the varying peripheral resistances in the vascular system (Fig. 2.**12**).

- *Arteries supplying a parenchymatous organ such as the brain* have a low peripheral resistance: the internal carotid artery, vertebral artery, superior thyroid artery, and (with reservations) the common carotid artery (to the extent that it supplies the internal carotid artery).
- *Arteries supplying muscle and skin* have a high peripheral resistance: the branches of the external carotid artery (facial artery, maxillary artery, superficial temporal artery), the subclavian artery, and (with reservations) the common carotid artery (to the extent that it supplies the external carotid artery).

Spectral analysis is useful in analyzing continuous wave as well as pulsed wave Doppler signals and provides semiquantitative numerical documentation of various parameters developed to describe the Doppler signal (Spencer and Reid 1979, Rittgers et al. 1983). Determining the peak frequency is the one most often used (Table 2.**1**). In addition, when several blood vessels are superimposed on one another, as

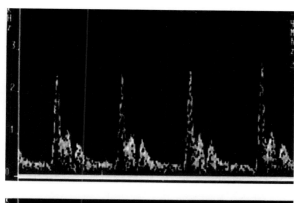

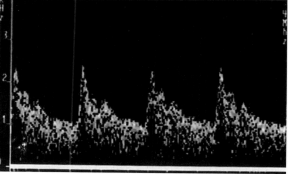

Fig. 2.**12** Examples of the effect of variations of peripheral resistances within the vascular territory on the form of the proximal Doppler spectrum
a External carotid artery waveforms reflect high peripheral resistance
b Internal carotid artery waveforms reflect low peripheral resistance

Peripheral resistance

High

Low

Table 2.**1**   The different parameters used for spectrum analysis in normal patients (transmission frequency 4 MHz), systolic/diastolic peak frequency (kHz), peak velocity (cm/s)

Author		CCA	ICA	ECA
Neuerburg et al. (1985)	(kHz)	1.95/0.43	2.08/0.78	2.58/0.42
Zbornikova and Lassvik (1986)	(cm/s)	75/20	85/40	80/20
Dauzat et al. (1991)	(cm/s)	67/17	50/13	–

happens particularly often at the carotid bifurcation and at the origin of the vertebral artery (Fig. 2.**13**), the different vascular segments can be evaluated separately from the spectrum.

■ Conventional and Color Flow Duplex Sonography

Initially, Doppler sonography made it possible to assess the flow relationships in the neck arteries functionally and hemodynamically (Planiol et al. 1972, Büdingen et al. 1976). Since the 1980s, blood vessels that are close to the surface have also been imaged using echotomograms (Baker et al. 1978, Hennerici 1983, Comerota et al. 1984, Marosi and Ehringer 1984).

Using the B-mode provides morphological information about the location and texture of the blood vessel and its contents (Fig. 2.**14**). When this method is combined with the Doppler mode in duplex sonography, it is possible to distinguish the various arterial segments, to display and assess the extent of stenoses. However, several conditions such as fresh thromboses that have the same acoustic impedance as blood could hardly be detected and evaluated until color flow duplex sonography was introduced (Middleton et al. 1988 a, b, Steinke et al. 1990 c) (Fig. 2.**15**).

Computer-assisted image processing can now provide a three-dimensional display of the different sectional planes. This facilitates the evaluation and quantification of findings encountered during the examination, particularly when there are unusual anatomical relationships and pathological conditions (Figs. 2.**14 d**, 2.**15 d**) (Meairs et al. 1995, Steinke et al. 1989, Guo et al. 1996).

Anatomy and Findings

Carotid System

▨ Anatomy

The extracranial *common carotid artery* originates from the aortic arch or the innominate artery, and divides into the internal carotid artery and the external carotid artery at different levels, usually at vertebrae C4/C5 (48%) or C3/C4 (34%), and only very rarely at a higher level, at C1, or significantly lower, at C2 (Krayenbühl et al. 1979). The structure of the *carotid bifurcation* is unique in the human vascular system. While the tunica media of the common carotid artery consists of elastic fibers in different layers and muscle cells, it is the muscle cells that predominate in the internal carotid artery. At the *carotid bifurcation,* the transition zone, which is not present in 6% of individuals, the hybrid structure of the media decreases, and the elastic fiber structure predominates. Four different sections of the *internal carotid artery* can be distinguished: cervical, petrous, cavernous, and cerebral. After entry, the internal course of the carotid artery is initially vertical for more than 1 cm, then horizontal until the apex of the petrous bone is reached. Since the artery is reduced to at least half of its diameter (3.0–3.5 mm) in the bony canal here, its pulsatile wall movements alter over a length of 25–35 mm. Stereomicroscopic investigations have shown that there is a close relationship between the carotid artery and the transbasal veins. The artery in the proximal section of the canal, mainly fixated by longitudinal fibers, cannot move in an extracranial direction, and it is only supplied by a few accompanying veins. Further distally, there is an extensive venous drainage system into the cavernous sinus, which shows clear changes in the volume and wall movements that are pulse-dependent. At the apex of the petrous bone, the internal carotid artery gives off branches to the middle ear (caroticotympanic artery) and to the pterygoid canal. When there is an occlusion at the external carotid bifurcation, these branches often keep the in-

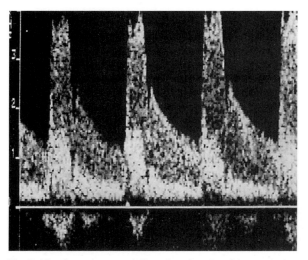

Fig. 2.**13**   Superimposed Doppler signals of two arteries can be differentiated from the spectrum

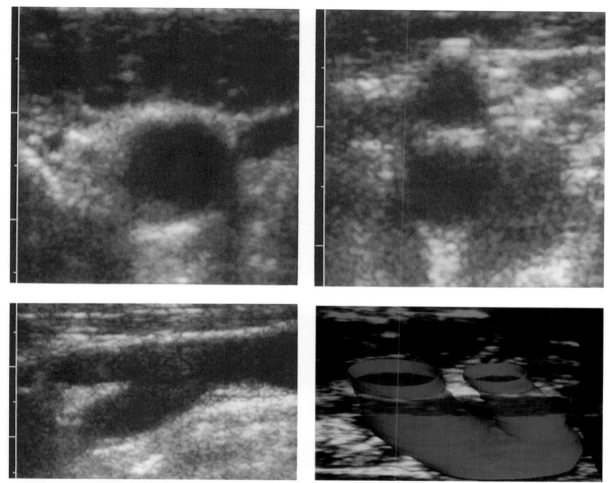

Fig. 2.**14**   Image from the carotid system using B-mode sonography **a** Cross-section of the common carotid artery **b** Longitudinal section of the common, internal and external carotid arteries **c** Cross-section above the bifurcation **d** Three-dimensional reconstruction of a carotid bifurcation, the internal and external carotid arteries

tracranial segment open to this point (Paullus et al. 1977). The C_3–C_5 segments are between 30 mm and 50 mm long, and are very variable in their diameters. Subsequently, the diameter narrows conically to 6 mm (C_5/C_6 segments) and then to 4 mm (C_1/C_2 segments) (Gabrielsen and Kreitz 1970). From proximal to distal, the most important branches of the internal carotid artery are (Hennerici et al. 1988):

– The ophthalmic artery
– The posterior communicating artery
– The anterior choroidal artery

The *external carotid artery* supplies the viscerocranium and, above the carotid bifurcation, divides into several branches (Salamon et al. 1968, Merland 1973). Its anastomoses with the branches of the internal carotid artery are very important under pathological conditions, and play a vital role in ultrasound diagnosis (on the anatomy of the ophthalmic artery, see pp. 33–34 above). In addition, there are many anasto-

moses between the branches of the external carotid arteries on both sides and also, under pathological conditions, between the deep neck arteries and the vertebral arteries. The latter are significant as collaterals, especially when there are proximal vascular occlusions of the common carotid artery and the subclavian artery. The main external branches of the carotid artery are the superficial temporal artery and the maxillary artery. Anastomoses formed with the middle meningeal artery and the fronto-orbital terminal branches of the ophthalmic artery are very important. Carotidobasilar anastomoses, which are persistent developmental disturbances of the fetal brain circulation, are rarely encountered. The best known of these is the primitive trigeminal artery.

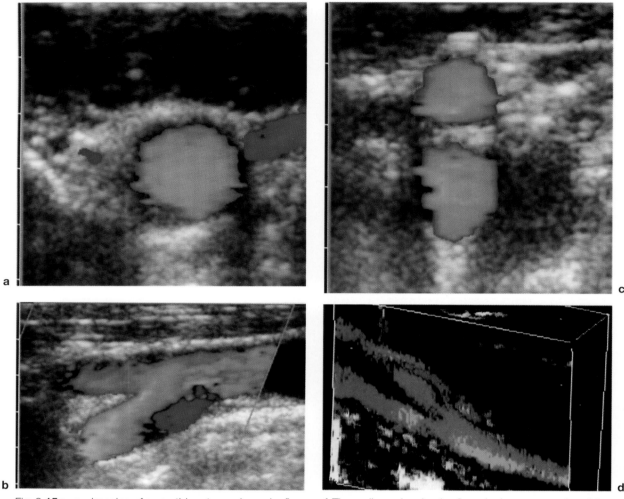

Fig. 2.**15 a–c**   Imaging of a carotid system using color flow duplex sonography (image sequence as in Fig. 2.**14**).

d Three-dimensional color flow duplex sonography of the extracranial carotid system (from Picot et al. 1993)

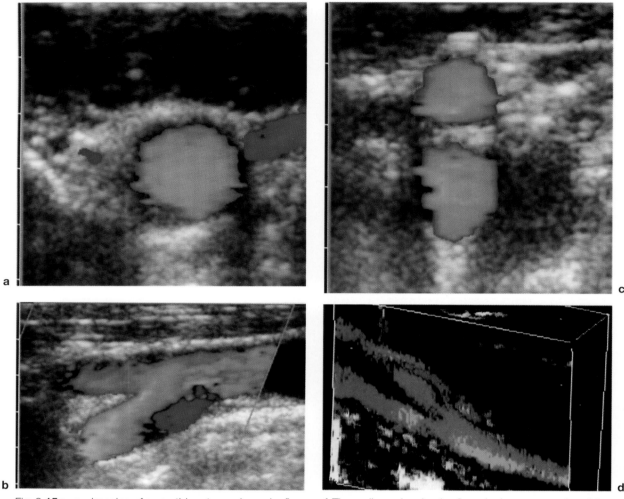

Common Carotid Artery

■ Continuous Wave and Pulsed Wave Doppler Sonography

The Doppler signal of the common carotid artery is similar both to that of the internal carotid artery, as a brain-supplying blood vessel, and to that of the external carotid artery, in its role as a vessel supplying muscle and skin. Compared to the internal carotid artery, the amplitude of its systolic signal is greater, and the diastolic flow velocity is smaller. In contrast to the external carotid artery, the common carotid artery has a smaller signal amplitude and a similar diastolic flow velocity. Especially during systole, the absolute values of the Doppler signals during a cardiac cycle and their amplitudes are strongly influenced by various factors. In addition to the incident angle of ultrasound application, the distance between the vascular segment being examined and the transducer (thick

neck, winding course), as well as age, are important factors. Younger people usually have a larger systolic signal amplitude, while the amplitude is usually lower in older patients, due to increasing wall rigidity. Occasionally, a wide, dilated blood vessel can show a high flow amplitude in systole if the blood pressure and the cardiac output are correspondingly high. The flow signal of the common carotid artery is acoustically associated with a rough sound (Reneman et al. 1985, Zbornikova and Lassvik 1986).

Signal frequency analysis is not as important here as it is with the internal and external carotid arteries. Stenoses usually lie at the origin, which is difficult to examine with CW Doppler ultrasound but may be possible with PW Doppler studies. They occur much less frequently than they do at the bifurcation.

The systolic peak frequency is lower than that of the external carotid artery, and is the same as or higher than the value obtained for the internal carotid artery (Table 2.**1**).

Conventional and Color Flow Duplex Sonography

Due to its course, which is extensive and close to the surface, it is very easy to image the common carotid artery. The wall structure has the characteristic double reflection, with a hypoechoic zone lying in between. This is termed the *intima–media thickness (IMT)*, and the measurement of this parameter is used in multicentric trials and in many standardized protocols as an indicator and prognostic parameter of atherosclerosis (progression vs. regression). Changes amounting to about 100 μm can be resolved with certainty (Hennerici and Steinke 1994). Normal values lie definitely below 1.5 mm (Riley et al. 1992) and, according to large series, range at 0.96 ± 0.19 mm in women and 1.04 ± 0.22 mm in men with risk factors of atherosclerosis. If there is associated coronary heart disease (CHD) or clinically manifest atherosclerosis, these values increase significantly (1.00 ± 0.22 mm and 1.10 ± 0.26 mm) (O'Leary et al. 1996). The anterior wall is usually more difficult to assess than the posterior wall, due to image distortion in the near field; to prevent image distortion, it helps to have a vein interposed between the artery and the transducer. Normally, the width of the blood vessel is between 6.3 mm and 7.0 mm (Marosi and Ehringer 1984).

Different methods of making quantitative measurements of the flow volume on the basis of duplex sonography have produced very contradictory results. Without exactly reconstructing the ultrasound planes from several pulsed wave Doppler components, it is not possible to correctly distinguish the laminar flow that takes place within the common carotid artery and the plug distortions near the arterial wall. Measurements using quantitative flow volumetry therefore have an error rate of between 10% and 40% (Müller et al. 1985, Eicke et al. 1994).

Using *color flow duplex sonography*, it is possible to improve the assessment of turbulence phenomena, flow separation zones and movement disturbances at the vascular wall, and variations in the hemodynamics, throughout two-dimensional sections of the blood vessel during each cardiac cycle, which is of diagnostic significance in patients with a suspected artery-to-artery embolism (Meairs et al. 1995, Hennerici et al. 1995).

Internal Carotid Artery

Continuous Wave and Pulsed Wave Doppler Sonography

Usually, the artery can be identified in its typical location by a soft, high-frequency sound and corresponding spectrum characteristics: the systolic signal amplitude is less steep, but the diastolic flow is significantly greater than that of the common carotid artery.

The transition from the common carotid artery to the internal carotid artery is almost always marked by a clear decrease in the signal amplitude and in the re-maining diastolic flow in particular. This may be due to a widening of the blood vessel within the carotid sinus, or may be caused by a change in the angle between the transducer and the blood vessel. Continuous wave Doppler sonography does not provide any clues about the complex hemodynamics of this region.

Depending on the specific equipment and the transducer used, the upper normal value for the systolic peak frequency ranges between 3000 and 4000 Hz at a transmission frequency of 4 MHz and < 120 cm/s flow velocity respectively. Near the carotid sinus, the peak frequency usually decreases markedly, the systolic window closes, and occasionally—depending on the incident angle at which the ultrasound is applied—a Doppler signal directed against the transducer is detected. These phenomena reflect the complex flow relationships within the carotid sinus. In case of hereditary absence of the carotid sinus (approximately 6–8%), these phenomena may be missed, however, this should not be interpreted as early arteriosclerotic plaque formation (Harward et al. 1986, Reneman et al. 1985, Steinke et al. 1990b, Hennerici and Steinke 1991, Steinke et al. 1992).

Conventional and Color Flow Duplex Sonography

It is usually possible to image the internal carotid artery at its origin, since the widened sinus is clearly recognizable. According to Marosi and Ehringer (1984), the width of the blood vessel at the sinus is 6.5–7.5 mm, and distal from this location it is 4.3–5.3 mm. IMT values range at 1.35 ± 0.64 mm and 1.57 ± 0.67 mm for women and men with the risk factor of atherosclerosis, and at 1.56 ± 0.70 and 1.87 ± 0.73 mm respectively in women and men with CHD and clinically manifest atherosclerosis, which is significantly associated with abnormal IMT values (O'Leary et al. 1996).

Depending on the size of the transducer used and the level at which the bifurcation is located, the artery can only be reliably depicted over a short distance in a more cranial direction. It should be noted that several longitudinal and cross-sectional planes are necessary and differentiation from the external carotid artery is made possible by both B-mode and Doppler mode studies (Fig. 2.**14**).

Dorsolateral access usually provides an adequate depiction of the course of the internal carotid artery. Although it is usually possible to depict its origin at the common carotid artery in a single longitudinal section, simultaneous display with the external carotid artery is only occasionally possible, when both dividing arteries are on the same plane.

Depiction of the complex flow changes within the carotid sinus and in the proximal segment of the internal carotid artery is considerably better using color flow duplex sonography (Schmid-Schönbein and Perktold 1995). Flow separation zones, variations in circulation, and cardiac cycle changes can be imaged in relation to the structural findings in at least two

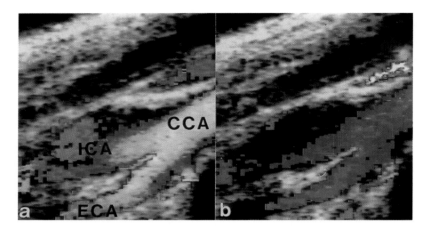

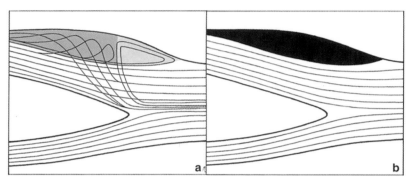

Fig. 2.**16** Schematic drawings and color flow duplex sonograms of normal hemodynamic phenomena in the carotid bifurcation during **a** systole and **b** diastole

planes (Fig. 2.**16**). It is only when the carotid sinus is absent that no flow separation zones are seen. Flow separation zones can usually be detected at the proximal origin of the internal carotid artery, and in almost one in two cases they are also found at the origin of the external carotid artery and, during systole, they can be seen as a zone of clearly altered flow direction (against the axial or wall-directed flow) (Steinke et al. 1990b). A color signal is also sometimes missing if there is a clear decrease in the flow velocity. Experimental results show that changes in flow are usually not found directly at the location that causes the flow separation, but within the carotid sinus at the origin of the external carotid artery, and that they are distributed in a horseshoe shape around the flow separation point (Zierler et al. 1987, Steinke et al. 1990b, Karino and Goldsmith 1985). The extent of the distribution of the flow separation is highly variable (Middleton et al. 1988b, Steinke et al. 1990b).

External Carotid Artery

■ Continuous Wave and Pulsed Wave Doppler Sonography

At its origin in the carotid bifurcation, the external carotid artery can be conclusively identified due to its characteristic whiplike sound, which is accompanied by a characteristic flow velocity curve. Since they supply the muscle and skin, most of its branches show a high systolic flow, with an almost complete absence of residual flow in diastole due to the high peripheral resistance in their vascular territories. Diastolic backflow, found in arteries of the extremities, is not observed. The superior thyroid artery, which proceeds in a caudal direction shortly after its origin, is an exception to these rules. As a vessel supplying the parenchyma, the superior thyroid artery has a flow signal that is similar to that of the internal carotid artery, due to the low vascular resistance within the thyroid gland. Thus, even in a completely normal situation, and particularly when there are loop formations, confusion with the internal carotid artery is likely. Using compression tests can clarify the situation in doubtful cases (p. 30).

Depending on the equipment and the transducer used, the systolic peak frequency lies at a maximum of 2.0–2.5 kHz at a 4 MHz transmission frequency (< 80 cm/s peak velocity) (Table 2.**1**). It is only within the carotid sinus that there is occasionally a reduction in the systolic window due to flow separation phenomena. The absence of such a reduction within the carotid sinus should not be interpreted as pathological, however.

■ Conventional and Color Flow Duplex Sonography

In B-mode, there is usually no difference between the origin of the external carotid artery and that of the internal carotid artery, although it is occasionally possible to use the origin of the superior thyroid artery to assist in the identification. Registration of the typical flow signals in the Doppler mode is more reliable. Since the orientation in relation to the internal carotid artery, which is usually found medially, varies, simultaneous display including the internal carotid artery is not always successful (Fig. 2.**6**).

In *color flow duplex sonography,* both branches of the common carotid artery can often be identified immediately by means of the color-coded flow signals within the side branches of the external carotid artery (Fig. 2.**16**). Further differentiation may be assisted by evaluating the flow dynamics in the two arteries. While the external carotid artery is usually displayed very briefly as a saturated color signal during the systole, the internal carotid artery has more constant flow characteristics during both systole and diastole. This reflects the differing vascular peripheral resistances in the two distribution areas. Application of energy-amplitude modes facilitates this separation further (Steinke et al. 1996).

In contrast to what might be expected from experimental results, significant flow separation and even flow reversal zones are found at the origin of the external carotid artery too. This is due to the configuration of the carotid sinus and the variable branching of the carotid bifurcation (Steinke et al. 1990 b) (p. 29).

Vertebral-Subclavian System

Anatomy

The vertebral artery forms the first branch of the subclavian artery. It rarely originates directly from the aortic arch. Its course proceeds in a cranial direction like a curve (V_1, prevertebral part). Near vertebra C5 or C6, it enters the costotransverse foramen and follows a perpendicular path toward the cranium (V_2, transverse part). After the C2 vertebra, it proceeds laterally and winds behind the lateral mass of the atlas (V_3, atlantic part). At this point, muscular branches originate, and form anastomoses with the branches of the external carotid artery. Continuing from the atlanto-occipital joint, it moves further cranially and proceeds into its subarachnoid part (V_4).

The anatomy of the subclavian artery and innominate artery is described on p. 156. Due to the bilateral origin of the vertebral artery from the subclavian artery, the proximal segment of the subclavian artery is also counted as an artery that supplies the brain.

Vertebral Artery

■ Continuous Wave and Pulsed Wave Doppler Sonography

The characteristics and formal analysis of the vertebral artery signal are similar to those of the internal carotid artery (i.e., an artery supplying the brain). With differing flow directions, depending on the position of the transducer, it is usually possible to detect both the afferent and efferent branches near the mas-

toid process (Fig. 2.**17 a, b**). Acoustically, the flow signal shows little frequency modulation, and is therefore clearly distinguishable from the occipital artery, which at this point is little more than a crossing terminal branch of the external carotid artery, or forms anastomoses with cervical collateral arteries (Figs. 2.**17 c, d**). Compressing the occipital artery proximally at the mastoid process helps avoid erroneous identification; in the case of the vertebral artery, the flow signal does not alter under these circumstances. It can sometimes be difficult to document the Doppler signal of the vertebral artery, particularly if spectral analysis is not being used, when the neck is thick, the blood vessel is quite deep-lying, or in the presence of variations in the caliber of the artery (hypoplasia). Rarely, particularly when the course of the carotid artery lies dorsal in patients with a very slender neck, the internal carotid artery may be mistakenly identified as the vertebral artery. In such situations, changing the displayed flow direction by slightly tilting the probe is usually a helpful diagnostic criterion. However, this still does not allow one to determine the flow direction in the vertebral artery, since there is no certainty as to whether it is the afferent or efferent branch of the vascular loop around the atlas that is being examined. In pathological cases, a compression maneuver involving the upper arm is therefore necessary. Bilateral differences in the flow velocity occur quite often. To differentiate between hypoplasia and aplasia or a stenotic vertebral artery, the blood vessel has to be examined with duplex sonography at different locations along its course.

In the neck region, the vertebral artery can only be examined discontinuously above C6, since the flow signal is obscured by the vertebral bodies. Until it enters the intervertebral foramen, it is usually possible to examine it with a posterolaterally positioned probe, especially when color flow duplex sonography is used. Doppler sonography often confuses it with collateral arteries from the thyrocervical trunk, from the subclavian artery, and directly from the aortic arch. In pathological cases such an evaluation is even more difficult because these collaterals often serve as anastomoses compensating for occlusion of the vertebral artery. Again, compression tests are helpful in differentiating these.

At its origin, the artery can be examined for several centimeters proximally (Fig. 2.**17 e**). The proximity of the vertebral artery to the common carotid artery, the thyrocervical trunk, the proximal subclavian artery, and numerous veins, makes it difficult to identify, especially in normal conditions (Fig. 2.**17 f**). Stenoses are usually easy to detect, due to their distinctive bruit. Proceeding from the origin of the vertebral artery at the subclavian artery, with the probe directed ventromedially and caudally, it is usually possible to record signals from the proximal segment (V_1). The flow signal is directed toward the probe, and the flow sound is identical to that obtained at the mastoid process (Fig. 2.**17 b, e**). With repeated tapping at the mastoid process, it is also possible to further identify the proximal section as a segment of

At a low ultrasound transmission frequency (2 MHz), flow velocities around 54.6 ± 16.9 cm/s are physiological when a supraclavicular access point for ultrasound application is chosen (Rautenberg and Hennerici 1988).

■ Conventional and Color Flow Duplex Sonography

It is possible to image the innominate artery using the supraclavicular or the suprasternal access points. The display of structural and flow-dynamic findings is better at low transmission frequencies. This vascular segment, along with the ascending aorta, can usually be displayed well using transesophageal access. The interpretation of the results is also improved when three-dimensional sectional planes are used (Rauh et al. 1996).

Evaluation

Optimal interpretation of the results of the examination is only possible when one observes the procedure oneself, taking into account the immediate acoustic and visual information and analyzing the documented findings carefully. Any interpretation that is made without knowledge of the specific circumstances associated with the examination of a patient can lead to significant misinterpretations. Graphic documentation provides a fair indication of the care exercised by the investigator and the extent of the examination, however, it cannot replace the subtle and direct experience of the individual investigator.

■ Continuous Wave and Pulsed Wave Doppler Sonography

The *analogue graphs* are usually only evaluated qualitatively, comparing the two sides. The end-diastolic flow within the common carotid artery is particularly important in bilateral comparisons. With optimal ultrasound application, variations of more than 15% in the values obtained are abnormal. Due to the slightly different courses of both common carotid arteries, it may therefore be necessary to repeat the side-to-side comparison at two or three levels, in order to avoid misinterpretations. By contrast, bilateral comparison of the measurements is less informative in the internal carotid artery, due to the more variable vascular course and the less standardized probe position in relation to the arterial axis. Asymmetries can be quite normal here. However, hypoplasia is extremely rare, in contrast to the vertebral system, where flow asymmetries are common (25–30%). In this context, it should also be noted that the flow velocity in a hypoplastically narrow blood vessel can be higher than that in a comparably normal or even pathologically narrow contralateral vertebral artery, while the flow velocity is more likely to be lower in a dilated vascular segment (Ries et al. 1996). It should also be noted that a hypoplastic vertebral artery often has a limited vascular territory, e.g., supplying the posterior inferior cerebellar artery instead of the basilar artery. In such cases, the flow is low even if the caliber of the blood vessel is normal. Better interpretation of the often difficult and inadequately classifiable hemodynamic findings can only be achieved using color flow duplex sonography (Trattnig et al. 1990/1993, Ries et al. 1997 b).

Formal analysis of the Doppler spectrum in the carotid and vertebral systems often makes it easy to determine which of the arteries in the neck is being depicted, particularly when compression tests and vibration maneuvers (i.e., tapping) are registered at the same time.

Only two parameters in the *Doppler spectrum* are usually evaluated:

- The systolic peak frequency
- The average mean frequency

In addition, the diastolic frequency is sometimes important for side-to-side comparison. Asymmetries in the spectrum have so far been difficult to detect.

Many algorithms have been proposed to calculate the width of the spectrum or the systolic window, particularly in the carotid system, including comparisons of preceding and subsequent vascular segments. These investigations mostly date from the period when the quality of echotomographic imaging was still relatively poor. The aim was to detect early forms of arteriosclerosis by calculating this parameter. However, systematic comparison of the structural and hemodynamic parameters shows that, even in completely normal situations, and particularly in younger patients with physiological flow displacements and separation phenomena, there may be limitations. In doubtful cases, an unusual result obtained when analyzing the spectrum should subsequently be checked using duplex sonography, which is the ultimate diagnostic arbiter.

■ Conventional and Color Flow Duplex Sonography

Duplex sonography has clearly simplified the evaluation of the results and made it easier to distinguish normal findings and variants within the vertebral artery system. This was made possible by introducing improved imaging capabilities for structural wall relationships, and especially after the introduction of simultaneous flow analysis in the color flow procedure. In general, a systematic evaluation of the parameters listed in Table 1.**5**, p. 24, is carried out under normal conditions, and the findings are carefully described and documented in the form of images or on videotape.

Sources of Error

Ultrasound Incident Angle

■ Continuous Wave and Pulsed Wave Doppler Sonography

In Doppler sonography, whether in analogue recording or spectral analysis, the most important source of error is still the unknown incident angle of ultrasound application.

Since the arterial course is very variable, and bilateral symmetry is not an absolutely reliable parameter, applying the ultrasound at too flat an angle may erroneously simulate a pathological flow acceleration,

or a decrease in flow, if the angle used is unfavorable (too steep or too flat) (Fig. 2.**19**). When the patient has a long, slender neck, this is easily possible in the submandibular region of the internal carotid artery.

Due to the almost perpendicular angle at which ultrasound is applied where the vertebral artery loops round the atlas, a lower flow velocity is registered here than is actually the case.

■ Conventional and Color Flow Duplex Sonography

Flow irregularity may be incorrectly represented by a change in color if the orientation of the blood vessel is not parallel to the probe. Actually, only the direction

Fig. 2.**19** Differences in Doppler waveforms during examination of the extracranial carotid system, **a** when following the course of the CCA and ICA with the probe position optimized, and **b** when holding the probe rigidly adjusted

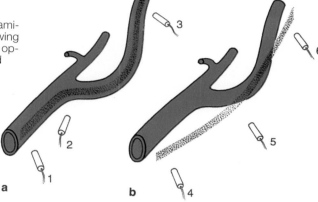

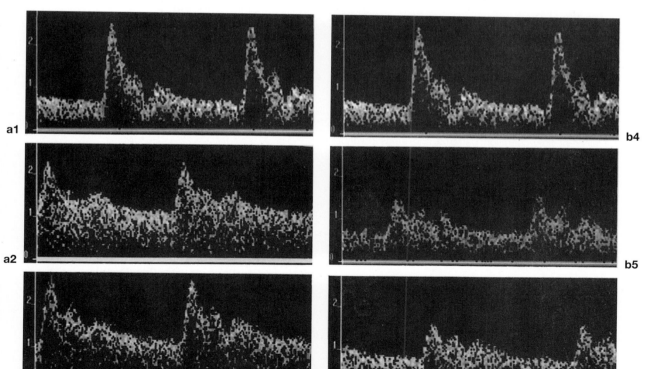

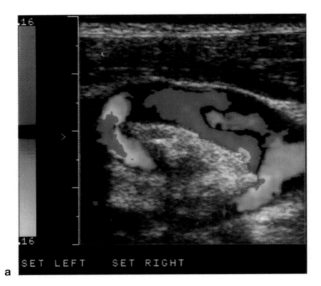

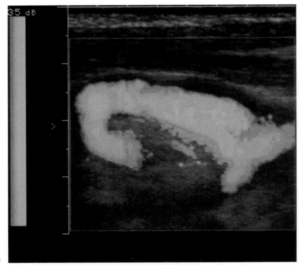

Fig. 2.**20 a** Velocity mode and **b** power energy mode using color flow duplex sonography of the carotid bifurcation and internal carotid artery with a severe kinking. Note the different flow-condition patterns and areas of undetectable flow velocity in the pulsatile velocity mode vs. continuous display of flow areas but without demonstration of flow direction in the power energy mode

of the vascular axis in relation to the transducer position has changed when this happens. This is a frequent source of misinterpretations, especially in loop formations (Fig. 2.**20**).

Cardiac Arrhythmia

Due to the varying filling phases of the cardiac cycle, absolute arrhythmia makes it more difficult to evaluate the flow spectra.

Vascular Width

At the same flow intensity, the flow velocity will be lower in a wider vessel, and higher in a narrower vessel. Therefore, the flow in a wide sinus can easily be disturbed, and may incorrectly suggest a pathological finding. Particularly in older patients, aneurysmatic dilatations are difficult to differentiate from a dilatative arteriopathy without duplex sonography.

Anomalies in Vascular Course

An atypical arterial course can make it more difficult to locate the blood vessels, and can simulate stenosis, kinking or coiling, or an occlusion. This is most often the case when the internal carotid artery originates medially (Fig. 2.**6**).

Venous Superimposition

Venous superimposition can disturb the assessment with continuous wave ultrasound and distort the signals as a result of opposed deflections. The venous noise often mimics a stenotic signal. By contrast, duplex sonography may be facilitated by venous superimposition.

Echogenicity

Fresh thrombotic material in blood vessels has the same acoustic impedance as blood. In the B-mode image, the blood vessel appears to be open. It is only with the Doppler signal that it is possible to determine whether the lumen of the blood vessel has narrowed or, when the signal is missing, has closed (Zwiebel et al. 1983) (Fig. 2.**21**). Missing pulsatility in the B-mode image and omission of color flow Doppler signals are characteristic.

If the sectional plane is selected unfavorably, a plaque may be bypassed, and the blood vessel be misdiagnosed as completely open. However, the opposite is also possible—plaque formation reaching into the lumen can be mimicked if wall segments of tortuous arteries are imaged diagonally. Examinations should therefore always be carried out in several planes, including both longitudinal and cross-sections, in order to avoid misinterpretations.

Diagnostic Effectiveness

■ Continuous Wave and Pulsed Wave Doppler Sonography

Directly assessing the arteries of the neck using ultrasound has significantly improved the noninvasive diagnostic capability of the vasculature in this region. It is now possible to differentiate the various vascular segments and confirm whether or not they are open.

Using direct, continuous wave Doppler sonography, stenoses exceeding 50% and occlusions in the carotid system can be excluded with a high degree of certainty. In the hands of experienced investigators, the *specificity* of the technique lies between 88% and 98% (Table 2.**8**) (de Bray et al. 1995). There is no increase in the reliability of diagnostic statements based on spectral analysis with regard to the exclusion of a pathological finding. Depending on which parameters are used to evaluate the spectrum, the number of false-positive results may even increase (Sheldon et al. 1983, Lally et al. 1984, Neuerburg-Heusler et al. 1985) (Table 2.**8**).

Fewer case numbers are available for the vertebral arteries. Due to the frequent variants, the specificity is lower, at 75–90% (von Reutern and Clarenbach 1980, Winter et al. 1987).

■ Conventional and Color Flow Duplex Sonography

The *specificity* of duplex investigations lies between 54% and 100% (Table 2.**9**) (de Bray et al. 1995). B-mode examinations on their own have a low validity, and are no longer adequate for today's investigations. Angiographic examination, which has been the accepted reference method for such measurements, can no longer serve as the gold standard (de Bray et al. 1995). For example, digital subtraction angiography, even using selective intra-arterial contrast medium administration, often fails to detect small changes in the walls of blood vessels. Comparative studies of duplex sonography and magnetic resonance angiography have recently been published (Riles et al. 1992, Mattle et al. 1991), and show a good correlation with normal findings. Flow irregularities that appear physiologically in the carotid bifurcation can produce artifacts in magnetic resonance angiography, which can often be identified by color flow duplex sonography if structural wall changes are absent.

The predictive value of the ultrasound method is significantly dependent on the quality of the examination procedure. Unfavorable conditions (e.g., an inability to study the arterial course in several sectional planes) can influence the specificity in individual cases. Also, isolated echotomographic imaging of the arteries of the neck, without additional examination of the hemodynamic findings using Doppler sonography, has a negative influence on the diagnostic predictive value (Ricotta et al. 1987), and should therefore be avoided. The advantage of color flow duplex sonography lies in the easy and quick simultaneous

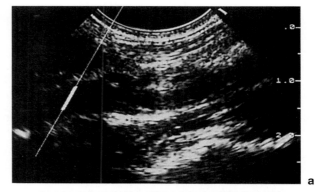

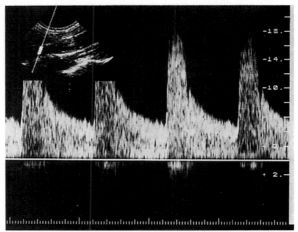

Fig. 2.**21**   Conventional duplex sonography
a B-mode image may be misdiagnosed as normal
b The stenosis can only be reliably demonstrated and graded using the frequency spectrum of the pulsed Doppler signal (sample volume 5 mm, transmission frequency 5 MHz). This is an extreme stenosis, with a systolic peak frequency of 18 kHz

display of morphological and hemodynamic parameters. The specificity of the method ranges between 86% and 100% (Steinke et al. 1990, Sitzer et al. 1993, Görtler et al. 1994), similar to that of continuous wave and pulsed wave Doppler versus angiography in the detection of carotid stenoses >50% (92–99% and 78–95% respectively) in experienced hands. It may be improved by the combined use of color flow velocity and signal amplitude analysis (*power imaging*) (Steinke et al. 1996) (Tables 2.**7** and 2.**10**).

Individual investigations show that the specificity in evaluating the vertebral arteries lies at about 83% (Ackerstaff et al. 1984). For color flow duplex sonography, these figures may well turn out to be higher.

Pathological Findings

Orbital Arteries

Principle

The ophthalmic artery and its terminal branches, the supraorbital and supratrochlear arteries (the lateral or medial frontal arteries), form a watershed with branches of the external carotid artery. Impediments to normal blood flow in this circulatory system, also termed the *ophthalmic artery anastomosis,* can lead to distortions in the flow equilibrium when they are accompanied by changes in the pressure relationships (Figs. 2.**22**, 2.**23**) (Melis-Kisman and Mol 1970, Müller 1971, Planiol et al. 1972, Büdingen et al. 1976). Using Doppler sonography, a variety of observations can be made via the supratrochlear, supraorbital, and ophthalmic arteries.

Findings

■ Continuous Wave and Pulsed Wave Doppler Sonography

Flow reversal. The direction of flow can change due to a decrease in pressure within the internal carotid artery (occlusion or high-grade stenosis). In the supratrochlear and supraorbital arteries, blood can be observed to move away from the probe—through the branches of the external carotid artery and the ophthalmic artery, blood flows into the carotid siphon and subsequently into the intracranial cerebral arteries. When the branches of the external carotid artery are *compressed,* this retrograde inflow is reduced or even reversed into the normal, physiological flow direction (Fig. 2.**22 d**).

Since the peripheral vascular resistance in cerebral arteries is lower than that in the vasculature supplying skin and muscle, the shape of the Doppler waveform also changes—the diastolic residual blood flow, in particular, often remains very high in the orbital arteries when flow reversal occurs.

No blood flow. Occasionally it is not possible to record a signal spontaneously, and often acoustic detection is not possible either (Fig. 2.**22 b**). When the branches of

Supratrochlear artery

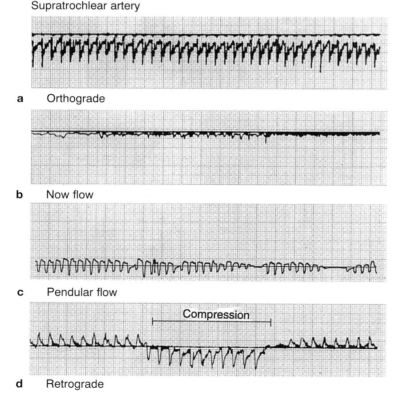

a   Orthograde

b   Now flow

c   Pendular flow

d   Retrograde

Fig. 2.**22**   Indirect Doppler sonography of the supratrochlear artery. The polarity of the Doppler equipment is adjusted to show orthograde blood flow (toward the probe) as a downward deflection
a   Orthograde blood flow: normal finding
b   No flow: suspected proximal obstruction
c   Pendular flow: suspected proximal obstruction (e.g., innominate artery, bifurcation stenosis)
d   Retrograde flow. Pathological finding indicating a high-grade obstruction of the internal carotid artery; when branches of the external carotid artery are compressed, the direction of flow reverses

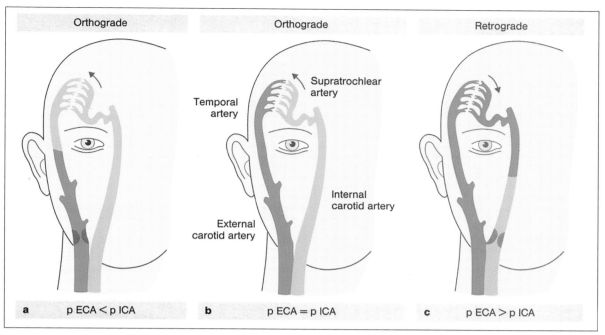

Fig. 2.**23** The collateral system and resulting blood flow directions in the ophthalmic artery **b** Normal findings (pressure in the internal and external carotid arteries is at equilibrium = orthograde blood flow in the orbital arteries)

a A pressure decrease in the external carotid artery = orthograde blood flow **c** A pressure decrease in the internal carotid artery = retrograde blood flow

the external carotid artery that form the anastomosis are compressed, physiological blood flow may result. This finding is often caused by a hemodynamically significant obstruction in the internal carotid artery (high-grade stenosis or occlusion), with insufficient collateralization through the ophthalmic artery anastomosis—e.g., in the presence of a stenosis of the extracranial carotid bifurcation, when pressure is reduced in both the internal and the external carotid arteries.

Alternating forward and backward (pendular) blood flow. When the fronto-orbital terminal branches of the ophthalmic artery are being examined, a rare observation is a breath-like audiosignal, corresponding to alternating forward and backward blood flow around the baseline on the recorded graph (Fig. 2.**22 c**). Compression of branches of the external carotid artery leads to a gradual return of blood flow to normal (watershed displacement). A high-grade obstruction in the innominate artery with significant reduction of pressure in the entire extracranial carotid system, is often responsible for this rare finding (Hennerici et al. 1981 b). Occasionally, a stenosis at the carotid bifurcation in the neck may also cause this phenomenon and in rare instances, hypoplastic orbital arteries (Keller 1961), or a variant in the origin of the ophthalmic artery, may be present (Vogelsang 1961).

Reduced physiological blood flow. A wide variety of reasons for asymmetric amplitudes (systolic or dia-

stolic, or both) are observed in side-to-side comparisons, with an increase in the flow signal when a branch of the external carotid artery participating in the collateral blood flow anastomosis is compressed; these differences are not significant in diagnosis. Flow obstructions in the extracranial and intracranial arteries may lead to such findings, as well as variations in the caliber of the ophthalmic arteries (Decker and Schlegel 1957, Di Chiro 1961). More often, these lateral differences are caused artificially by variations in the angle between the ultrasound probe and the vasculature.

Pathologically elevated nonpulsatile flow. Rarely, a marked, continuous sibilant audiosignal may be heard, e.g., in the vicinity of a significant stenosis of the neck arteries (Budingen and von Reutern 1993). If the flow direction is toward the probe, venous drainage through the orbital veins should be considered, as in the case of a cavernous sinus fistula involving the carotid artery. Increased pressure in the cavernous sinus can be caused by retrograde blood flow through the orbital veins, which subsequently passes through the jugular vein. The differential diagnosis should include large intracranial arteriovenous malformations, which generally produce a pathological retrograde nonpulsatile flow signal in the orbital arteries (von Reutern et al. 1977).

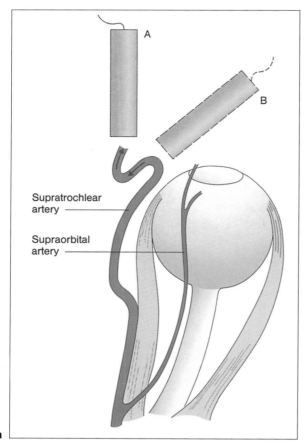

Supratrochlear artery

Supraorbital artery

a

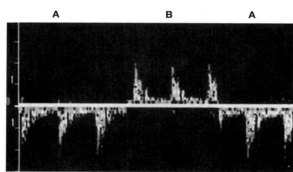

b

Fig. 2.**24a**   Loop formation in terminal branches of the orbital artery (**b** shows Doppler spectra from the supratrochlear artery) with changing flow direction, depending on the probe position
b Despite orthograde perfusion, blood flow away from the probe is seen in position B. Flow runs toward the probe only in A

■ Color Flow Duplex Sonography

Recently attempts have been made to examine branches of the ophthalmic artery in the orbit to identify small embolic events. If emboli travel into the retinal blood vessels, they may be demonstrated, e. g., by absent flow signals and hyperintense reflections sometimes producing echo-shadows in the central retinal artery (Hedges et al. 1993, Hedges 1995). In addition, color flow duplex sonography can differentiate between various mechanisms leading to amaurosis fugax attacks (Sergott et al. 1992) and support orbital blood flow measurements (Lieb et al. 1991 a/b).

Evaluation

The most reliable *indirect criteria* for assessing flow obstructions with hemodynamic significance in the internal carotid artery are, in *decreasing order of importance:*

- Retrograde blood flow
- Alternating forward and backward (pendular) blood flow
- No blood flow

in the main branch of the ophthalmic artery (usually the supratrochlear artery).

Asymmetries noted during side-to-side comparisons when the flow direction is orthograde can have several causes, and cannot be properly assessed until results are obtained from direct Doppler sonography of the neck arteries (Büdingen et al. 1976, Ackermann 1979, Trockel et al. 1984).

Sources of Error

Loop formation. Since, in a loop formation, compression tests in the external carotid artery branches affected are necessary to determine the actual direction of flow within the orbital arteries, a signal directed toward the probe can indicate both orthograde and retrograde flow (Fig. 2.**24**, Table 2.**2**). A positive compression effect with an accompanying *decrease* or *reversal* in the registered blood flow confirms *retrograde flow* in the orbital artery. Any *increase* in the flow velocity observed while compressing the branches of the external carotid artery indicates a physiological *orthograde* flow direction. If only one of the two components of the loop can be assessed, the compression test is indispensable to determine the flow direction.

Variants. Usually, the supratrochlear artery is a terminal branch of the ophthalmic artery, whereas the supraorbital artery is frequently hypoplastic. More than 80% of the time, the supratrochlear artery and the dorsal artery of the nose form the two terminal branches of the ophthalmic artery. In approximately 3% of cases each, the supratrochlear and supraorbital arteries, or all three of the above arteries, form the terminal branches. Only in about 5% of cases each does the supratrochlear or supraorbital artery alone form the terminal branch (Krayenbühl et al. 1979). High-grade obstructions to blood flow in the internal carotid artery, with an accompanying decrease in pressure, usually produce a clear flow change within the supratrochlear artery (*retrograde flow*, or *no flow*). In individual cases, however, the supraorbital artery is the main terminal branch of the ophthalmic artery, and the supratrochlear artery is hypoplastic. Retrograde blood flow within the hypoplastic supraorbital artery does not reliably indicate a significant carotid obstruction, but can also reflect altered flow relationships within the orbit in an intact carotid system. Therefore, only the orbital blood vessel that has the *strongest Doppler signal* supports the interpretation of any flow obstruction in the extracranial carotid system.

No signal. If no flow is observed for the supratrochlear or supraorbital arteries in the typical probe position, it may be due to a variant location, and the positioning of the probe may have to be changed. In this case, it is best to extend the search for the blood vessel by constant compression of the facial artery and/or the superficial temporal artery bilaterally, and to observe the effect when the vascular system of the external carotid artery is released again.

No compression effect. *Retrograde flow with a negative compression effect* can be caused by:

– A combined stenosis of both the internal and external carotid arteries
– A more proximal obstruction in the common carotid artery

Compression maneuvers in the branches of the contralateral external carotid artery are required, and these often confirm retrograde flow in the orbital arteries.

A negative compression effect can also be present:

– When the ophthalmic artery is collateralized via the maxillary artery and its largest terminal branch, the middle meningeal artery, which cannot be selectively compressed

A positive compression effect may also be absent in:

– Bilateral obstructions of the external carotid arteries.

Obstruction of the ophthalmic artery and its terminal branches. Rarely, proximal obstructions of these arteries can cause changes in the Doppler signals, in the absence of any obstructions within the extracranial carotid system. In such cases, it is best, in addition to the supratrochlear and supraorbital arteries, to locate and record the ophthalmic artery by placing the probe directly on the eyeball over the closed eyelid, or to evaluate its origin transorbitally using the pulsed wave Doppler (Spencer and Whistler 1986).

Tabelle 2.**2**   Interpreting the Doppler-sonographic findings in the orbital arteries

Flow direction related to the probe	Compression effect on flow	Verified flow direction
	Increase	Orthograde
	Decrease/ Reversal	Retrograde
	No reaction	None

Diagnostic Effectiveness

The specificity of indirect Doppler sonography lies between 88% and 97%, and—because of its simple and straightforward application—it can be used well as a rapid screening test (Büdingen et al. 1976, Neuerburg-Heusler 1984) rather than more complex ultrasound tests (Howard et al. 1996). In pathological cases, it serves as an additional way of evaluating the hemodynamic significance of an extracranial stenosis of the ICA. In addition, in cases of intracranial internal carotid artery obstruction, it may help to establish whether the process is proximal or distal to the ophthalmic artery.

However, examination of the orbital arteries alone when diagnosing obstructive lesions of the carotid artery is obsolete. Indirect test results support the final diagnosis only if combined with direct evaluation of the neck arteries, and if the findings are definitively pathological.

False-negative results, i.e., orthograde blood flow in the orbital arteries, also occur in hemodynamically marked flow obstructions involving the internal carotid artery if:

– There is good collateralization through the circle of Willis. In spite of a high-grade ICA obstruction, no retrograde flow or flow incidents occur (approximately 20% of cases) (Fig. 2.**25**).
– The anastomosis of the ophthalmic artery is not open, e.g., there is a high-grade external carotid artery stenosis ipsilaterally, and the contralateral external carotid artery does not contribute to collateral blood flow in the ophthalmic anastomosis. The pressure gradient remains unchanged in this case, since the pressure in both the internal and external carotid arteries decreases.

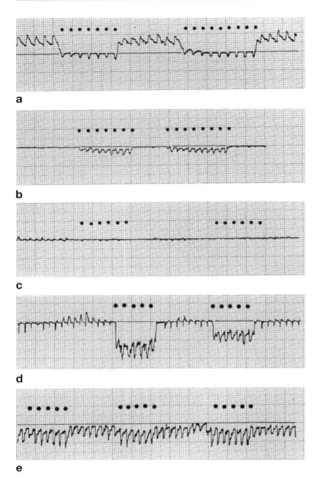

Fig. 2.25 Findings from supratrochlear arteries in five patients with an occluded ipsilateral internal carotid artery. This is an example of the different hemodynamic conditions that may be found in the watershed between the internal and external carotid artery branches (from Trockel et al., J Neurol Neurosurg Psychiatr 1984; 47: 43)

a Retrograde flow, with flow reversal when the branch of the ipsilateral external carotid artery is compressed (...)

b No flow, with orthograde blood flow developing after compression

c Variable retrograde and orthograde flow, with a small signal amplitude and no significant change under compression

d Alternating pendular flow, with clear orthograde flow when the branches of the ipsilateral external carotid artery are compressed

e Normal orthograde blood flow, showing an increase after compression

– The flow obstruction lies in the neck region proximal to the carotid bifurcation (common carotid artery or innominate artery). Here, the pressure gradient also remains unaffected.

False-positive results are extremely rare, provided that:

– Anatomical variants are taken into account.
– Compression maneuvers are performed completely and correctly. The most frequent mistake involves an artificial head movement while Doppler signals are being recorded, resulting in a simulated reduction in flow velocity.
– The possibility of an ophthalmic artery stenosis is taken into account.
– Intracranial arteriovenous malformations are not forgotten.

There is *no* reliable correlation between a change in flow relationships within the orbital arteries and the degree of stenosis in the internal carotid artery. Rather, the opening of the ophthalmic artery anastomosis and the resulting flow relationships reflect only the potential and effectiveness of collateralization to compensate for the obstruction in the carotid arteries.

Hemodynamic compensation through the anastomosis of the ophthalmic artery is only relevant to the perfusion of the brain in rare instances. Much more often, it is of no significance at all. In some cases, after a transcranial examination has been carried out, it may be possible to ascertain whether or not the ipsilateral middle or anterior cerebral arteries are being supplied through the collateral pathways of the ophthalmic artery (Fig. 2.26).

In summary, the verification of retrograde flow in the orbital arteries is of considerable diagnostic significance. It indicates a marked flow obstruction (severe stenosis or occlusion) in the preceding carotid system.

Vice versa, verified orthograde flow in the orbital arteries is conceivable with accompanying direct findings in the neck arteries that suggest a high-grade stenosis or occlusion in the distribution area of the internal carotid artery. In order to avoid misinterpretation, however, attempts should always be made to document the underlying collateralization pathways via the contralateral carotid system or the posterior circulation and the circle of Willis—i.e., supplementary intracranial ultrasound diagnostics are advisable.

Apparently abnormal but not necessarily pathological findings (e.g., alternating forward and backward blood flow, no flow, as well as clearly asymmetric orthograde flow) can be an important aid in locating an intracranial flow obstruction or pathological vascularization (e.g., combined with asymmetries in the common or internal carotid artery). In these cases, supplementary intracranial Doppler studies are also useful.

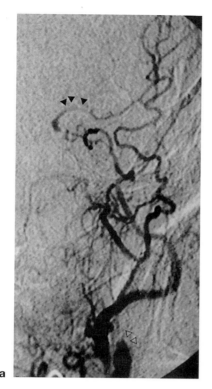

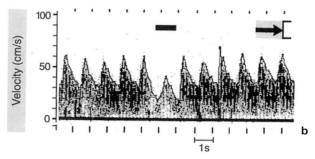

Fig. 2.26 Angiogram (**a**) and Doppler spectrum (**b**) in a patient with an occluded internal carotid artery (open arrows) and exclusive collateralization of the middle cerebral artery (closed arrows) via external carotid artery branches, with an incompletely developed circle of Willis. The flow velocity of the middle cerebral artery is clearly reduced (sample volume depth 50 mm, spontaneous peak velocity 60 cm/s) when the external carotid artery branches that feed the ophthalmic artery collateral are compressed (superficial temporal artery and facial artery, dark bar)

Carotid System

Principle

■ Continuous Wave and Pulsed Wave Doppler Sonography

Obstructions within the carotid system can be detected by continuous wave Doppler sonography when the narrowing of the lumen is at least 40%, and it is accessible to direct ultrasound application (Miyazaki and Kato 1965a, b, Keller et al. 1976a, von Reutern et al. 1976b, Spencer and Reid 1979, Reneman and Spencer 1979, Blackshear et al. 1980, Trockel et al. 1984, Humphrey et al. 1984, Beach et al. 1989).

Comparison with the angiographic findings in the literature is puzzling, due to heterogeneous criteria in estimating the degree of stenosis from image planes (Croft et al. 1980, Eikelboom et al. 1983, Langlois et al. 1983, Thomas et al. 1986, Widder et al. 1986b, Vanninen et al. 1994), and the variable procedures used to calculate the hemodynamic compromise from Doppler studies (Bladin et al. 1994, de Bray et al. 1995). Also, comparing Doppler findings with measurements of the degree of stenosis obtained from endarterectomy preparations (Arbeille et al. 1985), or from estimates of the original lumen using echotomographic findings (Terwey et al. 1984) has not yet provided a satisfactory solution. The same applies to interpretations of the degree of stenosis that are based

on angiographic findings, which usually involves an examination in two planes at best, representing a sketchy image of the real three-dimensional relationships. Similar problems are also encountered when evaluating the results from magnetic resonance angiography (MRA), which simulates a morphological image based on flow-sensitive relaxation phenomena (Huston et al. 1993, Laster et al. 1993). Even with a pathoanatomical approach, quantitative and reliable estimation of the degree of stenosis is difficult, due to the heterogeneous nature of arteriosclerotic processes and the structural artifacts that result from the processing of tissue (Picot et al. 1993).

Since the completion of trials that proved the effectiveness of surgery in symptomatic patients with stenoses of >70% in the carotid system (ECST 1991, NASCET 1991), the appropriate method of estimating the degree of stenosis has become a matter of controversy (Barnett and Warlow 1993, Alexandrov et al. 1993). When interpreting the extent of the stenosis angiographically, both the *local* and *distal degree of the stenosis* should be measured (Fig. 2.27). In addition, evaluating the proximal degree of stenosis seems to be useful, provided this calculation is related to the aforementioned criteria used in the multicenter trials. In individual cases, these results can also be further complemented with appropriate MRA and ultrasound diagnostic evaluations to determine the degree of stenosis. All of these methods have a similar predictive value, provided that appropriate care is taken when interpreting the results. Clinical decisions can be

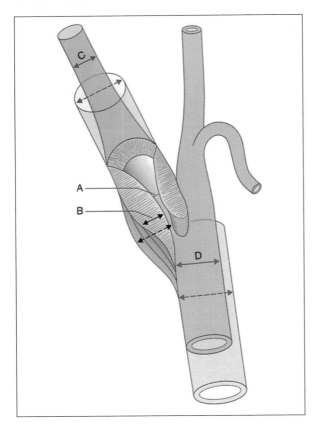

Fig. 2.**27**   Parameters used to calculate the degree of stenosis
A residual lumen, B plaque, C distal vascular diameter,
D standard lumen 3 cm proximal to the bifurcation
Local degree of stenosis (1–A/B) × 100%
Distal degree of stenosis (1–A/C) × 100%
Proximal degree of stenosis (1– A/D) × 100%

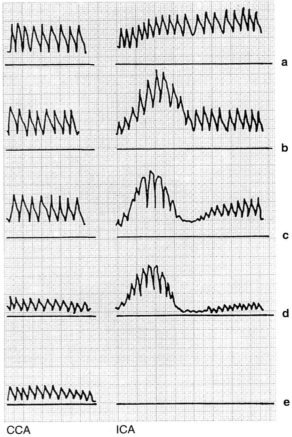

CCA   ICA

Fig. 2.**28**   Schematic drawing of continuous analogue-wave Doppler waveforms suggesting a low-grade stenosis at the origin of the internal carotid artery (**b**), a medium-grade stenosis (**c**), and a high-grade stenosis (**d**), as well as an occlusion (**e**), in comparison with a normal finding (**a**). To record the signal, the Doppler probe is continuously shifted from the common carotid artery (CCA) across the bifurcation into the internal carotid artery (ICA)

taken on the basis of noninvasive findings without the help of conventional angiography in the majority of patients (de Bray et al. 1995, Erdoes et al. 1996).

Hemodynamic values from ultrasound and MRA studies have their own classification, and should not be considered as synonymous with the "morphological degree of stenosis." An adequate safety margin is recommended for individual values (e.g. 50–60%, 60–80%) (Glagov and Zarins 1983).

Using five fundamental Doppler criteria (Table 2.**3**) (Neuerburg-Heusler 1984), it is possible to classify carotid obstructions according to the degree of stenosis, and to compare them with the angiographic results (Fig. 2.**28**) (Hennerici et al. 1981 a, von Reutern and Büdingen 1993, Trockel et al. 1984) (Table 2.**4**):

– Low-grade stenoses (40–60%)
– Medium-grade stenoses (60–70%)
– High-grade stenoses (around 80%)
– Subtotal stenoses (over 90%)
– Occlusions

Table 2.**3**   Important parameters for the evaluation of the degree of stenosis in extracranial cerebral arteries

– Changes in the audiosignal ("turbulences")
– Increase in the diastolic or systolic flow velocity, or both
– Decrease or loss in pulsatility, with diastolic flow acceleration and systolic deceleration
– Decrease in the diastolic or systolic flow velocity (proximal and distal to the obstruction)
– Missing Doppler signal

Table 2.**4**   Structural and hemodynamic criteria for the evaluation of the degree of stenosis in the internal carotid artery (after Widder); frequency data refer to 4 MHz transmission frequency

	I Nonstenotic plaques	II Low-grade stenosis	III Medium-grade stenosis	IV High-grade stenosis	V Subtotal stenosis
Angiographic estimates					
Local degree of stenosis	< 40%	40–60%	60–70%	ca. 80%	> 90%
Distal degree of stenosis	0	< 30%	≈50%	≈70%	> 90%
Dopplersonographic parameters					
Indirect criteria	No sign of flow obstruction			Ophthalmic artery: no flow or retrograde flow Common carotid artery: reduced flow	
Direct criteria					
Analogue waveform					
Near the stenosis	Normal	Audiosignal change, slight increase in local flow	Clear flow increase, pulsatility loss and systolic deceleration	Strong local flow increase with systolic deceleration	Variable stenosis signal with decreased intensity
Post-stenotic	Normal	Normal	Normal	Decreased systolic flow velocity	Difficult to find, strongly reduced signal
Spectrum analysis					
Near the stenosis	Normal	Spectral broadening	Spectral broadening with increasing intensity of the low-frequency component	Inverse frequency components in a broadened spectrum	Inverse frequency components in a broadened spectrum
Systolic peak frequency	< 4 kHz	> 4 kHz	4–8 kHz	> 8 kHz	Variable
End diastolic frequency	< 1.3 kHz	< 1.3 kHz	> 1.3 kHz	> 3.3 kHz	Variable
Systolic peak velocity (cm/s)	< 120	> 120	> 120	> 240	Variable
End diastolic velocity (cm/s)	< 40	< 40	> 40	> 100	Variable
Systolic ratio (ICA/CCA)	< 1.5	< 1.8	> 1.8	> 3.7	Variable
B-mode					
Proof quality	+ + +	+ + +	+ +	+	+
Findings	Small plaque extension		Medium-grade lumen constriction	High-grade lumen constriction	Maximum lumen constriction
Color flow mode					
Findings	No or local turbulences only	Long segmental systolic flow acceleration Color fading in systole only	Localized segmental systolic flow acceleration Color fading Turbulences Increased flow velocity in diastole	Extremely localized segmental high-grade flow acceleration, post-stenotic backflow components (mosaic pattern), secondary vortices Short segment of marked color fading or aliasing	

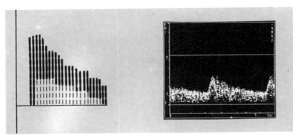

Normal findings (<40%)

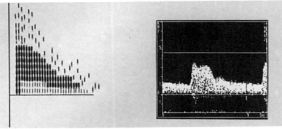

Low-grade stenosis (40–60%)

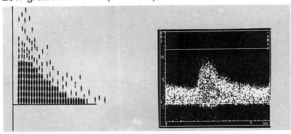

Medium-grade stenosis (60–70%)

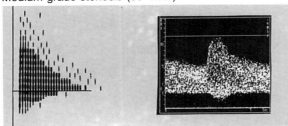

High-grade stenosis (ca 80%)

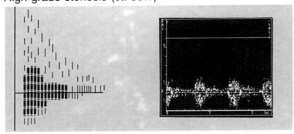

Subtotal stenosis (>90%)

Fig. 2.**29** Schematic drawings and FFT spectra showing continuous wave Doppler signals from the internal carotid artery for different degrees of stenosis (after Arbeille et al.)

This classification scheme only applies to unilateral, isolated obstructions of the internal carotid artery. Modifications are necessary for:

- Common carotid artery obstructions
- External carotid artery obstructions
- Processes involving several blood vessels
- Unilateral multiple stenoses at specific levels of the circulatory system ("tandem stenosis"), especially when these cannot be directly assessed with ultrasound (e.g., those in the proximal segment close to the aorta, or extending submandibularly until they reach an intracranial location)

Spectral analysis can be used to determine the degree of stenosis from the parameters summarized in Table 2.**5**:

Table 2.**5** Important parameters from the spectrum analysis for evaluation of the degree of stenosis in the extracranial cerebral arteries

- Systolic and diastolic peak frequency
- Average mean Doppler frequency
- Mode frequency band with the highest signal amplitude
- Systolic frequency window (spectral broadening)
- Negative frequencies

- The *systolic and diastolic peak frequency* can easily be measured in normal flow conditions; in pathological cases, problems can be encountered here. The peak frequency is usually used as the primary criterion for classifying the extent of a stenosis (Table 2.**8**, Fig. 2.**29**) (Spencer and Reid 1979, Norrving and Cronquist 1981, Hennerici et al. 1981, Humphrey and Bradbury 1984, Trockel et al. 1984, Johnston et al. 1986).
- The *average Doppler frequency* reflects a calculated variable (at least 30 different formulas are used in commercial instruments), which is insensitive to local flow changes near wall processes, and insensitive to variable intervals of sequential heart beats.
- The *distribution of amplitude-intensive frequency bands* differs at different degrees of stenosis; with increasing obstruction, a displacement toward the baseline results.
- The *systolic frequency window* describes the relationship between amplitude-weighted low frequencies and the frequencies of the overall spectrum. With an increase in the degree of flow obstruction, the percentage value of the frequency window decreases (Brown et al. 1982 a, Rittgers et al. 1983, Bandyk et al. 1985).
- *Negative frequencies* indicate flow components toward the probe, and reflect turbulence phenomena.

In a recent international Consensus Conference (de Bray et al. 1995), systolic peak Doppler shifts >4 KHz (f_o = 4 MHz, > 120 cm/s) to identify most ICA stenoses >50% in local diameter reduction and end-diastolic values >4.5 KHz (f_o = 4 MHz, >135 cm/s) to identify ICA stenoses >80% were considered to be the most reliable parameters for classification.

■ Conventional and Color Flow Duplex Sonography

B-mode. Various morphological and ultrasound tomographic studies have described a correlation with arteriosclerotic vascular processes (Comerota et al. 1981, Reilly et al. 1983, Hennerici et al. 1984, Cape et al. 1984, Zwiebel 1986, Hennerici 1987 a, Ricotta et al. 1987). It is useful to evaluate these changes in B-mode imaging, using suitable parameters (Table 2.**6**), and it is important to detect plaques in several longitudinal sections and cross-sections (Fig. 2.**30**). B-mode studies of the cerebral circulation alone, however, are often misleading and no longer represent the "state of the art." This should also be considered if IMT-measurements are included in multicenter trial protocols (see p. 39; O'Leary et al. 1996). Without reliable differentiation of ICA and ECA (in Doppler-mode or color-mode signal analysis), B-mode studies alone can cause significant errors.

Sonographic characteristics of plaques are

- *Echogenicity* (from anechoic to hyperechoic or echodense); this is a reliable, sufficiently reproducible parameter as evidenced by three independent studies (Geroulakos et al. 1994, Widder et al. 1990, Baud et al. 1996). "Echoic" can refer to anatomic structures ("isoechoic" with view to the sternocleidomastoid muscle and "hyperechoic" with view to the cervical vertebrae).
- *Texture* (from homogeneous to heterogeneous); classification schemes have been proposed but are not fully established (de Bray 1997).
- *Surface* (from irregular to excavated forms) characteristics are difficult to evaluate and often a matter of interpreter disagreement; the accuracy of diagnosis can be improved by additional color flow duplex sonography: recesses and vortices (> 2 mm in depth and length) are characteristic ulcer features (Sitzer et al. 1994, Steinke et al. 1996).
- *Evidence of embolic activity* (indirect signs like HITS = high intensity transient signals); several studies have shown a significant association with such findings and plaque activity as a potential predictor for stroke and TIA (Ries et al. 1996, Valton et al. 1995, Markus et al. 1994, Siebler et al. 1993).
- *Changes in plaque appearance* over time; this is the most valid predictor of associated stroke risks at present apart from the degree of stenosis (> 70% obstruction) (Hennerici and Steinke 1991, Sillesen et al. 1995).
- *Plaque motion* is an interesting, recently studied parameter; it is hypothesized that plaque motion resembles the risk of plaque distortion and rupture giving rise to cerebral embolism (Meairs et al. 1995, 1996).

Table 2.**6**   Criteria for evaluation of the B-mode image findings in the carotid artery according to Consensus Conference (de Bray et al. 1997)

Form and Course
   Regular
   Variation/anomaly
   Dilated
   Tortuosity
   Kinking/coiling

Vessel size
   External diameter
   Internal diameter
   Stenosis diameter
   Pulsatility changes (systolic/diastolic)
   Intima-media-thickness
   Vessel motion
     Transversal
     Axial
     Systolic–diastolic

Pathological changes
Location
   Common carotid artery (anterior wall/posterior wall, lateral or medial)
   Internal carotid artery (anterior wall/posterior wall, lateral or medial)
   External carotid artery (anterior wall/posterior wall, lateral or medial)
Form
   Circular
   Semicircular
Size
   Plaque diameter in longitudinal and cross-section at the broadest point
Surface
   Regular/irregular
   Flat, smooth
   Ulcerated
   Cavitation
Echo pattern
   Strength (strong/weak)
   Size (progressive/regressive)
   Echogenicity (anechoic/echogenic)
   Texture (homogeneous/heterogeneous)
   Ultrasound shadow (present/absent)
Plaque motion (uniform/discrepant)
Plaque volume
Embolic activity

Doppler mode. In continuous wave and pulsed wave Doppler procedures—e.g., in duplex systems—similar parameters can be used to describe the Doppler spectrum (Roederer et al. 1982, Blackshear et al. 1980, Jacobs et al. 1985, Robinson et al. 1988, Whiters et al. 1990, Moneta et al. 1993). The application of quantitative parameters from the Doppler spectra is particularly useful in patients with evidence of dynamic changes in the degree of stenosis. As in continuous wave Doppler sonography, the criteria used are mainly values for maximum and average frequencies. The spectral width is a less reliable indicator of minimal changes in the vascular wall.

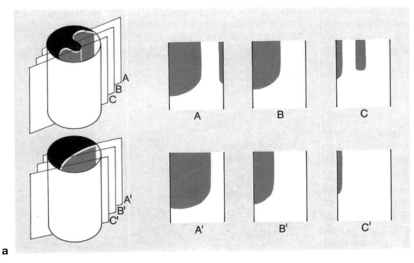

a

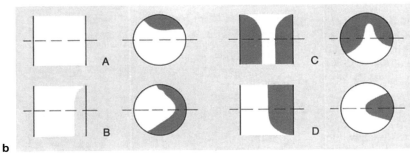

b

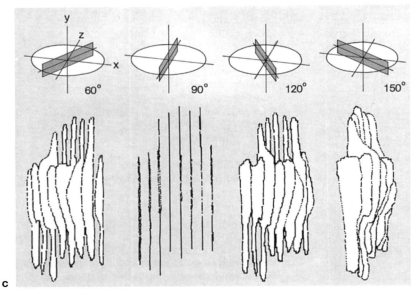

c

Fig. 2.**30**   The way in which the appearance of wall changes in B-mode depends on the location of the longitudinal or cross-sections (**a, b**), and ways of obtaining a three-dimensional reconstruction (**c**)

a   Sources of misinterpretation from inadequate evaluation of a single longitudinal section

b   Sources of misinterpretation from inadequate cross-sections with corresponding longitudinal sections

c   Three-dimensional reconstruction of a plaque overview from eight sequential longitudinal sections

Interpreting spectral parameters is carried out in the same way as in continuous wave and conventional pulsed wave Doppler sonography (Fig. 2.**31**). Parameters reflecting hemodynamic compromise induced by severe ICA obstructions are considered to be relevant (de Bray et al. 1995). The *systolic carotid ratio* should be recorded between the site of the stenosis and the CCA obtained 3 cm below the bifurcation: a value >1.5 determines stenoses >50%, and a threshold >4 determines stenoses of >70%.

Color flow Doppler mode. Since the early 1990s, studies have increasingly used color flow duplex sonography (velocity-guided and amplitude-guided analysis) to assess the value of a combined analysis of the echotomographic and associated hemodynamic changes in the presence of varying degrees of carotid stenosis (Polak et al. 1989, Londrey et al. 1991, Anderson et al. 1992, de Bray et al. 1995 b, Steinke et al. 1990). Steinke et al. (1992 a) carried out a systematic comparison between a multi-channel pulsed wave Doppler system and color flow duplex sonography, and found that detection of nonturbulent flow asymmetries caused by low-grade carotid plaques is more effective when the multi-channel pulsed wave Doppler method is used—which is technically much more complex, and in practice only available for experimental scientific research—than with color flow duplex sonography. However, turbulence can be detected by both methods in the vicinity of nonstenotic wall changes, with high specificity (90%) and adequate sensitivity (70%).

Systematic analysis of high-grade carotid stenoses (Steinke et al. 1992 b, Erickson et al. 1989, de Bray et al. 1995 b, Sitzer et al. 1993) in comparison with angiographic imaging yielded agreement in over 70% of cases in relation to the surface characteristics of plaques. Measurements of the degree of stenosis and cross-sectional measurements corresponded in 85% of cases. So far, evidence of specific hemodynamic patterns associated with various structural wall changes has not yet allowed any clinical correlations to be made in evaluating the malignant potential of such vascular processes. Using Doppler mode, anal-

Table 2.**7** Comparison of the distal degree of stenosis according to NASCET and the proximal degree of stenosis (CCA method) vs. CDFI. There is excellent agreement between the velocity mode and power energy mode of color-coded duplex sonography in separating >70% stenosis from <70% stenosis

Angiography (CCA Method)				
Velocity mode	<50%	50–69%	70–79%	80–99%
<50%	2	0	0	0
50–69%	0	3	0	0
70–79%	0	2	9	14
80–99%	0	0	0	12
	2	5	9	26
Amplitude mode				
<50%	1	0	0	0
50–69%	0	3	0	0
70–79%	0	2	3	11
80–99%	0	0	0	12
	2	5	8	23
Angiography (NASCET Method)				
Velocity mode	<50%	50–69%	70–79%	80–99%
<50%	2	0	0	0
50–69%	3	3	0	1
70–79%	1	5	8	6
80–99%	0	1	6	6
	6	9	14	13
Amplitude mode				
<50%	1	0	0	0
50–69%	3	5	0	0
70–79%	2	2	6	6
80–99%	0	1	7	15
	6	8	13	11

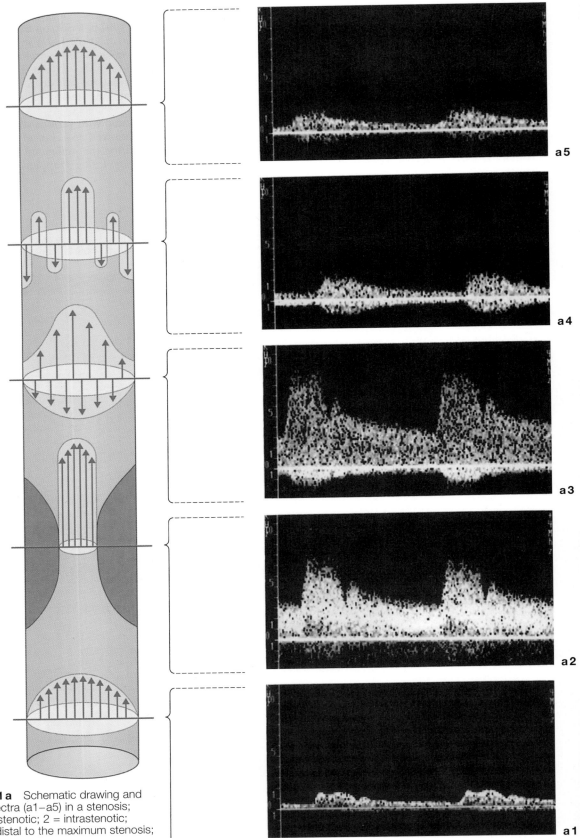

Fig. 2.**31 a**   Schematic drawing and FFT-spectra (a1–a5) in a stenosis; 1 = prestenotic; 2 = intrastenotic; 3, 4 = distal to the maximum stenosis; 5 = post-stenotic

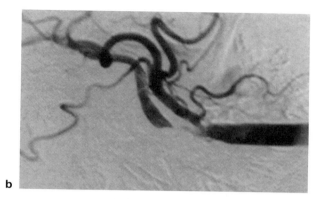

b

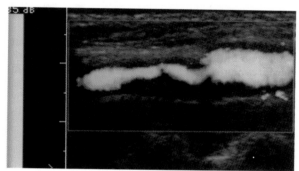

c

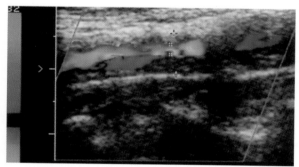

d

Fig. 2.**31 b–d** Comparison of color flow duplex analysis using velocity mode (**c**) and power energy mode (**d**) with regard to intra-arterial conventional angiography (**b**). The morphology of the stenosis and the delineation of the tightest lumen area is better by means of power energy than flow-velocity mode, although the degree of stenosis in longitudinal sections is generally underestimated by power energy mode due to the poor spatial resolution when compared to the velocity mode. In cross-sections this disadvantage is negligible

yses have been made of the intrastenotic and post-stenotic maximum peak frequencies, of the extent and location, as well as the spread, of the color saturation zone in correlation with "jet flow," and the spread and expression of the color reversal and color mixture zones (Table 2.**7**, Figs. 2.**31** and 2.**37**). Application of power energy modalities or echo-contrast (Sitzer et al. 1994) is more recent, and only limited results are available (Steinke et al. 1996). Both methods seem to provide additional information to the standard technologies: there has been an overall improvement in test reliability and validity, and both can be used in difficult conditions, e. g., high-grade stenosis and pseudo-occlusion of the ICA.

Findings

Common Carotid Artery

■ Continuous Wave and Pulsed Wave Doppler Sonography

Common carotid artery obstructions are rarer than those involving the internal carotid artery. They can also be detected using the Doppler-mode criteria mentioned above.

The results obtained are often the same as those described for the internal carotid artery (p. 64). The same applies to changes associated with dilatative arteriopathy, which is characterized by an increased vascular diameter, with decreased pulsatility and flow velocity. In cases of acute occlusion, pulsatility is completely absent. In the chronic stage, when the scar-like changes have produced a complete restructuring of the vascular lumen, it can be difficult even to identify the lumen of the common carotid artery.

As with internal carotid artery obstructions, stenoses of the common carotid artery show similar changes in the *Doppler spectrum* parameters mentioned in Table 2.**5**.

Increased Flow Velocity

Stenoses at the origin cannot be detected directly using CW Doppler alone. Nevertheless, in individual cases, their presence can be indirectly determined from changes in the audiosignal due to an increase in blood flow velocity. Using pulsed wave Doppler and an appropriate depth selection to record the Doppler signals, a flow velocity acceleration can be detected as positive proof (Fig. 11.**2**, pp. 344–345). A proximal stenosis at the common carotid artery can be differentiated from the innominate artery by demonstration of intact flow relationships in the ipsilateral and proximal subclavian and vertebral arteries.

Continuously elevated flow. In contrast to the internal carotid artery, a significant compensatory increase in flow within the common carotid artery cannot usually be detected when contralateral occlusions are present. However, when there are extensive intracranial shunt volumes (fistula formation, arteriovenous blood vessel malformation), an increase in the flow velocity can be detected within the entire neck region accessible to examination. In this case, it is quite possible to superimpose flow signal irregularities, both acoustically and in the spectrum waveform itself. Rarely, flow accelerations may be found, resulting from circular constrictions of the common carotid artery (e.g., in inflammatory vascular diseases, Fig. 2.**32 c**).

Decreased Flow Velocity

1. The most frequent and most significant pathological finding in the common carotid artery is decreased flow velocity caused by a *distal high-grade internal carotid artery obstruction* (internal carotid artery stenosis exceeding 80%, internal carotid artery occlusion, hemodynamically significant intracranial ipsilateral vascular obstruction).
2. The differential diagnosis should include a *dilatative arteriopathy* of the common carotid artery (Fig. 2.**32 b**, Fig. 11.**16**, pp. 368–369), which is often encountered at an advanced age (unilateral or bilateral), and causes a reduction in the flow signal similar to that seen in distal vascular processes with hemodynamic significance.
3. In a few cases, an acute obstruction of the middle cerebral artery, with poor collateralization through leptomeningeal anastomoses and near the circle of Willis (M_1 segment), can lead to a flow velocity reduction in the ipsilateral common carotid artery (Fig. 11.**15**, pp. 366–367).
4. Rarely, a high-grade common carotid artery stenosis at the origin of the artery is also present (Fig. 11.**2**, pp. 344–345).

Absent Signal

When the common carotid artery is occluded, no signal can be recorded. In contrast, signals from the external and internal carotid arteries may still be recorded distal to the bifurcation, because the most common collateral fills other branches of the external carotid artery from the ipsilateral vertebral artery via the occipital artery, which shows retrograde flow. If the internal carotid artery is open, intracranial circulation can also be maintained through this vertebral–occipital anastomosis. The important search for a flow signal within the internal carotid artery with intact circulation is much more successful using the duplex system under "visual control" than with the conventional Doppler technique. This is important, since in these cases surgical reconstruction of the carotid system is possible (Fig. 2.**32 a**).

■ Conventional and Color Flow Duplex Sonography

B-mode. The course of the common carotid artery in the neck region can be ideally demonstrated using a B-mode image, since this vascular segment can be displayed regularly and very reliably in several planes,

even when a large transducer is being used. Local wall changes in this region can therefore be detected, and their morphological structure can be described with more certainty than with angiography (Croft et al. 1980, Eikelboom et al. 1983, Hennerici et al. 1984).

Doppler mode. Flow changes in the common carotid artery can also be detected easily and regularly under visual control using conventional duplex sonography. By moving the sample volume, flow accelerations can be detected in the direct vicinity of plaque using sequential ultrasound application. Multi-channel pulsed wave systems have mainly been used in scientific studies and to calculate the blood volume because of their complex application modalities. Such systems have not acquired any significant diagnostic relevance in clinical use.

In the case of obstruction within the innominate artery (p. 86), duplex sonography can be particularly helpful when continuous wave Doppler sonography has registered only weak signals, or signals that are difficult to classify. If there is a clear reduction in the flow signal caused by a vascular process in the artery, and there are simultaneous changes in the right common or internal carotid artery, then it is almost always missed by continuous wave Doppler sonography. In this case, identifying any structural wall changes in B-mode, and locating the sample volume in this particular vascular segment, can be helpful in registering the mostly intermediate signals. When appropriate compression maneuvers are used, the hemodynamics of the compensatory collateral circulatory systems, which are often extremely complex, can be comprehended. Depending on the location of the watershed, various results may be obtained (Fig. 2.**46**).

Color flow mode. This procedure is particularly suitable for detecting abnormalities within the middle and distal common carotid artery. If the internal carotid artery is open, but has low flow velocities, a search for a flow signal under visual control is particularly important here. Often, the vertebro-occipital collateral—which is usually effective—is directly detectable, running via the ipsilateral vertebral artery, with retrograde perfusion of the external carotid branch, and ultimately passing through the bifurcation distal to the occluded common carotid artery into the internal carotid artery (Fig. 2.**32 a**).

Plaque formations occur particularly often in the common carotid artery. Fresh wall adhesions, which are marked by an absent flow signal and hypoechoic or absent structural wall changes in B-mode imaging, are by no means rare. Changes in the color signal can only occasionally be predicted from the structural image. Bizarre configurations may be accompanied by nearly laminar flow, while minimal plaque formations with a smooth surface structure can produce significant separation zones and flow turbulences. The extent to which clinically relevant information might be provided by these findings (e.g., differentiating between threatening and innocuous plaque formation) is not yet clear (Steinke and Hennerici 1992 b, Meairs et al. 1995, Hennerici et al. 1995).

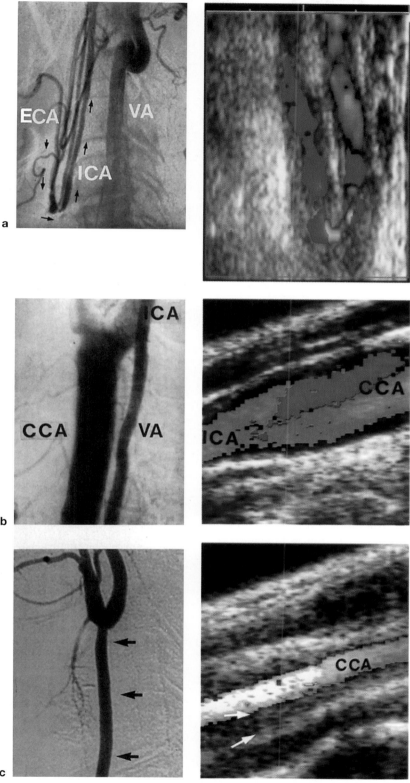

Fig. 2.**32** Diseases of the common carotid artery depicted in the angiogram and corresponding color flow duplex sonography. **a** Occlusion. **b** Dilatative arteriopathy. **c** Vasculitis with a small residual lumen and wall thickening

CCA Common carotid artery
ECA External carotid artery
ICA Internal carotid artery
VA   Vertebral artery

Internal Carotid Artery

Continuous Wave and Pulsed Wave Doppler Sonography

Increased Flow Velocity

Low-grade stenosis (40–60% lumen narrowing). A local increase in flow velocity is observed, with an acoustically intensified Doppler signal. The systolic–diastolic amplitude modulation is retained; pre-stenotically and post-stenotically, there are no changes in the Doppler waveforms; there are sometimes marked irregularities in the audiosignal (turbulence phenomena), comparable to fine or medium crepitating rales (Fig. 2.**29**).

Compensatory elevated flow. In the presence of hemodynamically significant contralateral flow obstructions, a similar Doppler signal intensification may eventually be recorded ipsilaterally, characterized by a *continuous* increase mainly in the diastolic flow velocity (rarely also in the systolic) throughout the internal carotid artery after the bifurcation. If the internal carotid artery can only be followed for a short distance (short neck, bifurcation located at a high level), it is impossible to differentiate between a low-grade stenosis or a compensatory elevated flow at the origin of the internal carotid artery.

Dilatative arteriopathy of the common carotid artery or hypoplasia of the internal carotid artery. Abnormal differences in the vascular caliber between the common and internal carotid arteries can cause similar changes in the Doppler signal, mimicking compensatory collateralization, or a mild stenosis at the origin of the internal carotid artery.

Intracranial arteriovenous malformation. Intracranial angiomas with a high shunt volume are sometimes recognizable due to a distinct increase in flow ipsilateral to the malformation, with a retained—or in rare cases also disturbed—amplitude modulation in the feeding extracranial cerebral arteries (i.e., the common and internal carotid arteries). In such cases, the Doppler signal is often reduced in the terminal branches of the ophthalmic artery, or even becomes retrograde when the blood is drained through these vessels into the malformation (Fig. 11.**17**, pp. 370–371).

Loss of Pulsatile Amplitude Modulation and Systolic Deceleration

Medium to high-grade stenosis (60–90% lumen narrowing). In medium to high-grade stenoses, with a lumen reduction exceeding 60%, the diastolic flow velocity at the stenosis is not only increased, but pulsatile amplitude modulation is lost in addition. When the transducer is moved along the obstruction, the analogue registration shows a systolic deceleration, also termed a "systolic peak reversal," with the average flow signal directed toward rather than away from the probe. Post-stenotically, the average flow velocity

again decreases, and significant turbulences in the audio signal appear (comparable to medium to coarse crepitating rales, crepitations, and rhonchi) (Fig. 2.**29**).

Extensive wall pulsations. A systolic deceleration may be imitated by extensive wall pulsations in vessels with abnormally high flow feed in the absence of any structural stenosis (e.g., "functional stenosis" in patients with intracranial arteriovenous malformations [Fig. 11.**17**, pp. 370–371]).

Kinking/coiling. Loop formations can also simulate a systolic deceleration. Misdiagnosis can be prevented when the exact physiological waveform is watched, particularly noting the normal audiosignal, which does not show any change in association with the changing flow direction.

Decreased Flow Velocity

High-grade and subtotal stenosis (more than 80% lumen narrowing). In high-grade and subtotal stenoses, the diastolic and systolic flow velocities decrease in both segments proximal and distal to the narrowing of the lumen. Post-stenotically, pulsatile amplitude modulation may be completely absent, can be replaced by an unmodulated, turbulent flow with pronounced audiosignal changes, and often reveals a marked systolic deceleration zone. High-grade stenoses show increased flow within the tightest vascular segment. In subtotal stenoses, these changes are frequently absent, and sometimes the Doppler signal is hardly audible, characterized by a quiet high-frequency sound, which is difficult to register because of the poor signal-to-noise ratio caused by the low ultrasound energy reflected by the few corpuscles passing (an acoustically fine, chirping sound) (Fig. 2.**29**).

Proximal and distal high-grade stenosis. Lumen narrowing in vascular segments near the aorta, in the submandibular or intracranial areas may be detected through a reduced flow velocity, with a retained or slightly reduced amplitude modulation. These are *indirect criteria* for the presence of a proximal or distal stenosis.

Slosh Phenomenon

A "carotid dissection" is usually characterized by a filiform stenosis, which originates submandibularly at the entrance of the internal carotid artery into the base of the skull, and reaches down to the bifurcation (Fig. 11.**8**, p. 353) (Fisher et al. 1978, Biller et al. 1986, Mokri et al. 1986, Steinke et al. 1989). Using Doppler sonography, a highly typical, but not pathognomonic signal is found, which the present authors were the first to describe as the "slosh" phenomenon in the internal carotid artery (Hennerici et al. 1989). In contrast to an intermediate flow signal, systolic flow components can be detected that are simultaneously directed in both orthograde and retrograde directions, indicating a significant increase in the peripheral resistance due to the pronounced distal stenosis in the absence of any flow velocity during the diastole. Some of these are pulsatile movement-induced artifacts caused by hemorrhage within the vascular wall, superimposed on a very low flow velocity signal. A similar flow signal, although usually without this characteristic systolic forward-and-backward (or "to-and-fro") blood flow, is seen in rare instances in which the main branch of the middle cerebral artery is acutely and completely occluded—in serious and extensive vasculitis, and in incomplete retrograde thrombosis of the internal carotid artery due to a siphon stenosis. Depending on the progression of the dissection, which is often followed by spontaneous recanalization after an average of four to six weeks, the slosh phenomenon shifts cranially, and the lumen opens again. During this process, various stenotic signs can be detected with Doppler sonography. Only rarely does the blood vessel remain completely occluded (Steinke et al. 1994).

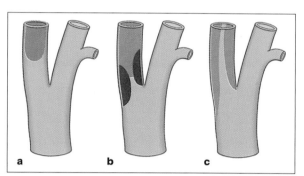

Fig. 2.33 Schematic drawings of different forms of carotid artery occlusion
a Retrograde thrombosis
b Cap-like occlusion due to local plaque formation
c A pointed, trailing occlusion; a typical finding in carotid artery dissections

Fig. 2.34 Comparison of individual parameters of the FFT ▷ Doppler spectrum analysis for various findings at the origin of the internal carotid artery. **a** Although the width of the *systolic window* decreases with increasing stenosis, overlapping values are so common that it is not possible to differentiate the underlying lesion only on the basis of this Doppler-sonographic criterion. **b** The systolic peak frequency continuously increases with increasing degree of stenosis but tends to decrease again if subtotal obstructions are analyzed. Thus, based on single values, the best differentiation possible allows us to distinguish < 50% from > 50% obstructive lesions, but not within those two classes. **c** The average mean frequency increases until a medium-grade stenosis is reached, and then decreases again. Like the peak systolic frequency, it does not provide sufficient discrimination between individual degrees of stenosis alone

Absent Signal

When the internal carotid artery is occluded at its origin, the flow signal is absent. In contrast to the changes described above, there is no positive criterion that can be used as an indicator of a flow obstruction.

Indirect criteria are therefore particularly important:

– All detectable Doppler signals have to be identified (e. g., most of the branches of the ECA by applying compression tests).
– Confusion with the vertebral artery, which often functions as a collateral, has to be avoided by following its course. If the signal is interrupted or cannot be traced back to the bifurcation, suspicion arises as to whether the ICA might be occluded.
– Possible variations in location have to be noted (medial location, loop formation).
– A filiform stenosis (i.e., pseudo-occlusion), often with only local flow acceleration, has to be looked for meticulously (Wernz et al. 1997). It is only rarely that the proximal stump is still open, or that an alternating flow signal extending in a submandibular direction can be detected in this location; this finding is characteristic of a dissection, as described above (Fig. 2.**33**).

As shown in Figure 2.**34**, the parameters given in Table 2.**5** vary depending on the degree of stenosis in internal carotid artery obstructions:

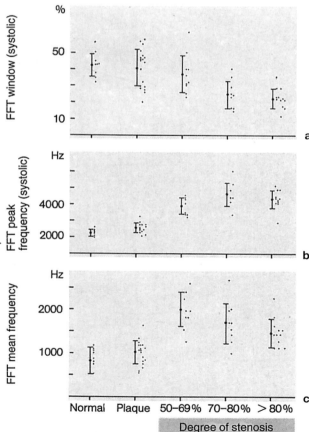

- The *peak frequency and velocity* (systolic and diastolic) increases continuously (from a lumen narrowing of 30% to around 90%) and then decreases again, corresponding to the transition from high-grade to subtotal stenosis.
- *The average flow velocity* also increases with the degree of stenosis (from approximately 30% to 90%), but is subject to less pronounced fluctuations.
- In high-grade stenoses (> 80%), it is useful to combine the values measured at different registration points (e.g., intrastenotic and prestenotic or poststenotic) in order to detect the change in the vascular resistance induced by the stenosis (Rittgers et al. 1983).
- *The frequency associated with the highest signal amplitude* decreases with an increasing degree of stenosis, and has a *negative frequency component*, especially in high-grade stenoses (over 75%).
- *The systolic window* decreases with increasing stenosis (from 30% to subtotal stenosis).

It is therefore not recommended to use *single parameters* to classify the degree of stenosis although most studies so far published on the topic have either used peak or mean frequency values (for review, see de Bray and Glatt 1995). Appropriate differentiation between normal results and evaluation of different degrees of stenoses (cf. Table 2.8) rather than a wide-range classification (i.e., < 50%, 50–80%, > 80%) requires the analysis of appropriate combinations of parameters for a reliable estimation (Hennerici et al. 1985a, Arbeille et al. 1985). This is especially important for follow-up studies (Daigle et al. 1988, Daffertshofer and Hennerici 1990, Ringelstein 1995, Young and Humphrey 1996).

■ Conventional and Color Flow Duplex Sonography

B-mode. The *normal finding* is an anechoic lumen, bordered by two light reflections. The echo line facing the lumen (intimal and subintimal portions) is separated from a second exterior interface reflection (medial and adventitial components) by a hypoechoic middle layer, i.e., "the intima–media contour," (see Fig. 2.**35a**; Pignoli et al. 1986, Poli et al. 1988). B-mode and color-coded duplex sonography of the CCA are being used increasingly to evaluate the intima–media thickness (IMT) (Glagov et al. 1995, D'Agostino et al. 1996, O'Leary et al. 1996). These values are used for the analysis of early atherosclerosis, its natural history, and changes induced by reduction of risk factors and pharmacological agents, e.g., in patients treated with special

Table 2.**8** Calculated reliability of conventional Doppler sonography versus intra-arterial angiography in internal carotid artery stenoses and occlusions

First author	Year	Arteries (n)	Specificity (%)	Degree of stenosis (%) 0–20	20–40	40–60	60–80	80–100	Occlusion (%)	Validity (%)	F	D
Barnes	1981	199	91	30			85		96	61	+	CW
Hennerici	1981	488	97	–		87	96	95	93	96	–	CW
Doorly	1982	190	79	85			91		90	84	+	PW
Rittgers	1983	123	89	28	62	77	88	63	–		+	CW
Neuerburg	1984	209	98	68			90		97	91	–	CW
Trockel	1984	431	96	34		81	93	86	96	85	–	CW
Hennerici	1985	28	89	68			–		–		+	CW
Fischer	1985	229	97	52	91		100		97	92	+	CW
Caes	1987	48	96	79		100		33	100	83	+	CW
Sillesen	1988	128	90	83			87		100		+	PW
Bornstein	1988	662	–	99					86	98	+	CW
Perry	1985	155	92		79		72	87	94	85	+	CW
Bandyk	1985	243	90		85		87	72	90	88	+	CW
Lindegaard	1984	432	94			96			97	94	+	PW
Floriani	1987	469	89		75		73	87	88	85	+	CW
Brown	1982	205	92			91			96		+	CW
Hames	1985	137	93			91			–		+	CW
Harward	1986	130	93			93			–	92	+	CW
Widder	1987	65	94			90			94	76	+	CW
Blackshear	1987	48	88			87			–		+	PW

F frequency analysis (+ yes, – no); D Doppler technique (CW continuous wave, PW pulsed wave Doppler)

diets, antihypertensive, or antilipid drug regimens. According to the literature, this parameter is rather useful in multicenter trials and can be applied reliably if the protocol exactly indicates where and when to measure the extracranial carotid system (e.g., the proximal and distal common carotid artery, the carotid bulb, the proximal and distal internal carotid artery bilaterally), etc. In addition, the diameter of the vessels and pulsatility variations have to be noted and be adapted for standardized values (Easton et al. 1994).

- *Flat plaques* are characterized by an increased broadening and restructuring of the normally hypoechoic middle layer. The inner or exterior interface reflections can also appear thickened (Fig. 2.**35a**). Although flat plaques represent the earliest detectable form of arteriosclerotic wall change, they are not always exclusively signs of a progressive development, but can also persist as the final stage of a regressive development (Hennerici et al. 1985 b, Hennerici and Steinke 1991) (Fig. 2.**35a**).
- If the wall change expands into the vascular lumen with increasing size, then a homogeneous echo structure is usually retained initially; the *surface structure* and the location have to be precisely evaluated by suitable longitudinal sections and cross-sections (Fig. 2.**35b, c**).
- With increasing size and morphological change, *complicated plaque formations* with a nonhomogeneous echo structure and surface irregularities can form (Fig. 2.**35c–e**). Hypoechoic and hyperechoic sections may alternate. Hypoechoic plaques at the carotid bifurcation are associated with a significantly higher risk of TIA and stroke (Sterpetti et al. 1988, Bock et al. 1993, de Bray et al. 1997). Sequential longitudinal sections or cross-sections then often show only individual aspects of a complex morphology. With advancing development, clear narrowing of the lumen and broadening of the atheromatous bed occur. During this process, bleeding *("hemorrhagic plaques")* and calcium deposits *("hard plaques")* may appear, which can cause an ultrasound shadow with significant energy loss (Fig. 2.**35f–h**). The validity of the test to identify hemorrhages within the plaque is still controversial: sensitivities and specificities reported range between 72% and 94% (79% and 89% respectively) (Reilly et al. 1983, Bluth et al. 1986, Widder et al. 1990).
- Ruptures in the plaque cap and *ulcerations* associated with them can appear at any stage of arteriosclerotic development, and cannot always definitely be detected in B-mode (Comerota et al. 1981, Sitzer et al. 1995, Valton et al. 1995). The presence of an ulcer in a tight stenosis (> 70% lumen narrowing) is associated with an increased risk of cerebral ischemia (Eliasziw et al. 1994, Sitzer et al. 1995). Recent attempts to improve the image quality and to display three-dimensional and even four-dimensional plaque reconstructions (including changes over time as an important indicator of stable or friable plaques) have been successful. Software programs may soon become available (Meairs et al. 1995, see Fig. 1.**12**).

Doppler mode. Depending on the extent of plaque formation, all the changes described above under continuous wave and pulsed wave Doppler sonography can also be seen in Doppler mode. It is usually easy to interpret flow changes in the vicinity of low-grade stenoses, and to assign flow accelerations and also sometimes flow irregularities. With medium and higher-grade stenoses, however, the primary concern of the Doppler mode in duplex sonography is the interpretation of hemodynamic criteria given above, since the image quality is often limited due to ultrasound shadows.

Color flow Doppler mode. Even under difficult examination conditions, color coding makes it easier to detect and grade medium to higher-grade stenoses in the carotid artery region than it is using the single-channel Doppler mode in echotomography. Usually, the flow acceleration, depicted as a color saturation, is immediately recognizable from the color-coded signal. Local or extensive sudden color changes, with artificially high flow velocities caused by aliasing, can be directly detected. The B-mode image only has a secondary role here, providing additional information by defining structural wall changes when these vascular processes, which are often calcified, are being examined. With low-grade plaque formations, there are usually changes near the carotid sinus, such as limited or completely absent flow separation phenomena, irregular color distributions near a detectable structural wall change in B-mode, or isolated color changes without any detectable morphological correlation. This is especially important as an indication of a fresh thrombosis, possibly with a high embolization potential. Cross-section imaging provides additional information about the actual degree of stenosis and is particularly useful if performed prior to surgery in the absence of conventional angiograms (Ringelstein 1995, de Bray and Glatt 1995, Steinke et al. 1996, Carpenter et al. 1996) (Fig. 2.**36**).

Color flow duplex sonography also provides better visualization of changes within a plaque, such as niche formation and surface irregularities, with or without accompanying flow changes, hypoechoic plaque formations, and ulcerations (Fig. 2.**37**). Ulcerated lesions show both recess ($\geq$ 2 mm in depth and $\geq$ 2 mm in length with a well-defined back wall at its base) and flow vortices (i. e., reversed flow without aliasing) within or at the level of the recess. Only one prospective study is available to date comparing CFDI and surgical specimen. It revealed rather low sensitivity and specificity values (0.33 and 0.76 respectively) (Kessler et al. 1991) that would be improved under better study conditions.

Table 2.**7** summarizes the sensitivity, specificity, and accuracy of the color flow Doppler mode in comparison to conventional Doppler sonography, on the one hand, and angiography, on the other.

If the system is accurate enough to detect slow flow velocities, then the color flow Doppler mode is also helpful in evaluating subtotal stenoses, pseudo-occlusions, and dissections (Steinke et al. 1990 a, Wernz et al. 1997).

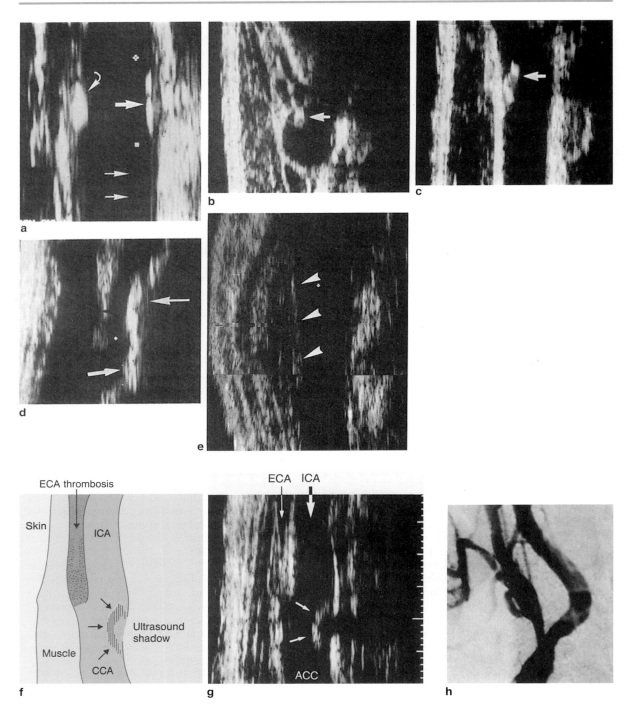

Fig. 2.35 Typical B-mode images of the extracranial carotid artery

a The normal two-layered wall contours (small arrows) contrast with changes in a healing ulcer (thick arrow). The ulcer shows a thickening of the interior echogenic layer, and also of the normally hypoechoic reflection zone lying underneath. In addition, a typical fibrous plaque formation can be seen as a hyperechoic zone without an ultrasound shadow on the opposite side (bent arrow)

b, c A "soft plaque" with penetration into the lumen from the common carotid artery's anterior wall in longitudinal section (**c**) and cross-section (**b**)

d A large extended ulcer from the distal common artery into the origin of the internal carotid artery (lower arrow), with a neighboring soft plaque in the internal carotid artery (upper arrow)

e Thrombotic aneurysm of the anterior wall of the internal carotid artery after surgery, containing heterogeneous echogenic structures

f–h Schematic drawing (**f**) to B-mode image (**g**) and corresponding angiogram (**h**) of a calcified plaque formation at the CCA's transition point into the ICA. There is clear echo shadow formation, with a smooth surface and minimal lumen constriction. The external carotid artery is segmentally thrombosed at its main branch, and presents a uniformly echogenic structure

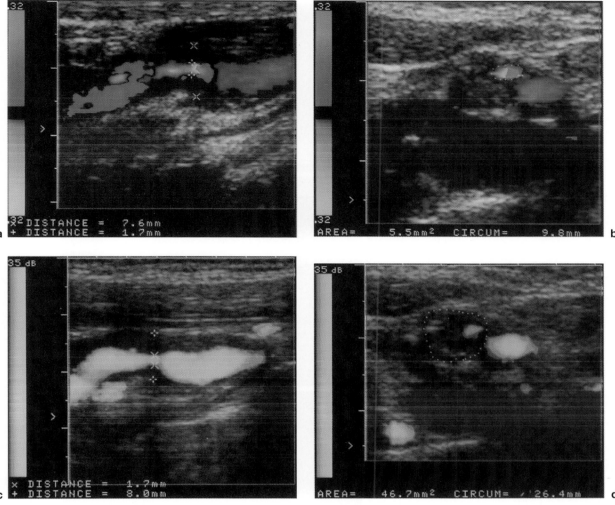

a × DISTANCE = 7.6mm
  + DISTANCE = 1.7mm

b AREA= 5.5mm² CIRCUM= 9.8mm

c × DISTANCE = 1.7mm
  + DISTANCE = 8.0mm

d AREA= 46.7mm² CIRCUM= 26.4mm

Fig. 2.**36** Color flow duplex sonogram of a severe carotid artery stenosis (>80%) extending from the bifurcation into the internal carotid artery, **a**, **b** velocity mode and **c**, **d** power energy mode. Displays are demonstrated in longitudinal (left) and cross-section (right). Pulse-dependent flow-velocity changes limit the illustration of the tightest area of stenosis, but restricted spatial resolution of the power energy mode may result in over-estimation of the degree of stenosis if its evaluation was made on longitudinal sections rather than cross-sections. The latter are more reliable and should be preferred for a quantitative analysis

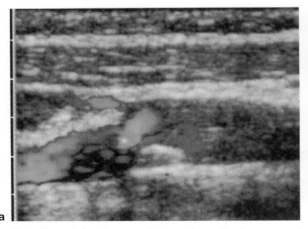

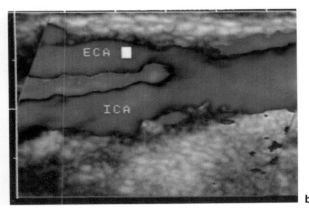

Fig. 2.**37** Various images of plaque formation by means of color flow duplex sonography

a Heterogeneous plaque with significant turbulences
b Ulcerated plaque

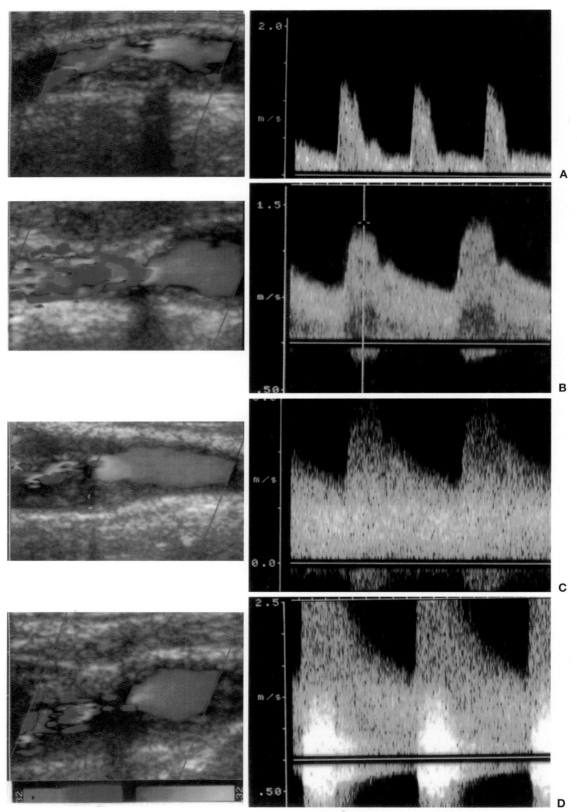

Fig. 2.**37 c**   Doppler spectra and velocity mode CDFI of different grades of ICA stenosis, (**A**) mild stenosis, (**B**) moderate stenosis, (**C**) severe stenosis, (**D**) subtotal stenosis (Steinke et al. 1994)

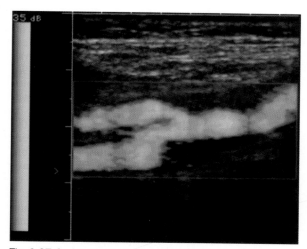

Fig. 2.**37 d** Hypoechoic, probably fresh thrombotic plaque at the posterior wall and extended isodense plaque at the anterior wall of the common carotid artery (power energy mode)

External Carotid Artery

■ Continuous Wave and Pulsed Wave Doppler Sonography

Isolated external carotid artery stenoses and occlusions are rare findings. They are more often seen in the context of bifurcation stenoses. Although the same Doppler-sonographic criteria apply as in the internal carotid artery, grading of the hemodynamic relevance is different.

Increased Flow Velocity and Systolic Deceleration

Since the blood vessel supplies both muscles and skin, i. e., tissue with a high peripheral vascular resistance, only stenoses with extensive lumen narrowing (> 70%) can be detected in the presence of systolic and diastolic flow acceleration, systolic deceleration, and irregularities in the audiosignal. Assessing the degree of stenosis is less reliable than in the internal carotid artery for these reasons and because of the inevitably larger alterations in the angle of ultrasound application for smaller-caliber vessels and the wide variations in origins and courses of the individual branches.

Due to the reduced peripheral vascular resistance, dural fistulas and angiomas with extracranial blood supply can cause significant increases in flow velocity at the origin and along the course of the feeding branches. This imitates all the criteria for high-grade stenosis (loss of pulsatility, systolic deceleration, extreme turbulences). Pronounced pulsations in the neck region are usually also found, which may cause artifacts in the Doppler signal.

Exceptions to this are conditions such as *extracranial–intracranial bypass* and to a lesser extent

ECA branches collateralizing occlusion of the internal carotid artery, in which the function of the artery changes. Subsequently stenoses have to be evaluated and classified according to the criteria mentioned for the internal carotid artery under these circumstances.

Decreased Flow Velocity

In contrast to occlusions of the internal carotid artery, obstructions of branches and even the proximal stem of the external carotid artery do not change the flow velocity in the proximal arteries, i.e., in the common carotid artery. Vascular stenoses of individual branches are rarities. Here, too, there is no indication of the distal vascular process at the origin of the external carotid artery. The process can only be identified directly within the stenotic area itself.

Absent Signal

As in the internal carotid artery, an occlusion of the external carotid artery or its individual branches is characterized by an absent Doppler signal. It is easy to overlook occlusions of the branches of the external carotid artery, since there are no indirect criteria for the diagnosis, in contrast to the situation in the internal carotid artery (no change of flow direction in the orbital, common, or internal carotid arteries). Often, only certain branches of the external carotid artery can be evaluated with ultrasound.

Spectral analysis is particularly helpful when external carotid artery stenoses are accompanied by internal carotid artery stenoses in bifurcation processes. Interpreting the audiosignal in such cases is difficult, and classification of different degrees of stenoses in the two branching arteries is sometimes impossible. In contrast, identification of the individual branches of the external carotid artery with pathological flow patterns is easily accomplished by compression tests (Fig. 11.**6**, pp. 350–351; Fig. 2.**13**).

There have been few systematic studies on the detection of stenotic vascular processes affecting the external carotid artery (Zbornikova et al. 1985 b). A sensitivity of 88% and a specificity of 94% was reported (for lesions of 50% or more) in 244 external carotid arteries examined by angiography and duplex sonography, with 21 stenoses of less than 50%, six stenoses of 50% or more, and two occlusions.

Absolute peak frequency values are not described, since the individual branches—even under physiological conditions, and depending on the peripheral vascular resistance—show wide and even extreme variations in the Doppler spectrum (e.g., superior thyroid artery and sublingual artery). In addition, it should be noted that obstructive vascular processes usually induce comprehensive vascular collateralization. Even occlusions of the proximal external carotid artery are perfused in a retrograde direction, partly also from the contralateral side, via branches of the subclavian and vertebral arteries, the thyrocervical trunk, and the neck muscle vasculature.

■ Conventional and Color Flow Duplex Sonography

B-mode. Usually, only the main branch of the external carotid artery can be adequately displayed in B-mode. The other branches can only be poorly imaged, due to:

- An unfavorable imaging angle
- Thin caliber
- Frequent loop formations
- Very variable anatomy

In many cases, it is not possible to identify any side branches in addition to the main branch itself. Consequently:

- It is not possible to differentiate between the internal and external carotid arteries using B-mode imaging alone.

Thus, additional identification of the flow signal using duplex system analysis is mandatory.

The wall changes in the external carotid artery are similar in structure to those described above for the internal carotid and common carotid arteries.

Color flow Doppler mode. It is easier to identify small, thin-caliber branches and loop formations with color flow duplex sonography than with conventional duplex procedures. Individual branches at their origin from the bifurcation are often displayed simultaneously (Fig. 2.**38**). In addition, flow signal changes can be visually recognized, e.g., saturation and separation phenomena, as well as local signal obliterations by thrombotic wall adhesions. This is especially significant when a stenosis is present at the bifurcation. With both continuous wave and conventional duplex sonography, it may be extremely difficult to distin-

guish terminal branches and differentiate the degree of stenosis in the external and internal carotid arteries independently. Retrograde blood flow through the external carotid artery, as in a proximal occlusion of the common carotid artery (Chang et al. 1995) (Fig. 2.**32a**), can also be easily recognized with color flow duplex sonography. Furthermore, complex processes involving several blood vessels can be depicted, and the collateral vascular network in the neck region can be identified (e.g., innominate artery obstruction with collateralization through the contralateral carotid and vertebral vascular systems). Local external carotid artery stenoses can easily be missed in conventional duplex and Doppler sonography in patients with "low flow" conditions, e.g., dilatative arteriopathies and subsequent arteriosclerotic processes, or high-grade stenoses near the aortic arch. If a vascular reconstruction has been carried out ("stump syndrome," Barnett 1978) in a case of occlusion of the internal carotid artery with an embolizing external carotid artery stenosis, or when new symptoms associated with a stenosis of the external carotid artery appear in patients with an extracranial–intracranial bypass, postoperative follow-up observations should include this method.

Evaluation

Important Pathological Situations

Internal Carotid Artery Occlusion

(Fig. 11.**5**, pp. 348–349)

Direct criteria:
- Absent Doppler signal, no color signal.
- Confirmed wall change, possibly with detection of a thrombus in the B-mode and color mode image respectively.

Indirect criteria:
- Retrograde flow direction in the orbital arteries (missing in 20% of cases).
- Reduction in the diastolic (systolic) flow velocity of the common carotid artery.
- External carotid artery serves as a collateral, with an altered waveform similar to that of the internal carotid artery despite a positive compression effect. Collateralization with a flow signal increase is possible in the vertebral artery system.

Internal Carotid Artery Stenosis

(Figs. 2.**39**, 11.**3**–11.**6**, pp. 346–351)

Direct criteria:
- Increase in flow velocity, systolic deceleration, and abnormal audiosignals near the stenosis, color saturation and color changes, depending on the degree of stenosis.

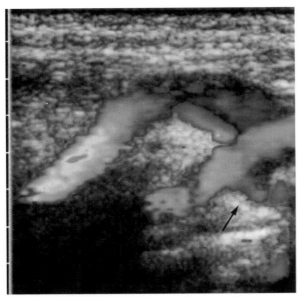

Fig. 2.**38** Color flow duplex sonogram of the carotid bifurcation showing the origin of the superior thyroid artery branching off the external carotid artery

- Post-stenotic audiosignal irregularities and, in color flow Doppler mode, a confirmed reduction in flow velocity.
- Wall changes in B-mode, associated with local flow change.

Indirect criteria:
- Post-stenotic normal or decreased flow velocity.
- Flow velocity changes in the orbital arteries.

The degree of obstruction is assessed using the combined parameters given in Table 3.**2**. *Subtotal stenoses* are often difficult to detect, and there is a risk of misdiagnosing an occlusion of the internal carotid artery. A specific examination using the color flow Doppler mode is often more successful than continuous wave Doppler sonography in detecting flow signals near a high-grade stenosis, since high-grade stenoses are often poorly imaged (Steinke et al. 1990, Steinke et al. 1996, Wernz et al. 1997).

Nonstenotic Wall Changes in the Carotid Artery

(Figs. 2.**35**, 2.**37**, 11.**3**, p. 346)

Direct criteria:
- Audiosignal irregularities, which can only be documented in the spectrum. For the beginner, recognizing these is difficult and overinterpretation is likely (nonexistent irregularities).
- B-mode and color flow Doppler mode are very important diagnostically in providing definite differentiation between various pathological processes, and distinguishing these from normal variants (e.g., wide or congenitally absent carotid sinus, caliber variations between the common and internal carotid arteries, "low flow phenomenon" in an extracranial carotid system aneurysm, and dilatative arteriopathy) (Figs. 2.**32**, 11.**7**, p. 352).

Indirect criteria:
- None.

Carotid Bifurcation Stenosis

(Fig. 11.**6**, pp. 350–351)

Direct criteria:
If both the internal and external carotid arteries are involved, it is necessary to classify the registered signals separately in the "carotid triangle" by:

- Carefully moving the Doppler probe
- Compressing or tapping the individual branches of the external carotid artery

in order to separately:

- Evaluate the degree of stenosis in each artery
- Identify the vascular processes

Analysis of the spectrum and application of color flow duplex sonography is recommended. These methods

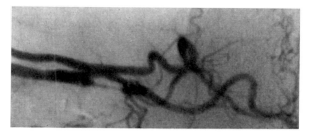
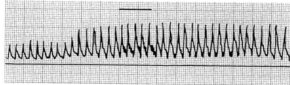
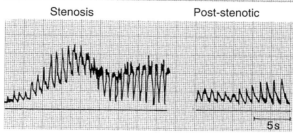

Fig. 2.**39** Doppler sonogram and angiogram in a high-grade stenosis involving the internal carotid artery, which is located medially. The intra-arterial digital subtraction angiogram cannot clarify whether the high-grade stenosis that is visible belongs to the external carotid artery or the internal carotid artery; in the Doppler sonogram, appropriate compression tests (–) show that the internal carotid artery is located medially and has a high-grade stenosis, while the external carotid artery, located laterally, appears completely normal

improve the reliability of the diagnosis in cases of multivessel pathology and associated flow irregularities.

Indirect criteria:
- Post-stenotically reduced flow, which can be assigned to the external or internal carotid arteries using compression tests.
- Changes in the flow relationships of the orbital arteries (in higher-grade stenoses).
- Prestenotically normal or decreased flow velocity in the common carotid artery is seen in high-grade internal carotid artery stenoses, but not in external carotid artery stenoses.

Sources of error:
- Local variants, e.g., internal carotid artery medial to the external carotid artery (Fig. 2.**39**).
- Superposition of two blood vessels (changing the neck and head position sometimes allows better assignment); color flow duplex analysis facilitates isolated Doppler signal registration.
- Loop formation in the internal carotid artery and individual branches of the external carotid artery.

Internal Carotid Artery Occlusion and Ipsilateral External Carotid Artery Stenosis

This situation often leads to misdiagnoses, since it is difficult to establish the absence of an internal carotid artery signal in the vicinity of an external carotid artery that has pronounced stenotic signs. In some cases, color flow duplex sonography helps, proceeding from the common carotid artery in a series of cross-sections that show no flow signal in the internal carotid artery.

Occasionally, the absence of a signal from the submandibular ICA may support the findings. In other cases, color flow duplex sonography (including power amplitude analysis) helps: in a series of cross-sections proceeding from the common carotid artery into the internal carotid artery, no flow signal can be detected. If the bifurcation is located high in the submandibular region, or there is a stenosis over an extended distance, the examination is sometimes subject to technical limitations due to the transducer size used. If surgery is planned, there is a broad range of indications for the use of magnetic resonance angiography and conventional angiography. In angiography, specialized examination techniques also have to be used to detect any pseudo-occlusion or total blockage.

External Carotid Artery Occlusion and Internal Carotid Artery Stenosis

In general, this situation does not present any diagnostic difficulties. To evaluate the degree of stenosis in the internal carotid artery, the criteria given above apply; there is no external carotid artery signal. Pathological flow changes in the branches of the ophthalmic artery are rare, due to the failure of ophthalmic artery to collateralize via the ipsilateral external carotid artery. Compressing the branches of the contralateral external carotid artery while recording Doppler signals from the ophthalmic artery and its branches is important here, because collateralization usually proceeds via the contralateral side.

Sequential Ipsilateral Stenoses in the Carotid System

(Fig. 11.**9**, pp. 354–355)

– If both stenoses lie in an area directly accessible to the Doppler probe, it is often possible to detect them.
– If the higher-grade stenosis is more proximal, then it is easy to miss the distal stenosis, because the Doppler-sonographic findings can be explained just by the changes in the flow relationships near the high-grade stenosis. This applies in all cases in which the distal stenosis is no longer directly accessible.
– If the higher-grade stenosis lies distally, and both stenoses lie in the directly accessible neck region, then they can usually be reliably detected. Even

when the higher-grade stenosis is not in the neck region that is directly accessible, indirect criteria often provide clues about the existence of a second stenosis.

Multivessel Disease

(Fig. 11.**5**, pp. 348–349)

Obstructions involving several arteries on both sides of the neck are more difficult to identify, due to:

– The possible presence of many different collateral mechanisms
– The greater difficulty of carrying out side-to-side comparisons
– The required modification of the criteria used to classify the degree of stenosis

Carotid Artery Dissection

(Fig. 11.**8**, p. 353)

There is a characteristic Doppler signal in many forms of carotid artery dissection. This "slosh phenomenon" may be found along the internal carotid artery in the neck region, depending on the extent to which the intramural hematoma formation, which usually starts submandibularly, reaches toward the carotid bifurcation in the neck and constricts the lumen. Depending on the extent to which the lumen is constricted and on the distance from the dissection area, there may be a sequence of stenotic signs ranging from high-grade obstruction to medium-grade or low-grade stenosis, all the way to completely normal findings. In a few cases, a complete occlusion after the bifurcation can be detected, which in B-mode typically resembles a long, pointed cap. In color flow Doppler mode, a flow signal is seen that reaches as far the tip of the cap, sometimes with a superimposed component showing retrograde flow.

Postoperative Findings

(Fig. 11.**7**, p. 352)

Control assessments are a frequent indication for ultrasound diagnosis, both in the early postoperative period to document the effectiveness of a procedure, and also later, to detect any re-stenoses, particularly in young patients (Valentine et al. 1996). Shortly after an operation, the examination can be difficult due to swelling of the neck, but continuous wave Doppler sonography can usually provide a reliable evaluation of the hemodynamic situation. The surgical technique used (e.g., patch) will determine whether a wide bifurcation can be displayed in B-mode. In color flow duplex sonography, well-developed flow separation zones occur (Steinke et al. 1991); occasionally, larger hypoechoic intravascular structures can be detected.

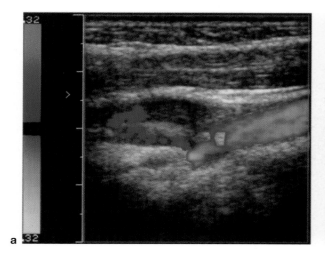
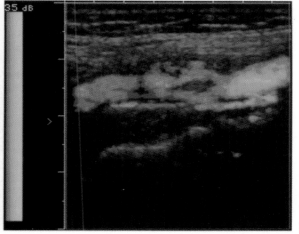

a   b

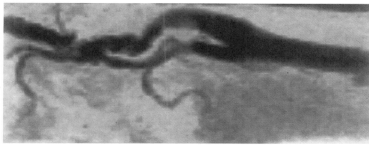

c

Fig. 2.**40**   Color flow duplex sonogram (**a**, **b**) and MR-an-giogram (**c**) of a bilateral carotid dissection in a patient with fibromuscular dysplasia (FMD) of the cerebral arteries. Note the better overview of flow vs. nonflow areas in the power energy mode (**b**) vs. the velocity mode (**a**). In contrast, the velocity mode nicely outlines *retrograde* flow components in the distal slow-flow territories of the ICA otherwise undi-agnosed by the power energy mode. Whereas the distal FMD-associated lesions are well demonstrated in the MRA and missed in CDFI, MRA fails to display the slow-flow ab-normalities in the proximal ICA

Sources of Error

Although there is usually no problem in the differen-tial diagnosis of isolated low, medium, and high-grade stenoses of the carotid arteries, mistakes in addition to those already mentioned are frequently caused by the following:

– Pseudo-occlusion versus total occlusion
– Proximal or distal stenoses outside the directly ac-cessible area
– Multiple obstructions of varying degree at several levels of the carotid system, both extracranial and intracranial

Misinterpretations can be avoided if the plausibility of findings from *all* arteries examined is carefully con-sidered, with particular attention to the following:

– Nonstenotic wall changes can be detected by con-ventional Doppler sonography only when directly accessible to ultrasound. Differential diagnosis from normal variants is only possible with sup-plementary duplex system studies.
– Flow signal changes in juveniles.
– Dilatative arteriopathy.
– Intracranial vascular processes, obstructions, fistu-las, and arteriovenous malformations.
– Analysis of altered collateral circulation.

– Confusion is possible between the internal carotid artery and the thyroid artery.
– Aneurysmal dilatations.
– Variations in location and course (a medial internal carotid artery, coiling, and kinking).
– Status after vascular surgery in the neck (scar formation, hematoma, edema, synthetic patch), or insertion of an extracranial–intracranial bypass.

A common misdiagnosis is the incorrect assumption of a stenosis (flow increase and audiosignal irregularities) in the presence of increased collateral flow, which is actually caused by a contralateral hemodynamically marked flow obstruction (80% stenosis or more, or occlusion). Furthermore, the degree of stenosis in arterial segments contralateral to a hemodynamically active stenosis is often overestimated, since the hemodynamic effects of the stenosis and the increased collateral flow are cumulative.

Since duplex sonography is able to provide structural confirmation in multiple planes of a slight or medium constriction of the lumen and depict the associated alterations in the vascular flow, appropriate classification of the degree of stenosis is usually possible in these cases. However, to assess the degree of the stenosis in high-grade stenoses of the internal carotid arteries is more difficult because of loss of image quality and signal-to-noise ratio as well as hemodynamic alterations induced in remote, either similarly diseased or structurally intact arterial segments. In these cases, evaluating the higher-grade stenosis should not only consider the local degree of stenosis and the circumscribed maximum constriction at one location, but should also take into account:

– The extent of the vascular process
– Collateral function in the vertebrobasilar system
– Possible cumulative effects due to stenoses that follow or precede the stenosis in question
– The current intracranial collateralization through the often (50%) insufficiently developed circle of Willis (Padget 1944)
– Collateralization through the external carotid artery anastomosis

If the vertebral artery collateralizes an occlusion in the ipsilateral carotid system (common or internal carotid artery), then the signal from the vertebral artery may be mistakenly interpreted as coming from the internal carotid artery, particularly in slender patients, since the blood vessels are projected on top of one other. Comparing this signal near the bifurcation to the vertebral artery signal at the posterior arch of the atlas usually allows the correct diagnosis to be made. Here too, extracranial detection of an internal carotid artery occlusion is possible using duplex sonography.

Proximal common carotid artery stenoses at the origin from the aortic arch can also cause changes in the Doppler curves in the neck region, preventing correct assessment of the degree of stenosis of the internal carotid artery at the bifurcation. It should be noted that in these cases, the ophthalmic collateral is not usually open, in spite of an intact external carotid artery. In the differential diagnosis, it is important to evaluate the collateral function in the remaining arteries supplying the brain: if there is no elevated blood flow in the contralateral carotid system or in both vertebral arteries, a careful search needs to be made for a proximal stenosis at the origin as an explanation. Intrathoracic pulsed wave Doppler sonography is often helpful here (Fig. 11.**3**, p. 346).

Figure 2.**41** provides a further summary of potential errors and interpretation problems associated with B-mode imaging of the common and internal carotid arteries (Hennerici 1987 b).

Diagnostic Effectiveness

■ Continuous Wave and Pulsed Wave Doppler Sonography

Using direct and continuous wave Doppler sonography, it is possible to exclude stenoses of over 50% and occlusions of the carotid vascular system with a high degree of certainty. With experienced investigators, the *specificity* lies between 88% and 98% (Table 2.**8**). Spectral analysis does not improve the diagnostic pre-

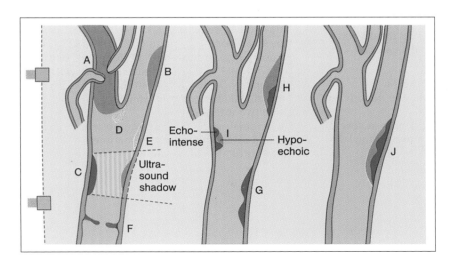

Fig. 2.**41** Potential sources of error in B-mode imaging of the extracranial carotid system
A   Carotid artery occlusion
B   A stenosis located too high
C   Wall changes are absent in the ultrasound shadow of a calcified plaque
D   Plaque in the bifurcation
E   Echolucent plaque
F   Buttonhole stenosis
G   Oblique surfaces on the plaque edges (no interposed ulcer)
H   Calcium deposit (no ulcer)
I   Superimposed hypoechoic structures (no ulcer)
J   Tandem plaques

dictive value of the method in excluding pathological findings.

In addition to reliable detection of obstructions in the extracranial carotid system, continuous wave and pulsed wave Doppler sonography can provide reliable estimates of the degree of stenosis (Table 2.**8**). However, the ultrasound investigation criteria and the reference methods used vary so widely in different studies that it is not possible to say whether the method can reliably detect a carotid stenosis of less than 70% in general. Therefore combined use of Doppler and imaging ultrasound techniques is current state-of-the-art practice (de Bray et al. 1995).

■ Conventional and Color Flow Duplex sonography

The specificity of a duplex investigation ranges from 47% to 100% depending on methodology, technical equipment, and expertise (Table 2.**9**). B-mode examinations alone are less valid, and no longer meet today's examination standards. X-ray angiography, which has usually been considered as the accepted reference method in making such assessments, can no longer be considered as the gold standard without further qualification (de Bray et al. 1995, Patel et al. 1995, Erdoes et al. 1996). Digital subtraction angiography, for example, often fails to detect small changes in the vessel walls, even when selective intra-arterial contrast medium is used (Eikelboom et al. 1983). Although still in its infancy, a few comparative studies of duplex sonography and magnetic resonance angiography have recently been published (Riles et al.

Table 2.**9** Calculated reliability of conventional duplex sonography versus intra-arterial angiography in internal carotid artery stenoses and occlusions

First author	Year	Arteries (n)	Specificity (%)	Sensitivity — Degree of stenosis (%) 0–20	20–40	40–60	60–80	80–100	Occlusion (%)	Validity (%)	Ulceration (%)	T
Abu Rahma	1987	200	71		51		81	79	94	71		D
Moore	1988	170	–		94		98		100	93,5	64	D
Mattle	1991	39	–		88		86			85		B
Comerota	1984	1723	–		87	72	66		64	80		B
Jones	1982	100	85			98			75	–		B
Terwey	1981	200	75		99		74		62	–		B
Wolverson	1983	97	47		83		78		36	69		B
Zwiebel	1983	393	95		72	52	21		91	62	44	B
Daiss	1984	488	89		50		85		87	66		D
Hames	1985	85	58		85		76	85	92	79		D
Hennerici	1984	193	76		88		100		71	83		B
Ratlif	1985	39	–		75	67	69	79	–	74		D
Zbornikova	1985	249	54	80	77		92		94	76		D
Ricotta	1987	1578	50		43		39	36	41	43	32	B
Langlois	1983	77	50	69	70		89		100	77		D
Cape	1984	586	93	64	59	71	68		54	84		B
Blackshear	1979	66	100	44	74		95		92	80		D
Rush	1985	20	81		75		92		100	–		D
Dreisbach	1983	101	94		84	76	83		94	–		D
Humphrey	1990	186	84		83	94	68		100	–		D
Burns	1985	500		71		38	26	6	83	–		D
Withers	1990	180		94		80	98		96	94		D
Lane	1982	304		80		76		–	–		D
Keagy	1982	171		94		89		94	93		D
Bornstein	1988	124		96				96	96		D

T: technique used (D: duplex mode, B: B–mode)

1992, Mattle et al. 1991, Huston et al. 1993) and showed good correlations (de Bray et al. 1995). Flow irregularities that appear physiologically in the carotid bifurcation can produce artifacts in magnetic resonance angiography. Duplex sonography using the color flow examination method is able to show that this apparent pathology represents normal flow separation phenomena, without any structural wall changes. Thus the two methods can complement each other.

The main advantage of color flow duplex sonography is that it can simultaneously display both the morphological and hemodynamic parameters, but there are still few data concerning the specificity of the procedure (Table 2.**10**) (de Bray et al. 1995/1996). Particularly when relationships are unclear and the examination conditions for other procedures are difficult, faster and more reliable assessment of the pathological relationships may be possible with this method.

In both in vitro and in vivo examinations, the sensitivity and specificity of high-resolution B-mode imaging (> 7.5 MHz) are very good for *nonstenotic plaques* (< 40% obstruction) (Table 2.**9**). Adequate differentiation of the various plaque forms, according to the criteria described above, is also possible. Using new methods that allow three-dimensional reconstruction from a series of sections (Steinke and Hennerici 1989, Meairs et al. 1995, Hennerici et al. 1995, Delcker et al. 1995), an outstanding image can be obtained that demonstrates the total extent and shape of the plaque formation. However, equipment manufacturers have been slow to provide the software required for this. As an atheroma expands and causes increasing stenosis, the quality of the image from various angles becomes poorer due to energy loss, even with sufficiently precise ultrasound. Uncomplicated plaques, usually consisting of an atheromatous seed, often still provide a good image, and the B-mode image is not significantly altered by fibrous caps covering a smooth muscle cell bed, or by collagen, elastin, or intracellular and extracellular lipids. However, *calcium deposits* and *surface irregularities*, which indicate progressive, complicated plaques, are unfavorable conditions for imaging in a high-resolution system. Whether this association with patients' individual prognoses can actually be scientifically established remains to be seen. At present, a large continental trial (ACSRS—asymptomatic carotid stenosis and risk of stroke) is being conducted in Europe that hopefully will answer this challenging question.

Plaque-associated flow changes and plaque movements can also be analyzed using color flow duplex sonography, and three-dimensional analysis is possible in hemodynamic investigations, although it is still somewhat experimental (Meairs et al. 1995, Hennerici et al. 1995) (Fig. 1.**12**). With appropriate clinical indications, a supplementary duplex-sonographic examination is therefore mandatory when the conventional Doppler sonogram or angiogram have normal or doubtful appearances—particularly signs of embolizing wall changes, since fresh thrombotic deposits that may have increased embolic potential, as well as niche and ulcer formation, can be detected fairly reliably using the color flow Doppler mode. The same applies to the detection of aneurysmal wall changes. Follow-up observations after obliterating carotid artery occlusions, and differentiation between dilatative vascular processes and obstructive ones are further important issues (Steinke et al. 1991, Bonithon-Kopp et al. 1996).

Combined use of continuous wave or pulsed wave Doppler sonography and color flow duplex sonography increases the diagnostic efficacy and validity of the classification of extracranial carotid disease (Vanninen et al. 1995, Howard et al. 1996, Carpenter et al. 1996). The advantage of the latter procedure, especially for the less experienced investigator, lies in the fact that it is easier to learn and provides both visual and acoustic access, in contrast to the traditional, rather one-sided acoustic investigation method. Significantly shorter examination times are possible, particularly in complicated high-grade vascular processes in which it is difficult to display the anatomy. Standardized research to determine the validity of this method using defined stenosis criteria in a reference procedure has not yet been carried out and probably never will be, since even without conventional angiography it is possible to establish the diagnosis of a carotid stenosis of more than 70% using the appropriate criteria for surgery (NASCET 1991, ECST 1991) (de Bray et al. 1995, Patel et al. 1995). Cervical bruit analysis, which has long been used in the US as a screening tool for the detection of carotid disease in patients at risk, is only predictive if combined with extracranial ultrasound techniques (Hennerici et al. 1981, Sauve et al. 1994).

In *high-grade and subtotal flow obstructions,* all of the available procedures may fail. It is difficult to differentiate a pseudo-occlusion from a complete occlusion, even when all possible hemodynamic and im-

Table 2.**10** Calculated reliability of color flow duplex sonography versus intra-arterial angiography in internal carotid artery stenoses and occlusions

First author	Year	Arteries (n)	Specificity (%)	Sensitivity — Degree of stenosis (%)					Occlusion (%)	Validity (%)
				0–20	20–40	40–60	60–80	80–100		
Steinke	1990	60	88	69		78	100	88	100	80
Polak	1992	41		91			85		100	88

aging methods, including angiographic procedures, are used. It can only be done when a weak flow signal is positively confirmed in an almost occluded blood vessel. It remains to be seen whether color flow duplex sonography will prove superior to the other methods in this instance as well, especially once algorithms for displaying slow flow signals, combined with new echo contrast media ("harmonic imaging," p. 3) have been developed (Steinke et al. 1996, Wernz et al. 1997).

Vertebral Artery System

Principle

There are fundamental differences in the diagnosis of vertebral artery obstructions compared to the extracranial carotid system:

- The course of the vertebral artery cannot be examined continuously.
- Flow obstructions are usually located at the origin, at the atlas loop, or in the intracranial segment.
- Variations in vessel caliber and hereditary anomalies occur more frequently here than they do in the carotid system.
- Subclavian artery flow obstructions affect Doppler signals from the vertebral artery.

While it is sufficient to record Doppler signals at a single point to provide an overview, examinations in several vascular segments have to be undertaken when there are pathological results, or in the context of specific problems (p. 21).

Findings

If the vertebral artery segment affected is accessible to direct examination with ultrasound, then the assessment of the degree of stenosis using Doppler sonography mainly follows the criteria discussed above in detail for the carotid artery system (pp. 53 ff). However, there are more problems here, because the origin of the vertebral artery can usually only be depicted in a single projection, and the degree of stenosis in the vascular course where it loops round the atlas cannot be reliably evaluated—often not even multiple projections are used. Further distally, and up to the point at which both vertebral arteries join the basilar artery, there are fluctuations in the vascular caliber. These variations can lead to evaluation difficulties when segmental or extended stenoses are involved. Similar problems, together with artifacts, result from subtraction angiography.

Some unique characteristics result from the

- Origin of the vertebral arteries from the subclavian artery

in addition to problems in interpreting processes involving several blood vessels, in:

- Unilateral tandem stenoses
- Proximal occlusions with distal collateralization

If hemodynamically relevant obstructions are present at the origin, then the flow relationships in the subsequent vertebral artery system change dramatically. The following may appear:

- Contralateral and ipsilateral decreases in flow
- Intermediate flow patterns
- Retrograde flow waveforms

As in the carotid system, the following parameters for spectrum analysis and for classifying the degree of stenosis are used:

- Systolic and diastolic peak frequency/velocity
- Average Doppler frequency/velocity
- Amplitude-intensive frequency/velocity band distribution

Depending on the development of collateral arteries, various steal forms are encountered (Vollmar 1975) (Fig. 2.**45**):

- Vertebrovertebral
- Caroticobasilar
- Externovertebral
- Caroticosubclavial

The additional unique trait of the vertebral artery system is the existence of many collateral circulatory pathways, which usually continue to support physiological blood flow through the basilar artery, even when the proximal or distal segment of either or both vertebral arteries is occluded. This is a significant difference in comparison to the carotid artery system, where stenoses at the origin of the internal carotid artery are compensated for less efficiently and only extracranially, via the ophthalmic anastomosis.

■ Continuous Wave and Pulsed Wave Doppler Sonography

Increased Flow Velocity

Stenoses. As in the carotid artery, *stenoses at the origin of the vertebral artery* (Fig. 2.**42**), or around the atlas loop, often show systolic or diastolic flow accelerations, which are sometimes accompanied by systolic deceleration and audiosignal irregularities. Near the *atlas loop*, care should be taken to avoid confusing a systolic deceleration with the physiological change in flow direction through the vascular loop itself (pp. 50–51). The Doppler signal from a systolic deceleration is sharper; spectral analysis shows pathological changes. Depending on the location of the stenosis, flow changes can be found in the proximal or distal segments, or in both vascular loop segments around the atlas. Clear audiosignal irregularities are usually found in the efferent loop segment. In the case of physiological vascular perfusion, the efferent segment is usually the distal vertebral artery, and in retrograde perfusion it may also involve the proximal vascular segment.

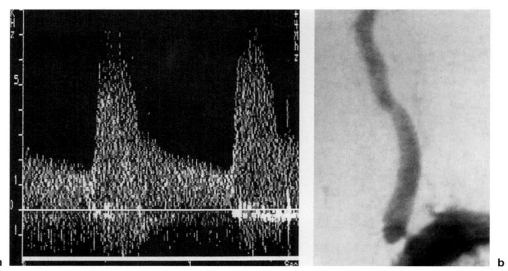

a   b

Fig. 2.**42 a**   Doppler spectrum and **b** corresponding angiogram of a high-grade stenosis at the origin of the vertebral artery

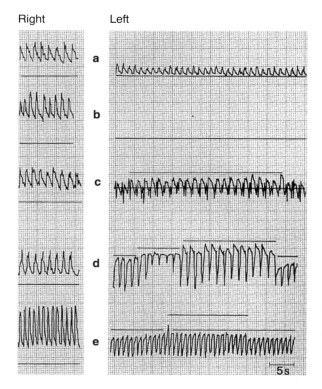

Right   Left

a

b

c

d

e

5 s

Fig. 2.**43**   Summary of abnormal findings in the left vertebral artery using continuous wave Doppler sonography
a Decreased flow velocity
b Absent signal but contralaterally elevated flow velocity
    (especially diastolic)
c Alternating flow around the baseline, without significant
    change during compression maneuver of the vascula-
    ture in the ipsilateral upper arm (–)

Collaterals. There may also be an increase in vascular flow velocity when:

– The contralateral vertebral artery is hypoplastic, aplastic, contains a high-grade stenosis, or is occluded (Fig. 2.**43 a, b**).
– The contralateral vertebral artery is perfused in a retrograde direction (Fig. 2.**43 e**).
– There is a significant carotid system obstruction.

Decreased Flow Velocity

Normal Doppler waveforms with clear side-to-side asymmetries (> 75% of the remaining diastolic flow) suggest various potential causes, provided that the examination excludes abnormal flow acceleration *at least at the origin and near the vertebral atlas loop* (Fig. 2.**43 a**):

– Unfavorable transducer location. This finding does not imply any pathology. It is the most common cause of a misdiagnosis.
– *Recommendation:* additional examination using color flow duplex sonography.
– Vertebral artery hypoplasia or dilatation. Color flow duplex sonography is helpful in distinguishing between a normal variant or dilatative arteriopathy.

d Systolic flow reduction above the baseline, with a significant change during ipsilateral upper arm compression
(–)
e Retrograde flow, without significant change during ipsilateral upper arm compression, apart from a small reactive, hyperemic increase in diastolic flow after release.
Contralaterally, there is a compensatory increase in
blood flow

- Flow obstruction in the intracranial segment of the vertebral artery (when there is unilaterally reduced flow velocity) or in the basilar artery (when there is bilateral flow velocity reduction).
- *Recommendation:* supplementary transcranial examination.
- Far less often than might be thought, a decrease in flow velocity results from changes in the neck and head position caused by partial or complete compression of the vertebral artery (neurological symptoms or subjective complaints are associated with this finding even more rarely), but misdiagnosis is frequent! From thousands of studies performed in our vascular laboratories, only three patients were found who suffered from sustained neurological deficits where both vertebral arteries were functionally obstructed (e.g., by reclination or inclination) in the absence of any collateralization through the circle of Willis from the carotid artery system.

Slosh Phenomenon

As in the internal carotid artery, an abnormal, usually low-amplitude flow signal, with a narrow systolic peak and at the same time a shorter backflow phase and a significantly delayed diastolic flow component, is found in vertebral artery dissections. This is not caused by intermediate blood flow, but is probably due to a combined intraluminal flow signal and secondary pulse waves that are superimposed on the Doppler signal by intramural hematoma formation. In contrast to carotid dissection, this phenomenon is encountered much less frequently in vertebral artery dissection (in approximately one-third of the cases). The diagnosis can usually be confirmed during the follow-up examination by a normalized flow signal after an interval of days to a few weeks. More often, unspecified findings are encountered in Doppler sonography when it is used to evaluate a vertebral artery dissection. Since angiographic confirmation of this clinical picture can be difficult, and experience shows that the intramural bleeding can be missed even with MRI and MRA, this condition is probably not diagnosed often enough (Steinke et al. 1994). Recent experience with color flow duplex sonography is promising (Sliwka et al. 1992, Bartels and Flügel 1996, Pfadenhauer and Müller 1995).

Intermediate Blood Flow

Different intermediate forms of flow velocity, also called pendular flow (alternating forward and backward flow) or systolic flow reduction (von Reutern and Pourcelot 1978), are found in the vertebral artery (Fig. 2.**43c, d**) when:

- A flow obstruction in the proximal subclavian artery or the innominate artery is detected. This is due to what is termed *latent subclavian steal phenomenon*, which can be a *temporary* phenomenon in hyperemia of the ipsilateral upper extremity (Fig. 11.**11**, pp. 358–359). A medium-grade stenosis

is often found to be the cause, and less frequently proximal subclavian artery occlusion.
- There are completely normal flow relationships in the proximal subclavian artery. This rare situation suggests *vertebral artery hypoplasia* and should be confirmed by duplex imaging (Fig. 2.**44**).

Absent Signal

This finding can point in several diagnostic directions:

- *Systematic error* (a frequent occurrence), e.g., due to incorrect or unfavorable probe placement or anatomically unfavorable conditions for conducting the examination. Supplementary duplex sonography is necessary.
- A *pathological finding,* e.g., an occlusion or segmental occlusion (aplasia) in the course of the vertebral artery. To differentiate this, duplex sonography also has to be used for an examination in all three locations (proximal, in the neck region, and at the posterior atlas arch). When there is a segmental occlusion, distal filling of the vertebral artery through collateral vasculature is frequently found, so that even when pulsed wave Doppler sonography is being used, evaluation of the intracranial segment of the vertebral artery from a transnuchal direction has to be carried out in order to ensure that an extensive extracranial occlusion is not missed. In contrast to proximal occlusion of the vertebral artery, a reduced flow signal is generally found in a distal occlusion. Stenosis of the vertebral artery is only rarely accompanied by an absent signal—at most, an extremely high-grade stenosis at the origin can cause a signal near the mastoid process, which can only be detected with great difficulty. When there is an absent signal, as in vascular hypoplasia, a compensatory increase in flow through the contralateral vertebral artery is the most reliable indirect indicator.
- A *flow signal that is only detectable acoustically.* This is often due to poor ultrasound examination conditions (unfavorable signal-to-noise ratio, deep location of the loop of the vertebral artery around the atlas, large distance between the examination point and the probe position). Supplementary examination using a pulsed wave Doppler or duplex sonography can be helpful.

Retrograde Blood Flow

Significant obstructions affecting the proximal subclavian artery (over 80% stenosis or occlusion) or the innominate artery often form a *permanent subclavian steal phenomenon,* with the arteries of the arm being supplied by the ipsilateral vertebral artery (Fig. 2.**43e**). The presence of retrograde flow can be demonstrated by a compression test (p. 31), which is accompanied by a clear change in the Doppler signal recorded from the vertebral artery. This effect is often clearer in the audiosignal than in the documented waveform changes. In normal conditions, no change,

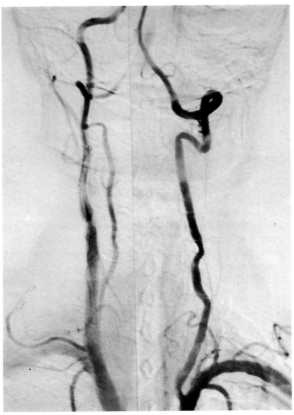

a

Fig. 2.**44a** Angiogram, **b**, **c** color flow duplex sonography, and **d**, **e** Doppler spectra (from the marked region) in a patient with a hypoplastic right vertebral artery. The lumen of the left vertebral artery is 4 mm (**c**), and the lumen of the right one is 1.1 mm (**b**). **d** Alternating flow in the right vertebral artery

or at most, only a slight flow reduction may occur during compression, followed by an increase of flow after the blood vessel is reopened. Comparisons with the angiographic findings have confirmed the excellent reliability of the method (100%) for the diagnosis of a permanent subclavian steal phenomenon (Keller et al. 1976b, Pourcelot et al. 1977, von Reutern and Pourcelot 1978, Hennerici and Aulich 1979, Büdingen and von Reutern 1993).

The various types of steal pathways are summarized in Figure 2.**45**.

■ Conventional and Color Flow Duplex Sonography

Duplex sonography is increasingly helpful in diagnosing the vertebral artery, and clearly improves the validity of the findings: this is however, mainly true for color flow duplex imaging, whereas conventional duplex scanning has only limited value. Areas of interest are:

– Segmental occlusions of the neck region.
– Stenoses before the vessel enters the intervertebral foramina.
– Depicting the lumen allows quantitative measurement of the diameter and direct evaluation of intra-arterial flow relationships (i.e., diagnosis of hypoplasia, stenosis, dilatative arteriopathy).
– Proximal dissections (V_1) of the vertebral artery in the neck. This issue has recently been raised by Bartels (1996), who observed dissections in V1/V2 segments far more often than hitherto expected and with reasonable sensitivity (> 90%): these findings from a small series need further confirmation, however.

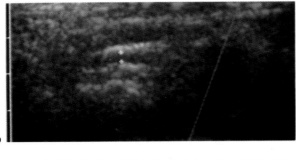

b

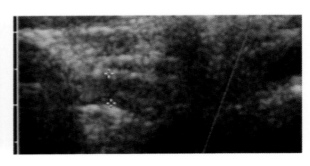

c

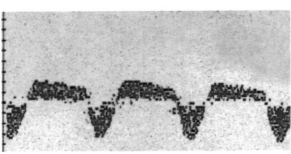

d

Right vertebral artery

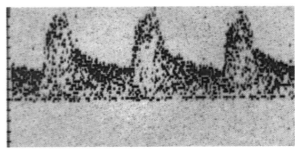

e

Left vertebral artery

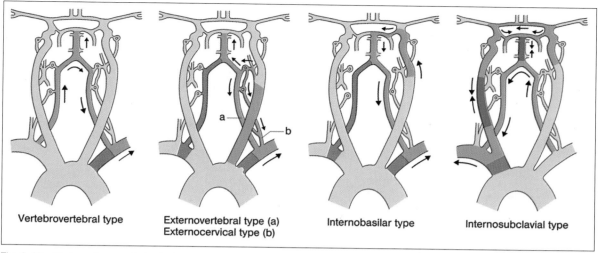

| Vertebrovertebral type | Externovertebral type (a) Externocervical type (b) | Internobasilar type | Internosubclavial type |

Fig. 2.**45** Various forms of steal phenomena associated with proximal supra-aortic flow obstructions

All of the parameters provide important indirect indications of proximal or distal flow obstructions.

Stenoses *in the neck region before the vessel enters the intervertebral foramina* are extremely rare, but can be better detected with color flow duplex sonography than with simple continuous wave duplex sonography.

Evaluation

Interpreting the Doppler-sonographic findings obtained is more difficult in the vertebral than in the carotid artery system. Combining different examination techniques and gathering data from several probe positions, evaluation mainly involves:

- Registering flow obstructions.
- Distinguishing between the different forms: proximal stenosis at the origin, proximal segmental occlusion, stenosis in the neck region, stenosis at the atlas loop, distal high-grade stenosis, distal segmental occlusion, complete occlusion, hypoplasia, aplasia, dissection (Table 2.**11**).
- A permanent or temporary subclavian steal phenomenon can be detected with great reliability if the flow relationships in the subclavian or the innominate arteries are observed and appropriate compression procedures are applied.
- When assessing the global circulation of the brain in processes involving blood vessels, in the anterior circulation, it is very important to know the collateral function of the vertebrobasilar arteries.

Sources of Error

Systematic errors occur when there is:

- A weak or missing signal
- A poor signal-to-noise ratio
- Unfavorable anatomy

- Venous superimposition
- Stenosis in hypoplastic segments with extremely slow flow velocity

Confusion is possible with:

- The inferior thyroid artery or the thyrocervical trunk and the subclavian artery, in proximal ultrasound application.
- The internal carotid artery in a slender neck (when ultrasound is applied at the arch of the atlas).
- External carotid artery branches (e.g., the occipital artery and other cervical arteries, when ultrasound is applied at the arch of the atlas).
- Other cervical arteries or open collaterals in rare types of steal phenomenon. The flow velocity in the cervical collateral vasculature can be so pronounced that the normal Doppler signals from the distal vertebral artery may be missed.

When only a selective examination is carried out at a single registration point:

- Stenoses at the origin, and *proximal* occlusions, may be collateralized via:
  - Thyrocervical trunk anastomoses; a normal signal results at the atlas arch.
  - External carotid artery branches into the distal vertebral artery; a normal signal is obtained at the atlas loop.
  - *Distal occlusion* of the vertebral artery; collaterals of the cervical arteries and the external carotid artery produce a normal signal proximally, and in some cases also distally (Table 2.**11**).

Various combinations of findings from pathological vascular processes in the distribution area of the vertebral artery are summarized in Table 2.**11**. Comparisons between various vascular examination methods are given in Table 2.**12**.

Table 2.**11** Summary of findings at various vertebral artery recording sites for different pathological conditions

	Doppler findings in the vertebral artery system				Special information (B-mode, CFDS)
	Origin	Neck area	Atlas loop	Intra-cranial	
Stenosis at the origin – High-grade – Low-grade	>2 ss <2 ss	↓ n	↓ n	↓ n	Detection is possible at the origin Detection at the origin is difficult
Stenosis at the atlas loop – High-grade – Low-grade	↓ n	↓ n	>2 ss <2 ss	↓ n	Detection is possible at the atlas loop Detection is possible at the atlas loop
Stenosis – Intracranial, high-grade – Intracranial, low grade	↓ n	↓ n	↓ n	>2 ss <2 ss	Normal lumen, possibly with a reduced color-coded signal
Occlusion (V0–V4)	∅	∅	∅	∅	Absence of color signal
Segmental occlusion (V0–V2, e.g., proximal dissection)	∅	∅	n	n	Cervical collaterals are detectable, occasionally wall lesion detectable at the site of dissection
Distal occlusion (V3, V4, e.g., distal dissection)	↓	↓	↓	∅	Normal lumen, reduced color signal in the neck region
Hypoplasia	↓	↓	↓	↓	Narrow lumen
Functional collaterals in carotid occlusion	↑	↑	↑	↑	Lumen often widened

n   Normal flow signal   ∅ No signal
↓   Reduced flow signal   ss: signs of stenosis (Tables 2.**3**, 2.**5**)
↑   Increased flow signal

Diagnostic Effectiveness

■ Continuous Wave and Pulsed Wave Doppler Sonography

Studies have shown that, compared to angiography, the diagnostic value of Doppler findings from the vertebral arteries is good if differentiated flow obstruction evaluation is not required (e.g., hypoplasia, stenosis, or occlusion). The sensitivity lies between 70% and 90%, and the specificity is reported to be over 95% (Keller et al. 1976 b, von Reutern and Pourcelot 1978, Büdingen and von Reutern 1993, von Reutern and Clarenbach 1980, Winter et al. 1987).

■ Conventional and Color Flow Duplex Sonography

More recently detailed studies have been carried out, mainly involving vertebral artery stenosis, dissection, and vessel variants. Ackerstaff et al. (1984) described a sensitivity for stenoses of more than 50% and up to 81%, and a specificity of 88% for conventional duplex systems, which are both higher (> 90%) if color flow imaging devices are used (Sliwka et al. 1992, Bartel and Flügel 1995, Pfadenhauer and Müller 1995).

Supra-Aortic System

Principle

The arterial segments that supply the brain close to the aortic arch (the proximal subclavian artery, innominate artery, and proximal common carotid artery) can only be evaluated to a limited degree with Doppler sonography, since they are poorly accessible and beyond the scope of continuous wave ultrasound. Usually, early atherosclerotic plaques in the proximal section are not detectable. Using a pulsed wave Doppler instrument, mild and moderate stenoses can be recorded, provided the sample volume is applicable intrathoracically to the origin of the arteries from the aorta (Fig. 11.**2**, pp. 344–345). High-grade stenoses and occlusions lead to changes in the Doppler signals in the neck arteries and can thus be identified indirectly by continuous wave and pulsed wave Doppler methods.

Table 2.**12**   Comparison of vascular examination methods

Method	Potential findings	Limitations and disadvantages
Pulse palpation	– Subclavian artery occlusion – Common carotid artery occlusion – External carotid artery occlusion	– Internal carotid artery not distinguishable – Side-to-side differences in the carotid pulse are often a methodological artifact
Auscultation	– Subclavian artery stenosis – Medium-grade carotid stenosis – Arteriovenous and dural fistula – Angioma	– The internal/external carotid artery cannot be differentiated – Flow sounds propagated from the heart or the subclavian artery may be misleading – Highest-grade and low-grade stenoses are often misdiagnosed – Collateral flow creates bruits in normal vasculature
Doppler sonography (indirect methods)	– Occlusion or stenosis >80% of the common and internal carotid arteries	– Locating the obstructions in the vascular system is not possible – Occlusion and stenosis cannot be differentiated – Stenoses <80% and bifurcation processes cannot be detected – False-negative results in approx. 20% of carotid obstructions >80%
Doppler sonography (indirect and direct methods) with spectrum analysis	– Carotid processes with lumen constriction >40% – Differentiation between occlusion and high-grade stenosis is often possible – Evaluating the vertebral artery system – Subclavian steal phenomenon – Quantitative evaluation of acoustic spectral phenomena	– Carotid stenoses <40% cannot be detected with certainty – Difficult to differentiate between vertebral artery aplasia-occlusion, and stenosis-hypoplasia – Only in a few cases, intracranial processes may be suspected based on indirect criteria
Conventional duplex sonography	– Low-grade carotid stenoses, plaques and large ulcerations – Evaluating vascular width, wall movement, and plaque structure	– Occlusions and subtotal stenoses are sometimes difficult to separate – Variants in anatomy of the carotid bifurcation and complex plaque structure reduces validity
Color flow duplex sonography	– Isodense plaques/thrombi vs. flowing blood – Subtotal stenosis vs. occlusion (i.e., pseudo-occlusion) – Small ulcers and dissections	– Lesions outside of the area directly examined can be easily missed – Artifacts and aliasing may cause misinterpretation

Findings

Subclavian Artery

■ Continuous Wave and Pulsed Wave Doppler Sonography

Obstruction Proximal to the Origin of the Vertebral Artery

Medium and high-grade stenoses cause a flow acceleration, with changes in the audiosignal and possibly a systolic deceleration, comparable to those changes associated with external carotid artery stenosis. Distal to the stenosis, a pathological waveform can be observed, with loss of the early diastolic backflow phenomenon (Fig. 11.**11**, pp. 358–359).

If the stenosis cannot be directly evaluated, or an *occlusion* is present, a reduced flow signal and an absent early diastolic backflow phenomenon is characteristic (Fig. 11.**12**, pp. 360–361). Since blood then flows through the vertebral artery in a retrograde direction and drains into the distal segment of the subclavian artery, the curves from the two blood vessels are similar in form and in the resulting audiosignal. It is only rarely that a temporary or permanent *subclavian steal phenomenon* (i.e., the hemodynamic pattern in the absence of typical signs or symptoms (Hennerici et al. 1988) fails to form—for example, when the vertebral artery originates directly from the aorta. The features of the subclavian steal phenomenon are described in detail on p. 81.

Even if no direct signs of stenosis are detected in the continuous wave Doppler sonogram, due to the inaccessibility of the site of stenosis far away from the close-up focus of the ultrasound area, a high-grade stenosis may still be present. This can usually be differentiated from occlusion by a careful supplementary examination using the pulsed wave method, with selective depth-gating.

Obstruction Distal to the Origin of the Vertebral Artery

- No steal phenomenon in the ipsilateral vertebral artery.
- If located in a segment directly accessible to ultrasound, reliable differentiation between stenosis and occlusion is possible.
- If located distally, there is often no change in the waveform of the proximal subclavian artery, or only indirect signs.

With spectral analysis, quite reliable quantification of the maximum systolic frequency is usually possible. It is only in high-grade stenoses that the signal is sometimes difficult to document.

■ Conventional and Color Flow Duplex Sonography

B-mode. Sometimes, the wall changes can be structurally detected from a supraclavicular direction using duplex sonography. Aneurysmal dilatations can be well documented.

Color flow Doppler mode. Diagnosing the subclavian artery in the supraclavicular fossa is easier using color flow duplex sonography, since it allows the artery to be identified more quickly in contrast to the neighboring vasculature, especially venous structures.

Innominate Artery

■ Continuous Wave and Pulsed Wave Doppler Sonography

As a result of compensatory collateral circulation, *hemodynamically significant obstructions* of the innominate artery cause a variety of changes in the extracranial system (Hennerici et al. 1981 a):

- In the orbital arteries, there is often a characteristic finding of a double-peaked physiological flow during systole, corresponding to a very characteristic breath-like audiosignal acoustically. When the arteries of the right upper arm are compressed, diastolic flow sometimes increases.
- In the ipsilateral internal and common carotid arteries, quite a variety of flow changes are seen in individual cases and even in the same patient on different occasions, depending on the location and lability of the watershed. *All* of these changes undergo modulation during the compression test (Fig. 2.**46**).
- The external carotid artery also often shows altered, but slightly less diverse, flow patterns, which are modulated in the compression test.
- In the ipsilateral vertebral artery, all physiological (rare), intermediate, or retrograde flow variations can be present.

As a peculiarity, the carotid and vertebral systems show rather *unstable hemodynamic behavior*, depending on systemic blood pressure, muscular activity, and changes in the arterial flow resistance of the right arm. Neurological signs and symptoms are seldom reported; rarely, symptoms may be provoked by physical stress, since a variety of compensatory mechanisms are available:

- The contralateral carotid artery
- The contralateral vertebral artery
- The circle of Willis

■ Conventional and Color Flow Duplex Sonography

B-mode. Due to the significant decrease of flow in the vascular territory affected, duplex sonography is extremely helpful in detecting structural changes that would otherwise have been missed because of the lack of associated flow disturbances, e. g., in the absence of any detectable velocity at rest, even a significant stenosis in the watershed region can remain undiscovered when using Doppler sonography alone. An additional B-mode examination is therefore particularly important if surgery is planned.

Color flow Doppler mode. The innominate artery and its branches can sometimes be documented, proceeding from a supraclavicular direction and using low-frequency probes such as those generally used in echo cardiography. By simultaneously displaying hemodynamic and structural parameters in the distal neck arteries, one can improve the results of both Doppler mode and B-mode studies.

Aortic Arch

■ Transesophageal Echocardiography

Changes in the aortic arch and in the heart, which may be a significant source of cerebral emboli, can be readily detected using transesophageal echocardiography (TEE) (Amarenco et al. 1992, 1994, 1995) (Fig. 2.**47**). Relevant works on echocardiography should be consulted in connection with the method itself and the interpretation of its findings (Tunick et al. 1991, Rubin et al. 1992, Horowitz et al. 1992, Amarenco et al. 1992, Jones et al. 1995, Rauh et al. 1996).

Evaluation

Continuous wave Doppler sonography is only of limited value in assessing vascular segments that are located intrathoracically (subclavian artery, innominate artery, and proximal common carotid artery) but they can usually be depicted using the pulsed wave Doppler method. Associated flow velocity changes in the distal territories of the extracranial carotid and vertebral artery systems can be well documented using all the available methods. They involve many hemodynamic changes.

Sources of Error

If severe flow reduction is present in obstructive lesions involving the innominate artery, it is often difficult to detect and assess additional structural changes in the distal territory. Occasionally, arterial signals are so distorted that they can be mistaken for venous signals. In the "low flow" watershed area, low and medium-grade stenoses, as well as high-grade ones, then fail to cause any abnormal changes in the Doppler signal, and are easily overlooked during the diagnosis. Even when angiography is used, such processes can be missed due to limited documentation in one single projection or inadequate concentration of contrast medium in the post-stenotic vascular segments. It is therefore mandatory to search deliberately for any as-

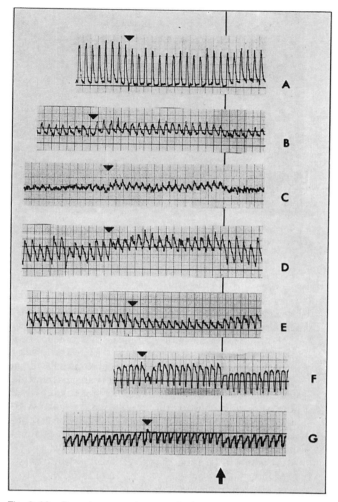

Fig. 2.**46** Doppler waveforms of the common carotid artery in seven patients with high-grade obstructions of the innominate artery. The arrows mark the compression (↓) and release (↑) of the right brachial artery in the upper arm. Physiological blood flow, directed toward the brain, corresponds to an upward curve deflection, retrograde flow to a downward deflection. Consistent with the hemodynamic relationships in the carotid artery system, the following variations are found, as well as transitional forms

A An apparently normal flow profile, the diastolic flow velocity of which changes during a compression test, reflecting a hemodynamic connection between the cerebral and the peripheral arterial systems

B Flow signal with reduced amplitude, **C** with only slightly modulated amplitude, and **D** with an exaggerrated modulated signal, with a significantly delayed diastolic decrease in the flow velocity. All of these have in common an increase of orthograde flow pattern during compression

E, F Early systolic retrograde flow, with a late systolic physiological flow signal and a renewed diastolic decrease in the flow velocity (intermediate flow); when the upper arm is compressed, physiological blood flow directed toward the brain can result (**E**), or, during the reactive hyperemic phase, a "carotid steal" phenomenon may appear (alternating flow) (**F**)

G A permanent "carotid steal" syndrome, with retrograde blood flow

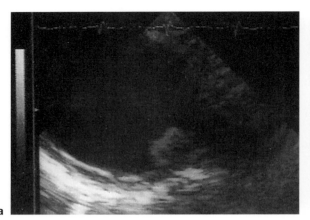

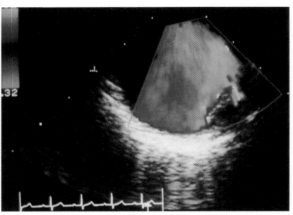

a

b

Fig. 2.**47 a** Transesophageal echocardiography of the aorta, with thrombotic plaque formation. **b** Transthoracic echocardiography depicting a thrombus near the wall of the left ventricle of the heart, using color flow duplex sonography

sociated atherosclerotic signs using a supplementary duplex system examination. Since MRA and helical CT share the same disadvantages, TEE of the aortic arch and the proximal branching cerebral arteries is often necessary (Rauh et al. 1996, Amarenco et al. 1992, 1994, 1995). In addition, otherwise unidentified important sources of cardioembolic stroke may be diagnosed in patients *without* evidence of artery-to-artery embolism or hemodynamic sources of cerebral ischemia (i.e., left atrial thrombus, atrial septal aneurysm with/without patent foramen ovale [PFO], ventricular aneurysm, global LV wall motion abnormalities, etc.).

Even high-grade stenoses or occlusions of the innominate artery can be missed by Doppler sonography when the examiner only expects highly abnormal flow curves, such as those shown in Figure 2.**46 C–G.** Less impressive findings, such as those shown in Figure 2.**46 A**, **B**, can then be misinterpreted prematurely as normal when functional diagnosis and upper arm compression are not carried out.

Diagnostic Effectiveness

High-grade flow obstructions in the innominate artery are rarer than stenoses or occlusions in the extracranial carotid artery system, the left subclavian artery, or the vertebral arteries (approximately 0.6% in the series examined by Hass et al. 1968, and also in our own study, Hennerici et al. 1981 b). Clinically, they are usually asymptomatic, just like subclavian artery stenoses and occlusions, but they can lead to significant changes in the extracranial and intracranial circulation including the subclavian steal phenomenon, and create a complex collateral circulatory network that can be analyzed sonographically in detail. More often, such lesions cause ischemic or embolic disturbances in the blood supply to the upper extremity. During a clinical examination in which *blood pressure*

in both upper arms should always be measured, such conditions can already be suspected. However, since measuring the blood pressure does not allow precise localization of the vascular process (e.g., location proximal versus distal to the origin of the vertebral artery), a Doppler-sonographic examination is indicated. Additionally coexisting flow obstructions in the carotid artery territory are sometimes difficult to assess, both with Doppler sonography and angiography. Duplex sonography is the most useful additional method here. Clinically, such vascular processes become active (transient or permanent neurological deficits) once the capacity of the collateral vasculature is limited by additional obstructions in other extracranial or intracranial arteries or when embolism occurs.

Due to the rarity of these findings and the limited angiographic display that is possible, hardly any comparative studies exist concerning the sensitivity and specificity of ultrasound diagnostics (Rautenberg and Hennerici 1988, Grosveld et al. 1988).

Local wall changes and low-grade or medium-grade stenoses are almost always missed in routine ultrasound studies if the examiner is not specifically informed about the grounds of suspicion. Studies in the intrathoracic zone are time-consuming and demanding and thus not included in routine services but may be useful on request.

3 Intracranial Cerebral Arteries

Examination

Special Equipment and Documentation

For the study of the intracranial basal cerebral arteries, a low frequency (1.5–2 MHz) pulsed wave Doppler system with high energy output is necessary. Continuous wave Doppler and conventional duplex systems cannot be used for this purpose. However, since 1990 color flow duplex analysis operating at 1.5–3 MHz emission frequency has been increasingly suggested.

Normally, a *hand-held probe* is used at various access points to examine the arteries at the base of the skull. Data from successive vascular segments are recorded under depth control. As there is no visual feedback about the relationships to the individual segments examined from the circle of Willis if only the Doppler spectrum is recorded, inexperienced examiners, in particular, find it difficult to classify the individual signals correctly solely on the basis of informations about depth and flow direction.

An earlier development in hand-held technology designed to circumvent these disadvantages was the *two-dimensional scanning system*. With this method, it was possible to display the position of the sample volume in two dimensions on a monitor and carry out a transtemporal examination on both sides, either simultaneously or successively, using two transducers in addition. Helmets incorporating ultrasound probes allowed changes in position of probes or sample volumes to be documented using voltage converters in a computer. This information was displayed on a monitor as a reference, and provided some views in the horizontal, lateral, and anteroposterior projections (Fig. 3.**10**). Although this system facilitated the documentation of recording sites of vascular segments, this procedure has not gained widespread acceptance.

Much more promising in this regard is the use of *color flow duplex sonography* to display the intracranial blood vessel system simultaneously with neighboring structures (Bogdahn et al. 1990, Bogdahn et al. 1993, Rosenkranz et al. 1993, Baumgartner et al. 1996, Ries et al. 1997, Baumgartner and Baumgartner 1996, Seidel et al. 1995) (Fig. 3.**7**).

As with helmet equipment, stable probe fixation for a longer period is required for *recordings that assess function and progress over time* (e.g., for *monitoring* purposes). In principle, simultaneous examination from both sides is also possible—for example, to carry out a comparative analysis of flow conditions in the right and left middle cerebral arteries.

Examination Conditions

Patient and Examiner

When the *intracranial arteries* are examined, it is necessary for the patient to be evaluated under controlled conditions of respiration. The patient should therefore lie relaxed in the examining room for some 20–30 minutes before the procedure takes place, or one should obtain constant end-expiratory P_{CO_2} values. Of course, the extracranial arteries can be examined during this time (Hennerici et al. 1987 b, see p. 25).

Conducting the Examination

The aim of the examination of the intracranial cerebral arteries is the documentation of the Doppler signals (i.e., PW Doppler only), or, increasingly, a display of flow velocity signals and adjacent tissue through the use of a duplex system from the same ultrasound access point. Anatomical orientation at the base of the brain is possible using the B-mode (for example, the mesencephalon can be depicted as a butterfly-shaped structure), and flow within the arteries at the base of the skull can be depicted with color application. After the sample volume has been placed in the color flow image, additional evaluation of the Doppler spectrum signals should follow. Application, probe position, and identification of the individual blood vessels are the same in both methods, as described below (Baumgartner et al. 1996, Klötzsch et al. 1996).

Ultrasound Application and Probe Position

Transtemporal Ultrasound

After applying contact gel, the probe is placed on the temporal region (Fig. 3.**1a, b**). The proper access point through the bone window (a thin-walled section of the skull) can be identified using slight, incremental changes in the incident angle and displacement (Fig. 3.**2b**). The required point is the one at which good probe coupling allows maximum flow signals to be recorded. In general, ultrasound energies between 75 and 100 mW/cm² are required.

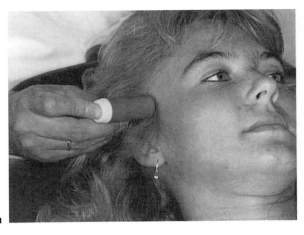

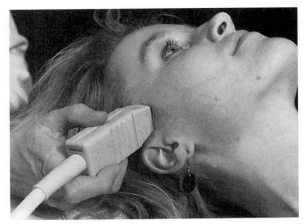

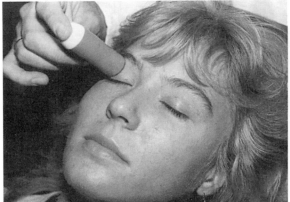

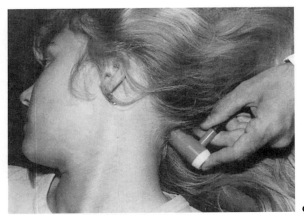

Fig. 3.**1a** Probe position in transcranial Doppler sonography to examine the anterior section of the circle of Willis. **b** Probe position in transtemporal duplex sonography. **c** Demonstrating the probe position for transorbital ultrasound application. **d** Transnuchal ultrasound application

Examination area. Anterior cerebral artery, middle cerebral artery, posterior cerebral artery, anterior communicating artery, posterior communicating artery, superior cerebellar artery, top of the basilar artery.

Transorbital Ultrasound

With the patient's eyes closed and after applying contact gel, the probe is placed without pressure above the eyeball and above the ophthalmic artery (Fig. 3.**1c**), or can be placed in the lateral canthus, where the contralateral anterior cerebral artery can also be assessed. The examination should proceed at the lowest intensity level (10–25 mW/cm^2). Since the angle between the ultrasound beam and the vascular axis is much more variable when the carotid siphon is being examined transorbitally than transtemporally, quantitative interpretation of the data obtained is limited.

Examination area. Siphon of the internal carotid artery, ophthalmic artery, anterior cerebral artery.

Transnuchal Ultrasound

The probe is placed in the midline or laterally underneath the inion, causing the ultrasound to enter mainly through the great foramen (Fig. 3.**1d**). The patient's head is inclined to the side; occasionally, the examination may be more successful with the patient seated than recumbent. Depending on the placement of the sample volume, an ultrasound energy of between 50 and 100 mW/cm^2 should be selected.

Examination area. Vertebral artery, proximal and middle basilar artery (posterior inferior cerebellar artery, PICA; anterior inferior cerebellar artery, AICA).

Ophthalmic Artery and Internal Carotid Artery (Siphon)

Probe position. Proceeding transorbitally, the flow signal from the ophthalmic artery is searched for at a depth of between 25 mm and 45 mm, and the ultrasound beam is adjusted along the course of the blood vessel to the carotid siphon (Fig. 3.**1c**).

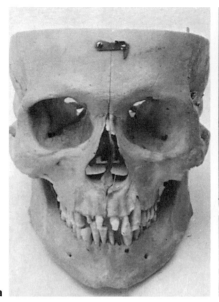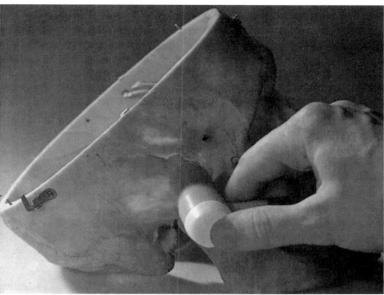

Fig. 3.**2a** Location of ultrasound windows for transorbital application, and **b** in the temporal region for transtemporal application (for ACA, MCA, and PCA recordings)

Identification. By continuously displacing the sample volume through the superior orbital fissure (Fig. 3.**2a**), the characteristic audiosignal of the internal carotid artery siphon can be obtained after approximately 60 mm (Fig. 3.**3c**). By slightly changing the incident angle of the applied ultrasound, it is usually possible to assess both the proximal part of the siphon (C_4 segment with flow toward the probe) and the distal part (C_2 segment with flow away from the probe). It should be noted that the flow direction changes as this is done.

Middle Cerebral Artery (MCA)

Probe position. To apply transtemporal ultrasound through the frontal bone window, the probe is placed at the lateral canthus, approximately 1 cm in front of the external acoustic meatus and above the zygomatic bone (Fig. 3.**2b**). The ultrasound beam is aimed in a slightly orbital direction (Fig. 3.**1a**). A slight frontal shift over the tragus is sometimes helpful.

Identification. At a depth of between 30 mm and 55 mm, the main trunk of the MCA (the M_1 segment) can be identified from the flow directed toward the probe, and the trunk can be followed for approximately 30 mm (Fig. 3.**3a**). Since the course of the distal branches (the M_2 segment) is almost perpendicular to the ultrasound axis, these blood vessels cannot be reliably evaluated. Color coding in the MCA is useful to distinguish it from other segments of the circle of Willis and to improve the identification of branches of the M_2 segment. This is particularly facilitated by additional use of echo contrast agents (Otis et al. 1995, Postert et al. 1997).

Anterior Cerebral Artery (ACA)

Probe position. To obtain the best possible signal, the transducer is placed for transtemporal ultrasound application through the frontal bone window, and it usually has to be inclined forward slightly. The direction of the ultrasound beam should be adjusted very flexibly, since there are many variants in this section of the circle of Willis (Padget 1944, Fisher 1965).

Identification. The anterior cerebral artery can be assessed at a depth of between 55 mm and 80 mm; the blood flow is normally directed away from the probe (Fig. 3.**3a**).When data are being recorded in deeper segments, it is sometimes possible to detect the contralateral anterior cerebral artery, showing blood flow directed toward the probe. Color flow information is useful for the identification of the ACA, although the course of the ACA is sometimes more difficult to follow than that of the MCA. After administration of echo contrast media, both A_1 segments and the anterior communicating artery can eventually be displayed (Postert et al. 1997).

Posterior Cerebral Artery (PCA)

Probe position. The probe is placed directly above the ear, with slight dorsal and caudal displacement, so that ultrasound can be applied transtemporally at the posterior bone window.

Identification. The posterior cerebral artery is usually detected at a depth of between 55 mm and 75 mm. In contrast to the middle cerebral artery, the signal cannot be followed in regions closer to the skull. There

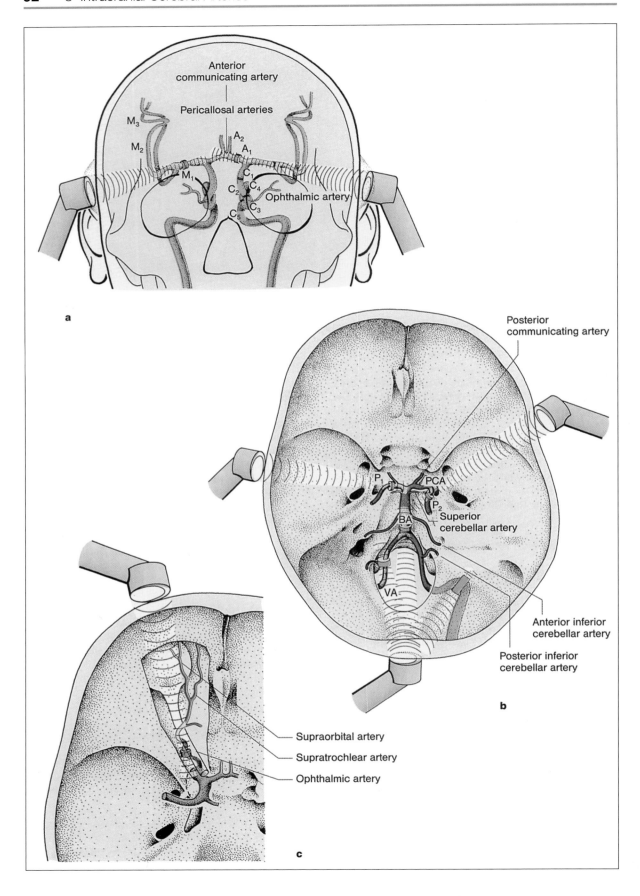

are two sections that can be distinguished from one another (Fig. 3.**3b**):

- *The proximal section* (P$_1$) usually has blood flow signals directed toward the probe, and can be detected at a depth of between 60 mm and 80 mm.
- *The distal section* (P$_2$) usually has the blood flow going away from the probe, and is found closer to the skull when the ultrasound probe is placed at a flatter angle (55–70 mm).

As with the ACA, it is possible to detect the contralateral PCA at greater depth when passing over the midline. It is sometimes possible to identify other blood vessels in this region, especially when using instruments with a smaller ultrasound beam or by color flow imaging, e.g., the superior cerebellar artery and the posterior communicating artery (Hennerici et al. 1987 b, Klötzsch et al. 1996).

With regard to the location of the butterfly-shaped mesencephalon, the P$_1$ segment, with blood flow toward the probe, and the P$_2$ segment can usually be depicted using color flow duplex sonography and distinguished from contralateral PCA segments (Bogdahn et al. 1993, Baumgartner et al. 1996). This is facilitated by the administration of echo contrast media (Postert et al. 1997).

Vertebral Artery (VA)

Probe position. Applying ultrasound nuchally, the vertebral arteries can be registered along their intracranial course in an adult at a depth of between 40 mm and 65 mm, and sometimes even deeper. To achieve this, the incident ultrasound angle has to be optimally adapted to the varying directions of these vascular segments (Figs. 3.**1d**, 3.**3b**). Occasionally, it may also be useful to displace the probe laterally away from the midline, especially when the vertebral arteries have loop formations.

Identification. Due to the incidence of unilateral hypoplasia (10–30%) and the variable course of the vasculature (Krayenbühl and Yasargil 1957), it may be difficult to classify the Doppler signals. This is especially so when there is pathology obstructing the blood flow in the posterior circulatory system, as well as when there are carotid artery obstructions resulting in the collateral circulation being taken over by the vertebrobasilar system. Uncertainties and mistakes in interpreting the flow signals can result if information about

the existence of vessel variants (e.g., from a conventional MRI or MRA) is lacking. Particular difficulties can be caused by the bidirectional Doppler signals that sometimes occur. At a depth of 40–65 mm, these should be classified as resulting from the intracranial loop of the vertebral artery, while at a depth of 60–90 mm, they can be attributed to simultaneous detection of the vertebral artery and the posterior inferior cerebellar artery (Fig. 3.**3b**). Due to the concomitant technical difficulties, compression tests applied to one or both vertebral arteries are not usually helpful. Furthermore, although both vertebral arteries can in principle be displayed along their full course from extracranial to intracranial, experience shows that consistent, continuous imaging of the vessels—even using color flow imaging—is not easy. Often, a conclusive judgment in such a situation cannot be reached until the true anatomical situation is known, as confirmed by angiography, magnetic resonance imaging (Mull et al. 1990, Röther et al. 1993), or helical CT studies (Gorzer et al. 1994, Katz et al. 1995, Baumgartner et al. 1995). However, this does not limit the value of this procedure as a way to follow the natural history of the arterial process.

Basilar Artery (BA)

Probe position. Using nuchal access, the origin of the basilar artery can almost always be detected at a depth of between 70 mm and 110 mm (Figs. 3.**1d**, 3.**3b**), but it can be only displayed along its full length in exceptional cases even if color flow duplex systems are used—whether the application of amplitude—or power energy mode and echo contrast media respectively can improve test validities has not yet been validated. Due to the variable position of the confluence of the two vertebral arteries, the proximal segment is likely to be encountered at various depths. However, since the basilar artery has an average length of 33 mm, the top of the artery is often not detectable (Hennerici et al. 1987b, Mull et al. 1990). In this case, transtemporal access is necessary.

Identification. It may be difficult to distinguish the basilar artery from the intracranial segment of the vertebral arteries and neighboring blood vessels:

- Due to unilateral hypoplasia (10–30%).
- Due to the variable distance between the basilar artery and the clivus (in 80% of cases this is less than 90 mm, in 20% of cases it is over 100 mm) (Krayenbühl and Yasargil 1957, Busch 1966).
- Due to variations in the distance to the transducer caused by the musculature of the nape of the neck.
- Due to the course of the basilar artery, which is rich in variants, sometimes horizontal or curved, and the other arteries in the posterior cranial fossa (when placing the sample volume incrementally, one should not expect to be able to follow the blood vessel continuously over its whole course; considerable gaps often occur).

◁ Fig. 3.**3** A diagram showing the relationship between the ultrasound axis and the vascular axis, for examining the large arteries at the base of the skull
  a Probe position and sample volume depth location for examining the MCA (left) or the ACA (right)
  b Probe position for examining the proximal segment (P$_1$ segment, left) and the distal segment (P$_2$ segment, right) of the PCA, with transtemporal ultrasound application, and the BA using transnuchal access
  c Application of transorbital ultrasound to the ophthalmic artery

Depth alone is therefore not a reliable parameter in identifying the basilar artery. Signals that are located

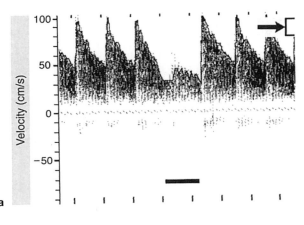

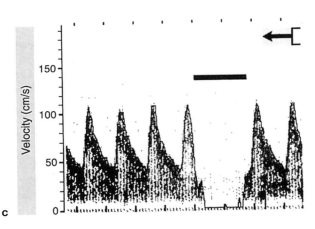

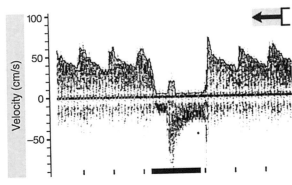

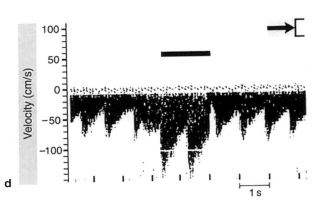

deeper than 80 mm from the transducer are usually from the basilar artery, but occasionally the confluence of the two vertebral arteries is not found until 90–110 mm, or cannot be determined at all.

Classifying the bidirectional signals that sometimes occur is also problematic. Although it is possible for such signals to originate from the large branches of the vertebral artery (posterior inferior cerebellar artery, PICA) or the basilar artery (anterior inferior cerebellar artery, AICA) (Fig. 3.**3b**), definitive classification of the signals has not so far been possible without knowledge of the anatomical situation, traditionally using angiography. A comparative examination using both MRA (or helical CT) and sonography has proved extremely helpful, and this will probably be the optimal combination in the future (Röther et al. 1993, Baumgartner et al. 1995).

Functional Tests

Compression Tests

Reliable compression tests can be carried out on the *common carotid artery*. Compressing the extracranial segments of the *vertebral arteries* at the mastoid process is difficult, and is often not complete due to the deep location of the blood vessels. Due to their intracranial confluence, only bilateral compression would make sense, but even with assistance from a second investigator, this is only occasionally satisfying and can be recommended for scientific reasons only.

Before compressing the common carotid artery, one should be certain:

- That light pressure on the blood vessel will not cause a significant decrease in the heart rate.
- That the compression will *definitely* take place below the carotid bifurcation (extracranial Doppler examination beforehand).
- That an area in which wall changes are evident will not be compressed (in order to avoid cerebral emboli, prior duplex studies are necessary).
- That compression is carried out only for single cardiac cycles, if repeated vascular percussion has not caused clearly noticeable pulse wave effects.

When the ipsilateral *common carotid artery* is compressed:

- The signal of the MCA is reduced (Fig. 3.**4a**).
- The signal of the ACA will reverse its flow direction (Fig. 3.**4b**) or will be reduced.

Fig. 3.**4** Doppler spectrum of the cerebral arteries near the base of the skull after ipsilateral compression of the common carotid artery (indicated by the bar)
a Flow velocity reduction in the MCA
b Flow signal reversal in the ACA
c Signal reduction in the PCA if originating from the carotid artery
d Signal increase in the PCA or the posterior communicating artery if originating from the BA

- The signal of the PCA will be reduced, if this vessel originates directly from the carotid artery and behaves like the middle cerebral artery (Fig. 3.**4c**).
- The signal of the PCA will be increased, if the vessel does not originate from the carotid artery and the posterior communicating artery is intact (Fig. 3.**4d**).
- The signal of the posterior communicating artery will increase, if good collateral function is possible (Fig. 3.**4d**).

Carotid compression tests may also be useful to assess the posterior circulation, especially if there is any pathological circulation in the vertebral arteries.

When compressing the upper arm arteries using the blood pressure cuff (e.g., in the subclavian steal phenomenon), the blood flow in the vertebral arteries, as well as in the basilar artery, may be either

- Reduced, or
- Elevated

depending on whether or not the artery being examined contributes to blood circulation in the brain.

During the reactive hyperemic phase after the blood pressure cuff has been released again, antagonistic effects may be observed, depending on the prevailing hemodynamics (see Figs. 11.**11**–11.**13**, pp. 358–364).

Normally, neither the vertebral artery nor the basilar artery show any reaction to this compression test.

Vasomotor Reactivity Test

Since CO_2 is a potent vasodilator in the arteries of the brain, examinations must not be conducted only under orthopneic conditions, but also under hypercapnic (CO_2 given via mask or during the breath-holding test) or hypocapnic (hyperventilation) conditions. Physiologically, an increase or decrease in the flow velocity with a corresponding change in the Doppler spectrum can then be found (Fig. 3.**13**).

The breath-holding test can be performed according to the Markus and Harrison (1992): After normal breathing of room air for approximately 4 minutes, patients are instructed to hold their breath after a normal inspiration. During the maneuver, the MCA mean blood velocity is recorded continuously. The mean blood velocity at the TCD display immediately after the end of the breath-holding period is recorded as the maximal increase of the MCA mean blood velocity (during breath-holding). This procedure is repeated after a rest of 2 to 3 minutes to allow mean blood velocities to return to their initial values. For the maximal MCA mean blood velocity increase and for the time of breath-holding, the mean values of both trials are used. A BHI (breath-holding test index) can be calculated as a percentage increase in MCA mean blood velocity divided by the seconds of breath-holding:

$$(V_{bh} - V_r/V_r) \cdot 100 \cdot s^{-1}$$

where V_{bh} is the mean blood flow velocity at the end of breath-holding, V_r the mean blood flow velocity at rest and per second s^{-1} of breath-holding.

Vasomotor reactivity can also be tested using parenteral acetazolamide (Piepgras et al. 1990). It should be clear that what both these procedures are testing is not, as is often claimed, the cerebral autoregulation, but rather changes in vascular width and arteriolar reactivity caused by metabolic or pharmacologic factors. Comparisons made between the blood pressure–dependent autoregulation capacity and limitations on this capacity under pathological conditions may therefore produce different results from those of vasomotor reactivity both in healthy volunteers and in patients.

Autoregulation Test

Changes in the perfusion of the brain can only be induced physiologically by extreme changes in the systemic blood pressure, causing the brain's arterioles to contract or dilate (so-called cerebral autoregulation). Although quantitative direct analyses of these events have not been successful, a tilt test can be used to observe changes occurring in the large arteries at the base of the brain, indirectly as a response to distal flow regulation (Fig. 3.**23**). Continuous documentation of the signal frequency shift and specific changes in the signal intensity may sometimes be superimposed by artifacts arising due to the manipulation of the head and the probe position, which may interfere with accurate measurement of the blood flow. Special attention is therefore given to achieving adequate fixing of the patient's head (Müller et al. 1991, Daffertshofer et al. 1991). To maintain a relatively constant head and probe position, other investigators have used changes in the blood pressure caused by manipulation of the blood flow through the lower extremity (Aaslid et al. 1989). However, due to the resulting discomfort, this method is not commonly applied to patients.

Vasoneural Coupling

Various experiments have shown that by physiologically activating certain brain regions, flow velocity changes can be observed in the large cerebral arteries at the base of the skull (e.g., activating the visual cortex by opening the eyes and presenting different visual stimuli, Fig. 3.**14**) (Aaslid et al. 1987, Conrad and Klingelhöfer 1989, Droste et al. 1989, Sitzer et al. 1992, Hennerici and Daffertshofer 1993, Becker et al. 1996). These studies are comparable to similar experiments using positron-emission tomography (PET) or fMRI (Kushner et al. 1988, Santosh et al. 1995). In contrast to the quantitative measurements obtained with PET, one should note that transcranial Doppler sonography (TCD) can only describe relative changes. In addition, there is greater topographical uncertainty using TCD, although it has the advantage that it provides better time resolution than PET, and can easily be repeated.

Transcranial Monitoring (TCM) and Echo Contrast Sonography

Due to its noninvasive nature and very high time resolution, TCD is also used as a *monitoring procedure* during surgery (e.g., examining the cranial circulation during carotid and cardiopulmonary bypass surgery, or angioma embolization). During surgical procedures and in patients with artificial heart valves, TCM has led to the detection of high-amplitude signals (HITS, *high-intensity transient signals*) (Hennerici 1995), which were initially interpreted as signs of cerebral emboli (Spencer et al. 1990). The true clinical significance of these findings, as well as of spontaneous HITS within the intracranial blood vessels distal to carotid stenoses, is at present still unclear (Hennerici 1994, Nabavi et al. 1996, Daffertshofer et al. 1996). There is much evidence to suggest that HITS may reflect real microembolic events, however, they are often due to causes that do not result in cerebral ischemia, such as microcavitations in artificial heart valves (i.e., transient irregularities in blood flow without formation of structural emboli, which may be distinguished from nongaseous emboli by means of oxygen inhalation (Dorste et al. 1997). Similar results were already described several years ago during intravenous and intraarterial injection of microscopic air bubbles (Rautenberg et al. 1987, Gerraty et al. 1996). On the other hand, it has become possible to detect potentially pathological embolization paths into the intracranial terminal vascular system by deliberately administering ultrasound contrast media. This has shown that there is reason to believe that paradoxical embolization through an open oval foramen of the heart, or through pulmonary arteriovenous shunts, may cause forms of cerebral ischemia that have so far remained etiologically unclear (Lechat et al. 1988, Schmincke et al. 1995, Zanette et al. 1996). Using contrast transesophageal echocardiography (TEE) and/or contrast TCD, it is possible to detect this situation with a high degree of sensitivity (Teague et al. 1991, Job et al. 1993, Nikutta et al. 1993) (Fig. 3.**24**). Finally, microemboli can be recorded in some patients, and these may turn out to be useful prognostic indicators in stroke-prone patients. This needs to be further clarified through prospective studies providing evidence about the prognostic relevance of HITS (Hennerici 1994) and consensus about the appropriate criteria to be used to distinguish artifacts from significant signals (van Zuilen et al. 1996, Markus et al. 1996).

Normal Findings

Principle

In *pulsed wave Doppler sonography,* the sample volume is usually positioned between 25 mm and 140 mm, proceeding in 1–5 mm increments (sometimes variable). Normal Doppler spectrum values refer to a healthy control group and vary for different arteries and selected vascular segments in different studies. Our own data (Table 3.**1**) correspond well with the results obtained by other groups (Aaslid et al. 1982, Arnolds and von Reutern 1986, Harders 1986, Büdingen and von Reutern 1993). Overall, they indicate a certain age dependence of flow velocities, which sometimes reach significance in the middle cerebral artery and show clearly lower average values for the posterior circulation when compared to the in-

Table 3.**1** Mean values and standard deviations of Doppler parameters obtained from spectrum waveforms recorded from 50 adult control subjects

Arteries		Peak systolic velocity (cm/s)	Mean velocity (cm/s)	Range of normal values (mean velocity) (cm/s)
ICA	Siphon (60 mm) (transorbital)	81.0 ± 16.1	52.3 ± 11.4	20–77
ACA	(70 mm) (transtemporal)	78.7 ± 19.1	48.6 ± 12.5	18–82
MCA	(50 mm) <60 yrs (transtemporal)	92.7 ± 15.3	58.1 ± 10.0	32–82
	>60 yrs	78.1 ± 15.0	44.7 ± 11.1	18–64
PCA	(60 mm) (transtemporal)	54.8 ± 14.6	33.6 ± 8.9	16–58
BA/VA	(75 mm) (transtemporal)	55.6 ± 14.5	33.9 ± 10.6	12–66

tracranial carotid region. It should be noted that mean values used to calculate the standard data might have been obtained from different arterial segments, due to interindividual variations in the arterial length and width. Also, it was not always possible to take into account the optimal flow signal in groups with a consistent location for the measuring point. Finally, topical asymmetries due to differing angles between the ultrasound and vascular axes also play a role and can be misleading if falsely interpreted from multi-gate pulsed Doppler instruments (Mess et al. 1996, Bäzner et al. 1996). According to our own comparative investigations, the size of the sample volume (1–5 mm) does not have a significant influence on the measurement values obtained (Rautenberg 1991). All these reasons explain the relatively large standard deviation in the reference values. Technical problems and artifacts caused by the methodology used also play a role (see Chapter 1). For the sake of clarity, only *the systolic peak velocity and the mean velocity* with their calculated averages and the standard deviations are given here. The lower and upper boundary values observed for the average flow velocity are also listed (Hennerici et al. 1987b).

Using *transcranial color flow duplex sonography*, individual vascular segments can be selectively examined under visual control and, after angle correction, velocity measurements can be made. From the examination of the extracranial carotid system, it was hoped that it would be possible to achieve quantitative rather than qualitative flow velocity analysis using transcranial duplex sonography, by correcting the incident angle in the duplex system. Eicke et al. (1994) have shown that the angle between the ultrasound and vascular axes of the intracranial arteries is indeed normally larger than previously assumed. They therefore reported significantly higher flow velocities after correction than those previously known from Doppler signal investigations alone. However, these results require additional evaluation, since there are various circumstances that restrict the ability of this procedure to be transferred analogously from the carotid or femoral system to the transcranial situation. For example, the S-shaped course of the middle cerebral artery, with its much smaller caliber, is significantly more curved, resulting in short segments lying close to one another within the same sectional planes but with different directions and vice versa (Mess et al. 1996). Orthogonal ultrasound application practically never occurs; transverse vascular sections are the rule. The still relatively wide transcranial ultrasound sections (ca 1 cm) compared to the small arteries at the base of the brain may therefore cause errors in calculating the corrected incident angle (Giller 1994). This is particularly the case when the value measured (and the corresponding angle correction) is large. However, in general, flow velocity calculations are more reliable with an angle correction, as introduced by transcranial duplex sonography. Whereas conventional duplex sonography is of limited use for transcranial application, color-coded flow velocity and signal amplitude mode analysis (power imaging) are increasingly being

Table 3.**2** Position of the sample volume and flow direction for identification of the cerebral arteries

Depth (mm)	Flow direction		Artery
60–70	←[	C_2 segment	ICA
	→[	C_4 segment	
60–75	←[		ACA
30–60	→[		MCA
55–80	→[	P_2 segment	PCA
	←[	P_1 segment	
70–110	←[		VA/BA

used for the examination of intracranial arteries (Baumgartner et al. 1996), and seem to be particularly advantageous when used in combination. If magnetic resonance angiograms are available, both noninvasive methods provide highly reliable information about the status and natural history of large-artery cerebral circulation which is not currently available otherwise (Röther et al. 1994).

The position and depth of the sample volume and the direction of the Doppler signal in relation to the ultrasound probe in the large basal arteries of the adult brain are given in Table 3.**2**. Due to the differing size relationships, significantly varying values can be expected in infants and small children.

Anatomy and Findings

Carotid Siphon—Ophthalmic Artery

Anatomy

After leaving the cavernous sinus, the ophthalmic artery is the first main branch of the internal carotid artery (ICA). Before the origin of the ophthalmic artery, the diameter of the ICA is 3.3–5.4 mm, and distal to this it is 2.4–4.1 mm; this is a physiological narrowing of the lumen, and does not represent an obstruction.

The ophthalmic artery has a diameter of 0.7–1.4 mm, and follows a course some 7.9 mm long that is intracranial, intracanalicular, and intraorbital (Krayenbühl et al. 1979).

Findings

Transorbital ultrasound. Applying low-energy ultrasound (2 MHz) to the *ophthalmic artery,* which is located at a depth of between 20 mm and 60 mm at the roof of the orbits, usually shows a low signal directed toward the probe (Fig. 3.**5a**).

At its origin in the *internal carotid artery,* one can sometimes register mixed signals in both directions (toward the ultrasound probe and away from it)

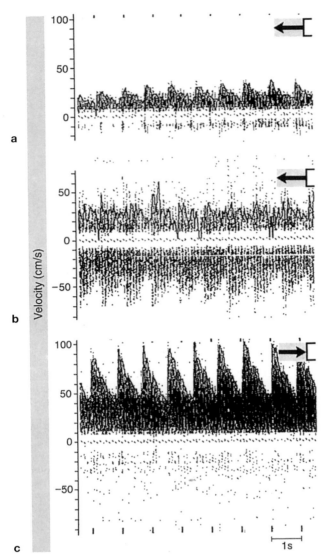

Fig. 3.**5** Doppler spectrum, **a** from a transorbital assessment of the ophthalmic artery, **b** from the origin of the ophthalmic artery at the C_3 segment of the internal carotid artery, and **c** from the proximal segment of the loop at the internal carotid artery (C_3/C_4 segment), from an 18-year-old woman with spontaneous dissection of the internal carotid artery. When using the low ultrasound frequency of the pulsed wave Doppler (2 MHz), the signal of the ophthalmic artery, with retrograde perfusion (←[), cannot be optimally registered, especially not at the depth of its origin at 65 mm (**b**). The proximal section of the C_3/C_4 segment of the internal carotid artery is documented as a positive or negative signal (note the difference in **b** and **c**), depending on the polarity of the flow direction display
→[Symbol for flow toward the probe
←[Symbol for flow in the probe's direction
All information given concerning the description of the Doppler spectrum also applies in the following figures
a Ophthalmic artery (depth 45 mm, peak systolic flow velocity 34 cm/s, mean flow velocity 20 cm/s)
b Ophthalmic artery (depth 65 mm, peak systolic flow velocity 38 cm/s, mean flow velocity 20 cm/s)
c Internal carotid artery, C_3/C_4 segment (depth 60 mm, peak systolic flow velocity 98 cm/s, mean flow velocity 60 cm/s)

(Fig. 3.**5b**). By moving the ultrasound probe slightly, or correcting the incident angle, a characteristic flow curve is recorded with a corresponding audiosignal in the C_3/C_4 segment of the ICA. The audiosignal is recognizable by its high diastolic flow velocity, with a slight systolic amplitude modulation (Fig. 3.**5c**). Color flow duplex sonography may be used for selected studies in patients with suspected retinal embolism (Hedges 1995).

Transtemporal ultrasound. The carotid siphon and the bifurcation (Fig. 3.**6b**) can also be displayed using the transtemporal bone window, orienting the probe caudally. Due to the angle between the ultrasound beam and the blood vessel, the flow velocities recorded are normally low. Only small flow accelerations or limited turbulences can be evaluated.

Normal Values

Since there is wide variation in the incident angle under which both the proximal (C_4 segment) and the distal components (C_2 segment) of the carotid siphon can be assessed using ultrasound, there is a wide range of normal values for healthy individuals, with no significant age dependence (Table 3.**1**).

Compression Test

When applying ultrasound transorbitally, it may be helpful in individual cases to distinguish between signals coming from the siphon segment and those coming from the posterior communicating branch. In these cases, compressing the proximal common carotid artery makes it possible to differentiate between these two vessels.

Middle Cerebral Artery

Anatomy

After its origin from the internal carotid artery, the middle cerebral artery (MCA) runs laterally along the wing of the sphenoid bone into the Sylvian fossa over a path 5–30 mm long (average 14–16 mm) (Herman et al. 1963, Jain 1964). The interior lumen of the blood vessel is 3–5 mm in diameter (M_1 segment: sphenoidal part). Usually, a large number of small arteries originate in this segment (the lenticulostriate arteries). Branching off in two or three segments at the M_1 level, distal long superficial arteries (M_3) support the superficial cortical and deep perforating arteries.

Findings

Using the temporal bone window, the main branch of the MCA (M_1 segment) can be examined step by step with ultrasound over a length of 30–60 mm (Fig. 3.**6a**). Since the artery usually divides into two or three main branches or into a double bifurcation at a distance less than 30 mm from the ultrasound probe, and since its

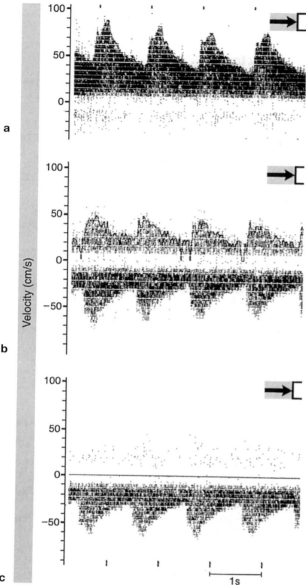

origin from the ICA is variable (usually 50–65 mm deep), normal values for the flow signals refer to a specific vascular segment (around 50 mm) (Tables 3.**1**, 3.**3**). It should be noted that significant changes in the Doppler frequencies are often only an expression of the loop formation in this region. Also, due to bending or crossing vascular segments, an impression of bi-directional flow may result. As a means of differentiating retrograde perfusion through leptomeningeal anastomoses, it is helpful to examine this region from several sample volume positions. The distal vascular segments (M_2 segment: insular part; M_3 segment: opercular part; M_4 segment: terminal part) are difficult to examine, due to the unfavorable incident angle of insonation. However, increasing experience with *color flow duplex imaging* indicates improved accessibility of M_1, M_2, and even M_3 branches with reasonable display, particularly if echo contrast media and/or amplitude energy modes are used (Kenton et al. 1996, Postert et al. 1997). Using color flow duplex sonography, the topographical course of the main branch of the middle carotid artery and its division in the M_2 segment can be clearly depicted. Furthermore, the T-bifurcation of the ICA and the anterior cerebral artery can often be displayed simultaneously, facilitating the evaluation of the anterior circle of Willis and avoiding misinterpretation (Fig. 3.**7**).

Normal Values

There is a clear age dependence in all of the Doppler spectrum values given.

Fig. 3.**6** Doppler spectrum from transtemporal evaluations of the middle cerebral artery (**a**), the T-junction of the internal carotid artery (**b**), and the anterior cerebral artery (**c**). All were obtained from a healthy control subject. At a typical sample volume position, the MCA shows flow toward the probe, and the ACA shows flow away from the probe. Near the bifurcation of the ICA, a flow signal in both directions is seen
a Middle cerebral artery (depth 50 mm)
b Internal carotid artery, T-junction (depth 70 mm)
c Anterior cerebral artery (depth 75 mm)

Table 3.**3** Calculated mean pulsatility indices (PI) (± one standard deviation), based on 50 control subjects, to compare Doppler spectra intra-individually and side to side

Arteries		Average PI	(± SD)
MCA	right	0.90	(0.24)
	left	0.94	(0.27)
ACA	right	0.78	(0.15)
	left	0.83	(0.17)
PCA	right	0.88	(0.23)
	left	0.88	(0.20)

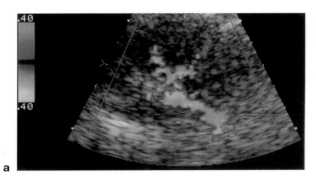

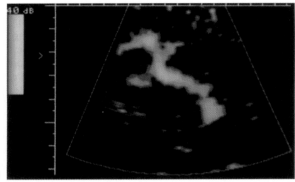

Fig. 3.**7** Transcranial color flow duplex sonograms (velocity mode **a** and power energy mode **b**) of the middle cerebral artery (M$_1$ stem) and its branches (M$_2$ segments)

Anterior Cerebral Artery and Anterior Communicating Artery

Anatomy

In the bifurcation of the internal carotid artery, the anterior cerebral artery (ACA) initially follows a horizontal path as a medial branch (A$_1$ segment = precommunical part = horizontal part) before it turns in a cranial direction as the pericallosal artery (A$_2$ segment) (Fischer 1938, Krayenbühl et al. 1979). The internal diameter of the ACA is 1–3 mm; a diameter of less than 1 mm indicates hypoplasia (Wollschläger and Wollschläger 1966):

- *Unilateral* in 4.1–14.0%
- *Bilateral* in 3.2% (Wollschläger and Wollschläger 1974, Krayenbühl et al. 1979)
- *Aplasia* in 0.7–1.1% (Tönnis and Schiefer 1959, Krayenbühl et al. 1979).

As the connecting link between both ACAs (anterior circle of Willis), the *anterior communicating artery* is the shortest cerebral artery (0.1–3.0 mm), and has significant variants (Fig. 3.**8**). The frequency of hypoplasia and aplasia is 0.5–1% and 6–8%, respectively (Sedzimer 1959, Krayenbühl et al. 1979).

Findings

After the T-bifurcation of the internal carotid artery has been identified (Fig. 3.**6b**), the ACA can be detected at 60–75 mm, with flow moving away from the probe (Fig. 3.**6c**). By increasing the depth adjustment, the contralateral ACA can sometimes be registered for a distance of 70–75 mm. It can be recognized by a change in the flow direction (toward the probe).

Flow signals from the *anterior communicating artery* cannot usually be documented in isolation. They are similar to those obtained from the posterior communicating artery, and reflect an intermediate position between the two A$_1$ segments, at a low signal amplitude (Fig. 3.**10**).

Color flow duplex sonography facilitates the depiction of the anterior circle of Willis. However, due to the often curving and winding arterial course of the vessel, continuous imaging in a single sectional plane is more difficult than in the middle cerebral artery if velocity modes only are used. Application of amplitude modes (power imaging) improves the validity of display remarkably (Kenton et al. 1996, Postert et al. 1997). Selective display of distal segments of this blood vessel that are not detectable with pulsed wave sonography alone is also possible.

Normal Values

Due to physiologically occurring variants in this segment, in particular, the normal values at a distance 70 mm from the probe (Table 3.**1**) fluctuate, and show no significant age dependence.

Compression Test

A compression test involving both the ipsilateral and the contralateral common carotid artery is important in the functional assessment of the anterior segment of the circle of Willis but can be dispensed with if color flow duplex imaging reveals good displays. Under ipsilateral compression, the flow signal reverses its direction; under contralateral compression, it increases if the anterior communicating artery is open. Following a compression-induced increase in the blood volume, signs of *functional stenosis* often appear in this small blood vessel. If a change in flow direction does not occur, or only a decrease in flow is observed, then either there is collateral blood flow taking place through the posterior communicating artery, or the anterior communicating artery is missing or hypoplastic.

Posterior Cerebral Artery and Posterior Communicating Artery

Anatomy

Phylogenetically and ontogenetically, the posterior cerebral artery (PCA) originates from the ICA. The connection to the basilar artery is a later development. Anatomical examinations in humans have shown that in 10–30%, the PCA originates directly from the ICA (Sunderland 1948, von Mitterwallner 1955). Both PCAs run as the terminal branches of the basilar artery around the brainstem (0.5–1.0 cm) (P_1 segment) until they join the *posterior communicating artery*, which is a branch of the internal carotid artery. In their further course (P_2 segment), they lead into the calcarine and parieto-occipital arteries.

Findings

The supratentorial segment of the posterior cerebral artery can be detected at a depth of 60–80 mm. The proximal segment (P_1 segment) usually lies slightly deeper, and its signal is directed toward the probe (Fig. 3.**9 a**). In the distal section (P_2 segment), the direction of flow is away from the probe (Fig. 3.**9 c**).

Doppler signals recorded under favorable ultrasound application angles should be compared to the calculated normal values (Table 3.**1**). The contralateral PCA can be evaluated at considerable depth (usually beyond 80 mm) while maintaining the specific angle used to examine the ipsilateral PCA. The flow direction reverses in relation to the ipsilateral P_1 segment, and is oriented away from the probe. Flow signals from the *posterior communicating artery* cannot usually be documented in isolation. They show an intermediate direction with low amplitude (Fig. 3.**10**). It is possible for the superior cerebellar artery to be confused with the posterior cerebral artery. The latter lies cranial to the superior cerebellar artery, and often shows signal modulation after the eyes are opened, due to changes in neuronal activation in the occipital lobe (Fig. 3.**14**).

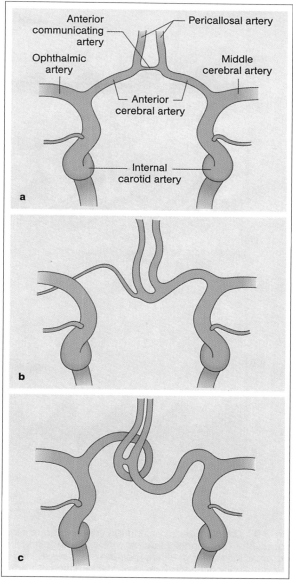

Fig. 3.**8** Variations in the anterior segment of the circle of Willis (according to Perlmutter and Rhoton)
a Symmetrical ACA
b Hypoplasia of the right ACA—the anterior communicating artery, as a continuation of the left ACA, belongs to the right pericallosal artery
c Significant winding of the ACA on both sides

In a *color flow duplex sonogram*, transtemporal depiction of the posterior circle of Willis and the posterior cerebral artery in relation to the upper brainstem is possible. However, complete anatomical imaging in a single sectional plane (P_1 and P_2 segments) is usually the exception, at least, if velocity modes are used exclusively. Amplitude modes, however, facilitate this display and improve the validity of insonation (Kenton et al. 1996 and Postert et al. 1997). The posterior communicating artery, important as a collateral blood vessel, can be located more easily under visual control.

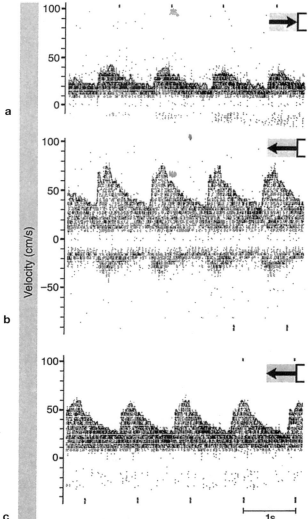

Fig. 3.**9** Doppler spectrum during transtemporal evaluation of the posterior cerebral artery (PCA), depicting the proximal segment (P_1) (**a**) and the distal segment (P_2) (**c**). In the deeper-lying P_1 segment, the flow direction is toward the probe. Due to the unfavorable conditions for ultrasound examination in this example, relatively low flow velocities are measured. At the transition point to the distal segment, the ultrasound angle is more favorable, the values are higher, and these alone are relevant for the quantitative evaluation. Sometimes, both flow signals (toward the probe and away from the probe) from the same depth can be registered at the bend of the blood vessel (**b**)

a Proximal segment P_1 (depth 80 mm, peak systolic flow velocity 38 cm/s, mean flow velocity 26 cm/s)
b Transition from P_2 to P_1 (depth 75 mm, peak systolic flow velocity 76 cm/s, mean flow velocity 50 cm/s in the P_2 segment)
c Distal segment P_2 (depth 75 mm, peak systolic flow velocity 60 cm/s, mean flow velocity 30 cm/s)

Normal Values

In a normal population, all of the values obtained from flow velocities recorded in the posterior circulation (PCA and basilar artery) are significantly lower than the values obtained in the anterior circulation (ACA and MCA).

Compression Test

When the ipsilateral common carotid artery is compressed, it is often possible to clarify whether the posterior cerebral artery originates directly from the ICA or from the basilar artery. In the first case, the flow velocity is reduced. The examination is best done in the P_2 segment. If the PCA is supplied through the basilar artery, and the posterior communicating artery is intact, an increase in flow velocity in the P_1 segment occurs. In this instance, care should be taken to examine the P_1 segment in order not to miss this collateralization (Fig. 3.**4**).

Vertebral Artery and Basilar Artery

Anatomy

The basilar artery, which has an average diameter of 4.1 mm and a length of 33.3 mm (Busch 1966), runs cranially from the junction of the two vertebral arteries at the clivus. It has many variants (Krayenbühl and Yasargil 1957, Busch 1966), due to embryological factors in its development and also, in older patients, as a result of arteriosclerotic, ectatic changes. Both the cranial bifurcation and the caudal junction are highly variable in location (Lang 1991) (Fig. 3.**11a**). The course of the artery winds repeatedly, and it is sometimes markedly elongated (the so-called megadolichobasilar artery). The vertebral arteries and the origins of the cerebellar arteries have a wide range of variations (Krayenbühl and Yasargil 1957). In many cases, a clear classification of the hemodynamic findings in the posterior circulation is only possible with topographical knowledge obtained from an angiogram, magnetic resonance angiography, or helical CT (Mull et al. 1990) (Fig. 3.**11b**).

Findings

Approaching from the occipital direction, one can assess the basilar artery in the midline or paramedially, and follow it for a longer distance by applying ultrasound through the great foramen or a thin bony access point at the posterior cranium. Near the loop formation of the vertebral arteries around the atlas, opposed signals similar to those made by crossing arteries, can be registered (PICA, AICA etc.) (Fig. 3.**12**).

When proceeding from the transnuchal direction, the vertebrobasilar area of distribution can be depicted using a *color flow duplex sonography*. However, the confluence of the vertebral arteries and

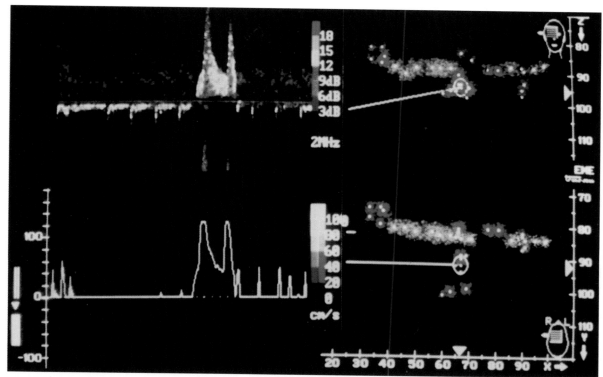

Fig. 3.**10** Display of a two-dimensional transcranial Doppler system. In the right half, the following structures are shown in a frontal section (top): the middle cerebral artery, the carotid siphon, the ipsilateral anterior cerebral artery and the contralateral anterior cerebral artery, the T-junction of the internal carotid artery, and the proximal middle cerebral artery. Red and yellow indicate flow toward the probe, and blue indicates flow away from the probe. In the corresponding lower half of the figure, the posterior communi- cating artery and posterior cerebral artery are shown in a horizontal section. Registering the Doppler spectrum (left) occurs on a selectable basis from the sectional images (−0), in this case for the posterior communicating artery. Spontaneously, an oscillating flow pattern is established in this blood vessel, which is significantly strengthened (for 1–2 cardiac cycles) as a signal toward the probe when the ipsilateral common carotid artery is compressed (from Hennerici et al. 1987)

the course of the basilar artery are often difficult to image, even if a power mode is used (Fig. 3.**7b**).

Normal Values

Normal values are comparable to those obtained for the PCA (Table 3.**1**). The data here do not show any age dependency either.

Flow signals directed toward the probe can sometimes come from branches of the vertebral artery (posterior inferior cerebellar artery, PICA) or the basilar artery (anterior inferior cerebellar artery, AICA) (Fig. 3.**10**).

Compression Tests

Compression tests of the vertebral arteries are difficult to carry out, due to the deep and protected site of these vessels and potential artifacts caused by motion. Although compressing the extracranial segment of the vertebral artery ipsilaterally or contralaterally causes a decrease or increase in the flow velocity of the intracranial segment, the flow signal of the basilar artery usually changes only slightly (reduction), or only under bilateral compression, which is not recommended for clinical use (Büdingen and Staudacher 1987).

Pathological Findings

Principle

Several parameters are selected from the spectra in transcranial pulsed wave Doppler sonography in order to evaluate pathological changes:

- Peak systolic (and possibly diastolic) velocity
- Mean flow velocity
- Spectral waveform (window width and signal amplitude distribution)

In addition, the following indices can be used:

- Pulsatility index (PI) (Table 3.**3**)
- Pulsatility transmission index (PTI):

$$PTI = \frac{PI_{ipsilat.}}{PI_{contralat.}} \times 100$$

The criteria listed below are used to classify the degree of stenosis.

Low-grade stenoses:
- A local, circumscribed increase in peak systolic and, if applicable, also diastolic velocity.
- Local increase in the mean flow velocity.
- Possibly a slight signal amplitude increase in the lower flow velocity range.

Medium and high-grade stenoses:
- A clear increase in the peak systolic and diastolic velocity.
- A clear mean flow velocity increase.
- Significant restructuring of the spectrum, with increased systolic backflow components.
- "Musical murmurs" (i.e., bands of discrete velocity at high signal amplitudes) may appear.
- Distal to the stenosis, there may be a clear reduction in the peak or mean flow velocity.

The criteria listed above do not allow differentiation between a *structural* and a *functional* arterial stenosis at the base of the brain: the latter is quite frequent in high-grade extracranial vascular processes with active intracranial collateralization through the anterior or posterior circle of Willis (Fig. 3.**18**).

Studies analyzing transcranial Doppler findings with angiographic and morphological results are reported in the literature (Lindegaard et al. 1986, Hennerici et al. 1987 a, Hennerici et al. 1987 b, Arnolds et al. 1986, Hennerici et al. 1992, Röther et al. 1994, Postert et al. 1997). These demonstrate, as do our own data summarized in Table 3.**4**, that the reliability of the method in detecting (i.e., sensitivity) or excluding (i.e., specificity) arterial processes at the base of the brain is comparable to that obtained with conventional angi-ography or MRA/spiral CTA with some over- and underestimates in low-degree and high-degree lesions. Therefore, hardly any results are yet available with regard to the validity of these methods in evaluating different *degrees of stenosis*. This is some extent due to the difficulty of evaluating arterial stenoses restricted by projection problems, loop formation, and often by the limitation of depicting the stenosis in only a single plane. Conventional angiography and MRA/spiral CTA share the same problems. These procedures are all mutually complementary (Fig. 3.**15**).

Table 3.**4**   Sensitivity, specificity, and validity of transcranial Doppler sonography compared to intra-arterial angiography in detecting intracranial stenoses in 467 patients (after Rautenberg 1991)

	MCA	ICA	PCA	VA	BA
True positive	24	21	10	19	9
True negative	438	445	455	442	453
False positive	3	–	–	1	–
False negative	2	1	2	5	5
Sensitivity (%)	92	91	83	79	64
Specificity (%)	99	100	100	99	100
Validity	99	100	100	99	99
Positive predictive value (%)	89	100	100	99	100
Negative predictive value (%)	99	100	99	98	99

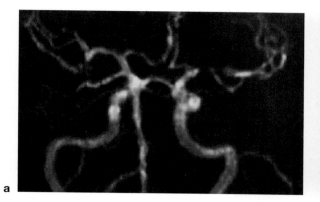
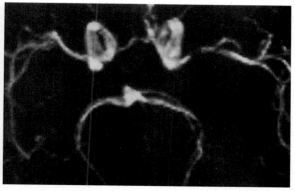

a

b

c

Fig. 3.**15** Magnetic resonance angiography (**a** antero-posterior, **b** axial projections) and **c** Doppler spectrum waveforms from a stenosis of the trunk of the middle cerebral artery. Depth 60 mm, peak systolic flow velocity 246 cm/s, mean flow velocity 158 cm/s

Findings

Stenoses

Criteria. Changes in the Doppler spectrum can be used to describe flow obstructions in the extracranial and intracranial cerebral arteries. In transcranial color flow duplex sonography, intracranial stenoses can sometimes be detected by a change in the color signal or only from guided spectrum analysis (Fig. 3.**16**). For routine use and in clinical trials, a classification recently proposed by Röther et al. (1994) may be useful (Table 3.**5**).

- The *peak flow velocity* (systolic and diastolic) initially increases with increasing stenosis, and decreases again in high-grade and subtotal stenoses.
- The *mean flow velocity* also increases with increasing stenosis, but is subject to less pronounced fluctuations.
- In significant high-grade stenoses, a clear mean flow velocity reduction may result, both prestenotically and post-stenotically.
- Medium and high-grade stenoses have *bidirectional signals,* i.e., negative flow components.
- With increasing stenosis, the low-frequency components increase (during systole in low-grade stenosis, during the entire cardiac cycle in high-grade stenosis) and the *systolic window* becomes smaller.

Table 3.**5** Classification proposal for MCA stenoses (modified from Röther et al. 1994)

0 = normal
I = mild stenosis
II = moderate stenosis
III = severe stenosis

MRA

0	no signal loss		
I	signal reduction	a =	< 50%
		b =	> 50%
II	complete signal loss	a =	limited to stenosis
		b =	including post-stenotic
III	signal loss including post-stenotic with signal rarefaction of distal vessels		

TCD

0	<140 cm/s
I	140–209 cm/s
II	210–280 cm/s
III	> 280 cm/s

DSA

0	no lumen reduction		
I	pallor of contrast medium		
II	lumen reduction	a =	< 50%
		b =	≥ 50%
III	subtotal stenosis with delayed distal filling and/or borderzone shift		

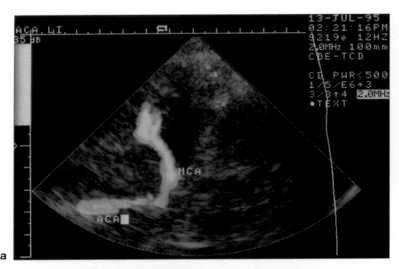

a

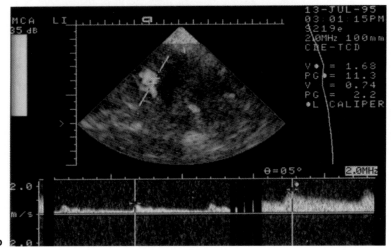

b

Fig. 3.**16** Power energy mode (**a**) of the transcranial color flow duplex sonogram of a proximal A_1 and a distal (M_2) branch stenosis in the cerebral arteries. In the M_1 segment (main stem), normal flow velocity values can be recorded. However, with spectrum analysis, the M_2-segment stenosis can be confirmed (**b**), although the power energy mode failed to disclose it: in the proximal anterior branch (right lower trace), the flow velocity is significantly increased as a sign of stenosis when compared to the posterior branch (left lower trace)

No single parameter seems to be able on its own to provide a reliable assessment of the degree of stenosis. Which combination of parameters correlates best with the morphological result has not yet been determined. Since evaluation of the "gold standard" (i.e., postmortem specimens) is unlikely to be achieved, correlations between different test procedures should be established from standardized studies using a priori determined criteria such as those already existing for extracranial measurements (de Bray et al. 1995, Baumgartner et al. 1994).

Location. Stenoses can be followed in the horizontal course of the MCA, i.e., they can be examined and documented prestenotically, intrastenotically, and post-stenotically. Near the intracranial ICA, the vertebral artery, the ACA, and the PCA, the examination is more difficult, as the vessels have a tortuous course, and only short vascular segments are optimally accessible to ultrasound. This is particularly the case when obstructive processes are to be followed (see Figs. 11.**14** and 11.**15**).

Change in the ascending arteries (e.g., the M_2 segment of the MCA) can be detected only if color flow imaging and spectrum analysis are used (Fig. 3.**16**).

Stenoses of the basilar and vertebral arteries can usually be detected quite well, provided they can be continuously examined prestenotically, intrastenotically, and post-stenotically. Due to the unique vascular course and specific positioning of the sample volume in relation to the transducer, only parts of the lesion can be assessed, and classification may not be possible. It is also difficult to differentiate between the two vertebral arteries intracranially, and to define the boundaries between the distribution areas of the vertebral artery and the basilar artery, which do not show any hemodynamic difference (Mull et al. 1990). Unless additional diagnostic methods are used, definite identification of flow obstructions is not possible.

Vascular changes within the intracranial segment of the carotid artery can be detected with combined transtemporal and transorbital ultrasound. A typical example is shown in Figure 11.**9**, pp. 354–355. Locating and detecting an intracranial carotid artery stenosis is more reliable than with continuous wave conventional Doppler sonography, which only indicates pathological findings associated with high-grade obstructions indirectly. Stenoses at the origin of the ophthalmic artery, which are extremely difficult to verify using extracranial continuous wave Doppler sonography, can be directly detected.

Using supplementary examination from the neck, it is sometimes possible to detect extracranial flow obstructions that are inaccessible to continuous wave Doppler sonography (Schwartz and Hennerici 1986 a, Ley-Pozo and Ringelstein 1990).

Occlusions

Occlusion criteria. Changes in the Doppler spectrum in intracranial arterial occlusions are characterized by:

- A locally absent signal, in spite of good study conditions (normal bony access point for ultrasound) and correct Doppler probe and sample volume position (clear signals from neighboring arteries with the identical probe position).
- Reduced flow velocity values in proximal vascular segments.

- Detection of collaterals and anastomoses with a flow velocity increase or flow reversal.

Location. Occlusions, often caused by emboli, are mainly found in the MCA. During acute ischemia, embolic material in the M_1 segment can fragment due to *spontaneous thrombolysis,* and subsequently cause distal branch occlusions (M_2 and M_3 segments). The following findings are therefore seen:

- In side-to-side comparisons, there is a clearly lower flow velocity at the transition point of the bifurcation of the internal carotid artery into the MCA.
- There is an absent signal toward the superior part of the cranium in the distal course of the vessel; at best, smaller anastomoses can sometimes be detected here, with reversed blood flow.
- A flow velocity increase in the ipsilateral ACA is sometimes also seen bilaterally.

When there is a *branch occlusion of the middle cerebral artery,* the following are also found (Fig. 3.17):

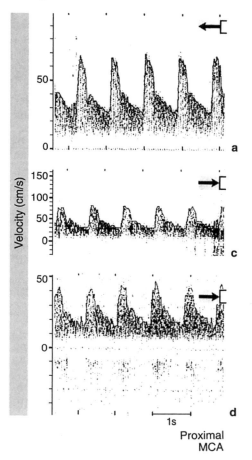

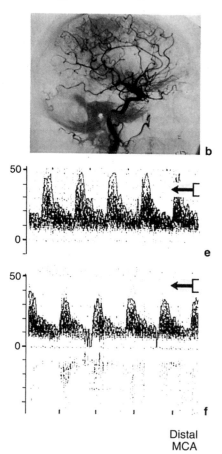

Fig. 3.**17** Selective angiogram of the internal carotid artery (**b**) and Doppler spectrum waveforms of the ACA (**a**) and MCA (**c–f**) in an occluded branch of the left middle cerebral artery. In the ipsilateral ACA, the Doppler spectrum waveform is normal (**a**), while in the proximal MCA (**c, d**), the systolic peak and mean flow velocities are reduced. Close to the brain surface (**e, f**), only arteries with inverse flow directions (away from the probe) and low flow velocities can be identified in leptomeningeal anastomoses here (from Hennerici et al., 1987)

a ACA (depth 70 mm, systolic peak flow velocity 66 cm/s, mean flow velocity 30 cm/s)
c Proximal MCA (depth 65 mm, systolic peak flow velocity 82 cm/s, mean flow velocity 40 cm/s)
d Proximal MCA (depth 50 mm, systolic peak flow velocity 48 cm/s, mean flow velocity 24 cm/s)
e Distal MCA (depth 40 mm, systolic peak flow velocity 48 cm/s, mean flow velocity 22 cm/s)
f Distal MCA (depth 30 mm, systolic peak flow velocity 36 cm/s, mean flow velocity 20 cm/s)

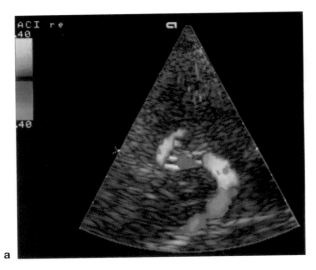

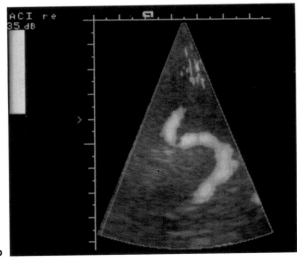

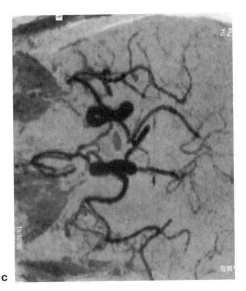

PCA, in the A_1 segment of the ACA without cross-flow).
- Retrograde flow in the proximal arteries at the base of the brain—e.g., retrograde flow in the distal MCA segments (M_2/M_3) to compensate for a proximal (M_1) occlusion of the main branch of the middle cerebral artery.

Collateralization via a *persistent primitive trigeminal artery* is difficult to detect. Theoretically, the signs of functional stenosis mentioned above could be found in the area between the anterior and posterior circulation. However, it is not possible to delineate any collateralization in the posterior communicating artery properly unless the morphological relationships are known.

Collateral vascular capacity. Although valid parameters are not yet available to estimate the hemodynamic capacity of the collateral vasculature, it can be approximated by comparing the pulsatility indices of the Doppler signals obtained from the bilateral corresponding cerebral arteries (so-called PTI values) (Lindegaard et al. 1985). Vasomotor reactivity (Widder et al. 1986 a) can be determined when there is a change in the end-expiratory P_{CO_2} or after administering acetazolamide, and also by evaluating the link between perfusion and neuronal metabolism (i.e., vasoneural coupling). These methods are based on the assumption that a vascular process has caused maximum vasodilatation in the terminal (arteriolar) system, so that a further increase in the flow velocity of the corresponding intracranial cerebral arteries can neither be attained by additional hypercapnia, nor by activating the vasoneural reserve. If there are no disturbances in neurological function, it can be assumed that the hemodynamic reserve capacity mechanisms have been exhausted, and that function is being maintained by metabolic compensation (e.g., increased oxygen extraction).

Evaluation

Dilatative Arteriopathy

With increasing age, arteriosclerosis may also show fusiform dilatations, the wall consistency of which may be difficult to distinguish histologically from sacciform aneurysms (Schwartz et al. 1993). Dilated and significantly elongated blood vessels can be found in all large arteries at the base of the brain, especially in the internal carotid artery and the basilar artery (so-called

◁ Fig. 3.**19** Transcranial color flow duplex sonograms (velocity mode **a** and power energy mode **b**) and MR-angiography **c** of a patient with severe dilatative arteriopathy—note the marked loops of the M_1 segment and the significant reflexes from the stiff anterior and posterior walls in the absence of any flow signal as a result of strictly perpendicular insonation

megadolichobasilar artery) (Boeri and Passerini 1964). The structural changes in the vascular wall and associated losses of elasticity lead to vascular dilatations (ectasias) that are accompanied by a reduced flow velocity. The following findings are characteristic (Figs. 3.**19**, 11.**16**, pp. 368–369):

– A significantly reduced mean flow velocity (less than the calculated average -2 SD)
– A reduced peak systolic velocity

Carotid Cavernous Fistula

Even with continuous wave Doppler sonography in the orbital arteries, a high-frequency flow signal can regularly be observed with a marked loss of pulsatility and flow directed toward the probe, reflecting the venous drainage of the cavernous sinus fistula through the orbital veins. Using pulsed wave Doppler sonography transorbitally or transtemporally, the fistula can sometimes be evaluated directly, and high-amplitude, high-frequency, nonpulsatile flow signals can be demonstrated. In addition, there are sometimes significant low-frequency components (machine noise), which decrease when the ipsilateral carotid artery is compressed. Depending on the shunt volume, abnormal results can also be detected above the neck arteries; however, these may also be completely absent (Fig. 3.**20**). During and after interventional therapy, closure of the fistula can be well documented with Doppler sonography and color flow duplex sonography.

Arteriovenous Malformation

Regardless of whether or not the diagnosis is made with other imaging procedures (CT, MRI, MRA, or conventional cerebral angiography), it is useful to carry out the following examinations using transcranial duplex sonography:

– Examine the hemodynamics in the arteries supplying the malformation
– Analyze the collateralization and the shunt size
– Recognize blood supply in the neighboring vasculature in time
– Observe the path of the blood flow before and after therapy (Hassler 1986, Schwartz and Hennerici 1986 b, Hennerici et al. 1987 a, 1992).

Characteristic changes in the vessels supplying the angioma (Fig. 11.**17**, pp. 370–371) are:

– Increased flow velocity
– Increased signal amplitude affecting slow flow velocities
– Bidirectional flow signals
– Significant audiosignal changes
– Clearly reduced pulsatility indices
– Disturbed autoregulation in hypercapnia and hypocapnia

In the neighboring arteries, the following are found:

– Decreased flow velocities
– Simultaneous registrations containing overlapping vascular spectra

Preoperatively, it is possible to differentiate the draining arteries using the criteria given above; postoperatively, reduced flow velocities are found in the vascular segments, which are usually still dilated and show disturbed CO_2 reactivity. Both parameters normalize with time.

Spasms and Aneurysms

One early use of intracranial sonography was to document and follow the clinical course of spasms in patients suffering from subarachnoid hemorrhage. Similarly, the sonographic diagnostic criteria described above for evaluating stenoses in arteries at the base of the brain can also be documented here. Depending on the extent of the bleeding, several neighboring vascular segments may be involved (Fig. 3.**21**).

Typically, flow change corresponds to the angiographic findings presented by a spasm. Observations of the clinical course have on several occasions shown that the risk of secondary cerebral ischemia can be adequately prevented by appropriate drug treatment (Aaslid et al. 1984, Harders 1986, Harders and Gilsbach 1987, Seiler et al. 1986).

The typical development of spasms between days 4 and 18, as well as the site and increase of flow velocities, determine whether it is necessary to carry out detailed follow-up examinations.

Due to the complex pathophysiology of vasospasms, potential false-negative and false-positive sonographic findings have to be taken into account. Despite angiographically detectable spasms, elevated intracranial pressure due to subarachnoid bleeding can cause increased peripheral resistance in both the arteriolar and capillary vascular systems, accompanied by a corresponding flow velocity reduction in the preceding large cerebral arteries that are accessible to ultrasound. This may lead to false-negative sonographic findings.

However, in other cases, substantial sonographic changes may be found without angiographic confirmation. These apparently false-positive results are generally interpreted as being caused by arteriolar constrictions appearing during subarachnoid bleeding. They cause a clear increase in peripheral resistance, and also induce increased flow velocity by simultaneously disturbing the arterial vasomotor reactivity.

In addition, it should be noted that routine use of calcium blockers in patients with subarachnoid hemorrhage might have had an influence on the value of sonographic findings. Laumer et al. (1993) detected a significant increase in intracranial flow velocity prior to secondary cerebral ischemia, which appeared in only three of their 11 treated patients.

Finally, although it is rarely significant, abnormally elevated flow velocities may be found in

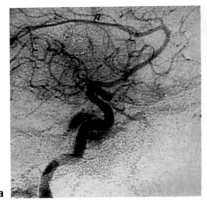

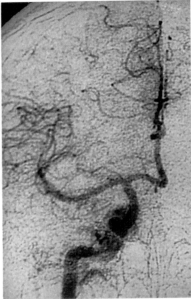

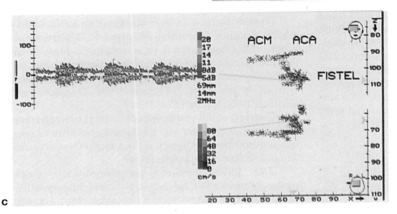

Fig. 3.**20** Angiograms in **a** lateral and **b** anteroposterior projections, and **c** transcranial Doppler sonography findings in a patient with carotid cavernous fistula using a two-dimensional scanning system (frontal and horizontal scans on the right). Ultrasound recorded from a depth of 69 mm demonstrates a pathological flow signal with low-frequency flow components (left)

patients who have not been diagnosed as suffering from subarachnoid hemorrhage. The extent to which functional spasms may be present (e.g., migraine), as in Prinzmetal's angina of the coronary arteries, remains unclear.

Apart from rare giant aneurysms, direct *detection of an aneurysm* as the bleeding source is not possible using TCD. In individual cases, transcranial color flow duplex sonography has been reported to facilitate direct imaging of an aneurysm (Wardlaw and Cannon 1996). Baumgartner et al. (1994) were able to detect 13 of 18 aneurysms sonographically, ranging in size between 6 mm and 28 mm, which had previously been demonstrated by angiography. Two thrombosed aneurysms and three ranging in size from 4 mm to 5 mm escaped sonographic detection.

Increased Intracranial Pressure and Cerebral Circulatory Arrest

An increase in intracranial pressure results after serious trauma, global hypoxia, and pronounced cerebral edema due to various causes. When the systemic blood pressure is constant, this leads to a decrease in intracranial circulation, with increasing peripheral resistance. Particularly in intensive care patients, the following changes in the Doppler spectra can be observed in short-term follow-up observations of the middle cerebral artery under increasing cerebral pressure (Fig. 3.**22**):

– Increased diastolic flow velocity.
– Change in the systolic/diastolic amplitude modulation.
– Appearance of a triphasic course with a diastolic backflow component, as seen physiologically in the body's peripheral arteries.

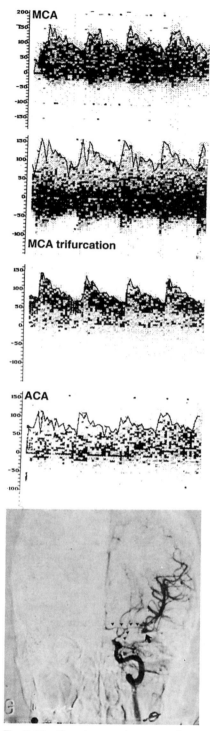

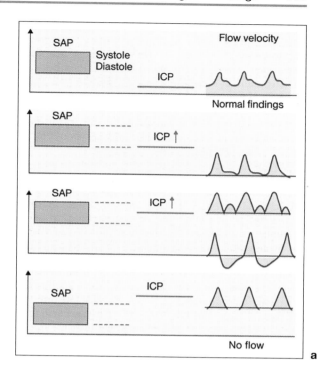

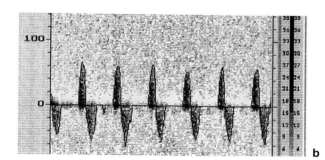

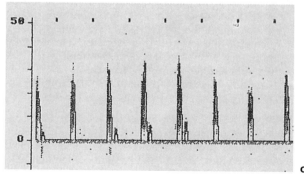

Fig. 3.**21** Doppler spectrum waveforms and corresponding carotid angiogram in a patient with a marked spasm (▲▲▲) of the intracranial T-junction of the carotid artery, the anterior cerebral artery (ACA), and the middle cerebral artery (MCA), after subarachnoid hemorrhage from a middle cerebral artery aneurysm (arrow). There is a flow acceleration in the entire area of both the MCA and the ACA at depths of 40–60 mm

Fig. 3.**22 a** Schematic illustration (**a**) of the relationship between systolic arterial pressure (SAP), intracranial pressure (ICP), and Doppler spectrum waveforms recorded from the intracranial cerebral arteries (according to Hassler et al. 1988). If the intracranial pressure exceeds the systolic blood pressure, flow velocity ceases, first during diastole and then completely. Examples of biphasic flow **b** and early systolic peaks **c** in a patient with intracranial arrest of cerebral perfusion

These changes can precede clinical symptoms and therefore allow early counterregulatory therapeutic measures to be taken. If these changes are fully developed in all large cerebral arteries, they indicate an intracranial circulatory arrest. In this situation, since the detectable flow waveform changes in the extracranial segments of the arteries are less pronounced and more difficult to interpret, false diagnoses were sometimes previously made when only an extracranial examination was carried out.

Thus a decrease in cerebral perfusion may be used as a supporting criterion for determining *brain death* if flow signals from both the extracranial and the intracranial cerebral arteries are modified in their course and if these changes are documented by an experienced investigator (von Reutern 1991). Criteria for brain death vary from country to country, and applicable national guidelines should be consulted; the position taken by the scientific advisory council of the Federal Council of Physicians in Germany requires at least two examinations that are at least 30 minutes apart to determine brain death. Similarly, according to the President's Commission on guidelines for determination of brain death in the US, confirmatory tests (like TCD, blood flow studies, radioisotope scanning) may be used in patients who fulfill the clinical criteria for brain death to shorten the observation period preceding organ harvest (President's Commission 1981). To prove cerebral circulatory arrest, the following findings must be documented bilaterally:

– Biphasic flow with equally developed anterograde and retrograde components, or early systolic peaks without systolic or diastolic flow in the middle cerebral arteries and the internal carotid arteries intracranially (50 cm/s), and also in the remaining intracranial arteries that are accessible to ultrasound, in the extracranial internal carotid arteries, and the vertebral arteries.
– Absent flow signals from the arteries at the base of the brain can only be interpreted as a definite sign of cerebral circulatory arrest if the same investigator has documented a signal loss in previously clearly detectable intracranial flow signals, and cerebral circulatory arrest can be demonstrated in the extracranial arteries supplying the brain.
– In infants less than six months old, Doppler sonography cannot be used to assess brain death.

Although false-negative results do not necessarily exclude a method from being used to assess brain death, false-positive findings would be fatal. This has been documented in patients with transient cerebral circulatory arrest, e.g., after SAH (Grote and Hassler 1988, Eng et al. 1993). It should therefore be noted that:

– A primary absence of signals from the arteries at the base of the brain cannot be considered a valid criterion of brain death—additional, repeat studies (after 30 minutes) are mandatory.

Intracranial Findings in Steal Phenomena

In the majority of patients with subclavian steal phenomenon, the direction of blood flow remains normal through the basilar artery. Approximately one-third of the patients show intermediate flow phenomena, and retrograde perfusion is a rarity, almost exclusively found in patients with multivessel disease, or anomalies in the circle of Willis, or both. About 40% of patients with subclavian steal phenomena show changes in the flow relationships within the basilar artery when the affected arm is compressed. In patients with innominate artery obstructions, complex hemodynamic changes can be detected in the intracranial vasculature, similar to those reported extracranially. The abnormal flow profiles in these patients are characterized by a muffled spectrum during the early systole, which is accompanied by a breath-like flow signal predominantly caused by a delayed systolic increase (Rautenberg and Hennerici 1988). A latent steal phenomenon can be detected in various intracranial blood vessels when the right upper extremity is compressed (Hennerici et al. 1988). Currently available studies suggest that latent and intermediate arterial flow changes at the base of the brain may act as a possible source of emboli. This revises the earlier pathophysiological view, according to which a hemodynamic perfusion disturbance was seen as the primary mechanism in this condition.

Functional Disturbances

Various functional disturbances can have a general influence on the Doppler signals in cerebral arteries, and their role as potential pathological mechanisms in neurological illnesses is under discussion.

Cerebral autoregulation disturbances. Examinations using a tilting table have recently been used in the diagnosis of syncope and cerebral ischemia of undetermined origin:

– In patients with serious peripheral or central autonomic nervous dysfunction (e.g., diabetic neuropathy, dysautonomia, multisystem degeneration).
– In patients without mechanisms of compensation for a significant reduction in flow parameters registered in the arteries of the brain.
– Due to endocrinal/metabolic failure of systemic blood pressure control.

More interesting is a different patient group, in whom this type of reduction was observable as a sign of isolated cerebral autoregulation disturbance (Fig. 3.**23**):

– In spite of stable systemic blood pressure regulation
– And with compensatory tachycardia

(Daffertshofer et al. 1991, Hennerici and Daffertshofer 1993).

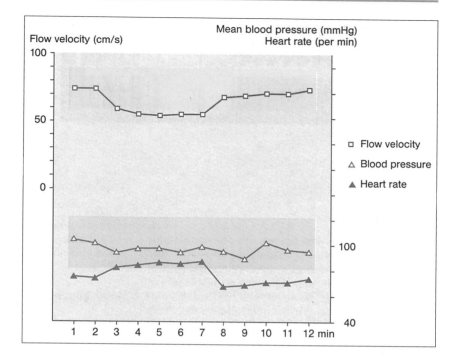

Fig. 3.**23** Findings in an orthostatic TCD study. In the
upright position (2–8 minutes)
the average flow velocity
of the middle cerebral artery
decreases despite normal
tachycardia and stable systemic blood pressure. This
condition indicates isolated
cerebral autodysregulation and
is only detectable using transcranial Doppler sonography

Monitoring HITS and Echo Contrast Media

Continuous monitoring of the flow relationships in
the middle cerebral artery has been carried out during
carotid endarterectomy (van Zuilen et al. 1995) and
during cardiopulmonary bypass surgery. Although it is
still not clear whether the complication rate associated with surgery can be reduced using this type of
monitoring (Hennerici 1993), these investigations led
to the interesting observation of *high-intensity transient signals* (HITSs) (Fig. 3.**25**). Similar signals have
also been detected during cerebral angiography of the
intracranial arteries at the base of the brain
(Rautenberg et al. 1987). HITSs are short, high-amplitude, unidirectional signals associated with a typical
chirping sound, which can appear at any time in the
cardiac cycle. In-vitro examinations have shown that
HITSs sometimes reflect the size and velocity of corpuscular intra-arterial particles. However, reliable
conclusions about the varying composition of the reflecting particles (e.g., atheromatous material, platelet
aggregate) from signals recorded are not currently
possible (Markus and Brown 1993). Neither the
causes nor the clinical significance of HITSs have
been sufficiently clarified yet, particularly since clinical symptoms are not associated with these phenomena (Hennerici 1994, Babikian et al. 1994). An interpretation involving microemboli might be plausible,
but using a synonymous term to describe these occurrences as "embolic signals" is not justified. Various
HITS frequencies have been reported in patients who
are potentially at risk from embolism (e.g., patients
with asymptomatic or symptomatic carotid artery stenoses and especially patients with artificial heart
valves) (Hennerici 1994). In these patients with artificial heart valves, 40,000–50,000 HITSs occur per day

without any clinically detectable neurological deficits—most likely the origin of these HITSs are harm-
less microcavitations rather than potentially harmful
microemboli. This has recently been confirmed by
Droste et al. 1997.

Intracranial emboli from extracranial vascular
processes or cardiac sources may fragment during
arterial passage or even after closure of distal arteries
through spontaneous lysis. To what extent this is of
any pathogenetic significance remains unclear. Automatic embolus detection software devices showed less
valid results when compared with the assessment of
experienced observers (Daffertshofer et al. 1996, van
Zuilen et al. 1996).

In TCD with contrast medium, the signal changes
that appear can be used diagnostically in the presence
of a patent foramen ovale (Lechat et al. 1988,
Schmincke et al. 1995). After an intravenous injection
of an echo contrast medium, paradoxical embolization paths can be spontaneously detected by observing microbubbles in cerebral arteries, or after using
the Valsalva maneuver to induce a right-to-left atrial
shunt. Depending on the particle size used, the reported sensitivity of contrast TCD may differ from
that seen with contrast TEE. With a diameter of less
than 5 μm, contrast TCD also detects small intrapul-
monary shunt formations; this error is lower with contrast medium particles larger than 20 μm (Di Tullio et
al. 1993, Nikutta et al. 1993) (Fig. 3.**24**). Although application of echo contrast media has been shown to
improve visualization of the cerebral arteries studied
by means of color flow duplex sonography, only vague
knowledge exists whether this can be used to improve
the validity of this test in pathological conditions. So
far, studies of cerebral veins have been repeatedly
performed with some success (see p. 12).

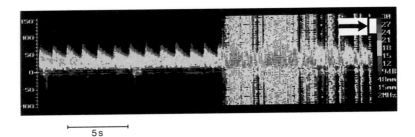

5 s

Fig. 3.**24** After injection of non-transpulmonary echo contrast medium into a cubital vein, numerous signals are detected with a high-amplitude audiosignal in the middle cerebral artery (ultrasound sample volume depth 48 mm), indicating activity of a paradoxical pathway through an atrial right-to-left shunt in a patient with a patent foramen ovale

Migraine

Follow-up investigations during, and in the intervals between, migraine attacks have not so far yielded any convincing and consistent examination results. The presence of local increases in flow velocities, as well as decreases, has contributed nothing to the diagnosis. More recent findings, however, obtained using complex stimulus patterns that stimulate almost the entire vascular territory of the PCA, including both the primary and secondary visual projection areas, indicate significant abnormal vasoneural coupling in almost all patients with migraine (with aura) who were examined, even when there was no current migraine attack. These results are modified when the patients are undergoing treatment. Confirmation of the findings and evaluations of these studies are not yet complete.

Sources of Error

Incorrect Anatomic Identification
Flow Accelerations

- "Functional stenosis" in a hypoplastic blood vessel
- A high shunt volume (arteriovenous malformation)
- Extracranial collateralization due to obstructive processes
- Intracranial collateralization due to obstructive processes
- Hyperperfusion after ischemia induced by arterial occlusion, cardiac arrest, hypoxia, spreading depression in migraine, meningitis, head trauma, etc.

Absent Signal

- When the sample volume is incorrectly placed, or there is a poor signal-to-noise ratio or unfavorable ultrasound window.

Inadequate Control

- Of orthopnea, hematocrit, hypoglycemia

Normal Variants

- Direct origin of the PCA from the carotid artery
- Pseudostenosis signs (e.g., collateral flow in the communicating arteries of the circle of Willis)
- Reduced flow velocity (e.g., basilar hypoplasia)

Miscellaneous

Knowledge of the extracranial vascular finding *is a prerequisite* for an adequate interpretation of the transcranial Doppler study.

In monitoring examinations, artifacts are a frequent problem. In contrast to so-called HITS, characterized by a short, high-intensity, unidirectional signal, artifacts caused by the probe are seen on both sides of the baseline (Fig. 3.**25**).

Diagnostic Effectiveness

The stenosis criteria available do not allow differentiation *between functional* and *structural* arterial stenosis of the intracranial cerebral arteries. In high-grade extracranial vascular processes, intracranial collateralization through the anterior or posterior circle of Willis often occurs. The extracranial vascular findings should always be known in order to avoid misinterpretations in such circumstances.

For methodological reasons, there have so far been hardly any findings concerning the validity of the evaluation of different degrees of stenosis, as is possible in extracranial Doppler sonography. As in other vascular regions, while a definite assessment of the degree of stenosis is quite possible on the basis of hemodynamic principles using Doppler sonography, angiographic evaluation of arterial stenoses of intracranial cerebral arteries is severely restricted by projection problems, loop formation, and the fact that the stenosis can often only be depicted on a single plane.

The literature only provides isolated information about the specificity of the procedure, since intracranial stenoses occur far more rarely than extracranial ones in Western communities. Information about the specificity in the posterior circulatory system is rare. According to our own investigations (Rautenberg et al. 1990), the specificity for the anterior and posterior circulatory systems of the brain lies be-

Fig. 3.**25** Doppler spectrum waveforms of the middle cerebral artery **a** Normal findings **b** Detection of a unidirectional high-intensity transient signal (HITS) **c** For comparison, an artifact is displayed for which baseline crossing signals are characteristic

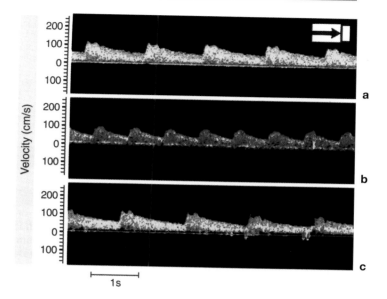

tween 99% and 100% (Table 3.**4**). Similar results (98%) were reported by Arnolds et al. (1986) and Ley-Pozo and Ringelstein (1990) in the anterior circulation.

In the anterior circulation, and when the normal Doppler spectrum values reported agree, reliable detection of vascular processes due to pathological phenomena is possible, as long as the vascular topography remains relatively standardized. In the posterior region, vascular variants and anomalies in the vessel course predominate; continuously applying ultrasound at the specific depths that are of interest is often difficult. Using transcranial Doppler sonography and MRA, spiral CTA in combination (Röther et al. 1993, Neff et al. 1997) appears to be promising. With its ability to display the vascular anatomy, especially in the posterior circulatory system, MRA, spiral CTA provides significant additional information. The detailed additions to the static examination technique described are useful in documenting functional changes. The diagnostic usefulness of transcranial color flow Doppler sonography is attractive, however, cannot be conclusively evaluated (Seidel et al. 1995).

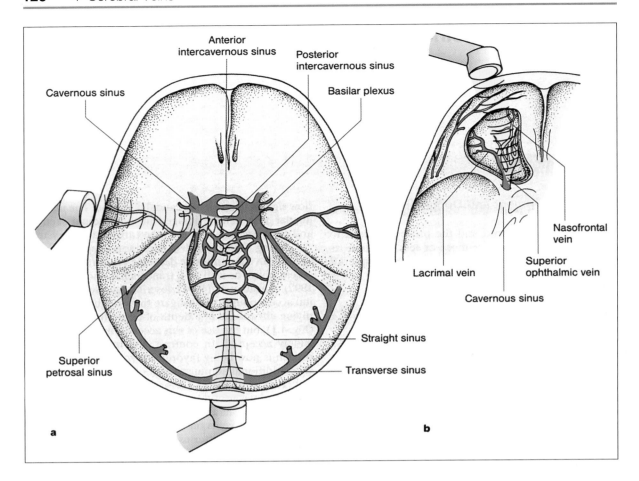

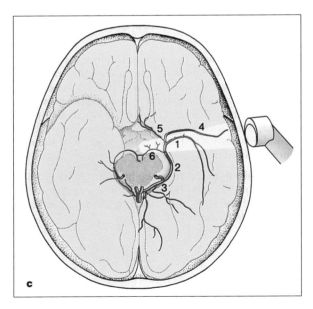

Fig. 4.**1**   Schematic drawings of:
a Basal sinuses
b Periorbital veins
c Basal cerebral veins.
Basal vein of Rosenthal: anterior segment (1), middle segment (2), posterior segment (3); deep middle cerebral vein (4), anterior cerebral vein (5), peduncular vein (6)

Fig. 4.**2** Display of Doppler spectra during
the Valsalva maneuver
a Internal jugular vein
b Intracranial vein

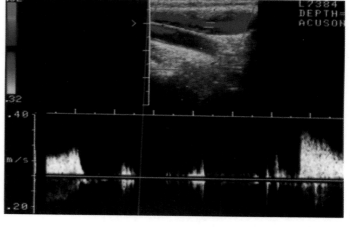

Normal Findings

Principle

■ Continuous Wave and Pulsed Wave Doppler Sonography

Continuous wave and pulsed wave Doppler sonography can provide a general overview of the jugular vein.

■ Conventional and Color Flow Duplex Sonography

The *B-mode* allows the morphological relationships in the jugular veins to be evaluated; the *pulsed wave Doppler mode* provides selective analysis of the venous flow signals without disturbance from arterial superimpositions. In *color flow duplex imaging,* it is possible to detect small veins that are otherwise hardly noticeable, such as the vertebral vein, in addition to the jugular venous system (Fig. 2.**18**, p. 43).

Anatomy and Findings

Cerebral drainage proceeds via superficial (external) and deep (internal) veins (Fig. 4.**1 a, b**) (for details see Andeweg 1996). Both venous systems drain into the sinuses, and further drainage is provided by the internal jugular vein and the vertebral vein. The intracranial basal veins (anterior cerebral vein, deep middle cerebral vein, basal vein of Rosenthal, vein of Galen) form a circuit around the midbrain, comparable to the arterial circle of Willis. Additional, quantitatively smaller, venous drainage is provided by

anastomoses between the sinuses, diploic veins, and emissary veins into the external jugular vein. There is further venous drainage via the cavernous sinus. Venous valves are rarely found in the neck and head region; the intracranial veins have no valves.

Extracranial Veins—Internal Jugular Vein

■ Continuous Wave and Pulsed Wave Doppler Sonography

The internal jugular vein is located lateral to the common and internal carotid arteries. Usually, venous return flow from the upper venous system occurs passively, due to the inspiratory pressure decrease within the thorax. A biphasic progression can be observed in the Doppler sonogram, with dominant systolic and less developed diastolic flow (Baumgartner and Bollinger 1991). Acoustically, a predominantly unmodulated signal, which is affected by breathing, swallowing, and the Valsalva maneuver, is found (Fig. 4.**2**). The venous signals can easily be eliminated by applying light proximal or distal pressure with the hand.

■ Conventional and Color Flow Duplex Sonography

In longitudinal section, the internal jugular vein is usually located lateral to the common carotid artery (Fig. 4.**3**). The vein is easily compressed. Depending on the direction of the pressure applied, a peaked venous image is produced below the sternocleidomas-

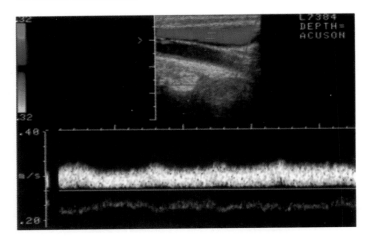

Fig. 4.**3** Display of the jugular vein using color flow duplex sonography (normal findings)

toid muscle (Steinke and Hennerici 1989). In some cases, venous valves can be depicted in the proximal venous segment. Sonographically, typical vascular wall pulsations can be detected. The maximum venous distension occurs during expiration, and the minimum venous diameter is reached at the end of inspiration (Fig. 4.**4a, b**). Fluctuations in the caliber of the vessels due to respiration are superimposed by higher-frequency, weak pulsations from the venous walls, which are caused by the heart. During the Valsalva maneuver, the vein dilates and the pulsations decrease. Often, streak-like internal echoes can be displayed, caused by slow-flowing blood. During deeper inspiration, the vascular lumen collapses, and retrograde blood flow simultaneously increases.

An experimental method for determining total cerebral blood flow uses flow measurements in the internal jugular veins (Müller et al. 1988).

Color-coded flow signal imaging makes it easier to differentiate between veins and arteries. Even small-caliber veins, such as the vertebral veins, can usually be detected in the extracranial V_2 segment (Fig. 2.**18**). In color flow duplex sonography, imaging of the vertebral vein when the vertebral artery is missing is a criterion for vertebral artery occlusion.

Intracranial Veins

■ Pulsed Wave Doppler and Color Flow Duplex Sonography

The scope of ultrasound diagnosis in the intracranial veins has long been underestimated and is still limited. Although venous signals can often be identified, with a breath-like, band-shaped, and hardly pulsatile signal during transcranial examination of the arterial system, systematic examination of the cerebral veins and sinuses with reproducible results is limited to the basal vein of Rosenthal, the deep middle cerebral vein, the lateral transverse sinuses, and—with some reservations—the rectus sinus, carvernous sinus, and inferior sagittal sinus, even with color flow duplex sonography (Hennerici 1990, Valdueza et al. 1996). As shown in Fig. 4.**1c**, the basal vein of Rosenthal can be insonated via the transtemporal window adjacent to the P_2-segment of the posterior cerebral artery. Venous signals adjacent to the middle cerebral artery may derive from the deep middle cerebral vein. Pulsatile flow signals can also be recorded from the straight sinus and the transverse sinuses from ipsi- and contralateral transtemporal windows, with a typical modulation effect following the Valsalva maneuver. Ultrasound

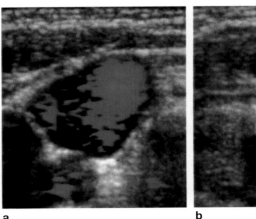
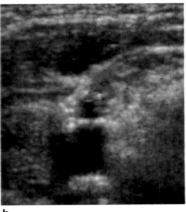

a   b

Fig. 4.**4** Varying dilatation of the internal jugular vein during:
a Expiration
b Inspiration

contrast media that pass through the lungs are used with color flow transcranial duplex sonography (Bogdahn et al. 1995, Ries et al. 1997). Although a conclusive assessment of the usefulness of the procedure is not yet possible, this seems to be a suitable procedure for follow-up studies in patients with transverse sinus thrombosis and to differentiate acute/chronic thrombosis against hypoplasia/aplasia, particularly if combined with magnetic resonance phlebography (Figs. 4.**5**, 4.**9**) (Mattle et al. 1991).

Evaluation

The evaluation of findings in the cerebral veins is primarily qualitative. For transcranial recording of the basal vein of Rosenthal and the deep middle cerebral vein in adults, mean flow velocities from 4–17 cm/s (mean: 10.1 ± 2.3 cm/s) have been reported.

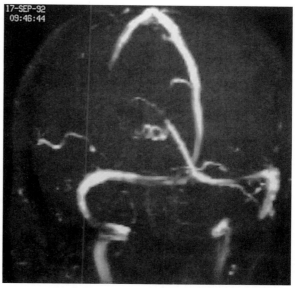

Fig. 4.**5** Noninvasive imaging of the intracranial venous system using magnetic resonance phlebography

Pathological Findings

Principle

Since pathological findings here are rare, ultrasound diagnosis in the cerebral veins is less important than diagnosis in the veins of the extremities. Extracranially, the most frequent pathological finding is thrombosis of the jugular vein, often due to iatrogenic manipulation. Findings obtained from the intracranial cerebral veins provide information concerning direct and remote obstructions in the cerebral veins, sinus, and plexus.

Findings

Extracranial Veins

Jugular Vein Thrombosis

■ Continuous Wave and Pulsed Wave Doppler Sonography

Continuous wave and pulsed wave Doppler sonography are of limited value in detecting or excluding processes involving the jugular vein. When there is clinical suspicion of jugular vein thrombosis and an absent flow signal using continuous wave Doppler, sonography, a supplementary examination with duplex sonography should be carried out.

■ Conventional and Color Flow Duplex Sonography

In thromboses of the internal jugular vein, the B-mode image shows a nonpulsatile, dilated blood vessel that has no detectable flow signals (Fig. 4.**6a**). Depending on the age of the thrombosis, irregular internal echoes that correspond to thrombotic material may be detected. The blood vessel cannot be compressed by exerting pressure with the probe. The Valsalva maneuver does not produce any further dilatation of the vascular lumen. A flow signal is not detectable, either with pulsed wave Doppler sonography or in color flow imaging (Fig. 4.**6b, c**). It is important to ensure that very low flow velocities are also detected and represented in color-coded form.

If venous valves will not close, it may indicate long-standing tricuspid valve insufficiency or a prior jugular vein thrombosis (Brownlow et al. 1985).

Damaged venous valves, vascular wall thickening, and also floating thrombi have been detected in patients who have undergone frequent jugular venipuncture procedures (Krünes et al. 1992).

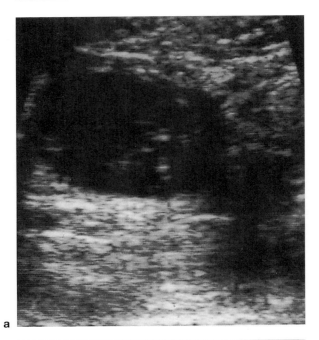

a

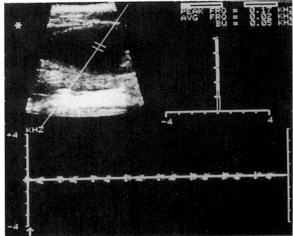

b

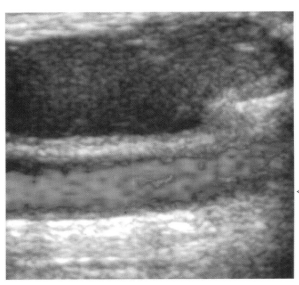

c

Ectasias and Aneurysms

These can be detected without difficulty using *conventional* or *color flow duplex sonography*. In contrast to the findings with thrombosis, the Valsalva maneuver causes dilatation, and the vein is compressible (Fig. 4.**7 a, b**).

Venous Congestion

When there is venous congestion, the disturbed venous drainage leads to decreased flow and venous dilatation (Steinke and Hennerici 1989).

Due to the close proximity of the cervical lymph nodes to the jugular vein, enlarged lymph nodes associated with metastatic processes may lead to displacement or compression of the jugular vein.

Arteriovenous Fistulas

Extracranial arteriovenous fistulas are very rare. Sonographically, the arterial and venous pathological findings involve increased flow velocities, and a volume increase in the afferent arteries and efferent veins (Fig. 4.**8 a, b**).

Intracranial Veins

Sinus Thrombosis

Noninvasive diagnosis of a sinus thrombosis using *transcranial Doppler sonography* is not reliable if not supported by concomitantly performed MR venography or conventional cerebral phlebography studies. When using transcranial Doppler sonography in the acute phase, a flow delay through the arterial branch, resulting from the obstructed venous drainage, may indicate a sinus thrombosis. In addition, flow velocity may be increased in the venous collateral vessels (Wardlaw et al. 1994, Valdueza et al. 1995, Ries et al. 1997). Normalization after recanalization of dural sinus thrombosis has also been monitored. *Transcranial color flow duplex sonography* using ultrasound contrast media can improve the diagnosis and facilitates imaging in detecting otherwise indetectable low-flow signals hidden by cerebral arteries (Becker et al. 1994). Systolic peak flow velocities with symmetrical appearance of the blood flow spectra show no signifi-

◁ Fig. 4.**6** Internal jugular vein thromboses imaged **a, b** using conventional duplex sonography, and **c** with color flow duplex sonography. A maximally dilated blood vessel without pulsation and with partially hyperechoic material is seen (**a, c**). Inside the vein, pulsed wave Doppler sonography (**b**) and color imaging (**c**) do not detect any flow signals. In the color flow image, the common carotid artery, with a normal flow signal, is seen dorsal to the vein

Fig. 4.**7 a, b**   A color flow duplex sonogram showing an ectatic internal jugular
vein

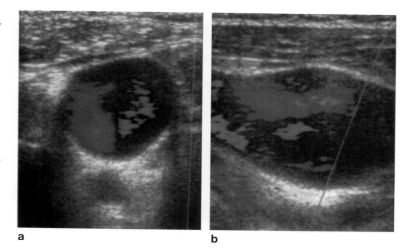

a   b

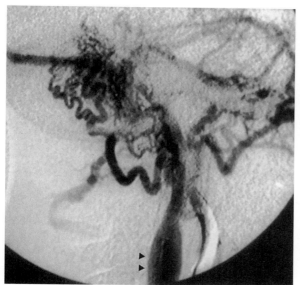

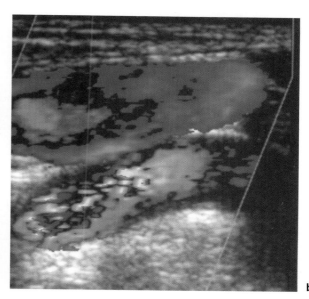

a   b

Fig. 4.**8 a**   Angiogram and **b** color flow duplex sonogram
in a patient with a dural fistula. Significant turbulences add
signal-void flow distortion areas located distally within the
internal jugular vein (**a** arrowheads, **b** upper blood vessel).
There are abnormally high flow velocities in the common
carotid artery as a result of the arteriovenous shunt

cant side-to-side differences. Velocities range from
15–20 cm/s (mean 17.5 ± 1.9 cm/s). In the hypoplastic
or partially blocked transverse sinus, velocities were
significantly reduced (mean 9.4 ± 4.0 cm/s) but increased contralaterally (mean 28.4 ± 6.5 cm/s) (Ries et
al. 1997). If combined with magnetic resonance phlebography, both noninvasive methods add complementary information and may help to avoid conventional
cerebral phlebography (Fig. 4.**9**).

Cavernous Sinus Fistula

Carotid cavernous fistulas consist of a connection between the internal carotid artery and the cavernous
sinus or the dural branches of the external carotid
artery. The typical finding using *continuous wave or*
pulsed wave Doppler sonography at the medial canthus is a hissing, high-frequency, machine-like sound,
with flow directed toward the probe. This signal is
caused by drainage through the orbital veins, in which
the blood flow is normally directed from exterior to
interior. Fistulas between external carotid artery
branches and the cavernous sinus often show normal
findings in extracranial Doppler sonography. In these
cases, *transcranial Doppler sonography* can produce
diagnostically significant findings by displaying the
fistula directly (Fig. 3.**20**, p. 118).

Arteriovenous Malformations

Although venous angiomas and cavernomas usually
escape sonographic detection, the venous drainage of

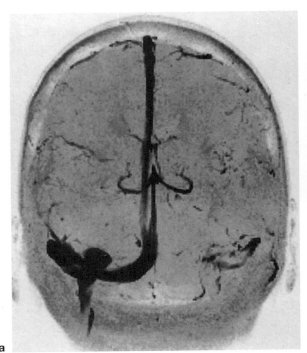

a

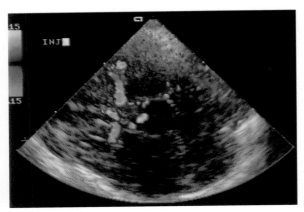

b

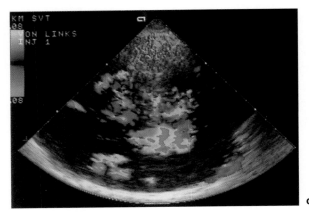

c

Fig. 4.**9 a**   A thrombosis in the transverse sinus and ipsi-lateral jugular vein, imaged noninvasively in a magnetic resonance angiogram. Corresponding transcranial duplex-analysis demonstrates normal flow in the contralateral transverse sinuses after echo-contrast medium administration (**b**: before, **c**: after injection)

arterial angiomas can usually be imaged well using *transcranial Doppler sonography,* especially when there is a high shunt volume.

Evaluation

The evaluation includes the spontaneous flow signals, the way in which they change after the Valsalva maneuver, and an assessment of vascular compressibility.

Sources of Error

Extracranial application. Exerting too much surface pressure with the probe often leads to errors when examining the extracranial venous system.

When there is an occlusion or high-grade stenosis of the innominate artery, marked hemodynamic changes occur distally in the distribution area of the carotid artery. Signals from the common carotid artery sometimes have a breath-like, venous quality, and confusing these with venous signals is a common mistake.

Intracranial application. The major disadvantages are difficulties in identifying and monitoring cerebral veins and sinus in older patients, nearby high-flow signals (arteries, AVM, shunt volumes, dural fistulas,

etc.) or total occlusion in hidden segments of the highly variable venous system.

Diagnostic Effectiveness

There are no data on the sensitivity and specificity of the method, since very few cases have been reported in the literature (Albertyn et al. 1987, Bloching et al. 1989, Terwey et al. 1981, Valdueza et al. 1995, Wardlaw et al. 1994, Ries et al. 1997). However, it can be assumed that using ultrasound to detect a jugular vein thrombosis is highly accurate. Duplex sonography is increasingly being used to clarify vascular relationships before a central venous catheter is placed. *Ultrasound-guided puncture* can reduce the number of incorrect punctures, and the technique is therefore likely to be particularly helpful in high-risk patients (Mallory et al. 1990).

Diagnosing intracranial sinus thromboses with ultrasound is more difficult and should always be supplemented by other imaging studies.

5 Peripheral Arteries

Systolic Blood Pressure Measurement

Examination

Special Equipment and Documentation

Unidirectional Doppler equipment is usually used for this examination method. The equipment reproduces the Doppler signal acoustically (with the so-called pocket Doppler), but does not allow any measurement of flow direction. However, bidirectional Doppler equipment, which does detect the flow direction, can also be used for the examination.

Measurements are made using standard blood pressure equipment. The cuff width must be appropriate for the placement area (e.g., upper arm, distal lower leg 13 cm; thigh 15–18 cm).

A *transmission frequency* of 8–10 MHz should usually be used to localize the Doppler signal at the arteries of the foot (posterior tibial artery or dorsal artery of the foot, or both), and the radial artery. A transmission frequency of 4–5 MHz may be more effective for deeper arteries (popliteal artery, brachial artery).

It is obligatory to keep a written numerical record of the measured pressure values, describing the measurement area (cuff position) and the arteries measured (probe position).

Examination Conditions

Patient and Examiner

The examination is carried out with the patient *reclining*. A preliminary resting period, lasting about 5–10 minutes in healthy individuals, is important. In pathological cases, the resting period should be extended to 15–20 minutes, since the poststenotic pressure decreases with the slightest exertion, and only returns to its original value slowly. The examiner should be in a comfortable position, possibly sitting.

Conducting the Examination

The blood pressure cuff is placed round the selected measurement area avoiding any external pressure. If possible, the pulse is first palpated at the artery to be measured, and contact gel is then applied. Using the Doppler ultrasound probe, the examiner searches for the optimal acoustic signal. Prior to inflating the cuff, the examiner should stabilize the probe by supporting it with his finger, hand, and/or lower arm. The patient must keep the leg or arm that is being measured still (Fig. 5.**1**).

The measurement procedure is the same as used in the conventional Riva-Rocci method. The cuff is inflated to above the systolic pressure until the Doppler signal disappears. The cuff is then *slowly* deflated (2–5 mmHg/s) until the signal is heard again. The pressure reading at this moment on the manometer scale is noted, and it corresponds to the systolic arterial pressure *under the cuff.*

It is important to distinguish between the cuff and probe positions. The pressure measured corresponds to the pressure in the artery or arteries underneath the cuff. Independently of the cuff location, the Doppler signal is usually recorded at the ankle (posterior tibial artery or dorsal artery of the foot, or both) (Fig. 5.**2**).

Note: Measurement area – cuff position
Measured artery – probe position

If the Doppler signal can no longer be detected at very low pressures, the patient's upper body can be raised. The hydrostatically elevated pressure will then be within the measurable range, and the difference to the ankle pressure while reclining has to be deducted from the measured value (1.0 cm H_2O $\geq$ 0.736 mmHg).

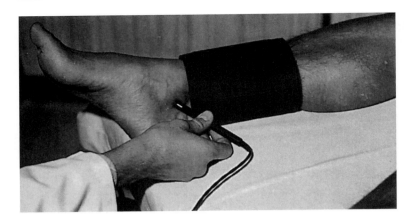

Fig. 5.**1**   Systolic ankle pressure measurement using the ultrasound Doppler method. Distal lower leg is the cuff position, the posterior tibial artery the probe position

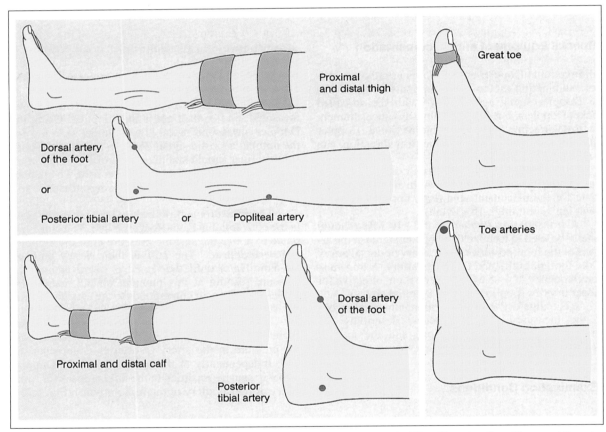

Fig. 5.**2**   Measurement areas and arteries for measurement at the lower extremity. The different cuff positions and the arterial regions that are suitable for measuring with the continuous wave Doppler probe are shown

Examination Sequence

Pelvic and Leg Arteries

Measurement of Pressure at Rest

The systolic blood pressure in the lower extremities is always compared to the systemic pressure. The undisturbed arm pressure is used as the reference pressure. This value should be determined directly before or after measuring the leg pressure, using either the ultrasound Doppler method or the conventional stethoscope technique. The comparative studies carried out by Thulesius (1971) showed a very good correlation between the two measured values. A prior clinical examination (palpation, auscultation) and comparison of the blood pressure values on each side must be carried out to exclude a vascular obstruction in the arm (subclavian artery, brachial artery).

The preferred area for leg measurement is the distal lower leg; the arteries measured are the posterior tibial artery or the dorsal artery of the foot, or both. All prior flow obstructions can be detected here, and it is also a good location for placing the blood pressure cuff without too much interference from the musculature.

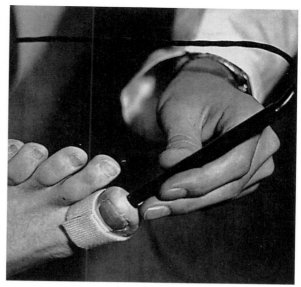

Fig, 5.**3**   Toe pressure measurement showing the special cuff at the basal joint of the toe and the position of the Doppler probe (8–10 MHz)

Measurement of Pressure Post Exercise

Defined active exercises that exert the leg muscles are used to differentiate borderline values. Simple exercise tests that have proved useful are tiptoeing and knee bending, which can be used without any difficulty (Köhler et al. 1972). More precisely defined exercise is possible using the treadmill ergometer (Cachovan and Maass 1984, Mahler 1990).

For patients in whom exercise testing is not possible, reactive hyperemia after arterial compression may be used (Fronek et al. 1973, Mahler et al. 1975).

The following exercises are recommended as standards for the different *exercise tests:*

- Tiptoeing 40 times
- Knee bending 20 times
- Walking 200 m, 120 steps/minute
- Treadmill 100 m, 3.2 km/h, at a 12.5° gradient
- Hyperemia: three minutes of arterial compression above the knee with a wide blood pressure cuff

Measurement of Segmental Pressure

Measuring the pressure at several locations in the thigh and lower leg is only necessary if the ankle pressure is pathological, and the obstruction level in the vascular system needs to be established. The usual *cuff placement locations* and corresponding cuff widths are:

- Arm 13 cm
- Proximal thigh 15–16 cm
- Distal thigh 15–16 cm
- Proximal lower leg 15–16 cm
- Distal lower leg 13 cm
- Great toe (proximal joint) 2–3 cm

Care should always be taken to ensure that the blood pressure cuffs are of the appropriate width and length for the specific placement area (Fig. 5.3). The arteries of the ankle (posterior tibial artery, or dorsal artery of the foot) are usually the ones measured. However, some authors (Franzeck et al. 1981) do recommend placing the probe close to the cuff (e.g., popliteal artery for thigh measurements).

Measurement of Toe Pressure

Measurement of the systolic toe pressure is carried out using a special cuff (width approximately 25 mm), and the result is compared to the distal lower leg pressure. Sensing devices that can be used include the ultrasound Doppler probe, a strain gauge, or photoplethysmography. This important measurement is made more difficult by the fact that ordinary blood pressure equipment does not allow pressure determination in small cuffs without special adaptation. In a practical clinical setting, one should preferably use special equipment that fully automates the simultaneous blood pressure measurements in the distal lower leg and toe, using a mercury strain gauge, and can also display the information graphically.

Upper Extremity Arteries

In the upper extremities, measurements of the systolic blood pressure using the ultrasound Doppler method are rarely made, since conventional blood pressure measurement with the stethoscope is simpler and produces identical results (Thulesius 1971). Alternatively, a blood pressure cuff can be placed round the upper arm or forearm, and the radial artery or ulnar artery can be used as the artery to be measured by the Doppler probe (Fig. 5.**4**).

Measurement of Digit Pressure

Using special blood pressure cuffs (width 20–25 mm), it is also possible to measure the systolic blood pressure of the fingers. The cuff is placed around the distal or middle portion of the finger, and distal to this a high-frequency ultrasound Doppler probe is used to locate the arterial signal (Fig. 5.**5**). However, it is more common to use plethysmographs, which provide suitably small cuffs, automatic inflation and deflation, with the aid of a strain gauge (Gundersen 1972, Nielsen et al. 1973).

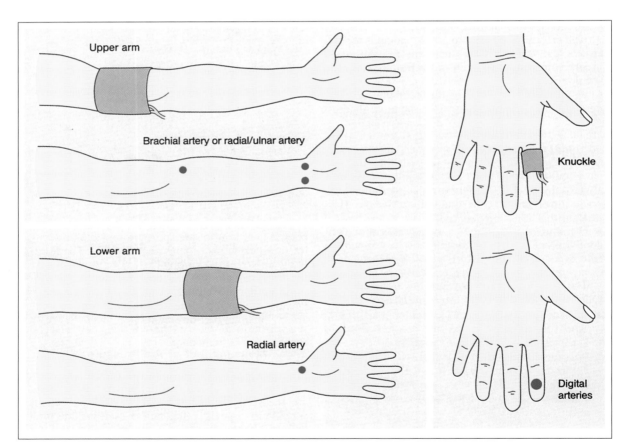

Fig. 5.**4** Measurement areas and arteries for measurement in the upper extremity. Proper placement of the blood pressure cuff and the most favorable positions for registering the Doppler signals are shown

Fig. 5.**5** Measuring the digital arterial pressure at the hand showing the position of the cuff and the ultrasound Doppler probe (8–10 MHz)

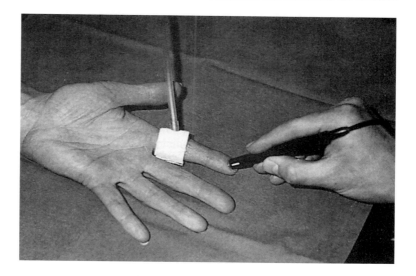

Normal Findings

Principle

The most important parameter for demonstrating undisturbed arterial circulation in the extremities is obtained by measuring the blood pressure in the peripheral arteries. Using the ultrasound Doppler method, it has also become possible to detect the flow in blood vessels in which Korotkoff sounds can no longer be auscultated (Strandness et al. 1966, Schoop and Levy 1969).

The procedure used (p. 135) corresponds to the conventional Riva-Rocci arterial blood pressure measurement, with the Doppler probe replacing the stethoscope.

However, only the systolic blood pressure can be measured, not the diastolic pressure. When the suprasystolic obstacle to blood flow is released, the ultrasound waves from the Doppler probe meet the peak flow of the blood that is running freely again. In contrast to the Korotkoff sounds, the signal does not cease during the diastolic phase. A special characteristic of pressure measurement in the distal lower leg, termed *ankle pressure measurement,* is that the normal peripheral systolic pressures are approximately 10–20 mmHg higher than the systolic arm pressure.

Wetterer et al. (1971) showed that "pressure pulses at a greater distance from the heart appear later than the central pulse, and differ from it both in their course and pressure amplitude." The systolic pressure increases substantially in a peripheral direction, while the diastolic pressure, and also the mean pressure, decrease slightly (Fig. 5.**6a, b**). This amplitude increase is called "amplification." It is explained by the back-and-forth movement of the waves in the arterial system, which when superimposed on one another, create a so-called "standing wave" (Busse 1982). These interesting phenomena, which are important in the evaluation of blood pressure measurements, are caused by wave reflection. Reflection occurs wherever wave resistance increases; in the arteries of the extremities, this happens peripherally. All of the longer arterial branches are individual elastic systems with their own variable reflection factors, which can show wide differences, corresponding to the particular vascular tone.

Findings

Pelvic and Leg Arteries

Measurement of Pressure at Rest

When the vascular system in the leg is patent, the peak systolic ankle pressure, as mentioned above, is higher than the systolic arm pressure.

Schoop and Levy (1969) and Bollinger et al. (1970) thus demonstrated that, on average, peripheral pressures in the leg were 10 mmHg higher than the arm pressures. Thulesius (1971) found in one study, however, that these pressures were the same; this may have been due to measuring the peripheral Doppler pressures while the cuff was still inflated.

Measurement of Pressure Post Exercise

After physical exertion (ergometry, tiptoeing, knee bending, treadmill), the peripheral systolic pressure increases in proportion to the elevated systemic pressure. A peripheral pressure decrease is usually not noted in healthy individuals. If it does occur, it is slight and brief (30–60 s) (Mahler et al. 1976, Ludwig et al. 1985) (Fig. 5.**7**).

Fig. 5.**6a**   Direct and simul-

Pathological Findings

Principle

Proximal to an occlusion, the arterial pressure does not increase, since excluding a small subsidiary circulatory system does not have any significant effect on the systemic blood pressure (Rieger 1993).

Distal to an occlusion, the *resting* pressure decreases, depending both on the number and length of the occlusions and on the extent of the collateral circulatory system (Schoop 1974). In contrast, the blood flow at rest is only detrimentally affected by extensive occlusive processes involving several levels. The *critical pressure level* is important here. This term refers to the postocclusive pressure that still allows undisturbed blood flow through the capillaries. It lies at 40–50 mmHg, sufficiently high above the venous pressure (20–25 mmHg) to maintain the arteriovenous blood flow (Rieger 1993) (Fig. 5.**9**).

All forms of *exercise* (walking, treadmill, tiptoeing, knee bending) result in vasodilatation. The increased blood requirements of the musculature are met by decreased peripheral resistance and increased flow volume. This occurs because of a pressure decrease due to the increased friction. Reactive hyperemia following suprasystolic arterial compression also has the same effect: the systolic blood pressure

decreases, and the flow velocity increases. The ankle pressure is inversely proportional to the flow velocity (Sumner and Strandness 1969, Köhler et al. 1972, Bollinger et al. 1973, Mahler 1990) (Fig. 5.**10**).

Findings

Pelvic and Leg Arteries

Measurement of Pressure at Rest

When resting values are being estimated, the *absolute value* of the systolic blood pressure corresponds approximately to the *circulatory reserve*. This is the value that most reliably indicates whether an extremity is endangered. The benchmark values given in Table 5.**1** have proved significant.

The pressure difference between the systemic and peripheral pressures provides information about the hemodynamic *compensation* for occlusions or stenoses (Table 5.**2**). It indicates the pressure decrease in an individual case, and shows a marked dependence on the systemic pressure, so that this value cannot be used as an absolute indicator of the seriousness of disease (Rode and Schütz 1977). The pressure difference is often also called the pressure gradient. However,

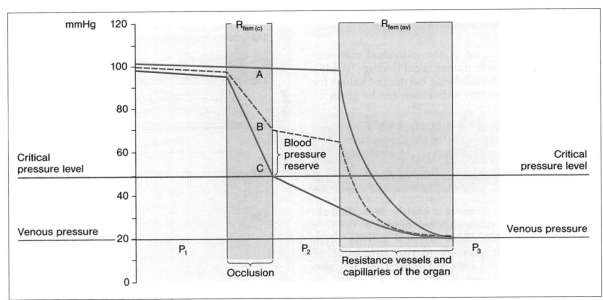

Fig. 5.**9**   A diagram showing the critical pressure level and the blood pressure reserve under normal and pathological conditions, in relation to the mean arterial pressure (adapted from Rieger). The course of the pressure in the lower extremities is shown:
A In a normal situation with a retained blood pressure reserve AC
B In an arterial occlusion with reasonable compensation and still adequate pressure reserve BC

C In several occlusions with no pressure reserve
P₁ Preocclusive pressure
P₂ Postocclusive pressure
P₃ Venous pressure
R$_{fem (c)}$ = Collateral resistance
R$_{fem (av)}$ = Organ resistance

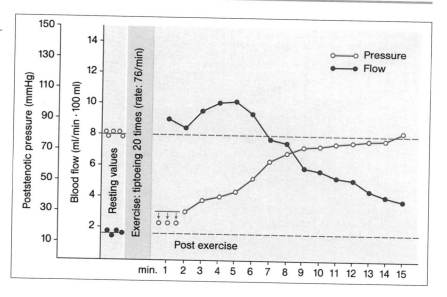

Fig. 5.**10**   A graph showing pressure and flow following exercise (muscular work) (adapted from Köhler). In a patient suffering from vascular disease (femoral artery occlusion), the postocclusive systolic pressure (measured with the ultrasound Doppler method) decreases following exercise, dropping into the area that is not measurable and only returning to its initial value after 15 minutes of rest. The length of the recuperation time indicates poor compensation. By contrast, the perfusion clearly increases following exercise, and has not yet reached its resting value after the same length of time has elapsed

the terms "pressure gradient" and "pressure difference" are not synonymous. "Pressure gradient" refers to a pressure difference over a specified vascular length.

The *ankle/arm index,* i.e., the pressure ratio between the peripheral and systemic pressures, is preferred at many clinical centers. However, this theoretical number provides no information about the systemic pressure or the absolute peripheral pressure, so that the value is only of limited use as a single parameter. The same ratio can be based on very different peripheral values, for example (Table 5.**3**).

On the other hand, the ankle/arm index has good predictive value in distinguishing a healthy patient from a sick one (Ouriel and Zarins 1982). It is well suited to the evaluation of large patient groups, and for follow-up studies (Yao et al. 1969).

Measurement of Pressure Post Exercise

When the ankle pressure borders on pathological values (pressure 0–5 mmHg below the systemic pressure), exercise may reveal low-grade stenoses that do not become hemodynamically significant until muscular exertion (Yao 1970, Thulesius 1978).

Using the various exercises (tiptoeing 40 times, or knee bending 20 times), it can be established whether the pelvic or thigh flow obstruction is the hemodynamically more active one in a two-level occlusion (Neuerburg et al. 1989). Looking at different types of exercise tests, Mahler (1990) showed that tiptoeing (30 times) is almost no different from treadmill ergometry, while hyperemia following arterial compression causes a slightly smaller pressure decrease (Fig. 5.**14**).

Table 5.**1**   Ankle pressure values in relation to the clinically significant circulatory reserve in peripheral arterial disease

Absolute pressure in the distal lower leg (mmHg)	Circulatory reserve
> 100	Good
≈ 80–100	Satisfactory
≈ 50–80	Moderate
< 50 (critical pressure level)	Insufficient

Table 5.**2**   The pressure difference between the systemic and peripheral systolic pressures, and its significance in indicating the hemodynamic compensation level in peripheral arterial disease

Pressure difference Δp arm/ankle (mmHg)	Hemodynamic compensation
< 20	Very good
≈ 40	Good
≈ 60	Satisfactory
≈ 80	Moderate
≈ 100	Poor

Table 5.**3**   Although the ankle/arm index (ratio) is highly dependent on the systemic systolic blood pressure, it simplifies following the clinical course in peripheral arterial disease

Systemic pressure (mmHg)	Peripheral pressure (mmHg)	Ratio
220	110	0.5
160	80	0.5
120	60	0.5

Measurement of Segmental Pressure

If the ankle pressure is pathological, i.e., lower than the arterial systemic pressure, the occlusion or stenosis level can be located using segmental blood pressure measurements (Fronek et al. 1978, Rutherford et al. 1979, Neuerburg et al. 1989).

The usual cuff placement positions are described on page 135. The pressure difference between the arm and the cuff placement area indicates an interposed vascular system obstruction. In addition, the pressure differences between the cuff locations on a leg are evaluated in longitudinal section.

The pressure difference between the measurement areas makes it possible to diagnose and locate obstructions that follow in tandem (Figs. 5.**11**, 5.**12**).

Measurement of Toe Pressure

In patients suffering from endangiitis and diabetes, it is also important to examine the arteries of the foot, which are often affected on an isolated basis. Using toe pressure measurement (p. 135), the absolute toe pressure, as well as the pressure difference between the lower leg and the toe, are examined. This detects any additional occlusions in the forefoot region, which may be missed using ankle pressure measurement only. In a pathological case, the pressure difference is more than 20 mmHg. One of our own studies yielded a value of between 45 and 80 mmHg (Fig. 5.**13**) (Knoblich et al. 1986). Absolute toe pressures of less than 30 mmHg indicate an ischemic endangerment of the foot or the lower leg. Bowers et al. (1993) set a threshold value of 40 mmHg, and in the follow-up they found that 19 of 56 patients (34%) with stable claudication had a clear deterioration in their condition.

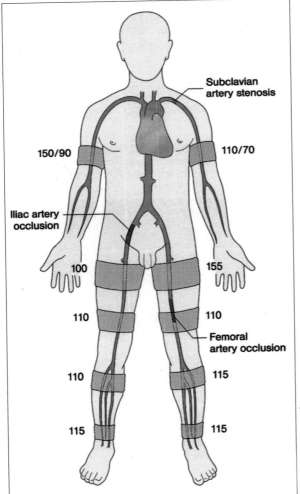

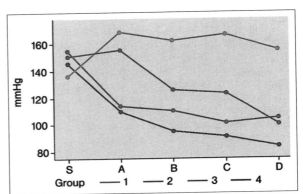

Fig. 5.**12** Segmental pressure measurement at various locations in the lower extremities. Cuff placement at the proximal thigh (A), the distal thigh (B), the proximal lower leg (C), and the distal lower leg (D) in relation to the systemic pressure (S)
Group 1 Normal results
Group 2 Isolated pelvic occlusions
Group 3 Isolated thigh occlusions
Group 4 Multilevel occlusions
When applying segmental pressure measurement, it can be seen that isolated pelvic level occlusions (group 2) are clearly distinct from isolated thigh occlusions (group 3). However, multilevel occlusions (group 4) can only be differentiated from isolated iliac artery occlusions by an additional peripheral pressure decrease

◁ Fig. 5.**11** Segmental pressure measurement showing the pressure differences between the individual cuff placement areas in relation to various occlusion locations. The model shows:
– A left subclavian artery stenosis
– A right iliac artery occlusion
– A left-sided, distal occlusion of the femoral artery
Using the ankle pressure alone as the measurement method would not make it possible to distinguish the level of an occlusion

Upper Extremity Arteries

Measurement of Pressure at Rest

In patients who are suspected of having arterial occlusive disease, bilateral comparison of the blood pressure values should always be carried out. Clinically unremarkable subclavian artery stenoses or occlusions can occasionally be detected; auscultating above the supraclavicular fossa can provide an additional indication of stenosis. Flow sounds detected above the supraclavicular fossa must always be clarified. Initially, it needs to be established whether or not the sound is actually coming from the heart. In addition, measuring the blood pressure can provide a general orientation regarding the hemodynamic significance of the finding. Blood pressure differences are not classified as pathological until they exceed 20 mmHg. It is by no means rare to encounter blood pressure differences in healthy individuals; these are due to nonsimultaneous blood pressure measurements. In particular, there is a time difference between the measurements taken on each side.

Measurement of Pressure Post Exercise

Here, too, an exercise test can be used to help distinguish between slight side-to-side blood pressure differences. The test is usually conducted using push-ups or dumbbells. However, this type of study is carried out much less frequently here than in the lower extremities.

Measurement of Digit Pressure

Studies done by Hirai (1978), as well as our own studies, have shown that fingers with segmental arterial occlusions in only one of the two arteries have pressures in the normal range. When there are occlusions in both arteries, or their afferent arteries, the pressures are pathological. On average, pressures clearly decrease by approximately 60 mmHg.

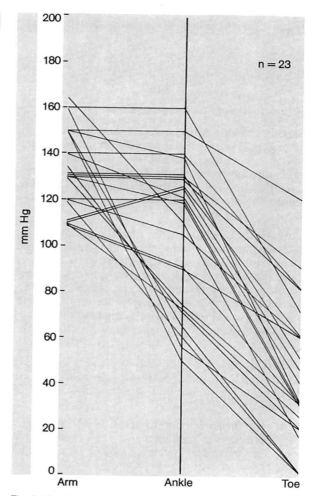

Fig. 5.**13**   The pressure difference between the ankle and the great toe in patients with isolated or combined forefoot or great toe occlusions, or both. Preceding vascular system obstructions are present in some cases

Evaluation

Evaluating the measured *resting pressures* is dealt with in different ways. In the Anglo-American countries, the ratio from the systolic leg and arm pressures is usually calculated, while in the German-speaking countries, measuring the absolute values is preferred. As described in Table 5.**4** on page 145, the individual measurements can provide different information about the hemodynamic situation:

- *Absolute value:* circulatory reserve
- *Pressure difference:* hemodynamic compensation
- *Ratio* (ankle/arm index): degree of severity

Each of these evaluation modalities offers important information, and enables the examiner to classify the extent of arterial circulatory disturbance.

The values obtained can be entered on data sheets provided for the purpose, or can be plotted graphically.

Sources of Error and Diagnostic Effectiveness

Sources of Error

Methodological Sources of Error

Resting period. An adequate resting period for the patient, lasting 10–20 minutes before measurements are taken, has to be observed (the patient can also rest in a sitting position). When there are poorly compensated flow obstructions, the resting period should be longer, in order to avoid measuring *erroneously low* values.

Cuff width. Sometimes, one may forget to adapt the cuff width and length to the placement area. If the same blood pressure cuff used at the arm and the distal lower leg is applied to the thigh, *erroneously high* values may be measured. In the thigh, wider and longer cuffs should be used than those used for the arm and distal lower leg (p. 135). Conical cuffs improve the applied pressure. At the digital arteries, narrow cuffs have to be used.

Deflation. Releasing the cuff pressure too quickly (> 5.0 mmHg/s) causes *erroneously low* values. Care must be taken to maintain a deflation velocity of around 2.0 mmHg/s.

Arm–leg measurement interval. The time difference between the measurements should not be too long. Intra-individual systemic blood pressure fluctuations can occur.

Distance between the measurement area and the probe position. If there are *extensive* occlusions located at several levels between the cuff position and the probe position—and only if this is the case—the blood flow through the collaterals may be slightly delayed. *Erroneously low* values result (Franzek et al. 1981). In such cases, therefore, it is preferable to measure an artery lying near the cuff.

Subclavian obstruction. If subclavian stenoses or occlusions are missed, the systemic blood pressure values measured are erroneously low; when peripheral pressures are also low, this may give a false impression of a patent vascular system.

Flow velocity in the arteries measured. If the blood flow velocity in the arteries that are being measured is too low, it is not possible to receive a Doppler signal. This phenomenon occurs at pressures below 30–40 mmHg (p. 133).

Other Sources of Error

Media sclerosis. In individuals suffering from diabetes, media sclerosis, which occurs frequently and can lead to *erroneously high* values, should be considered. It should be suspected when the ankle pressure exceeds the arm pressure by more than 50 mmHg, or when the Doppler signal cannot be suppressed using a 300 mmHg cuff pressure. In such cases, one must use procedures that register the pulse instead (i.e., oscillography).

Edema. In a solid edema, especially a lipedema, adequate arterial compression may fail. *Erroneously high* values result.

Hypertension. When the systemic pressure is pathologically elevated, the absolute poststenotic values present a picture that is too favorable *(erroneously high* pressure). Since there is no linear relationship between a change in the systemic pressure and the peripheral pressure (Matsubara et al. 1980), the measurement should always be repeated after the systemic pressure has normalized.

Measuring the digital artery pressure. Measurement of the digital artery pressure is not capable of detecting segmental occlusions affecting individual finger arteries (Hirai 1978). The same is true of arterial toe pressure measurement. These pressures are clearly decreased when compared to the ankle pressure only if occlusions are also present in the afferent arteries of the forefoot.

Measurement of pressure post exercise. After physical exertion to evaluate both extremities, two examiners have to carry out the examination simultaneously, or there has to be an adequate rest period between the measurements of the right and left sides. The leg that has the lower resting pressure should be measured first, because the recovery time post exercise is otherwise too long in pathological cases.

Diagnostic Effectiveness

Peripheral systolic pressure measurement, which is relatively simple methodologically, provides a great deal of useful information.

Decreased ankle pressure confirms the *diagnosis* of occlusive arterial disease. This is extremely useful in differentiating a variety of leg complaints.

The *absolute value* provides information about the available circulatory reserve, while the *pressure gradient* reflects the degree of hemodynamic compensation.

Finally, *pressure measurement at different levels of the body* allows the level of a flow obstruction to be identified (Table 5.**4**).

The value of systolic leg pressure measurement lies in the fact that it is simple to use, time-saving, and cost-effective, with easily reproducible results (Carter 1969, Grüntzig and Schlumpf 1974).

When a resting pressure is detected that is higher than the undisturbed arm pressure, this finding excludes any hemodynamically significant vascular system stenoses below and proximal to the blood pressure cuff. With borderline results (arm and ankle pressure with the same magnitude), an exercise test should be completed to allow better discrimination (Yao et al. 1969, Carter 1972, Mahler 1990). In diabetics with media sclerosis, it is sometimes not possible to measure the peripheral systolic blood pressure, or only an erroneously high result is produced. This value can therefore not be used to exclude other diagnoses.

In contrast, it is not possible to differentiate between an *occlusion* and a *stenosis*. For this purpose, simple auscultation with a stethoscope at an appropriate location (pelvis, groin, adductor canal, popliteal region) is still used when duplex equipment is not available. However, recognizing and evaluating the sounds of stenosis requires some practice. In addition, *locating* an occlusion or a stenosis within a particular vascular level is not possible, nor is it possible to establish the *length of the occlusion*.

According to Rutherford et al. (1979), the *sensitivity* in detecting isolated lesions is 93% (n = 103). According to this study, occlusions at several levels are more difficult to locate, even using segmental measurements; the sensitivity amounts to only 86%. Ouriel and Zarins (1982) were able to demonstrate a 99% overall sensitivity for detecting occlusive arterial disease in the legs; further results are shown in Table 5.5. Wall changes such as plaques and ulcerations—the early forms of arteriosclerosis—are not detectable, since they do not cause any decreases in pressure, even during exertion.

The consensus of the reports in the literature is that the *specificity* of peripheral systolic pressure measurements in most studies is very high (Table 5.5). In a study involving 108 angiographically normal lower extremities, the peripheral systolic pressure was higher than the arm pressure in 83% of cases, the same in 15%, and a lower ankle pressure was detectable only in 2% (Köhler and Krüpe 1985).

Doppler pressure measurements are significant, because these noninvasively measured systolic pressures have the same value as pressure measurements obtained using more invasive procedures (Bollinger et al. 1976, Köhler and Lösse 1979). *Measured digital artery pressures* are a different matter. According to

Table 5.4 The importance of systolic pressure measurement in arterial disease

- Detecting or excluding obstructions
- Circulatory reserve (remaining peripheral pressure)
- Hemodynamic compensation (pressure difference)
- Degree of severity (pressure index)
- Location (pressure measurement at different segmental levels)

our own results and information from the literature, these show a somewhat larger scatter range, which can be explained by methodological problems (Gundersen 1972, Nielsen et al. 1973, Knoblich et al. 1986).

Differentiating between normal and pathological groups is markedly improved using *exercise tests* (Thulesius 1971, Ludwig et al. 1985). The same applies to the use of hyperemia following suprasystolic arterial compression (Fronek et al. 1973, Mahler et al. 1975), although this has a somewhat smaller effect on the pressure decrease than physical exertion (Fig. 5.14).

Table 5.5 The validity of segmental pressure measurements at rest using the ultrasound Doppler method, in stenoses of more than 50% and arterial occlusions at the pelvic and thigh levels

Vascular region	Rutherford (1979) (n=205)		Flanigan (1981) (n=95)		Lynch (1984) (n=345)		Neuerburg (1989) (n=144)		Moneta (1993) (n=151)	
	Sensi-tivity (%)	Speci-ficity (%)	Sensi-tivity (%)	Speci-ficity (%)	Sensi-tivity (%)	Speci-ficity (%)	Sensi-tivity (%)	Speci-ficity (%)	Sensi-tivity (%)	Speci-ficity (%)
Isolated aortoiliac	75	97	79	56	97	50	60	96	59	86
Isolated femoropopliteal	60	97	–	–	89	68	88	96	85	53
Aortoiliac + femoropopliteal	78	97	–	–	55	–	80	96	–	–

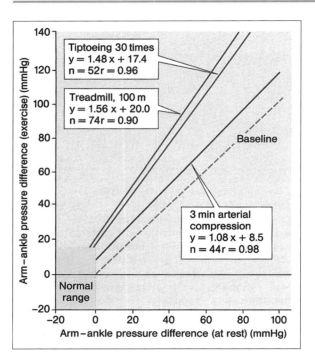

Fig. 5.**14** Various forms of stress tests that can confirm the diagnosis of arterial obstructions based on borderline pressure differences at rest (adapted from Mahler). Following suprasystolic cuff occlusion and arterial compression (three minutes), a 20 mmHg pressure difference increases to 30 mmHg. After the patient has tiptoed 30 times, it increases to 46 mmHg, and after 100 m on the treadmill, it increases to 51 mmHg

Doppler and Duplex Sonography

Examination

Special Equipment and Documentation

■ Continuous Wave and Pulsed Wave Ultrasound Doppler Equipment

Examinations of the flow velocity in the arteries supplying the extremities are carried out using bidirectional Doppler equipment, with a transmission frequency of 4–5 MHz. Ultrasound application in the superficial ankle arteries (posterior tibial artery, dorsal artery of the foot) may sometimes be better with an 8–10 MHz probe. The flow velocity waveforms are obtained either by zero-crossing, or as frequency spectra. Due to the greater complexity of the equipment, exclusive use of pulsed wave Doppler sonography has not become widespread for routine examinations. It is only employed in special research situations.

■ Conventional and Color Flow Duplex Systems

For the vasculature of the extremities, linear scanners with a 5.0–7.5 MHz B-mode transmission frequency are preferred (Fig. 1.**15**, p. 18). These allow clear demonstration of arteries located relatively close to the surface.

In contrast, in the pelvic region, sector scanners (transmission frequency 3.5–5 MHz) may be easier to use, especially near the groin, as they have a smaller transducer, which also provides better depth resolution (Fig. 1.**14**, p. 17). It is essential to use pulsed wave, continuous wave, or color flow Doppler mode ultrasound (2.5–5 MHz), since evaluating the hemodynamics in the arteries of the extremities is of primary importance, while pure B-mode imaging provides information about vascular wall deposit morphology and plaque location, which is of secondary importance only. However, the imaging of true or false aneurysms, and determining their form and location, are best accomplished with B-mode imaging.

When documenting the findings, important regions should always be recorded, such as the femoral bifurcation, the origin of the superficial femoral artery, the deep femoral artery, and also the popliteal artery. Pathological findings, such as stenoses or occlusions, should be depicted separately. Color flow Doppler imaging is no substitute for precise measurement of the flow velocity using the pulsed wave Doppler mode (pp. 23, 24) or formal analysis of the Doppler velocity waveform.

Examination Conditions

Patient and Examiner

The patient is examined at the pelvis or leg while lying down. As in blood pressure measurements, he or she should have rested for some 10–20 minutes beforehand. This is also important when measuring the flow velocity in healthy individuals, since incorrect pathological results can thereby be avoided.

Ultrasound is applied at the groin with the patient in a supine position, and at the popliteal region with the patient prone, or in a supine position with the knee bent.

With duplex systems, the examination should also be carried out while the patient is lying down. The pelvis, thigh, and lower leg should also be examined in supine position, while the popliteal artery is easier to evaluate in prone position (Figs. 5.**15**, 5.**16**).

It is easier to assess the subclavian artery with the patient in a semi-sitting position; ultrasound is applied from a posterior, i.e., cranial direction. The examiner should be seated during all procedures to facilitate stable positioning of the transducer or probe. The arm arteries are most effectively located in an abducted arm resting on a supporting surface (Fig. 5.**20 b**).

Conducting the Examination

■ Continuous Wave and Pulsed Wave Doppler Sonography

Ultrasound is applied to the arteries at the same locations used to palpate the pulse. In the lower body, these are: the groin, the popliteal region, the lateral margin of the lateral malleolus, and the dorsum of the foot. In the upper body, the subclavian artery is mainly searched for via a supraclavicular access point, and the radial artery is detected at the pulse palpation point.

Initially, the pulse should be felt manually at the locations to be investigated, and after contact gel is applied, the arterial signal should be located with the Doppler probe. While this is being done, the probe is shifted with minimal movements, medially and laterally, proximally and distally. Acoustically, an attempt is made to obtain the signal with the loudest volume and highest frequency. It should be displayed on the monitor as a maximum amplitude curve, but without any venous superimposition.

The Doppler frequencies are documented as an analog curve, or as a frequency spectrum (pp. 10, 11). With spectral analysis, automatic calculation of indices is usually possible. Quantitative measurement of the maximum systolic velocity with continuous wave Doppler sonography is limited, since the incident angle of the ultrasound beam cannot be determined. The maximum frequency is therefore analyzed, and a qualitative formal analysis of the Doppler waveform is usually acceptable. In contrast to the carotid vascular system, stenoses are seldom directly located in the peripheral arteries. These arteries, which lie too deep anatomically, cannot be continuously evaluated with the probe.

■ Conventional and Color Flow Duplex Sonography

B-mode. Imaging of the peripheral arterial vascular system in B-mode concentrates on the pulsating arterial wall, which is more hyperechoic than the venous wall. When the anatomical course is known, the arteries are usually displayed in longitudinal section. The cross-sectional view can sometimes be useful in locating the blood vessels. In a specialized investigation, only specific segments are usually examined. Since exact description of plaque morphology and location is not diagnostically decisive, the number of sectional planes does not have the same significance as in the carotid arteries. Instead, *locating* and *quantifying* vascular system obstructions in order to plan appropriate treatment is more important here.

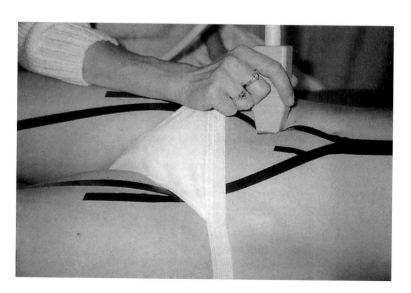

Fig. 5.**15** Applying ultrasound to the pelvic vascular system with a sector scanner from a duplex system. The position of the transducer is appropriate for examining the external iliac artery at its origin. Depending on the artery being examined, the position of the transducer is displaced along the indicated lines, which are shown here as an example

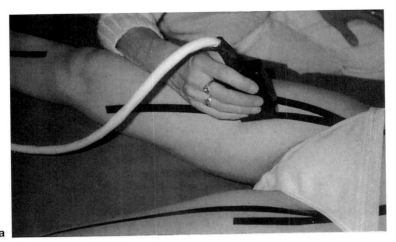

a

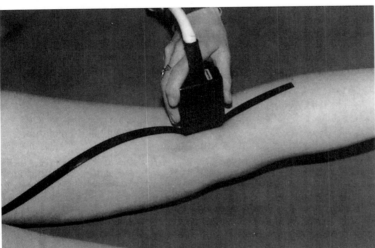

b

Fig. 5.**16 a**   Applying ultrasound to the superficial femoral artery with a linear scanner at the transition zone from the proximal to the middle segment.
b   The linear scanner of a duplex system in a longitudinal plane above the popliteal region, to evaluate the popliteal artery. A sectional plane is marked from the dorsal side to depict the middle section of the superficial femoral artery as far as the trifurcation. Particularly in the femoropopliteal transition zone, care must be taken to apply ultrasound continuously, also when the patient changes position

Doppler mode. The Doppler sample volume serves to measure the flow velocity and detect all types of disturbed flow, from laminar to turbulent flow. Systematic examination of the total lengths of vessel (pelvis, thigh, lower leg) using the pulsed wave Doppler mode requires considerable time. The specific area of interest can be identified using prior simple examination procedures (segmental pressure measurements, or directional continuous wave Doppler sonography, or both). Within the area of interest, the hemodynamic effect of stenoses, or the occlusion length, can then be quantified through systematic examinations with the Doppler mode.

Color flow Doppler mode. The color-coded display of blood flow has greatly facilitated the examination of the arteries of the extremities. Arterial flow is usually represented in red, and venous flow in blue. It is necessary to note any regions with lighter coloring (increased flow velocity) and to identify gaps in the vasculature. These are additionally identified through outflow into thin collateral blood vessels. The color flow Doppler mode is important for imaging the lower leg arteries, which without this method would be extremely difficult to display and classify topographi-

cally. Even when color flow duplex sonography is used, however, exact measurement of the flow velocity in the regions of interest has to be carried out using the pulsed wave Doppler mode.

Examination Sequence

Pelvic and Leg Arteries

■ Continuous Wave and Pulsed Wave Doppler Sonography

The most important location to examine is the groin region, since cuff methods, i.e., pressure measuring procedures, cannot be used in the pelvic vascular system. The pulse of the common femoral artery is relatively easy to find. This is where the Doppler signal is registered.

Ultrasound is preferably applied to the popliteal artery, with the patient in a prone or lateral position. It is often difficult to achieve a favorable incident angle.

The Doppler flow curves from the posterior tibial artery and the dorsal artery of the foot, relatively small-caliber blood vessels located near the surface, are somewhat more difficult to detect, and are therefore recorded less often.

■ Conventional and Color Flow Duplex Sonography

To image the complete *pelvic and leg vascular systems* unilaterally or bilaterally, the examination begins at the distal aorta, with a search being made for aneurysmal dilations or plaques in cross-section and longitudinal section. The iliac vasculature is followed distally to the bifurcation of the femoral artery. Care should be taken to note wall deposits in B-mode imaging, as well as stenoses or occlusions using color flow or pulsed wave Doppler (Fig. 5.**15**).

The ultrasound examination can also start at the common femoral artery, proceeding proximally to the aortic bifurcation. If the pulsed wave Doppler sample volume is gradually included, this procedure has the advantage that it can identify any suspicion of preceding flow obstructions in the vascular system of the common femoral artery. However, due to the deep location of the pelvic vasculature, especially in patients with obesity or meteorism, continuous imaging of the vascular course may be difficult or impossible.

Subsequently, the examination of the *vascular system of the thigh* also starts at the groin. The transducer position, in a ventrodorsal sagittal plane, allows imaging of the common femoral artery and the bifurcation area initially, and in longitudinal section, the superficial femoral artery and the deep femoral artery can be depicted (Fig. 5.**24**). Along its distal course, the femoral artery is increasingly examined in a sectional plane, moving from ventromedial to dorsolateral (Fig. 5.**16a**).

Near the femoral bifurcation, the origin of the deep femoral artery must be displayed without fail; stenoses of the origin often cannot be seen in angiography. Indirect examination methods do not allow successful differentiation between the superficial femoral artery and the deep femoral artery (Fig. 5.**31**). Following the deep femoral artery further distally is often difficult, since the artery changes course in a dorsal direction. A critical location for the superficial femoral artery is its transition into the popliteal artery in the adductor canal. Here, the patient's position has to be changed from supine to prone; the artery at this point is more deeply entrenched in the muscles.

The distal femoral artery, or the proximal segment of the popliteal artery, is examined with ultrasound from a dorsal direction (Fig. 5.**16b**). The *popliteal region* at the posterior aspect of the knee can be readily evaluated with ultrasound. Further distally, ramification into the three lower leg arteries (trifurcation) occurs. Usually, this can still easily be imaged. Imaging of the superficial femoral artery and the popliteal artery should be done in longitudinal section, so that flow velocity measurements can be made with angle correction.

The distal *lower leg arteries* are easier to follow starting from a distal direction (ankle level) and pro-

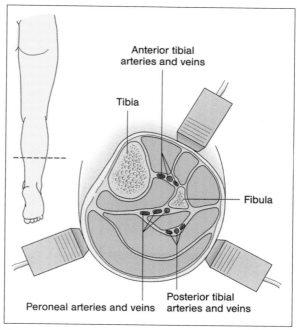

Fig. 5.**17** An anatomical cross-section through the lower leg from the proximal to the middle section, showing the three arteries: the anterior tibial artery, the peroneal artery, and the posterior tibial artery, with the corresponding veins accompanying them. The transducer positions at which ultrasound should be applied to the vascular bundles are indicated

ceeding proximally. However, precise topographical knowledge (Fig. 5.**17**) is a necessary precondition for successfully distinguishing between the three arteries (posterior tibial artery, anterior tibial artery, and peroneal artery). Color flow Doppler mode makes it much easier to follow the arteries proximally (Fig. 5.**18a, b**). Here, as in other arterial segments, vascular system stenoses and occlusions are easier to detect by the usual signs (color lightening, color interruption) (Fig. 11.**35**, p. 406). Quantitative measurement of the flow velocity is possible using the adjusted Doppler sample volume.

In the groin and popliteal regions, attention should be given to the aneurysms that accompany dilative arteriopathy, which are not infrequent.

The groin region requires special attention, since iatrogenic vascular damage (hematomas, false aneurysms, and arteriovenous fistulas) occurs here most often after catheter interventions. Due to the wider-lumen access channel that is used, these complications are being observed with increasing frequency after coronary angioplasty (Steinkamp et al. 1992).

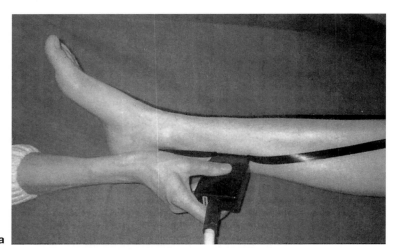

Fig. 5.**18**   The lower leg arteries evaluated along their course using a linear scanner, **a** Posterior tibial artery at the distal lower leg. **b** Anterior tibial artery in the middle section of the lower leg

a

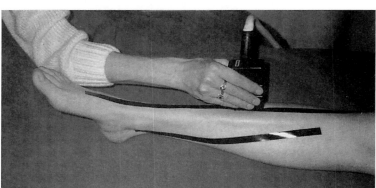

b

Fig. 5.**19**   The way in which to hold the probe and the measurement position for registering velocity waveforms when applying continuous wave Doppler sonography at the proximal digital arteries

Upper Extremity Arteries

■ Continuous Wave Doppler Sonography

The subclavian artery can be defined as an artery supplying the extremities and also, via the origin of the vertebral artery, as an artery supplying the brain. The data registration area is described on page 32.

The other arteries of the arm (axillary artery, brachial artery, radial artery, ulnar artery) are rarely directly examined with ultrasound; the Doppler flow curves are obtained at the usual palpation points.

The digit arteries at the fingers can also be located, and appropriate flow velocity curves can be registered (Fig. **5.19**). However, this cannot detect or exclude segmental occlusions with certainty. Measuring the pressure should be the preferred method (p. 136).

■ Conventional and Color Flow Duplex Sonography

Using a side-to-side comparison of the measured pressure and continuous wave Doppler sonography at the upper extremities, precisely quantified information concerning the patency of the vascular system or the presence of flow obstructions can be obtained. The most important region for applying duplex sonography is the origin of the subclavian artery, which can be imaged well with a little practice, and with appropriate patient positioning (Fig. 5.**20 a**). Since the proximal subclavian artery and the innominate artery can be classified as arteries supplying the brain—via the vertebral artery located near the origin—the corresponding section above should be referred to (pp. 41–43).

Another important region is the passage of the subclavian artery between the clavicle and the first rib. Occasionally, this is mechanically constricted, leading to functional or permanent stenoses (Fig. 5.**38**, p. 169). Due to the osseous structures, it is not easy to image this region, even with duplex sonography. The further courses of the subclavian artery, the axillary artery (Fig. 5.**20 c**), and the brachial artery, and the two forearm arteries, the radial artery and ulnar artery (Fig. 5.**20 b**), are continuously imaged only when there is a clinical suspicion of vascular stenoses and occlusions. The evaluation of hemodialysis fistulas is particularly important, however. These can be searched for with high-resolution transducers in the upper arm and forearm. Quantitative measurement of the flow rate using the adjusted Doppler mode is important, and this can be done with duplex sonography, since both the incident angle and the vascular diameter can be determined (p. 175).

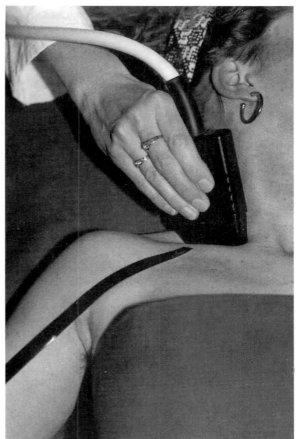

a

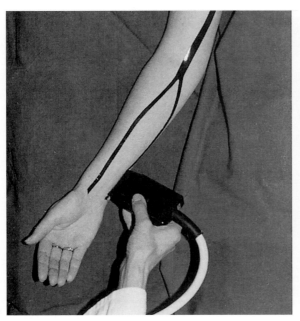

b

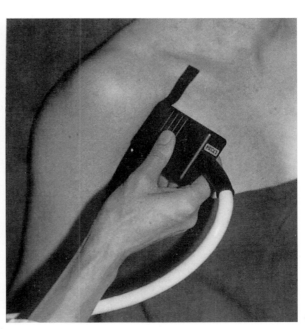

c

Fig. 5.**20** Positions for applying ultrasound with a linear scanner of a duplex system at the upper body. Applying ultrasound **a** to the supraclavicular subclavian artery, **b** to the ulnar artery, and **c** to the axillary artery

Normal Findings

Principle

■ Continuous Wave Doppler Sonography

At rest. The normal flow pulses in the extremities are triphasic, with a steep systolic ascent (acceleration), a narrow peak, a quick descent (deceleration), and a backflow component in the early diastole that is approximately one-fifth of the height of the systolic forward flow. In the late diastole, a short forward flow peak is also registered (Fig. 5.21 a).

The formation of this characteristic wave phenomenon is basically explained by a primary pulse wave being reflected in the periphery, proceeding retrogradely through the arterial system, and being reflected again at the aortic valve, which has in the meantime closed (backflow component). The wave then again proceeds peripherally, and there causes the second forward peak (dicrotism, "ping-pong phenomenon") (Busse et al. 1975).

The form of the flow pulse in the extremities is determined by the high peripheral resistance present in the terminal vascular regions (skin, muscles) under resting conditions. In normal circumstances, the flow velocity curves must show the same, typically triphasic, formal elements, at the usual registration points—groin, popliteal region, and ankle, as previously described. Occasionally, the amplitude height decreases distally, since the flow velocity decreases when passing parallel connecting blood vessels with decreasing diameter (Rieger 1993).

Post exercise. In a healthy individual, the systolic pressure briefly decreases post exercise, and the flow velocity increases significantly, especially in its diastolic component (decreased peripheral resistance), due to the higher blood supply required by the muscles. However, it returns to approximately normal values within one minute (Figs. 5.7, 5.21 b) (Sumner and Strandness 1969, Mahler et al. 1976).

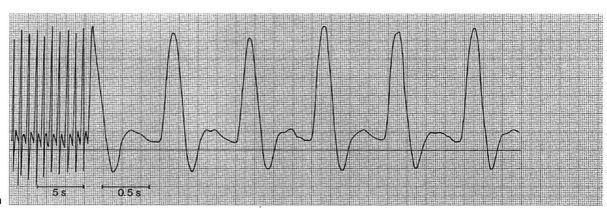

a

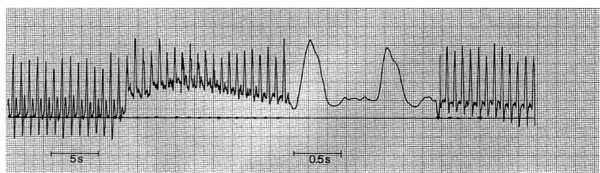

b

Suprasystolic compression   ↑ Blood pressure cuff release

Fig. 5.**21**   A flow velocity waveform in a healthy individual **a** at rest, and **b** after suprasystolic arterial compression, registered from the common femoral artery above the groin. At rest, there is a characteristic tricyclic curve with a steep systolic increase, early diastolic back-flow, and diastolic forward flow. During suprasystolic arterial compression, the backflow component deepens, and a decreased diastolic flow while resting is seen, due to increased peripheral resistance. After the cuff is released, the hyperemic phase results in a clearly increased diastolic flow velocity, due to decreased peripheral resistance

■ Conventional and Color Flow Duplex Sonography

B-mode. Under physiological conditions, the B-mode serves to determine the vascular diameter and course. It is possible to exclude aneurysmal dilatations and atherosclerotic wall deposits, or thrombotic occlusions.

Doppler mode. The criteria for evaluating flow velocity curves, which in duplex sonography are usually obtained with the pulsed wave Doppler sample volume and are displayed as a frequency analysis, correspond to the formal analysis using continuous wave Doppler sonography, mentioned above. However, quantitative measurement of the maximum velocity, mean velocity, and other parameters is also possible, since the angle between the Doppler beam and vascular axis, as well as the diameter of the vessel can be calculated (p. 159).

Color flow Doppler mode. The color flow Doppler mode makes it easier to locate arteries, follow their course, and in normal cases to detect a patent arterial vascular system. The vascular course and diameter deviations, such as positional anomalies and hypoplasias, are more easily recognized with this method than with pure B-mode imaging.

Findings and Anatomy

Classifying peripheral arterial disease by anatomical region is based on the Ratschow location types (Heberer et al. 1974).

- Pelvic level (pelvic type)
- Thigh level (thigh type)
- Calf and foot arteries (peripheral type)

The occlusion level determines not only the clinical symptoms, but also requires differentiated therapy. We have retained this tried and tested classification system, even for normal findings. Of course, there is often so-called *multilevel disease*.

Pelvic Level

Anatomy

The *distal abdominal aorta* belongs to the pelvic level. At the fourth lumbar vertebra, and at an angle of between 60° and 80°, the aorta branches into the common iliac arteries. After nearly 5–6 cm, these divide into the *external iliac artery* and the *internal iliac artery*. With its large ventral and dorsal main branches, the latter supplies the viscera in the small pelvis, and contributes to the collateral circulation in pathological cases. The external iliac artery follows a course through the small pelvis, without giving off any branches. Shortly before its entry into the lacuna of vessels, it gives off two smaller blood vessels: the *inferior epigastric artery* and the *deep circumflex iliac artery* (Fig. 5.**23**).

■ Continuous Wave Doppler Sonography

Due to inadequate penetration depth and insufficient scope for arterial identification in the small pelvis, it is not possible to apply ultrasound to the pelvic vasculature using continuous wave Doppler sonography.

A normal triphasic Doppler waveform obtained from the common femoral artery at the groin *indirectly* confirms the patency of the pelvic vascular system (Fig. 5.**21 a**).

■ Conventional and Color Flow Duplex Sonography

B-mode. If the preceding examinations using Doppler pressure measurement or continuous wave Doppler sonography, or both, have excluded significant vascular system obstructions, only the pelvic vascular system is normally examined with ultrasound, in order not to miss any aneurysms or to acquire morphological information. According to measurements made by several authors, the sonographically measured diameter of the external iliac artery is 0.80 cm (Table 5.**6**). For the common iliac artery, radiological anatomical measurements on average give a figure of 0.83 cm (left side) and 0.89 cm (right side) (Luzsa 1972, p. 294).

Doppler mode. The Doppler mode is only used for specialized investigations. According to Jäger et al. (1985), the flow velocity in the external iliac artery is on average 119.3 cm/s (Table 5.**6**).

Color flow Doppler mode. It is easier to locate the arteries using color flow Doppler mode, especially at the origin of the internal iliac artery (Fig. 5.**22**). After femoral artery punctures, it is important to exclude a false aneurysm or a hematoma.

Thigh Level

Anatomy

Below the inguinal ligament, the external iliac artery is called the *common femoral artery* (not according to anatomy textbooks, but among angiologists, radiologists, and vascular surgeons). This arterial segment is classified as belonging to the thigh level, and in its short, stretched course near the surface, it is particularly well accessible to all types of examination.

Approximately 4.0 cm below the inguinal ligament, the artery branches into the *superficial femoral artery* and the *deep femoral artery*. The deep femoral artery supplies the thigh muscles, and shortly after the origin gives off two larger branches, the *medial and lateral circumflex femoral arteries* (Fig. 5.**23**). At adductor canal level, the terminal branches of the deep femoral artery collateralize with smaller thigh branches from the superficial femoral artery, and form the most important collateral circulation for femoral artery occlusions proximal to the adductor canal. The superficial femoral artery continues as the *popliteal artery*, which begins with its proximal segment at the

Table 5.**6** Results from several studies measuring the vascular diameter and the maximum systolic flow velocity ($V_{syst.}$), including the standard deviation (SD), in normal individuals using duplex sonography

Artery	Jäger et al. (1985)	Jäger et al. (1985)	Luska et al. (1990)	Hatsukami et al. (1992)	Sacks et al. (1992)	Strauss et al. (1989) Karasch et al. (1991 a)
	Diameter ± SD (cm/s)	V_{syst} ± SD (cm/s)	V_{syst} ± SD (cm/s)	V_{syst} ± SD (cm/s)	V_{syst} ± SD (cm/s)	V_{syst} ± SD (cm/s)
External iliac	0.79 ± 0.13	119.3 ± 21.7	102	98 ± 17.5*	112.0 ± 49.0	–
Common femoral	0.82 ± 0.14	114.1 ± 24.9	104	80 ± 16*	90.0 ± 41.0	–
Proximal superficial femoral	0.60 ± 0.12	90.8 ± 13.6	99	73 ± 10	89.0 ± 23.0	–
Distal superficial femoral	0.54 ± 0.11	93.6 ± 14.1	–	56 ± 12	74.0 ± 21.0	–
Deep femoral	–	–	–	64 ± 15	–	60.0 ± 14.0 (Strauss et al. 1989)
Popliteal	0.52 ± 0.11	68.8 ± 13.5	62	53 ± 17*	59.0 ± 12.0	89.7 (Karasch et al. 1991 a)

$V_{syst.}$ = maximum systolic velocity
SD = standard deviation
* Mean value of proximal and distal measurement

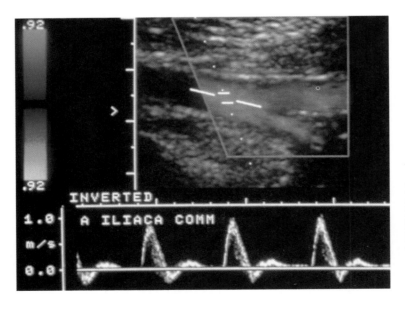

Fig. 5.**22** Color-coded image of the distal common iliac artery, showing its branching into the larger-caliber external iliac artery and the smaller-caliber internal iliac artery, and also displaying the corresponding tricyclic flow velocity waveform. The consistently intense color indicates a uniform mean flow velocity within the vascular bifurcation

lower margin of the adductor canal, and reaches to the lower margin of the popliteal muscle. The popliteal artery belongs to the femoropopliteal segment, and in the knee region gives off several small branches—*genicular arteries* and *sural* arteries—which together form the *articular rete of the knee.*

■ Continuous Wave Doppler Sonography

Hemodynamically significant flow obstructions in the thigh region are again *indirectly* excluded by applying ultrasound to the popliteal artery. A triphasic Doppler curve obtained above the popliteal artery indicates that the subsequent vascular system is patent. In slender extremities, it is also possible to apply ultrasound continuously to the superficial femoral artery using the continuous wave probe. However, due to methodological problems, this form of examination has not gained widespread use as a standard.

■ Conventional and Color Flow Duplex Sonography

B-mode. Using B-mode imaging alone to depict the femoral artery and the deep femoral artery is not recommended, since the exclusion of vascular system stenoses is not morphologically certain enough with this method, and it may be difficult to differentiate the vasculature. According to Jäger et al. (1985), in normal individuals the diameter of the common femoral artery is 0.82 cm, and that of the proximal superficial femoral artery is 0.60 cm, with its distal segment giving an average figure of 0.54 cm (Table 5.**6**).

However, the B-mode does allow the exclusion of wall changes, ectasias, aneurysms, and also hematomas after vascular punctures.

Doppler mode. Since the bifurcation of the deep femoral artery is a preferred location for arteriosclerotic stenoses, and can often not be assessed with sufficient certainty using angiography, the Doppler mode should be used for the identification and evaluation of this vascular segment. In addition, course variations in the femoral bifurcation can be detected, especially those affecting the origin of the deep femoral artery, the many branches of which may originate directly from the superficial femoral artery (Fig. 5.**23**). Evaluating the entire length of the vascular segment (20–30 cm) using conventional Doppler mode is very time-consuming, and can only be recommended when a specialized investigation requires it. By contrast, the popliteal artery, which lies near the surface, is easy to examine from a dorsal direction.

The flow velocity in normal individuals has been examined by several authors (Table 5.**6**). In the common femoral artery, Jäger et al. (1985) measured the peak systolic velocity as 114.1 cm/s; Sacks et al. (1992) arrived at a figure of 90.0 cm/s. For the proximal superficial femoral artery, the results obtained by Jäger and Sacks varied only minimally (90.8 cm/s versus 89.0 cm/s).

Strauss et al. (1989) measured the flow velocity of the deep femoral artery at its origin in healthy individuals as 60 cm/s ± 14 cm/s on average.

In a separate study using comparative measurements by two examiners using two separate duplex apparatuses, the mean flow velocity in the popliteal artery was determined to be 89.7 cm/s (Karasch et al. 1991 a). Jäger and Sacks obtained values of only 68.8 cm/s and 59.0 cm/s, respectively.

Color flow Doppler mode. Using color coding, the thigh level is especially easy to evaluate, due to the extended course of the femoral artery. The entire distance can be rapidly followed to the adductor canal. In this region, however, a change in position is usually necessary, and continuously displaying the vascular course is somewhat more difficult. With practice, it can be reliably evaluated. In contrast, the popliteal artery, which is located near the surface, is easily displayed from a dorsal direction (Fig. 5.**24**).

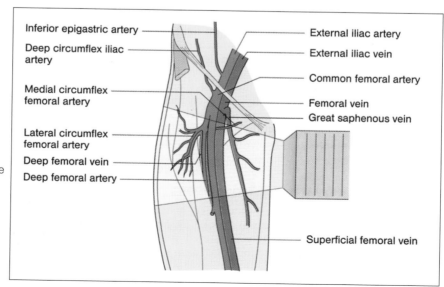

Fig. 5.**23** The anatomy of the common femoral artery and vein, and the division of the common femoral artery into the superficial femoral artery and deep femoral artery with its side branches, the medial femoral circumflex artery and the lateral femoral circumflex artery. Distal to the inguinal ligament, the femoral vein has an arched junction with the great saphenous vein

Inferior epigastric artery
Deep circumflex iliac artery
Medial circumflex femoral artery
Lateral circumflex femoral artery
Deep femoral vein
Deep femoral artery

External iliac artery
External iliac vein
Common femoral artery
Femoral vein
Great saphenous vein

Superficial femoral vein

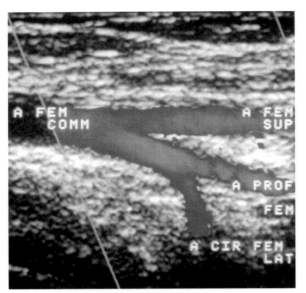

Fig. 5.**24**   A color-coded image showing the femoral bifurcation in longitudinal section, including the origin of the superficial femoral artery and the deep femoral artery with its side branch, the lateral femoral circumflex artery

Calf and Foot Arteries

Anatomy

The first branch originating from the popliteal artery is the *anterior tibial artery,* which proceeds deeply downward between the lower leg extensors and, between the two malleoli, supplies the foot as the dorsal artery of the foot. One of its terminal branches makes a turn and forms the five dorsal metatarsal arteries. After the origin of the anterior tibial artery, the popliteal artery is called the *tibiofibular trunk* (Fig. 11.**1h**).

The *posterior tibial artery* is the largest terminal branch, which proceeds downward in stretched form between the soleus muscle and the deep flexors. Behind the medial malleolus, it supplies the sole of the foot with its terminal branches, the tibial plantar artery and the fibular plantar artery, and then forms the main afferent path to the deep plantar arch. The *peroneal artery* runs parallel to the posterior tibial artery, and in the ankle region joins with the malleolar rete and the fibular rete.

■ Continuous Wave Doppler Sonography

The lower leg arteries are clearly identified in the ankle region with the continuous wave probe. Doppler waveforms can be successfully registered from both the posterior tibial artery and the dorsal artery of the foot. In this region, measuring the ankle pressure, however, is better than directly registering the pulse.

■ Conventional and Color Flow Duplex Sonography

Conventional duplex sonography has rarely been used to examine these arteries, and has only been attempted in order to answer specific questions at highly specialized medical centers. Color flow duplex sonography has made it practical to image and evaluate this vascular region, but when the findings are normal, it does not form part of the routine program used to exclude certain diagnoses. Exact topographical knowledge of the way in which the lower leg arteries are arranged within the muscles (Fig. 5.**18**) is definitely required for proper evaluation and differentiation of these arteries (Fig. 5.**25 a, b**).

Upper Extremity Arteries

Anatomy

As arteries supplying the extremities, the *subclavian arteries* are classified as belonging to this level; in their proximal segment, they also count as arteries supplying the brain. The left subclavian artery originates directly from the aortic arch. The right subclavian artery originates from the *innominate artery,* which after 3–5 cm divides into the *common carotid artery* and the *subclavian artery.* Through its large side branch, the *vertebral artery,* the subclavian artery becomes both an artery supplying the extremities and also one supplying the brain (Fig. 11.**1d**).

Distal to the lower margin of the clavicle, the artery is called the *axillary artery,* which can have varying locations, depending on the position of the arm. After giving off several branches supplying the muscles, the axillary artery, on entering the arm, is termed the *brachial artery* (Fig. 11.**1g**). The *deep brachial artery,* which is joined by the radial nerve, originates medially at the upper arm. Approximately 1.0 cm below the elbow, the brachial artery divides into the *radial artery* and the *ulnar artery.* On the radial side, the ventral side of the forearm, the radial artery becomes the so-called pulse artery, and in the palm it forms the *deep palmar arch.* The main afferent path to the *superficial palmar arch* proceeds through the *ulnar artery.* Both palmar vascular arches are, one below the other, connected with the contralateral artery and give off vasculature to the fingers. Each finger has two blood supplies through the *proper palmar digital arteries* (Hafferl 1969, p. 784).

■ Continuous Wave Doppler Sonography

The flow waveform of the subclavian artery obtained with Doppler sonography shows the same triphasic components as every blood vessel that supplies the extremities. A high systolic forward flow is seen, with a steep increase and a quick decrease. In the early diastole, a backflow component is detected, and in the late diastole, a short forward flow is seen. Using Doppler sonography, the same waveforms can be registered at

Fig. 5.**25 a** A color-coded image showing the anterior tibial artery with the concomitant veins in longitudinal section. **b** Applying ultrasound with a linear scanner at distal lower leg level and depicting the corresponding flow velocity waveform

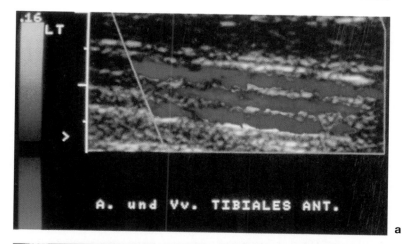

a

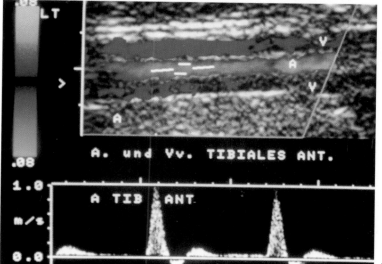

b

the brachial artery palpation points, shortly above the bend of the elbow, at the radial artery, and the ulnar artery. The same indices can be used as with Doppler waveforms from the lower extremities. However, these evaluations do not have the same significance as in the lower extremities, since level-oriented diagnostics do not play a significant role here, and the quantitative value that is usually used is bilateral comparison of arm blood-pressure measurements.

■ Conventional and Color Flow Duplex Sonography

B-mode. The origin of the subclavian artery from the aortic arch is easy to display when the patient is suitably positioned (pp. 150, 151). Continuing to apply ultrasound to the artery is difficult beneath the clavicle, due to ultrasound signal loss. The transition zones into the axillary artery and the brachial artery, and also into the ulnar artery and the radial artery in the forearm, can be readily followed using B-mode sonography to exclude pathological structures (Fig. 5.**26 a, b**).

Doppler mode. Using Doppler mode, the same flow tracings can be obtained as with continuous wave Doppler sonography. Since the vasculature lies near the surface, only transducers with a high transmission frequency (over 5 MHz) can be used at the extremities.

Color flow Doppler mode. Displaying the entire arterial course is significantly quicker with the color flow Doppler mode. This can become important incidentally in intensive care medicine, since the subclavian artery or the brachial artery can serve as a guide for easier identification of the accompanying veins. Before a hemodialysis shunt is placed, the arterial component of the vessel should be examined.

Evaluation

■ Continuous Wave Doppler Sonography

The assessment of a normal flow waveform from the extremities is usually only qualitative, following the clear formal criteria given above (Fig. 5.**29**).

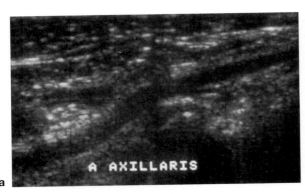

a

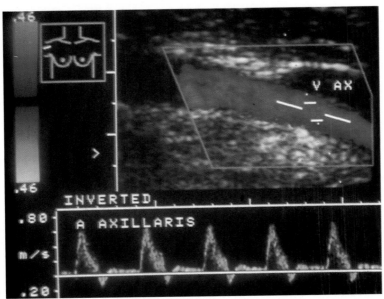

b

Fig. 5.**26**   Applying ultrasound to the axillary artery.
a   A composite presentation of the axillary artery in longitudinal section using B-mode.
b   The registered frequency spectrum from the axillary
artery shows a tricyclic form, typical for an arterial velocity waveform of an extremity. The positive diastolic component is less pronounced than in the lower extremities

However, quantitative measurement of different parameters increases the precision of the information, especially when pathological findings have to be excluded.

Different indices have been developed to measure the flow pulses. Among these, the *pulsatility index* (PI), in particular, has gained widespread acceptance, following Gosling and King (1974). For this calculation, the height between the positive systolic peak and the negative diastolic backflow is measured, and this quantity is divided by the average flow velocity (mean) (Fig. 5.**27**).

$$\text{Pulsatility index (PI)} = \frac{\text{Height of the maximum forward and reverse velocity}}{\text{Mean velocity}}$$

The advantage of the PI lies in the fact that it is relatively independent of the angle between the probe and the vascular axis. This is so because a potential error affects both the numerator and the denominator to the same degree. An index of around five indicates a normal value.

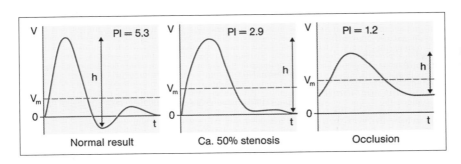

Fig. 5.**27**   The pulsatility index (PI) (adapted from Gosling) in both normal and pathological flow pulses within the extremities. The PI is calculated from the height (h) of the maximum forward and reverse flow, divided by the mean flow velocity (V_m)

The *"damping factor"* (DF) is the ratio between the proximal pulsatility index and the distal pulsatility index:

$$\text{Damping factor} = \frac{\text{PI}_{\text{proximal}}}{\text{PI}_{\text{distal}}}$$

It is a sensitive parameter for detecting intervening intraluminal stenoses—even low-grade ones. Normally, a value of one is calculated (FitzGerald and Carr 1977).

The *"transit time"* is sometimes measured. This expression refers to the pulse wave propagation from one registration point to the next. This value allows the stiffness of the arterial wall to be assessed. However, the Doppler waveform has to be simultaneously registered at two different points, and possible time delays in the pulse increase have to be calculated (Humphries et al. 1980).

Fronek et al. (1976) measured the pulse increase and decrease times, and divided them by the peak velocity (Fig. 5.**28**). The most favorable parameters for discrimination purposes were the peak velocity and the deceleration (Baker et al. 1986).

Cobet et al. (1986) and Scharf et al. (1988) examined, both in vitro and in vivo, the pulsatility index (PI), the resistance index (RI) (pp. 454, 455), and the systolic half-life (SHL) in relation to different degrees of stenosis and varying peripheral resistances. They found that the SHL reacted with the greatest sensitivity to all proximal and distal stenotic changes.

Finally, mention should be made of the indices developed using the *Laplace transformation*, which were introduced by Skidmore and Woodcock (1980).

The practical use of such indices usually depends on the ability to evaluate the flow pulses digitally. Windeck et al. (1992) compared four indices, namely the pulsatility index, the damping factor, the systolic time delay index, and the "height/width index." The latter index showed the greatest sensitivity (92.5%) in diagnosing preceding vascular system obstructions, both before and after angioplasty. An index which they developed themselves, the *curve broadening index*, which uses a complicated measuring procedure to calculate both the pulse increase and decrease times at certain curve points, even reached a sensitivity of 100%.

■ Conventional and Color Flow Duplex Sonography

Evaluating the pelvic and leg arteries over their long vascular distance is complicated, particularly because of the deep arterial location within the small pelvis. The course of the external iliac artery, which disappears into the depths, is a particularly critical point. Several authors describe the imaging of the vascular system of the thigh in the adductor canal as being difficult; the positional change required at the transition into the proximal popliteal artery further complicates the continuity of the examination.

Displaying the lower leg arteries without using color provides little information.

B-mode. The vascular diameter can be determined in B-mode. In addition, the image is useful for excluding

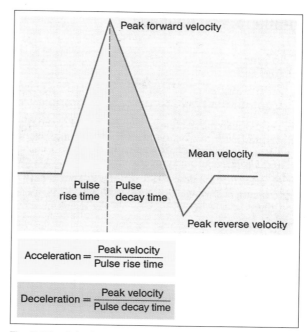

Fig. 5.**28** Criteria for measuring a normal artery velocity signal in the extremities (adapted from Fronek). The parameters of acceleration and deceleration were calculated from the measured peak systolic velocity, and the pulse increase and pulse decrease times. These modified criteria are also used in other indices

true aneurysms and, especially after femoral artery punctures, for excluding a false aneurysm or hematoma, or an A-V (arteriovenous) fistula. In the popliteal region, a search can be made for popliteal cysts.

Doppler mode. With the integrated pulsed wave Doppler mode, the same parameters can be evaluated as when using continuous wave Doppler sonography. As in all vascular regions, it is important to note the angle to the so-called vascular axis, which should not exceed 60°, and which should be kept at approximately the same value when following the course of an axis and making quantitative measurements. The advantage in doing this lies in the ability to convert the Doppler frequencies into velocities by a known ultrasound angle. This means that the comparison of the measured values is independent of the transducer transmission frequency selected, which often fluctuates due to the varying depth of the peripheral vasculature. In this procedure, the B-mode transmission frequency and the Doppler transmission frequency need not necessarily have the same high or low value. Evaluating the Doppler frequency or flow velocity, as well as the advantages and disadvantages of each of these measurements, are discussed in Chapter 1, pages 3–4 (Phillips et al. 1989).

Color flow Doppler mode. The color flow Doppler mode makes it easier to detect the vasculature in all regions. Excluding vascular system stenoses or occlusions, pathological structures, or aneurysmal dilatations, can be done quickly and reliably.

Pathological Findings

Principle

■ Continuous Wave Doppler Sonography

At rest. Due to the decreased peripheral resistance, the flow rate distal to *occlusions* largely remains stable. The flow pulses are similar in form to those found in regions of physiologically decreased organ resistance (e.g., cerebral arteries, renal arteries). The backflow component is absent, the amplitude height is lower, and there is a relative increase of the diastolic component. (Busse 1982) (Fig. 5.29).

Stenoses can rarely be directly detected at the typical registration points (groin, popliteal region) in the arteries supplying the legs. Iliac artery stenoses are usually located deep in the pelvic region, which is not accessible to CW Doppler sonography. In addition, a su-

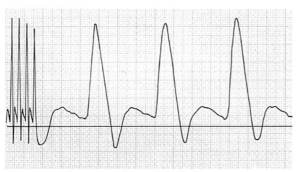

Patent vascular system (normal finding)

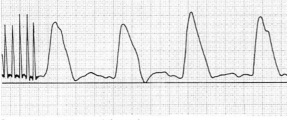

Stenosis ca. 50% pelvic region

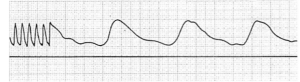

Occlusion or stenosis > 80% pelvic region

Fig. 5.**29** The Doppler waveform from the common femoral artery obtained with continuous wave Doppler sonography at the groin in a patent vascular system. The triphasic waveform that is typical for an artery in the extremities is seen. In medium-grade pelvic stenoses, the amplitude height decreases, and the flow reversal disappears. Postocclusively, a monophasic waveform with a low amplitude height and a clear diastolic flow is recorded (decreased peripheral resistance). The pulse rise and the pulse decay times are clearly lengthened

perficial femoral artery stenosis, usually found in the adductor canal, often cannot be located with the usual continuous wave ultrasound probes (4–5 MHz transmission frequency). However, when applying ultrasound to a superficial stenosis (e.g., in the common femoral artery, beneath the inguinal ligament), the noise phenomena and the flow pulse characteristics are identical to those obtained in the easily accessible distribution area of the carotid artery: the flow velocity increases, and what is termed "systolic peak reversal" occurs (Fig. 11.**22**, pp. 380–382).

Obtaining more exact information about the location of occlusions and stenoses, differentiating between the two, and determining the length of the occlusion, require the use of duplex sonography.

Post exercise. This examination is not of very great importance in pathological cases, since even in healthy individuals, a hyperemic reaction, i.e., a flow velocity increase, especially in the diastolic component, can be observed after exercise. According to Fronek et al. (1973), it amounts on average to 226% after a four-minute suprasystolic compression. In pathological cases, the flow velocity increase is lower, and amounts to under 80%. The time needed to return to normal flow pulses is longer than in healthy individuals (Sumner and Strandness 1969, Fronek et al. 1973).

■ Conventional and Color Flow Duplex Sonography

B-mode. The opportunity to carry out morphological evaluation of vascular wall deposits, the vascular diameter, and the vascular course in B-mode imaging has made an outstanding contribution to diagnosis. Previously, aneurysmal dilatations could only be suspected through palpation, and the total extent of them could not be recognized using angiography (Fig. 11.**27**, pp. 390–391). Hematomas and cysts, e.g.. Baker's cyst in the popliteal region, are now accessible noninvasively, and can be distinguished from aneurysms.

Doppler mode. It is essential to use pulsed wave Doppler sonography when formally analyzing the Doppler waveform, qualitatively or quantitatively determining the stenosis grade, calculating indices, differentiating the vasculature, and evaluating the peripheral resistance.

Color flow Doppler mode. As mentioned above (p. 149), it was only with the introduction of the color flow Doppler mode that it first became practicable to examine the long vascular segment from the aorta to the distal lower leg, with targeted placement of the Doppler sample volume. Locating vascular system stenoses, differentiating occlusion and stenosis, and determining the length of an occlusion can all be carried out quickly and reliably. Arteriovenous fistulas and the inflow and outflow of blood to and from a pseudoaneurysm can also be diagnosed without difficulty (Fig. 11.**28**, pp. 392–393).

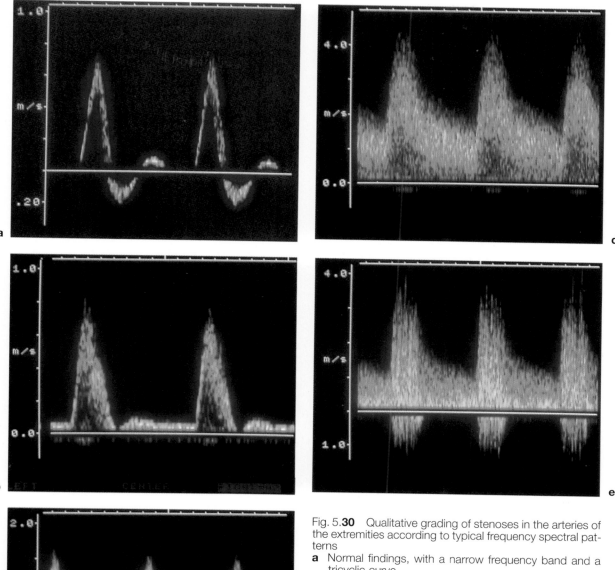

Fig. 5.**30** Qualitative grading of stenoses in the arteries of the extremities according to typical frequency spectral patterns

a Normal findings, with a narrow frequency band and a tricyclic curve

b Flow disturbances in minimal wall deposits with a diameter reduction of less than 25%. Slight spectral broadening is observed. However, the systolic window is open

c Low-grade stenosis. The systolic window has dissolved, but there is not yet any flow velocity increase, or only a minimal one. The backflow component is missing

d The image is the same as that for a stenosis of more than 50%. The maximum systolic velocity is clearly elevated (3 m/s). The peak diastolic velocity is also elevated, and includes spectral broadening

e High-grade stenosis, over 80%. The peak velocity is close to 4 m/s, and there is systolic reverse flow component and a clear amplitude increase at low frequencies

Grading of Stenoses

In the peripheral arteries, the graduation of the stenoses is subject to different hemodynamic criteria from those in the carotid stream, since the peripheral resistance is different, and above all because a series of consecutive stenoses, or occlusions and stenoses, complicates the hemodynamic assessment (see below) (Table 5.**7**).

Principle

Using frequency spectrum analysis, it is possible to carry out qualitative and quantitative evaluations.

Qualitative evaluation. Even in the peripheral arteries, it has proved valuable to carry out an approximate assessment of the severity of the constriction of the flow using a qualitative determination of the alteration of the waveform. The Doppler curve scheme proposed by Jäger et al. (1985) has gained widespread acceptance. The method uses the pulsed wave Doppler mode of a conventional duplex apparatus (Fig. 5.**30**).

Quantitative evaluation. As in the carotid stream, grading a stenosis by measurements of the peak systolic velocity serves as a quick way of obtaining quantitative information concerning the peripheral arteries. The rule of thumb is that a doubling of the starting velocity indicates a stenosis greater than 50% (Jäger et al. 1985, Moneta et al. 1993). Other authors have determined an optimal threshold value of 200 cm/s (Cossman et al. 1989) or 180 cm/s (Ranke et al. 1992). Since the avarage velocity, e.g. in the femoral artery, is 90 cm/s, a stenosis of over 50% is assumed at a velocity of 180 cm/s. However, this is valid for isolated stenoses only (Allard et al. 1994). Additional parameters for frequency analysis such as end-diastolic velocity (EDV) or mean frequency are used less often.

Since measurements of the absolute velocity do not take into account distal or proximal stenoses or occlusions, the following physical laws (Neuerburg-Heusler and Karasch 1996) have proved their value alongside the familiar pulsatility index and resistance index.

Principle of the Continuity Equation

The continuity equation states that the volume flow in all consecutive sections of the vascular system remains constant. This means that the flow velocity in regions of reduced diameter (stenoses) must increase in order to maintain the volume flow, which is calculated according to the following formula:

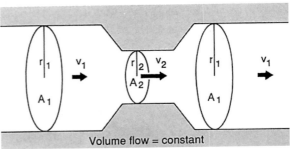

Volume flow = constant

Principle of the law of continuity. Due to variation in the mean velocity of flow (*v*), the volume flow (Q) at each point in the vessel remains the same even when the area of the vessel's cross-section (A) changes. An increase in the flow velocity can therefore be used to calculate the extent of the reduction in the cross-section and the degree of stenosis (adapted from: K.-J. Wolf, F. Fobbe 1993).

Table 5.**7** Comparison of the arterial systems supplying the extremities and the brain

Parameter	Peripheral arteries	Carotid arteries
Peripheral resistance	High	Low
Region supplied	One transporting artery	Several transporting arteries
Vascular length	Long	Short
Exertion test	Muscular work Hyperemia	CO_2 test
Collaterals	Variable, depending on location	Pre-formed circle of Willis
No. of stenoses	Mainly affect several levels	Often isolated stenoses
Significance of stenoses	Severity Only higher-grade stenoses effective	Morphology Lower-grade stenoses also cause embolism

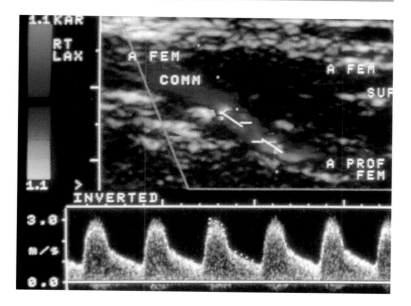

Fig. 5.**31** Stenosis at the origin of the deep femoral artery in a case of superficial femoral artery occlusion. The systolic velocity in the stenosis jet is 3.15 m/s, and the maximum diastolic velocity is 0.84 m/s. Aliasing is seen, with a recognizable color change proceeding through white. The Doppler waveform shows a monophasic course, with clear spectral broadening and a high diastolic flow velocity

$$Q = A \times v$$

Q = Volume flow
A = Vessel cross-sectional area
v = Mean flow velocity

Since according to the continuity equation	$Q_1 = Q_2$
are equal, then	$A_1 \times v_1 = A_2 \times v_2$
or	$\dfrac{v_1}{v_2} = \dfrac{A_2}{A_1}$

1 = prestenotic,
2 = intrastenotic

The flow velocities thus behave in an inverse proportion to the vessel's cross-section.

When the flow velocity in a narrowing is compared according to the law of continuity with the prestenotic flow velocity, then in purely arithmetical terms it must be possible to use the above formula to calculate the reduction in the cross-sectional area at the point of narrowing. This velocity ratio has been used with considerable success as a way of grading stenoses, and it is known as the peak velocity ratio (PVR).

The use *of the peak velocity* in the equation is based on the experimental studies carried out by Johnson et al. (1989) that demonstrate that the intrastenotic spectral peak velocity correlates well with cross-sectional area (r = 0.93), while the intensity-weighted mean velocities showed only a poor correlation. In cardiology, however, it is more usual to use the mean (or averaged) velocity to calculate the valvular cross-sectional area.

Principle of the Bernouilli Equation

In addition to measurement of the reduction in the cross-section in the stenotic area, measurement of the pressure loss caused by the stenosis is also important.

This value is calculated according to the Bernouilli equation, known as the law of the maintenance of constant energy in the movement of fluids:

$$E = P + {}^1\!/_2\, p\, v^2 + p\, g\, h$$

P	= potential energy
${}^1\!/_2\, p\, v^2$	= kinetic energy
p g h	= gravitational energy
p	= density of the fluid
v	= velocity
g	= gravitational constant
h	= height

In a *stationary liquid,* the static or hydrostatic pressure is constituted by the gravitational pressure (P) and the influence of gravitational forces (p g h). In a *flowing* liquid, part of the total energy is given off with the movement of the liquid, and is described as kinetic energy (${}^1\!/_2\, p\, v^2$).

When flow velocities increase in the stenosis (see the continuity equation), kinetic energy is required, which is used up at the expense of the static pressure. Since the sum total of the energy remains constant, and since the gravitational energy does not change when the vessel region being measured remains at the same height, the gravitational pressure must fall so that the same total is derived from the three parameters.

The Bernouilli equation thus states that the energy level in front of the stenosis and at the stenosis is the same:

$$P_1 + {}^1\!/_2\, v_1^2 + p\, g\, h = P_2 + {}^1\!/_2\, p\, v_2^2 + p\, g\, h$$

If the Bernouilli equation is then simplified by factors that stay constant or are negligible, the pressure gradient can be calculated as follows:

$$P_1 - P_2 = \Delta p = 4 \times (v_2^2 - v_1^2)$$

Assuming that the intrastenotic velocity (v_2) is high and that the prestenotic velocity (v_1) is low—i.e.

below 1 m/s—the latter can be ignored as a subtrahend. With further simplification, the formula thus becomes:

$$\Delta p = 4 \times v_2{}^2.$$

This formula has already been in use for a considerable period to calculate the pressure loss in cardiac valvular stenoses. In the area of the peripheral arteries, the principle has already been tested in vivo and in vitro in several studies (p. 184), and has shown satisfactory correlation coefficients of r = 0.8 to 0.9. However, a recent study by Legemate et al. 1993 comparing the measurement of the pressure gradient according to the Bernouilli equation with the calculation of the stenosis grade according to the velocity difference ΔPSV (peak systolic velocity), showed that the use of the Bernouilli equation gave a markedly lower correlation quotient of r = 0.62 than the measurement of the velocity difference ΔPSV at r = 0.72.

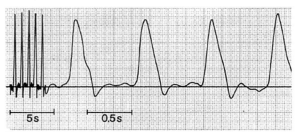

Iliac artery stenosis < 50 %

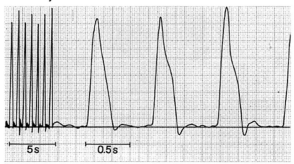

Proximal occlusion of the femoral artery

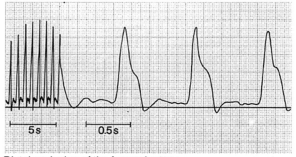

Distal occlusion of the femoral artery

Fig. 5.**32** Analogue waveforms of the common femoral artery in an iliac artery stenosis of less than 50%, with subsequent femoral artery occlusions. In each case, the backflow component is reduced, while the systolic forward flow is little affected. It is not possible to distinguish between preceding stenoses and subsequent occlusions qualitatively

Findings

Pelvic Level

Prior to the introduction of duplex sonography, the diagnosis of pelvic occlusions and stenoses was questionable, because cuff methods were not applicable in this region, and neither pressure measurements nor oscillographic examinations were possible. Even continuous wave Doppler sonography was only able to provide indirect information about pelvic vascular system stenoses. Differentiating between the proximal and distal site of the lesion had always presented difficulties. Clinically, only palpating the pulse at the groin and auscultation were practicable for orientation purposes.

Even if isolated pelvic occlusions occur in only 10.2% of the cases, two-level iliofemoral occlusions are still very frequent. According to Vollmar (1982), one-third of peripheral occlusive processes occur in the aortoiliac segment. However, 82% of these are combined with peripheral occlusions. Individual evaluation of the hemodynamic effect of an obstruction at each level is important, due to the availability of lumen recanalization techniques.

■ Continuous Wave Doppler Sonography

Registering Doppler waveforms from the groin is a practicable and fast method of indirect examination, although it only provides information about higher-grade stenoses of the pelvic vascular system.

It is possible to assess the extent of the aortoiliac obstruction approximately from the form of the flow pulse. In the subsequent vascular system, stenoses of less than 50% cause no change in the recording form. When the systolic peak is unchanged with regard to both the amplitude height and the steepness rate of the increase, only a reduced backflow component results. Qualitatively, this waveform image is no different from that obtained in a patent pelvic vascular system with subsequent femoral artery occlusions (Fig. 5.**32**).

By contrast, preceding stenoses of more than 50%, and occlusions, both have a clearly decreased amplitude height and reduced steepness in the systolic rate of increase, along with a delayed diastolic decrease, without any backflow component.

A decreased Doppler waveform is predominantly caused by obstructions in the preceding vascular system; subsequent occlusions only decrease the back-flow component (Neuerburg et al. 1981).

■ Conventional and Color Flow Duplex Sonography

B-mode. Due to the deep location of the arteries and the often unfavorable ultrasound application angle, continuous imaging of the vascular wall structure is not always possible. It is more important to locate the plaques and then determine, using Doppler mode, whether or not a hemodynamically significant stenosis is present. Occlusions and stenoses are most often encountered near the aortic bifurcation, the common iliac artery, and at the origin of the external iliac artery.

Imaging of the internal iliac arteries is not always possible, and this is reserved for specialized investigations concerning vascular impotence, or collateralization of an iliac artery flow obstruction (Fig. 5.**33**).

Doppler mode. The Doppler mode displays the flow velocity curves with frequency analysis. Usually, the groin region is examined first, to see whether the pelvic vascular system is patent or obstructed. Qualitative evaluation of the form of the recording is sufficient (Fig. 5.**30**). When a definite or suspected indication of preceding pelvic obstruction is encountered, the stenosis grade must be determined, both *qualitatively,* using formal analysis of the Doppler waveform and also *quantitatively,* by measurements (pp. 161–164). A few authors (p. 184) have attempted to calculate the stenosis grade using the pressure gradient, as in echo-cardiographic measurements. The results were encouraging.

Color flow Doppler mode. Locating the deep pelvic arteries, with their markedly curved course, has become very much easier with the addition of color. Using color flow duplex sonography to display the pelvic vascular system has made angiographic controls largely superfluous following recanalization techniques such as angioplasty, stent implantation, and bypass operations.

Thigh Level

Femoral artery occlusion is the most frequent manifestation of obstructive arterial disease in the arteries of the extremities (approximately 48%). It usually originates at the adductor canal, where a mechanical physiological arterial narrowing caused by the tendon of the great adductor muscle is thought to serve as the initiating mechanism for vascular wall lesions. In this region, stenoses also occur quite often. Occasionally, the occlusion extends to the origin of the deep femoral artery, which with its strong flow hinders appositional growth. Since femoral artery occlusions and stenoses are readily accessible to angioplastic techniques, noninvasive diagnosis in this region is particularly important.

Evaluating the common femoral artery before and after using the catheter in coronary or peripheral angioplasty is vital. *Beforehand,* plaques and stenoses are located, and unintentional plaque ablation is avoided; *after* these procedures, dissections, pseudoaneurysms (Fig. 5.**42**), and hematomas can be detected.

Preferred locations for *stenoses* are the bifurcation, with the origins of the superficial femoral artery and deep femoral artery, and also the adductor canal (Fig. 5.**34a, b, c**). The most important collateral in this region is the deep femoral artery, which as a blood vessel capable of significant throughput can compensate well for occlusions above the adductor canal.

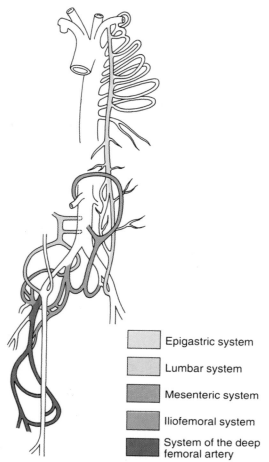

Epigastric system

Lumbar system

Mesenteric system

Iliofemoral system

System of the deep femoral artery

Fig. 5.**33** Different collateral systems for bypassing aortoiliac vascular system occlusions and superficial femoral artery occlusions (adapted from Vollmar)

■ Continuous Wave Doppler Sonography

Detecting vascular system stenoses in the course of the superficial femoral artery is usually only done indirectly, by applying ultrasound to the popliteal artery. In the process, both qualitative and semiquantitative assessment of the magnitude of the hemodynamic effect are possible. Examining the deep femoral artery is possible neither with clinical palpation, nor with continuous wave Doppler sonography.

■ Conventional and Color Flow Duplex Sonography

B-mode. The arterial images allow detection and structural evaluation of plaques and, most importantly, of aneurysms (pp. 170–171) the preferred locations of which in the extremities are the common femoral artery and the popliteal artery (Figs. 11.**26**, 11.**27**, pp. 387–391).

Pseudoaneurysms are primarily located near the common femoral artery and in the proximal segments of both the superficial femoral artery and the deep femoral artery (Figs. 11.**28**, 11.**29**, pp. 392–395).

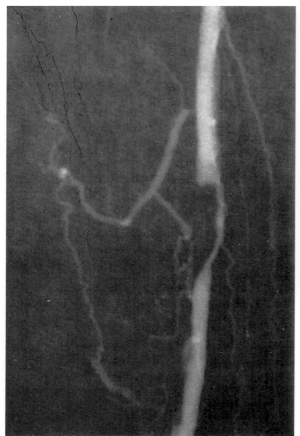

a

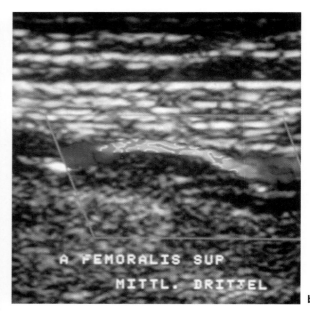

b

Fig. 5.**34** A threadlike stenosis in the middle third of the superficial femoral artery. **a** Imaging using selective angiography, and **b** the corresponding image in the duplex color flow mode. There is marked aliasing in the stenosis canal. **c** The Doppler flow waveform shows a peak systolic velocity of 4.36 m/s with spectral broadening, and a diastolic velocity of 1.03 m/s

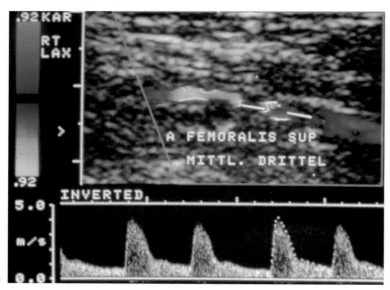

c

Doppler mode. The qualitative and quantitative evaluation of stenoses using the Doppler mode is carried out as described above. Most importantly, at the femoral bifurcation, it becomes possible to differentiate stenoses as belonging to either the superficial femoral artery or the deep femoral artery, since the Doppler waveform with the B-mode image can be specifically registered from both of the originating blood vessels. By contrast, the origin of the deep femoral artery is often depicted in superimposed form when using angiography (Strauss et al. 1989). In 1989, Strauss et al. demonstrated that a femoral artery occlusion at the origin causes the flow velocity in the *patent* deep femoral artery to double (maximum resting flow velocity 60 cm/s ± 15 cm/s, with superficial femoral artery occlusion 142 cm/s ± 44 cm/s). With both a combined femoral artery occlusion and deep femoral artery stenosis at the origin, the velocity in stenoses of more than 50% amounted to at least a twofold or threefold increase over the original value of the flow velocity (on average 255 cm/s ± 60 cm/s).

Color flow Doppler mode. Continuous application of ultrasound to the femoral artery vascular system, a preferred location for atherosclerosis, has been very much simplified through color coding. Stenoses can be quickly recognized by their lighter color, so that, guided by the color, exact placement of the pulsed wave sample volume becomes easier. Femoral artery occlusions can be located by the color interruption and by the originating collateral vasculature; the occlusion length can be reliably measured (Karasch et al. 1993) (Fig. 5.**49**, p. 185, Fig. 5.**35 a, b**). Even stenoses in tandem, or occlusions, are rarely missed with color flow duplex sonography. Difficulties can arise when the subsequent flow velocity is too small for adequate coding. Additional findings are given in the following sections (pp. 169–178).

Calf and Foot Arteries

Arterial occlusions in this region are found in 18.4% of leg occlusions. The anterior tibial artery is affected most frequently, the fibular artery least frequently. Isolated arterial occlusions of the foot are clearly less frequent, and occur predominantly in diabetics and in those suffering from endangiitis. Noninvasive diagnosis of these is only possible in summary form, by measuring the pressure at the toes. Magnification angiography has to be used to confirm the diagnosis. Combined with thigh occlusions, the distal arteries are often affected, forming a large group of the combined occlusive types (26%).

■ Continuous Wave Doppler Sonography

It is not difficult to register the flow recording from the posterior tibial artery, the anterior tibial artery, and sometimes also from the peroneal artery, at the ankle. However, pressure measurements are used more often in this region. With these relatively nar-

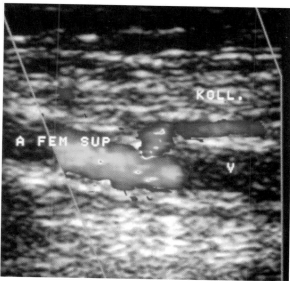

a

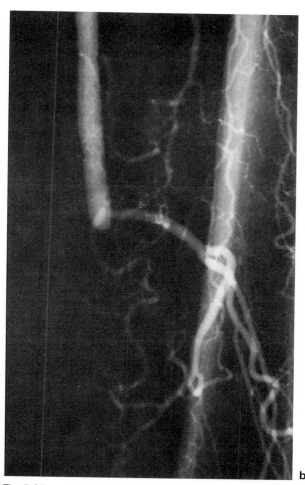

b

Fig. 5.**35 a** A color flow image of a femoral artery occlusion in the distal third of the artery. The branching collateral shows strong coloring. The blue color at the origin of the collateral reflects a flow direction change. **b** The corresponding angiogram shows the occlusion, with the branching, vigorous collateral

row-caliber arteries, imaging of the Doppler waveform may be somewhat imprecise, since the optimal angle cannot always be easily adjusted.

■ Conventional and Color Flow Duplex Sonography

In this region, conventional duplex sonography plays a subordinate role, since finding the arteries without the benefit of color coding can be tedious and time-consuming.

Color flow Doppler mode. With color flow duplex sonography, it has become possible to image all three lower leg arteries. However, the examination is only carried out as part of a specialized investigation, since even with color coding, the time needed is quite considerable, and the topographical classification is not simple. The procedure requires some experience (Fig. 5.**36 a, b**). Practiced examiners, however, report short examination times and good results (Langholz et al. 1991, Larch et al. 1993). In the foot arteries, color coding allows rapid measurement of the flow velocity, so that from this value indirect indications of the collateral circulation in lower leg arterial occlusions can be obtained.

Upper Extremity Arteries

Atherosclerotic wall changes in the arteries of both upper extremities are very much less frequent than in the arteries supplying the legs. Stenoses or occlusions occur frequently only at the origin of the subclavian artery from the thoracic aorta. According to Vollmar (1982), 16% of supra-aortic arterial occlusions are subclavian artery occlusions. According to the joint study, the left subclavian artery is affected in 14.9% of cases, and the right subclavian artery in only 9.2% of cases (Hass et al. 1968; Fields et al. 1970).

When there are hemodynamically significant stenoses in this region, the vertebral artery serves as a collateral supplying the arm, and this flow reversal within the vertebral artery, which is also important for the cerebral circulation, ("subclavian steal") is discussed in Chapter 2 (pp. 79–84).

Afferent collateral flow through the vertebral artery is usually more than sufficient to supply the axillary artery, the brachial artery, and the forearm arteries, so that symptoms such as arm claudication are not noticed until higher levels of exertion are applied (carrying heavy objects, or working above head level).

When there are embolic occlusions affecting the digital arteries of the hand, it is important to look for aneurysms in the subclavian artery. These can sometimes occur due to a neurovascular compression syndrome (p. 169). Arteriosclerotic stenoses in the arteries of the arm are a rarity. Inflammatory causes, such as Takayasu's arteritis, or arteritides of varying pathogenesis, predominate (Fig. 11.**36**).

The ultrasound examination has additional significance in detecting occlusions after intra-arterial punctures or catheter manipulations, or in testing the hemodialysis shunt output.

■ Continuous Wave Doppler Sonography

Suspected *subclavian artery occlusions* or *stenoses* are initially indicated by the conventional blood pressure measurement, using bilateral comparison. However, Doppler waveforms registered in a supraclavicular location may also reliably reflect stenoses at the origin of the subclavian artery. Formal analysis can classify these according to the degree of stenosis in the same way as in the arteries of the leg. Nevertheless, only indirect indications of the vascular system stenosis are obtained. Due to the inadequate penetration of the continuous wave Doppler probe, distinguishing between an occlusion and a stenosis is rarely successful. In this case, it is recommended to use the pulsed wave Doppler system or, simply, a stethoscope.

The *axillary artery and the brachial artery* are rarely examined with continuous wave Doppler sonography, because atherosclerotic occlusions and stenoses are a rarity in this region. Obstructions are more commonly caused by vasculitis, or are due to catheter placements.

In principle, the velocity recording from the *digital arteries of the hand* can be registered using Doppler sonography. However, continuous tracing of the course of the palmar digital arteries is not reliably possible.

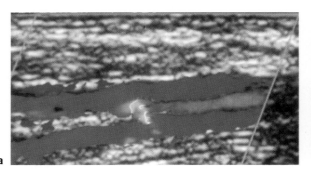

a

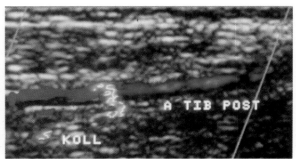

b

Fig. 5.**36**   A posterior tibial artery occlusion in the distal lower leg.
a Refilling through a collateral. The artery follows a course between the accompanying veins, while proximal to this (on the left in the image), the absence of an arterial band can be observed. **b** Imaging in a different plane to verify the collateral, which shows a high flow velocity, with aliasing

■ Conventional and Color Flow Duplex Sonography

B-mode. Imaging of the subclavian artery, which originates on the left from the aortic arch and on the right from the innominate artery, is adequate with proper patient positioning and practice, although direct evaluation of this region is not simple. It is important to detect aneurysms distal to the clavicle, which may be due to a mechanical stenosis, and sometimes appear as a poststenotic dilatation (p. 170).

Doppler mode. Under visual B-mode control or using color, the sample volume is used to quantify the stenoses. The velocity curves are evaluated using the criteria applied in the arteries of the extremities (p. 179). It is often sufficient to use indirect criteria in the poststenotic or postocclusive area, since stenoses of the *subclavian artery* vascular system are quantitatively assessed by measuring the arm pressure. Imaging of the subclavian steal phenomenon is discussed in Chapter 2, pages 79–84.

The Doppler mode is also very important in quantitatively determining the flow rate through a *hemodialysis shunt* in the arm. The output of the shunt is very significant (p. 175). Conventional duplex sonography is rarely used in the arm or in the digital arteries of the hand, which have a narrower lumen.

Color flow Doppler mode. Again, color flow Doppler mode simplifies the imaging of the arteries and allows faster diagnosis in specialized investigations. A special area of application for color flow duplex sonography is for the fast detection of stenoses or aneurysmal dilatations in hemodialysis shunts. The importance of these is discussed separately below (Fig. 5.**37**).

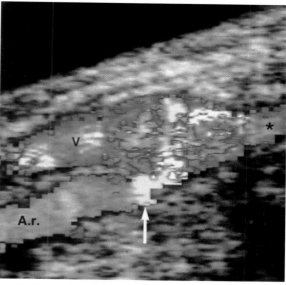

Fig. 5.**37**   An insufficient Brescia–Cimino dialysis shunt in the left lower arm. The radial artery (A. r.) serves as the afferent blood vessel for the shunt. A stenosis at the anastomosis (white arrow) can be recognized from the lumen narrowing and the high flow velocity (light red/ white color coding). Within the shunt vein, the color coding is blue, due to flow in the opposite direction. In the segment of the radial artery that is distal to the anastomosis (*), retrograde flow is seen, due to the shunt vein's low peripheral resistance and steal from the ulnar artery with a patent palmar arch. (Illustration courtesy of Dr. P. Landwehr, Institute and Polyclinic for Radiological Diagnostics, University of Cologne)

Special Sets of Findings

Neurovascular Compression Syndrome of the Shoulder Girdle

In the shoulder girdle, the subclavian artery, along with the subclavian vein and the brachial plexus, as a vascular nerve bundle, have to pass through three naturally narrow points, and this can lead to functional arterial stenoses.

When there is continuous compression of the subclavian artery, organic changes with poststenotic dilatation, culminating in an aneurysm, may also occur. Peripheral emboli in the digital arteries of the hand often indirectly indicate the pathogenesis (Fig. 11.**36**, pp. 407–408).

Figure 5.**38** shows the subclavian artery passing through osseous and muscular structures, and indicates the narrow points. Clinically, the costoclavicular syndrome causes functional or organic damage in 80% of cases.

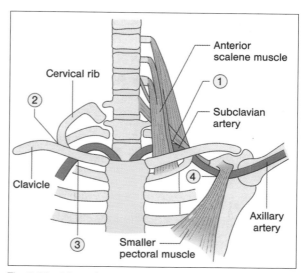

Fig. 5.**38**   Muscular and osseous points of narrowing that the subclavian artery has to pass in its course. The compression syndromes are defined by the location of their origin (adapted from Kappert)
1 Anterior scalene syndrome
2 Cervical rib syndrome
3 Costoclavicular syndrome
4 Hyperabduction syndrome

Examination and Findings

Indirect clinical criteria for a stenotic vascular narrowing can already be obtained from bilateral comparative blood pressure measurement, unilateral pulse weakening, and auscultation of a stenosis bruit in provocation position.

During the ultrasound examination, the continuous wave Doppler probe, as well as the transducer from duplex sonography equipment are used to locate the subclavian artery above the clavicle and distal to it.

A narrow point itself can rarely be displayed in the B-mode image, since the bone structures obstruct ultrasound penetration. Imaging aimed at detecting aneurysms or poststenotic dilatations is very important, as is the location of thrombotic components in an aneurysm, which is best done using color coding. Since ablation of small emboli occurs occasionally, occlusions affecting the digital arteries of the hand that are unclear in their origin always require a search to be made for subclavian aneurysms.

Aneurysms

Arterial aneurysms are differentiated according to etiological, morphological, and clinical criteria (Table 5.**8**). They play a significant role, not only in the abdomen, but also in the peripheral arteries. Both arteriosclerotic true aneurysms and iatrogenic false aneurysms are found (Fig. 5.**39**).

Clinically, suspicion of an aneurysm in the lower extremities is mainly raised by the acute appearance of claudication or ischemia, while subclavian aneurysms are mostly recognized by digital atheroembolisms. In a multicenter study of popliteal aneurysms, Varga et al. (1994) found in 125 symptomatic popliteal aneurysms that there was claudication in 58% of cases, ischemia in 56%, and atheroembolism of the limbs in 11%. According to a recent study by Mönig et al. (1996), infrapopliteal aneurysms have been reported in the world literature in only 15 cases so far.

True Aneurysm

The arteriosclerotic *true aneurysm* in the peripheral arteries is preferentially located in the common femoral artery and in the popliteal segment. It usually develops from a preceding dilatative arteriopathy (Hepp and Pallua 1991, Loeprecht and Bruijnen 1991). The low incidence of aneurysms in these locations (0.003–0.007%) is analyzed in detail in a study by Lawrence et al. (1995), but no connection with atherosclerotic risk factors was found. However, the late appearance of the condition in patients aged over 65 is notable. By contrast, the subclavian artery aneurysm usually forms from a poststenotic dilatation subsequent to a proximal mechanical stenosis, which is caused by the compression of narrow points in the shoulder girdle area, as shown above.

Vascular graft aneurysms are due to a wall insufficiency in venous grafts or in synthetic prostheses.

Findings

Pathological arterial dilatation is recognizable in *sonographic B-mode imaging,* and the size and form of the aneurysm can be measured (Fig. 5.**40).** The distinction between thrombosed lumen and patent lumen is important. The patent lumen can be most effectively displayed and evaluated with color flow.

In the periphery, the greatest risk is represented by aneurysmal thrombosis and the acute arterial occlusion associated with it, while the danger of rupture in this arterial region is minimal. The thrombotic deposits serve as an source of embolization into the acral regions of the upper and lower extremities. The subclavian aneurysm is a particularly frequent source of emboli.

False Aneurysm

While traumatic aneurysms predominated earlier, the injury pattern has now shifted toward iatrogenic vascular lesions. Due to the wide variety of diagnostic and therapeutic procedures that involve introducing catheters into the common femoral artery, the number of pseudoaneurysms in this region has significantly increased. An additional cause of false aneurysms is wound dehiscence, which is seen not infrequently after procedures involving prosthesis placement.

Table 5.**8**   Classification of aneurysms

Etiotogical	Morphological	Clinical
Congenital	**True**	Ruptured
Arteriosclerotic	Saccular	Closed
Syphilitic	Fusiform	
Traumatic	Fusiform/saccular	Patent lumen
Mycotic	**Dissecting**	Thrombosed lumen
Poststenotic	**False** *(spurious)*	Partially thrombosed lumen
Wall insufficient	*also:*	
Iatrogenic	Pseudoaneurysm	

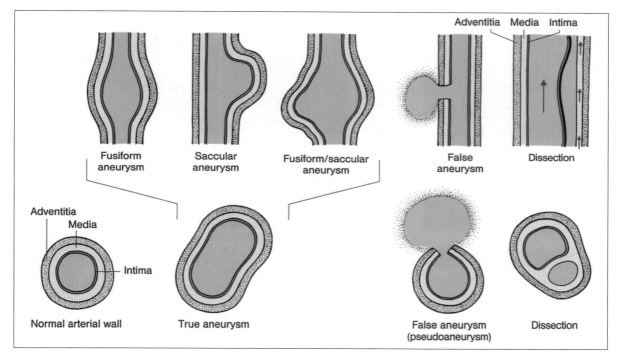

Fig. 5.**39**   The various types of arterial aneurysm in longitudinal section and cross-section. A true aneurysm signifies arterial wall sacs in which all of the wall layers are retained, and which have various configurations. The false aneurysm results from iatrogenic interruption of all of the wall components, forming an undefined varying lumen in the perivascular tissue. Dissection refers to iatrogenic or spontaneous media splitting, creating a double lumen that can also cause aneurysmal arterial expansion

Findings

While the morphological diameter and size of an aneurysm can be determined with the *B-mode image,* and thrombotic components can be recognized, the *Doppler mode* registers a characteristic Doppler waveform, which is marked by the "to-and-fro" sign (Abu-Yousef et al. 1988). During systole, the arterial blood is pressed like a jet (forward flow) into the aneurysmatic cavity, and in the diastole then flows reverse again into the originating artery (backward flow) (Mitchell et al. 1987) (Fig. 5.**41**). The Doppler waveform shows a pronounced pendular flow. The reverse flow component, which is broader and is present throughout the entire diastole, is clearly distinct from the brief backflow (dip) in normal arteries.

In the *color flow Doppler mode,* the aneurysmatic neck or stalk is easy to identify, and the phenomenon of forward and backward flow can be impressively demonstrated in real time (Fig. 5.**42**).

When there is wound dehiscence, the original aneurysmal defect is often broader, and can lead to large, pulsating hematomas.

Compression Therapy

Although pseudoaneurysms, especially those with small volume, thrombose partially and spontaneously (Paulson et al. 1992), the cautious waiting approach, even with regular follow-up examinations, involves an inherent risk of aneurysmal rupture (Buchholz et al. 1991). Using color-guided compression of the aneurysmal stalk by the transducer, Fellmeth et al. (1991), were able to detect a thrombosis in 27 of 29 cases. In the meantime, these results have been confirmed by Sorrell et al. (1993), Do et al. (1993), and also in our own patients. While compression is being applied, regular assessment during the interval using color flow Doppler mode proves helpful in determining whether or not the femoral artery is being compromised. Otherwise, there is a danger of a thrombosis in the main blood vessel.

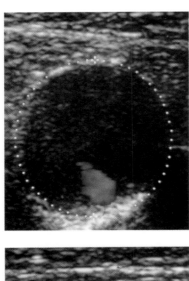

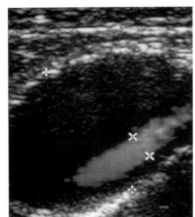

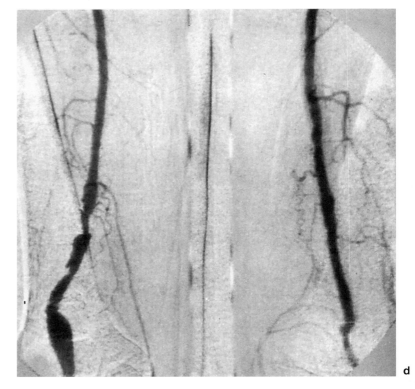

Fig. 5.**40** A large popliteal artery
aneurysm, **a** in cross-section, **b** in
longitudinal section, and **c** in longitudinal section with the corresponding
frequency spectrum. The popliteal
artery has a predilection for peripheral
aneurysms. In this case, the lumen can
be seen in both cross-section and
longitudinal section, with minimal blood
flow. The form of the flow waveform is
relatively normal. The flow velocity at
40 cm/s is very low.
d Corresponding angiograms

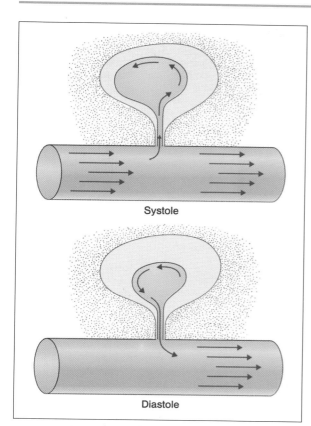

Fig. 5.**41** A schema of false aneurysm. During systole, there is flow into the aneurysm, caused by damage to the arterial wall, which burrows into the perivascular tissue. During diastole, blood is again pressed out of the aneurysmatic sac. In the neck of the aneurysm, flow in the opposite direction results

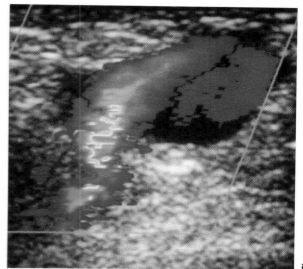

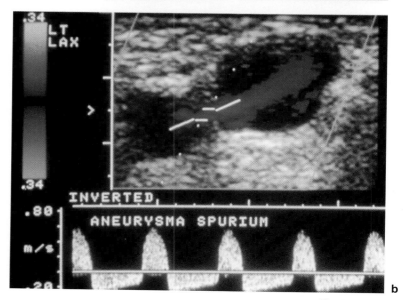

Fig. 5.**42** A false aneurysm seen in color flow mode shown in longitudinal section, near the left groin, **a** A summation image with aliasing in the aneurysmatic stalk, **b** During systole, there is forward flow into the aneurysm.

Fig. 5.**42 c** ▷

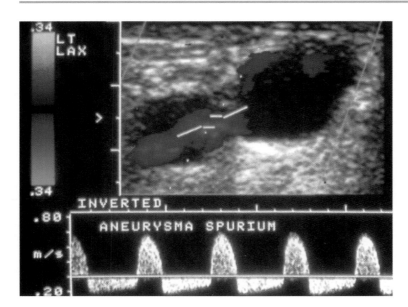

Fig. 5.**42 c** During diastole, there is flow in the opposite direction. The Doppler flow curves show the typical "to-and-fro" sign, which is marked by a long backflow phase registered during diastole

Table 5.**9** The origin, morphological structure, and functional effect of arteriovenous fistulas

Etiology	
Congenital (10–15%)	Predominantly in the head region, hemangiomas
Acquired (80–90%)	
Traumatic	Penetrating vascular injuries
Iatrogenic	Caused by surgery, catheter manipulations
Spontaneous	Aneurysmal invasion, neoplasms
Therapeutic	Hemodialysis shunt, postoperative temporary shunt
Morphology	
Single	With/without pseudoaneurysms
Multiple	With/without pseudoaneurysms
Function	
With/without effect on the cardiovascular system	Increase (pulse frequency, cardiac output-per minute), heart failure, venous ectasias, growth in length

Arteriovenous Fistulas

Principle and Etiology

An arteriovenous fistula is defined as a combined arterial and venous injury at corresponding locations. A distinction is made between congenital and acquired fistulas (Table 5.**9**).

The acquired forms are especially significant for ultrasound diagnosis in the extremities. *Traumatic* fistulas appear after gunshot or knife wounds, while *iatrogenic* fistulas occur after surgery, mainly following orthopedic operations (Karasch et al. 1991b, Seifert et al. 1987) and after *diagnostic or therapeutic catheter manipulations.*

In a consecutive series of observations following cardiac catheterization, Steinkamp et al. (1992) reported an 18–20% occurrence of arteriovenous fistulas.

Pathophysiologically, the short-circuited circulatory system leads to a significant decrease in the flow resistance and to a volume loss from the arterial system into the venous low-pressure system. A pulse frequency increase and vasoconstriction in the peripheral circulation cause increased cardiac output, decreased blood pressure, and cardiac hypertrophy, which can lead to cardiac decompensation. These potential consequences are to be taken into account particularly in hemodialysis shunts (p. 175).

Findings

Localizing the fistula canal can begin clinically. A thrill can be observed on palpation, and during auscultation, a pronounced, continuous systolic–diastolic sound (machinery murmur) is detected.

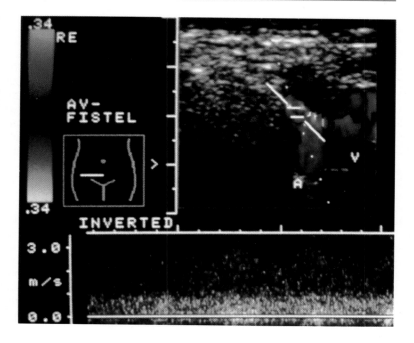

Fig. 5.**43** An arteriovenous fistula near the right groin, presented in transverse section. The arteriovenous channel is shown with a high, only slightly modulated flow, proceeding from the arterial into the venous vascular bed. The velocity is over 3 m/s. This fistula represents an iatrogenic arteriovenous shunt following manipulation with a catheter

In the *B-mode image,* the canal between the artery and the vein can be seen. Often, this canal undergoes aneurysmal dilatation and additional pseudo-aneurysmal dilatations and windings in the afferent artery and the afferent vein, which contains arterial blood (Fig. 5.**43**).

Using *Doppler mode,* a turbulent, elevated flow velocity is often found in the fistula canal (shunt). In the arterial segment proximal to the fistula, there is an increase in systolic velocity and an appearance of diastolic blood flow (Strano et al. 1995). In the venous segment, there is arterial blood flow directed toward the heart (flow reversal). When the ultrasound application angle and the vascular diameter are known, the flow rate can be calculated.

The *color flow Doppler mode* makes it easier to locate the fistula canal, and the arterial backflow into the vein becomes dynamically visible. Of course, color coding allows the examiner to recognize a change in the flow direction toward the heart, opposite to the arterial inflow.

Hemodialysis Shunt

Noninvasive morphological and functional evaluation of dialysis shunts has become increasingly important.

A distinction is made between internal arteriovenous shunts, which are formed with the body's own tissue, and prosthetic shunts with interposed vascular grafts (Wolf and Fobbe 1993). The oldest and best-known fistula between the radial artery and the cephalic vein is the Brescia–Cimino internal shunt. This is initially placed as far distally as possible near the hand, and is not moved proximally until it has become occluded. After the internal shunt has failed, or

when there are unsuitable venous relationships, a synthetic implant is interposed.

Qualitative detection of an increase in the flow velocity in the shunt to the point of turbulence is possible even with *continuous wave Doppler sonography.*

Conventional and color flow duplex sonography has expanded the diagnostic process, both morphologically and functionally. The shunt itself, the origin of stenoses in the sutured area of the venous or arterial connection, and also aneurysm formation or thromboses, can be morphologically detected (Middelton et al. 1989 b) (Fig. 11.**39**).

Shunt volumes can be measured using the Doppler sample volume. This quantity is determined by electronically measuring the length of the vascular radius (r) and the mean flow velocity (V_m).

Flow rate (ml/min) = Vmean (cm/min) $\times$ r^2 (cm^2) $\times$ π

Wittenberg et al. (1993) were able to show that several investigators repeatedly measuring the flow rates achieved an acceptable degree of reproducibility, with an 11% error rate. The following threshold values have become accepted as representing the norm (Kathrein et al. 1989):

– PTFE fistulas	614 ± 242 ml/min
– Brescia-Cimino fistulas	464 ± 199 ml/min
– Mean shunt volumes	514 ml/min

Since flow rates that are too low (less than 200–300 ml/min) can be corrected by widening the shunt, and those that are too high can be surgically reduced to eliminate the cardiac consequences, measuring the output is clearly important.

Temporary Shunts

Postoperatively, temporary shunts are often created in order to increase the flow into the arteries or veins operated on. This is often done after an operation for a fresh venous thrombosis, to prevent the high tendency toward re-thrombosis (Schmitz-Rixen et al. 1986).

Monitoring Lumen Recanalization Techniques

Monitoring lumen recanalization measures after various angioplasty procedures, and evaluating stent implantations (Fig. 5.**44**) is as important as the follow-up examination after vascular surgery. Morphological monitoring, and measurements of the flow velocity in the patent arterial segment, in the interposed prostheses, and also in the bypasses, make it possible to recognize complications and potential reocclusions at an early stage.

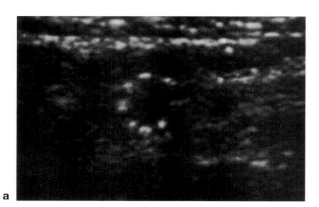

a

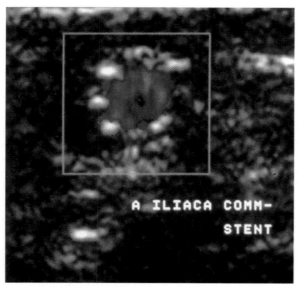

b

Fig. 5.**44** A stent in the common iliac artery, shown in transverse section.
a A B-mode image showing the metal rods in cross-section. **b** Color flow Doppler mode, showing the patent lumen with freely flowing blood

Angioplasty

After percutaneous transluminal angioplasty, remaining wall deposits and partial dissections can be detected with *B-mode imaging.*

Using *Doppler mode,* it is possible to evaluate vascular system stenoses and flow disturbances both qualitatively and quantitatively. Regular noninvasive follow-up exams are possible (p. 185).

Implantation of a stent after angioplasty can be depicted with excellent quality in the B-mode image, with or without color coding. Adequate spreading of the stent in relation to the regular vascular diameter can be well documented in the B-mode image, and undisturbed flow or stenotic narrowing can be quantitatively measured with the Doppler mode (Fig. 5.**45 a, b**).

It is not only in the immediate and later checking of angioplasty measures that the use of duplex sonography is of value. The lumen can also be opened under visual ultrasound guidance using a catheter that is incorporated into the duplex scanner (Cluley et al. 1993). Katzenschlager et al. (1996) showed that PTA was even possible without an integrated catheter using color duplex ultrasound guidance.

Vascular Prostheses and Bypasses

Follow-up observations of vascular prostheses and bypasses have already been carried out intensively for a long time using various ultrasound procedures. A bypass occlusion rate of approximately 30% within five years is considered normal (Bandyk 1993). Quantitative measurement of the flow velocity and qualitative formal analysis of the Doppler waveforms indicate possible bypass stenoses or aneurysms, as well as any reduced output that may reflect an impending reocclusion.

Various studies have provided values that can be used to predict pathological changes.

As early as 1985, Bandyk et al. were able to demonstrate using *continuous wave Doppler sonography* that bypasses have differing flow velocities depending on their location. The femoropopliteal bypass showed a flow velocity of 90 cm/s; the femorotibial bypass yielded a flow velocity of 68 cm/s on average. A decrease in the peak systolic velocity to less than 45 cm/s was interpreted as being a certain sign of impending reocclusion. In evaluating 379 venous bypasses, Mills et al. (1990) were also able to demonstrate that a velocity of less than 45 cm/s indicated a preceding stenosis with a lumen obstruction of over 50%.

In a later study using *conventional duplex sonography,* Bandyk et al. (1988) were able to show that formal analysis of the flow velocity recordings can indicate preceding or following stenoses (Fig. 5.**46 a, b**). A monophasic configuration indicates a preceding or following stenosis with a lumen obstruction of 50–75%. Pronounced pendular blood flow indicates a distal stenosis.

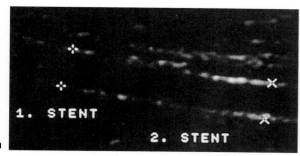

a

b

Fig. 5.**45a** A cross-section of the superficial femoral artery, showing stents that were placed one after the other in an overlapping manner. **b** The color flow Doppler mode detects a stenosis with a velocity of 3 m/s at the second stent.

In a study carried out by Sladen et al. (1989), in which both decreased flow velocity (less than 45 cm/s) and elevated velocity (higher than 300 cm/s) were used as stenosis indicators, it was shown that of 15 stenoses with a lumen obstruction of over 80%, only six showed a poststenotic flow velocity of less than 45 cm/s.

In a larger, prospective study by Gooding et al. (1991), the predicted value, less than 40 cm/s *reduced* peak systolic velocity, could only be confirmed with a 33% sensitivity. However, the *elevated* peak systolic velocity was a more informative parameter.

In stenoses, Polak et al. (1990a) observed a peak velocity of between 117 cm/s and 225 cm/s. Absent diastolic backflow could also be used to indicate a subsequent stenosis. In a larger study by Buth et al. (1991), color flow duplex sonography was used in follow-up observations to examine five continuous wave Doppler waveform parameters in 116 grafts. It was shown that a velocity index formed from the velocity

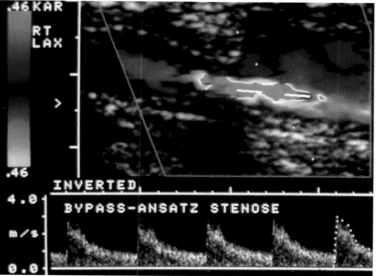

a

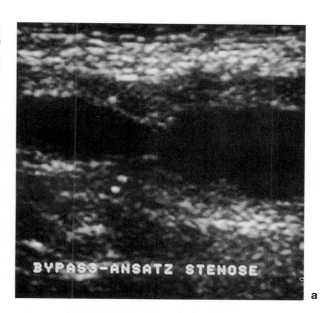

b

Fig. 5.**46** A bypass stenosis in the right groin, shown in longitudinal section. **a** In the B-mode image, the wall deposits that have led to lumen narrowing at the bypass junction are clearly seen. **b** In color flow Doppler mode, the critical area, with pronounced aliasing, is seen. The adjusted Doppler sample volume shows a monophasic curve with a maximum systolic flow velocity of 3.06 m/s

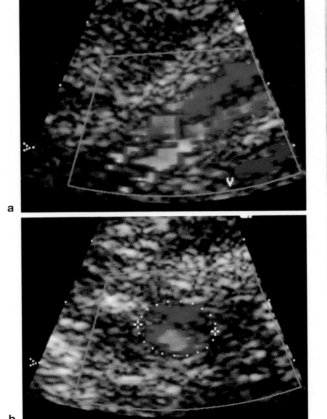

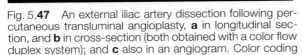

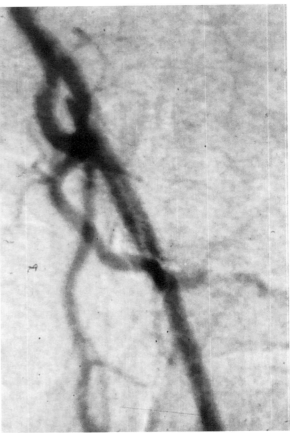

Fig. 5.**47** An external iliac artery dissection following percutaneous transluminal angioplasty, **a** in longitudinal section, and **b** in cross-section (both obtained with a color flow duplex system); and **c** also in an angiogram. Color coding displays the two filled lumens parallel and alongside one another in longitudinal section and also in cross-section (V= external iliac vein). The angiogram confirmed the findings obtained with duplex sonography

ratio in a normal bypass and a stenotic bypass, with a threshold value of 0.65, identified all stenoses with a lumen obstruction of over 40%. Taylor et al. (1992) also found that the velocity quotient had the greatest predictive value. However, they set their threshold value at over 2.0.

In addition to diagnosing stenoses and predicting impending bypass occlusions, it is important to search for pseudoaneurysms at the bypass insertion point.

Arterial Dissection

In the arteries of the leg, including the pelvic region, iatrogenic dissection is not infrequently observed after angioplasty, and can be clearly documented using color flow duplex sonography (Fig. 5.**47**).

Spontaneous dissections, as encountered in the aorta and also in the carotid arteries, are rare in the arteries of the extremities.

Iatrogenic dissections can cause occlusions, or can spontaneously regress without obstructing the lumen; both of these phenomena can be documented in their course using duplex sonography (Murphy et al. 1991).

Evaluation

■ Continuous Wave Doppler Sonography

As mentioned earlier (pp. 157–159), many indices have been developed that use quantitative values obtained from the measurement recording segments to determine a correlation with the extent of the stenosis.

The threshold value of the *pulsatility index* (PI) lies at around five, and often shows much higher values (up to 10 or 12). In a comprehensive study (of 175 aortoiliac segments) comparing the results to those from intra-arterial blood pressure measurements, Johnston et al. (1983) demonstrated that 5.5 as a threshold value provided the smallest error range in detecting aortoiliac obstructions. In poorly compensated occlusions, the index can sink to a value of one, proximal to the registration point. In our own study (Neuerburg et al. 1991), the pulsatility index values obtained applying ultrasound to the common femoral artery were:

- In normal findings: PI of 8.5 ± 3.5
- Isolated iliac occlusions: PI of 2.8 ± 1.6
- Combined pelvic femoral artery occlusions: PI of 2.3 ± 1
- Isolated subsequent femoral artery occlusions still retained normal values: PI of 6.3 ± 2.6

Only preceding vascular system obstructions are detected with this index; subsequent occlusions are not evaluated.

In practical clinical use, qualitative evaluation of the waveform is sufficient (Fig. 5.**29**), since many results that complement one another are available.

In contrast to the pressure, which is not detrimentally affected by subsequent vascular system obstructions, the Doppler waveform is already altered in the prestenotic segment. The reverse flow component lessens, due to decreased peripheral resistance in the collateral or peripheral vasculature, or both. This recording is not distinguishable from poststenotic recordings, with minimal vascular system stenoses (less than 50%) proximal to the registration point (Fig. 5.**32**).

When occlusions and stenoses of more than 50% occur in isolated form, registering the flow pulses segmentally makes it possible to assign them to specific vascular levels. Occlusions covering several levels are difficult to classify topographically (Fig. 5.**48**).

Examining the circulation to the superficial and deep palmar arterial arches can be important in vascular injuries, and especially when placing a dialysis shunt. The functional clinical *Allen test* has a 13% error rate, and cannot be used in patients suffering from posttraumatic circulatory shock (Fronek 1989). The *palmar test* described by Ruland et al. (1988), which uses the Doppler probe, provides better results.

In this study, the most reliable criterion was flow reversal in the distal radial artery after proximal occlusive compression of the radial artery at the margin of the brachial muscle. Equally practicable was testing the flow direction in the radial artery of the thumb after compressing the radial artery at the typi-

cal pulse palpation point. In contrast, the flow velocity increase in the ulnar artery or radial artery during occlusive compression of the opposite artery produced too wide a variation in the values obtained.

■ Conventional and Color Flow Duplex Sonography

B-mode. The interpretation of wall deposits and plaques follows the same criteria as those discussed above for the carotid artery vascular system (p. 57). Plaques are characterized according to their location, size, and form, as well as their surface, echo structure, and echogenicity (Table 1.**5**, p. 24). However, as mentioned previously, due to the wide variety of possible wall deposits, the characterization of plaques in the long, stretched vascular bed of the extremities is not as significant as it is in the carotid arteries. The possible peripheral microembolization consequences are less important here than they are in the sensitive cerebral parenchyma.

The B-mode is especially useful in detecting aneurysms. They can be precisely located and also classified according to their form and size. Even the thrombotic component of aneurysms can usually be readily detected in the B-mode image. Vascular dilatations, hematomas, and cysts can also be evaluated in B-mode.

Doppler mode. Until now, predominantly *qualitative* Doppler waveform evaluations from the peripheral arteries have been used to estimate the degree of stenosis (Fig. 5.**30**). They make it possible to classify approximately the extent of vascular system stenoses. The following assessment criteria, adapted from Jäger and Landmann (1994), are recommended:

Normal findings: Triphasic Doppler curve with narrow frequency band.

1–25% stenosis: Triphasic curve shape, discrete widening of the frequency spectrum, flow velocity not increased.

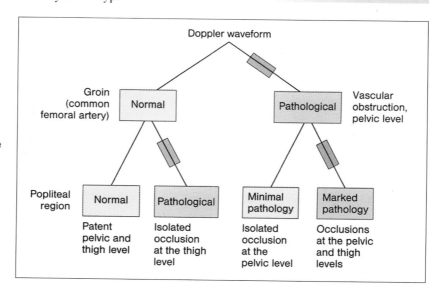

Fig. 5.**48** Determining the level of hemodynamically significant vascular system obstructions using directional continuous wave Doppler sonography. Formal analysis of the flow velocity waveforms at the usual registration points allows qualitative assessment of the presence of a high-grade stenosis or occlusion proximal to the registration point. Comparative evaluation of the Doppler waveforms at two typical registration points allows a topographical diagnosis

25–50% stenosis: Triphasic curve shape, broadening of the frequency spectrum, flow velocity increased >*30%*.

50–75% stenosis: Loss of reverseflow component, so-called systolic window filled, flow velocity increased some 100–200%.

75–99% stenosis: High diastolic flow, strong-amplitude signals around the zero line under the systolic peak. Usually an increase in flow velocity >300% (only filiform stenoses produce a reduction in the velocity again).

Stenosis	V (cm/s)	Quotient
> 50%	120	1.4
> 70%	160	2.0
> 90%	180	2.9

These curves are constructed by plotting the sensitivity of the test against its specificity at various thresholds. Usually the value of the optimal relationship between test sensitivity and specificity is the one chosen (Sumner et al. 1993).

However, these values are lower than in most other studies, in which stenoses of more than 50% indicate a threshold value of 180 cm/s for peak systolic velocity, and a ratio of 2.5. This agrees with in vivo results obtained by Whymann et al. (1993), who obtained a figure of 2.67 for the peak velocity ratio (PVR) for stenoses of more than 50%.

Color flow Doppler mode. As mentioned above, the color flow Doppler mode has become an indispensable aid in locating stenoses and determining the length of an occlusion. Color flow Doppler sonography can also be used to detect a false aneurysm following femoral artery punctures. Pendular flow in the leak can be impressively depicted using color, and compression therapy can be applied specifically under color control (pp. 170–174).

Of course, it is also possible to calculate the pulsatility index with pulsed wave Doppler mode. However, this does not have the same value as in continuous wave Doppler sonography, since imaging techniques make it possible to measure the stenotic area directly.

Quantitatively, the peak systolic velocity values and the ratio of the maximum velocity within and 2 cm proximal (or distal) to the stenosis are usually used, the peak velocity ratio (PVR). Using ROC curves, Sacks et al. (1992) determined the following threshold values in stenoses of more than 50%:

Sources of Error and Diagnostic Effectiveness

Sources of Error

■ Continuous Wave Doppler Sonography

Incident angle of ultrasound. Independent of the evaluation method used (analogue registration or spectral analysis), the most important error source in continuous wave Doppler sonography is the fact that the ultrasound application angle is not known. At an acute incident angle of less than 30° a flow acceleration is simulated: at an angle of more than 60°, the amplitude height of the Doppler waveform is reduced; at a 90° angle, no signal is registered.

Identifying the arteries. Properly assigning the artery signal can pose difficulties; not every signal that is located belongs to the artery sought. The anatomical course, the neighboring arteries, the side branches, and possibly also the collaterals, have to be known.

In regions in which peripheral compression maneuvers are possible (oscillating vibration, suprasystolic arterial compression), identification is easier (e.g., subclavian artery, brachial artery). In the thigh, for example, it is not possible to apply ultrasound separately to the superficial femoral artery and the deep femoral artery.

Vascular diameter. The vascular diameter is not known, so that the flow velocity can be slightly disturbed or somewhat slower in a wide or pathologically enlarged blood vessel. In a narrow blood vessel, the flow velocity increases.

Hematoma, obesity. Particularly in the groin region, applying ultrasound can be more difficult due to a hematoma (following angiography), or in cases of massive obesity.

Superimposed veins (popliteal region) can also complicate the identification of the arteries.

■ Conventional and Color Flow Duplex Sonography

Methodological Sources of Error

Transducer/transmission frequency. Choosing an unsuitable transducer or an inadequate transmission frequency can have a significant adverse effect on the examination quality.

The transmission frequency, adjusted for the depth of the ultrasound application, has to be observed without fail (see the section on examination, p. 146).

Sectional plane. When an unfavorable section is used, plaques may be missed, or an incorrect diameter value may be measured (Fig. 2.**30,** p. 58).

Echogenicity. Thrombotic material in the vasculature can have the same acoustic impedance as the blood. In the B-mode image, the blood vessel or the aneurysm appears to be patent. It is only when using the Doppler mode (with or without color coding) that stenoses, occlusions, or thrombotic components can be detected. It is also important to note vascular wall pulsations.

Doppler mode. Positioning the Doppler mode at the necessary angle of less than 60° can present problems, especially in the pelvic region. The vasculature does not have a stretched course here. Guiding the probe is more difficult, and not as easily varied as it is in the extremities.

Color flow Doppler mode. The following fundamental problems associated with this procedure should be noted.

Flow Velocity

A mean flow velocity is color-coded, and this only provides qualitative and semiquantitative information.

Angle Dependence

When using color, the angle of the many pulsed wave Doppler sample volumes used cannot be adequately adjusted for changes in the vascular course. This means that different color intensities do not always indicate stenoses, but may also be due to an alteration in the incident angle of the ultrasound in relation to the vascular axis.

Color Change

When changing the true flow direction in relation to the transducer, the sudden color change proceeds through black. In the aliasing phenomenon, the color change proceeds through white, since it takes place at high frequencies.

Perivascular Color Artifacts

This phenomenon entails color clouds consisting of multicolored pixels, which can be projected onto the tissue surrounding stenotic vessels. Widder (1993) describes this as a "confetti effect." Middelton et al. (1989a) discovered that these artifacts vary with the heartbeat, and are most prominent during systole (less noticeable during diastole). The authors assume that turbulent flow conducts the vibrations into the perivascular tissue, and there causes Doppler shifts that are reproduced in color. They mainly observed these phenomena in arteriovenous fistulas, hemodialysis shunt stenoses, and also in renal artery stenoses, pelvic stenoses, and stenoses in the leg arteries.

Other Sources of Error

Rest period. An adequate rest period is necessary, both in healthy patients and in pathological cases. After physical exertion, the flow velocity increases while the peripheral resistance decreases, and the resulting monophasic curve simulates a preceding vascular system obstruction.

Noting positional anomalies. Positional anomalies, which are especially frequent at the origin of the deep femoral artery and its branches, must be known. A distal deep femoral artery occlusion may be missed. When there is a superficial femoral artery occlusion, the deep femoral artery, which then assumes a steep position, is often interpreted as being the superficial femoral artery, and its side branches are viewed as the deep femoral artery. Loop formations affecting the pelvic arteries should not be misinterpreted as an occlusion. The points of origin of the lower leg arteries also have variants. Specifically, the anterior tibial artery can originate from the popliteal artery at varying levels.

Classifying space-occupying lesions. When using *B-mode imaging* alone, it is not easy to differentiate between a hematoma, a false aneurysm, and an arteriovenous fistula. Complications occurring after procedures in the groin region, which are being carried out more and more often, can only be verified using differential diagnosis with the Doppler mode, or even better using the *color flow Doppler mode.* Depicting the fistula canal (arteriovenous fistula, false aneurysm), in particular, is much easier when the inflow into the narrow channel can be color-coded.

Recognizing and distinguishing occlusion and stenosis. A stenosis can be falsely deduced:

- When a collateral is stenotic at its origin, and is characterized as being the main blood vessel.
- When the grading of a stenosis proceeds according to the morphological image, which almost always proves to be misleading.

Grading the stenosis is made more difficult by preceding or subsequent vascular system obstructions. If a stenosis is apparent in the area of the bifurcation, the peak velocity ratio (PVR) can only be calculated using the peak flow velocity distal to the stenosis.

Measuring occlusion length. The occlusion length can be *underestimated* when proximal or distal collaterals in the vascular course simulate a patent vascular system.

The occlusion length can be *overestimated:*

- When plaques in the proximal or distal occlusive segment extinguish the color through ultrasound shadows; and/or
- When a significantly reduced postocclusive flow velocity can no longer be color-coded.

Diagnostic Effectiveness

■ Continuous Wave Doppler Sonography

With directional continuous wave Doppler sonography it is possible to register flow velocity curves—the visual, *qualitative evaluation* of which makes it possible to assess the presence of preceding or subsequent vascular system obstructions.

Registering the so-called *groin pulse curve* is particularly important, since cuff methods cannot be used in the pelvic region, and quantitative pressure measurements are therefore not possible.

Segmental flow velocity curves are valuable in *diagnosing the level* of vascular system stenoses. The examination can be carried out more quickly than segmental pressure measurement.

A normally configured triphasic flow pulse in the groin (common femoral artery), or at other registration points, excludes the presence of a hemodynamically active flow obstruction in the preceding afferent arteries.

The grade of preceding vascular system obstructions can be estimated by formal analysis (Fig. 5.**32**). Using visual interpretation in a prospective study, Walton et al. (1984) demonstrated an 87% sensitivity in occlusions and stenoses with a lumen obstruction of more than 50% in the aortoiliac region. By contrast, stenoses of less than 50% cannot be distinguished from subsequent femoral artery occlusions using formal analysis (Neuerburg-Heusler et al. 1981).

According to comparison of several indices completed by Humphries et al. (1980), all individually evaluated *quantitative* measurements that were employed, i.e., parameters such as the pulsatility index, the "damping factor," the pulse wave duration time, and the pulse increase time, failed to distinguish between normal and low-grade pathological findings. By contrast, FitzGerald and Carr (1977), combining several indices, were able to obtain a sensitivity of 93% in detecting obstructions in 187 arterial segments.

When examining the aortoiliac level alone, Nicolaides et al. (1976) arrived at a sensitivity of 86%. For the pulsatility index, Johnston et al. (1984) measured a sensitivity of 96% when no subsequent vascular system obstruction was present. When the femoropopliteal area was also affected, the sensitivity dropped to 83%. According to this study, in which several indices were compared, the pulsatility index is the value that has the best hit rate and a high specificity of 98%.

Our own results evaluating the PI (pulsatility index), obtained from different sets of findings, showed that for isolated stenoses in the pelvic arteries, a sensitivity of 80% is possible, while for occlusions at two levels a sensitivity of 100% could be obtained. Subsequent femoral artery occlusions in a patent pelvic vascular system could only be suspected with a sensitivity of 41%. The total overall specificity was 88%. In this study, the pulsatility index of 5.0 was used as the threshold value between normal and pathological findings (Neuerburg et al. 1989).

Using 4.0 as the threshold value, Thiele et al. (1983) attained a sensitivity of 94% and a specificity of 82% for isolated aortoiliac obstructions. When there were occlusions at two levels, the corresponding values were 99% and 45%, respectively.

Conventional and Color Flow Duplex Sonography

In contrast to the arteries supplying the brain, many noninvasive procedures for diagnosing vascular system obstructions in the arteries supplying the extremities have already been available for a long time. These allow both qualitative and quantitative evaluation of the hemodynamic effects of vascular system stenoses. Locating occlusions and stenoses with respect to the vascular level is possible both with clinical methods and with equipment. Segmental pressure measurements using the ultrasound Doppler method (p. 135) and directional Doppler sonography (see above) have increasingly come to predominate.

A gap existed with regard to conclusive *differentiation* between *occlusion* and *stenosis, localization* of occlusions and stenoses within the vascular levels, and *measuring the length* of an occlusion. An additional weak point shared by all previous procedures was differentiation between isolated pelvic occlusions and *combined* pelvic and thigh *occlusions*.

Conventional duplex sonography, and particularly the color flow method, have now closed this gap. Although *conventional duplex sonography* with careful arterial scanning already made it possible to differentiate between occlusion and stenosis, confirmed the location of vascular system obstructions, and allowed semiquantitative stenosis grading (Jäger et al. 1985, Kohler et al. 1987), this method required considerable time to carry out, and this was an obstacle to its widespread use.

Using *color flow duplex sonography* has made following the arteries easier in varying degrees. Color-guided sample volume placement in the stenotic areas has simplified the identification of the highest-velocity region within the stenosis. In addition, the color interruption has made it possible to recognize occlusions and determine their extent more quickly.

Various studies have confirmed the validity of examinations using duplex sonography, both with and without color (Table 5.**10**). Comparisons between conventional and color flow duplex sonography provide recognizable, but not marked, improvements in the findings, due to the color flow Doppler mode. Jäger et al. (1985) and Kohler et al. (1987a) were able to obtain very good results using conventional duplex sonography (sensitivity 77% and 98%, respectively, specificity 81% and 92%, respectively). They used a simplified grading process with spectral analysis patterns (Fig. 5.**30**) that is still very useful for a rapid diagnosis.

More comprehensive studies using color flow duplex sonography, such as those carried out by Cossman et al. (1989), Koennecke et al. (1989), Landwehr et al. (1990 a), and Polak et al. (1990), produced sensitivity figures of between 87% and 97%, and had an outstanding specificity lying between 95% and 99%. Increasingly, the validity of the method for individual vascular segments when using duplex sonography is also being assessed (Table 5.**11**). A limiting factor is that almost all studies evaluate stenoses and occlu-

Table 5.**10**   The validity of angiographically controlled studies using conventional and color flow duplex sonography to diagnose various grades of stenosis and occlusions of pelvic and leg arteries

Authors	Patients (n)	Segments (n)	Sensitivity (%)	Specificity (%)	Threshold value of measurement procedure
Conventional duplex sonography					
Jäger et al. (1985)[1]	30	338	77	98	} Spektrum qualitativ
Kohler et al. (1987)[1]	32	393	82	92	} PSV ↑ >100%
Color flow duplex sonography					
● **Stenoses >50% + occlusions**					
Cossman et al. (1989)[2]	61	629	87	99	PSV > 200 cm/s
Koennecke et al. (1989)[3]	53	344	97	96	PSV ↑ 30–100%
Landwehr et al. (1990)[3]	52	132	92	99	qualitativ
Polak et al. (1990)[1]	17	238	88	95	PSV ↑ >100%
Legemate et al. (1991)[4]	61	918	84	96	PVR ≥2.5
Ranke et al. (1992)[1]	62	121	87	94	PVR >2.4
	62	121	66	80	PSV ↑ >180 cm/s
Strauss et al. (1993)[1]	598	1460	88	86	PSV ↑ >100%
● **Occlusions**					
Cossman et al. (1989)[5]	61	560	81	99	–
Legemate et al. (1991)[5]	61	105	91	99	–
Karasch et al. (1993)[5]	94	150	98	–	–

[1] Common femoral artery obstructions to and including the distal popliteal artery >50%
[2] Iliac artery obstruction to and including the trifurcation >30%
[3] Iliac artery obstructions to and including two lower leg arteries >20%
[4] Aorta to and including the popliteal artery >50%
[5] Femoropopliteal occlusions
PSV = peak systolic velocity; PVR = peak velocity ratio

Table 5.**11** The validity of angiographically controlled studies using conventional and color flow duplex sonography to diagnose stenoses of more than 50% and occlusions of different arterial segments

| Arteries (segments) | Conventional duplex | | | | Color flow duplex | | | | | | | |
| | Jäger (1985) (n = 338) | | Kohler (1987) (n = 393) | | Cossman (1989) (n = 629) | | Koennecke (1989) (n = 344)* | | Legemate (1991) (n = 918) | | Moneta (1992) (n = 286) | |
	Sensitivity (%)	Specificity (%)	Sensitivity (%)	Specificity (%)	Sensitivity (%)	Specificity (%)	Sensitivity (%)	Specificity (%)	Sensitivity (%)	Specificity (%)	Sensitivity (%)	Specificity (%)
Obstructions >50%												
Iliac	81	100	89	90	81	98	96	98	79/92[2]	94/96[2]	89	99
Common femoral	56	46	67	98	80	100	–	–	57	98	76	99
Superficial femoral	76	97	84	93	91	96	98	100	73/100	96/98	87	98
Deep femoral	86	100	67	81	80	100	–	–	71	94	83	97
Popliteal	80	100	75	97	90	100	88	98	75	93	67	99
Lower leg	–	–	–	–	–	–	95/94[1]	92	–	–	90	93/92[1]
All segments	77	98	82	92	87	99	97	96	84	96	–	–

[1] Anterior tibial artery/posterior tibial artery
[2] Common iliac artery/external
[3] Proximal superficial femoral artery, middle, distal
* Obstruction of all grades

sions that are over 50%. In differentiating stenoses of less than 50% from those of more than 50%, there are certain methodological inaccuracies involved, both when using angiography and also when applying duplex sonography.

The diagnostic validity of Duplex sonography increasingly has to be measured against another noninvasive examination procedure, namely magnetic resonance arteriography (MRA). However, investigations completed by Mulligan et al. (1991) and Baumgartner et al. (1993) show that duplex sonography is clearly superior. Correlation with the angiographic findings produced the following results:

	Mulligan et al. (1991)	Baumgartner et al. (1993)
MRA	71%	81%
Duplex sonogram	93%	88%

Stenosis Grading

Currently, there are still no unanimously accepted criteria for grading a stenosis; this contrasts with the extremely exact stenosis grade determination in the cerebral arteries (Widder et al. 1986). In general, a doubling of the maximum systolic velocity is interpreted as indicating a stenosis of more than 50% (Moneta et al. 1993). By contrast, Leng et al. (1993) believed that one can assume a stenosis of more than 50% only when the peak velocity has undergone an increase of over 300%. Allard et al. (1991), who evaluated many velocity waveform parameters in 379 segments, arrived at a sensitivity of only 50%, with a specificity of 98%.

In a paper submitted by Sacks et al. (1992), an attempt was made to distinguish between 50%, 70%, and 90% stenoses by evaluating 558 segments using a quotient formed from the flow velocity 2.0 cm proximal to the stenosis and the flow velocity within the stenosis itself the so-called *peak velocity ratio* (PVR). The results obtained (stenosis over 50%: 71% sensitivity, stenosis over 70%: 80% sensitivity, stenosis over 90%: 85% sensitivity, with a specificity of between 93% and 97%) are not yet comparable to those from the carotid artery vascular system. Ranke et al. (1992) evaluated the sensitivity and specificity of stenosis grading in 10% increments between 20% and 90%, and found sensitivities between 84% and 96% with specificities ranging from 75% to 97%. They examined the peak systolic velocity within the stenosis and also the peak velocity ratio in relation to the velocity both in the proximal and the distal segment (Table 5.**10**). However, as mentioned above, angiographic evaluation of the degree of a stenosis as a gold standard is problematic, since imaging of the pelvic and leg regions is usually only practicable in a single plane.

In a new study by Smet et al. (1996), various parameters for the Doppler curve (PSV, PSV ratio, PSV difference, EDV) were investigated in 112 aortoiliac segments in comparison with arteriographic diameter reduction and invasive blood pressure measurement. The best result was achieved with the PSV ratio in detecting aortoiliac stenoses of more than 50%, with a threshold value of 2.8, giving a sensitivity of 86% and a specificity of 84%. Arteriographic stenoses of more than 75% were measured by a PSV ratio of 5, although with a lower sensitivity of 65% and a higher specificity of 91%.

A different method of measuring the stenosis grade is to *calculate the pressure gradient* according to the Bernoulli equation (analogous to measuring the pressure gradient at the aortic valve). However, this equation can only be used in larger arteries, such as the iliac vascular system. Kohler et al. (1987b) experimentally found a close correlation with invasive blood pressure measurements, although this could not be reproduced in clinical studies they conducted.

By contrast, Langsfeld et al. (1988) were able to attain a correlation coefficient of r = 0.9 in 11 patients. They believe that this method provides the best correlation, especially for higher-grade stenoses with a peak velocity of between 2 m/s and 4.5 m/s. In 28 patients, Strauss et al. (1993a) were able to attain a correlation coefficient of r = 0.8 when comparing the mean pressure decrease through stenoses. Legemate et al. (1993) evaluated the sensitivity and specificity of this procedure with 81% and 88% values, respectively, for hemodynamically significant stenoses. In addition to measuring the pressure gradient according to Bernoulli, they also compared the difference between the maximum systolic velocity in the stenosis and the lowest maximum velocity proximal or distal to the stenosis. For this parameter, they were able to obtain the same sensitivity and specificity, namely 81% and 88% respectively. In this study, the correlation to the invasively measured pressures was r = 0.62 for the Bernoulli measurement and r = 0.72 for the maximum velocity measurement.

It seems that the final word has not yet been said on the topic of stenosis grading, however, the most recent studies indicate an increasing tendency to grade stenoses according to the peak velocity ratio.

Determining Occlusion Length

In some studies, separate evaluation of the *differentiation between stenosis and occlusion* is not carried out. However, particularly in relation to therapeutic lumen recanalization techniques such as angioplasty, information regarding the exact location and length of an occlusion is very significant. In some studies, detecting occlusions was successful with a sensitivity of 81–98% and a specificity of 99% (Table 5.**10**).

A more recent study of ours, which evaluated 100 extremities in 94 patients, showed that determining the *femoral artery occlusion* length with color flow duplex sonography gave a correlation coefficient of r = 0.95, fully comparable to the results obtained with angiography (Karasch et al. 1993) (Fig. 5.**49**). Segmental classification of the location of the occlusion was also successful in 95% of the cases. It is expected that purely diagnostic angiographic procedures will increasingly lose their importance, or that they will only be selectively used in combination with angioplasty.

Aneurysms

A further area of application for duplex sonography is in recognizing and differentiating *aneurysms;* differentiating a true aneurysm from a false aneurysm is clearly successful with color coding (pp. 170–171). The incidence *of false aneurysms* after diagnostic and therapeutic catheterization procedures has been examined in various larger studies using color flow duplex sonography. In 1120 patients following cardiac catheterization examinations, Moll et al. (1991) found a false aneurysm in 4% of the cases. After diagnostic catheterization procedures, Habscheid and Landwehr (1989) discovered pseudoaneurysms in 5% of the cases. After 144 coronary angioplasties, eight aneurysms (6%) were diagnosed by Kresowik et al. (1991). When examining symptomatic patients in whom femoral artery lesions were suspected, Gross-Fengels et al. (1987) discovered 13 aneurysms in 30 patients (43%), and Steinkamp et al. (1992) found 46 aneurysms in 132 patients (35%).

The sensitivity and specificity of duplex sonography in detecting a false aneurysm are reported in an earlier study by Coughlin and Paushter (1988) as 94% and 97%, respectively. Helvie et al. (1988) detected pseudoaneurysms in 60 retrospectively examined patients, with a sensitivity of 95% and a specificity of 94%.

Ultrasound-guided compression closure of pseudoaneurysms, first introduced by Fellmeth et al. (1991), is now in widespread use, although in some studies (Kresowik et al. 1991, Johns et al. 1991), spontaneous thromboses without any subsequent complications were observed. The success rate of compression therapy was found by Sorrell et al. (1993) to be 91%.

In a prospective long-term study, Hajarizadek et al. (1995) achieved an overall success rate of 95%. During an 18-month follow-up period, there were no late recurrences or significant changes in ankle brachial pressure. Even with uninterrupted anticoagulation, Dean et al. (1996) achieved a satisfactory success rate of 73%.

Angioplasty

Changes in vascular diameter and in the vessel walls are increasingly being examined post angioplasty directly and in long-term studies using the duplex procedure. In the study by Mewissen (1992), the open lumen rate after 6–12 months in patients with normal findings after angioplasty was 84%, while only 15% with a velocity ratio ≥ 2.0 after PTA showed no occlusions or residual stenoses. In the study by Sacks et al. (1994), the open lumen rates with a velocity ratio of less than 2 and with a velocity ratio of 2 or more were 54% and 74%, respectively. Henderson et al. (1994) showed that in 98% of 49 lesions, a significant increase in lumen diameter was seen immediately after angioplasty. The overall diameter of the vessel, including the vessel wall, was increased in 87.7%. The mean change in the lumen diameter was 2.1 mm, and the

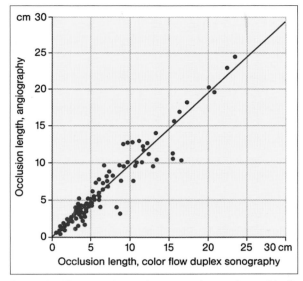

Fig. 5.**49** The correlation between the angiographically measured and sonographically (color flow duplex mode) measured length (in cm) of an occlusion in the superficial femoral artery or the popliteal artery, based on data from 98 lower extremities. y = 0.96 × + 0.27; correlation coefficient r = 0.95; significance level p = 0.001

mean change in the overall vessel diameter was 1.6 mm; after six months, these figures were 1.1 mm and 1.2 mm, respectively.

In a recent study by Spijkerboer et al. (1996) on iliac stenoses, no significant difference was found between PTA results with "residual stenoses" ≥ 2.5 PVR) and those of 2.5 PVR or less, although the boundary value was set fairly high. A large proportion of the residual stenoses showed significant improvement within one year of PTA—an observation that had already been made in angiographic and clinical checkups before the duplex period. Baumgartner et al. (1996) also confirmed that a residual stenosis of less than 50% after endovascular treatment was not associated with an increased risk of restenosis during the follow-up period, but they found that a hypoechogenic endoluminal wall thickening immediately after endovascular procedures appears to be predictor of restenosis.

Vascular Prostheses and Bypasses

The patency of bypasses and vascular prostheses, and also the prediction of *bypass reocclusions,* have been examined in several studies using simple continuous wave Doppler sonography. Bandyk et al. discovered in 1985 that a reduction of the maximum systolic velocity to less than 40 cm/s indicates an impending bypass occlusion. In addition, they found in a later study (1988) that a transformation of the triphasic Doppler waveform into a monophasic configuration, when related to a decreased velocity, indicates proximal or distal segmental stenoses between 50% and 75%. The staccato configuration of the Doppler waveform with low physiological flow and pronounced reverse flow, was viewed as being an additional criterion associated with distal bypass stenoses.

Grigg et al. (1988) viewed a very low velocity ratio of 1.5 as being the threshold value for detecting stenoses of more than 50% in bypasses. In 14 patients with 92 graft segments, Polak et al. (1990 a) documented a 95% sensitivity and a 100% specificity for detecting and locating stenoses using color flow duplex sonography; a peak velocity ratio of more than 2.0 served as the threshold value.

In a prospective study involving 54 patients with 62 bypasses, Trattnig et al. (1992) were able to detect a bypass stenosis by purely qualitative optical means, using the lightening of the color coding. These angiographically controlled cases produced a sensitivity of 92% and a specificity of 100%, but without determining the stenosis grade.

More *recent follow-up studies* conducted by Idu et al. (1993) and Mattos et al. (1993), using color flow duplex sonography, also showed that the increased flow velocity criterion within a stenosis is a more important parameter for indicating imminent bypass occlusions than decreased velocity. Mattos et al. (1993) were able to demonstrate that flow velocity doubling, or a peak velocity ratio of more than 2.0, requires a bypass revision. They also showed that the patency rates of revised stenoses in comparison with stenoses that had not undergone revision were 88% and 63%, respectively.

A similar observation was made by Idu et al. (1993) in a follow-up study over five years. In evaluating stenoses of more than 50% with color flow duplex sonography, it was shown that bypass occlusions occurred in 57% of the unrevised stenoses, and only in 9% of the revised grafts. With B-mode image and color coding, other postoperative bypass complications, such as aneurysms, hematomas, and seromas can also be quickly diagnosed (Fig. 5.**50**).

The studies mentioned show that color flow duplex sonography has clearly improved the potential for noninvasive diagnosis of bypass stenoses and impending reocclusions, and that a threshold value for the peak velocity ratio at 2.0 currently appears to be the most reliable parameter for indicating stenoses of more than 50%. Since early bypass revision clearly produces an increased patency rate (Mattos et al. 1993), systematic follow-up bypass evaluations using duplex sonography appear to be indicated postoperatively.

Upper Body Arterial Occlusive Disease

This is clearly less important, since those suffering from peripheral vascular disorders show an affected upper body region in only 5–10% of cases (Edwards and Porter 1993). In the *large arteries,* subclavian artery stenoses and subclavian artery occlusions occur most frequently. The significance of these is discussed along with the cerebral arteries in Chapter 2, pp. 86–88 above. Takayasu's syndrome or other inflammatory vascular diseases can sometimes cause extensive brachial artery stenoses, or other arterial stenoses in the arm (Fig. 11.**38**, pp. 412–413).

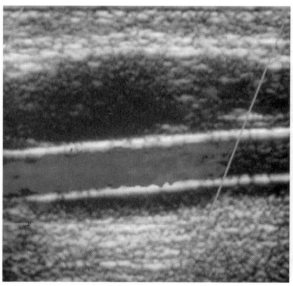

a

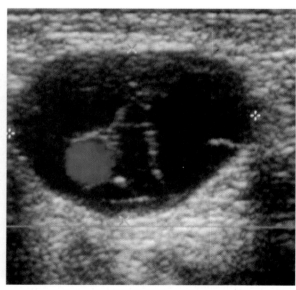

b

Fig. 5.**50** Color flow duplex sonograms of a femoropopliteal bypass, **a** in longitudinal section and **b** in transverse section near the proximal thigh. Around the vascular prosthesis, a hypoechoic to anechoic and partially septated, space-occupying lesion can be seen, with a maximum cross-sectional diameter of 31.5 mm. This corresponds to a postoperative seroma. In the differential diagnosis, a prosthetic infection or an aneurysm following suture dehiscence should be considered

When there are disturbances in the circulation of the *small arteries of the hands,* a distinction needs to be made between a functional Raynaud syndrome and acrovascular occlusions. The latter have so far been diagnosed more often using pulse-registering procedures, or by measuring the pressure in the digital arteries of the hand, and cannot yet be the subject of validation studies using duplex sonography. However, high-resolution transducers allow the imaging of individual digital arteries. As shown by Hübsch and Trattnig (Frühwald and Blackwell 1992), it is easier to detect arteriovenous malformations and hemangiomas in the digital arteries of the hand than occlusions in the small arteries.

Hemodialysis Shunt

The examining of hemodialysis shunts in the *upper extremities* has become very important, especially since the introduction of color flow duplex sonography clearly increased the precision of the assessment. The most important complications are shunt stenoses, thromboses, pseudoaneurysms, and excessive shunt volume.

Tordoir et al. (1989) studied the diagnosis *of stenoses* using the peak systolic frequency, and determined a 10 kHz threshold value for implants and an 8 kHz value for Brescia–Cimino fistulas. Compared to results obtained with angiography, they attained a sensitivity of 79% and a specificity of 84% for Brescia–Cimino fistulas, and figures of 92% and 84% for implant fistulas.

Due to the turbulences that are always present, Kathrein et al. (1989) held that quantifying stenoses was not as important as measuring the shunt volume. They were of the opinion that a revision was not necessary until the shunt volume had fallen to less than 200–300 ml per minute. In a study involving 264 segments in 66 shunts, Landwehr et al. (1990 b) arrived at a sensitivity of 91% and a specificity of 96% for locating and diagnosing vascular system stenoses. However, they quantified the stenosis flow rate (n = 96) before percutaneous transluminal angioplasty (PTA) of the shunt in only 17 cases. Before PTA, the volumes were less than 200 ml/min, and after it they were around 500 ml/min.

Thromboses and *pseudoaneurysms* are further complications in addition to stenoses. Middelton et al. (1989 b) were able to detect these with sensitivities of 87% for stenoses, 100% for thromboses, and 95% for pseudoaneurysms. Dousset et al. (1991) also confirmed the simplified diagnosis of shunt complications when using color flow duplex sonography.

6 Veins

Examination

Special Equipment and Documentation

■ Continuous Wave and Pulsed Wave Ultrasound Doppler Equipment

Unidirectional ultrasound Doppler equipment can be used to provide a general diagnostic orientation. Bidirectional Doppler equipment is preferable, however, as it can indicate the flow direction, and can graphically register and document the findings. Pulsed Doppler systems do not have a significant place in routine clinical practice.

The transmission frequency varies according to the region being examined: 4–5 MHz at the groin and the popliteal artery, 8–10 MHz at the ankle. For documentation purposes, it is recommended that flow directed toward the heart should be adjusted to indicate an upward deflection, and that a slow paper speed should be selected.

■ Conventional and Color Flow Duplex Systems

To display the deep veins of the extremities, a 5 MHz linear scanner is preferable. For the vena cava and the iliac veins, a 3.5–5.0 MHz sector scanner provides more effective depth resolution. When the color flow Doppler mode is being used, care should be taken to assure that the flow is coded blue, and that the color scale is selected to detect low flow velocities ("slow flow technique"). Otherwise the color display of the slow venous flow will not be adequate.

There is a limiting factor in the fact that the color coding setting in some color flow duplex equipment simultaneously determines the deflection of the Doppler spectrum direction, so that separate directional change is not possible. In these cases, the preferred documentation type has to be chosen— either the blue venous flow coding, usual in angiology, or the recommended recording format, using the polarity that results in signals directed toward the probe having an upward deflection.

Examination Conditions

Patient and Examiner

The patient's position changes according to the region being examined and the clinical objectives. The examiner should be in a sitting position if possible, in order to ensure stable holding of the probe or transducer with low contact pressure. The procedure described below, which was developed for continuous wave Doppler sonography and is also generally valid for duplex sonography, is recommended (Table 6.**1**).

Deep Pelvic and Leg Veins

For appplications of ultrasound to the *iliac, common femoral,* and *superficial femoral veins,* the patient should be in a supine position, and a slight elevation of the upper body is advantageous, as it causes better venous filling in the extremities (Fig. 6.**1**).

For examining the *popliteal vein,* the prone position is preferable; the ankles should be slightly elevated using a suitable support. The patient's position can be varied depending on the objectives of the examination and the clinical findings. Examining the popliteal vein is also possible with the patient in the supine or lateral position, or sitting (Fig. 6.**2**). The supine position is recommended mainly for bedridden and elderly patients, for whom the prone position is difficult. Due to the improved display, the *lower leg veins* should be examined while the patient is seated, or with the leg held lower.

Superficial Leg Veins

For examination of the *great saphenous vein* with its branches, and the *small saphenous vein,* the patient should be standing (Fig. 6.**3**). If necessary, the examination can also be carried out with the patient supine. The leg being examined should not be carrying any body weight. In contrast, the *perforating veins* are better examined with the patient seated (Fig. 6.**6**).

Table 6.**1** Examination conditions for applying ultrasound to the venous system of the lower extremity using continuous wave Doppler sonography

Venous system	Patient position	Application region	Provocation maneuver
Deep veins			**Compression**
Iliac veins	Supine	Groin	Thigh
Femoral vein	Supine	Groin/thigh	Thigh (Valsalva maneuver)
Popliteal vein	Prone (with the ankle supported on a pillow)	Popliteal region	Calf
Lower leg veins	Standing or seated	Ankle	Sole of the foot, possibly lower leg
Superficial veins	Standing	Along anatomical course	**Valsalva maneuver**
Great and small saphenous veins			
Perforating veins			**Compression**
Dodd–thigh	Seated	Above the anatomical region or "blow-out"	Distal/proximal, using tourniquets
Boyd–proximal lower leg			
Cockett–distal lower leg			

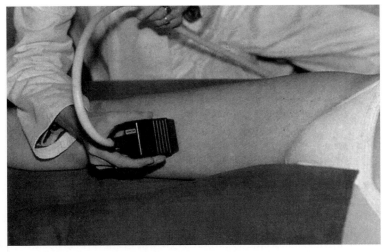

a

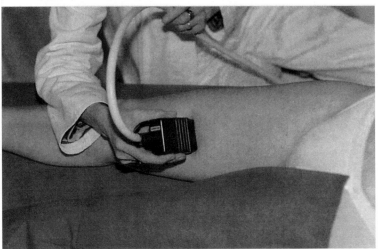

b

Fig. 6.**1** Examination of the superficial femoral vein, with the patient supine, in transverse section with the transducer of a duplex apparatus. Compression maneuver to exclude or detect a thrombosis **a** without, **b** with compression

Upper Extremity Veins

The patient should be in the supine position. In this position, the subclavian vein, axillary vein, and brachial vein can be located; the examiner sits next to or behind the patient. The deep and superficial veins of the forearm can also be examined with the patient seated (Talbot 1986).

Conducting the Examination

Pelvic and Leg Veins

■ Continuous Wave Doppler Sonography

The spontaneous venous signal is characterized by respiration-dependent modulations, which sound like the howling of wind, or roaring waves. Usually, the vein is found directly medial to the artery, at the typical palpation points of the arteries of the extremities. In the popliteal region, the vein lies dorsal to the artery, which means it is closer to the transducer when the patient is in the prone position (p. 200).

The various respiration and provocation tests are summarized in Table 6.**2**.

Spontaneous Respiration (S-sounds)

This term describes the observation and recording of the spontaneous venous flow. The phasic increase and decrease of the venous flow during quiet respiration is detected acoustically, and an appropriately modulated flow directed toward the heart is registered.

During inspiration, the blood flow in the veins of the lower extremities is minimal; on expiration, blood flows more intensely toward the heart. In the upper extremities, the opposite sequence of events occurs. During inspiration, blood flow into the thorax increases, and on expiration it decreases (Fig. 6.**12**).

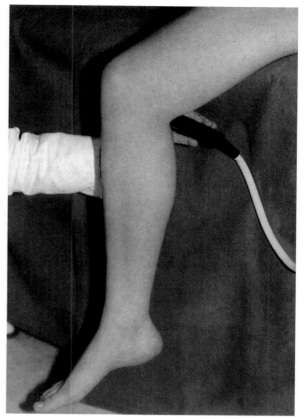

Fig. 6.**2** Applying duplex transducer to the popliteal vein at the posterior aspect of the knee, in transverse section, with the patient sitting

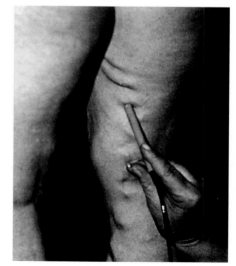

Fig. 6.**3** Using continuous wave Doppler sonography to examine the great saphenous vein at the distal thigh, with the patient standing. The clear, varicose dilatation of the superficial vein is easily recognized

Table 6.**2** Functional testing of venous flow while registering Doppler curves

Spontaneous signals
S-sounds (spontaneous sounds)
  Quiet respiration

Provoked signals
A-sounds (augmented sounds)

  Maneuvers during respiration
  – Deep respiration
  – Valsalva's maneuver

  Manual signal amplification
  – Distal compression/decompression
  – Proximal compression/decompression
  – Modulation
  – Tourniquets

Augmented Sounds (A-sounds)

Increased Respiration

During deepened but not forced respiration, an almost complete cessation of blood flow occurs in the late inspiratory phase. Expiration causes an increase in venous flow directed toward the heart. In the presence of venous thrombosis, continuous flow is registered, due to changed pressure-flow relationships (pp. 207–208).

Valsalva's Maneuver

The Valsalva maneuver involves exerting strong pressure while the patient's mouth is closed after deep inspiration. This causes increased thoracic and abdominal pressure and inflow congestion into the right heart. The maneuver causes the centripetal venous flow to cease (Fig. 6.**4**). When the venous valves, especially those of the groin, are insufficient, a centrifugal reflux of varying duration occurs in the extremities, due to the marked abdominal pressure elevation. Immediately after the maneuver has been completed, an elevated, reactive flow towards the heart is registered (Fig. 6.**17b**).

Valsalva's maneuver should be practiced with the patient first, since patients often do not exert the required sudden lower abdominal pressure in the optimal way.

Manual Signal Amplification

Signal augmentation, the so-called A-sound (augmented sound), refers to rapid manual compression executed distal or proximal to the registration area, or both.

Sometimes, veins peripheral to the popliteal artery are difficult to locate. When compression is applied *distal* to the registration point, the venous signal is amplified, making it easier to locate the venous flow. Normally, brief flow peaks toward the heart are registered, due to distal compression. If the valves are insufficient, peripheral reflux results from the release of the compression.

Proximal compression is used, as is Valsalva's maneuver, to test valvular function. Sufficient valves produce a cessation of blood flow during compression, and subsequently, during decompression, cause increased flow toward the heart (Fig. 6.**5**). Insufficient valves produce distal reflux.

Venous modulation is a variation of signal augmentation. It plays a role in locating the *superficial* veins when using the continuous wave Doppler probe. Within the anatomical course of a superficial vein, testing is usually carried out distal to a registration point to see whether the signals are propagated with quick, repetitive modulation (Straub and Ludwig

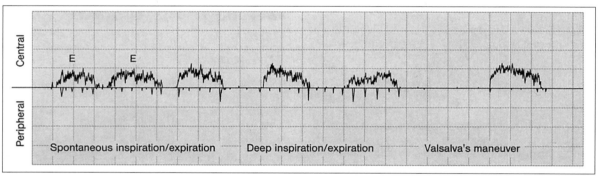

Fig. 6.**4** A recording of the continuous wave Doppler waveform in a healthy individual at the popliteal vein during spontaneous respiration, deep (forced) respiration, and the

Valsalva maneuver. Venous flow cessation is seen, due to elevated intra-abdominal pressure following forced respiration into the abdominal cavity

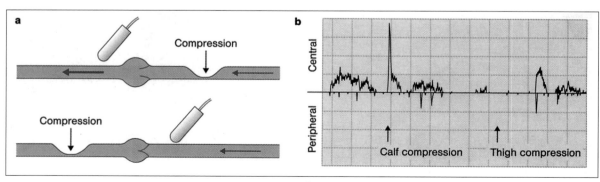

Fig. 6.**5** Distal and proximal compression maneuver (A-sounds) in a normal patient
a The venous valve is open during distal compression, and closes during proximal compression to prevent retrograde flow into the periphery

b The Doppler waveform of the popliteal vein shows peak flow during distal calf compression and flow cessation during proximal thigh compression, with a subsequent, brief peak flow toward the heart

1990). Using modulation, which usually occurs at an anatomically known location, the course of a vein or varix can be easily followed, and its patency can be evaluated.

Testing for Insufficiency of Perforating Veins

With the patient seated, the dangling leg is examined. Proximal and distal to the suspected area (blow-out, or known anatomical region), a tourniquet is placed to compress the superficial varices. Subsequently, the tissue distal or proximal to the tourniquet, or both, is compressed. Insufficient perforating vein valves cause a reflux that flows out through the deep veins via the perforating vein (flow toward the probe), subsequently producing internally directed, retrograde suction (flow away from the probe) (Bjordal 1981). Proximal compression is only effective when there are no valves or insufficient valves between the compression area and the perforating vein. The tourniquet obstructs the propagation of the venous flow through the epifascial varices. If the perforating veins are intact, then neither distal nor proximal compression will subsequently result in flow toward the probe (Fig 6.**6**).

■ Conventional and Color Flow Duplex Sonography

B-mode. In a transverse section, the veins are displayed in a round or oval cross-section, which increases by about 50–100% when the Valsalva maneuver is applied.

In longitudinal section, the venous lumen is anechoic, and the wall cannot be distinguished consistently. There is no arterial pulsation. However, there is visible modulation due to respiration. A patent vein can be completely compressed when pressure is applied with the transducer, while the accompanying pulsating artery cannot be compressed by the same pressure (6.**13**). In the presence of an acute thrombosis, the diameter of the vein is usually clearly larger than that of the artery, and compression is not effective, or only partially so. Sometimes the thrombus can be recognized by its greater echo intensity. In longitudinal section, the head of a thrombus may be clearly distinguishable from the patent vascular lumen (Fig. 6.**21**).

Doppler mode. With the adjusted Doppler mode, reactions to respiratory maneuvers, to deliberate pressure exertion, and also signal augmentation are ex-

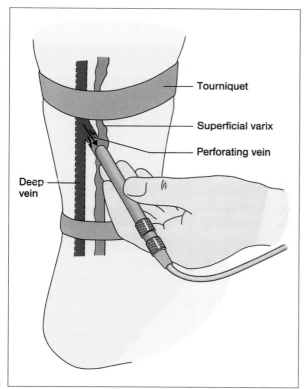

a

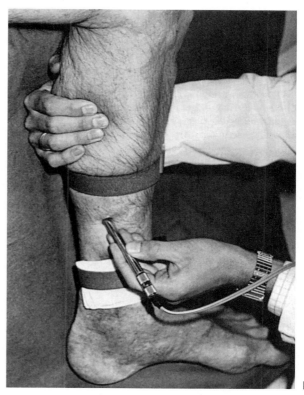

b

Fig. 6.6 Testing for insufficiency of the perforating veins
a The probe is placed in the direction of the perforating vein. During distal or proximal compression, venous flow is registered through the insufficient valve of the perforating vein toward the surface and approaching the probe. The superficial varices have to be compressed with tourniquets so that epifascial venous flow is not incorrectly registered

b Conducting the examination with the patient sitting. Proximal compression is used here, because the patient has a postthrombotic syndrome, with insufficient deep venous valves. If the deep venous system is intact, distal compression should be employed. However, this usually results in weaker signals, since only the foot is available for compression

amined in the same manner as previously described for continuous wave Doppler sonography. In addition to the morphological image, functional parameters describing the venous flow can be obtained (Fig. 6.**23**).

Color flow Doppler mode. When diagnosing a thrombus, the location and extent of the thrombotic material are more easily recognized with this method than in a pure B-mode image, since thrombus, especially if it is fresh, has the same echogenicity as blood. In addition, partially thrombosed thrombi, surrounded by flowing blood, or partially recanalized venous segments, can be clearly detected (Fig. 6.**22**).

Respiratory maneuvers, such as cessation of inspiratory flow, and reactions during the Valsalva maneuver and compression maneuvers, can be detected with certainty using the color reproduction dynamics. Venous reflux during the Valsalva maneuver can be recognized as a reversal flow with a color coding change. Turbulent flow patterns are sometimes also present (Fig. 11.**44 d**). Diagnosing insufficient valves in this way is more certain and faster than when using the Doppler mode. As in the arterial system, locating the lower leg veins is significantly easier with color coding. The direction of blood flow in the perforating veins can be detected very easily.

Upper Extremity Veins

■ Continuous Wave Doppler Sonography

Spontaneous Respiration (S-sounds)

Respiratory modulation is observed and registered in the same way as in the lower extremities. However, the phenomena observed are the opposite of those in the legs: during inspiration venous flow increases, and on expiration it decreases (Strandness and Sumner 1972) (Fig. 6.**12 b**).

Augmented Sounds (A-sounds)

Increased Respiration

Venous flow modulation cannot be increased as significantly as it can in the lower extremities. During expiration, the near-cessation of blood flow does not occur.

Valsalva's Maneuver

As in the lower body, the test using forced respiration causes a blood flow cessation and subsequent, typical flow acceleration. If an incomplete thrombotic occlusion is present, only a minimally increased flow velocity can be obtained after Valsalva's maneuver.

Manual Signal Amplification

As in the lower extremities, an increased flow velocity (peak flow) can be acquired by a quick compression distal to the probe position. Proximal compression tests the valve function.

■ Conventional and Color Flow Duplex Sonography

B-mode. As in the lower extremities, the most important criterion in pure B-mode sonography is the *compressibility* of the deep veins (axillary vein, brachial vein), which allows their patency to be evaluated. A patent vein is very easily compressed. Care must therefore be taken to use an externally supported light touch when guiding the transducer; this is particularly true when the epifascial veins are being imaged. Noncompressible segments, which usually have a dilated lumen in comparison with the artery, indicate thrombosis or thrombophlebitis. In addition, the latter is often recognizable through increased intraluminal echoes.

Doppler mode. As described above under continuous wave Doppler sonography, the Doppler mode can be used to examine venous flow specifically, with its phasic modulation and its reactions to provocation tests. The veins should be imaged in longitudinal section, so that a favorable incident ultrasound angle can be adjusted for the Doppler sample volume. Particularly in regions in which venous compressibility is uncertain or cannot be assessed at all, as in the subclavian artery below the clavicle, it is indispensable to register Doppler waveforms to confirm the diagnosis (Longley et al. 1993).

Color flow Doppler mode. The color-coded display of the venous system is also helpful in this region, especially for veins in the forearm that have a narrow lumen. It is important here to test the blood flow throughput in the venous branch of a hemodialysis shunt and to carry out qualitative or quantitative flow measurements (further details on p. 175).

Examination Sequence

The recording of values from the examination points should be carried out proceeding from proximal to distal, beginning on the asymptomatic side, and side-to-side comparisons should be made. Pelvic veins and lower leg veins are often evaluated with ultrasound proceeding in the opposite direction.

Pelvic and Leg Veins

■ Continuous Wave Doppler Sonography

Pelvic veins can not be assessed with the CW-Doppler probe. The examination starts at the common femoral artery in the *groin*. The inguinal pulse is felt first and then, after contact gel is applied, the probe makes contact to search for the typical, whiplike arterial signal. The probe is then displaced slightly in a medial direction until the low noise of the femoral vein, modulated by respiration, is heard. Using the various maneuvers described previously, tests for patency and valve function are carried out. If the vein is completely open, the venous signal is affected by respiratory fluctuations. Distal to thromboses, there is a continuous flow (Fig. 11.**40**, pp. 416–417).

When the venous valves are intact, the Valsalva maneuver causes a complete cessation of blood flow, or at the most a very brief reflux, which means that the blood in the femoral vein does not come to a standstill until slightly distal to the groin, due to a valve ("early incompetence"). If the groin valves are insufficient, a continuous low noise and distal flow during pressure exertion occurs.

Differentiation between reflux in the great saphenous vein and the femoral vein is achieved by placing a light tourniquet. This compresses the great saphenous vein, so that the return flow either ceases, or can only proceed through the femoral vein (Fronek 1989).

Similarly, the *popliteal vein* is located and evaluated at the posterior aspect of the knee. The venous signal here is more difficult to hear spontaneously than in the groin, since the angle of the probe to the vessel is often too steep. The signal can be amplified by compressing the calf.

In routine practice, one can often dispense with evaluating the posterior tibial vein and the dorsal vein of the foot, since the signals can usually only be detected by foot compression (A-sounds).

The superficial veins are easy to follow using Doppler sonography. The junction of the main collecting vein, the *great saphenous vein,* can often be located medial to the femoral vein, approximately 9 cm below the inguinal ligament. Clinical assessment follows the vein along its anatomical course (Fig. 6.**3**), and the segment that shows reflux is evaluated with successive Valsalva maneuvers or proximal compression.

In valvular insufficiency, the course of the *small saphenous vein* can also be identified when the flow is modulated by repeated compression of the distal segments.

Insufficient *perforating veins* are usually only examined in their typical anatomical location (Fig. 6.**6**) when a clinically suspicious region is noticed (hyperthermia, redness, blow-out) or chronic venous insufficiency of unclear etiology is present.

■ Conventional and Color Flow Duplex Sonography

As with continuous wave Doppler sonography, the examination usually starts in the *groin region,* where the femoral vein is easy to find directly medial to the femoral artery. Initially, the course of the vein is followed distally, and after only a few centimeters (approximately 5–9 cm) the junction of the great saphenous vein with the common femoral vein can be identified (Fig. 6.**14**). Further distally and dorsally lies the junction of the deep femoral vein. The superficial femoral vein is examined with ultrasound proceeding from an anterior direction, both parallel and medial to the superficial femoral artery, all the way to the adductor canal, until the vein passes dorsally in front of the superficial femoral artery and the popliteal artery.

Proceeding from a dorsal direction, the *popliteal vein* is examined with ultrasound at the posterior aspect of the knee, where it is easily located and close to the transducer (Fig. 6.**2**). The division into the three lower leg veins is usually located easily distal to the interarticular space of the knee (48%). However, it should be borne in mind that the lower leg veins are found in pairs accompanying the corresponding artery, and that the division can also be encountered as a variant located near the popliteal vein, but above the interarticular space (40%) (May and Nissl 1966).

Experienced examiners can follow the *lower leg veins* in transverse or longitudinal section, or both (Fig. 6.**7**). Sometimes, in normal findings, the veins are not displayed in a completely continuous form. In contrast, a thrombosed, dilated vein is easily recognized, and its lack of compressibility confirms the diagnosis of lower leg venous thrombosis (Habscheid and Wilhelm 1988).

Examining the *pelvic veins,* which can be done either initially or at the end of the examination sequence, depending on the clinical objectives, also starts at the groin, where the femoral vein becomes the external iliac vein above the inguinal ligament. In its distal segment, the vessel can be displayed well and easily compressed. However, when it bends, proceeding deeper into the body, its compressibility is no longer guaranteed. Although the ease with which the veins can be displayed clearly depends on the extent of a patient's obesity, it is always difficult to follow the external iliac vein along its further course, or to follow the common iliac vein to the vena cava within the deeper regions of the body. The vena cava is searched for on the right of the umbilical region.

The epifascial veins (great and small saphenous veins) are continuously followed as with continuous wave Doppler sonography; the search usually proceeds in a proximal to distal direction.

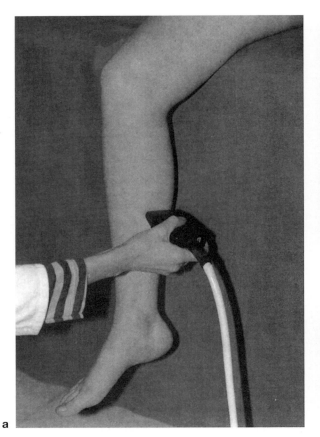

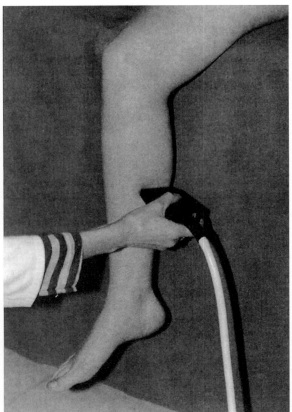

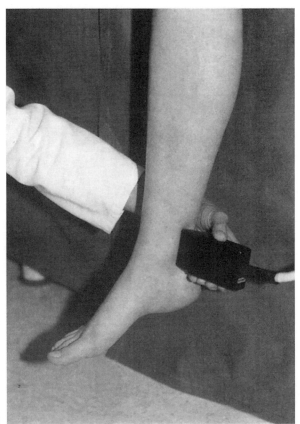

Fig. 6.**7** Applying duplex transducer to the posterior tibial veins in the loosely hanging lower leg. Transverse section **a** without compression, **b** with compression, **c** in longitudinal section posterior to the medial malleolus

Upper Extremity Veins

■ Continuous Wave Doppler Sonography

It has proved useful to start by carrying out side-to-side ultrasound comparison of the subclavian vein, and conducting the various maneuvers to test the patency of the vessels as described above. The probe is placed in a supraclavicular location, or below the clavicle if the artery covers the vein at that position.

Depending on the nature of the clinical situation, i.e., usually when a thrombosis is suspected, the axillary vein in the armpit and the brachial vein slightly above the bend of the elbow can be examined with Doppler sonography.

Evaluating patency after the placement of indwelling venous catheters sometimes requires assessment of the cephalic vein. The anatomical course of this vessel can be followed using the modulation test. Examinations of arteriovenous dialysis shunts in the upper arm or forearm can also be conducted with continuous wave Doppler sonography, which provides an overall orientation (p. 175).

■ Conventional and Color Flow Duplex Sonography

Usually, the veins of the upper extremity are followed in an ascending manner and in cross-section, starting from the bend of the elbow. This allows imaging of both the *brachial vein,* which is paired, and also the superficial veins, namely the *basilic vein* and the *cephalic vein.* For the superficial, delicate veins, the contact pressure applied with the transducer must be very light. Imaging these veins and testing their patency is very important in connection with the placement of permanent venous catheters (in intensive-care medicine). Imaging the *axillary vein* as far as the clavicle can sometimes be difficult, while the *subclavian vein* can be displayed both below and above the clavicle. Below the clavicle, the vein lies above the artery, and can be more easily depicted here than in its supraclavicular location, where it lies underneath the artery. The clavicle itself extinguishes the ultrasound signal, so that it is not possible to follow the course of the subclavian vein continuously (Fig. 6.**8a**).

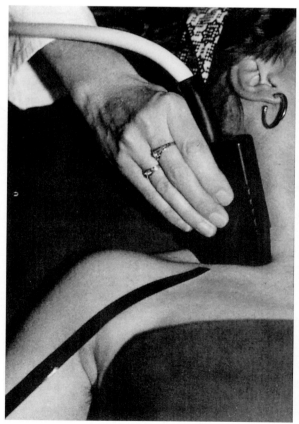 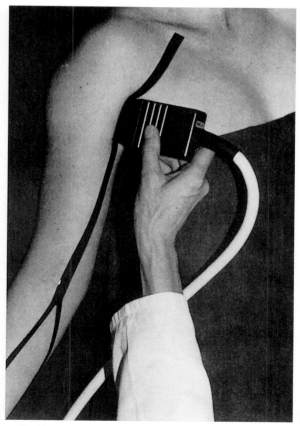

a · b

Fig. 6.**8** Applying duplex transducer to the subclavian vein **a** and the axillary vein **b** in the shoulder region, with the linear scanner of a duplex apparatus. The venous course is marked in black on the skin, providing a guide for moving the transducer

Normal Findings

Principle

In a reclining position, the return flow of venous blood is predominantly modulated by respiration. The pressure difference between the periphery and the right atrium amounts to approximately 15 mmHg. Flowing from the legs to the heart, the blood passes through two "closed containers," the abdominal and the thoracic cavities (Sumner 1984).

During *inspiration,* the diaphragm moves caudally, and the abdominal pressure increases. It exceeds the venous pressure in the legs, and the venous flow toward the heart stops. Simultaneously, thoracic pressure decreases, so that blood from the upper body region is moved or suctioned into the thorax (Bollinger et al. 1970) (Fig. 6.**9**).

On *expiration,* the process is reversed. The pressure within the abdominal "container" sinks, and the blood flows from the lower extremities, in which the pressure is now higher than in the abdomen, into the abdominal cavity. Flow into the thorax decreases. Bollinger et al. (1968) termed this process the "abdominothoracic two-phase pump."

During standing and walking, the venous pressure increases hydrostatically by approximately 60–80 mmHg. However, since an identical pressure increase occurs in the arterial system, the arteriovenous pressure gradient remains the same. In addition, during walking, the pump-like action of the muscles comes into play, and this can lead to a significantly increased blood flow toward the heart. In this process, the muscles are the energy source, and the veins are the bellows. The blood, which has been propelled toward the heart by the muscles, is prevented from returning through the venous valves (Sumner 1984) (Fig. 6.**10**). During rest, however, the venous valves are constantly open, both when the body is reclining or when standing. They only close following an acute increase in intra-abdominal pressure (coughing, straining) and if there is a sudden change of body position (Schoop 1988, p. 21).

Special provocation maneuvers and compression tests can be used to gather diagnostically useful information concerning venous function (pp. 191–193).

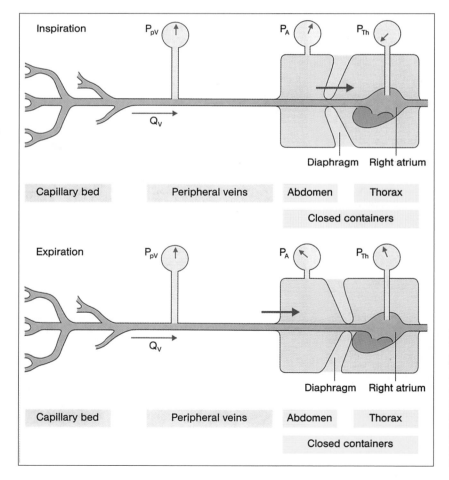

Fig. 6.**9** Physiology of venous flow toward the heart when the patient is lying down, showing the principle of abdominal and thoracic containers (according to Sumner), with changing pressure relationships that are respiration-dependent

P_{pV} Peripheral venous pressure
P_A Abdominal cavity pressure
P_{Th} Thoracic cavity pressure
Q_V Venous outflow

Fig. 6.**10** Physiology of the
pump function provided by the
muscles, which actively propel
blood back to the heart when the
body is in a standing position
(adapted from Sumner). *Resting:*
at rest, the deep and superficial
venous valves are open. The
valves of the perforating veins are
closed. *Contraction:* during
muscular contraction, the deep
and superficial veins increasingly
empty in a central direction. *Relaxation:* during muscular relaxation, blood is sucked from the superficial veins through the open
valves of the perforating veins
into the deep venous system,
and is guided toward the center

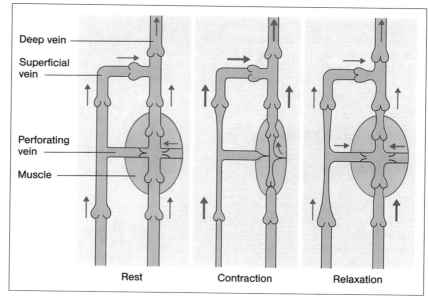

■ Continuous Wave and Pulsed Wave Doppler Sonography

Continuous wave Doppler sonography has proved to
be an important noninvasive procedure for venous diagnosis, and it can provide information on functional
venous flow parameters. *Pulsed wave Doppler sonography* is not generally used as the only examination
method, since determining venous depth is not the
primary aim. With increased use of duplex sonography equipment, examinations using the pulsed wave
Doppler mode are often carried out as part of duplex
sonography.

■ Conventional and Color Flow Duplex sonography

Imaging procedures have the advantage of allowing
the venous course, which is often very variable, to be
followed, and pulsed wave Doppler signals from the
vascular lumen being examined can be registered
specifically under B-mode imaging control. A precise

knowledge of the vascular anatomy, including the
many variations in the courses of the veins that are
often seen, is an important and necessary precondition for correct classification of the findings.

The *B-mode* provides almost continuous display
of the venous vascular system, including both the deep
and superficial veins of the upper and lower extremities, as well as the pelvic and abdominal regions. An
evaluation of the venous lumen, excluding the presence of thrombotic components, and venous dilatation following the Valsalva maneuver, can be used to
exclude thromboses. The most important test for assessing the patency of a vein is still mechanical compression with the transducer, which is successful in a
genuinely patent vein when slight downward pressure
is applied (Fig. 6.**11**).

With the integrated *Doppler mode,* spontaneous
venous flow and its phasic course can be examined.
Venous valve sufficiency, both in the deep and in the
superficial venous system, can be confirmed using the
Doppler mode with the Valsalva maneuver or proximal compression.

Fig. 6.**11** Compression testing to
examine the patency of the deep
veins (adapted from Habscheid)
a With no external pressure applied, the veins normally appear to be approximately the
same size. The arterial wall can
be identified by its more intense reflections
b During compression, the vein
collapses, while the size of arterial lumen remains unchanged
c In a pathological case, the
freshly thrombosed vein,
without any compression, is
larger than the artery. During
compression, it does not
change its diameter at all, or
only very slightly

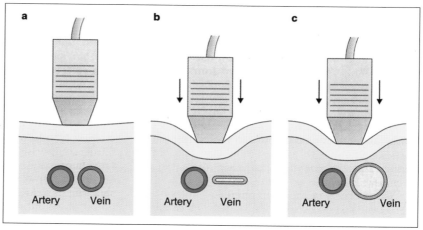

Color flow Doppler mode makes it easier to locate the veins, with color intensity and flow direction coding. It is particularly useful in anatomically difficult regions such as the lower legs, where three pairs of veins have to be located. In the pelvic region as well, however, color coding is extremely helpful in locating the deep common iliac veins (Fig. 6.**20**).

Anatomy and Findings

Pelvic and Leg Veins

When dealing with the veins of the extremities, it is useful to divide the venous system into deep and superficial systems as well as a system of perforating veins. This differentiation is reflected in the discussion of the normal and pathological findings that follows.

Deep Venous System

Anatomy

The deep veins of the lower extremity and the pelvis follow a course within a vascular sheath parallel to the arteries, and have the same names as the arteries.

The deep veins of the foot, i.e., the dorsal veins of the foot and the medial and lateral plantar veins, which are all paired, are connected to the superficial rete at the dorsum and the sole of the foot through the perforating veins. They drain the blood into the veins of the lower leg, specifically the *posterior and anterior tibial veins* and *the fibular veins,* which also accompany the three corresponding arteries in pairs. They sometimes divide into three or four blood vessels. In the proximal segment, they fuse; their junction with the popliteal vein occurs either distal to the knee joint (46%), at joint level (9%), or proximal to it (40%) (Weber and May 1990, p. 25).

The *popliteal vein* follows a dorsal course at the posterior aspect of the knee above the corresponding artery, and has a diameter of about 0.8 cm. After it enters the adductor canal, the popliteal vein becomes the superficial femoral vein and joins the deep femoral vein 2–7 cm below the inguinal ligament. From this point onward, it is termed the *common femoral vein,* according to both the angiological and radiological nomenclature. Even the superficial femoral vein shows variations from the normal unpaired form (62%). In 21% of cases, there is a division in the distal segment; in 13%, multiple veins are found. A completely paired superficial femoral vein only occurs in 3%.

Above the inguinal ligament, the femoral vein becomes the *external iliac vein* (diameter 1.2–1.4 cm), which follows a course parallel to the corresponding arteries, and unites with the internal iliac vein. It then continues as the *common iliac vein* (diameter 1.6–1.8 cm) into the inferior vena cava.

The junction of the common iliac veins with the inferior vena cava occurs to the right of the spine. For this reason, the left common iliac vein, with a length of 7.5 cm, is longer than the right common iliac vein, which is approximately 5.5 cm long. The right common iliac vein is located dorsal to the corresponding artery, while the left common iliac vein lies medial to the artery. Shortly before its junction, it is crossed by the right common iliac artery and pressed against the body of the fifth lumbar vertebra. At this point, the so-called *venous spur* can form (May 1974, p. 168)—a membrane that projects into the lumen and obstructs it. The spur is important, since it represents a thrombogenic obstruction to blood flow.

The number of venous valves varies. The most important are the subinguinal valve of the common femoral vein, located just below the inguinal ligament, and the valve of the popliteal vein, located at the level of the interarticular space of the knee. The superficial femoral vein and the common femoral vein have between one and six valves. Fairly consistently, a valve is found directly below the junction of the deep femoral vein, as well as in the adductor canal. Many valves are found in the lower leg veins at approximately 2.2 cm intervals.

A special group of veins located in the calf muscles are termed *muscle veins.* These are particularly important due to their tendency to form localized thrombi. They can be seen as indirect perforating veins to some extent but are usually classified as belonging to the deep veins.

Two groups are distinguished: the *gastrocnemic veins* and the *soleic veins.* The soleic veins, which form three intramuscular longitudinal branches, are considered by Dodd and Cockett to be a broad, valveless venous sinus, which, like the gastrocnemic veins, flow into the popliteal vein.

Continuous Wave Doppler Sonography

The Doppler-sonographic examination of the deep veins proceeds from a proximal to a distal orientation, and uses the favorable areas for applying ultrasound described in Table 6.**1**. Functional evaluation of venous flow provides information about the venous segment located caudal to the probe. Two important aspects need to be evaluated: the *patency* and the *venous valve functioning* of the undisturbed vein. When the venous flow is genuinely free, the blood flow usually is phasic, decreases during inspiration, and increases again toward the heart during expiration (p. 198). If the findings are unclear, the spontaneous flow is tested during deep respiratory excursions (Fig. 6.**12**).

When no spontaneous flow can be registered, as may happen at distal registration points, especially at the ankle (posterior tibial veins), an A-sound, namely a flow peak (Fig. 6.**5**) directed toward the heart, is elicited by rapidly compressing the forefoot distally. Increased blood flow can be produced indirectly even during compression proximal to the registration point. On compression, the flow initially ceases if the valves are intact. However, after decompression, an overshooting flow directed toward the heart follows. In addition, A-sound propagation indirectly indicates the patency of the intervening venous segment.

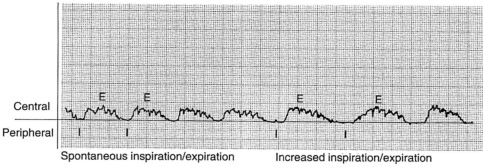

Central

Peripheral

Spontaneous inspiration/expiration   Increased inspiration/expiration

a   Femoral vein

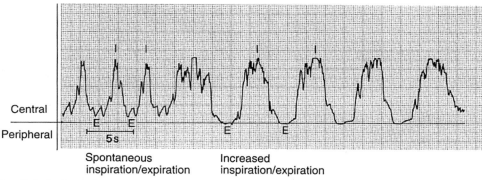

Central

Peripheral   5s

Spontaneous   Increased
inspiration/expiration   inspiration/expiration

b   Subclavian vein

Fig. 6.**12**   The graph shows physiological venous blood flow in the lower body, femoral vein (**a**), and in the upper body, subclavian vein (**b**) using continuous wave Doppler sonography

a   In the femoral vein, the spontaneous venous flow almost stops during inspiration, and flows centrally only during expiration. During forced inspiration and expiration, the modulation of the venous curve becomes even more pronounced

b   The flow behavior in the subclavian vein is precisely the opposite. During spontaneous respiration, the marked, centrally directed flow peaks can be registered in inspiration. During expiration, the venous flow decreases. While forced inspiration and expiration is going on, the flow ceases briefly during expiration

Usually, the Valsalva maneuver is used to *assess valvular function.* An analysis of the curve can be used to ascertain whether the flow comes to a standstill, because outflow from the extremities ceases due to the elevated proximal pressure, and reflux does not occur when the valves are sufficient. After the maneuver, an elevated flow velocity toward the heart is registered for a short time (overshoot) (Köhler and Neuerburg 1978).

■ Conventional and Color Flow Duplex Sonography

B-mode. Imaging of the vein allows evaluation of the venous lumen, the diameter of the vessel, and its compressibility, and some authors therefore consider functional flow examinations to be unnecessary (Sullivan 1984, Habscheid and Wilhelm 1988).

However, this only applies to testing the patency in regions that can be directly compressed by the transducer (Fig. 6.**13**).

This is not necessarily the case in the pelvic region. In addition, examining valve function with the B-mode alone is usually not possible. Depicting the venous valves and their functioning is only feasible with very good imaging quality (Fig. 6.**15**). However, in most regions, a general overview excluding a deep

venous thrombosis can be confirmed by an experienced examiner using so-called compression sonography in transverse section (Table 6.**5**, p. 224).

Doppler mode. As described above for continuous wave Doppler sonography, the Doppler mode is used to register spontaneous flow modulations, as well as the provocation maneuvers that assess valvular function, both under visual control (Fig. 6.**23 a, b**).

Color flow Doppler mode. The color flow Doppler mode makes it possible to locate the veins more quickly, also in the lower leg and pelvic regions, and can therefore exclude venous thromboses with a greater degree of certainty. However, the distinction between respiration-modulated flow and continuous flow, as seen in thromboses, can only be reliably made with the Doppler mode. Continuous color representation in the venous segment being examined can exclude a thrombosis as well (Fig. 6.**13 c**). Valsalva maneuvers can also be recognized using the color flow Doppler mode. Reflux is observed when there is a color change. With color flow Doppler sonography, it may often still be possible to detect venous flow in the lower legs without using provocation maneuvers when a venous signal can no longer be detected by continuous wave Doppler sonography.

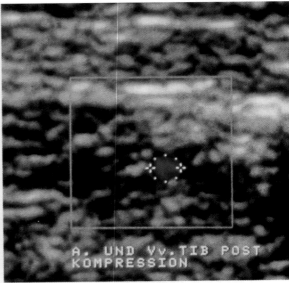

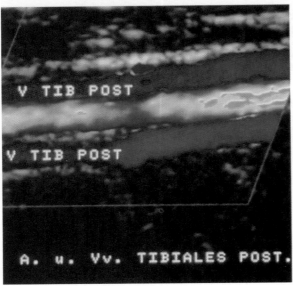

Fig. 6.**13**   Compression testing in the posterior tibial veins using color flow duplex sonography

a   Transverse view of the posterior tibial artery and the posterior tibial veins at ankle level. The arterial and venous lumina are approximately the same size. Arterial blood flow is coded red, and venous flow is shown in blue

b   During compression, only the arterial lumen can still be recognized. The accompanying veins are compressed by contact pressure exerted by the transducer

c   Longitudinal view. The posterior tibial artery is accompanied by the paired posterior tibial veins. Due to the low pulse repetition frequency of the color flow Doppler mode, the venous color coding appears saturated and intense; aliasing is seen in the artery. The sectional plane corresponds to that in Fig. 6.**7 c**

▨ Superficial Venous System

▨ Anatomy

The superficial and epifascial veins lie above the fascia within the subcutaneous fatty tissue and, in contrast to the deep veins, have no accompanying arteries along their course.

The most important superficial vein is the long *great saphenous vein,* which originates from the superficial veins at the medial margin of the dorsum of the foot, and continues to the lower leg in front of the medial malleolus. Here it follows a course along the medial side and ascends in to the thigh dorsolateral to the medial condyle of the femur. It proceeds through the saphenous opening into the deeper regions, and approximately at the level of the junction of the deep femoral vein—often somewhat proximal to this—

curving from medioventral, joins the common (or superficial) femoral vein. The confluence has a characteristically arched, chiasmatic form (Fig. 6.**14**). The distal diameter of the great saphenous vein is 0.4–0.5 cm, and at the proximal junction point, which is often dilated into a funnel shape, it can range up to 2.0 cm. Approximately ten valves are encountered along its course. In 73% of cases, a single trunk of the great saphenous vein is present; in 27%, it is paired. Near its junction, there is a confluence of several veins: the superficial gastric vein, the external pudendal vein, the circumflex vein, the superficial medial femoral vein, the superficial lateral circumflex femoral vein, and the superficial circumflex iliac vein. This region has been described as a "venous star" because of the star-shaped form of this venous junction.

The second important collecting vein is the *small saphenous vein,* which originates from the lateral margin of the dorsum of the foot, and proceeds dorsal to the lateral malleolus to the dorsal lower leg. It continues onward and, approximately at mid-calf, runs through the fascia between the heads of the gastrocnemic muscle into the deeper regions, and, usually at knee-joint level, flows into the popliteal vein. The junction point is very variable: in 79% of cases, it lies 3–5 cm above the interarticular space of the knee joint; in 14%, it lies at interarticular space level or up to 3 cm above it; in 1%, it is located distal to the interarticular space. In 50% of cases, a single, unpaired vein is found, and in approximately one-third, the vein is paired (Weber and May 1990, p. 37). The number of valves is between six and 12.

■ Continuous Wave Doppler Sonography

Examining the *great* and *small saphenous veins* is only done when valvular insufficiency or thrombophlebitis is suspected.

If the valve in the region of the confluence of the great saphenous vein is found to be sufficient, no further examinations are undertaken unless a trunk varicosis, which can be caused by insufficient perforating veins, indicates distal insufficiency. The *small saphenous vein* is also only examined in varicosis, phlebitis, and chronic venous insufficiency.

■ Conventional and Color Flow Duplex Sonography

In general, the superficial veins can be displayed with high-resolution, high-frequency transducers. However, examining them is not normally part of routine diagnostic procedures (Fig. 6.**14**).

The superficial veins in the lower body are often used as veins for bypass procedures. Testing the unobstructed patency of a vein and excluding paired segments and varicose abnormalities is important in these cases, especially when the great saphenous vein is used as an in-situ bypass. Since they are removed preoperatively, the valves do not have to be sufficient.

▨ Perforating Veins

▨ Anatomy

The blood vessels that are termed *perforating veins* connect the superficial, epifascial veins and the deep, subfascial veins, while connections *within* the region of the superficial or the deep veins are made by the *communicating veins* (Weber and May 1990, p. 40). The perforating veins, which transport blood from the superficial venous system into the deeper regions, are of clinical significance. Reverse flow is normally prevented by valves.

The perforating veins are often referred to by the proper names of the anatomists who first described them. They proceed through all the vascular regions of the extremities. Of the 95 groups described by van Limborgh (1963), only 18 groups, the majority of

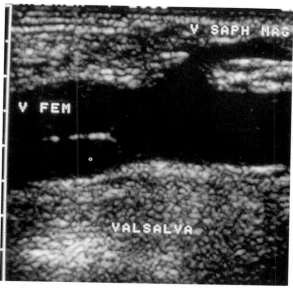

Fig. 6.**14**   B-mode image showing the femoral vein with the origin of the great saphenous vein during a Valsalva maneuver (normal results). There is a closed valve in the femoral vein during the Valsalva maneuver, and a chiasmatic shape at the junction with the great saphenous vein, in which the return flow ceases during the Valsalva maneuver, as does that of the femoral vein

which are located in the lower leg, have clinical significance (Weber and May 1990, p. 349).

In the *lower leg*, eight groups are distinguished, including the medial, lateral, and posterior perforating veins, with Cockett veins I, II, and III having the greatest clinical significance.

The *Cockett* perforating veins lie in a defined region above the sole of the foot, and connect the great saphenous vein with the posterior tibial veins distally:

Cockett I:	7 ± 1 cm
Cockett II:	13.5 ± 1 cm
Cockett III:	18.5 ± 1 cm

The group termed the *Boyd* veins is also well known. It lies approximately a hand's breadth below and medial to the knee joint, and is important because of its potential to supply the great saphenous vein in case of insufficiency.

The veins of the *thigh* are also divided into three main groups: the medial, lateral and posterior perforating femoral veins. The clinically most important blood vessels here are the *Dodd perforating veins*, which are located at the medial distal thigh, and connect the great saphenous vein to the femoral vein.

■ Continuous Wave Doppler Sonography

The method used to evaluate the perforating veins is described above in the section on examination (p. 193). Functional testing of these veins is not usually important. The perforating veins do not have any clinical significance until valvular insufficiency appears.

■ Conventional and Color Flow Doppler Sonography

The perforating veins can also be located in B-mode images, so that functional tests can be carried out using the *Doppler mode* with definite sample volume placement. The *color flow Doppler mode* helps to locate the connecting veins more quickly, and is useful in evaluating the flow direction (Fig. 11.**46**, pp. 424–425).

Upper Extremity Veins

Deep Venous System

Anatomy

The ulnar and radial veins collect blood from the deep palmar arch, and course in pairs alongside the corresponding arteries. They unite slightly above the elbow joint, and flow into the *brachial veins,* which may be either paired or singular.

The confluence of the brachial veins with the *axillary vein,* which is only paired in 1% of cases, is at the level of the inferior margin of the greater pectoral muscle. The superficial *cephalic vein* joins 2–3 cm distal to this point. The axillary vein contains a valve, and is approximately 3–5 cm long, with a diameter of 1.3 cm (0.8–1.9 cm). After passing below the clavicle, the axillary vein is called the subclavian vein.

The *subclavian vein* is separated from the artery of the same name by the anterior scalene muscle, and, on the right, flows into the innominate artery. On the left side, the vein, in contrast to the artery, is termed the left innominate vein after the junction with the internal jugular vein. The terminal valve is located in the venous angle; its functioning is influenced by the intrathoracic pressure.

■ Continuous Wave Doppler Sonography

In the upper extremities, including the subclavian vein, the behavior of venous flow during respiration is the opposite of what it is in the lower body region. During inspiration, pressure in the thorax decreases, and the venous blood flows toward the heart. During expiration, the thoracic pressure increases, and the flow decreases, or even ceases during deep respiration (Fig. 6.**12b**).

Forced respiration can also be used to assess valvular sufficiency in the upper extremities. Examinations using proximal compression are not usually done, and are not clinically relevant. However, quick distal compression, which produces A-sounds to help locate the veins, can be useful in this region.

■ Conventional and Color Flow Duplex Sonography

As in the lower extremities, the veins of the arm can be followed from distal to proximal using *B-mode,* predominantly in transverse section, and their compressibility can be evaluated at the same time. The subclavian vein is examined proximal and distal to the clavicle. Since compression is not always possible at this location, the normal collapse of the subclavian vein after quick, deep inspiration is tested (Gooding et al. 1986). In B-mode imaging, definite intravasal identification of a central venous catheter can be tested (Möllmann et al. 1987). Subclavian vein punctures, and particularly axillary vein punctures, are much easier to do under visual control using B-mode imaging (Taylor and Yellowless 1990).

Using the *Doppler mode,* phasic flow during normal respiration can be observed, registered and, if necessary, provocation maneuvers can be used (Fig. 6.**27a**).

Color flow duplex sonography allows faster documentation of the six important upper arm veins (the paired brachial veins, cephalic veins, and basilic veins) and the four forearm veins, i.e., the paired radial veins and ulnar veins. Continuous color intensity makes it possible to exclude thromboses.

Superficial Venous System

Anatomy

The most important collecting veins of the superficial veins of the arm are the cephalic vein and the basilic vein.

The *cephalic vein* proceeds from the radial margin of the dorsal venous rete of the hand, and follows a course diagonally above the radial margin of the forearm to the cubital fossa. From there, it continues in a stretched manner medially along the upper arm, 2–3 cm below the clavicle and into the axillary vein. It has six to ten valves.

The *basilic vein* starts from the ulnar margin of the dorsal venous plexus, and travels onward to the bend of the elbow, where it is joined by the cephalic vein. From there it continues into the medial bicipital sulcus and, approximately in the middle section of the upper arm, flows into the medial brachial vein at some depth. It contains four to eight valves, and may be paired, mainly in the forearm. There are two additional small branches that pass superficially on the volar side of the forearm—the median antebrachial vein and the median cubital vein—which flow into the cephalic vein and the basilic vein somewhat above the bend of the elbow.

■ Continuous Wave Doppler Sonography

Doppler sonography is normally only used to examine the larger collecting veins of the upper arm, the cephalic vein and the basilic vein, if there is any doubt about their patency prior to placing venous catheters. Assessing valvular insufficiency has no clinical role in the upper body.

The modulation test (p. 192) is recommended when the course of superficial veins is being followed.

Conventional and Color Flow Duplex Sonography

As in the lower extremity, the superficial veins can be located using high-resolution, high-frequency transducers in *B-mode* imaging, with a water buffer if necessary. Examining these veins can be important prior to the placement of a hemodialysis shunt to confirm unobstructed vascular patency.

In *Doppler mode,* quantitative flow volume measurements can be made. These are important in the measurement of hemodialysis fistula output, and are discussed above under "Pathological Findings" in Chapter 5 (p. 175).

The *color flow Doppler mode* is helpful in these investigations, since it makes it easier to follow the branching, superficial venous network, and facilitates the imaging of dialysis fistulas.

Perforating Veins

Although the *upper extremities* also have many communicating veins, they have no clinical significance, and are therefore not specifically named and located.

Evaluation

Patency Examination

Continuous Wave Doppler Sonography

Initially, the behavior of the venous blood flow is evaluated acoustically. As it is affected by respiration, the signal modulation is tested at rest and during deep inspiration and expiration. Flow cessation during the Valsalva maneuver is also examined. For overall orientation purposes, acoustic evaluation of the Doppler signals is sufficient.

It is advisable to register the venous flow in analogue curve form, or as a Doppler frequency spectrum, in order to provide more precise documentation of the individual examination maneuvers (p. 189), and especially of follow-up examinations. Quantitative information is not usually provided in venous Doppler sonography. It is usually only the presence or absence of a reaction to respiratory or compression maneuvers that is of interest.

Conventional and Color Flow Duplex Sonography

When using *"compression sonography,"* which involves compressing veins with the B-mode or duplex transducer, venous compressibility is preferably tested in transverse section. If the vein, in contrast to the accompanying artery, can be fully compressed, then venous thrombosis in the region being examined can be excluded with great certainty.

After the vein has been located in *B-mode imaging,* respiration-dependent signal modulation at rest and during deepened inspiration and expiration is observed using the *Doppler mode.*

In *color flow Doppler mode,* unobstructed venous patency is already detectable, without any compression, from the continuous color intensity both in cross-section and longitudinal section.

These tests mainly apply to the deep venous system, but they can also be used for the superficial veins.

Evaluating Valvular Function

Continuous Wave Doppler Sonography

In normal findings, an examination is carried out at the usual registration points to assess whether or not flow cessation occurs during the Valsalva maneuver. This confirms that definite valve closure has taken place. Alternatively, valve closure can also be provoked by proximal compression of the lower abdominal veins, or of the veins along the course of the extremity.

Conventional and Color Flow Duplex Sonography

In *B-mode,* the valves and their function can often be displayed morphologically. However, *Doppler mode* is usually also used to ensure a definite assessment of flow cessation. Assessment with *color flow Doppler mode* is even easier. During the Valsalva maneuver or proximal compression, no color change should take place (Fig. 6.**15**).

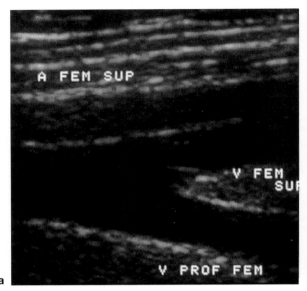

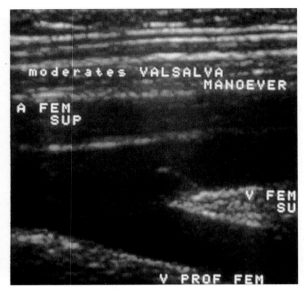

a

c

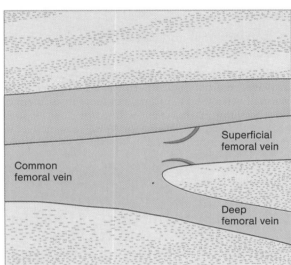

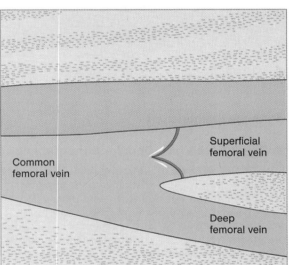

b

d

Fig. 6.**15** A longitudinal sonographic section through the bifurcation of the femoral vein. A few centimeters distal to the arterial vascular bifurcation (**a, c**), the confluence of the superficial femoral vein and the deep femoral vein is seen, forming the common femoral vein; the corresponding diagrams are shown below (**b, d**). The venous lumen is almost anechoic. In the accompanying superficial femoral artery, which is located ventrally, sparse reflections are recognizable.

During expiration (**a, b**), the venous lumina have a normal diameter. Shortly below the confluence of the deep femoral vein, a venous valve in the proximal segment of the superficial femoral vein is open; the cusps of the valve lie flat on the venous wall.

A moderate Valsalva maneuver causes an increased venous caliber. The cusps of the femoral vein valve have noticeably closed (**c, d**)

Pathological Findings

Principle

The division of the venous system into deep and superficial parts, as well as a system of communicating veins, is reflected in the different clinical pictures seen in pathological cases.

The most frequent disease of the deep venous system is acute *venous thrombosis,* with shorter or longer deep venous obstruction. After spontaneous or drug-induced lysis, this usually leaves behind damaged venous valves. Subsequent clinical symptoms such as congestion, hyperpigmentation, secondary varices, and ulcers may appear, and this characteristic clinical picture is termed the *postthrombotic syndrome.*

According to Kerr et al. (1990a), deep and superficial venous system thromboses occur above the knee in 51% of cases, in 32% below the knee, and in 17% within the superficial veins. This retrospective data summary was carried out on the basis of 1084 consecutively studied extremities with acute venous thrombosis, which were examined using duplex sonography. In order of decreasing frequency, of the 3169 venous segments, 509 were popliteal vein thromboses (above the knee), 475 were located in the superficial femoral vein, 425 in the posterior tibial vein, 418 in the common femoral vein, and 314 thromboses were located in the great saphenous vein. A study using phlebography by Schmitt (1977) provided percentage figures for thrombosis frequencies in a group of 383 patients (Fig. 6.**16**).

In superficial veins, *valvular insufficiencies* with their accompanying sequelae, varicose venous dilations, are important. In addition, local inflammatory reactions such as *thrombophlebitis* are not uncommon in normal veins, or veins dilated due to varicosis.

Valvular insufficiency of the communicating veins, *termed, perforating vein insufficiency,* also leads to drainage disturbances and secondary varicose dilatations affecting the superficial veins.

All of these symptoms are caused by characteristic *morphological* and *functional* changes, which can be detected with ultrasound imaging procedures and diagnosed in their hemodynamic effects using Doppler devices.

■ Continuous Wave and Pulsed Wave Doppler Sonography

In the case of acute *venous thrombosis,* it is not possible to register a signal above the thrombotic occlusion itself. Distal to the occlusion, the venous pressure increases, and is higher than the intra-abdominal pressure during inspiration as well. Flow cessation or flow reduction during inspiration no longer occurs. Instead, depending on the quality and extent of the collateral circulatory system, a continuous trickle or an elevated flow toward the heart is seen (Fig. 6.**17a**) (Partsch and Lofferer 1971, Thulesius 1978).

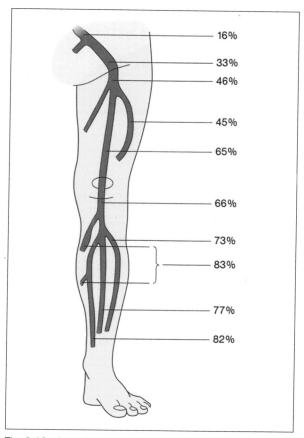

Fig. 6.**16**   Location of thromboses and the frequency with which different venous segments are affected, expressed as percentage distributions (adapted from Schmitt, in J. Weber, R. May, *Functional phlebology,* Stuttgart: Thieme, 1990)

During the postthrombotic phase, the spontaneous flow can again become phasic. Since the valves are damaged following thrombosis, deep venous *valvular insufficiency,* which can be detected with Doppler sonography, occurs subsequently. This means that during the Valsalva maneuver, *reflux* can be heard and documented peripherally (Fig. 6.**17b**).

In the *superficial venous system,* continuous wave Doppler sonography can be used easily to detect valvular insufficiencies, which are recognized by a distinct reflux toward the periphery during the Valsalva maneuver. This occurs on proximal compression, as well as during the deflation phase following distal compression.

■ Conventional and Color Flow Duplex Sonography

It is possible to detect *deep venous thrombosis* using the B-mode component of duplex equipment, as well as with dedicated B-mode equipment. This is at its

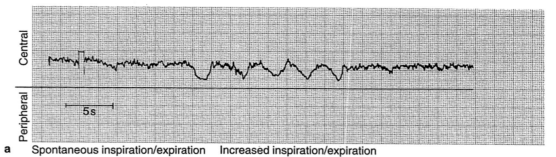

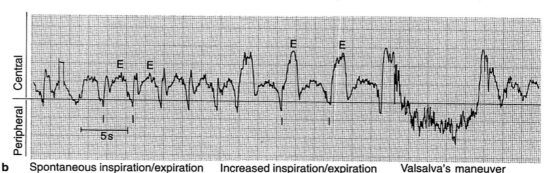

a   Spontaneous inspiration/expiration   Increased inspiration/expiration

b   Spontaneous inspiration/expiration   Increased inspiration/expiration   Valsalva's maneuver

Fig. 6.**17**   Pathological Doppler flow recordings registered from the femoral vein at the groin
a Venous flow during a pelvic venous thrombosis. During spontaneous respiration, there is a continuous, relatively high rate of flow without any modulation due to respiration. During forced inspiration and expiration, respiratory fluctuations are recognizable, although they are clearly less than those obtained in normal findings. During inspiration, a centrally directed flow remains
b Reflux during a Valsalva maneuver. In patients with insufficient groin valves, retrograde flow into the deep or epifascial thigh veins occurs when pressure is exerted. This retrograde flow lasts until the reflux is stopped by a functioning valve, or until the maneuver ends. With continuous wave Doppler sonography, it is not possible to differentiate on the basis of the waveform whether the reflux is in the superficial femoral vein or in the great saphenous vein. When duplex apparatus is not available, this has to be decided by compressing the great saphenous vein with a tourniquet

most reliable in regions that are readily accessible to compression with the transducer. The thrombosed vein shows no reaction on compression, or only a minimal one, and when compared to the normal vein it often contains scattered echo patterns (Fig. 6.**18a, b**). Further size increase during a Valsalva maneuver is not possible.

Dilatation, wall thickening, and internal structures within the venous lumen that are seen in the B-mode image of the superficial veins also suggest *thrombophlebitis* (Fig. 6.**25**).

In *Doppler mode,* reflux phenomena accompanying valvular insufficiency can be clearly classified under visual control as belonging to the superficial or deep veins, and can also be followed along their course (Markel et al. 1992).

In *color flow Doppler mode,* thrombosis can be recognized by a color gap. Marginal flow in partial thromboses or during recanalization can be recognized with certainty (Fig. 6.**18c**), and valvular insufficiencies can be detected by a color change during the Valsalva maneuver (change of flow direction).

Findings

Pelvic and Leg Veins

Deep Venous System

As described above, a deep venous thrombosis of the leg represents the most common and serious disorder of the deep venous system. It may give rise to pulmonary embolism and chronic sequelae, in particular, chronic venous insufficiency as a result of a post-thrombotic syndrome.

Thrombotic Processes

■ Continuous Wave Doppler Sonography

It is decisive here to observe whether the examination is carried out at the typical registration points in the groin and popliteal region, proximal or distal to the thrombotic occlusion, or whether the probe is lying above the thrombotically occluded vein. In the latter case, no signal is registered, or minimal or increased continuous flow, depending on the vascular diameter,

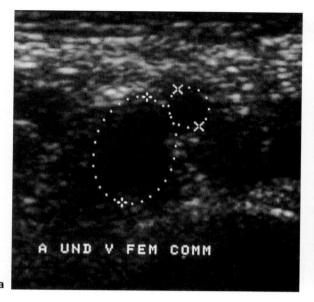
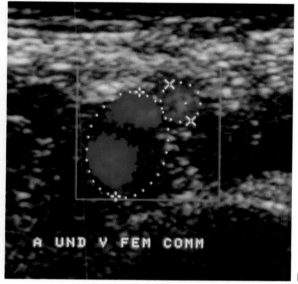

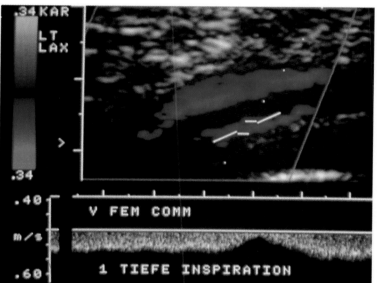

Fig. 6.18 A partially recanalized common femoral vein thrombosis

a Transverse section at the left groin. The common femoral artery and the common femoral vein are shown, without compression. The vein is clearly dilated

b The same image after adding color flow Doppler mode. Only with color coding does the partial venous recanalization become clear

c The common femoral vein in longitudinal section, clearly displaying two recanalized canals at the margins of the vascular lumen. In the pulsed wave Doppler mode spectrum, discrete respiratory modulation can be seen. During deep inspiration, the continuous venous flow while resting decreases. For technical reasons, the spectrum representing physiological venous flow was, exceptionally, registered below the baseline

is detected from the collaterals running parallel to the occluded region. Peripheral to the occlusion, continuous flow is the clearest criterion for flow obstruction. Proximal to the occlusion, the flow is still predominantly modulated by respiration (Fig. 6.**19 a-c**) (Strandness 1990, p. 169).

There is a characteristic reaction to *distal compression,* which causes an abrupt flow cessation or a very low flow peak following compression when the probe lies distal to the occlusion (Fig. 6.**19 d**). The compression maneuver therefore causes a reaction opposite to that seen under free flow conditions, in which compression results in increased flow (A-sounds) being registered toward the center. When the probe is placed proximally, no flow reaction to *distal* compression is detectable, or only a very limited one (Fig. 11.**40 c**, p. 417).

The clearest appearance of these phenomena results when ultrasound is applied above the groin, and

they become somewhat less noticeable toward the periphery. In the Anglo-American literature, proximal and distal compression is thought to be highly useful (Sigel et al. 1972, Barnes et al. 1976, Sumner and Lambeth 1979, Strandness and Thiele 1981). However, these phenomena should not be overvalued. In the German medical literature, Partsch (1976) and Wuppermann (1986) found no convincing diagnostic improvement when additionally using signals elicited by compression.

■ Conventional and Color Flow Duplex Sonography

B-mode. Diagnosing a deep venous thrombosis in the femoral and popliteal veins has become the domain of pure B-mode imaging diagnosis, and is termed *compression sonography.*

This procedure was first described by Talbot (1982), who formulated the most important criteria for B-mode imaging diagnosis (Table 6.**3**). These observations have proved to be reliable, and have been confirmed by many studies (Table 6.**5**) (Fig. 11.**42**, p. 419).

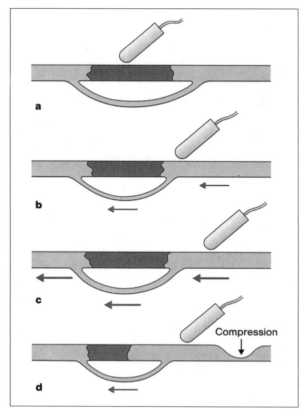

Fig. 6.**19** Doppler signals in case of a deep venous thrombosis and the way in which they depend on the location of the Doppler probe, the quality of the collateral blood supply, and the effect of distal compression (adapted from Strandness and Thiele)
a Probe above the thrombosis: no signal
b Probe distal to the thrombosis, poor collateralization: continuous low velocity flow
c Probe distal to the thrombosis, good collateralization: continuous high velocity flow
d Doppler probe distal to the thrombosis, compression peripheral to the probe: low flow peaks directed toward the heart

Table 6.**3** Criteria for detecting deep vein thrombosis using B-mode imaging with compression testing

- Dilated, thrombosed venous lumen
- Absent or clearly decreased compressibility
- Stationary internal echoes within the thrombosed lumen
- Missing additional dilatation during the Valsalva maneuver

Compression testing is not possible in the pelvic region (Fig. 6.**20**), where there is no resistance to compression. In addition, evaluating the lower leg veins solely with B-mode sonography requires good anatomical and topographical knowledge, practical experience, and commitment on the part of the examiner. It has been shown that thrombosed and dilated venous segments, which accompany the lower leg arteries as soft, tissue-like, noncompressible cords, are a good criterion for diagnosing thromboses, and are easily located (Fig. 6.**21**). It is more difficult to follow the three patent paired veins continuously, however, and many authors believe this is not necessary (Krings et al. 1990, Habscheid and Landwehr 1990, Yucel et al. 1991, Herzog et al. 1991). The variable quality of B-mode imaging (depending on older or newer equipment) has a significant effect on whether these smaller-caliber veins can be located.

Doppler mode. Using Doppler mode, altered flow criteria, as described above, can be detected, and reactions to compression under visual control can be specifically elicited in the deep and superficial venous segments. Comparison with the contralateral side is essential here.

Color flow Doppler mode. The color flow Doppler mode is particularly helpful in detecting marginal flow phenomena in partially thrombosed veins (Fig. 6.**22**). Reliable detection of thromboses is possible even when it is difficult to locate the veins (pelvic veins and lower leg veins) and compression sonography cannot be used, or only inadequately.

Valvular Insufficiency

Various studies (Strandness et al. 1983, Markel et al. 1992) have shown that valvular insufficiency develops after deep venous thrombosis at varying time intervals, depending on the recanalization process (p. 228). It does not usually appear until five or six months later, and signs of insufficiency are primarily seen in the valves below the knee, documented by reflux during the Valsalva maneuver or proximal compression.

■ Continuous Wave Doppler Sonography

Valvular insufficiency in the deep conducting veins is already detectable with continuous wave Doppler sonography. Whether the refluxes occur in the deep or superficial veins has to be differentiated.

A reflux that is registered at the *groin* can flow into the superficial femoral vein or great saphenous vein. When the great saphenous vein is compressed with a tourniquet, reverse flow into the epifascial veins is avoided, and assignment to the deep venous system becomes easier.

Similarly, when Doppler sonography is being used to examine the popliteal region, reflux can be differentiated by obstructing blood flow to the great saphenous vein and the small saphenous vein in the distal lower leg. If the reflux persists, then it involves the deep veins. If the reflux ceases, then it was located in the superficial, epifascial venous system.

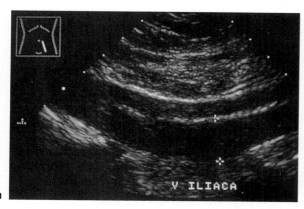

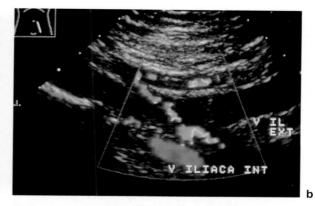

a

b

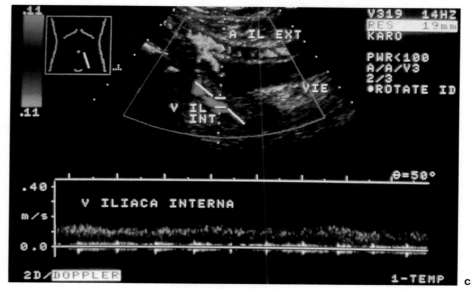

c

△
Fig. 6.**20** An external iliac vein thrombosis
a External iliac artery and external iliac vein in longitudinal section, in an image obtained with the sector scanner from a color flow duplex apparatus. The artery lies ventrally; dorsally, the dilated external iliac vein is recognizable
b Refilling of the common iliac vein via the internal iliac vein (blue color coding). The external iliac artery, with the origin of the internal iliac artery, is located close to the transducer (red-yellow color)
c Registering the Doppler frequency spectrum in the internal iliac vein. Continuous flow with a following thrombosis is seen. There is no color coding in the thrombosed external iliac vein

▷
Fig. 6.**21** Posterior tibial vein thrombosis. The color-coded B-mode image shows the thrombus, its head can be readily recognized in the patent lumen (arrow). Close to the transducer, the muscles are seen

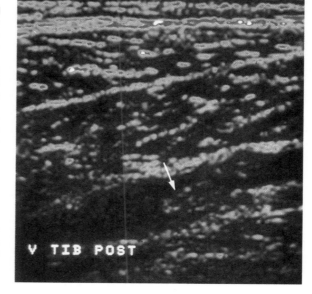

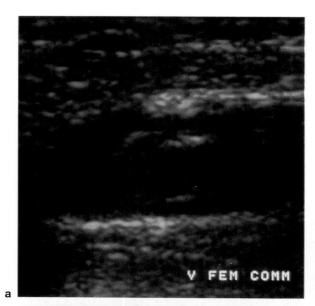

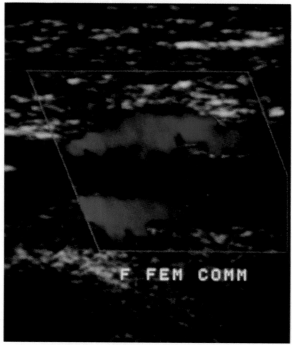

Fig. 6.22 A parietal thrombus in the common femoral vein
a A longitudinal section through the common femoral vein, in color-coded B-mode image. In the B-mode image alone, it is not possible to assess whether one is dealing with a parietal thrombus (attached to the wall)
b With color coding, bilateral marginal flow can be seen in the longitudinal section

In fresh thrombosis, valve closure does not occur during the Valsalva maneuver, but the centripetal blood flow is generally preserved. Despite the elevated abdominal pressure (20–80 mmHg), venous pressure in the leg is still higher than the abdominal venous pressure.

When the venous thrombosis is fresh, the Valsalva maneuver should only be used if absolutely necessary, since there is a danger that the increased pressure may cause thrombi to be released, with subsequent pulmonary embolism (Kriessmann et al. 1990, p. 67).

■ Conventional and Color Flow Duplex Sonography

B-mode. With good local resolution in B-mode, larger venous valves can be recognized and their failure to close during the Valsalva maneuver can be observed (Talbot 1986). However, usually not all the valves can be depicted, and demonstrating their insufficiency in particular is normally impossible.

Doppler mode. Like continuous wave Doppler sonography, the Doppler mode allows evaluation of the reflux reaction, its duration, and, with systematic scanning, the length of the reflux area.

Color flow Doppler mode. In color flow Doppler mode, the reflux can be recognized by a change in the color coding. At first, the changed flow direction can also lead to turbulences, which can be detected as differently colored pixels. When venous flow toward the heart is color-coded as blue, the reflux is coded red, for example. The examination can be completed in different specific regions of the deep venous system, and the endpoint of the reflux area can be identified visually.

▭ Superficial Venous System

Detecting insufficient valves in the great saphenous vein and the small saphenous vein is clinically very important, and they can be reliably diagnosed even with continuous wave Doppler sonography. Usually, the *valvular insufficiency* can be suspected due to a superficial varicosis. Clinically, taking the case history clarifies whether primary or secondary varicosis is involved. The latter often occurs in the postthrombotic syndrome, which can be confirmed with duplex sonography. The deep venous system should therefore always be examined, and valve closure in the deep conducting veins should be tested. *Thrombophlebitis*, i.e., epifascial vein inflammation, can also be verified with duplex sonography using high-resolution transducers (Fig. 6.25).

Valvular Insufficiency

■ Continuous Wave Doppler Sonography

In valvular insufficiency, *spontaneous flow* in the superficial veins is slightly pendular, and contains centrifugal and centripetal components during increased respiration. The response to the Valsalva maneuver is a long *reflux,* with a flow peak that is directed toward the periphery and followed by overshooting flow directed toward the heart (Fig. 6.**17b**). The length of time of the reflux depends on the number of insufficient valves and whether the Valsalva maneuver is applied effectively. An examination is conducted throughout the anatomical course of the main epifascial veins to assess whether partial or complete valvular insufficiency is involved, and particularly to determine how much of the vein the reflux affects.

According to Hach et al. (1977), varicosis of the great saphenous vein consists of four stages, which differ from one another in the length of the reflux that is observed:

- *Stage I:* Valvular insufficiency in the confluence region and in the two venous valves located distal to it (proximal thigh).
- *Stage II:* Valvular incompetence in the confluence region and in the femoral region of the great saphenous vein (distal thigh).
- *Stage III:* During the Valsalva maneuver, venous reflux to and beyond the knee (proximal lower leg).
- *Stage IV:* Valvular insufficiency all the way to the ankle.

Assessing valvular insufficiency in the *small saphenous vein* using Doppler sonography has not become as important as it is in the great saphenous vein, because the confluence region has significant anatomical variants, and the course of the small saphenous vein in the deeper regions is not easy to follow. Without using B-mode, it is also difficult to distinguish whether a reflux in the popliteal region is coming from the popliteal vein or the small saphenous vein. Compressing or modulating the small saphenous vein behind the lateral malleolus is helpful in the identification process.

■ Conventional and Color Flow Duplex Sonography

B-mode. In the B-mode image, the dilated confluence regions of the great saphenous vein and the small saphenous vein can be measured, and their diameters can be compared to those of the main veins, i.e., the femoral vein and the popliteal vein. Bork-Wölver and Wuppermann (1991) showed that these values correlate closely to measurements of vascular diameters obtained from phlebography. B-mode imaging can also evaluate the suitability of the great saphenous vein for use as a bypass vein by excluding any residual inflammatory changes, paired veins, or aneurysmal dilatations (Ruoff et al. 1987).

Doppler mode. Functional evaluation of the reflux area is completed in the same way as with continuous wave Doppler sonography. Careful placement of the pulsed wave sample volume allows clear differentiation between the superficial and deep veins (Fig. 6.**23**).

Color flow Doppler mode. It is much easier to follow the course of a reflux phenomenon using color flow Doppler sonography. The exact spot can be observed at which the reflux color change can no longer be detected (Fig. 11.**44**, pp. 422–423).

▬ Collateral Veins

In the case of thrombotic deep venous occlusions, drainage of blood from the region takes place via collateral veins, with elevated pressure distal to the occlusion, in the direction of the heart.

Following May (1974, p. 141), drainage through the great saphenous vein is termed first-order collateral circulation; drainage through the small saphenous vein and the femoropopliteal vein into the veins of the buttocks is termed second-order collateral circulation; and drainage through the small saphenous vein and the superficial veins at the medial thigh that flow into the pudendal plexus is termed third-order collateral circulation. The continuous, usually elevated, drainage through these collateral veins can be clearly identified with a duplex scanner, preferably a color flow Doppler.

A spontaneous form of Palma's operation is also known. In a unilateral pelvic venous occlusion, this leads the venous flow suprapubically through the external pudendal veins to the contralateral side (Fig. 11.**41**, p. 418).

When there is venous occlusion of the pelvic vascular system, Palma's operation uses the contralateral great saphenous vein, which is dissected free from the surrounding tissue and anastomosed suprapubically distal to the thrombotic occlusion with either the common femoral vein or the superficial femoral vein. The patency of a spontaneous Palma or of Palma's operation, and the flow direction in each case, can be confirmed with continuous wave Doppler sonography. However, it is better to evaluate patency and valvular insufficiency using conventional or color flow duplex sonography.

Regular examination of valvular sufficiency in the collateral veins is recommended. At the beginning of a thrombotic occlusion, the valves are usually still intact, but in chronic occlusion, valvular insufficiency results and secondary varicosis develops (Kriessmann et al. 1990, p. 76).

Thrombophlebitis

Initially, thrombophlebitis of the superficial veins is diagnosed on the basis of clinical symptoms such as reddening, hyperthermia, swelling, and pain in the affected epifascial venous segment. It has been shown that high-resolution transducers can determine the ex-

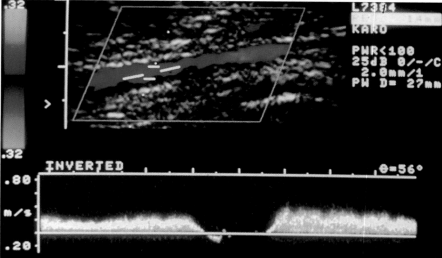

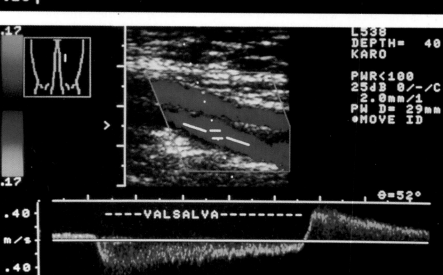

Fig. 6.**23** Valsalva maneuver in the proximal third of the superficial femoral vein

a An example with sufficient valves. In the Doppler frequency spectrum, valve closure during the Valsalva maneuver is recognized by a flow cessation

b Pathological finding during a Valsalva maneuver. Here, a long reflux is registered. After the Valsalva maneuver is finished, the spectral analysis shows reactively increased flow toward the heart

tent of the thrombosis and can occasionally locate thrombus growth in the patent lumen of the deep draining veins (femoral vein or popliteal vein) (Fig. 6.**24**). Lutter et al. (1991) were able to confirm that, especially when it affects the great saphenous vein, thrombophlebitis is associated with complications in 35% of cases. These entail either pulmonary embolism or deep venous involvement.

■ Conventional and Color Flow Duplex Sonography

In *B-mode,* the vein sometimes appears to be irregularly configured, with a thickened wall. As in the deep veins, the lumen cannot be compressed. Depending on the echogenicity of the thrombus, a stronger reflection intensity in the lumen is noticeable even without compression (Fig. 6.**25**).

Using *Doppler mode,* the absent venous flow can be documented (Ruoff et al. 1987). With *color flow Doppler sonography,* Schönhofer et al. (1992) were

able to demonstrate that thrombi partially adhering to the walls could be superimposed on phlebitis in the great saphenous vein. In individual cases, these thrombi can lead to pulmonary embolisms (p. 225).

■ Perforating Veins

Valvular *insufficiency of the perforating veins* is clinically significant. It can lead to drainage disturbances and to the development of varicose malformations in the epifascial veins. Whenever the causes of venous congestion and ulcers are unclear, the examination of these veins should not be omitted.

■ Continuous Wave Doppler Sonography

A search is made for areas in which there is a gap in the fascia at anatomically typical locations. These are often recognized by redness or a "blow-out" (a protruded, limited, livid varix loop from a fascial opening). After sealing off the varices with tourniquets

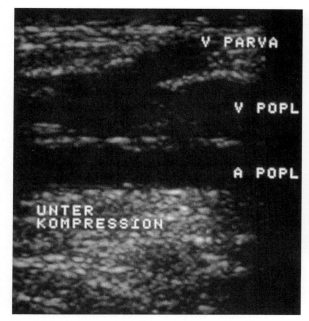

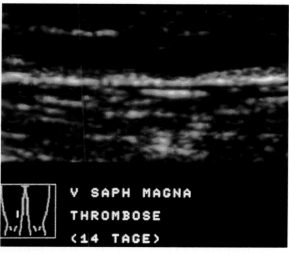

Fig. 6.**25** Great saphenous vein thrombophlebitis in the distal third of the thigh. The dilated superficial vein is seen, with a thickened wall and "frozen" valvular cusps. The lumen does not decrease under compression

Fig. 6.**24** A sonographic B-mode image showing a popliteal vein thrombosis and a small saphenous vein thrombosis. The popliteal vein lies in the popliteal region dorsal to the popliteal artery, close to the transducer. Compressing the veins is not successful, confirming superficial and deep venous thromboses

(see the examination section on p. 193), distal compression, preferably at the foot or shortly above the examination area, is applied, causing flow from the interior to the exterior, toward the probe. When the compression is released, suction toward the interior can be recognized acoustically, and is also registered (Fig. 6.**26**).

■ Conventional and Color Flow Duplex Sonography

In *B-mode imaging,* the usually winding perforating veins can be depicted. The *Doppler mode* documents an unphysiological flow direction in the epifascial veins during compression, and retrograde suction during decompression. If a *color flow* duplex system is used, tourniquets are not necessary. The outward blood flow into the epifascial veins can be recognized by the same color coding as is seen in the deep venous flow toward the center (Fig. 11.**45 c, d,** p. 425).

Fig. 6.**26** The Doppler waveform in an insufficient perforating vein. During compression distal to the registration point, a reflux signal is obtained, directed toward the probe. When the compression is released, suction directed physiologically (toward the heart) occurs in the lower leg veins (physiological = toward the deep venous system, retrograde = flow directed toward the surface)

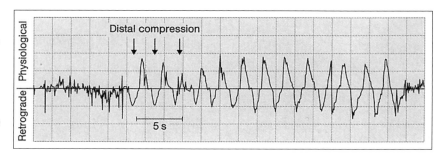

Upper Extremity Veins

Deep Venous System

Thrombotic Processes

Spontaneous venous thromboses of the arm, involving the axillary vein and the subclavian vein, occur relatively rarely (approximately 2% of all thromboses). However, they can arise acutely following physical exertion, and are usually related to a neurovascular compression syndrome (p. 169). They are therefore described as *"exertion induced thromboses."* Deep venous thrombosis in the upper extremities is also termed *Paget–von Schroetter* syndrome (Hübsch et al. 1988). Central venous catheters are a not uncommon source of thrombosis, and this type of thrombosis is being diagnosed with increasing frequency in patients in intensive-care units (Weissleder et al. 1987, Horattas et al. 1988).

■ Continuous Wave Doppler Sonography

As in the lower extremities, the venous flow distal to the occlusion, shows an absence of respiratory modulation and continuous flow toward the heart, while no venous flow can be registered directly above the thrombotic segment itself.

Due to good collateral blood flow, compression maneuvers that are applied proximal and distal to the thrombotic occlusion are less effective for differentiation purposes. The Valsalva maneuver usually produces an extensive flow cessation, but only results in a slight increase in flow in comparison with the healthy side afterward (Kriessmann et al. 1990, p. 81).

■ Conventional and Color Flow Duplex Sonography

The criteria for diagnosing a thrombosis are the same as those in the lower extremities (p. 210).

B-mode. Compression testing using a B-mode or duplex transducer can be applied to the veins of the upper extremities in the same way as in the veins of the leg. In areas where this is not possible, such as in the subclavian vein below the clavicle and at its junction with the superior vena cava, the thrombosed vein can be recognized by its dilated lumen, with stationary texture patterns. The Valsalva maneuver does not cause an additionally increased lumen; also, the "sniff test" (rapid panting) does not cause venous collapse (Gooding et al. 1986) (Fig. 11.**47**, pp. 427–429).

Doppler mode. As with continuous wave Doppler sonography, examinations with the Doppler mode also assess venous flow under visual control (Fig. 6.**27 a, b**).

Color flow Doppler mode. The color flow Doppler mode allows veins to be identified quickly, can detect occlusions or partial thromboses that are surrounded by flowing blood, and allows follow-up evaluations of the progress of recanalization after deep venous thromboses.

Superficial Venous System

In the superficial venous system of the upper extremities, the most important blood vessels in which are the cephalic vein and the basilic vein. Primary or secondary varicosis is extremely rare. Welch et al. (1994) report three cases of unilateral segmental varicosis in the arm not caused by a malformation and without accompanying varicoses in the lower extremities. In congenital venous dysplasias (Klippel-Trénaunay syndrome), superficial varicosis is one of the typical signs (p. 218).

Superficial veins have become more important in intensive-care medicine with the introduction of central venous catheters and venous indwelling cannulas. In addition, they are used to place hemodialysis fistulas (p. 175). After longer-term placement in the body, venous catheters cause phlebitis, which is readily clinically recognizable by redness and pain, and does not always require additional clarification using sonography.

■ Continuous Wave Doppler Sonography

The superficial veins can be located with the pen-shaped probe from continuous wave Doppler equipment, and their patency can be evaluated. This can sometimes be important prior to venous punctures, or prior to the introduction of venous catheters. Assessment of valvular sufficiency has not yet formed part of these examinations.

Secondary reticular collateral veins often appear in the upper arm and in the shoulder region after subclavian and axillary artery thromboses. Varicose malformation, however, is rare.

■ Conventional and Color Flow Duplex Sonography

B-mode. Using B-mode sonography, the superficial veins can be examined with a high-resolution transducer, with a water buffer if necessary, and their patency can be confirmed, or inflammatory changes and thrombotic displacements can be recognized. B-mode is important for measuring the vascular diameter, which is necessary to calculate the flow volumes. B-mode imaging has become indispensable for monitoring the location of a central venous catheter.

Doppler mode. Doppler sonography registers a flow velocity waveform, and the computer equipment calculates a value for the average flow velocity. With this value and the size measured in B-mode, the flow rate of the hemodialysis shunt can be determined (p. 175).

Color flow Doppler mode. Imaging of the superficial veins has been made very much easier with color coding. Fast orientation is provided by using the color flow Doppler mode to diagnose the functional capacity of a hemodialysis shunt, and the presence of venous stenoses or venous thrombotic displacements. For quantitative measurements, additional use of the Doppler mode is necessary.

Fig. 6.**27** Color-coded B-mode image, showing the subclavian vein (blue) and the corresponding frequency spectrum of the venous flow

a *Normal findings.* Fluctuations during respiration can be seen. They decrease during deep expiration, while the flow velocity increases on inspiration. The double pulse of the venous flow due to tricuspid valve closure is easily recognized in the analytical spectral depiction

b *Subclavian vein thrombosis.* Continuous venous flow distal to a deep thrombosis of the subclavian vein. Respiratory fluctuations are not transmitted; the flow velocity is relatively high, indicating good collateral drainage

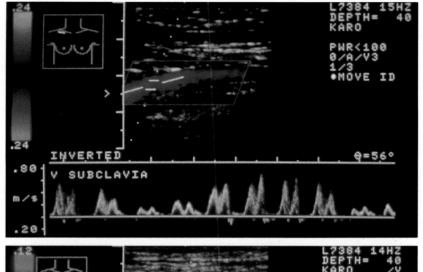

a

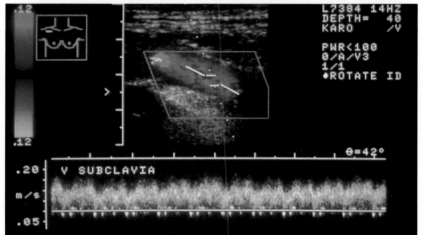

b

Special Sets of Findings

Congenital Venous Dysplasias

Congenital vascular malformations are more important in the venous system than they are in the arterial component of the circulatory system. There are a number of characteristic clinical syndromes, and the symptoms of these should be mentioned here briefly. Since combined forms occasionally occur, Bollinger (1979, p. 244) recommends that specific naming of the syndromes should be avoided. Malan (1974) recommends a new different classification. Here, we will only list these clinical manifestations that are still well known briefly, focusing on the atypical venous changes that are connected with them and which can be confirmed using ultrasound diagnosis (Rutherford 1989). The extent of these changes is sometimes only recognized using imaging procedures. For additional information, the appropriate textbooks can be referred to (Bollinger 1979, Vollmar et al. 1989), and the specialist literature can be consulted (Langer and Langer 1982, Voss 1984, Vollmar et al. 1989, Belov et al. 1989).

Klippel–Trénaunay Syndrome

(Diepgen et al. 1988)

- Unilateral, atypical varicosis
- Soft-tissue and bone hypertrophy affecting one extremity
- Nevus flammeus
- Deep venous system anomalies:
  - Valve agenesis
  - Aneurysmal venous transformations

F.P. Weber Syndrome

(Leipner et al. 1982)

- Secondary varicosis
- Soft-issue and bone hypertrophy affecting one extremity
- Arteriovenous fistulas
- Capillary hemangiomas

Servelle–Martorell Syndrome

(Martorell and Monserrat 1962, Servelle 1949)

- Atypical varicosis
- Skeletal hypoplasia

- Cavernous hemangiomatosis affecting one extremity
- Deep venous system anomalies:
  - Valve agenesis
  - Aneurysmal venous transformations

In more detail, the following sets of findings should be noted and, as far as possible, clarified using ultrasound diagnostics.

Varicosis

A unilateral, atypical varicosis of the extremities must be examined with regard to its extent and its blood supply source. The examiner should check whether an arteriovenous fistula is supplying the varicosis, whether valvular agenesis involving the deep venous system is present, or whether apparent monomelic gigantism points to the above clinical syndromes (Karasch et al. 1991).

Hemangiomas

Capillary hemangiomas, such as nevus flammeus, should be differentiated from *cavernous* hemangiomas.

Nevus flammeus forms part of the classic clinical picture of Klippel–Trénaunay syndrome. The nevus is a capillary dysplasia located below the epidermis, with no tendency to grow further. It is usually congenital in origin, and can often regress. However, regression is rarely observed in the Klippel-Trénaunay syndrome. Since the clinical diagnosis is clear, Doppler or duplex sonography is not currently used to evaluate this condition. The high-resolution transducers (20 MHz) re-

cently introduced in dermatology may be able to provide new morphological details.

Cavernous hemangioma consists of soft, venous cavernous structures that are filled with blood and lined with endothelium, in which chronic, intravasal clotting occurs. They are classified as tumors, and duplex sonography can be used to determine their depth and extent before any radiotherapy that may be required, and also in subsequent follow-up examinations (Handl-Zeller et al. 1989).

Arteriovenous Fistulas

The presence of arteriovenous fistulas in angiodysplasias suggests a possible F.P. Weber syndrome if the resulting secondary varices or aneurysmal structural disturbances appear unilaterally, and are related to increased growth of one extremity during childhood. The diagnosis of arteriovenous fistulas using ultrasound is described on pages 174–175 (Fig. 6.28a, b) (Partsch et al. 1975).

Venous Hypoplasia and Aplasia

In 50% of cases involving the clinical syndromes listed above, Vollmar et al. (1989) report finding a shorter or longer hypoplastic or aplastic segment of the deep venous system, with an accompanying increased resistance to drainage.

So far, it has only been possible to detect these malformations using phlebography, but it should now also be feasible using color flow duplex sonography. In superficial varices that have an atypical course, the deep veins should always be examined to determine their course and the diameter of their lumen.

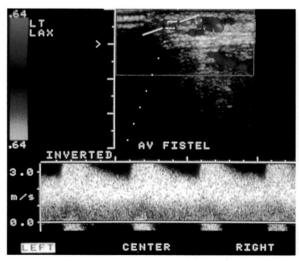

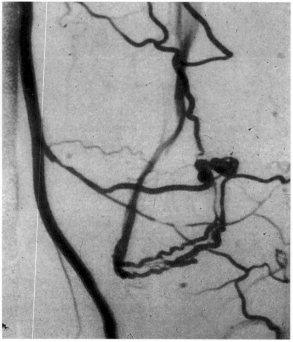

Fig. 6.**28a** An in-situ arteriovenous fistula in a bypass of the great saphenous vein. The Doppler spectrum shows the fistula's high flow velocity of more than 3 m/s, similar to a high-grade stenosis with turbulence, **b** In the radiographic image of the same phase, the venous bypass, containing arterial flow, and a simultaneously filled vein are seen

Venous Valve Agenesis and Hypoplasia

Doppler and duplex procedures have made it much easier to diagnose congenital valvular insufficiency. It has been shown that the syndromes listed above are associated with valvular agenesis, more often than was previously suspected. This means that the developing varicosis should not just be interpreted as manifesting congenital venectasias, but should also be viewed as reflecting a secondary varicosis due to venous insufficiency. Partial and complete agenesis must be differentiated from each other. Very often, the valves of the deep main branches and the connecting veins are missing (Bollinger et al. 1971).

Venous Aneurysms

More extensive imaging diagnosis (Vollmar et al. 1989) has been able to show that the Klippel–Trénaunay syndrome and Servelle–Martorell syndrome are often accompanied by venous aneurysms, and that they also show deep venous anomalies. The anomalies are almost always seen in these syndromes (Klippel–Trénaunay 96%, Servelle–Martorell 100%). According to the above study, venous leg angiodysplasias were seen in 80% of the cases, and arm angiodysplasias in only 20% of the cases. Aneurysmal transformations from normal venous structure appeared in half of the cases. Percentage figures for so-called cylindrical ectasias or aneurysmal transformations are not given. In venous graft aneurysms, Karkow et al. (1986) indicate a 50% venous dilatation as a criterion for detecting the presence of aneurysms.

Secondary Arteriovenous Fistulas

Secondary arteriovenous fistulas are discussed on pages 174–175 (arteries of the extremities). It is important to examine the venous component contributing to the fistula using Doppler, or, even better, color flow Duplex sonography. For example, stenoses in the vein forming the fistula are often encountered in hemodialysis shunts, and thrombi in the fistula may be found, or a combination of the two (Nonnast-Daniel et al. 1992).

Acquired Venous Aneurysms

In contrast to the arterial sector, acquired venous aneurysms are rarely observed. They appear to be related to congenital wall weakness in primary varicosis, elevated volume and pressure loading, and also due to acquired arteriovenous fistulas (Vollmar et al. 1989). In addition, Friedman et al. (1990) suggest trauma and inflammatory processes as further factors causing venous aneurysms in the upper extremities. The observations made by Loose and Drewes (1983), based on 600 cases, showed that approximately 30% of the aneurysms occur in the lower extremities, and only 4.2% in the upper extremities. It is to be expected that noninvasive investigation of the veins using B-mode imaging will in the future be able to identify clinically silent aneurysms more often. It is also important to detect aneurysmal dilatation in bypass veins (Karkow et al. 1986).

Evaluation

To begin with, the criteria described for normal findings (p. 205) should also be considered when pathology is encountered. Beyond that, the following aspects are important or noteworthy.

Thrombotic Processes

■ Continuous Wave Doppler Sonography

Continuous flow through the collateral veins should be documented; the speed of the flow reflects the effectiveness of the collateralization. In uncertain findings, compression and decompression maneuvers should be carried out, and the results should be registered (p. 209). Note that respiration-independent, characteristic continuous flow, or flow that is only very slightly respiration-modulated, can only be detected peripheral to a thrombotic occlusion, or parallel to it.

■ Conventional and Color Flow Duplex Sonography

Compression sonography is important in diagnosing thromboses, and can be documented with the *B-mode imaging* component of duplex equipment, or using dedicated B-mode imaging equipment without a Doppler component. It is very important to determine the location and the length of the thrombotic segments. The following criteria should be evaluated and documented:

- Venous display with the accompanying artery in transverse section
- Display without and with compression
- The distal and proximal ends of the thrombus, if its echogenicity is different from that of flowing blood

In a *postthrombotic syndrome,* or if there is a suspicion that venous thromboses have occurred in the past, the irregular and thickened wall structures should be clearly recognizable in longitudinal section after recanalization. Compressibility is decreased, but not absent.

Uncertain B-mode findings, and a need to establish whether the margins of the thrombus are still surrounded by flowing blood, are important reasons for additional use of the *Doppler mode.* An absence of spontaneous flow, or an absence of modulation by respiration, confirms the diagnosis of fresh thrombosis, especially in regions that are usually not accessible to compression sonography (e.g., the pelvic region).

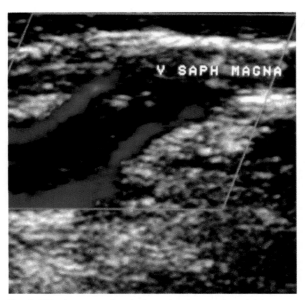

Fig. 6.**29** A recanalized thrombosis of the right great saphenous vein, showing vascular dilatation and hypoechoic, intraluminal structures. In the color flow Doppler mode, marginal flow can be recognized visually

In *color flow duplex sonography,* an interruption of the color and imaging of the collateral veins clearly indicates the presence of a *thrombosis.* Results obtained from Doppler sonography are declining in importance due to the easy observation of the dynamic flow processes this method provides. Additional use of compression sonography to confirm the diagnosis is recommended. It should be noted in particular whether the thrombi are still surrounded by flowing blood (color), or have adhered to a vessel wall (Fig. 6.**29**).

Thromboses of the superficial veins, so-called *thrombophlebitis* or varicophlebitis, can be examined with high-resolution transducers using the same criteria as those for deep venous thromboses (Fig. 6.**25**).

Valvular Insufficiency

■ Continuous Wave Doppler Sonography

Valvular insufficiency of the *deep conducting veins* is examined using the Valsalva maneuver. If necessary, the superficial veins can be compressed. A short reflux to the first valve is not pathological. The duration time of the reflux reflects the length of the valvular insufficiency, but it also depends on the intensity of the Valsalva maneuver.

In the *superficial veins,* the distal endpoint of the reflux can be established by repeated testing along the course of the great saphenous vein. Wuppermann et al. (1981) were able to show that the measured length of the reflux area obtained with Doppler sonography and the length of the reflux measured by phlebography correlate exactly.

When evaluating *insufficiency of the perforating veins,* it is important to locate the characteristic regions.

The Cockett perforating veins, located on the medial lower leg, are affected preferentially, especially those at a defined region above the sole of the foot (p. 203). During compression, forward flow toward the probe and return flow on decompression can be recognized acoustically and registered (Straub and Ludwig 1990, Wuppermann 1986).

■ Conventional and Color Flow Duplex Sonography

Testing for valvular insufficiency is largely a functional examination, in which morphological assessment of the venous lumen plays a lesser role. However, even in *B-mode imaging,* so-called frozen valves can be detected if the resolution is good. These show no reaction to proximal compression, and in particular do not close (Zwiebel 1992, p. 309) (Fig. 6.**25**).

With *Doppler mode,* evaluating reflux during the Valsalva maneuver or compression maneuvers is carried out as described above. Assigning reflux phenomena to the individual venous regions (deep venous system, superficial venous system, connecting veins) is simpler and more reliable than with continuous wave Doppler sonography.

The fastest and most impressive way of detecting valvular insufficiency is using the *color flow Doppler mode.* The color change that occurs during the Valsalva maneuver clearly indicates the changed flow direction (Fig. 11.**44 c, d,** p. 422). In a purely visual manner, the length of the color change along the course of the veins shows the extent of the valvular insufficiency in all three venous regions (deep, superficial, and connecting veins). Since this is a dynamic process, the static documentation provided by color flow duplex images can only reflect this flow reversal inadequately. Theoretically, one could code the color differently in the same location. Videotape recording is a suitable form of documentation.

Recent research has sought to identify the anatomic distribution of venous reflux and to quantify venous reflux time (Weingarten et al. 1993, van Ramshorst et al. 1994). A total reflux time of 9.66 seconds was predictive of ulceration.

Welch et al. (1996) considered segmental reflux to be present if the valve closure time was greater than 0.5 seconds, and system reflux was considered to be present if the sum of the segments was greater than 1.5 seconds. Van Bemmeln et al. (1989) report that the examination was conducted with the aid of distal cuff deflation, with the patient in the supine position, with pressures of 80 mmHg in the thigh and 100 mmHg in the calf.

Sources of Error and Diagnostic Effectiveness

Sources of Error

In addition to the many *objective* findings that can adversely affect diagnostic certainty, there are also *subjective* limitations, as in all ultrasound investigations, due to insufficient knowledge and practice on the part of the examiner.

The distinction between false-positive and false-negative results presented below is based on the fact that false-positive results restrict the specificity of the method, i.e., the certainty that it will be capable of recognizing normal findings as being normal, while false-negative results affect its sensitivity, i.e., its ability to confirm pathological findings with a high degree of probability.

False-Positive Results

Thoracic respiration is often found in young women and asthenic individuals. The respiratory pressure fluctuations are minimal in this type of respiration, and venous return flow is almost continuous. However, during deep inspiration, or when the patient is practicing abdominal respiration, it is possible to provoke a flow cessation, and this does not succeed if a thrombosis is present (Bollinger et al. 1970).

Congenital aplasia or hypoplasia of the venous valves. This finding can lead to misinterpretation of a venous valvular insufficiency when using continuous wave Doppler sonography. Depending on the quality of the B-mode image, the venous valves cannot always be recognized with certainty.

Missing venous signal. Due to minimal flow velocities, which are often present in the venous system, venous signals can sometimes not be detected with Doppler sonography. In addition, color coding requires the use of a special mode of the equipment, the so-called slow flow technique.

Excessive contact pressure exerted by the transducer may result in the compressed vein not being detected.

External venous compression. When it is impossible to use B-mode imaging to confirm the diagnosis, external compression from a tumor or a cyst may be incorrectly interpreted as being due to a thrombosis.

Edema, painful swelling. Extreme edema may clinically simulate a thrombosis, and complicate the examination as a result of mechanically more difficult compression and also as a result of the painful swelling itself. Occasionally, local edemas conceal a hematoma (Fig. 6.**30**), a muscle tear, or a Baker cyst, which can only be differentiated using B-mode imaging (Fig. 6.**31a, b**) (Cronan et al. 1988, Borgstede and Clagett 1992).

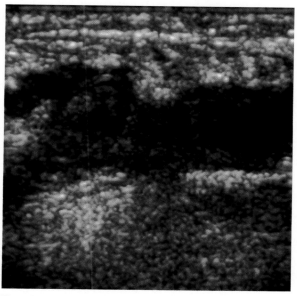

Fig. 6.**30** A color-coded B-mode image, showing a hematoma in the gastrocnemic muscles, which can be recognized as an irregularly echogenic, predominantly hypoechoic, space-occupying lesion in the muscle. Clinically, an unclear lower leg swelling was observed, the sonographic examination confirmed the hematoma

False-Negative Results

Nonocclusive thrombi. In nonocclusive thrombi, respiratory modulation may remain unchanged, so that these thrombi may escape detection if B-mode imaging is not used, and especially if the color flow Doppler mode is not used.

Good collateralization of a thrombotic occlusion. If there is good collateralization, the respiratory flow modulation may appear to be completely normal when functional test are conducted in the Doppler mode.

Parietal thrombi, or thrombi covering a short distance. These can be missed using B-mode sonography if an additional functional examination with Doppler ultrasound is not used to assess the venous flow.

Unsuccessful provocation maneuvers. Compression with the transducer can fail, especially in longitudinal section, if the vein is caught at a tangent.

The *Valsalva maneuver,* which tests the capacity for dilatation, may be carried out incorrectly. Patients often exhale deeply instead of breathing into the abdomen, or they may be afraid of urination or flatulence.

The distal or proximal *compression maneuver* may not be executed above the afferent vein, or may be more difficult due to a painful edema.

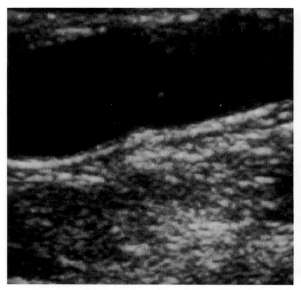
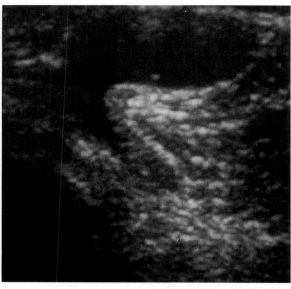

a

b

Fig. 6.**31** A Baker cyst in the right popliteal region, **a** in longitudinal section and **b** in transverse section. This finding represents a differential diagnosis from popliteal vein thrombosis, particularly in the longitudinal section. In the transverse section, the extension of the cystic structure, which has a connection to the synovial space, can be seen

Multiple veins. If the clinical presentation indicates a suspected thrombosis, a search should always be made for possibly paired veins, which are mainly encountered in the popliteal and the distal femoral region. However, the clinical manifestation may often be very obscure, since the second parallel vein functions as a collateral (Quinn and Vandeman 1990).

Perforating veins. Failing to detect insufficient perforating veins when a "blow-out" or fascial gap is not recognizable. Inadequately excluding epifascial veins when testing for insufficiency of the perforating veins.

Unfavorable patient position. The information yield of the findings can be improved if care is taken to note the patient's changing positions when ultrasound is applied to the veins (Table 6.**1**, p. 190). Proper positioning can enhance sufficient venous filling, and can also make it easier to move the transducer.

Diagnostic Effectiveness

■ Continuous Wave Doppler Sonography

Although continuous wave Doppler sonography has been replaced in some areas by B-mode sonography (for diagnosing thrombosis), conventional and color flow duplex sonography, it is still used quite correctly for preliminary diagnosis and for specific investigations.

In areas such as the pelvic region, where compression testing using real-time B-mode is difficult or impossible, functional evaluation of venous flow using simple continuous wave Doppler sonography, or the pulsed wave Doppler mode of a duplex apparatus, is still important. Validation of this method of detecting or excluding *venous thromboses in the pelvis and thigh* was carried out by many authors in the era before duplex sonography (Table 6.**4**). Overall, good results, especially with regard to the method's specificity, were attained; the sensitivity is subject to wider fluctuations, and depends on both the vascular level and the grade of the lumen obstruction. For the *lower leg* veins, Barnes et al. (1976) and Sumner and Lambeth (1979) obtained a sensitivity of 94% and 91%. Flanigan et al. (1978) did not attain a comparable value (sensitivity 36%).

Testing *valvular function* is another role for Doppler sonography. However, assignment to specific regions, the deep or the superficial veins, is only possible with aids (tourniquets). Excluding insufficient valves at the origin in the groin region was successfully achieved by Hach et al. (1977), with a specificity of 91%. In the same study, the sensitivity was 93.6%, while Sigel et al. (1972), for example, only reported a figure of 76%. In 1981, Wuppermann et al. were able to demonstrate that the length of the reflux area and the phlebographically determined reflux length correlate precisely. This is important in classifying the stage of a superficial varicosis, according to Hach et al. (1977) (p. 213).

Examining the perforating veins using continuous-wave Doppler sonography is a difficult procedure. There have been no studies to validate this procedure, to our knowledge.

Table 6.**4** The validity of phlebographically controlled studies in the diagnosis of deep venous thrombosis of the leg using **continuous wave Doppler sonography**

First author	Year	Phlebographies (n)	Sensitivity (%)	Specificity (%)
Grüntzig	1971	42	92	93
Sigel	1972	248	85	91
Strandness	1972	57	79	95
Yao	1972	50	87	94
Holmes	1973	71	100	85
Barnes	1976	122	96	92
Dosick	1978	160	99	92
Sumner	1979	75	95	89
Hanel	1981	183	91	92
Bendick	1983	140	54	90
Sigel	1986	121	86	88

■ Conventional and Color Flow Duplex Sonography

Thrombotic Processes

In 1982, Talbot listed criteria for detecting acute deep venous thrombosis using *duplex sonography:*

- Absent compressibility
- Increased cross-sectional diameter
- Increased intraluminal reflection intensity
- No vascular caliber variation during respiration maneuvers
- No Doppler signal

These criteria were confirmed in a small number of cases (n = 11, n = 23) in the earliest studies, carried out by Rhagavendra et al. (1984) and Sullivan et al. (1984), and are still valid today. Both of the groups mentioned only used a real-time B-mode apparatus, without a Doppler mode. Rhagavendra et al. (1984) considered that the echo pattern and the absence of caliber variation to be more important than the absent compressibility. They achieved a sensitivity of 100, and a similar specificity. In the same year Sullivan et al. (1984) published a study involving 215 extremities. In 23 phlebographically controlled results, the sensitivity and specificity values in the lower body region were 92% and 100%, respectively (Table 6.**5**).

In a large study conducted by Dauzat et al. (1986), which evaluated 145 patients with 100 thromboses, B-mode sonography as well as continuous wave Doppler sonography were used (sensitivity 94%, specificity 100%). In 1987, further studies by Aitken and Godden (1987) and by Cronan et al. (1987) using B-mode sonography followed. Cronan et al. (1987) were able to show that assessing *venous compressibility as the sole diagnostic parameter* was sufficient.

In the same year, five years after Talbot's (1982) pioneering research, the first additional studies using *duplex sonography* appeared: Appelman et al. (1987), Elias et al (1987), Langsfeld et al. (1987) and Vogel et al. (1987). In a study including 430 patients, Elias

achieved a high sensitivity of 98% and a specificity of 95%. Despite the use of duplex sonography, continuous wave Doppler sonography was initially applied in this study, so that, in contrast to pure B-mode studies, evaluating the iliac veins (n = 57) and the inferior vena cava (n = 47) was also possible. Elias et al. (1987) first described the examination of the *lower leg veins* using so-called compression sonography under controlled conditions in 1987. They detected thromboses in this area with a 91% sensitivity (84 of 92 cases). Habscheid and Wilhelm (1988) were able to attain a sensitivity of 89% and a specificity of 100% in 77 extremities that had undergone phlebography, and contained 29 lower leg thromboses (Table 6.**8**).

In the years 1988 to 1991, many further studies followed using the different ultrasound procedures (Tables 6.**5**, 6.**6**). Some of these only used B-mode compression sonography; some used duplex sonography or B-mode combined with continuous wave Doppler sonography (Krings et al. 1990). Parallel to this, the first studies carried out using *color flow duplex sonography* also produced good, although not significantly better, results (Table 6.**7**).

Overall, the indicated sensitivity values for conventional and color flow duplex sonography usually lie between 90% and 95%; the specificity lies between 95% and 100%. The results obtained depend not only on the methodologically differing access and evaluation criteria, but are primarily influenced by whether or not findings obtained from the pelvic and lower leg veins are included. This is shown in Tables 6.**5** and 6.**7**. Since the disease prevalence and clear field tables were not provided in all studies, we have not listed results for the predictive value.

In a study published by Heijboer et al. in 1993, a very high 94% positive predictive value was attained for a pathological finding (of 89 pathological B-mode results, 84 were confirmed through phlebography) using compression sonography, and the significance of repeat examinations (on the second and eighth day) was shown.

Becker et al. (1989) published a critical analysis summarizing 15 studies conducted between 1984 and

Table 6.**5** The validity of phlebographically controlled studies in the diagnosis of deep venous thrombosis of the leg using **B-mode sonography**

First author	Year	Patients/ phlebographies (n)	Thromboses (n)	Sensitivity (%)	Specificity (%)
Sullivan	1984	23/ 23	12	100	92
Rhagavendra	1986	20/ 20	14	100	100
Aitken	1987	46/ 42	16	94	100
Cronan	1987	51/ 51	28	89	100
Gaitini	1988	45/ 45	23	87	91
Habscheid	1988	146/146	127	94	97
O'Leary	1988	53/ 50	25	88	96
Rollins	1988	63/ 46	35	87	98
Lensing*	1989	220/220	77	91	99
Habscheid*	1990	238/301	153	96	100
Herzog*	1991	113/101	57	88	98
Langholz	1991	64/ 64	25	76	88

* Including calf vein thromboses

Table 6.**6** The validity of phlebographically controlled studies in the diagnosis of deep venous thrombosis of the leg using **duplex sonography**

First author	Year	Patients/ phlebographies (n)	Thromboses (n)	Sensitivity (%)	Specificity (%)
Dauzat[1]	1986	145/145	100	94	100
Appelman	1987	112/112	52	96	97
Elias[2]	1987	430/854	268	98	95
Langsfeld	1987	58/ 65	24	100	78
Vogel	1987	54/ 54	25	91–94	100
Barnes[2]	1989	78/309[3]	14	86	97
Killewich*	1989	47/ 50	38	92	92
Stapff*,[1]	1989	49/ 50	26	96	71
Krings	1990	–/235	–	93–100	96–99
de Valois	1990	180/101	61	92	90
Wright	1990	84/ 71	34	91	95
Comerota	1990	103/ 72	44	96	93
van Ramshorst	1991	117/120	64	91	95

* Including calf vein thromboses
[1] B-mode sonography + continuous wave Doppler sonography
[2] Additional continuous wave Doppler sonography
[3] Preoperative and postoperative

1988, which diagnosed deep venous thromboses using B-mode sonography alone (8) and B-mode sonography including Doppler sonography (7). They indicated the many notable sources of methodological error that can appear when constructing and executing such studies.

For *lower leg thromboses,* more recent studies (Table 6.**8**) indicate lower sensitivity and specificity values than for the thigh and pelvic regions, but with increasing experience, satisfactory diagnostic information can also be obtained here, especially with the application of color flow Doppler mode. In some studies, the certainty of identifying the different lower leg ar-

teries has been listed separately (Grosser et al. 1991). Lohr et al. (1991) demonstrated in a prospective study that in 75 patients with isolated calf vein thrombi, the thrombi in 24 patients (32%) propagated, and 46% of those into the popliteal or large veins of the thigh.

Mattos et al. (1996) showed in a retrospective study that imaging of three paired calf veins is feasible in 94% of cases and that calf-vein thrombosis is present in two-thirds of limbs with documented acute symptomatic thrombosis of the deep veins.

More attention still needs to be given to the study of *floating thrombi* in the deep and superficial veins, which, according to a study by Voet and Afschrift

Table 6.**7**   The validity of phlebographically controlled studies in the diagnosis of deep venous thrombosis of the leg using **color flow duplex sonography**

First author	Year	Patients/ phlebographies (n)	Thromboses (n)	Sensitivity (%)	Specificity (%)
Persson*	1989	264/ 24	16	100	100
Fobbe	1989	103/129	58	96	97
Foley	1989	47/ 47	19	89	100
Fürst	1990	75/102	39	95	99
Grosser*	1990	180/180	154	94	99
Rose*	1990	69/ 75	32	79	88
Schindler	1990	97/ 94	54	98	100
Langholz	1991	116/116	65	100	94
van Gemmeren	1991	114/141	74	96	97
Mattos	1992	75/ 77	32	100	98
Schweizer[1]	1993	78/ 70	70	98	100

* Including calf vein thromboses
[1] Intensified by using contrast medium

(1991), cause *pulmonary embolisms* with an incidence of 9%.

Several research teams have examined the application of duplex scanning in patients with suspected pulmonary embolism. Killewich et al. (1993) showed that, in the case of angiographically confirmed pulmonary emboli, color flow duplex sonography registered a positive finding in only 44% of cases. In 664 patients with suspected pulmonary embolism, Matteson et al. (1996) detected a confirmed deep venous thrombosis in only 13% of the cases and consider ventilation/perfusion scanning to be the more important diagnostic measure.

Nicholls et al. (1996) studied venous thromboemboli in a new way. Transcranial Doppler monitoring is an established means of identifying emboli in the arterial circulation. Nicholls et al. extended this technique to identify emboli in the venous circulation. Of 60 patients with deep vein thrombosis, embolism was demonstrated in 43%. In patients taking heparin, the embolism disappeared within 72 hours.

Deep venous thrombosis affecting the *shoulder and arm veins* is a rare occurrence, and is described in the literature as having a frequency of between 1% and 2% of all thromboses (Huber et al. 1987). If newer investigations, which primarily focus on catheter-related thrombosis in the superior thoracic aperture (Habscheid et al. 1992), are taken into account, a higher incidence of upper body venous thromboses may be suspected. The validity of (color flow) duplex sonography in diagnosing venous thromboses in the shoulder–arm venous region is very satisfactory. A more comprehensive study involving 99 phlebographies completed by Knudson et al. (1990) arrived at a sensitivity of 78% and a specificity of 92%. In 14 phlebographies, Falk and Smith (1987) attained a 100% sensitivity. Habscheid and Landwehr (1992) reproduced this result in 17 phlebography patients, with both a 100% sensitivity and 100% specific. The sensitivity may decrease if proximal subclavian thromboses

cannot be directly visualized, and only the absence of a color signal and not the Doppler spectra are used to make the diagnosis.

In summary, it can be said that the indication for using duplex sonography in the veins of the upper body should be broadly interpreted, most importantly in patients with central venous catheters, and unclear complaints such as arm swelling and arm pain. Owing to the difficulty of applying ultrasound to the junction of the subclavian vein and at its course beneath the clavicle, a hemodynamic evaluation with the Doppler mode should always be done additionally. With regard to the causes of thromboses in the upper extremities, subclavian vein and axillary vein regions, the following data are provided by Hill and Berry (1990) from 40 subclavian thromboses and Kerr et al. (1990) from 85 thromboses (Table 6.**9**):

The formation of thromboses is strongly influenced by the underlying illness in the various patient groups. Usually, catheter thromboses involve extremely ill patients who are often receiving substances intravenously that damage the venous wall.

All the enthusiasm for using imaging techniques to examine the veins should not lead us to forget that the diagnosis of venous thromboses in the upper extremities already showed good results with continuous wave Doppler sonography. For example, a study conducted by Tower et al. (1981) attained a sensitivity as high as 100%, and an equally good value for the specificity. The same applies to the detection of venous thromboses in the lower extremities using continuous wave Doppler sonography, as shown in Table 6.**4**.

Postthrombotic Venous System

Noninvasive follow-up evaluations after deep venous thrombosis have been carried out with continuous wave Doppler sonography and as well as with other measuring procedures, such as invasive and noninvasive venous pressure measurement or venous occlu-

Table 6.**8** The validity of phlebographically controlled studies using B-mode, duplex and color flow duplex sonography in diagnosing the level of deep venous thrombosis

First author	Year	Patients/phlebo-graphies (n)	Iliac vein Sensi-tivity (%)	Iliac vein Speci-ficity (%)	Common femoral vein Sensi-tivity (%)	Common femoral vein Speci-ficity (%)	Superficial femoral vein Sensi-tivity (%)	Superficial femoral vein Speci-ficity (%)	Popliteal vein Sensi-tivity (%)	Popliteal vein Speci-ficity (%)	Lower leg Sensi-tivity (%)	Lower leg Speci-ficity (%)
B-mode sonography												
Habscheid	1988	104/146	–	–	100	99	95	99	95	99	89	100
Habscheid	1990	238/301	–	–	100	100	97	99	97	100	94	99
Herzog	1991	13/101	78	98	–	–	100	100	98	98	60	97
Duplex sonography												
Dauzat[1]	1986	145/145	Sensitivity 100 →		Sensitivity 100 →			Sensitivity 94 →			62	–
Vogel	1987	54/ 54	Sensitivity 91 →		91			Sensitivity 94 →			–	–
Elias	1987	430/854	Sensitivity 100 →		100			Specificity 98 →			91	96
Stapff	1989	49/ 50	100	100	–	–	100	95	100	89	92	50
Krings[1]	1990	235/235	93	99	100	96	96	99	97	98	93	96
Color flow duplex sonography												
Fürst	1990	75/102	–	–	95	99	93	97	90	99	72	100
Rose	1990	69/ 75	100	100	→		Sensitivity 92, Specificity 100 →				73	86
Langholz	1991	116/ 65	100	100	–	100	99	94	100	96	89	91
Yucel	1991	45/ 45	–	–	–	–	–	–	88	–	88	96
Grosser	1991	325/325	Sens. 95–100, Spec. 99–100 →			100	Sens. 94–100, Spec. 99–100 →			→	93–100	99–100
Mattos	1992	75/ 77	–	–	94	100	100	98	85	92	94	81
Schweizer[2]	1993	78/ 70	100	100	–	–	96	100	100	100	95	100
Söldner	1993	84/103	Sens. 95–100, Spec. 98 →			→	95	98	97	100	93	97

[1] Or B-mode sonography + continuous wave Doppler sonography
[2] Intensified by a contrast medium

Table 6.**9** Etiology of thromboses in the subclavian vein and axillary vena regions

Thromboses in the upper extremities	Hill (1990) n = 40	Kerr (1990) n = 85
Catheter thrombosis	32%	61%
Carcinoma	23%	31%
Compression syndrome	45%	6%
Other causes	–	2%

sion plethysmography. B-mode or duplex sonography has introduced a new dimension to this process, in that it allows the measurement of the extent of the thrombus, and the reduction in size both in length and width. In addition, subsequent valvular insufficiency can be assigned chronologically and an indication of the age of the thrombus can be derived from its echogenicity and the caliber decrease during the subsequent stage. Also, echographic criteria for distinguishing between a fresh venous thrombosis and a rethrombosis in postthrombotic changes have been compiled.

The Strandness group (Strandness et al. 1983, Killewich et al. 1985) has been concerned for a considerable time with comprehensive long-term studies of postthrombotic venous changes. With the assistance of duplex sonography, Killewich et al. (1989a) showed that, in 53% of patients, all of the venous segments had opened after 30–90 days, while a small group (19%) showed an expansion of the thrombosis and an occlusion time lasting up to 180 days. In an additional group (14%), thrombosis growth recurred after seven days, but after this a tendency toward recanalization within 90–180 days was also observed.

Cronan and Leen (1989) observed 58 patients with a deep venous thrombosis over 16 months. The study was retrospective, and showed that 75% of veins in patients with a single femoral vein thrombosis or a popliteal vein thrombosis were open at the follow-up examination. Patients who had a lengthy occlusion involving both the femoral vein and the popliteal vein had open veins in only 33% of cases. In a prospective study lasting six months, Murphy and Cronan (1990) also confirmed the unfavorable course of lengthy thromboses in 46 patients suffering from acute venous thrombosis affecting the femoral and popliteal region. After three months, they observed that only 30% of the patients had undergone recanalization; in contrast, Killewich et al. (1989b) had found this in 53% of cases. After 180 days. Murphy and Cronan (1990) still found abnormal findings in 48% of the extremities; Killewich et al. (1989b) found residual effects in only 14% of their patients.

In a study by Mantoni (1991) involving 49 patients who were observed for one year (no specific examination dates were given), 36 extremities were without pathology and 73% of findings were normal, while 13 patients still showed residual effects. A very close connection between the extent of the throm-

boses and the tendency toward incomplete recanalization was also found in this study.

Sonographic measurement of the *age of venous thromboses* was carried out by Fobbe et al. (1991) and van Gemmeren et al. (1991). Fobbe sonographically examined 23 legs before venous operations. The echo structure of the histologically confirmed fresh thrombi (n = 17) was predominantly homogeneous and hypoechoic (76%), and the venous wall was not clearly defined in 11 patients. The remaining veins had a heterogeneous echo structure and distinct wall boundaries. The correlation of the *vascular quotient,* i.e., the venous diameter/arterial diameter quotient, proved to be significant. Both in the patients who underwent surgery and in 78 patients with thrombosis who did not receive surgery, the value at a thrombosis age of less than ten days was more than 2, and at a thrombosis age of more than ten days it was less than 2.

Richter et al. (1992) quantitatively analyzed thrombus echogenicity in 45 patients with 65 venous thromboses; they were not able to detect a relationship between echogenicity and the age of the thrombus. Similar results were obtained by Appelman et al. (1987) and Murphy and Cronan (1990). The latter also concluded that a decreasing venous diameter in the different vascular segments correlated better with the age of the thrombus than the echo structure did. As an example, the diameter of the superficial femoral vein during the first week amounted to 8 mm, and after five weeks, it was on average only 6 mm. O'Shaughnessy and FitzGerald (1996) investigated the aging process of a fresh thrombosis in 100 patients in a sequence of five investigations over 1 year and established that calf-vein thrombi resolve quickly with few long-term effects. The organization of thrombi above the knee was seen to be influenced by the age and physical condition of the patient.

Differentiating between a fresh initial thrombosis and an acute *re-thrombosis* in a previously damaged venous system has only been examined in a few studies. Rollins et al. (1988) found an acute venous thrombosis in 63 legs from a total of 76 examined extremities. An initial thrombosis was present in 15 cases, and an acute re-thrombosis was detectable in 48 instances. They were successful in detecting a chronic venous thrombosis, as distinct from an acute venous thrombosis, with a sensitivity of 94% and a specificity of 95%. Fobbe et al. (1991) and Langholz and Heidrich (1991) also focused on acute occlusions and postthrombotic changes, without separately validating the results. Since clinical indications concerning completed thromboses are often imprecise, a great deal of experience is required to make this differentiation successfully solely on the basis of the B-mode image. In a long-term study of 43 patients with thromboses, Gaitini et al. (1991) also discussed this problem, and recommended that ultrasound findings should be obtained about 6–12 months after venous thromboses, in order to be able to differentiate later acute re-thromboses from postthrombotic ultrasound results.

A recent study by the Strandness group (Meissner et al. 1995) refers to the differentiation of rethrombosis, propagation, and new contralateral thrombi. In a prospective examination of 204 extremities, re-thrombosis occurred in 31%, propagation in 30%, and new thrombi in 6% within one year. Unfortunately the study does not indicate if clinical symptoms were involved in these findings.

An earlier investigation conducted by Krupski et al. (1990) demonstrates propagation in as many as 38% of the 24 patients with deep venous thrombosis, even though all had undergone clinically controlled anticoagulation treatment. The study claims the clinical outcome did not differ in the initial presentation.

From these studies, it can be concluded that involutional developments in the thrombus, caused by retraction and fragmentation, fibroblast ingrowth, and intrathrombotic lysis, are often already complete within three months. They are certainly complete by the time six months have elapsed. Veins that are not patent at this time usually remain chronically occluded, or only partially recanalized. Data concerning the percentage of veins that still show residual effects 180 days after a thrombosis vary between 14% (Killewich 1989 b) and 48% (Murphy and Cronan 1990).

Valvular Insufficiency

Locating and quantifying venous reflux during the Valsalva maneuver or a compression maneuver has taken on a new dimension with duplex sonography. Retrograde venous flow can be reliably classified as belonging to the deep or superficial veins. Problems arise when *quantifying* the causal respiratory or compression maneuvers. Hirschl and Bernt (1990) applied the Valsalva maneuver with a specific pressure level (40 mmHg) and a specified time interval (10 s). Van Bemmelen et al. (1989) examined the reflux resulting during distal cuff deflation; the compression intensity and its duration were predetermined. Vasdekis et al. (1989), from the Nicolaides group, quantified the reflux automatically according to its velocity and quantity (ml/s). As in studies undertaken by Bork-Wölwer and Wuppermann (1991), Lindner et al. (1986), Marshall (1990) and also Hirschl and Bernt (1990), the diameters of the insufficient veins were also evaluated in Vasdekis' study.

Special attention has been given to *measuring the reflux during postthrombotic status*. In a comprehensive follow-up study conducted by Markel et al. (1992), involving 123 patients following a deep venous thrombosis, 17% of 106 patients who had had sufficient valves during the initial examination showed valvular insufficiency even during the first week. After one month, this figure had risen to 40%, and after a year, 69% had demonstrable valvular insufficiency, which developed in the recanalized veins in 40–50%. The veins predominantly affected were the common femoral vein (53%), the superficial femoral vein (44%), the popliteal vein (59%), and the posterior tibial vein (33%). In contrast, the reflux prevalence in the control group without deep venous thrombosis was only 6%. In a long-term study following deep venous thrombosis, Lindner et al. (1986) also found valvular insufficiency to the extent of 63%.

However, they were not able to find a clear correlation between the hemodynamic status of the extremities and the valvular insufficiency.

Van Ramshorst et al. (1994) compared reflux duration in 21 healthy individuals and 27 patients. They found a threshold of 0.5 seconds between the reflux duration of normal veins and the incompetent vein segments that showed values two or more times longer than the normal values of reflux duration.

Using duplex sonography, Hirschl and Bernt (1990) and Marshall (1990, 1991) examined the *insufficiency of the conducting veins* following chronic venous insufficiency, which also occasionally appears without a preceding thrombosis.

The available venous function parameters, such as diameter, area measurements, and also flow velocities, showed a large standard deviation in the study by Hirschl and Bernt (1990). No correlation could be found with the levels of seriousness of the manifest venous insufficiency.

In contrast, Labropoulos et al. (1994) found that the severity of chronic venous insufficiency in 217 legs increased in limbs with combined superficial and deep venous distal reflux. Reflux confined to the deep veins alone is less harmful. Shami et al. (1993) recorded similar findings in a smaller study.

Vasdekis et al. (1989) tested refluxes in *symptomatic varicosis* in 47 extremities, and confirmed that a reflux of more than 10 ml/s correlates closely to skin damage and ulcers. In cases that demonstrated no correlation, they postulated that additional factors (insufficient perforating veins, poor pump function in the calf muscles, ankle joint immobility) needed to be considered. Van Bemmelen et al. (1990) also examined valvular insufficiency in 42 patients with ulcerations, and were also able to detect extensive valvular insufficiency related to skin lesions.

The best way to quantify venous reflux is still a matter of debate. Weingarten et al. (1996) found that total limb reflux time correlated significantly with a specific air-plethysmographic variable. A total reflux time greater than 9.66 seconds was predictive of venous ulceration in this study.

In contrast, Rodriguez et al. (1996) found that duplex-derived valve-closure times do not correlate with the magnitude of reflux and should not be used to quantify the degree of reflux.

Despite the methodological problems associated with the quantitative measurement of reflux, purely *qualitative* evaluation of venous valvular insufficiency and also evaluation of the location and extent of reflux has proved to be significant, both prognostically and in relation to the etiology of chronic venous insufficiency. The *pathogenesis of chronic venous insufficiency* should be identified as being due to one of the following:

- Deep venous thrombosis
- Conducting vein insufficiency
- Perforating vein insufficiency
- Primary varicosis

Insufficient Perforating Veins

The diagnosis of *insufficient perforating veins* has been revived with the use of color flow duplex sonography. The insufficiency can be clearly demonstrated through the changed flow direction within the perforating vein on distal compression. Franzeck et al. (1993) and Stiegler et al. (1994) produced comprehensive validating studies. However, Franzeck et al. (1993) confirmed that there was still a significant discrepancy between clinically suspected venous insufficiency cases (n=62), those ascertained using continuous wave Doppler sonography (n=37), and those detected using color flow duplex sonography (n=20). Using color flow duplex sonography, Stiegler et al. (1994) compared the preoperatively diagnosed cases of insufficient perforating veins (n=334) in 94 patients with primary varicosis to the intraoperative findings. The success rate with which the previously diagnosed and marked insufficient perforating veins were surgically confirmed amounted to 96%. By contrast, the perforating veins detected in 31 phlebograms could only be surgically confirmed in 65% of cases. The study shows that preoperative localization of insufficient perforating veins using color flow duplex sonography can produce a significant improvement in their targeted surgical removal.

Thrombophlebitis

The imaging of superficial venous thrombotic inflammations using B-mode sonography has only begun to attract attention recently, since the clinical diagnosis of thrombophlebitis is usually quite clear (Skillman et al. 1990). However, the first larger studies organized by Pulliam et al. (1991) and Lutter et al. (1991) show that the inflammatory thrombotic processes in the superficial veins are occasionally combined with complications and thrombus growth into the deep venous system. In a large retrospective investigation involving 12,856 venous examinations using duplex sonography, Lutter et al. (1991) found 186 patients with primary superficial phlebitis. Of these, 57 patients (31%) had a complication. In eight instances (4%), a pulmonary embolism occurred; in 28% (n=53) an expansion of the thrombosis into the deep venous system was observed. In 34% of cases, great saphenous vein phlebitis was associated with complications; small saphenous vein phlebitis yielded complications in 16% of the instances. Isolated local phlebitis led to complications in only 8% of the occurrences.

The following predisposing factors should be noted: age >60 years, female gender, prior deep venous thrombosis, bed rest, and systemic infection.

In a group consisting of 20 superficial thrombophlebitis cases, Pulliam et al. (1991) discovered a subgroup of six patients who developed progressive phlebitis involving the deep venous system. In five individual cases with great saphenous vein varicophlebitis, Schönhofer et al. (1992) and Yucel et al. (1992) both discovered appositional thrombus growth, evident in color flow duplex sonography. Schönhofer et al. (1992) observed a free-floating appositional throm-

bus in the common femoral vein, and were able to follow the course of the loosened thrombotic material with centripetal drainage without noting any signs of pulmonary embolism.

The examples described, some of which are still only case studies, support the value of the diagnostic procedure in acute extensive thrombophlebitis, especially when it involves the great saphenous vein. The utmost vigilance, together with closely-spaced ultrasound follow-up examinations, are required to recognize appositional growth into the patent lumen of the common femoral vein, and have it removed surgically if required. The connection between thrombophlebitis and deep venous thrombosis is also discussed in a more recent investigation. In this study, Skillman et al. (1990) were able to detect a deep venous thrombosis (without complications) in five of 42 patients suffering from thrombophlebitis (12%). In a prospective study, Jorgensen et al. (1993) discovered a deep venous thrombosis in ten of 44 patients with thrombophlebitis (23%). However, it could only have been transmitted onward in three of these patients. Predisposing factors for the combined appearance of thrombophlebitis and a deep venous thrombosis could not be identified in this study. In a recent large study, Chengelis et al. (1996) reviewed the records of 263 patients with superficial venous thrombosis. Thirty (11%) patients showed progression, with the most common site being progression of the disease from the greater saphenous vein into the common femoral vein (70%). Chengelis et al. recommend anticoagulation therapy in proximal great saphenous vein thrombosis.

Special Sets of Findings

Congenital venous dysplasias are not observed often enough for validating studies of their examination with duplex sonography to be available. However, there is no disputing the immense diagnostic benefit which the imaging procedures have provided, together with functional Doppler-sonographic examinations of venous flow. Color coding of the Doppler shift has made the diagnosis of unusual anatomical formations, e.g., arteriovenous fistulas or aneurysms, much easier. The same is true of *secondary arteriovenous fistulas* and *acquired venous aneurysms*. The latter are receiving increasing attention in the literature. Ritter et al. (1993) published an overview examining the frequency and distribution of venous aneurysms in the literature and in their own patients (n= 152 of 1000 phlebograms). Hach (1985) found great saphenous vein aneurysms (10.9%) in a trunk varicosis of the superficial veins. But venous system aneurysms can also occur in other body regions. According to Ritter et al. (1993), who evaluated 152 phlebographically confirmed cases, the aneurysms were located as follows: 65% in the lower extremity, 14% in the upper extremity, and the remainder at the head, neck, thorax, and abdomen.

Weber and May (1990, p. 358) emphasize that the origin of an aneurysm cannot be deduced with certainty from the phlebographic or histological images, or from its form (fusiform or saccular). The same is true, of course, for B-mode images. Further examination of the venous system is necessary to classify aneurysms as being due to congenital angiodysplasias or to secondary local wall degenerations (usually based on a varicose symptom complex or bypass dilatations).

It should be mentioned that the genesis of an unclear pulmonary embolism can be traced back to a venous aneurysm (Aldridge et al. 1993). However, according to this study, only 22 symptomatic venous aneurysms affecting the popliteal vein have so far been described in the literature throughout the world.

Cost-Effectiveness

The issues of quality assurance and cost-effectiveness have been playing a significant role internationally for some years now, especially with regard to new diagnostic procedures. This applies particularly to ultrasound procedures, since there is no inherent limit to the number of times these noninvasive procedures can be applied.

In his presidential address, Kempinski (1994) cited the example of the Medicare carrier for Michigan that became frustrated with its inability to determine which venous duplex scans were indicated and simply refused reimbursement for all of them.

The following considerations should govern all ultrasound examinations: *which examiner can* conduct the examination effectively, *which procedure* will provide a diagnosis most economically, and *how often* will follow-up examinations be necessary? An increasing number of publications discuss these problems in association with duplex examinations conducted for suspected deep vein thromboses.

Poppiti et al. (1995) compared the accuracy of a limited B-mode compression technique (two sites per limb) with a complete color flow duplex examination for the detection of proximal deep vein thrombosis. The sensitivity for B-mode technique was 100%, the specificity 98%. Color flow duplex sonography was considered the gold standard. Strothman et al. (1995) investigated the clinical necessity of contralateral venous duplex scanning in patients with unilateral symptoms. They concluded that unilateral scanning would result in improved cost-efficiency for vascular laboratories. Fowl et al. (1996) analyzed the indications for duplex scans retrospectively in 2993 cases: 74% of all results were completely normal, while only 13% of the scans detected acute proximal deep vein thrombosis. He claimed this diagnostic method is being inappropriately used.

These exemplary studies demonstrate the necessity of better-defined guidelines for ordering venous duplex scanning. It will, however, still be difficult for the director of a vascular laboratory to refuse to carry out investigations prescribed by a general practitioner or clinic physician.

7 Abdominal Arteries

Examination

Special Equipment and Documentation

Classic *continuous wave Doppler sonography* on its own, without an additional B-mode image, is not used in investigations of the abdominal organs. In contrast to the neck region and the peripheral arteries, the anatomical classification of individually registered signals by assigning them to the various abdominal arteries is uncertain. In addition, the penetration depth provided by the usual pen-shaped probes is not usually sufficient to detect the deep retroperitoneal vasculature reliably. However, there is some clinical value in ultrasound studies of subcutaneous vascular prostheses (e.g., axillofemoral and axillobifemoral bypasses).

Duplex and color flow Doppler sonographic examinations of the abdominal cavity are therefore highly dependent on the clear display of the tissues and vascular structures provided by B-mode sonography. This ensures a proper anatomical and morphological orientation. Usually, sector or vector transducers with a B-mode frequency of around 3.5 MHz allow the display of more extensive arterial segments in each image. Multifrequency transducers, which do not require the transducer to be changed, have proved useful in providing a lower transmission frequency, usually around 2 MHz, to increase the penetration depth in obese patients (Figs. 1.**14**–1.**17**, pp. 17–20). Because of the lower penetration depth, a 5-MHz transmission frequency is adequate only in slim patients for the imaging of intra-abdominal vascular structures.

Three-dimensional reconstructions of two-dimensionally produced B-mode ultrasound sections provide a more sculptural representation of the arteries, and particularly of their position in relation to one another and to surrounding structures (cf. Fig. 8.**3**, p. 283). At present, the advantage of this type of imaging lies in its ability to document the findings more clearly, rather than in providing any fundamentally new diagnostic insights. Potential advantages might include the greater reproducibility of findings that the method provides, more exact volume measurements (e.g., in aortic aneurysms) and, with digital storage of the image data, the ability to select freely and reproduce at will various sectional planes from a particular block of tissue. The latter facility would allow an experienced investigator to reexamine special findings at a later date, without depending on the fixed two-dimensional documentation provided by the initial examiner.

Transgastric ultrasonography can be applied additionally during intraoperative procedures in patients with unfavorable transabdominal sonographic conditions and may be useful for evaluating atherosclerotic disease of the abdominal aorta, renal arteries, and mesenteric arteries (Kenn et al. 1996).

Qualitative and quantitative evaluation of flow within the abdominal arteries require continuous wave or pulsed wave Doppler modes. In the smaller arteries, the pulsed wave Doppler mode shows better results than continuous wave Doppler sonography because it can identify the location and depth of the registration point more precisely. On the other hand, the continuous wave Doppler mode can measure the high flow velocities associated with high-grade and extremely high-grade stenoses more precisely.

B-mode is often used on its own in abdominal sonography to detect arteriosclerotic wall plaques and aneurysms of the abdominal aorta and the pelvic arteries.

Documentation of the results follows the criteria given in Chapter 1 (pp. 20–24).

Examination Conditions

External influences are of decisive significance in precise duplex-sonographic studies, and not only during complex investigations or in difficult examination conditions (obesity, gaseous superimposition). Assuming that the equipment has been optimally adjusted, it is necessary to have a room that can be darkened to ensure good reproduction of the examination findings on the monitor. It must also be possible to provide comfortable positions for both the patient and the examiner during the examination. Otherwise, especially during longer examinations, the difficulty of acquiring sectional images may have a negative influence on the quality of the findings.

Calculating and allowing sufficient time for the examination is an absolutely necessary prerequisite. The time required is not only decisively dependent on the examiner's experience and the individual examination conditions presented by the patient, but is also affected by the result of the investigation. Pathological results may require further examinations.

Patient and Examiner

The examination is usually carried out with the patient in supine position. Under special circumstances (pregnancy, obesity) or to display intrarenal arterial segments, it may be better to place the patient in a lateral position. The patient should always be in a relaxed position, since any tension in the abdominal muscles increases the distance from the transducer to the abdominal arteries. A suitable support placed under the knees helps to relax the abdominal muscles.

In elective, non-emergency examinations, the patient should have fasted for at least eight hours before the procedure, since gas and food remnants in the stomach and intestines complicate the examination. In addition, carminative medications can be administered on the day preceding the examination. Both measures help to improve the examination conditions, but they are by no means indispensable for obtaining a good sonographic image of intra-abdominal structures.

If the abdominal arteries are superimposed by the intestinal wall and intestinal contents, continuous gentle pressure with the transducer for several minutes can displace these structures and significantly improve the examination conditions.

Conducting the Examination

■ Continuous Wave and Pulsed Wave Doppler Sonography

Due to the insufficient penetration depth and inadequate resolution of the usual pen-shaped probes, *continuous wave* Doppler sonography is not used on its own in the abdominal region. *Pulsed wave* Doppler sonography is only used in the context of duplex sonography.

■ Conventional and Color Flow Duplex Sonography

Presented below is a basic description of the various examination techniques in the organs and arteries, including appropriate positioning of the transducer. Owing to the number of intra-abdominal blood vessels that can be specifically evaluated with color flow duplex sonography, several specialized examination modalities are also included in the description of individual arteries in the section on normal findings—particularly since the individual sonographic components involved in the B-mode, Doppler mode, and color flow Doppler mode require the use of special sections.

B-mode. The *abdominal aorta* has to be searched for in sagittal and several transverse sections. In both sectional planes, the transducer is placed directly distal to the xiphoid process, and is guided distal to the level of the navel (Fig. 7.**1**).

Depending on the examination conditions in the individual patient, transabdominal access can proceed from medial, right, or left paramedian orientations. The latter access point is often the most favorable one, since the abdominal aorta is located on the left, ventrolateral to the spinal column, and the distance to the transducer is therefore shorter.

A complete examination assesses the entire intra-abdominal course of the abdominal aorta, from the subphrenic segment proximal to the celiac trunk as far as the aortic bifurcation.

To measure the lumen, an attempt should be made to display the cross-sectional diameter continuously throughout the entire course in transverse sections. When measuring the cross-section, particular care must be taken to adjust an exact transverse plane, in order to avoid diagonal vascular sections. This is especially important when following the course of an aortic aneurysm and attempting to detect possible progression in the vascular wall dilatation.

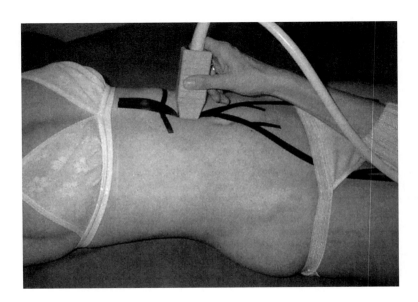

Fig. 7.**1**   The correct transducer position in duplex sonography for a paramedian examination of the infrarenal segment of the abdominal aorta, proceeding from the left to provide a sagittal section. The guidelines show the path for imaging of the abdominal aorta, with the origins of the renal arteries and large pelvic arteries (external and internal iliac arteries) as far as the groin

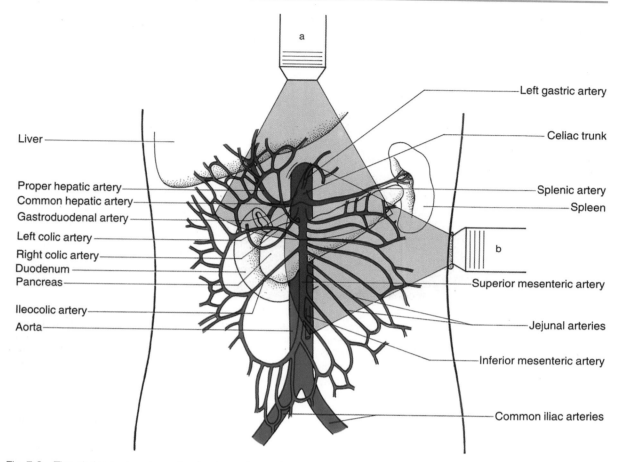

Fig. 7.2   The abdominal aorta, celiac trunk, and superior and inferior mesenteric arteries with their branches. Example transducer positions are shown. **a** Transverse, **b** sagittal sections

The origins of the large abdominal branches near the aorta can also be observed (Fig. 7.2) using median to paramedian sagittal and transverse sections, which are described in more detail below. In normal conditions, the amplification *(gain)* is adjusted so that the arteries only show weak intraluminal reflections, while the veins show no reflections.

Transabdominal examination of the *uterine artery* should be carried out when the patient's bladder is full, to take advantage of the natural water buffer placing the vascular structures in a more favorable position for the ultrasound.

Doppler and color flow Doppler mode. In *all the abdominal arteries,* it is important to note that Doppler examinations are optimized, or may even only be possible, when the Doppler signal meets the vascular flow direction at the smallest possible incident angle. This means that in addition to the sectional planes used in B-mode examinations, transducer positions need to be selected for duplex examinations that allow imaging of the target arterial segment at the most acute angle possible between the vascular axis and the transducer.

In color flow mode, this allows optimal coding of the mean flow velocity and flow direction. In addition,

using the Doppler sample volume (gate), frequency spectra should be registered at different locations within the blood vessel, again at an acute angle, in order to obtain representative information about the bloodstream.

In contrast to the strict requirements for maintaining appropriate sectional planes when measuring the vascular caliber (cf. Fig. 7.25, p. 255), the relatively large distance between the *abdominal aorta* and the transducer, and the vessel's course ventral to the spinal column, makes it necessary to turn the transducer onto the plane of the longitudinal vascular axis. Then the examiner will achieve optimal color coding of the aortic lumen as well as angle-corrected measurements of the flow velocity.

Identification of the *celiac trunk* and its branches, the common hepatic artery, the splenic artery, and the left gastric artery, is successful in transverse and sagittal sectional planes, with the transducer positioned several centimeters below the xiphoid process (Fig. 7.3). The *common hepatic artery* and the *splenic artery* are examined along their course in a targeted fashion, using transverse sections adapted to the vascular axes. Adequate imaging of the left gastric artery is not always possible. Near the hepatic arteries, the course of

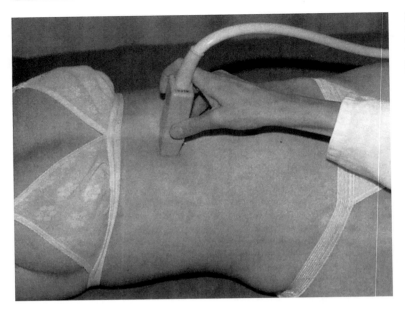

Fig. 7.**3**   The transducer position for duplex-sonographic imaging of the celiac trunk and its branches in transverse section. The transducer approaches from a subxiphoid direction. In comparison to Figure 7.**4**, showing transducer placement for examining the renal arteries, the transducer here is positioned directly below the xiphoid process, relatively far cranially

the common hepatic artery and its branching into the left and right proper hepatic arteries can be examined.

The proximal segment of the *superior mesenteric artery* can usually be followed for a longer distance in transverse sections, as well as in median or paramedian sagittal sections. Velocity measurements are possible in longitudinal sections near the origin, and in the proximal vascular segments (Fig. 7.**13a, b**). For specific questions, comparative measurements of the superior mesenteric artery both before and after a specific stimulating meal are advisable (Aldoori et al. 1985, Sieber et al. 1992).

The *renal arteries* are examined near the aorta from beneath the xiphoid process (Fig. 7.**4**) using an adapted transverse section, or in a paramedian location along their longitudinal axes. Subsequent measurement of the intrarenal velocity spectra can be carried out from a dorsolateral or, more rarely, from a ven-

trolateral orientation to the kidneys (Fig. 7.**5**). In addition to the much simpler evaluation of the segmental arteries and the interlobular arteries provided by color flow duplex sonography, this transrenal access also facilitates the detection of the distal components of the renal arteries that are near the hilum (Fig. 7.**6**).

In addition to the use of sonographic B-mode imaging to detect plaques (Fig. 11.**55e, f**, p. 443), the evaluation of the renal arterial system requires the registration of several frequency spectra along the course of each renal artery, including the intrarenal components and those near the hilum. A sonogram of the renal parenchyma is obligatory, including the presentation of the kidney's size, form, the relationship between the parenchyma and the renal pelvis, and its echogenicity (Otto et al. 1981).

In kidney transplants that are located near the surface nonanatomically, special care should be taken

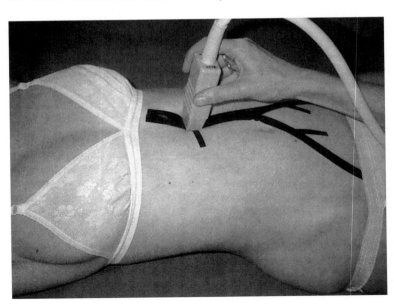

Fig. 7.**4**   Positioning the transducer for duplex-sonographic imaging of the proximal segments of the renal arteries near the origin (transverse section)

Fig. 7.**5**   Positioning the transducer for duplex-sonographic examination of the renal parenchyma in longitudinal section, with the intrarenal vasculature and the distal renal artery

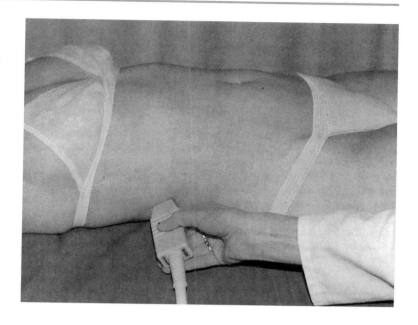

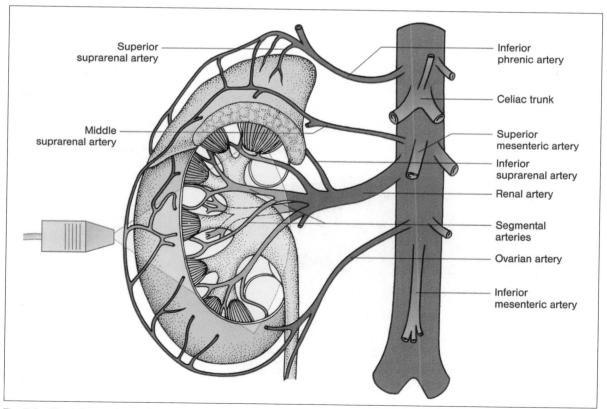

Fig. 7.**6**   The intrarenal vasculature and renal artery, showing potential collateral vessels during obstructions of the organ artery (inferior phrenic artery, suprarenal artery, ovarian artery). The collaterals can rarely be displayed with duplex sonography if the renal artery remains patent. Increased arterial caliber due to collateral function may result in possible confusion with the main artery

not to exert pressure on the transplant with the transducer because this can lead to decreased arterial inflow and elevated peripheral resistance.

For flow velocity measurements, the *inferior mesenteric artery* is best displayed near its origin in sagittal sections proximal to the aortic bifurcation. It can often be imaged simultaneously with the superior mesenteric artery (Fig 7.**19**).

The *lumbar arteries* are sometimes depicted using color flow Doppler sonography in a transverse section as small arterial branches originating dorsolateral to the aorta. Ultrasound imaging is not always successful, and their examination with duplex sonography has therefore only acquired limited clinical significance. However, it is important to be able to recognize them when identifying other abdominal arteries, and when an aortic flow obstruction is collateralized through these blood vessels, which then have a larger caliber.

In the lower abdominal arteries, clinically interesting vessels include the internal iliac artery in men, particularly in relation to clarifying vascular impotence, as well as the ovarian arteries and the uterine vasculature in women. The *internal iliac artery* can be displayed at its origin by proceeding transabdominally in an almost sagittal plane from ventromedial to dorsolateral, initially following the course of the external iliac artery (Fig. 7.**7**). The artery often turns off in a dorsomedial direction proximal to the dorsal convex arch formed by the vascular system of the iliac artery (Fig. 5.**22**, p. 154).

In favorable examination conditions, the *ovarian artery* can be depicted within the ovarian suspensory ligament in order to register Doppler frequency spectra from the vessel. Transabdominally, the *uterine artery* can be examined with ultrasound in a transverse section, supravaginally at the level of the cervix. If possible, women patients being examined should have full bladders.

Although transabdominal access is basically possible for displaying the female pelvic vasculature (Burns 1988), several examiners routinely apply ultrasound transvaginally. According to a large international survey dating from 1990, these procedures were carried out in 90% of cases by obstetricians or gynecologists (Bernaschek et al. 1991). The examinations are closely related to gynecological or obstetric examinations, and are carried out for *specific* purposes deriving from the clinical indication (Kurjak et al. 1991). Values measured transabdominally for different flow parameters are basically comparable to the transvaginal results (Funk et al. 1992).

Facilities for imaging the fetomaternal vasculature to monitor pregnancy and the development of the embryo and fetus can only be referred to here in passing (Fig. 7.**42 a, b,** p. 277). The methodologies of these applications are in a constant process of further development, and extensive studies of them are available elsewhere; the specialist gynecological and obstetric literature should be consulted.

Color Doppler energy mode. When examining the abdominal arterial circulation with the color flow Doppler energy mode, the same examination conditions apply to patient positioning and sectional planes as in B-mode and Doppler-sonographic procedures. The special advantages of this amplitude-weighted procedure lie in the fact that it can display even slower flow in even smaller blood vessels. However, it requires particularly steady and almost static preliminary fixation of the sectional plane to avoid movement artifacts. Due to the physiological displacement of the abdominal organs during respiration, the examination is usually carried out with halted inspiration. After the other sonographic procedures have provided preliminary information about the region being examined, the amplitude-weighted color flow mode can be used for *specific* color coding of the small

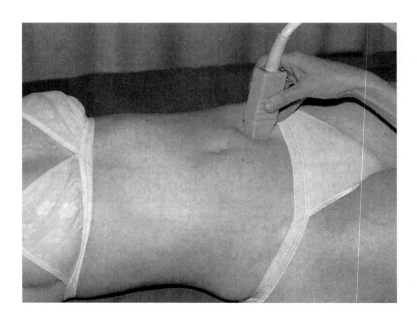

Fig. 7.**7** Transducer placement for duplex-sonographic imaging of the internal iliac artery in longitudinal section. Compare this with Figure 7.**1,** showing the sectional planes for examining the internal iliac arteries

intraparenchymatous vasculature in the liver, kidneys, and spleen (Fig. 7.**18**). For quantitative evaluation of the flow velocity, the pulsed wave Doppler mode, also using color, is used in combination with the color flow Doppler energy mode.

New *ultrasound contrast agents* can pass the lungs without impairment after peripheral intravenous application and amplify reflected sound waves. The amplification produced by appropriate concentrations of the contrast agent markedly improved visualization of blood movement in the abdominal arteries (Leen and McArdle 1996, de Jong and Ten-Cate 1996). The use of *harmonic imaging* with ultrasound contrast agents may also give additional flow information and improve the feasibility and diagnostic value of color Doppler ultrasound by lowering the threshold of detectable velocity in the abdominal arteries and enabling detection of smaller vessels (Burns 1996). Future developments with modification of ultrasound equipment should increase the capabilities of these agents to improve imaging as well as Doppler sensitivity (Goldberg et al. 1994).

Examination Sequence

There is no rigid sequence of examination steps for the abdominal arteries. Displaying the abdominal aorta in either a sagittal or transverse ultrasound plane is usually the first step, however, as it provides an intra-abdominal orientation for the arterial vasculature.

The further course of the examination depends on its purpose. If there is no special indication, then routine analysis of frequencies from the various upper abdominal arteries will not be particularly useful, in view of the significant time investment it requires. Specific clinical objectives are more likely to decide the further course of the examination. The sequence of individual examination phases for the renal arteries, including the intrarenal components and the other abdominal arteries, is described above.

For special investigations, transabdominal ultrasound examinations can be supplemented by endoscopic procedures (endoscopic color Doppler sonography), which cannot be discussed in detail here (Frazin et al. 1994, Keen et al. 1996, Koito et al. 1996). This technique, when integrated with fiberoptics for guidance, could provide a method for abdominal vascular ultrasound with less tissue-imposed attenuation, providing high resolution imaging and allowing additional structure recognition. This has potential applications for studying patients with disease of the abdominal aorta or its branches.

Normal Findings

Principle

■ Conventional and Color Flow Duplex Sonography

B-mode. Real-time sonography provides the examiner with a morphological image of the abdominal organs and vasculature. Particular note should be made of the location of the arteries, their diameter, their wall thickness and structure, and also any vascular pulsations.

Along with duplex and color flow sonography, real-time B-mode imaging provides an indispensable sonomorphological and anatomical basis for all subsequent and supplementary color flow Doppler or frequency spectrum analyses.

Doppler mode. Depending on the specific *peripheral resistance* of the organ supplied, the *vascular caliber*, and the absolute *flow volume,* the frequency spectra of the abdominal arteries show various differences. These are described in the individual sections on specific arteries under "Findings" below. For example, in the abdominal aorta, the diastolic flow velocity proximal to the origin of the upper abdominal arteries is higher than in its distal segment.

Moreover, the flow velocities within the individual arteries are affected by *respiratory, digestive, hormonal,* and *pharmacological* influences. For example, the resistance indices in the abdominal arteries increase during inspiration, and increase even more during the Valsalva maneuver (Kamps et al. 1992, Iwao et al. 1996). After eating, or after intravenous injection of various peptide hormones (e.g., gastrin, glucagon, secretin, cholecystokinin), the systolic and end-diastolic flow velocity, as well as the vascular caliber of the superior mesenteric artery and celiac trunk, show a dose-dependent increase (Lilly et al. 1989, Sieber et al. 1991), or the quick postprandial increased flow may be blocked (Kooner et al. 1989). In the uterine artery and the ovarian artery, hormonal and pharmacological effects lead to menstruation-related and pregnancy-dependent changes in the flow pattern (Zaidi et al. 1996, Coppens et al. 1996).

To calculate the *flow rate* (ml/min), the mean flow velocity within the artery is multiplied by the vascular cross-section (cf. p. 282).

Color flow Doppler mode. Time-saving examinations of particularly *small* vessels within the abdominal cavity, which are difficult or even impossible to depict using conventional methods are specific indications for the use of color flow Doppler mode.

Since most abdominal arteries hardly ever follow a course along a straight axis, or only for a short distance, the color coding of this vasculature is determined by the frequent relative changes in the incident angle of the color flow Doppler. The use of a sector or vector transducer already involves an additional angle-dependence of the color coding, depending on the location of the vascular segment within the ultrasound window. This dependence is also present in blood vessels that have a straight course. Due to a change in the direction of flow relative to the transducer, this results in quickly changing color coding (red/blue) over short vascular distances. Quantitative measurement of the flow velocity in the abdominal arteries always requires *angle-corrected measurements* obtained from the frequency spectrum. Due to the visually detectable flow direction, this is an easier and particularly secure method when the arterial wall cannot be completely depicted by sonography.

Due to the uniform color coding of similar average flow velocities, a vascular region in which the mean velocity deviates from that of the remaining vascular system, i.e., has a lighter color, is clearly noticeable. Within this arterial segment, frequency analysis using the *pulsed wave* Doppler mode must be carried out. This *selective application* reduces the examination time significantly in comparison to conventional duplex sonography.

Detecting slow flow velocities, such as those seen in the arteries of the intestinal wall, using color flow duplex sonography is at present only occasionally possible. Further developments will show the extent to which flow phenomena can be reliably detected in these small vessels in order to distinguish between hyperemic and ischemic intestinal wall lesions.

Color Doppler energy mode. The amplitude-weighted color flow mode makes it possible to increase the sensitivity to *slow* flow velocities that can be displayed in color by a further order of magnitude. In comparison with Doppler and color Doppler procedures, the color coding in the energy mode depends significantly less on the incident angle of the ultrasound waves. The appropriate field of application for the method follows from this: identifying even smaller vessels than was previously possible with color flow Doppler sonography, and color-coding of arterial circulation that cannot be evaluated with ultrasound when it is applied at a small (Doppler) incident angle, and is therefore not adequately color coded, or not at all. However, it should be noted that *no differentiation of the flow direction* is possible with the color flow Doppler energy mode, nor can quantitative velocity measurements be carried out. A distinction between abdominal arteries and abdominal veins is therefore not possible with color coding alone.

Anatomy and Findings

Exact anatomical knowledge of the spatial relationships in the abdominal vasculature is a fundamental requirement for examining the abdominal arteries (and veins) using (duplex) sonography. In the section on anatomy that follows, the vascular anatomy can only be described in a general way to provide overall orientation. For more detailed information, anatomical and radiological atlases should be consulted (e.g., Luzsa 1972).

Abdominal Aorta

Anatomy. The *descending aorta* passes through the aortic hiatus at the level of the twelfth thoracic vertebra into the abdominal cavity, and depending on body size and other factors measures approximately 16–25 mm in diameter (Kremer et al. 1989). It proceeds somewhat to the left of the sagittal median plane and ventral to the vertebral bodies in a stretched manner and in a caudal direction, to branch into the common iliac arteries at the level of the fourth lumbar vertebral body.

■ Conventional and Color Flow Duplex Sonography

B-mode. The abdominal aorta is displayed as a hypoechoic to anechoic vascular band with delicate wall structures ventral to the vertebral bodies. The cross-sectional diameter decreases slightly as it proceeds from its subdiaphragmatic segment to the bifurcation.

Doppler mode. The flow velocity curve of the aorta has a triphasic signal. The early diastolic backflow component is developed to varying degrees, depending on the peripheral resistance. In the cranial aortic segments proximal to the origins of the visceral and renal arteries, a greater diastolic flow velocity, with a lower early diastolic backflow, can be registered than in the distal aortic segments, which have a pronounced backflow and absent or decreased diastolic flow (Fig. 7.**8 a, b**).

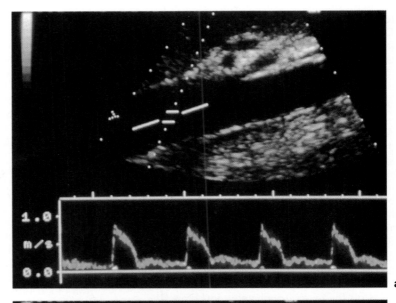

a

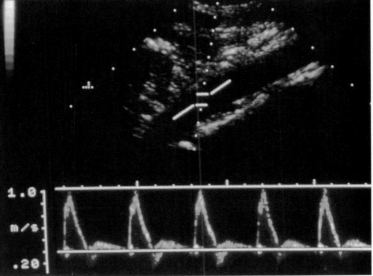

b

Fig. 7.**8**   A duplex sonogram of the abdominal aorta, **a** *proximal* to the origin of the celiac trunk and **b** in the *distal* segment after the origin of the celiac trunk and the superior mesenteric artery, inferior mesenteric artery, and renal arteries, along with the corresponding Doppler frequency spectra. In the proximal segment, there is confirmation of clear holodiastolic flow, while distally a triphasic frequency spectrum can be registered, corresponding to the greater peripheral resistance in the *subsequent* vascular system

Overall, lower flow velocities are present in the abdominal aorta than in the peripheral arteries, due to the relatively wide vascular caliber of the aorta. The flow velocities lie between 70 cm/s and 140 cm/s. The *mean* flow velocity lies between 40 cm/s and 70 cm/s.

However, flow velocities of a maximum 140–160 cm/s can still be physiological in delicate, rather narrow-caliber abdominal arteries. In these arterial segments, local circumscribed zones with a *relatively* increased velocity should then be given particular attention.

Color flow Doppler mode. Proximal aortic segments are displayed in a sectional plane tilted cranially, and the distal sections, including the bifurcation, are shown in an ultrasound plane that is tilted caudally. As previously described, uniform arterial color coding with normal vascular caliber excludes the presence of a flow obstruction.

When using a sector or vector transducer, arterial segments that are examined with the color flow Doppler at an acute incident angle show a higher calculated Doppler shift in their color flow. This is indicated by the display of lighter colors, although a higher flow velocity is not necessarily responsible for this.

Celiac Trunk and Branches

Anatomy. The *celiac trunk,* measuring 0.6–1.3 cm in diameter, is the first substantial subdiaphragmatic arterial branch given off by the aorta, in the midline or somewhat to the left of it, in a ventral direction. At the level of the twelfth thoracic to the first lumbar vertebral body, the celiac artery branches into the common hepatic artery, the splenic artery, and the left gastric artery. The *common hepatic artery* turns to the right in the hepatoduodenal ligament, and divides into the *gastroduodenal artery* and the *proper hepatic artery.* At the liver hilum and after the origin of the *right gastric artery,* the latter divides into a right branch, which gives off the *cystic artery,* and a left branch.

With a diameter of 0.4–0.8 cm, the splenic artery proceeds cranially to the splenic vein, which has a large caliber and runs dorsal or cranial to the pancreas for a distance of 7–14 cm peripherally, dividing into 6–30 terminal branches at the hilum of the spleen. The third and smallest branch of the celiac trunk, the *left gastric artery,* runs to the lesser gastric curvature in the gastropancreatic fold, and anastomoses there with the right gastric artery.

■ Conventional and Color Flow Duplex Sonography

B-mode. The celiac trunk can be clearly depicted in sagittal sectional planes cranial to the origin of the superior mesenteric artery (Fig. 7.**9 a**), and also in a transverse sectional plane with branches into the arteries (Fig. 7.**9 b**). Anatomical variants are very likely to be encountered (Wenz 1972). In such cases, precise differentiation and anatomical classification of the ar-

teries is only possible by displaying the vascular course over a longer distance and in this way identifying the organ system to which it belongs.

Doppler mode. The flow direction of the celiac artery is directed ventrally toward the transducer. Influenced by the actual peripheral resistance in the subsequent vasculature (proper hepatic artery, left gastric artery), the normal frequency spectrum has a monophasic profile (Fig. 7.**9 d**) and its systolic flow velocity is around 138 cm/s ± 99 cm/s, with a generally low end-diastolic velocity (Mallek et al. 1990 b). In a study carried out by Qamar et al. (1985) involving 42 test subjects, the flow volume was found to be 703 ml/min ± 24 ml/min (471–1017 ml/min). No significant age-dependent or sex-dependent differences were detected. Deep inspiration decreased peak systolic and end-diastolic velocities of the celiac artery origin. Proximal to distal Doppler velocities of normal celiac artery origins were comparable (Geelkerken et al. 1996).

Postprandially, within the first 90 minutes after the ingestion of 300 ml of a mixed nutrient solution, Moneta et al. (1988) registered a slightly increased systolic, end-diastolic, and mean flow velocity, and also flow volume around 20 ± 6%, 24 ± 9%, and 18 ± 4%, and also 18 ± 4%. The increase observed was not significant in only seven of the test subjects. Qamar et al. (1985) described an increased flow volume of around 38% (n = 12) directly after introducing 400 ml of a mixed nutrient solution containing 1700 kJ (405.5 kcal).

According to data provided by Sato et al. (1987), and Nakamura et al. (1989), the mean flow velocity in the *splenic artery* lies between 18.7 cm/s ± 4.2 cm/s and 34.6 cm/s ± 9.5 cm/s, and that of the *common hepatic artery* is 31.3 cm/s ± 13.3 cm/s. The flow profiles of the splenic artery and the common and proper hepatic arteries, as arteries supplying the parenchyma, show a relatively high diastolic level and a low Pourcelot resistance index value (Figs. 7.**10**, 7.**11**). The splenic arterial blood flow remains unchanged after ingestion of food (Iwao et al. 1996), whereas the postprandial resistance index in healthy subjects increases by about 42% (Joynt et al. 1995). The resistance index value in the hepatic artery lies between 0.6 and 0.7 (Kamps et al. 1992). The *left gastric artery* also has a relatively low peripheral resistance. In a Doppler spectrum, this is documented as holodiastolic flow, with an end-diastolic velocity of around 10–20 cm/s (Fig. 7.**12 a**).

Color flow Doppler mode. The origin of the celiac trunk, which has a relatively large caliber, can be easily displayed in color flow Doppler sonography by proceeding with transverse and sagittal planes that are oriented from a subxiphoid to a cranial direction, and adjusted so that the vascular axis is projected onto the transducer (Fig 7.**9 c, d**). As in the case of the superior mesenteric artery, the color coding contains lighter-colored zones at the origin, and especially in the sagittal plane. These represent relatively higher frequency shifts caused by incident angle variations, and do not in themselves indicate stenosis.

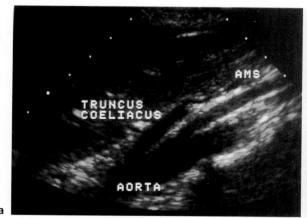

a

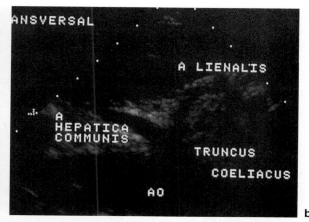

b

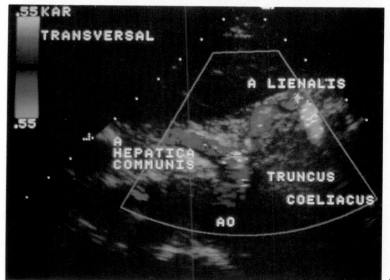

c

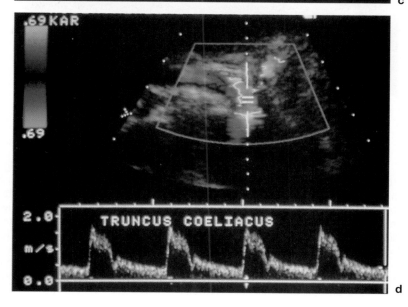

d

Fig. 7.**9** **a** The celiac trunk and superior mesenteric artery in sagittal section, and **b** the celiac trunk as it branches into the common hepatic artery and splenic artery in a transverse sectional plane. **c** A color-coded display of the flow in the celiac trunk and the proximal segments of the common hepatic artery and splenic artery. **d** A frequency spectrum from the distal section of the celiac trunk proximal to the origin of the common hepatic artery and the splenic artery, with a maximum systolic flow velocity of 166 cm/s and a Pourcelot resistance index of 0.74 (AO aorta, SMA superior mesenteric artery)

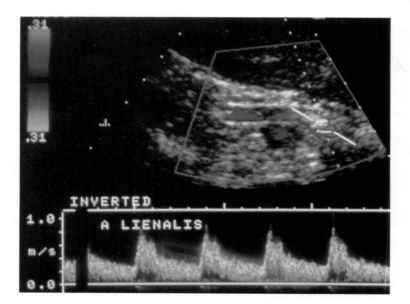

Fig. 7.**10**   A color flow duplex sonogram of the splenic artery and a frequency spectrum obtained from the distal vascular segment, approximately 6 cm after its origin. A color change in the sample volume can be seen, which is due to aliasing, caused by an angle-related higher frequency shift during the color flow Doppler mode. There is no pathological flow acceleration. Maximum systolic flow velocity 92 cm/s, end-diastolic flow velocity 37 cm/s, Pourcelot resistance index 0.60

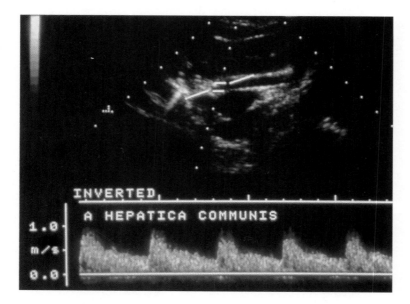

Fig. 7.**11**   A Doppler frequency spectrum from the distal common hepatic artery. Maximum systolic flow velocity 98 cm/s, end-diastolic flow velocity 43 cm/s, Pourcelot resistance index 0.56

In favorable examination conditions, the origin of the *gastroduodenal artery* from the common hepatic artery can be displayed, and the vessel can be followed over several centimeters. Using the color flow Doppler mode, the *proper hepatic artery* can also be evaluated more easily, and in many patients it can only be assessed with this method. The left gastric artery likewise shows a relatively low peripheral resistance that can be documented in a Doppler spectrum as having holodiastolic flow and an end-diastolic flow velocity of circa 10–20 cm/s (Fig. 7.**12a, b**). The intrahepatically highly ramified branches of the hepatic artery follow a course through the liver parenchyma spatially close to the branches of the portal vein (Fig. 8.**7a**, p. 286). Using

color flow Doppler sonography, the identical flow direction of the hepatic artery and portal vein can be recognized within the parenchyma by their identical color coding (Fig. 8.**7b**, p. 286). The pulsatility index of the hepatic artery is 0.61–1.36 (mean 0.98); the resistance index lies between 0.59 and 0.70 (McGrath et al. 1992, Kamps et al. 1992).

In measurement values obtained from 30 test subjects, McGrath et al. (1992) were also able to register frequency spectra from the *cystic artery*. The pulsatility index was 0.54–1.8 (mean 1.04). Due to the difficult incident angle correction involved, measurement of the absolute flow velocities was found to be uncertain.

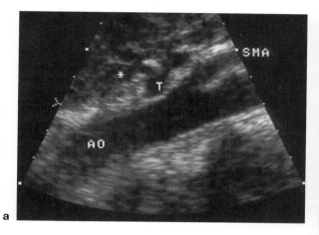

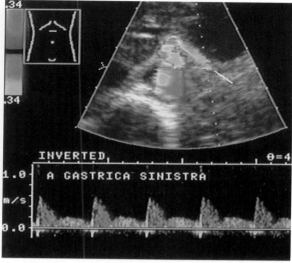

Fig. 7.**12** **a** Sonogram of the origin of the left gastric artery (*) from the celiac trunk (T) in transverse section (AO = abdominal aorta, SMA = superior mesenteric artery). **b** Subxiphoid transverse section, showing blood flow in the abdominal aorta (cross-section, blue color coding), with the origin of the celiac trunk proceeding ventrally. From the celiac trunk, the left gastric artery proceeds in a left lateral direction (red color coding). The frequency spectrum shows a profile with continuous, relatively low diastolic flow. The maximum systolic, mean, and end-diastolic flow velocities are 62 cm/s, 28 cm/s, and 14 cm/s, respectively

Superior Mesenteric Artery

Anatomy. The *superior mesenteric artery* originates ventrally from the aorta 0.5–2.0 cm below the celiac trunk and behind the pancreas, approximately at the level of the first lumbar vertebral body. It passes through the pancreatic notch ventral to the inferior part of the duodenum and between the two mesenteric layers, where it gives off side branches in a curved manner, which progress to the ileocecal region. The artery, which is easy to evaluate with ultrasound in its proximal segment, has a diameter of 0.8–1.4 cm at this point, and provides the arterial blood supply for the jejunum, the ileum, the ascending colon, and the transverse colon as far as the left flexure of the colon.

■ Conventional and Color Flow Duplex Sonography

B-mode. In a sagittal longitudinal section, the course of the superior mesenteric artery ventral and almost parallel to the aorta can be detected (Fig. 7.**13 a, b**). In a transverse sectional plane, the mesenteric artery can be seen in cross-section dorsal to the splenic vein and ventral to the aorta, which is also depicted in a single cross-section. Depending on the digestion, the cross-sectional diameter in healthy individuals is stated to be 0.5–0.8 cm (Jäger et al. 1986, Sato et al. 1987, Schäberle and Seitz 1991). Occasionally, and depending on the examination conditions, it is possible to image the proximal sections of the jejunal arteries using adapted transverse sections (7.**38 b**).

Doppler mode. The *frequency spectrum of* the superior mesenteric artery also depends on the patient's phase of digestion. After *fasting*, the flow profile is triphasic, with a maximum systolic velocity of 97–157 cm/s (Table 7.**1**), a small backflow component, and a low end-diastolic velocity of approximately 8–25 cm/s, with a mean velocity of 15–35 cm/s (Jäger et al. 1986, Sato et al. 1988, Nakamura et al. 1989, Bowersox et al. 1991, Sabbá et al. 1991, Sieber et al. 1991). The flow volume is 517 ml/min ± 19 ml/min (250–890 ml/min, n = 70); the pulsatility index (n = 82) is given as 3.57 ± 0.11 (1.83–6.78). Neither of these values shows any age-dependent or sex-dependent differences (Qamar et al. 1986 a, b).

Ingestion of food. Jäger et al. (1986) report a maximum circulatory increase 45 minutes after a defined meal (4,200 kJ) with an increased maximum systolic flow velocity to a value of 189.1% ± 50.7% and an increased end-diastolic velocity to a value of 350% ± 175%. This correlated with an increased vascular diameter. Following nonstandardized *food ingestion,* Schäberle and Seitz (1991) measured an increased maximum systolic flow velocity at a magnitude of 196 cm/s ± 25 cm/s and increased end-diastolic flow velocity to a value of 47.5 cm/s ± 8.3 cm/s. In neonates, in whom the velocities are generally low, the dependence of the flow acceleration on food ingestion was confirmed, with accompanying decreased peripheral resistance (resistance index: fasting 0.85, postprandial 0.73) (Leidig 1989).

After the ingestion of 500 ml of liquid nutrients (3053 kJ), Sieber et al. (1991) registered a maximally increased vascular diameter (after 60 minutes) from 6 mm ± 0.3 mm to 6.7 mm ± 0.2 mm, and an increased mean flow velocity (after 30 minutes) from 29.8 cm/s ± 5.1 cm/s to 68.4 cm/s ± 11.5 cm/s. This corresponds to an average increase of around 12% (5–26%) or

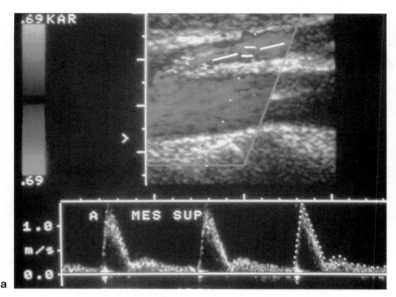

a

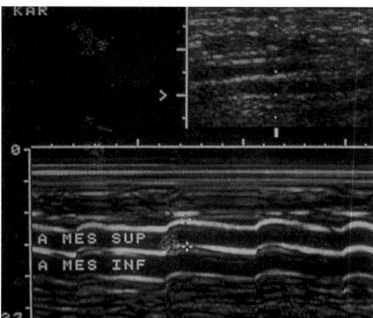

b

Fig. 7.**13** **a** Sagittal section through the abdominal aorta, with the ventral origin of the superior mesenteric artery shown in the color-coded image. Aliasing is seen near the origin of the superior mesenteric artery. It is due to a steeper incident angle of the color flow Doppler beam. The flow direction is directed toward the transducer. Maximum systolic flow velocity 138 cm/s, end-diastolic flow velocity 13 cm/s, mean flow velocity 41 cm/s, Pourcelot resistance index 0.91, Gosling pulsatility index 3.03. **b** In M-mode, the walls of the aorta and the superior mesenteric artery are seen, with pulsatile and respiration-dependent movements. The diameter of the superior mesenteric artery is 7.4 mm

130% (31–245%), with the increased velocity showing significantly large scattering. Due to the increased end-diastolic velocity, in particular, the postprandial values calculated for the resistance index and pulsatility index are constantly lower. Ten minutes after a mixed nutrient meal containing 2860 kJ (683 kcal), Qamar et al. (1986 b) found a decreased pulsatility index amounting to 46% in a series of measurements (n = 15).

At least under experimental conditions, the *composition* of the nutritional components influences the time of maximum hyperemia. After ingestion of 400 ml of a mixture consisting of carbohydrates, fat, and protein (1675 kJ, 400 kcal), the increased circulation in 12 normal individuals amounted to 64%, 60%, and 57% with maximum values taken at 15, 30, and 45

minutes. Consuming only a liquid of identical volume without any nutritional components does not cause any changed flow parameters in the superior mesenteric artery (Qamar and Read 1987).

Color flow Doppler mode. Using color flow Doppler mode, the origin of the superior mesenteric artery typically shows a lighter color than the remainder of the blood vessel (Fig. 7.**13 a**). As described above, this corresponds to a relatively higher regional Doppler shift, which is related to the ultrasound incident angle, and does not in itself indicate the presence of a stenosis. In its further course, the uniform flow velocity results in homogeneous color coding of the blood vessel. Example Doppler frequency spectra should regularly be registered and documented from the origin, as well as from the trunk of the artery itself.

Table 7.**1** Normal values for duplex-sonographic measurements in various abdominal arteries; maximum systolic (V_{max}), end-diastolic (V_{dias}) and mean flow velocity (V_{mean}), resistance index (RI) and maximum cross-sectional diameter, based on data in the literature and our own measurements

Arteries	V_{max} (cm/s)	V_{dias} (cm/s)	V_{mean} (cm/s)	RI	Diameter (mm)
Abdominal aorta	50–120	–	–	–	15–25
Celiac trunk[1–3, 19]	100–237	23–58	45–55	0.66–0.82	6–10
Splenic artery[4, 5, 16]	70–110	–	15–40	–	4–8
Common hepatic artery[3–5]	70–120	–	20–40	–	4–10
Left gastric artery	45–80	10–20	20–40	–	–
Gastroduodenal artery[6]	13–29	–	–	–	2–5
Superior mesenteric artery[7–11]	124–218*	5–30*	15–35*	0.75–0.9*	5–8
Renal artery[12–16]	60–180	20–65	–	0.6–0.8	5–8
Inferior mesenteric artery[16]	108–155	5–20	–	0.8–0.9	3–7
Ovarian artery[17]	56–148**	–	–	–	–
Uterine artery[18]	–	–	–	0.8–0.9**	–

*Influenced by the stage of digestion, **dependent on the menstrual cycle, [1]Bowersox et al. (1991), [2]Jäger et al. (1992), [3]Sieber and Jäger (1992), [4]Nakamura et al. (1989), [5]Sato et al. (1987), [6]Uzawara et al. (1993), [7]Jäger et al. (1986), [8]Sato et al. (1988), [9]Sabbá et al. (1991), [10]Bowersox et al. (1991), [11]Schäberle and Seitz (1991), [12]Karasch et al. (1993b), [13]Hoffmann et al. (1991), [14]Schäberle et al. (1992), [15]Sievers et al. (1989), [16]our own measurements, [17]Feichtinger et al. (1988) stimulated cycles, [18]Kurjak et al. (1991), [19]Moneta et al. (1988)

Renal Artery and Intrarenal Arteries

Anatomy. Bilaterally, the *renal arteries* originate approximately 1.5 cm caudal to the superior mesenteric artery at the level of the first and second lumbar vertebrae, almost at a right angle to the abdominal aorta (Fig. 7.**14**). In many cases, the vasculature proceeds ventrally, and reaches the end organs in a convex, curving fashion. In side-to-side comparisons, the somewhat longer right renal artery (3.5–7.0 cm; left renal artery 2.5–5.0 cm) proceeds dorsal and lateral to the inferior vena cava. In the hilum, the vessel divides into a dorsal and ventral main branch that usually give off five *segmental arteries* in the renal sinus. After further division in the parenchyma, the *interlobar arteries* proceed to the cortex and, between the medullary rays, supply the glomeruli as the *interlobular arteries* and afferent vessels. Multiple renal arteries are not uncommon (Lippert and Pabst 1985).

■ Conventional and Color Flow Duplex Sonography

B-mode. Under favorable examination conditions, the renal arteries can be followed distally. starting from their origin at the abdominal aorta for a varying distance (Figs. 7.**15**, 11.**55e, f,** p. 443). Their bilateral diameter is approximately 0.5 cm. It should be noted that the right renal artery often originates somewhat ventrally to the abdominal aorta, while the contralateral renal artery often proceeds dorsally from the aorta.

Usually, the right kidney is somewhat smaller than the left one, and the kidney size is slightly larger in males than in females. A series of measurements obtained by Emamian et al. (1993) in 665 patients with

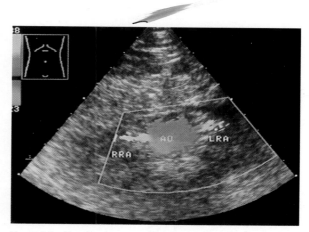

Fig. 7.**14** Proximal transverse section of the upper abdomen at the level of the origin of the renal arteries. With color flow Doppler sonography, it is an exception for the origins of both renal arteries to be displayed together here in one sectional plane. Normally, the right renal artery originates distally to the origin of the left renal artery from the aorta

healthy kidneys yielded a sonographically determined length and width of the right kidney with average values of 109 mm and 57 mm, compared to 112 mm and 58 mm for the left kidney (Table 7.**2**). Increasing age is associated with a slight decrease in kidney size.

Doppler mode. The normal *frequency spectrum* presents the typical form of a flow pulse, with low subsequent vascular resistance (Fig 7.**16**). In evaluating the circulation, the normal maximum systolic flow velocity is given at 100–180 cm/s (Avasthi et al. 1984, Ferretti et al. 1988, Hoffmann et al. 1991, Schäberle et al. 1992, Karasch et al. 1994). The end-diastolic flow

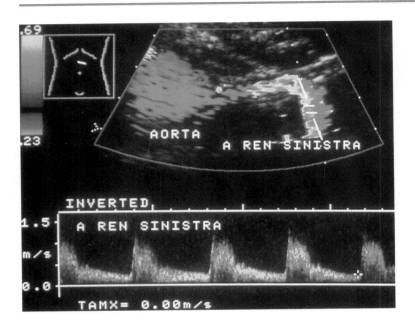

Fig. 7.**15** Sonogram of a left renal artery, with its origin from the abdominal aorta, in a subxiphoid transverse section through the upper abdomen. The curving arterial course, proceeding dorsolaterally, can be seen

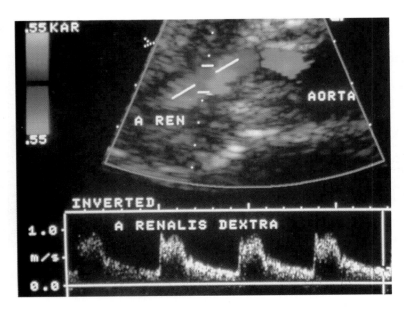

Fig. 7.**16** An upper abdominal transverse section at the origin of the right renal artery from the abdominal aorta. Doppler sample volume in the proximal segment of the right renal artery: normal Doppler frequency spectrum, with an open systolic window

Table 7.**2**   Normal values for sonographically measured kidney sizes (length × width) in 307 female and 358 male patients with healthy kidneys (Emamian et al. 1993)

	Right kidney (10th and 90th percentiles)		Left kidney (10th and 90th percentiles)	
	Length (mm)	Width (mm)	Length (mm)	Width (mm)
Normal group (n = 665)	109 (98–122)	57 (51–64)	112 (101–123)	58 (51–65)
Men (n = 358)	112 (101–124)	59 (53–66)	115 (104–126)	60 (53–68)
Women (n = 307)	107 (95–120)	56 (50–61)	110 (99–121)	56 (49–62)

velocity is 25–50 cm/s, and the Pourcelot resistance index, which is especially useful in the intrarenal arteries, lies in the range 0.5–0.75 (Sievers et al. 1989, Karasch et al. 1993 b). The side-to-side difference in the resistance indices should be less than 12%. Table 7.**3** summarizes the normal values for the maximum systolic and end-diastolic flow velocities and also the resistance indices in the renal artery and intrarenal segmental arteries, obtained from measurements made in angiographically normal-appearing arterial renal vascular systems (Karasch et al. 1995 a). In the same group of normal individuals, it was shown that the end-diastolic flow velocity in the renal arteries and intrarenal arteries decreases, so that an age-dependent increase in the intrarenal resistance indices can be expected (Table 7.**4**). The normal values apply in children after the first completed year of life, and the diameter of the renal artery shows an age-dependent

increase here as well. The maximum systolic and mean flow velocity do not show any age-dependent fluctuations (Grunert et al. 1990). The resistance index increases with increasing age (Schwerk et al. 1993), and also in various diseases (hypertension, diabetes mellitus, urinary obstruction, renal compression).

Some authors have also measured the quotient represented by the maximum velocity in the renal artery and the maximum velocity in the abdominal aorta, and have determined a maximum value of 3.0–3.5 for this (Berland et al. 1990, Hoffmann et al. 1990). Using this as the sole criterion for stenosis has proved inadequate in our experience because the quotient can be influenced, among other things, by the pathological aortic flow velocities that occur in aneurysms, stenoses, or occlusions.

With regard to the physiological reaction of the renal circulation during a normal pregnancy, various

Table 7.**3** Normal values for the maximum systolic (V_{max}) and end-diastolic flow velocity (V_{dias}), the resistance index and renal/aortic ratio (RAR) int the proximal renal artery, and the resistance index in the intrarenal segmental arteries, measured in 133 patients with 181 angiographically normal-appearing renal arteries. The follwing values are indicated: mean ± one standard deviation (SD), the 95% confidence interval (mean ± twice the SD), and the absolute minimum and maximum values as upper and lower boundaries

	Mean ± SD	95% confidence interval	Upper and lower boundaries
Renal artery			
V_{max} (cm/s)	126.5 ± 32	62.5 ± 190.5	51 –195
V_{dias} (cm/s)	36.6 ± 11.9	12.8 ± 60.4	11 – 71
Resistance index (RI)	0.71 ± 0.07	0.56 ± 0.84	0.50– 0.88
RAR	1.8 ± 0.8	0.2 ± 3.4	0.5 – 4.6
Intrarenal arteries			
Resistance index	0.68 ± 0.070	0.54 ± 0.82	0.50– 0.86
Right-left difference	4.5%	–	0.0 – 12%

Table 7.**4** Age-related normal values for the maximum systolic (V_{max}) and end-diastolic (V_{dias}) flow velocity in the renal artery, and for the resistance index (RI) in the proximal renal artery and intrarenal segmental arteries. Values are given as the mean ± one standard deviation, with the minimum and maximum values (upper and lower boundaries/range) (Karasch et al. 1995 a)

Age (y) Test subjects (n)	21–30 (5)	31–40 (4)	41–50 (17)	51–60 (80)	61–70 (28)	71–80 (9)
Renal artery						
V_{max} (cm/s)	113± 23	148± 45	126± 29	134± 33	127± 30	116± 23
Range	90–137	88–193	76–168	51–225	71–164	87–143
V_{dias} (cm/s)	41± 5	46± 10	41± 12	40± 13	32± 11	22± 7
Range	33– 48	35– 58	24– 53	12– 95	14– 64	14– 33
RI	0.63±0.07	0.68±0.06	0.67±0.061	0.70±0.06	0.74±0.07	0.80±0.06
Range	0.53–0.71	0.60–0.74	0.50–0.74	0.37–0.84	0.57–0.91	0.68–0.89
Segmental Arteries						
RI	0.63±0.088	0.65±0.05	0.67±0.04	0.67±0.06	0.72±0.05	0.78±0.07
Range	0.54–0.73	0.58–0.70	0.59–0.73	0.56–0.81	0.61–0.85	0.70–0.87

values have been obtained. Sohn et al. (1988) described a decreased peripheral resistance within the renal vascular system, along with a lower intrarenal resistance index. By contrast, Sturgiss et al. (1992) were not able to observe any significantly changed flow patterns in the renal arteries of pregnant women.

Color flow Doppler mode. Using color flow Doppler mode helps significantly to *identify* the renal arteries and to follow their course. This becomes more important the poorer the examination conditions are for B-mode sonography. Color is also beneficial in the middle arterial segment in slim patients. The reliability of imaging, which involves registering several Doppler frequency spectra from the proximal and middle segment of the renal arteries, lies at approximately 88%. Intrarenal frequency spectra can usually be registered in healthy individuals (Karasch et al. 1994). Using *echo contrast media* further improves the imaging of the renal arteries, and can be useful in difficult sonographic examination conditions.

A complete examination of both renal arteries with duplex sonography takes about 20–90 min (Berland et al. 1990), whereas the evaluation with color flow imaging can be done within 30 minutes (Karasch et al. 1993a).

Color imaging of the *intrarenal vascular segments* also enables a fast location of individual arteries and provides targeted spectral analysis in the terminal vascular system (Fig. 7.17). According to measurements taken by Grunert et al. (1990), the intrarenal resistance indices only show slight differences, both in the kidney as a whole and in side-to-side comparisons. In 143 angiographically normal-appearing renal vascular systems, the intrarenal resistance index measured was 0.69 ± 0.064. The value measured in the segmental arteries was somewhat lower than that recorded in the segment of the renal artery near the origin (0.78 ± 0.079) (Karasch et al. 1993b). The average side-to-side difference in 34 angiographically normal renal arteries was 4.5% (0.1–12%), measured intrarenally. Figure 7.18 shows the intrarenal flow image using color flow Doppler energy mode. In the selected color window, covering the lower pole of a right kidney, the intrarenal perfusion is shown in yellow. The fan-shaped spreading of the vasculature, which proceeds from the renal hilum to the renal cortex, is clearly recognizable. However, in this color-coded image, only the circulation is coded, without further differentiation of the flow direction, so that it is not possible to distinguish between arterial and venous signals in the image.

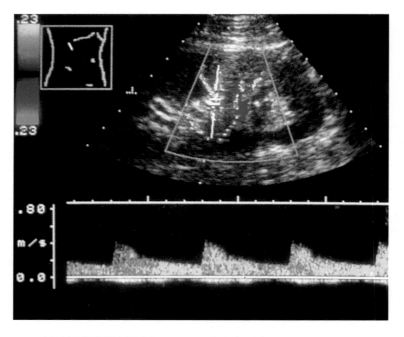

Fig. 7.**17** A color flow Doppler sonogram of the intrarenal arteries and veins, using a dorsolumbar section. The arterial flow directed toward the transducer is coded red, while the opposite flow direction in the intrarenal veins is shown in blue. The pulse repetition rate is preselected at a relatively low level on the apparatus, since relatively slow flow velocities predominate in both vascular systems

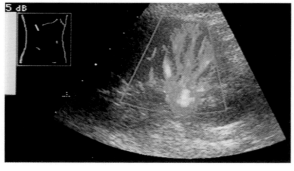

Fig. 7.**18** An oblique dorsolumbar section displaying the right kidney. In the color Doppler energy mode, extensive blood flow is seen in the lower segment of the kidney. The structure of the vascular tree between the cortex and the hilum can be clearly recognized

Inferior Mesenteric Artery

Anatomy. The *inferior mesenteric artery* originates from the ventral abdominal aorta, 5–6 cm proximal to the bifurcation, to the left of the midline, at the level of the third and fourth lumbar vertebrae. With its branches (left colic artery, sigmoid arteries, superior rectal artery) it provides the blood supply for the descending colon distal to the left flexure of the colon, the sigmoid colon, and the proximal sections of the rectum. Its proximal diameter is approximately 0.25 cm (0.1–0.5 cm).

Conventional and color flow duplex sonography. Depending on the examination conditions and underlying disease, the origin and the trunk of the inferior mesenteric artery can be displayed in 56% to more than 80% of cases in a median sagittal section, ventral to the abdominal aorta (Kathrein et al. 1990, Mirk et al. 1994). The superior mesenteric artery is sometimes depicted simultaneously in its ventral course, and for a distance it proceeds caudal and parallel to the inferior mesenteric artery (Fig. 7.**19 a, b**). This may be a possible source of confusion when identifying the arteries. If necessary, the arteries should be followed to their origin.

According to several measurements, the mean cross-sectional diameter is 0.4–0.7 cm, and the mean maximum systolic flow velocity amounts to 130 cm/s (Table 7.**1**).

Lumbar Arteries

Anatomy. As parietal branches of the abdominal aorta, four *lumbar arteries* normally proceed bilaterally behind the great psoas muscle, and dorsal to the quadrate muscle of the lumbar vertebrae, to the abdominal muscles.

Conventional and color flow duplex sonography. The frequency spectrum corresponds to the triphasic curve of an arterial image in the extremities, with subsequent high peripheral resistance in contrast to arteries supplying the parenchyma. This is particularly important when examining the renal arteries, since the difficult B-mode examination conditions mean that confusion with the renal arteries is possible—although the lumbar arteries, when compared to the renal arteries, appear smaller in caliber.

Ovarian Artery

Anatomy. The thin-caliber *ovarian arteries* originate from the ventral aorta below the renal arteries and proceed caudally, anterior to the psoas muscle. They reach the ovary from the margin of the small pelvis, within the suspensory ligament of the ovary. Branches are given off to the tubal extremity of the ovary, to the ampulla, and connections to branches of the uterine artery are made within the ovarian arcade.

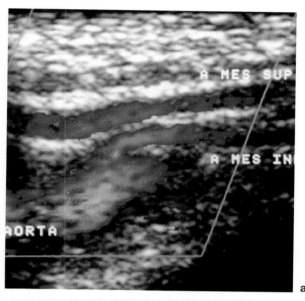

a

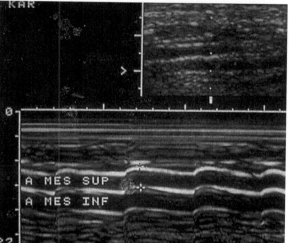

b

Fig. 7.**19** **a** A right paramedian sagittal section at the level of the origin of the inferior mesenteric artery from the abdominal aorta, showing the parallel and ventrally located course of the superior mesenteric artery. **b** An M-mode image of the superior and inferior mesenteric arteries, with pulsatile movements. The diameter of the superior mesenteric artery is 4.4 mm

Conventional and color flow duplex sonography. The frequency spectrum of the ovarian artery has menstruation-dependent flow patterns. Prior to ovulation, low-frequency signals with a clear diastolic forward flow are detectable in *both* ovarian arteries (Fig. 7.**20 a**). During the remainder of the cycle, a vascular spectrum indicating high peripheral resistance is observed. The resistance index decreases from a preovulatory value of 0.5–0.6 to 0.4–0.5 (Kurjak et al. 1991).

According to examinations carried out by Schurz et al. (1990) in 21 women with normal menstrual cycles, hormonal stimulation causes the pulsatility index (which can be used as a parameter to indicate cycle-dependent circulatory changes) to decrease from a

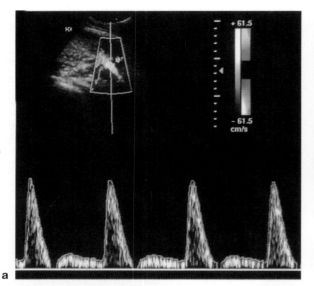

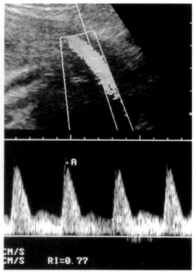

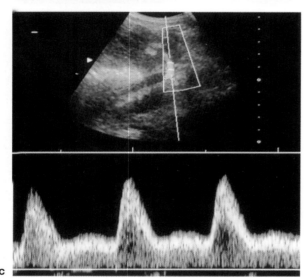

Fig. 7.**20**   A transabdominal image of the ovarian artery (**a**) and the uterine artery (**b**) in a section adapted to the vascular axes at the level of the small pelvis. (**c**) A uterine artery in a pregnant woman. Note the increase in the systolic and diastolic velocity and the decrease of the resistance index compared to the nonpregnant woman in (**b**).
Illustration from Dr. Bald, Klinik für Frauenheilkunde und Geburtshilfe, University Clinic of Cologne

preovulatory value of 1.5 ± 0.6 to 1.1 ± 0.4 on ovulation. Similar values were also reported by Taylor et al. in 1985.

Feichtinger et al. (1988) were not able to confirm these changes when using transvaginal measurements. Instead, they asserted that a modified resistance index, in which the relationship between the maximum systolic and end-diastolic flow velocity is determined, had greater predictive value. In their measurements, which were in each case conducted one day prior to (n = 35) and one day after (n = 46) induced ovulation, the maximum systolic flow velocity in the ovarian artery was 80.8 cm/s ± 40.2 cm/s and 104.2 cm/s ± 44.0 cm/s respectively.

Uterine Artery

Anatomy. The *uterine arteries* originate from the anterior visceral branch of the internal iliac artery, and proceed along the lateral wall of the small pelvis both caudally and ventrally. At the base of the uterine ligament, they reach the lateral margin of the uterus at cervix level. Here the main branch of the arteries continues upward at the uterus in a curved fashion. Their branches anastomose with the contralateral side, and one terminal branch (the ovarian branch) is connected to the ovarian artery (ovarian arcade).

Conventional and color flow duplex sonography. Bilaterally, the uterine arteries usually show equivalent circulation (Long et al. 1989, Kurjak et al. 1990 b). However, like the internal iliac arteries, they also show menstruation-dependent circulatory fluctuations, and in particular an increased flow velocity prior to ovulation and during the luteal phase. The resistance index of approximately 0.8–0.9 decreases slightly as ovulation approaches (Kurjak et al. 1991). According to measurements made by Santolaya-Forgas (1992), when the endometrial thickness increases over the course of the cycle, the pulsatility index should gently decrease (Fig. 7.**20b**). The pulsatility index changes are said to be more pronounced on the ovulation side (n = 16, 20 cycles). During a normal pregnancy, increased diastolic circulation and decreased peripheral resistance (Fig. 7.**20c**) commence even after a few days (Feichtinger et al. 1988, Sohn and Stolz 1991).

Evaluation

■ Conventional and Color Flow Duplex Sonography

B-mode. Measurements of the arterial lumen using B-mode imaging have to apply ultrasound in different planes to be able to determine the true vascular diameter of the abdominal arteries, particularly in the abdominal aorta, in order to recognize aneurysmal dilatations and determine the flow volume. Using the M-mode procedure makes it easier to ascertain the precise boundaries between wall structures and the internal lumen (Fig. 7.**19b**).

Especially in longitudinal examinations, it is important to define the cross-section unambiguously. In diameter measurements using the "leading edge method," which should always be carried out in the sagittal plane, the true diameter of the lumen as well as the thickness of *one* wall needs to be included in the defined cross-section. For that reason, a brief description of the structures measured is helpful when reporting the findings. With careful examination, reliable size indications are possible for the abdominal arteries in three planes.

Doppler mode. Assessing the frequency spectra registered by continuous wave or pulsed wave Doppler sonography can be subdivided into *qualitative* and *quantitative* evaluations, the criteria for which are described above (p. 161 ff.).

In the *abdominal aorta,* under physiological conditions, it is almost always possible to register a pure Doppler frequency spectrum that is free of artifacts, and also to determine quantitative measurements.

When measuring flow velocities based on frequency spectra obtained from smaller abdominal arteries, particular care must be taken to ensure that precise angle corrections are calculated. Table 7.**1** lists normal values for various measurements obtained with duplex sonography for the arteries described.

In interpreting the parameters measured, it is very important to take into account the peripheral resistance in the subsequent vascular bed. This resistance influences the flow dynamics in the supplying artery. Arterial resistance is subject to effects from the nervous system, which in the individual organ regions are affected by digestion, menstruation, and particu-

larly by any preexisting or accompanying diseases (arteriosclerosis, arterial hypertension, diabetes mellitus, or renal insufficiency of varying etiology).

Color flow Doppler mode. Color flow Doppler sonography makes duplex examination of many abdominal arteries possible—e.g., the cystic artery and the gastroduodenal artery—which can only be evaluated by conventional means very time-consumingly, if at all (McGrath et al. 1992, Uzawa et al. 1993). Color coding, however, allows a more targeted and definitive recognition and localization of the smaller abdominal arteries.

By visually displaying the course of thin-caliber arteries (e.g., interlobar arteries of the kidney), *angle-corrected* flow velocity measurements are also possible in these vascular segments for the first time.

In contrast to the normal vasculature, color flow Doppler signals make it possible to detect and differentiate anomalies along the vascular course—e.g., multiple blood vessels, accessory arteries, and pathological vascular processes such as aneurysms or independent vascular neoplasms. The extent of tumor vascularization can also be determined (Chapter 10).

Although normal values have not yet been determined for all of the smaller arteries at present, and pathological patterns of findings have not yet been standardized in large patient groups, it seems certain that normal values, clinically relevant indications for an examination, and pathological sets of findings will also be defined and recognized for these vascular regions in the future.

Color Doppler energy mode. Although the potential areas of application for the color flow Doppler energy mode are currently still in the process of development, it is already clear that the procedure will be mainly used to detect very slow flow, and flow in very small blood vessels, which has so far only been partially successful with frequency-dependent color flow Doppler sonography. When the findings are normal, consideration should be given to initially detecting and assessing the regional perfusion in parenchymatous organs such as the liver, spleen, kidneys, and perhaps also the placenta.

Pathological Findings

Principle and Methodological Peculiarities

■ Conventional and Color Flow Duplex Sonography

B-mode. With B-mode sonographic imaging, it is possible to image anomalies of arterial location and arterial course (Loyer and Egli 1989), as well as pathologically dilated arterial segments. Information can be gathered about the extent of ectasias and aneurysms from cross-sectional measurements in different planes, and these values can also be used in the relevant follow-up observations (Lederle et al. 1988, Cronenwett et al. 1990, Kremer et al. 1989).

In addition to their most frequent location in the infrarenal aorta, true aneurysms can appear in almost all of the abdominal arteries (overview in Elliot and Ashley 1991). The anatomical relationships between these vascular dilatations and arterial branches or neighboring organ systems can be assessed with sonography. Confirmation of an aneurysm in the abdominal aorta requires examination of the pelvic and popliteal arteries for possible vascular dilatations as well.

In addition, more echogenic structures within the vascular lumen can be recognized due to their stronger ultrasound reflection when compared to blood. This makes sonographic imaging of morphological wall changes of varying pathology possible (Ludwig et al. 1989, Pignoli et al. 1986), as they occur in various types of aneurysm, including wall deposits (plaques) and also parietal thrombi (Figs. 7.**22**, 7.**24a, b**, 11.**26a–c**, p. 387, 11.**48b–d**, pp. 430–431).

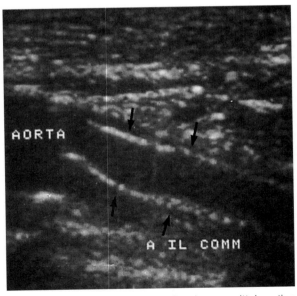

Fig. 7.**21**   A B-mode sonogram showing a sagittal section at the level of the distal abdominal aorta, with the origin of the right common iliac artery. An expansion stent (arrows) is seen in the lumen of the common iliac artery. With its proximal component, it reaches as far as the distal aorta

As a postoperative observation method, B-mode imaging is useful to detect many incipient clinical problems after vascular surgery (Fig. 7.**26a**, 7.**27a**). It can recognize soft-tissue hematomas, abscesses, prosthetic infections, and anastomotic aneurysms. After interventional procedures, imaging and locating vascular supports (stents) is possible (Fig. 7.**21**).

In the descriptions of the specific vascular findings, the contribution of B-mode imaging is discussed in the context of ultrasound vascular diagnosis under the heading of *duplex sonography.*

Doppler mode. To confirm the presence of a stenosis, spectral analyses have to be undertaken at various registration points along a blood vessel, in order to assess the flow dynamics in the artery and document the zones of pathologically elevated or decreased velocity.

If the B-mode image shows pathological findings (plaques or ultrasound shadows emanating from the vascular wall near the transducer, suddenly increased vascular caliber), then the region must be examined with particular care to identify flow accelerations or flow disturbances with the assistance of the Doppler signal. The *degree of stenosis* in a particular artery can be estimated using the following quantitative flow parameters:

- Elevated maximum systolic flow velocity
- Decreased or absent early diastolic backflow velocity
- Elevated end-diastolic flow velocity
- Elevated mean flow velocity
- Decreased pulsatility and resistance index

Due to its more precise local resolution capabilities, using the pulsed wave Doppler is preferable in the abdominal cavity. If there are high-grade arterial narrowings that cause a marked increase in the intrastenotic flow velocity, integrated *continuous wave* Doppler signal processing is useful. The pulsed wave Doppler has limitations in relation to depth, and due to the Nyquist limit when attempting to detect higher frequencies or velocities. Precise measurement of a very high maximum frequency is only possible with the continuous wave Doppler mode (Figs. 11.**51d**, p. 438; 11.**52b**, p. 439; see also p. 4ff).

Color flow Doppler mode. With the color flow Doppler mode, visual identification of zones of elevated flow velocity, which are typically encountered near stenoses, or disturbed flow patterns near plaques or in the lumen of aneurysms is possible (Fig. 11.**49b**, p. 433). An experienced examiner with a knowledge of the equipment-specific color adjustments will be able to make a rough estimate of the degree of stenosis on the basis of the color pattern. However, this cannot replace the need to register and evaluate a frequency spectrum. Criteria indicating a pathological finding with the color flow Doppler mode are:

- Changed color intensity (lightness)
- Aliasing
- Color change (changed direction)
- Color elimination
- Color band with a tapering width
- Separation phenomena

Occluded arteries do not show any color signals within the vascular lumen (Fig. 7.**30**). Directly in front of the occlusion, a slower, physiologically directed flow is occasionally coded in the blood vessel, and this flow then breaks off, or is not constantly displayable. Directly prior to the occlusion, the flow still goes through a color change, which passes through the baseline. This corresponds to flow components that are being reflected at the occlusion itself.

Color Doppler energy mode. The color flow Doppler energy mode is especially useful for detecting segmental perfusion disturbances due to occluded terminal arteries (e.g., renal infarction). It is helpful to adjust the color parameters relevant to the examination in a normally perfused organ segment or, when examining the kidneys, in the contralateral parenchyma.

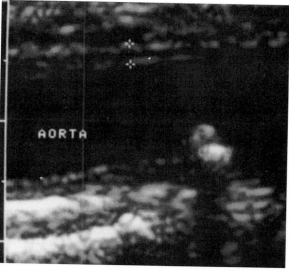

Fig. 7.**22** A sagittal longitudinal section through the distal abdominal aorta. At the dorsal wall, hyperechoic plaques can be detected as a narrow ultrasound shadow. These wall deposits were not hemodynamically active in the way that a stenosis would be

Findings

Since routine examination of all of the abdominal arteries is not feasible due to the time required, the examiner should be given all the relevant clinical findings concerning the patient before the start of conventional or color flow duplex sonographic investigations. An unambiguous clinical objective allows targeted collection of information on findings and therefore speeds up the examination.

Abdominal Aorta

B-mode. Within the overall context of vascular diagnosis, B-mode sonography can identify aneurysmal vascular wall dilatations, particularly in the abdominal aorta, and can postoperatively monitor vascular prostheses with regard to their anastomotic width and potential complications (hematomas, seromas, infections, and stenoses).

In addition to vascular caliber measurements, the imaging of plaques in the aortic lumen is also clinically interesting. These plaques are potential embolic sources, and may cause peripheral arterial occlusions in some patients, which are usually acute events (Fig. 7.**22**). Three-dimensional reconstructions of aortic aneurysms can clarify their spatial extent and determine the positional relationships of the aortic branches to the aneurysm itself.

▮ Aneurysms

B-mode. In sonography, *aneurysms* are classified morphologically as follows (Fig. 7.**23**):

- True aneurysm
- Dissecting aneurysm
- False aneurysm

A *true aneurysm* is a local, clearly defined vascular dilatation of varying etiology, with a cross-sectional diameter larger than 2.5–3.0 cm (Kremer et al. 1989) or a diameter increased by a factor of at least 1.5 in comparison with the proximal aorta (Cronenwett et al. 1985). Further differentiation of aneurysm forms as:

- Fusiform
- Saccular
- Fusiform/saccular

is possible (Fig. 7.**23**). This classification has clinical relevance, due to the differences in the prognosis, treatment, and surgical techniques associated with each form (Vollmar 1982).

In addition to determining the absolute size, form, and location of the aneurysm, it must be ascertained whether additional abdominal blood vessels—particularly the renal and mesenteric arteries—or the iliac vasculature also participate in the aneurysmal dilatation. This is important when planning surgery (Torsello et al. 1991).

According to Vollmar (1982), 95–96% of abdominal aneurysms develop in the infrarenal segment, and they primarily develop when there is underlying arteriosclerosis. A sonographically detectable involvement of the renal arteries is only seen in 4% of cases.

If an aneurysm is diagnosed in the abdominal aorta, a search must be carried out for additional aneurysms in other vascular segments of the periph-

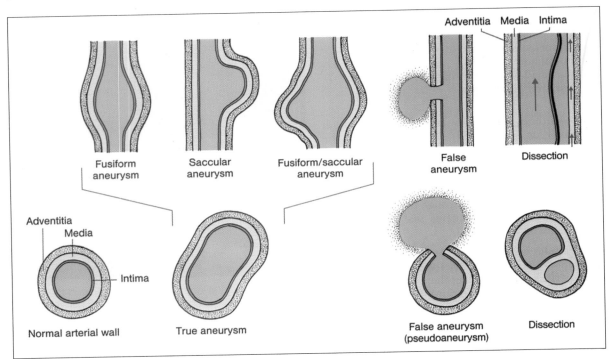

Fig. 7.**23**   Various types of aneurysm in longitudinal section and cross-section

eral and supra-aortal arteries, since the co-occurrence of an aortic aneurysm and wall dilatations in other arteries can be as high as 80% (Vollmar 1982, Hennerici and Steinke 1991).

Special attention has to be paid to what are termed *inflammatory aneurysms of the abdominal aorta,* because of their association with higher complication and mortality rates that are clinically and prognostically relevant (Kniemeyer et al. 1991). According to Gaylis et al. (1989), this special form comprises 5–15% of infrarenal aortic aneurysms. It is characterized by an inflammatory adventitial reaction, of unclarified etiology, which can be detected sonographically as a predominantly ventral hypoechoic external layer around a usually arteriosclerotically altered aortic wall. In addition to the required differentiation from uncomplicated arteriosclerotic aneurysms, Müller-Schwefe et al. (1991) list the following sonographic differential *diagnoses:*

- (Covered) perforated aneurysm of the abdominal aorta
- Dissecting aneurysm of the abdominal aorta
- Ormond's disease
- Retroaortal fibrosis
- Giant-cell arteritis
- Xanthofibrogranulomatosis
- Horseshoe kidney
- Malignant lymphoma
- Other retroperitoneal tumors

In inflammatory and fibrotic diseases of the abdominal aorta or retroperitoneal tissue (Fig. 7.**24a, b**), attention should be given to possible simultaneous in-

volvement of the inferior vena cava. A curved or loop-shaped *elongation* of the abdominal aorta without any vascular dilatation should be distinguished from a true aneurysm (Paes and Vollmar 1991). Particularly when there is a combination of an aneurysm and an elongation, measurements of the maximum cross-sectional diameter have to take the true anatomical situation into account, since incorrectly large diameters might otherwise result (Figs. 7.**25a, b**, 11.**48b**, p. 430).

Rupturing of an aneurysm of the abdominal aorta can occur with perforation of neighboring organs ("covered"), or into the open abdominal cavity. If the hematoma is at least partially encapsulated by intra-abdominal structures, the rupture will be detected sonographically from the presence of a hypoechoic to anechoic para-aortal space-occupying lesion. In addition, the detection of freely flowing fluid in the abdominal cavity, accompanied by appropriate clinical symptoms, may indicate a ruptured blood vessel.

A *dissecting aneurysm* is the result of the splitting of the vascular wall in the aorta, and it is usually associated with vascular dilatation. The split forms two lumina, through which blood flows. According to De Bakey (1965), three types are classified in vascular surgery:

Type I:	Dissection beginning in the proximal segment of the ascending aorta and extending to the distal abdominal aorta and possibly the aortic bifurcation.
Type II:	Dissection only in the ascending aorta.
Type III:	Dissection proceeding distally from the left subclavian artery and, in some instances, reaching the distal aorta.

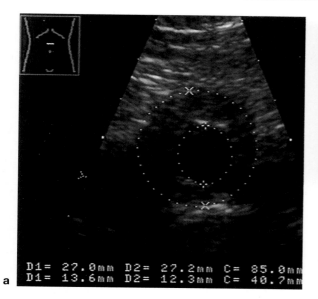

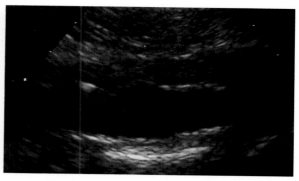

a

D1 = 27.0mm D2 = 27.2mm C = 85.0mm
D1 = 13.6mm D2 = 12.3mm C = 40.7mm

b

Fig. 7.**24** The abdominal aorta in a patient with retroperitoneal fibrosis (Ormond's disease). In the transverse section (**a**) and sagittal section (**b**), a broad, hypoechoic, periaortal border 5–8 mm long can be seen. It is delimited from the vascular lumen by a hyperechoic, thin structure corresponding to the original wall layers

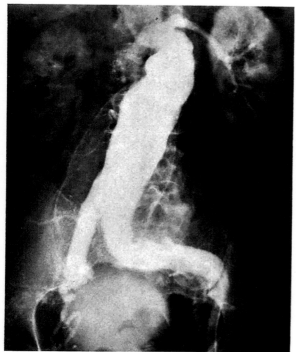

a

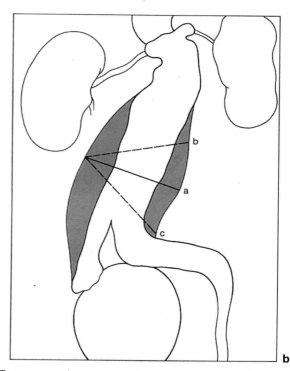

b

Fig. 7.**25 a** An angiogram showing a pronounced aneurysm of the abdominal aorta and common iliac arteries, with a vascular system elongation.

b To measure the cross-sectional diameter correctly, an adapted sonographic section vertical to the vascular axis (a) should be used, since sections deviating from this (b and c) produce incorrectly large values for the diameter

Types I and III can be suspected sonographically in the abdominal segment due to the intraluminal vascular wall structures (Chow et al. 1990). They can also be diagnosed with duplex sonography, which can identify the presence of two lumina with a separating membrane, through which blood flows at varying speeds and with temporal displacement (Giyanani et al. 1989, Thomas and Dubbins 1990). The exact location in the aortic arch and thoracic aorta can be displayed using transesophageal echo cardiography.

An aortic *dissection*—with or without an aneurysmal dilatation—can also be recognized by two lumina with asymmetrical blood flows (Fig. 7.**26**). In the B-mode image, a swinging membrane within the aortic lumen can be seen.

In general, *false aneurysms* of the aorta only occur rarely. The etiology is usually traumatic. The diagnostic criteria in duplex sonography correspond to those used for the peripheral arteries (p. 170).

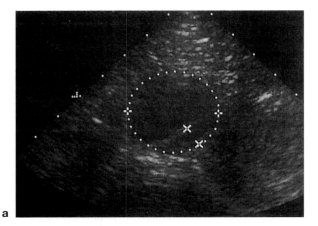

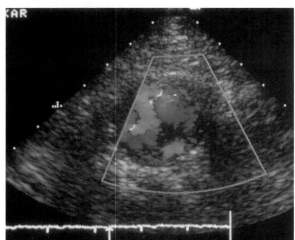

Vascular Prostheses

In the abdominal aorta, the bifurcation prosthesis (Y-prosthesis), which uses an iliac or femoral anastomosis, is the most important. Ultrasound examinations can evaluate the morphological status directly after various vascular operations, such as thrombendarterectomies, vascular patches, and placement of vascular prostheses (Figs. 7.27–7.29). They are suitable for follow-up assessments and evaluating complications after surgical interventions (Gross-Fengels et al. 1988, Zwicker et al. 1988, Imig et al. 1991). Imaging of the following phenomena is possible:

– Anastomotic aneurysms
– Prosthetic aneurysms
– False aneurysms
– Lymphoceles
– Hematomas
– Prosthetic infections

Fig. 7.**26** A dissection of the distal abdominal aorta. **a** Transverse sectional plane. In the left dorsolateral segment, an intraluminal membrane can be seen floating inside the vessel during the real-time imaging. The total vascular diameter is 29 x 30 mm. The false lumen shows a diameter of 7 mm in this sectional plane. **b** In the color flow Doppler mode, the second lumen is distinct from the remaining blood vessel, with the dissection membrane forming the boundary. **c** Using color flow Doppler sonography in a sagittal section, two lumina, with blood flowing through both of them can be presented

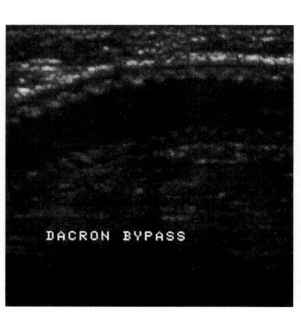

Fig. 7.**27** Sonogram of a Dacron bypass, with a markedly wave-shaped prosthetic wall in longitudinal section. The prosthetic material is easy to distinguish from the surrounding tissue, and it can be differentiated from wall deposits in the original vasculature due to its regular structure

Fig. 7.**28** With color flow Doppler sonography and a triphasic frequency spectrum, a normal, uniform, mean flow velocity of 23 cm/s is seen. Maximum systolic flow velocity 87 cm/s, early diastolic backflow velocity 17 cm/s, Gosling pulsatility index 4.5. Color flow Doppler sonography provides good recognition of the contours of the bypass wall

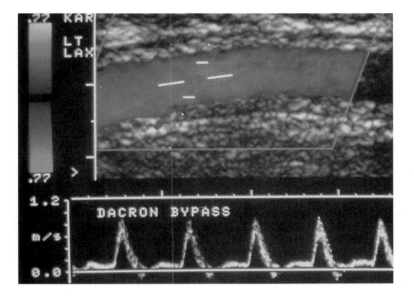

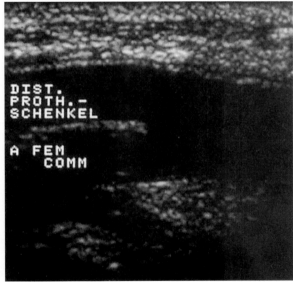

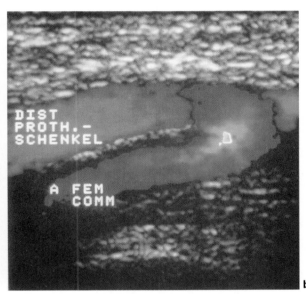

Fig. 7.**29** **a** A longitudinal duplex-sonographic section through the left groin, showing a femorally placed bypass anastomosis from an aortobifemoral Y-prosthesis. **b** In color flow Doppler mode, the physiological flow direction in the prosthesis (toward the transducer) is coded red. The original common femoral artery shows retrograde flow in the anastomosis, and is therefore coded blue. Aliasing of the color change occurs due to the more acute incident angle of the color flow Doppler beam in relation to the flow direction

Stenoses and Occlusions

Doppler mode. *Stenoses and occlusions.* Generalized arteriosclerosis is often accompanied by hyperechoic plaques causing ultrasound shadows, which protrude into the vascular lumen but do not have any hemodynamic effect on the wide blood vessel. Areas that have massive wall deposits also need to be analyzed by the Doppler component to determine whether a stenosis is present. This can be recognized by a locally increased flow velocity. When there is a maximum systolic flow velocity of more than 200 cm/s, the presence of a hemodynamically effective stenosis can be as-

sumed. Since flow obstructions predominate in the *distal aorta,* the examiner must always check whether the stenosis has progressed to involve the common iliac artery. If stenoses in the abdominal aorta occur further proximally, e.g., as a rare form of aortic coarctation, particular attention should be given to the location of the flow obstruction to the mesenteric and renal arteries (Karasch et al. 1995).

Occlusions of the abdominal aorta, which are also mainly located distal to the origin of the renal arteries and usually involve arteries at the pelvic level, can be recognized by the following:

As in aortic stenoses, occlusions are usually classified as belonging to the pelvic form of peripheral arterial occlusive disease, and are therefore discussed in this chapter (p. 163).

Vascular prostheses. Due to the often relatively large diameter of the prosthesis, the flow velocity is often lower than it is in normal blood vessels. Hendrickx et al. (1991) indicate a value of 40.5 cm/s ± 12.6 cm/s for the maximum systolic forward velocity, 9.2 cm/s ± 4.0 cm/s for the mean flow velocity, and for the Gosling pulsatility index distal to the prosthesis a value of 5.6 ± 2.3. (The patient group consisted of 21 rosthetic segments, involving Y-bypasses and aortofemoral or iliacofemoral bypasses).

Postoperatively disturbed flow can be detected near the sutures and distal to anastomoses on the basis of:

- Spectral broadening with low-frequency components
- Characteristic bubbling sounds
- Slow flow velocity in large-caliber vascular segments

With the Doppler mode, the information provided by B-mode imaging can also be expanded to quantify *anastomotic stenoses* and to evaluate postoperative changes in the vascular caliber between the original blood vessel and the prosthetic insertion point, with regard to their hemodynamic effectiveness (Gritzmann et al. 1988). A diagnosis of a false aneurysm can be confirmed with Doppler sonography, which detects pendular flow within the canal of the fistula (p. 170 ff; Figs. 5.**41**, 5.**42 a–c**; Fig. 11.**28 c**, p. 392).

Bypass stenoses, which usually form at the insertion point of the prosthesis, can be detected by a locally increased maximum systolic velocity. One can often recognize wall structures or plaques sonographically, as they protrude into the vascular lumen. It is also important to evaluate the poststenotic flow when estimating the grade of the stenosis.

When the subsequent peripheral vascular system is patent, the frequency spectrum in Dacron prostheses is usually triphasic (Fig. 7.**28**). The early diastolic backflow component is absent in polytetrafluoroethylene (PTFE) prostheses.

Color flow Doppler mode. *Aneurysms.* Color coding of the flow has the advantage that it allows hypoechoic wall material (thrombi) to be distinguished visually from the colorfully imaged aortic lumen, through which blood is still flowing (Bluth et al. 1990) (Fig. 11.**48**, p. 430). Unfortunately, in very hypoechoic thrombi in particular, this definitive sonographic differentiation is not always possible. It is obligatory to image the blood vessel in at least two planes when using color flow Doppler sonography.

When the color coding is adjusted to detect slow velocities, it becomes possible to identify disturbed flow visually, particularly on the margins of the vascular wall dilatation.

Stenoses and occlusions. An *aortic stenosis,* which usually forms in the distal segment of the blood vessel, can be visualized by color flow Doppler sonography due to the flow acceleration that occurs in the stenotic lumen. Targeted frequency analysis gives a value for the maximum flow velocity within the stenotic jet, and the stenosis can then be additionally quantified in the light of functional considerations.

No color signals can be registered from the region of an *aortic occlusion* (Fig. 7.**30**). Particularly when occlusions have existed for some time, arteries that under physiological conditions show little circulation, and are therefore also rarely identified, will often present a larger vessel diameter (Fig. 7.**37**, p. 265). Collateral arteries, which can form a significant bypassing circulatory system due to changed pressure and flow relationships, are: the superior and inferior mesenteric arteries, the superior and inferior epigastric arteries, the median sacral artery, and the lumbar arteries (Fig. 5.**33**, p. 165).

Vascular prostheses. Occasionally, the arterial lumen in the region of bypass anastomoses and vascular patch operations has a larger caliber than the preceding and subsequent vascular system. This does not in itself represent a pathological finding. If a stenosis forms in this type postoperative ectasia, and if the color flow Doppler mode is not available, then the narrow maximum jet can only be detected within the dilated vascular bed by carefully examining the entire prosthetic region. With color coding, visual detection of the maximum stenosis is possible at a glance. Figure 5.**46 a, b,** p. 177, shows a typical example.

Coarctation of the abdominal aorta. In contrast to classical subisthmic coarctation of the proximal descending aorta with a narrowing between the origin of the left subclavian artery and the region of the ligamentum or ductus arteriosus representing 98% of the reported coarctation cases, the rare connatal form of the abdominal coarctation (2%) is normally located in the subdiaphragmatic part of the aorta. The morphological intensity can be divided in hypoplastic (or even atretic) and segmental groups. The former includes cases with the small lumen of the aorta in a longer segment, and the latter refers to short stenosis, which can be suprarenal, juxtarenal (interrenal), or infrarenal.

The most common type (35%) is the juxtarenal, short segmental coarctation followed by the diffuse hypoplastic (21%) and suprarenal hypoplastic (15%) types. The infrarenal localization, either hypoplastic or segmental, is described in 15% and the suprarenal segmental type was found in 13% (Ben-Shoshan et al. 1973). With duplex sonography, hemodynamically significant stenosis of the coarctation can be diagnosed, and involvement of the renal arteries in the constricted segment of the abdominal aorta can be evaluated (Karasch et al. 1995).

Celiac Trunk and Branches

B-mode and Doppler mode. Sonographic detection of aneurysms in the *upper abdominal arteries* is possible. A local dilatation of the vascular wall is seen, and the registered flow phenomena show clearly disturbed arterial signals, or pendular flow patterns emanating from an anechoic or hypoechoic cystic space-occupying lesion (Fröhlich et al. 1988, Fransen et al. 1989, Endress et al. 1989).

Stenoses of the celiac trunk and its branches are recognized by a locally circumscribed increased flow velocity (Mallek et al. 1990b). Particularly careful consideration should be given to the often curved arterial course, so that a stenosis is not incorrectly assumed when an acoustically and quantitatively recordable higher Doppler frequency is encountered at one of the vascular curves. Only the exact angle-corrected measurement of the flow velocity allows correct estimation of the flow relationships when showing the vascular course by sonography or color flow Doppler mode (Fig. 11.**52**, p. 439). A maximum systolic flow velocity higher than approximately 200–400 cm/s indicates a celiac trunk stenosis. Figures 7.**31a–c** show a high-grade stenosis of the celiac trunk near the origin. With B-mode sonography, a clear, but not always detectable poststenotic dilatation of the celiac trunk can be seen in this patient (Fig. 7.**31**). This can be seen as the morphological correlate of the functionally effective stenosis.

During the acute phase of *hepatitis*, the peak systolic and end-diastolic velocity of the *hepatic artery* measured in 15 patients with hepatitis was larger than those in normal arteries (n= 15). During the recovery phase, these indexes of the hepatic artery decreased significantly to the control levels. The resistance indexes related to vascular resistance in the hepatic artery during the acute phase were significantly less than those in normal arteries, and they increased significantly to the control levels during the recovery phase (Tanaka et al. 1993). In patients with *severe liver disease,* the postprandial increase of the resistance index is significantly reduced (Joynt et al. 1995).

Color flow Doppler mode. Detecting smaller arterial *aneurysms* in the upper abdomen is also possible with color flow duplex sonography, due to the presence of eddy-like zones in a locally dilated arterial segment. Since these zones, which often occur next to one another, show varying flow directions and flow velocities, a colorful pattern with a broad range of color coding is seen (Warshauer et al. 1991). All of the criteria mentioned above for the diagnosis of an aneurysm also apply.

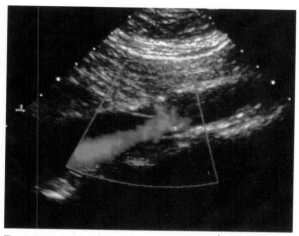

Fig. 7.**30** A color flow Doppler sonogram showing a distal segment occlusion of the abdominal aorta proximal to the bifurcation. The light color caused by aliasing in the two lumbar arteries originating above the occlusion reflects the high flow velocity in these collateral blood vessels

Pseudoaneurysm of the *cystic artery* is extremely rare and only a few cases have been reported in the world literature. However, color Doppler sonography can detect a false aneurysm ot the cystic artery in the gallbladder as a cause of hemobilia and upper gastrointestinal bleeding (Nakajima et al. 1996).

Arteriovenous fistulas between the abdominal arteries and veins can be recognized with duplex and color flow Doppler sonography from the twofold to threefold increase in the flow velocity in the fistula canal. The frequency spectrum documents a massively disturbed flow, with separation phenomena and eddy-like formations. As in the peripheral vasculature, dilatation, elongation, or loop formation in the afferent artery, and ectasia in the draining vein, may occur (Cantarero et al. 1989). If the shunt connection persists for a longer period, arterial and venous aneurysms may develop because of the elevated flow volume. In the venous component, an arterialized flow profile is registered (Stiglbauer et al. 1988, Endress et al. 1989, Hausegger et al. 1991; see the section on abdominal veins, pp. 299–300).

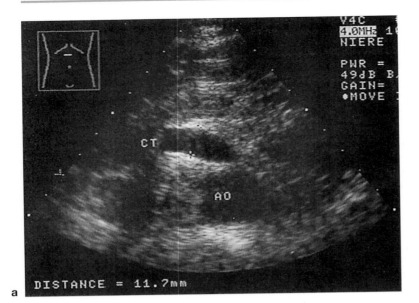

a

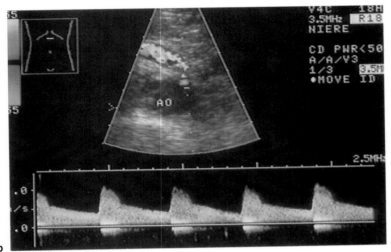

b

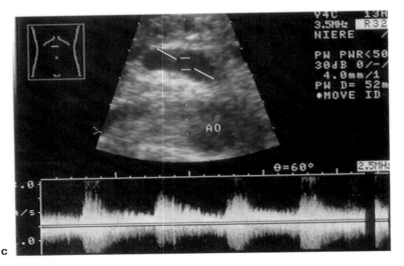

c

Fig. 7.**31** A stenosis of the celiac trunk (CT) at its origin. **a** With B-mode sonography, arterial dilatation can be seen after the origin from the abdominal aorta. **b** With color flow Doppler sonography, a narrow flow lumen can be seen near the origin of the celiac trunk; Doppler sonography indicates a high flow velocity in this region. **c** In the poststenotically dilated vascular segment, the eddy-like movements of blood flow can be heard and recognized from the frayed contour of the frequency spectrum, with retrograde components below the baseline. The maximal systolic velocity was 250 cm/s

Arcuate Ligament Compression Syndrome

B-mode and Doppler mode. The rare *ligamentous compression syndrome of the celiac trunk (arcuate ligament compression syndrome)* occurs predominantly in young females. An increased flow velocity is seen during expiration and a decreased flow velocity during inspiration, near the origin of the celiac trunk (Münch and Jäger 1988, Feindt et al. 1992). In our experience, this can be detected using color flow duplex sonography (Fig. 11.**53**, p. 440). The pathophysiological mechanism is still unclear. Ligamentous structures with a connection to the diaphragm or enlarged celiac ganglia, or both, might be possible causes (Kernohan et al. 1985, Williams et al. 1985, Reilly et al. 1985, Sieber and Jäger 1992).

Superior Mesenteric Artery

B-mode and Doppler mode. A monophasic curve form in the Doppler frequency spectrum, with a maximum systolic flow velocity higher than 220 cm/s, indicates a hemodynamically effective *stenosis* of the superior mesenteric artery when the patient has fasted and there are no celiac trunk abnormalities.

In eight patients with an angiographically confirmed superior mesenteric artery stenosis, Bowersox et al. (1991) measured a mean maximum systolic flow velocity of 299 cm/s ± 40 cm/s and an end-diastolic velocity of 78 cm/s ± 11 cm/s, with loss of the triphasic signal character in the Doppler frequency spectrum. In the *early dumping syndrome,* and particularly within the first 15 minutes after the ingestion of food, intestinal hypermotility and a notably increased flow volume, which is detectable with duplex sonography, occur in the superior mesenteric artery (Aldoori et al. 1985, Snook et al. 1989).

Intestinal hyperemia, which may accompany gastroenteritis, results in an increased flow volume, which is due to elevated flow velocity in a dilated blood vessel with increased caliber (Münch and Jäger 1988).

Color flow Doppler mode. Due to the acute incident angle (less than 30°) when applying color flow Doppler sonography, pronounced aliasing occurs near the origin of the superior mesenteric artery when there are hemodynamically effective stenoses, usually located proximally and having an arteriosclerotic etiology (Fig. 7.**32**).

Renal Artery and Intrarenal Arteries

■ Conventional and Color Flow Duplex Sonography

B-mode. A hypoechoic to anechoic space-occupying lesion with a connection to the renal artery indicates the presence of *aneurysmal vascular wall dilatations* of the renal arteries, both in children (Bunchman et al. 1991) and in adults (Martin et al. 1989). A massively disturbed flow with low flow velocities can usually be registered by the Doppler signal. Often, the separation phenomena can be followed as far as the distal arterial segment, which has a normal caliber again, and in which the velocity increases.

In B-mode sonographic imaging, unambiguous plaques in the renal arteries indicate the presence of a *stenosis* (Fig. 7.**33a, b**). However, the imaging of these is not always successful. It is helpful for the examiner to be familiar with the various etiological forms of stenoses found in the renal arteries, which have varying predilection points, and also with the various age-related and sex-related distributions.

Arteriosclerotic stenoses (80–90% of all forms of renovascular hypertension; high age at first manifestation; occurring in women significantly less often than men) mainly arise near the origin, but intrarenal segments may also be affected. On the other hand, these often

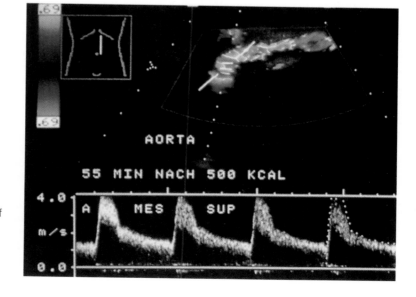

Fig. 7.**32** A color flow duplex sonogram showing a superior mesenteric artery stenosis in a sagittal, right medial, upper abdominal section, with aliasing in the stenosis jet. In a patient suffering from abdominal angina, the following parameters show elevated values 55 minutes after oral ingestion of a 500-ml liquid meal (500 kcal): maximum systolic flow velocity (408 cm/s), end-diastolic flow velocity (150 cm/s), and mean flow velocity (228 cm/s). The Pourcelot resistance index is 0.63

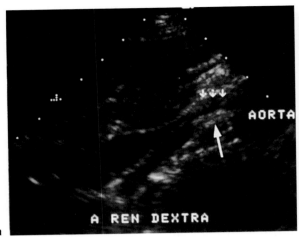
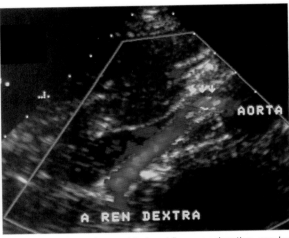

Fig. 7.33 a Sonogram of the abdominal aorta, with the origin of the right renal artery. The renal artery has a plaque at its origin, which is protruding into the vascular lumen from a dorsal direction. **b** In color flow Doppler mode, blood flows around the plaque. No local flow acceleration can be seen. The color change from orange to blue is due to the changed flow direction in the renal artery relative to the position of the sector transducer (Fig. 11.**56 a-f**)

multiple fibromuscular stenoses may be preferentially located in the middle and distal segment of the renal arteries (10–20% of all forms of renovascular hypertension; congenital; already symptomatic in children; first manifestation at a young age; occurs in women significantly more often than in men; histological classification into fibrous stenosis of the intima, fibromuscular stenosis of the media, and periarterial stenosis; McCormack et al. 1966). Possible rarer causes that may be considered are diseases of the aorta (dissecting aneurysm, syphilitic aortitis, congenital coarctation), renal arteritides (in obliterating endangiitis, periarteritis nodosa), or trauma of the renal arteries (Kaufmann et al. 1992). Renal artery stenoses can also be detected in these diseases with duplex sonography.

Doppler mode. Frequency spectra, which are registered using the pulsed wave or continuous wave Doppler mode, indicate the presence of a *stenosis,* which can be quantified when:

- The *angle-corrected* maximum systolic flow velocity is locally greater than 180 cm/s.
- The end-diastolic flow velocity increases.
- Clear spectral broadening appears.
- The flow velocity and the resistance index decrease in distal arterial segments and intrarenally (resistance index less than 0.5) (Handa et al. 1986, Patriquin et al. 1992, Özbek et al. 1993, Schwerk et al. 1994).

Figure 7.**34** shows a typical example.

If the renal arteries are clearly identified with B-mode imaging sonography, an arterial segment *occlusion* can be diagnosed when:

- There is no flow signal registered from the vascular lumen.
- In side-to-side comparisons, there is a clearly lower intrarenal flow velocity (Fig. 7.**35 a, b**).
- In side-to-side comparisons, a clearly lower value is calculated for the resistance index.
- The kidney appears to be smaller (less than 9 cm), without other causes for this being present (Hoffmann et al. 1990).

Failure to detect a signal alone, without being able to display the artery morphologically, is an uncertain sign. If duplex sonography indicates possible renal artery occlusion, an attempt should be made to register the refilling of the renal artery or of the accessory renal arteries. In our own patients, intrarenal arterial flow signals could not be registered in four of eight patients with renal artery occlusions. Even using the color flow Doppler mode did not improve the situation. In the remaining four patients, the intrarenal Pourcelot resistance index gave an average of 0.54 on the occluded side. This was on average 26% (13–36%) lower than in the contralateral kidney (Karasch et al. 1994).

In various diseases involving the kidney, of varying etiology, which functionally restrict renal filtration capacity, an *increased* intrarenal resistance index due to a decreased diastolic flow component has been reported (de Toledo et al. 1996). According to examinations conducted by Patriquin et al. (1989), including 17 children, and a case study of a transplantation complication reported by Jansen et al. (1990), the resistance index in an *acute hemolytic uremic syndrome* correlates in this sense with the renal functional capacity. However, Gleeson et al. (1992) maintain that no predictions concerning the long-term prognosis of renal function are possible in this clinical syndrome.

Fig. 7.**34** A renal artery stenosis demonstrated with color flow duplex sonography. The frequency spectrum shows an elevated systolic flow velocity (250 cm/s) and end-diastolic flow velocity (50 cm/s)

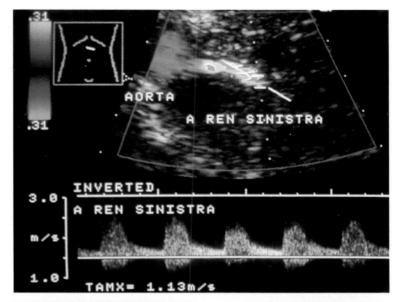

Fig. 7.**35** Doppler frequency spectra from intrarenal segmental arteries, showing **a** a normal-appearing renal vascular system on the left, and **b** a right renal artery occlusion, with reduced flow velocities and a decreased Pourcelot resistance index when compared side to side postocclusively.
Left: the maximum systolic and end-diastolic flow velocities are 53 cm/s and 17 cm/s, respectively, with a resistance index of 0.68; *right:* the maximum systolic and end-diastolic flow velocities are 31 cm/s and 12 cm/s, respectively, with a resistance index of 0.45

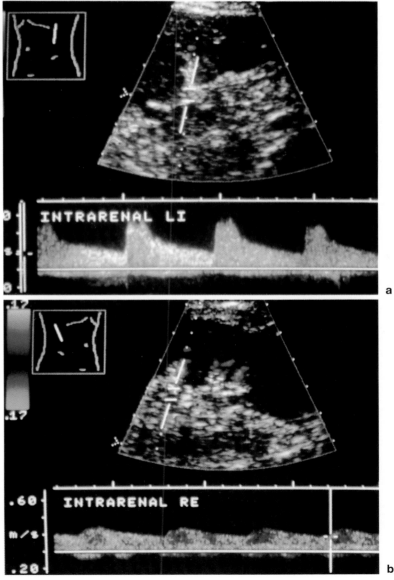

Platt et al. (1989, 1992) report measurement of the intrarenal resistance index in differentiating between obstructive and nonobstructive dilatations of the renal calices. When an obstruction was not present, patients with dilatations of the renal calices had a resistance index no higher than 0.7 (n = 7); when there was an obstruction, values higher than 0.7 were measured (n = 14). After placement of a percutaneous nephrostoma, the resistance index fell to values under 0.7 in all controlled patients (n = 10). In transplanted kidneys, an elevated resistance index may indicate a urinary obstruction (Flückiger et al. 1990).

The intrarenal resistance index is also used to evaluate the renal circulation during the course of pregnancy. According to Sohn and Fendel (1988), pregnant women with edema–proteinuria–hypertension (EPH) gestosis (n = 12) show a significantly increased resistance index, reaching values around 0.75, while in normal pregnancies (n = 52) low resistance index values of around 0.64 are measured.

Color flow Doppler mode. Although the proximal and distal portions of the renal arteries can be depicted using B-mode imaging in favorable examination conditions, hemodynamically effective *stenoses* in the middle segment of the renal arteries, in particular, may be easier to diagnose due to their lighter color signals or the presence of aliasing (Fig. 7.**36**), and they can be quantified using the pulsed wave Doppler mode.

Extremely high-grade stenoses, which usually cause aliasing in the stenosis jet, are often only recognizable through a brief flashing of the color signals within the stenosis (Fig. 11.**54**, p. 441). However, they can be suspected on the basis of a clearly disturbed flow with separation phenomena in the vascular segment that immediately follows.

Intrarenal arteriovenous malformations can be recognized on the basis of a focally disturbed flow pattern, with high flow velocities of up to 180 cm/s and low resistance indices (0.32–0.52). When color-coded, they show light, irregular shades of color with aliasing

(Takebayashi et al. 1991). It is occasionally possible to depict vascular segments that have a curved course. Depending on the magnitude of the arteriovenous shunt volume, elevated flow velocities with pulse-synchronous flow patterns can also be measured in the draining veins.

In addition to the possible occurrence of bleeding or infections, *complications following punctures* of the renal parenchyma include the development of arteriovenous fistulas or false aneurysms (Vassiliades and Bernardino 1991). Both of these complications can be detected easily with color flow duplex sonography, due to the massively disturbed flow with relatively high flow velocities in the fistula, and also because of the pendular flow in the false aneurysm.

Inferior Mesenteric Artery

Conventional and color flow duplex sonography. A stenosis of this visceral artery can be detected through the elevated maximum systolic flow velocity and, in the case of a higher stenosis grade, it can also be recognized by the increased end-diastolic flow velocity. To our knowledge, quantitative measurements from larger groups of patients are not yet available.

In a group of 63 patients with acutely infectious intestinal diseases, such as ulcerative colitis and Crohn's disease, Kathrein et al. (1990) were able to show that the maximum systolic, the end-diastolic, and the mean flow velocities increase when the disease is florid, while the resistance index decreases. During remission of the disease these parameters normalize again.

Mesenteric Steal Syndrome

Conventional and color flow duplex sonography. Preformed collateral arteries exist between branches of the *inferior mesenteric artery* and the *internal iliac arteries,* and they are also present between the inferior and superior mesenteric arteries themselves (Fig.

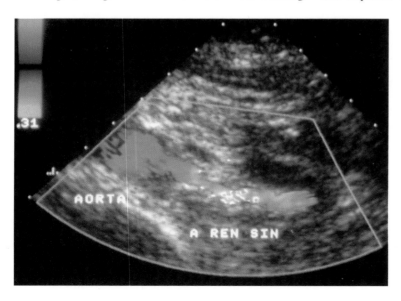

Fig. 7.**36**   A left renal artery stenosis after the origin from the aorta, demonstrated by color flow Doppler sonography through locally circumscribed aliasing, which reflects the elevated mean flow velocity in this segment. The maximum systolic flow velocity in the stenosis was 289 cm/s

7.**37**). When there is gradual obstruction of the common iliac artery, these collaterals can develop into vasculature of relatively substantial caliber to supply the lower extremity with blood (Fig. 5.**33**, p. 165). When this bypassing circulatory system is in operation, the external iliac artery and the subsequent vascular system receive blood from the distribution area of the inferior or superior mesenteric artery through the internal iliac artery, which shows *retrograde* flow. Particularly during exercise-induced hyperemia, this can lead to an insufficient intestinal blood supply, with resulting colicky abdominal pains, which typically cease when the leg muscles are rested *(mesenteric steal syndrome)*. The collateral arteries appear as vessels with a lumen of above-average size and high circulation, with an atypical location. They consist of inferior mesenteric artery branches, or more rarely superior mesenteric artery branches, and can be continuously followed with duplex sonography as far as their junction with the internal iliac artery.

Lumbar Arteries

Conventional and color flow duplex sonography. Pathological findings such as stenoses or occlusions in individual lumbar arteries usually cannot be diagnosed with transabdominal sonography, even with color flow Doppler sonography. Particularly when an artery is not displayed, it cannot be determined with certainty whether its absence is due to an occlusion, or is simply due to factors associated with the examination.

When individual lumbar arteries or the *median sacral artery* assume collateral function due to high-grade flow obstructions in the distal abdominal aorta or the common iliac arteries, it causes increased circulation and therefore increased caliber in these vessels. In this altered hemodynamic situation, the partial imaging of the lumbar arteries in transverse sectional planes is possible near their origin in the preocclusive segment of the aorta (Fig. 7.**30**) or the median sacral artery.

Ovarian Artery and Uterine Artery

Conventional and color flow duplex sonography. It may be possible to evaluate ovarian functional status by studying pathological deviations from the normal changes in the flow pattern of the *ovarian artery* associated with the menstrual cycle. Schurz et al. (1990) were able to demonstrate that women with polycystic ovaries (n = 10) had higher pulsatility index values during a stimulated cycle on the day of follicle rupture, without any preovulatory decrease (4.4 ± 1.1 compared to 1.1 ± 0.4 in the control group; n = 21). These higher values correlated with a lower conception rate.

Absent changes in the resistance during anovulatory cycles have also been described in the *uterine artery* (Kurjak et al. 1991). In these transvaginal measurements, absent end-diastolic flow in several women was accompanied by primary infertility.

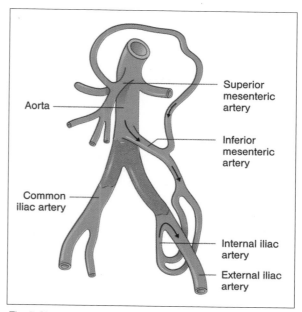

Fig. 7.**37** Potential paths of collateralization during common iliac artery occlusions that can be detected with duplex sonography. Refilling of the external iliac artery can occur via an ipsilateral internal iliac artery, with retrograde blood flow. The blood supply can be provided by connections to the contralateral internal iliac artery (light blue) or through branches from the distribution area of the inferior and superior mesenteric arteries (dark blue). A steal syndrome in the intestinal arteries, with accompanying relative ischemia, can present clinically as a mesenteric steal syndrome. In rare instances—usually in bilateral obliterations of the iliac arteries—steal through the contralateral internal iliac artery may be responsible for erectile dysfunction when body weight is placed on the leg (p. 315)

Organ Transplants

Kidney Transplants

Conventional and color flow duplex sonography. The following structures or parameters in a transplanted kidney can be displayed and evaluated using color flow duplex sonography:

- Perirenal soft-tissue structures (edematous organ swelling, hematoma, seroma, abscess)
- Anastomotic region (vascular stenosis, anastomotic aneurysm, false aneurysm)
- Transplanted artery and vein (stenoses, occlusion)
- Segment-specific organ circulation (qualitative)
- Intrarenal resistance indices (Pourcelot, Gosling)
- Intrarenal vascular complications (arteriovenous fistula, false aneurysm)

In addition to the important findings that can be diagnosed solely with B-mode imaging, evaluating the arterial inflow and the venous outflow is also very important, since it allows the detection of vascular kinking, stenoses, and occlusions in the anastomotic vasculature (Fobbe et al. 1989, Reuther et al. 1989, Jurriaans and Dubbins 1992, Krumme et al. 1994).

Depending on the operative technique used, some patients postoperatively exhibit separation movements caused by irregularities in the contour alignment within the sutured area of the anastomotic vessels, with an acoustically and visually detectable flow disturbance. However, this alone does not indicate a hemodynamically effective stenosis due to suturing. An *anastomotic stenosis* can be assumed when the registered maximum systolic flow velocity exceeds 150–200 cm/s, shows a local increase of more than 50%, or if the velocity related to the iliac artery increases beyond a factor of 2.6–3.0 (Leichtman et al. 1989, Soper et al. 1989, Deane et al. 1990, Alvarez et al. 1991). In a study with 109 patients, a peak systolic velocity of ≥ 2.5 cm/s in the transplanted renal artery had a sensitivity of 100% and a specificity of 95% for the detection of renal artery stenosis. Although a significant difference in pulsatility index, resistance index, acceleration index, and acceleration time was recorded from the intrarenal vessels in the angiographically normal and stenosed groups with Doppler, these measurements were less useful as discriminating diagnostic tests (Baxter et al. 1995). To detect *stenoses* that are due to *kinking*, it is useful to examine the transplanted organ while the patient is standing because these stenoses may become hemodynamically significant only in certain positions.

In arteries of various sizes in a transplanted kidney with normal circulation, the *Pourcelot resistance index* is less than 0.71, and shows a gentle decrease toward the periphery (Fig. 7.**38**). Within the range 1.**12**–1.**26**, the *Gosling pulsatility index* does not show any significant differences at various locations within the renal arterial system (Schwaighofer et al. 1987, Rigsby et al. 1987). Changes in the resistance index and pulsatility index that have been described in connection with individual postoperative complications after transplantation can be calculated using the Doppler frequency spectrum. However, the true significance and the diagnostic predictive value of these indices in assessing the etiology of transplant-related complications remains controversial (Mallek et al. 1990 a, Perchik et al. 1991, Meyer et al. 1990, Pelling and Dubbins 1992).

A reduced diastolic flow velocity associated with an increase in the intrarenal resistance index can be detected in:

- Acute rejection reactions
- Chronic rejection reactions
- Urinary obstruction (ureteral obstruction)
- Arteriosclerosis of the vasculature
- Acute tubular necroses

(Taylor et al. 1987 a, Arima et al. 1989, Don et al. 1989, Harris et al. 1989, Flückiger et al. 1990, Van Leeuwen

et al. 1992). In some transplants, diastolic flow reversal may occur in the context of an acute rejection (Oh et al. 1989).

As they appear to be related to complete or partial renal infarctions, *occlusions of the transplant arteries and segmented arteries* show either regionally limited or completely absent flow in the intrarenal vasculature (Fig. 11.**55**, p. 442) (Taylor et al. 1987, Alvarez et al. 1991, Grenier et al. 1991).

Arteriovenous fistulas formed after percutaneous punctures of kidney transplants can be recognized by focal massively disturbed flow, with irregular Doppler signals above and below the baseline and a high intrarenal flow velocity, or frequency due to a high arteriovenous pressure gradient (Taylor et al. 1987 a, Sievers et al. 1989 a, Hübsch et al. 1990, Plainfosse et al. 1992, Gainza et al. 1995, Branger et al. 1995).

Renal venous thrombosis can be located not only using criteria from B-mode sonography (p. 298) or by confirming absent venous flow intrarenally and in the renal vein, but also on the basis of altered arterial Doppler signals from the ipsilateral renal artery and the intrarenal arteries (Fig. 11.**58**, p. 447). During systole, a steeply increased and then decreased flow velocity is observed, followed by *retrograde* flow during the entire diastole (Krumme et al. 1993, Trattnig et al. 1993).

Liver, Pancreas and Bone-Marrow Transplants

In addition to focal and generalized changes in the liver parenchyma, duplex and color flow Doppler sonography can detect vascular complications following liver transplantations Abad et al. 1989). In a group of 90 patients after liver transplantation, Barton et al. (1989) reported a total of 14.4% sonographically identified and angiographically verified postoperative vascular complications. The following were detected:

- Stenoses and occlusions of the hepatic artery (5.5%)
- Thromboses, stenoses, and occlusions of the portal vein (5.5%)
- Stenoses and partial thromboses of the inferior vena cava, both suprahepatic and infrahepatic (6.6%)

After combined pancreas and kidney transplantations, a postoperative Doppler frequency spectrum with a relatively high end-diastolic component in the *intraparenchymatous pancreatic vasculature* may be found, the Pourcelot resistance index of 0.50–0.58 indicates a low peripheral arterial resistance.

In the case of an acute rejection reaction, case studies have observed a reduced diastolic flow velocity with an additional end-diastolic decrease, which in turn leads to an increased resistance index at values of 0.70–0.76 (Patel et al. 1989, Pöllmann et al. 1989).

Fig. 7.**38** **a** A color flow Doppler sonogram of a transplanted kidney (NTX) in the right inguinal fossa, simultaneously showing the ipsilateral pelvic vasculature. **b** The intrarenal segmental arteries, with flow toward the probe, are color-coded red. The frequency spectrum from the intraparenchymatous arteries shows a normal flow pattern. The maximum systolic, end-diastolic, and mean flow velocities are 45 cm/s, 15 cm/s, and 26 cm/s, respectively (resistance index 0.67, pulsatility index 1.15). **c** On the contralateral side, the remaining cystic kidney presents an intrarenal flow profile with decreased systolic, end-diastolic, and mean flow velocities of 30 cm/s, 15 cm/s, and 5 cm/s, respectively and also a relatively high peripheral resistance index at 0.83

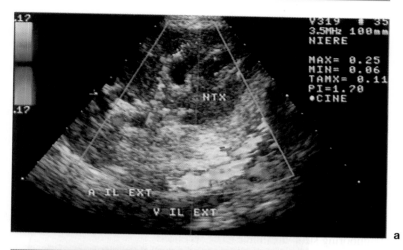

a

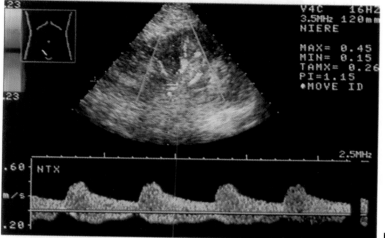

b

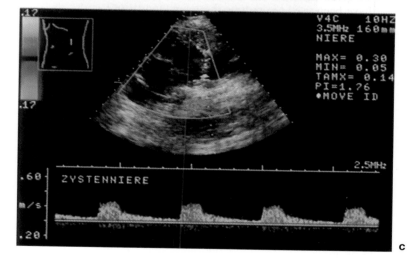

c

After *bone-marrow transplants*, liver damage in the form of occlusive venous disease can be recognized from nonspecific signs such as ascites, thickening of the gallbladder wall, circulatory disturbances in the portal vein, and also an elevated resistance index in the hepatic artery of 0.81 (0.75–0.87, n = 19) in comparison to a value of 0.69 (0.58–0.76) in a healthy control group (n = 20) (Herbetko et al. 1992).

Evaluation

■ Conventional and Color Flow Duplex Sonography

B-mode. Although true aortic aneurysms are quantitatively by far the largest group of all the *aneurysm types* that occur in the abdominal cavity, rarer aneurysms of the visceral arteries can also be diagnosed with B-mode sonographic imaging (Ugolotti et al. 1984, Fransen et al. 1989, Grech et al. 1989, Grün et al. 1989, Song et al. 1989, Namieno et al. 1991, Verma et al. 1991).

For all aneurysms, the exact location and morphology, size measurements, and relationship to neighboring structures have to be recorded.

In favorable examination conditions, it is also possible to clearly demarcate *vascular wall structures* in small abdominal arteries, so that plaques can be detected by B-mode imaging sonography. The use of intravascular and intraluminal ultrasound probes with high resolution to recognize vascular wall structures (plaques, dissections), although it is bound to remain restricted in its scope, may provide additional morphological and functional information in certain groups of patients (Pandian 1989).

Doppler mode. Using frequency analysis according to the criteria mentioned above allows the recognition of flow changes that are due to nonstenotic wall deposits (Labs et al. 1990), and also the grading of a stenotic lumen according to its hemodynamic effectiveness.

Additional parameters used for qualitative and quantitative assessment of the circulation (the examples given here are various measurable quantities that have been used to describe the renal circulation) are: modified resistance indices (Marchal et al. 1989); the *Gosling pulsatility index* (Oh et al. 1989, Alvarez et al. 1991); the *pulsation flow index* (PFI), derived from the quotient of the area beneath the maximum frequency curve and the product consisting of the maximum systolic flow velocity and the pulsation time (Sievers et al. 1990); and the *peripheral circulation index,* which is a ratio of the intrarenal circulation quantities to the iliac flow volumes (Nishioka et al. 1989). The use of these is discussed in the individual sections dealing with the specific organ arteries.

The significance of parameters measured with Doppler sonography in the differential diagnosis of parenchymatous organ disease cannot yet be determined.

Color flow Doppler mode. Using color flow Doppler sonography has significantly expanded the possible areas of application for abdominal examinations. In addition to locating and identifying small-caliber arteries, the results obtained with color flow Doppler sonography provide decisive visual feedback, which shows a circulatory disturbance throughout the entire area of a sectional plane.

Eddy-like circulatory disturbances detectable by their color coding can be observed near aneurysms of the upper abdominal arteries. These correspond to the subordinate flow within the aneurysmatic sack, and make it easier to differentiate cysts (Bunschman et al. 1991). A low pulse repetition frequency should be selected for the color flow Doppler mode in order to detect the (sometimes very slow) velocities displayed by these intra-aneurysmal currents. Rare findings, such as false aneurysms in the arteries of the abdominal wall, can de distinguished with certainty from abscesses or hematomas on the basis of the arterial flow (Gage et al. 1990).

Sources of Error and Diagnostic Effectiveness

Sources of Error

Obesity

Particularly in obese patients with an android lipid distribution pattern, the examination can be complicated by intervening cutaneous and mesenterial lipid layers to such an extent that the assessment of some individual abdominal arteries becomes impossible, since the vasculature cannot be adequately displayed. Under unfavorable conditions and in rare instances, it may even be difficult to evaluate the large pelvic arteries completely and directly with ultrasound.

Meteorism

The superimposition of intestinal gas on intra-abdominal structures complicates the ultrasound examination, because ultrasound waves are completely reflected at interfaces with gas. In these situations, it can be helpful to apply moderate pressure with the transducer on the area of interest, since it is sometimes possible to displace the gas-filled intestinal loops. Occasionally, it may be better to schedule a new examination after the flatulence has been eliminated by appropriate preventive measures (diet or fasting, bodily exertion prior to the examination, carminative medications).

Movement Artifacts

When examining the small abdominal arteries, pulsations in larger vascular branches or positional changes due to respiration may sometimes make it difficult to register an extensive artifact-free frequency spectrum continuously. Sometimes, a representative frequency spectrum which can then serve as the basis for further analyses can only be registered in brief, respiration-dependent examination phases. To register frequency spectra that are artifact-free, it is sometimes useful for the patient to be in an inspiratory, expiratory, or average respiratory resting position.

Peristalsis

Peristaltic waves, which occur intermittently during every sonographic examination of the abdomen, cause an extensive color display resembling the color noise obtained when using amplification that is too high. This can be explained by the registration (and color-coding) of the intestinal wall movements. During a peristaltic wave, the arterial color flow Doppler signals are superimposed, and for a brief moment cannot be evaluated. Similarly, in the Doppler mode an acoustically recognizable crescendo and decrescendo of bubbling, hissing signals is produced, which can be seen in the spectrum as a broad frequency band with "frayed" fine notches above and below the baseline (Fig. 7.**39**).

Cardiac Arrhythmia

Flow within the abdominal arteries is also determined predominantly by the cardiac output. In addition to heart failure, cardiac arrhythmias in particular can complicate the registration and calculation of reliable flow velocities, since the registered Doppler frequencies can change from systole to systole. Due to the different filling times for the left ventricle, the stroke volume varies, as does the end-diastolic flow velocity. The latter decreases during the course of the diastole, due to significant fluctuations in the length of the diastole caused by the start of a renewed cardiac contraction. Both changes also influence the calculation of

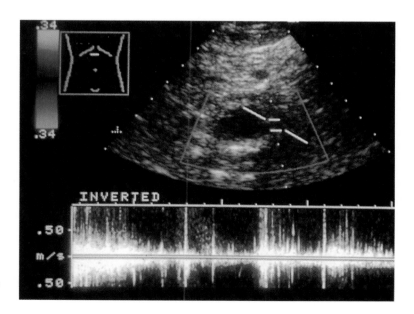

Fig. 7.**39** A Doppler spectrum registering peristaltic intestinal wall movements

the resistance index and pulsatility index. As an example, Figure 7.**40 a** shows the registration of several Doppler frequency spectra from the superior mesenteric artery in a patient suffering from intermittent absolute arrhythmia. The systolic and end-diastolic flow velocities, varying from one cardiac cycle to the next, can be recognized. After a rhythm spontaneously establishes itself, uniform frequency spectra can again be evaluated for a longer period (Fig. 7.**40 b**).

Planes and Angles for Applying Ultrasound

B-mode. To avoid tangential sections with incorrect measuring points, the measurement of values obtained from the arteries requires strict adherence to defined sonographic sectional planes. As a standard practice, an artery whose caliber is to be determined should be depicted in a cross-section perpendicular to the long vascular axis and in a longitudinal section,

possibly with a large internal lumen (Figs. 7.**23**, 7.**25 a, b**, 11.**48 c**, p. 431):

– Imprecisely positioned *cross-sections overestimate* the arterial caliber.
– Imprecisely selected *longitudinal sections* can *underestimate* the vascular caliber when measuring the vascular wall components either close to or far away from the transducer.

Doppler mode. The influence of the incident ultrasound angle is very important in the abdominal arteries. Often their retroperitoneal location deep within the abdominal cavity restricts the ability to position the transducer in relation to the vascular axis. Particular care must therefore be taken to ensure that only frequency spectra with a properly corrected angle adjustment are used to determine quantitatively measured values (maximum frequency, maximum

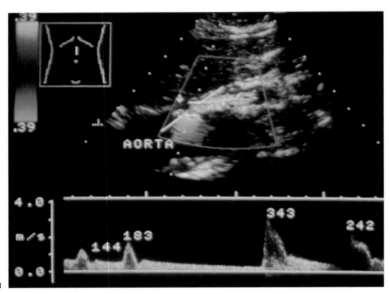

a

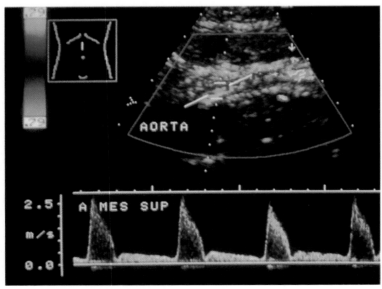

b

Fig. 7.**40** **a** A longitudinal section through the upper abdomen, showing the proximal superior mesenteric artery and the registration of several Doppler frequency spectra in a patient with intermittent absolute arrhythmia. From one cardiac cycle to the next, the significantly varying systolic and end-diastolic maximum velocities can be seen. The systolic maximum velocity (144–343 cm/s) was calculated for four cardiac cycles, and can be determined from the spectrum. In the region near the origin of the artery, a hyperechoic plaque is observed (arrow). **b** After a steady rhythm has been established, consistently shaped frequency spectra can be registered. From these, maximum, end-diastolic, and mean flow velocities of 282 cm/s, 20 cm/s, and 86 cm/s, respectively, can be calculated. A side branch of the superior mesenteric artery (arrow) is displayed using color flow Doppler sonography. Due to a low-grade stenosis, the maximum flow velocity is elevated in this patient, who had fasted

velocity). Incorrectly low angle adjustments erroneously produce a flow velocity that is too low; incorrectly large angle corrections produce a flow velocity that is too high.

Color flow Doppler mode. Since the abdominal arteries do not usually follow a course along a straight line
in the sectional plane at which ultrasound is applied,
the ultrasound waves can impinge on an artery at
different angles during its curving course. The various
frequency shifts resulting from this lead to a nonhomogeneous color display, with varying shades of color
and color saturation, although no concrete flow velocity change is present in the arterial segment to which
the ultrasound is being applied (Figs. **7.10, 7.13**).

At an obtuse incident angle, these physical limitations can lead to poor or absent color coding (darker
colors or black); at an acute ultrasound angle, *aliasing*,
with a single or multiple color change through white,
may result (Mitchell 1990).

Color Flow Doppler Parameters

Evaluating the displayed color patterns can be ren-
dered impossible by *unfavorable color adjustments*.
Frequent potential errors include:

- Specifying an inadequately low velocity (pulse
  repetition frequency) for the color adjustment,
  with resulting aliasing or multiple aliasing.
- Having almost a right angle between the color
  flow Doppler plane and the vascular axis, with
  resulting poor to absent color coding of the arterial lumen.
- Incorrect color filter selection, with reduced
  color output.
- Undercontrolled or overcontrolled color amplification, with poor color display or color noise.

Color Doppler energy mode. Since the amplitude
mode cannot determine the velocity or direction of
flow, it is possible to confuse arterial and venous flow.
This must be borne in mind particularly when examining intraparenchymatous arteries of the kidney and
the spleen.

Positional Anomalies and Variants

There are a large number of variants in the anatomy
and location of the abdominal arteries. Identification
is possible on the basis of the typical flow patterns of
the individual arteries (triphasic or monophasic signal, high or low peripheral resistance, flow velocity).
In addition, anatomical classification of an artery as
belonging to a particular endorgan can also assist in
the identification. Simple modulation tests like those
used in the peripheral arteries or the arteries supplying the brain are not available.

In addition to positional anomalies, arteries that
are *multiple,* or *accessory* arteries such as those found
in the diagnostically relevant *renal vascular system*
(pole arteries) (Lippert and Pabst 1985), complicate
the complete evaluation of an organ's arterial system
(Berland et al. 1990, Hoffmann et al. 1990). For that

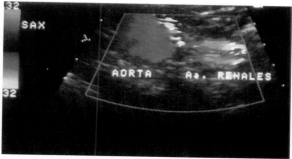

Fig. 7.**41** The abdominal aorta (color-coded blue) is
shown with color flow Doppler sonography, with the origin
of two renal arteries on the left. Both arteries originate from
the aorta at approximately the same level, so that they can
both be imaged in a single sectional plane. The flow direction of the renal arteries is coded red in this section, since it
is directed away from the transducer. Doppler sonography
measured the systolic and end-diastolic flow velocities in
the two renal arteries as 130 cm/s and 35 cm/s, respectively

reason, the possibility of more than one renal artery
has to be kept in mind. Thus, the abdominal aorta has
to be systematically examined for additional arteries
that may be supplying the kidney. Figure 7.**41** shows
an example of a dual arterial blood supply, in this case
a left kidney supplied by two renal arteries.

Particularly when there are simultaneous
aneurysmal dilatations and possible partial thrombosis, an *elongation* of the arteries requires precise location and delimitation of the vascular course (Paes and
Vollmar 1991). Careful adjustment of the vascular
axes in several planes is a prerequisite for measuring
the vascular caliber (see the section on planes and angles for applying ultrasound, above).

After organ transplantations with anatomical or
extra-anatomical vascular anastomoses, it can be difficult to identify transplant arteries when the ultrasound examination capability is limited as a result of
postoperative scarring and signal superimposition
from other vasculature (pelvic vasculature in kidney
transplants).

Venous Superimposition

Venous signals that are superimposed on the arterial
flow curves complicate the assessment of the
frequency spectra that are registered. Due to the complex anatomical structures in the abdominal cavity,
the appearance of arterial and venous flow in the
same direction should be expected, in addition to arterial and venous flow in opposite directions.

Depending on the location and the specific vascular segment, venous and arterial signals with a similar
flow direction may become superimposed on one
another. This applies both to visual (color flow Doppler) and acoustic registration, as well as to the
frequency spectrum. As an example, the distal left

272   7 Abdominal Arteries

renal vein after it crosses the abdominal aorta and right renal artery can be mentioned (Fig. 8.**23a**, p. 303).

In the part near the hilum, the left renal vein, which courses next to the renal artery with an opposite flow direction, can be distinguished from its accompanying artery due to its different color coding, and also by registering the frequency spectrum in the opposite direction.

Filiform Stenoses

Extremely high-grade flow obstructions that have only a thread-like fast flow jet or an intrastenotically decreased flow velocity can sometimes be clearly recognized with duplex sonography only after extensive and careful examination of the overall arterial caliber. In particular, filiform stenoses with a reduced maximum flow velocity can be inadvertently missed if the sensitivity of the Doppler frequencies during the examination is not preselected and evaluated for both the high-frequency and low-frequency ranges (high or low pulse repetition frequency).

Although this hemodynamic situation can also be a problem in color flow Doppler sonography, using the color flow Doppler mode in such cases provides visual support for the identification of the remaining narrow lumen of an extremely high-grade stenosis through which the blood is still flowing.

Occlusions

As a sole criterion, the simple inability to detect a Doppler frequency or color signal in the abdominal region does not yet justify the conclusion that there is a vascular occlusion. Instead, possible positional variants must be considered in the less than ideal conditions for ultrasound application that may apply (postoperative, obesity, intestinal gas superimposition). In addition, an absent image in the abdominal vasculature may also be due to technical factors associated with the examination itself.

A diagnosis of arterial occlusion can therefore only be made after:

– *Clearly identifying* the artery over a longer distance, *and*
– Observing the absence of a Doppler frequency signal

and only provided that the necessary preconditions— i.e., correctly adjusted equipment adapted to the examination conditions (amplification, filter selection, appropriate scale for the color and frequency display, pulse repetition frequency)—have been fulfilled!

At least in the case of extremely high-grade vascular obstructions, diagnosis *without* differentiating between a stenosis or an occlusion becomes more secure the larger the number of additional criteria that are made available to assess the entire vascular system of an organ.

Indirect criteria indicating an extremely high-grade narrowing of the lumen are:

– Clearly reduced maximum systolic flow velocity distal to the suspected obliteration, with accompanying flow disturbances
– Clearly lower end-diastolic and mean flow velocity postocclusively
– Reduced resistance index in the subsequent vascular system
– Clear bilateral differences in the resistance index and flow velocity parameters (e.g., in the renal arteries)

The relevance of the perfusion differences detected while diagnosing and quantifying obstructions of the abdominal arterial system by color flow Doppler sonography is not yet clear.

Collateral Vasculature

Hemodynamically effective stenoses or occlusions of individual organ arteries or the abdominal aorta can be compensated for by preformed collateral systems, which at least partially take over the blood supply to the affected vascular system (Fig. 7.**6**). The increased circulation velocity causes luminal widening in the collateral arteries (Schoop 1974), which leads to improved display in duplex sonography, and may even allow a blood vessel to be displayed for the first time (Fig. 7.**30**). This provides a potential for morphological confusion between the collaterals and the original blood vessel, and can also give rise to difficulties in differentiating the functional findings. A high flow velocity in a collateral can mistakenly be interpreted as a stenotic main blood vessel.

Diagnostic Effectiveness

Abdominal Aorta

Aneurysms

Sonography is considered to be the noninvasive method of choice in detecting an aneurysm of the abdominal aorta, and also in conducting the follow-up assessments that are required (Zoller and Stapff 1989). However, special attention needs to be given to the reproducibility of cross-sectional measurements (sectional planes, definition of method, wall delimitation, interindividual examination variability) during follow-up observations (Crawford and Hess 1989).

In the measurement of the sonographic size of intra-abdominal space-occupying lesions (n = 255), Hess et al. (1988) indicate interindividual variations averaging 2 mm ± 6 mm. In tumors larger than 50 mm, 16% of differences in the values measured were already over 12 mm (mean 7.1 mm ± 5.2 mm). These results emphatically demonstrate the examination-dependent variability of quantitative size determinations.

In a *dissecting aneurysm,* diagnosing the two lumina through which blood flows is possible with duplex and color flow Doppler sonography (Giyanani et al. 1989, Thomas and Dubbins 1990). With an appropriate indication, additional diagnostic procedures (transesophageal echocardiography, computed tomography, magnetic resonance imaging, angiography) can subsequently be applied in a targeted manner.

Vascular Prostheses

Y-prostheses, aortic prostheses, and unilateral bypasses can usually be clearly imaged using duplex and color flow Doppler sonography (Gritzmann et al. 1988, Hendrickx et al. 1991). Evaluating proximal Y-anastomoses and aortic prostheses is complicated in patients who have recently undergone surgery and in obese individuals, since the required abdominal wall compression with the transducer is sometimes not tolerated due to pain.

Additional use of the color flow Doppler for extensive imaging of the aortic lumen through which blood is flowing may make it easier to distinguish between normal and pathological findings, since it allows hemodynamically effective stenoses, occlusions, and hypoechoic wall deposits to be excluded more quickly.

Celiac Trunk and Branches

Celiac trunk stenoses can be excluded with high diagnostic certainty, since this relatively short arterial segment is usually easily accessible to ultrasound. In approximately 93–98% of the examined cases (Uzawa et al. 1993), it is possible to depict the flow in the *common hepatic artery* and *splenic artery* with color flow Doppler sonography. In patients with liver cirrhosis (see Chapter 8) and older individuals, intrahepatic segments of the hepatic artery can be imaged particularly clearly, due to the increased flow volume and larger caliber of the artery in comparison to the portal vein (Grant et al. 1989).

Even the smallest truncal branch, the *right gastric artery,* is accessible to an experienced examiner 20–27% of the time using *conventional* duplex sonography.

In measurements involving 41 patients, Uzawa et al. (1993) were able to depict the *gastroduodenal artery* with duplex sonography 66% of the time, and in 98% of patients, when using color flow duplex sonography.

The *cystic artery* can be displayed 70% of the time using color flow duplex sonography (McGrath et al. 1992).

Since duplex sonography makes a significant contribution to the differential diagnosis of *cystic space-occupying lesions* in the upper abdomen, every anechoic or hypoechoic tumor of unclear origin must first be examined by duplex sonography to exclude the presence of a true or false aneurysm, before proceeding to puncture or surgery (Fröhlich et al. 1988,

Fransen et al. 1989, Paes and Vollmar 1991, Warshauer et al. 1991).

At present, we are only aware of studies with small patient numbers that have attempted to establish the value of conventional and color flow duplex procedures in the celiac trunk and its branches (Mallek et al. 1990b, Pierce et al. 1990). However, these confirm the expectation that there will be useful and sensitive applications of duplex and color flow duplex sonography to diagnose stenotic processes in this vascular region as well. In a group of 38 patients, Harward et al. (1993) obtained a sensitivity of 100% (30/30) and a specificity of 88% (7/8) in diagnosing stenoses of more than 50% using duplex sonography when assessing their correlation with results obtained from intra-arterial angiograms. Postprandially increased circulation influences the measurement of the flow velocity much less here than it does in the superior mesenteric artery (Nicholls et al. 1986, Moneta et al. 1988).

Ligamentous and high-grade stenoses can sometimes present problems of differential diagnosis as against arterial occlusion, when stenoses of both the celiac trunk *and* the suprior mesenteric artery are present.

Superior Mesenteric Artery

The superior mesenteric artery can be evaluated with duplex sonography in 82–89% of examinations (Mallek et al. 1990b). With color flow Doppler sonography, this value can even rise to 98% (Uzawa et al. 1993). The proximal segments of the jejunal arteries, the smaller branches of the superior mesenteric artery, cannot be consistently displayed in all examinations with duplex sonography.

Several examiners have carried out quantitative measurements of normal values for various circulatory parameters in smaller patient groups (Table 7.**1**). When interpreting flow velocity waveforms, the time that has elapsed since the patient's last meal, and the quantity and composition of the meal, have to be taken into account, since higher flow velocities and flow volumes in this mesenteric artery are to be expected postprandially.

In patients who had fasted and for whom upper limits (relatively high ones) for the maximum systolic flow velocity *and* the maximum end-diastolic velocity were set at 300 cm/s and 45 cm/s, respectively, Bowersox et al. (1991) calculated a 100% *sensitivity* and a 92% *specificity* for measurements made with duplex sonography. Using an absent triphasic waveform in the Doppler spectrum as the *sole* criterion resulted in a specificity of only 46% in their group of 24 patients.

With a 100% *sensitivity,* the most definite parameter for recognizing a hemodynamically effective arterial stenosis—apart from the absent triphasic form of the Doppler frequency spectrum—is an end-diastolic flow velocity higher than 45 cm/s. Using a maximum systolic flow velocity higher than 300 cm/s as the sole criterion for predicting an effective stenosis only yields a sensitivity of 63%.

Harward et al. (1993) detected or excluded hemodynamically effective stenoses of more than 50% in examinations (n = 38) using conventional duplex sonography, with a 96% (24/25) *sensitivity* and a 92% (12/13) *specificity* in comparison with results obtained from infra-arterial angiography.

Patients with a clinical suspicion of *abdominal angina* should be examined with (color flow) duplex sonography to determine if there are any vascular changes in their mesenteric arteries (Jäger et al. 1984, Flinn et al. 1990, Muller 1992, Arienti et al. 1996). Elderly patients with other manifestations of arteriosclerosis and the characteristic triad of postprandial abdominal pain, food aversion, and weight loss very often have stenoses in the celiac and mesenteric arteries (Järvinen et al. 1995). Stenoses or occlusions can be detected in the trunk and in the first 6–10 cm of the superior mesenteric artery. The distal arterial segments are usually not accessible to examination. Initial studies have shown that it is possible to evaluate the results of revascularization operations by means of duplex sonography (patency of bypasses or reopened blood vessels) and to conduct relevant follow-up observations in the superior mesenteric artery (Sandager et al. 1987, Flinn et al. 1990). In patients with acute abdominal pain, intestinal ischemia is a differential diagnosis. Doppler sonography may also be a feasible method for detecting *acute intestinal ischemia* due to proximal superior mesenteric artery occlusion. In a group of 770 patients with emergency admission to a sonography, five of six surgically or arteriographically confirmed occlusions of the superior mesenteric artery were correctly detected by Doppler sonography (Danse et al. 1996).

When interpreting the measurement results from individual arteries, the flow relationships in the other upper abdominal arteries always have to be taken into account. Functionally effective collateral systems exist between the branches of the celiac trunk, the superior mesenteric artery, and the inferior mesenteric artery. If there are local vascular system obstructions in one arterial segment, these may cause compensatory increases in flow volumes and flow velocities in the rest of the mesenteric vasculature.

Renal Artery and Intrarenal Arteries

The ability to *depict* the renal arteries is 84–100% (Handa et al. 1988, Distler et al. 1992). In our own prospective investigations using color flow duplex sonography, which yielded 370 findings confirmed by angiography, the ability to evaluate the renal arteries in their proximal and middle vascular segments was found to be 88% overall (right renal artery 90%, left renal artery 85%). The imaging ability was higher in women, at 96%, than in men (84%), and the intrarenal blood vessels could be displayed with certainty in all of the patients (Karasch et al. 1994).

In addition to the ability to recognize *aneurysms* and *arteriovenous malformations* in the renal arterial system (Bunchman et al. 1991, Takebayashi et al. 1990), detecting or excluding a vascular stenosis is particularly interesting.

Duplex and color flow duplex sonography of the renal vascular system have acquired special significance in the differential diagnosis of treatable forms of hypertension (Dawson 1996). With these procedures noninvasive detection of a renal artery stenosis is possible with a *sensitivity* of up to 97.5% (Table 7.**5**).

The prevalence of renovascular hypertension is reported to be 1–4% among individuals suffering from hypertension in general, and who have not been preselected in any manner (Kaufmann et al. 1992). At specialized medical centers with selected groups of patients who already have manifest arteriosclerosis in other vascular segments, the use of a noninvasive screening procedure to recognize a renal artery stenosis appears sensible, particularly since a significantly higher prevalence is expected in these patient groups (Swartbol et al. 1992, Karasch et al. 1994).

In our own investigations to detect arteriosclerotic stenoses and occlusions of the renal arteries using a 180 cm/s threshold value for the angle-corrected maximum systolic flow velocity, we were able to attain a sensitivity of 92.7%, based on 277 renal arteries in comparison with the angiographic results, and a sensitivity of 94.4% in comparison with 80 conventional angiographic catheter examinations (Karasch et al. 1993 c). The positive predictive value in this group was 0.98; the negative predictive value was 0.89 (Table 7.**5**).

Experienced examiners are able to exclude a renal artery stenosis with a *specificity* of 73–98.5% (Avasthi et al. 1983, Ferretti et al. 1988, Hoffmann et al. 1990, Spies et al. 1990; cf. Table 7.**5**). This is above the level attained by intravenous digital subtraction angiography (Schörner et al. 1984). In our own group of patients, a specificity of 96.2% was attained in comparison with 80 conventional catheter angiographies.

Our own recent experiences with patients suffering from *fibromuscular dysplasia* show that, with a sensitive examination, these stenoses, which are located mostly in the middle and distal segments of the renal arteries, can also be recognized by color flow Doppler sonography. Studies to confirm this in larger patient groups are not yet available for this relatively rare disease.

Multiple and accessory renal arteries and pole arteries may represent a diagnostic gap in the arterial renal vascular system, since they may not be displayed unless the examiner carries out a targeted and systematic search for these vascular variants (Spies et al. 1995).

Exclusively evaluating the intrarenal flow patterns to detect, exclude, or grade a stenosis (Özbek et al. 1993) has not yet had its validity established. Recent studies suggest that a prolonged acceleration time of more than 0.12 s is associated with a high likelihood of significant stenosis of a renal artery (Baxter et al. 1996). Although there were also significant differences for intrarenal peak and end-diastolic velocities between both groups, in this study of 120 kidneys there was no significant difference in in-

Table 7.**5**   Limits for the maximum flow velocity (V_{max}) and the renoaortal index (RAR), the percentage successfully imaged, and the sensitivities and specificities in detecting angiographically controlled stenoses of the renal arteries using duplex and color flow duplex sonography

Authors	Blood vessels/ stenoses (n)	V_{max} (cm/s)	Imaging capability (%)	Sensitivity (%)	Specificity (%)	Reference angiography
Duplex sonography						
Avasthi 1984	52/26	100	84	89	73	i.a.
Norris 1984	86/12	–	94	83	97	?
Rittgers 1985	84/12	–	90	83	97	?
Kohler 1986	43/?	RAR > 3.5	90	91	95	?
Ferretti 1988	104/27	>100	84	100	92	conv.
Handa 1988	40/10	–	100	100	93	?
Jäger 1988	88/?	–	84	97.5	94	?
Taylor 1988	58/14	RAR > 3.5	–	84	97	?
Hillmann 1989	40/10	–	98	100	93	?
Zoller 1989	113/13	–	100	86.7	99	i.a. DSA
Strandness 1990	58/14	RAR > 3.5	–	84	97	?
Hoffmann 1991	85/64	> 180	87	95	90	?
Schäberle 1992	91/44	> 140	–	86	83	i.a., conv.
Color flow duplex sonography						
Berland 1990	29/7	>100	58	0	37	i.a.
Desberg 1990	55/?	>100	69	0	79	i.a., conv.
Spies 1990	87/21	–	80	94.4	98.5	conv.
Breitenseher 1992	41/8	> 120	56	17	89	i.a.
Distler 1992	100/?	–	70	88.9	95.6	i.a.
Schulte 1992	86**/42	–	75	93	92	i.a.
Karasch 1993 c*	277/109	>180	88	92.7	89.8	conv., i.a., i.v.
	80/54	>180	–	94.4	96.2	conv.
	43/24	> 180	–	91.7	84.2	i.a., DSA
	154/31	> 180	–	90.3	89.4	i.v., DSA
Spies 1995	268/42	–	73	93	92	i.a.

*   Includes five renal artery occlusions differentiated according to various angiographic procedures as the reference method.
** 86 patients with no data concerning the renal arteries.
Conv. = conventional angiography; i. a. = intra-arterial conventional or subtraction angiography; i. v./i.a. DSA = intravenous/ intra-arterial digital subtraction angiography

trarenal pulsatility or resistance index noted between the angiographically confirmed stenosed and normal arteries.

Percutaneous transluminal renal angioplasty is a safe and effective treatment for arteriosclerotic non-ostial renal artery stenoses. In patients with renal artery stenoses involving the ostium, stent placement may be an effective interventional treatment after unsuccessful balloon angioplasty with good early and long-term results (Blum et al. 1997). In these patients, in particular, and also after reconstructive vascular surgery, duplex and color flow Doppler sonography are suitable for assessing the success of the procedure and carrying out follow-up observations (Edwards et al. 1992, Karasch et al. 1993 a, Tullis et al. 1997).

The prognostic significance of the preinterventional resistance index as a predictor of clinical success of percutaneous transluminal angioplasties or operations for renal artery stenosis is still uncertain. First experience may point out, that a low preinterventional intrarenal resistance index correlates with clinical failure in subsequent treatment of hypertension by renal revascularization, whereas a higher RI correlates with better postinterventional results (Frauchiger et al. 1996).

Furthermore, progression of atherosclerotic renal artery stenosis is relatively common. In a patient group including 36 men and 40 women with a mean follow-up of 32 months, Zierler et al. (1996) report a cumulative incidence of progression from below 60% up to or above 60% stenosis of: 30% at 1 year, 44% at 2 years, and 48% at 3 years. Therefore, duplex scans should be performed regularly in patients with known renal artery stenosis.

Inferior Mesenteric Artery

Stenoses and *occlusions* of the inferior mesenteric artery are often symptom-free, on the one hand, but the classification of clinical abdominal symptoms as deriving from this arterial segment is often uncertain, on the other. The examination of the mesenteric artery has therefore not yet had very much clinical significance. Imaging of the inferior mesenteric artery is possible approximately 60–70% of the time (Sieber and Jäger 1992 a). In the future, color flow duplex sonography should result in faster and more definitive ultrasound examinations of the inferior mesenteric artery. It should then regularly be possible to evaluate this vascular distribution area.

A stenosis can be assumed when there are flow velocities higher than 180–200 cm/s in the trunk of this artery. It should also be mentioned that autopsy statistics indicate that previously clinically asymptomatic stenoses of the mesenteric blood vessels are found in 50–70% of autopsies, so that the assignment of gastrointestinal symptoms to specific findings that is possible with duplex sonography should take precedence over the clinical aspects of a case.

Inflammatory diseases of the small and large intestines can lead to a reactive and uniform increase in the flow velocity in the inferior mesenteric artery (Kathrein et al. 1990).

Lumbar Arteries

Since stenoses or occlusions of the lumbar arteries usually remain clinically asymptomatic, there is rarely any indication for a specific examination of these arteries. In addition, imaging of these thin-caliber blood vessels with their far distal location is not usually possible when there are physiological flow relationships in the abdominal aorta and the large pelvic arteries. Nevertheless, it is important to be able to differentiate them from other small abdominal arteries, to avoid any unnecessary confusion.

In stenoses or occlusions in the distal abdominal aorta or the pelvic arteries, the lumbar arteries can act as collateral vasculature, and due to the larger circulation they may present with a significantly increased caliber. As a result of the morphological formation of the collateral system, there is a potential to assess the hemodynamic significance of a flow obstruction in the large arteries indirectly.

Ovarian and Uterine Arteries

Imaging of the ovarian artery is successful in circa 80% of cases. According to Skodler et al. (1991), the uterine artery can be depicted 83–94% of the time. The circulation in both of the *ovarian arteries* changes during the menstrual cycle, and reflects the measured plasma estradiol levels in a cycle-dependent manner. These values are used during in-vitro fertilization. In this context, they can be used as parameters to evaluate an undisturbed cycle and to monitor ovulation (Feichtinger et al. 1988, Schurz et al. 1990).

Because of the anatomy, most flow measurements in the female lower abdominal arteries are carried out transvaginally. The endosonographic access is better than transabdominal ultrasound application for displaying and differentiating anatomical structures within the small pelvis, and has therefore become the primary method of vascular diagnosis in this region (Deutinger et al. 1989a, 1989b, Tessler et al. 1989, Thaler et al. 1990, Kurjak and Zalud 1990b, Bernaschek et al. 1991).

The main emphasis of noninvasive diagnostic procedures in the diagnosis of the *uterine vasculature* is on measuring circulatory relationships during pregnancy. However, further differentiation of pregnancy-independent pathological findings is also possible in the small pelvis using both duplex and color flow Doppler sonography. The initial results suggest a possible application in identifying vascular changes in space-occupying lesions (see Chapter 10).

Rarely, arteriovenous malformations of the uterus and its appendages can also be confirmed with duplex sonography (Musa et al. 1989, Jain et al. 1991), and in case of interventional therapy color Doppler controls are recommended (Flynn and Levine 1996).

Fetomaternal Vasculature

Since the end of the 1970s, Doppler-sonographic measurements have increasingly been used in obstetric diagnosis. Due to the improved detail resolution in more recent equipment, the larger intrauterine fetal blood vessels can be depicted and morphologically evaluated with *B-mode* (Fig. 7.**42a**) and *color flow Doppler sonography* over a longer distance (Fig. 7.**42b**). This allows targeted and reproducible registration of frequency spectra using Doppler sonography in individual uteroplacental and fetal blood vessels in order to quantify the flow velocity (Hastie et al. 1988). However, the standard application of the procedure and the general acceptability of the value of the various measurements are not yet finalized (Hüneke et al. 1991).

Examinations carried out in groups of female patients, which can only be briefly mentioned here, indicate instances in which spectral analyses from uterine and fetal blood vessels can be usefully employed in monitoring a pregnancy (Sohn et al. 1991, Kudielka et al. 1992, Jaffe and Woods 1996, Barth et al. 1996, Tekay and Jouppila 1993). Color flow duplex sonography can be useful in the identification and evaluation of structural abnormalities of the umbilical cord (Fig. 7.**43**). Abnormalities in cord size, degree of coiling, attachment, and position can have important implications for the outcome of the pregnancy. Structural abnormalities of the umbilical cord such as single umbilical artery, knots, cysts, and tumors may be associated with fetal distress or malformations (Dudiak 1995). Overviews of the possible areas of application

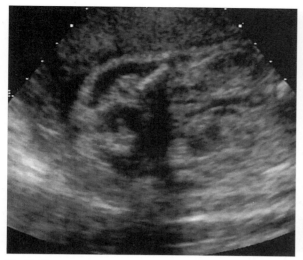 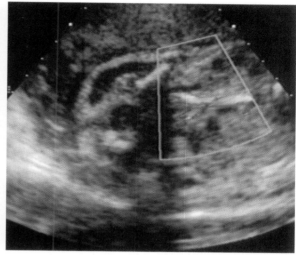

Fig. 7.**42** **a** A B-mode sonogram of the abdominal aorta (arrow) of a fetus in the 26 th week of pregnancy, with the vessel branching into the large pelvic blood vessels.

b Flow coding in color flow Doppler sonography makes it easier to recognize the branching of the iliac vasculature into the external and internal iliac arteries on both sides

for Doppler sonography in gynecology and obstetrics are provided by Hill et al. (1988), Grab et al. (1992) and Sohn et al. (1993 c).

Because of the methodological difficulties of the examination itself, and safety aspects with regard to the duration and magnitude of the energy emitted, the use of duplex sonography on the fetus basically requires special training. This diagnostic method is best left to experienced medical centers.

Organ Transplants

Kidney Transplants

Following kidney transplantations, the implanted organs have to be monitored particularly intensively. Various vascular and parenchymatous complications (anastomotic stenosis, rejection reactions of various etiologies, cyclosporin A toxicity, acute tubular necrosis, viral and bacterial infection, glomerulonephritis, urinary obstruction) can endanger the success of the transplantation, and have therapeutic consequences for both the patient and the physician. Color flow duplex sonography is an established noninvasive diagnostic procedure for many aspects of the investigations required, and is successful in 80.0–98.4% of the examinations (Ward et al. 1989).

Duplex sonography is now firmly established as the method of choice for recognizing a *vascular stenosis within the anastomotic segment* or a renal artery thrombosis, since the extra-anatomical superficial location of the transplant makes this region easily accessible to examination (Plainfosse et al. 1992, Krumme et al. 1994). When detecting occlusions and stenoses in transplant arteries, procedures using duplex sonography have a specificity of between 75% and 100%, and a sensitivity of 100% (Taylor et al. 1987 a, Deane et al.

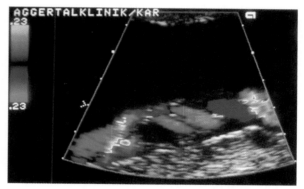

Fig. 7.**43** An umbilical cord is shown in utero with color flow Doppler sonography. Arterial flow is coded red; venous flow is coded blue

1990, Grenier et al. 1991). An early postoperative examination of the transplant vasculature is useful to provide secure classification of any complications that may appear later in comparison with the original vascular status.

Many groups have devoted a great deal of time and effort to establishing duplex and color flow duplex criteria that might indicate an impending rejection reaction or functional restriction of the transplant (Rifkin et al. 1987, Fobbe et al. 1989, Don et al. 1989, Evans et al. 1989, Leichtman et al. 1989, Wan et al. 1989, Kelcz et al. 1990, Rigauts et al. 1990, Grenier et al. 1991, Perchik et al. 1991, Deane 1992). Measuring the *Pour-celot resistance index* is the most widespread criterion. Using relative flow velocities *independent* of the ultrasound angle of the Doppler beam, the resistance index allows evaluation of the peripheral resistance. It is particularly useful in conventional duplex sonography, which does not allow visual display of the small intrarenal vasculature.

An elevated resistance index or pulsatility index corresponds to increased resistance in the renal arteries, and this can occur in: acute, primary vascular, or cellular and chronic rejections, urinary obstruction, arteriosclerosis, and acute tubular necrosis. However, elevated values can also appear after external transplant compression. Among these differential diagnoses, further distinction is not clearly possible without relevant clinical information (including biopsy) (Rigsby et al. 1987, Pozniak et al. 1988, Meyer et al. 1990, Pelling and Dubbins *1992). Intraindividual follow-up observations* over time increase the sensitivity of resistance measurements (Van Leeuwen et al. 1992). Hollenbeck and Grabensee (1994) claimed an increased probability of a rejection reaction due to an elevated resistance index several days before the histological diagnosis. However, a low resistance index (less than 0.7) does not necessarily exclude an acute rejection reaction (Drake et al. 1990).

In other possible complications, such as glomerulonephritis, cyclosporin A toxicity, or *Cytomegalovirus* infection, the resistance index is not significantly elevated, so that it cannot be used to distinguish between pathological and normal findings (Flückiger et al. 1990). Independently of the above complications, transplant artery stenoses can change the flow pattern in the renal vascular system, and must therefore also be taken into consideration in interpreting the measurement values (Rigsby et al. 1987).

The color flow Doppler mode has proved superior to the exclusive use of conventional duplex sonography for the detection of complications after organ punctures, such as arteriovenous fistulas or false aneurysms (Hübsch et al. 1990).

Liver and Pancreas Transplants

Postoperatively, arterial and venous transplant vasculature can be evaluated with duplex sonography, and stenoses, occlusions, arteriovenous fistula, pseudoaneurysms, and thromboses of these blood vessels can be diagnosed (Barton et al. 1989, Hall et al. 1990, Kubota et al. 1990, Snider et al. 1991, Propeck and Scanlan 1992, Keener et al. 1995, Worthy et al. 1994). In particular, pathological findings that do not develop until the postoperative phase (acute thromboses, arterial occlusions, altered flow velocity parameters) can be reliably recognized over time, provided that it was possible to assess the original status immediately after the transplantation. The sensitivity of detecting a hepatic artery thrombosis with duplex sonography is given at 92%–100% (Flint et al. 1988, Pinna et al. 1996).

According to Longley et al. (1988), acute rejection reactions cannot be detected with certainty using an elevated resistance index and accompanying low flow velocities in the hepatic artery. If percutaneous pancreas or liver graft biopsy is necessary, it can be performed with color flow Doppler ultrasound localization (Klassen et al. 1996). The value of Doppler frequency spectral analysis to diagnose a *transplant rejection* cannot yet be conclusively evaluated. First observations suggest that a decreased splenic-vein velocity can be a sign of a thrombosis of the pancreatic transplant (Nghiem 1995).

A further application area for color flow duplex sonography is in the recognition of occlusive venous disease after *bone-marrow transplantations* on the basis of an elevated resistance index in the hepatic artery (Herbetko et al. 1992).

8 Abdominal Veins

Examination

Special Equipment and Documentation

■ Continuous Wave and Pulsed Wave Doppler Sonography

In examinations using Doppler sonography, *continuous wave* and *pulsed wave Doppler systems* are primarily used to determine the flow relationships in the veins of the groin, in order to acquire information indirectly about the patency of the pelvic veins and the inferior vena cava. Due to their relatively large distance from the outer surface of the abdomen, the inferior vena cava and the pelvic veins, as well as the veins of the abdominal organs (renal vein, splenic vein, hepatic veins) generally cannot be evaluated with the ultrasound probe. Even if venous signals can be identified in a particular instance, assigning these signals to specific organ veins is at best inadequate, due to the lack of orientation.

A narrow area of application for continuous wave Doppler sonography is in examining the *umbilical veins* to detect the presence of any flow, since blood does not usually circulate in these vessels under physiological conditions.

■ Conventional and Color Flow Duplex Systems

Due to its higher local resolution and the lower flow velocities in comparison with the arterial vascular system, the *pulsed wave* Doppler mode is mainly used to assess the abdominal veins in *duplex-sonographic examinations*. The device's integrated Doppler mode should allow the expected low flow velocity to be registered, using freely selectable (wall) filters that can be adjusted in small steps, and allowing low-frequency components to be detected without filtering out a broad frequency band in the lower frequency range.

In addition to the equipment required to examine the abdominal arteries (p. 231) with *color flow Doppler sonography*, adequate imaging of abdominal venous flow patterns requires powerful and reliable color coding for very low flow velocities, so that flow patterns in the small veins of the organs and marginal flow in thromboses can still be detected in color-coded form—especially in connection with thrombolysis (Seitz and Reuss 1992). To achieve this, optimal conditions must be provided by appropriate technical ad-justment of the equipment for frequency registration, both in Doppler and in color flow Doppler sonography:

- Low velocity scale (pulse repetition frequency) for the Doppler and color flow Doppler modes
- Low (wall) filter adjustment
- Adequate signal amplification
- Slow recording speed

Examination Conditions

The same recommendations for the external examination conditions apply here (positioning of the patient, examination position, patient preparation) as for the examination of the abdominal arteries using duplex and color flow Doppler sonography (p. 231).

Patient and Examiner

Special attention should be given to placing the patient in a relaxed position, usually supine, since any tension in the abdominal muscles causes an increase in intra-abdominal pressure, which affects the venous flow relationships. For additional relaxation of the abdominal muscles, it may be helpful to ask patients to raise their legs, bending the hip and knee joints and placing their feet on the examination table. Normal, relaxed breathing should be ensured.

Conducting the Examination

■ Continuous Wave Doppler Sonography

To detect blood flow in the umbilical veins, the examiner places the probe in the area of the umbilicus and tries to register venous signals from these collateral veins by optimizing the direction and incident angle of the ultrasound.

■ Conventional and Color Flow Duplex Sonography

Several general examination steps used in imaging the abdominal veins are described below. Special section planes and adjustments for B-mode imaging and Doppler-sonographic examinations are explained in the section on normal findings, following a short discussion of the anatomy.

B-mode. The *inferior vena cava* can be imaged in left lateral, paramedian longitudinal sections of the upper abdomen and in right paravertebral, subcostal transverse sections next to the abdominal aorta. Special attention should be given to *respiration-dependent variations in caliber.* When the patient is standing, the increased hydrostatic pressure causes increased filling of the intra-abdominal segments of the vena cava and its branches, with an accompanying increase in the vascular caliber.

The *hepatic veins and portal vein* are imaged in a right oblique subcostal section, or when transducer contact has been made, in a median or right paramedian orientation (Fig. 8.1). In B-mode sonography, special attention should be given to the caliber of these veins.

The middle and distal segments of the *splenic vein* are displayed in transverse section, and the proximal segment can be depicted in a left dorsolumbar section, or transsplenically.

The *superior* and *inferior mesenteric veins* can be evaluated using an oblique paramedian sagittal section, proceeding from left or right.

Doppler and color flow Doppler mode. The sectional planes used for Doppler and color flow Doppler examinations are basically the same as those used in B-mode sonography (Fig. 8.2). To provide as small an incident angle as possible for the pulsed ultrasound waves—making it much easier to detect a frequency shift with the slower flow velocities in the abdominal veins—the position of the transducer and the section have to be selected in such a way that the vascular axes are displayed almost longitudinally, with targeted use of the duplex procedure to optimize the registration conditions.

To measure the flow rate, exact images of the vessels in two planes and an incident Doppler beam angle of less than 60° are basic requirements (Sabbá et al. 1990 b). Absolute values for the volume as well as flow rates related to the patient's body weight (ml/min/kg) should be given (Kawasaki et al. 1990).

When measuring the flow velocity and the absolute flow rate—requiring calculation of both the vascular caliber and cross-section—care must be taken not to exert pressure on the abdominal veins with the transducer. Particularly with vessels that lie near the abdominal wall (splenic vein, left renal vein), pressure can cause a reduction in the vascular cross-section, with an accompanying increase in the flow velocity, mimicking an artificial stenosis.

In color flow duplex sonography, the best procedure is to use B-mode sonographic imaging first to adjust for the best possible sectional plane, especially when examining the hepatic and portal veins, and then use the color flow Doppler mode to assess the flow direction and velocity in a targeted fashion.

In order to detect an obstruction of the inferior vena cava, it is essential to register Doppler frequency–time spectra from both *pelvic veins* (p. 194). A hemodynamically effective drainage obstruction in the inferior vena cava can also be recognized from the altered Doppler and color flow Doppler signals in the region of both external iliac veins and common femoral veins.

Color Doppler energy mode. The sectional planes differ only slightly from those used in the color flow Doppler mode: the dependence on the incident Doppler beam angle is significantly lower, however, which often enables a more extensive color coding of the abdominal veins. The color Doppler energy mode is usually used in conjunction with pulsed wave or color flow Doppler modes to determine the flow direction in the abdominal veins and to identify possible collateral vessels in pathologic cases.

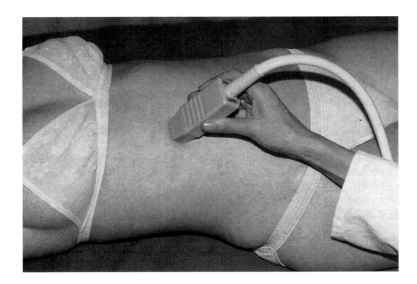

Fig. 8.**1**   The correct transducer position for imaging the hepatic veins from a subcostal orientation using duplex sonography. The sectional plane applied has to be properly optimized according to the vascular axes of the veins being examined

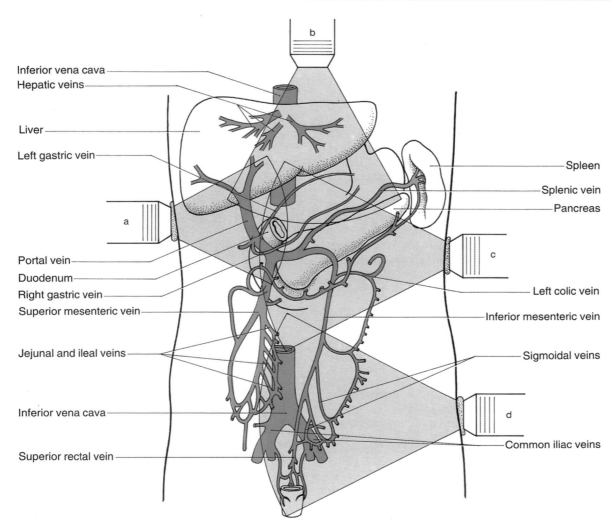

Fig. 8.**2** The abdominal veins. Example transducer positions are shown (a-d), and these have to be varied to provide the best image of the abdominal veins corresponding to the vascular axes being examined; a, c, and d = sagittal sections; b = transverse section

Examination Sequence

The intra-abdominal veins (and arteries) are leading structures that can provide orientation in many abdominal sonographic studies. There is no standard examination sequence for targeted morphological and functional imaging of the abdominal veins. The examination sequence is determined by clinically relevant questions and indications for duplex sonography.

To exclude a thrombosis of the vena cava, for example, it is necessary not only to carry out a B-mode ultrasound search for thrombotic material in the inferior vena cava and venous collateral systems, but also to give particular attention to the respiratory modulation of the vascular caliber and the flow velocity—with side-to-side comparisons of the pelvic veins and leg veins. In addition, the effect of various provocation maneuvers (Valsalva, thigh compression) on venous

flow has to be taken into consideration. An ultrasound assessment of the intra-abdominal segment alone would provide an incomplete examination.

In the case of a renal cell carcinoma, carrying out a targeted search for a *thrombosis of the renal vein* or a *tumorous protrusion,* which may extend into the caval vein, is particularly important (Tsushima et al. 1993). For that reason the clinical indication has a decisive influence on the sequence of the examination of the veins or venous systems being imaged.

Normal Findings

Principle

The organs that are directly or indirectly responsible for the digestion of food physiologically share an interconnected drainage area, the flow dynamics of which (direction, velocity, cardiac and respiratory modulation of the blood flow) are determined by the intravenous pressure relationships. These in turn are subject to external influences (e.g., examination position, physical exertion) (Ohnishi et al. 1985 c).

Venous blood from substantial portions of the unpaired abdominal organs—the small and large intestine, spleen, pancreas, and stomach—flows into the liver via the portal vein, which is the common final blood vessel before the liver. This represents approximately 70% of the hepatic flow rate; the hepatic artery contributes approximately 30% of the liver's blood supply.

The velocity and rate of flow in the portal vein and superior mesenteric vein are influenced by the patient's digestion status (Pugliese et al. 1987, Gaiani et al. 1989, Alvarez et al. 1991) as well as the respiratory status (Abu-Yousef 1992). They are also affected by pharmacological (Alvarez et al. 1991, Gaitini et al. 1991) and hormonal (Bolondi et al. 1990) changes, which become significant when interpreting the findings obtained. Specific descriptions of the qualitative and quantitative findings obtained with duplex sonography are given below under the individual abdominal veins.

The renal veins constitute a separate venous drainage area, which passes directly into the inferior vena cava. However, it is possible for this area to come into contact with other drainage areas (e.g., in a splenorenal shunt) as a result of earlier collateral pathways (see the section on pathological findings below, p. 295).

■ Conventional and Color Flow Duplex Sonography

B-mode. B-mode sonography allows direct imaging of the abdominal veins and their anatomical relationships with the organ systems, so that they can be evaluated in terms of their caliber, caliber fluctuations, wall structure and, if relevant, internal structures. For reasons of space, the sonographically relevant anatomical aspects of the abdominal veins cannot be discussed here at any length. For a more comprehensive and detailed discussion of the topic, the specialist literature on abdominal sonography should be consulted (Meckler et al. 1992, Weiss and Weiss 1990, Rettenmaier and Seitz 1990, 1992).

Doppler mode. The pulsed wave Doppler mode can be used to detect flow in the abdominal veins and evaluate it in relation to its direction, velocity, and flow pattern. In contrast to the quantitative measurements used in the abdominal arteries, flow patterns in

the veins have so far mainly been described qualitatively. Recently it has become possible to estimate normal values for quantitative measurements here as well. These figures still have to be validated with larger groups of patients, but they do provide an initial basis for conducting comparative examinations.

However, even in earlier years, the renal and portal veins were an exception to this. For these blood vessels, *the flow* rate—indicated in ml/min or relative to the body weight as ml/min/kg—has long been in widespread use for diagnostic evaluation of the Doppler frequency spectra (Burns et al. 1987, Patriquin et al. 1987, Kawasaki et al. 1991, Abu-Yousef 1992).

The calculation is made using the formula:

$$\text{Flow rate} = V_{mean}\ \frac{\pi d^2}{4}$$

V_{mean} is the average venous flow velocity, and d represents the diameter of the vein.

The normal values for various quantities measured are listed for each of the abdominal veins in Table 8.1. In addition, the flow rate measured with duplex sonography, combined with the intravasally measured pressure values, can be used to calculate further quantities. For example, *resistance in the portal vein* is defined as the quotient consisting of the sum of the pressure in the portal vein, minus the pressure in the renal veins, divided by the flow rate. This can be used for further quantification in cases of portal hypertension, for example (Moriyasu et al. 1986 a).

When measuring the flow velocity in the portal vein and the mesenteric veins, any digestive and pharmacological effects also have to be taken into account, as in the region of the superior mesenteric artery (p. 243) (Sabbá et al. 1991).

Color flow Doppler mode. Color-coding the direction and velocity of flow provides a quick overview of the flow patterns in the abdominal veins. In the liver region, for example, the distinction between flow directed toward the liver (*hepatopetal* or *hepatocentral* flow) and *hepatofugal* flow can only be recognized with venous color coding.

Particularly in unfavorable examination conditions, the color flow Doppler mode may help to detect smaller veins. To achieve the optimal quality here and use the procedure as a quality standard, it is necessary to be able to detect very low flow velocities and display them in color-coded form.

Color Doppler energy mode. The use of the color Doppler energy mode to display abdominal vessels is discussed in Chapter 7 on abdominal arteries (p. 236). Its advantages for the abdominal veins lie in the signal's low dependence on the incident beam angle, which enables a uniform coding of even those vessels lying on an unfavourable angle to the ultrasound, and in the display of the slower velocities that can occur in the abdominal veins. Because the power- or color Doppler energy mode does not indicate the flow

Table 8.1 Normal values for duplex-sonographic measurements obtained in the various abdominal veins (in adults). Values for maximum flow velocity (V_{max}), average flow velocity (V_{mean}), flow rate, and maximum cross-sectional diameter, are given from the literature and our own measurements

Abdominal veins	V_{max} (cm/s)	V_{mean} (cm/s)	Flow rate (ml/min/kg)	Diameter
Inferior vena cava [1,2]	44–118	5–25		15–30
Hepatic veins [1,3]	16– 40(–61) [1]	–		–
Portal vein [3-8,11,12]	15– 30*(–88) [1]	12–18*	10–21*	7–15*
Splenic vein [2,7,9,10,13]	9– 30	5–12	2– 5	5–10
Renal vein	18– 33	10–20		4– 9
Segmental veins	10– 30	–	–	1– 4
Superior mesenteric vein [2,7,13]	8– 40*	9–18*	3–9*	4–13*
Inferior mesenteric vein	–	–	–	3– 7

* Influenced by the phase of digestion; cf. Table 8.4. [1] Herbetko et al. (1992); [2] Weiss and Weiss (1990); [3] Abu-Yousef (1992); [4] Bolondi et al. (1984); [5] Meifort et al. (1990); [6] Kawasaki et al. (1990); [7] Ohnishi et al. (1987a); [8] Sabbá et al. (1991); [9] Ohnishi et al. (1987b); [10] Ohnishi et al. (1989b); [11] Leen et al. (1993); [12] Moriyasu et al. (1986b); [13] Zoli et al. (1986b).

direction, it is most commonly used in conjunction with the color flow Doppler mode.

Anatomy and Findings

The descriptions below of the B-mode and Doppler-sonographic *normal findings* for each of the abdominal veins are preceded by short anatomical introductions, describing only the most important blood vessels that can be imaged with conventional or color flow duplex sonography.

Inferior Vena Cava

Anatomy. After the confluence of the *common iliac veins,* the *inferior vena cava* proceeds at the level of the fourth and fifth lumbar vertebrae in a paravertebral orientation on the right next to the aorta, and continues in a cranial direction, passing through the vena caval foramen into the thorax. Proximal to the *right ovarian vein* or the *right testicular vein (internal spermatic vein), renal veins, right suprarenal vein, lumbar veins* and *phrenic veins,* two to five short *hepatic veins* join the inferior vena cava below the diaphragm. The distal diameter of the caval vein is approximately 18 mm, at the level of the renal veins it is 24 mm, and above the hepatic veins it can be up to 30 mm in the middle respiration range.

■ Conventional and Color Flow Duplex Sonography

B-mode. Sonographically, the inferior vena cava can be displayed in a median to paramedian section, paravertebrally next to the aorta on the right, with an oval to round *respiration-dependent* diameter (15–25 mm), observed in transverse section. In a sagittal section, it is possible to image the vein from the confluence of the common iliac veins as far as the junction of the he-

patic veins. Its wall is delicate, and cannot be delimited uniformly. The lumen is free of reflections, and shows a subdiaphragmatic inspiratory reduction in caliber and expiratory increase in diameter. It is compressible, mainly in slim patients. The cardiac-modulated flow dynamics (described under the Doppler mode, below) correspond sonographically to a two-phase reduction in caliber, also termed "double beat." The Valsalva maneuver causes the vein to dilate to its maximum extent.

Due to the complex nature of the vein's embryological development, anatomical variants can be expected in 1.5–7% of ultrasound examinations (e.g., left-sided or paired inferior vena cava, periaortic ring formation in the hepatic veins) (Pierro et al. 1990, Kubale 1992, Brüggemann et al. 1993). Figure 8.3 shows a three-dimensional representation of the infe-

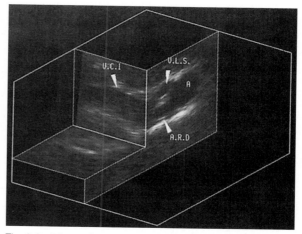

Fig. 8.3 Three-dimensional representation of the inferior vena cava (arrow **VCI**) and the abdominal aorta (**A**) at the level of the confluence of the left renal vein (arrow **VLS**) and the branching off of the right renal artery (arrow **ARD**). The close topographical relationship of the vessels is recognizable in the visual display. *(Illustration from PD Dr. Ludwig, University Clinic of Bonn)*

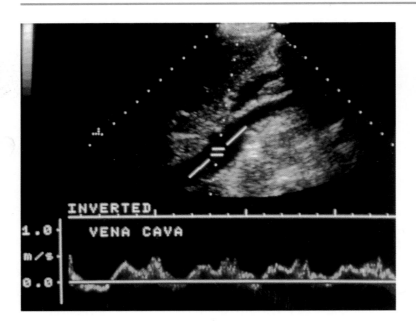

Fig. 8.**4** A longitudinal section of the inferior vena cava distal to the confluence of the hepatic veins, from a left paramedian orientation, and the Doppler frequency spectrum, showing a double-peaked profile indicating expiratory increase and inspiratory decrease in the cardiac-modulated flow velocity

rior vena cava and the abdominal aorta at the level of the confluence of the left renal vein.

Doppler mode. Corresponding to the sonographic "double beat" image, the *Doppler spectrum* shows an acceleration of flow directed toward the heart at the beginning of the systolic ventricular contraction, and a second, slightly lower, peak when the tricuspid valve opens (Fig. 8.**4**). After the phasic increase in the double-peaked inflow into the right atrium, a decrease in the blood flow follows during atrial contraction. Cardiac modulation remains, with an inspiratory decrease and expiratory increase in the flow velocity usually being seen (Smith et al. 1985).

Color flow Doppler mode. Using color flow Doppler sonography, flow in the inferior vena cava, which is modulated by respiration and the cardiac cycle, can be imaged in color-coded form by making steep ultrasound contact with the blood vessel. This allows an evaluation of the vessel's patency even when it is not (completely) compressible.

Hepatic Veins

Anatomy. Two to five *hepatic veins* collect blood from the central veins and pass through the liver sagittally along the segmental boundaries from the lower margin in a dorsal direction, joining with the inferior vena cava directly below the diaphragm.

■ Conventional and Color Flow Duplex Sonography

B-mode. With the sectional plane tilted cranially, and proceeding subcostally from the right costal arch or from an intercostal orientation, the hepatic veins can be followed sonographically from the periphery of the liver to their confluence with the inferior vena cava.

Their caliber increases steadily and homogeneously until the junction with the inferior vena cava. The course proceeds in a straight line and is stretched; confluent veins meet at an acute angle.

Doppler mode. Since the hepatic veins are anatomically and hemodynamically closely connected with the subdiaphragmatic caval vein, the frequency spectra also show a cardiac double-peaked flow acceleration. The first peak velocity (ca. 22–39 cm/s), caused by the movement of the tricuspid valve toward the apex of the heart, is separated by a short phase of retrograde flow (filling of the right atrium) from the slightly lower second peak velocity (ca. 13–35 cm/s), which is caused by the opening of the tricuspid valve and the filling of the right ventricle (Fig. 8.**5**). The atrial contraction that follows leads to a phase of clearly reduced flow, no flow, or flow reversal (Abu-Yousef 1992).

Within the parenchyma, an *expiratory* increase follows an *inspiratory* decrease in flow (Rabinovici and Navot 1980 provide an overview of the pressure and flow relationships in various abdominal veins in the rabbit model). This flow modulation decreases during the Valsalva maneuver while the flow direction is maintained, and there may even be a brief cessation of blood flow.

Color flow Doppler mode. With color coding, it is technically easy to image the respiration-modulated flow in the hepatic veins, which proceeds from the periphery of the liver to the caval vein. The sectional planes used require a fairly acute incident angle between the color flow Doppler beam and the vascular axis, allowing uniform color coding of the flow.

Color Doppler energy mode. Amplitude-weighted imaging of the flow in the hepatic veins leads to an intensive coloring of the vessels. The advantage of power mode imaging is that even the slower velocities can be coded, allowing representation of smaller hepatic veins as well (Fig. 8.**6**).

Fig. 8.**5**   A Doppler spectrum in a hepatic vein, showing two peaks and cardiac modulation. The second flow peak is slightly smaller than the first

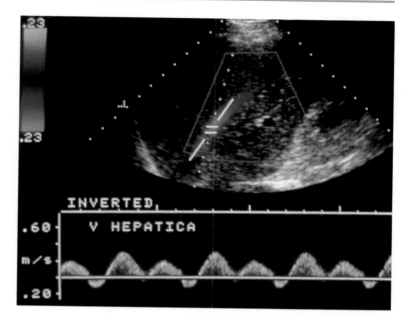

Portal Vein

Anatomy. The liver receives blood from extensive areas of the intestines, stomach, pancreas, and spleen, via the portal vein with its sources, the *superior mesenteric vein, splenic vein,* and *coronary vein of the stomach (right and left gastric veins, prepyloric vein)* and directly or indirectly the *inferior mesenteric vein.* The trunk of the portal vein is located dorsal to the head of the pancreas, and proceeds over 3–9 cm with a diameter of 7–15 mm behind the superior horizontal part of the duodenum in the hepatoduodenal ligament. It is accompanied by the proper hepatic artery (left ventrally) and the common bile duct (right ventrally) as far as the hilum of the liver. Intrahepatically, the portal vein divides into a right and left main branch, with approximately 6–10 smaller branches that supply the corresponding liver segments in turn, and which pass together with branches of the hepatic artery and the bile ducts all the way to the periphery of the liver.

■ Conventional and Color Flow Duplex Sonography

B-mode. The trunk of the portal vein, along with its afferent sources and branches proceeding into the individual segments, can be observed consistently along its entire length for 5–8 cm using paramedian, transcostal, or subcostal dorsolumbar sections. The maximum diameter of the portal vein is 7–15 mm (Table 8.**1**); it is affected by respiration, and has an oval lumen. In the liver parenchyma, branches of the portal vein can be distinguished from hepatic veins due to their more echogenic wall (Fig. 8.**7a**).

Doppler mode. Portal vein flow is hepatopetal, and is usually not pulsatile. The average flow velocity is approximately 15 cm/s ± 3 cm/s (Bolondi et al. 1984,

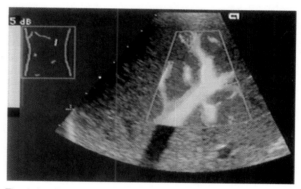

Fig. 8.**6**   Color Doppler energy mode display of the hepatic veins. The confluence of smaller veins to a larger vein, which can also be visualized in B-mode, is easily recognizable. The smaller vessels could not be seen in B-mode

Seitz and Kubale 1988); the maximum velocity is 15–88 cm/s (Kawamura et al. 1983, Saito et al. 1983, Meifort et al. 1990, Herbetko et al. 1992). The flow rate is given as 10–21 ml/min/kg body weight (Moriyasu et al. 1986b, Ohnishi et al. 1987a, 1989b, Kawasaki et al. 1990; Table 8.**1**). However, in 45 test subjects, Brown et al. (1989) measured a decrease of 26% during a standing examination (662 ml/min ± 169 ml/min, compared to 864 ml/min ± 188 ml/min). After correction of the flow rate in relation to body weight, no sex-related differences were evident.

The *Doppler frequency spectrum* is less modulated than in the hepatic veins, and due to respiration-dependent organ compression is also affected by an inspiratory decrease and an expiratory increase in flow velocity (Rabinovici and Navot 1980, Smith et al. 1985; Fig. 8.**8**). During the Valsalva maneuver, a band-like flow profile can be registered.

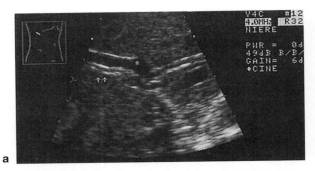

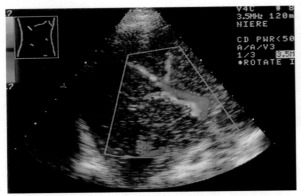

Fig. 8.**7** **a** B-mode sonographic view from the right flank: intrahepatic branches of the portal vein (*) and an accompanying proper hepatic artery (→ →). The close, at times parallel course of the two vessels is visible
b With the same sectional view as in **a**, the color Doppler sonographic display shows the flow in the branches of the portal vein and the hepatic artery in an overview. The common flow direction leads to identical color coding (red). The higher flow velocity in the hepatic artery can be recognized through the aliasing phenomenon, while the portal vein shows a consistent color

Postprandially, an increase in the venous diameter of approximately 14% occurs; the average flow velocity increases by ca. 24–49% (Gaitini et al. 1989, Meifort et al. 1990). The flow rate increases to 189% ± 7%, reaching its maximum value approximately 30 minutes after food ingestion (355 kcal, liquid) (Sabbá et al. 1991).

Anatomically preformed portocaval anastomoses via esophageal veins, umbilical veins, and hemorrhoidal veins do not show any measurable blood flow in physiological pressure and flow conditions, but they can become diagnostically significant in case of prehepatic, intrahepatic, or posthepatic venous congestion (Fig. 8.**19**).

Color flow Doppler mode. With color coding, the direction and velocity of flow in the portal vein can be detected visually. Depending on the sonographic sectional plane used, the flow direction in large areas of the liver will be opposite to that in the hepatic veins. Branches of the interhepatic portal veins show the same flow direction as branches of the hepatic artery that accompany them (Fig. 8.**7b**).

Splenic Vein

Anatomy. The splenic vein is formed from the confluence of three to five venous branches in the hilum of the spleen. It proceeds along the dorsal side of the pancreas, caudal to the splenic artery, and joins with the *superior and inferior mesenteric veins* to form the *portal vein.* It is 7–17 cm long, and has a diameter of ca. 9 mm (range 5–10 mm); it shows only minimal respiratory fluctuations. Additional afferent venous pathways are found in the short gastric veins, the left gastroepiploic vein, and numerous small veins of the pancreatic tail and the duodenum.

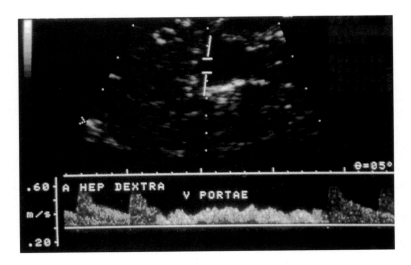

Fig. 8.**8**   A Doppler frequency spectrum of the portal vein and right hepatic artery, showing the same flow direction toward the transducer. Due to the respiratory displacement of the liver and the close proximity of the blood vessels, the spectra of the arterial and venous flow are registered with minimal respiratory displacement of the measured volume

■ Conventional and Color Flow Duplex Sonography

B-mode. In subcostal transverse sections, the segment of the splenic vein located ventral to the superior mesenteric artery and the abdominal aorta can be distinguished from the pancreas due to an absence of internal echoes. In most patients, compression can be used to assist in diagnosing a thrombosis. By means of the same transducer position, the left renal vein, as well as various other abdominal arteries can be displayed (Fig. 8.9a).

Doppler mode. During quantitative measurement of the various flow parameters in this vein, which shows hepatopetal blood flow, special care has to be taken not to compress the vessel, which lies relatively close to the abdominal wall, by exerting too much contact

pressure with the transducer. The resulting artificial stenosis may produce altered flow patterns. The frequency spectrum is band-shaped, and only minimally affected by respiratory modulation (Fig. 8.9b). The flow velocity is approximately 5–12 cm/s; the flow rate ranges from 120 ± 47 ml/min to 231 ± 19 ml/min (Ohnishi et al. 1987a, 1987b and 1989b, Zoli et al. 1986b).

Color flow Doppler mode. Due to the convex, curving course of the splenic vein in relation to the transducer, blood flow toward the probe is seen when using a transverse section in segments near the hilum, while in segments near the portal vein, flow away from the probe is registered. This physiological "color change" is divided by an area in which the incident angle of the color flow Doppler beam meets the vein at right angles, resulting in no color coding (Fig. 8.10).

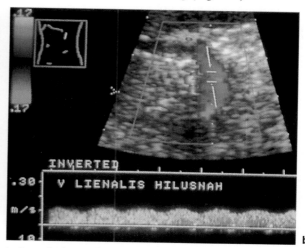

a

Fig. 8.**9** A sonographic section along an almost transverse body axis at the level of the splenic vein in its preaortic segment (**a**). Simultaneous sections of the left renal vein (3) directly ventral to the abdominal aorta (1) and the right renal artery (2). The superior mesenteric artery (4) is seen dorsal

b

to the splenic vein (5). **b** The Doppler frequency spectrum in the splenic vein near the hilum is not modulated by respiration (transducer approach from an oblique left lateral dorsolumbar flank section)

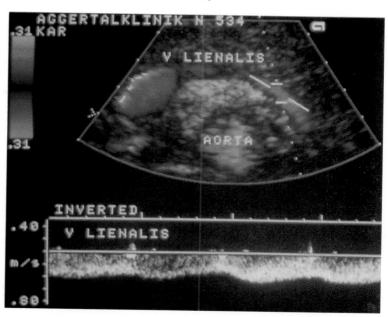

Fig. 8.**10** Color flow duplex sonogram of the splenic vein. Flow in the vein running from the splenic hilum toward the transducer is coded red. Flow away from the transducer into the venous confluence is coded blue. In the distal splenic vein, the frequency spectrum shows little respiratory modulation

Renal Vein and Intrarenal Veins

Anatomy. The renal vein forms from the fusion of numerous *arcuate veins* and *interlobar veins,* and then proceeds from the hilum of the kidney, usually ventral to the renal artery, to the inferior vena cava. While the right renal vein reaches the inferior vena cava almost at a right angle after some 2.5 cm, the left renal vein, which is approximately 8 cm (6–12 cm) long, crosses the aorta ventrally, dorsal to the superior mesenteric artery, after joining with the left testicular vein and the left suprarenal vein. Multiple manifestations of the renal veins appear in 19% of cases on the right side and approximately 4% on the left.

■ Conventional and Color Flow Duplex Sonography

B-mode. Sonographic examination of the renal veins is limited to the area from the hilum of the kidney to their confluence with the vena cava (Fig. 8.9 a). Intrarenal veins cannot usually be morphologically imaged.

Doppler and color flow Doppler mode. Using color flow duplex sonography and oblique dorsolumbar sections, the renal venous system can be depicted from the roots of the arcuate veins, through the interlobar veins, as far as the confluence with the inferior vena cava. As when examining the renal arteries, the left renal vein needs to be shown over a longer distance in a transverse section.

Intrarenally, the Doppler spectrum has a frequency band that is modulated by respiration (Fig. 8.11). As one proceeds further toward the caval vein, the flow patterns can be influenced by the characteristic atrial modulation of the inferior vena cava. A clear intrarenal cardiac flow modulation is sometimes also seen, as described above in connection with the hepatic veins (Fig. 8.12).

Attention should be given to potential pairing of the renal vein, which occurs significantly more often on the right than on the left side (Wenz 1972), as well as to a retroaortic course of a single or paired left renal vein, or the presence of a periaortic venous ring (Fig. 8.13).

Mesenteric Veins

Anatomy. The *superior* and *inferior mesenteric veins* and their contributing veins accompany the mesenteric arteries and transport blood from the arterial distribution area of the superior mesenteric artery and the drainage area of the right gastroepiploic vein to the portal vein. The trunk of the *superior mesenteric vein* (diameter up to 13 mm) passes on the right next to the artery, and crosses the inferior horizontal part of the duodenum to join the splenic vein on the dorsal side of the pancreas.

The *inferior mesenteric vein* (diameter about 8 mm) takes up the sigmoidal veins and left colic veins, and then anastomoses via the superior rectal vein and the venous rectal plexus, through the internal pudendal vein, with the vena cava. The vein also lies next to the inferior mesenteric artery, and joins the splenic vein or the superior mesenteric vein behind the pancreas.

■ Conventional and Color Flow Duplex Sonography

As the second major vein contributing to the portal vein, the *superior mesenteric vein* is relevant both clinically and sonographically. Frequency spectra registered from mesenteric veins with physiological blood flow have a band-like flow profile, which is not influenced to any great extent by respiratory modulation, and is directed toward the portal vein (Fig. 8.14). The flow velocity is 9–18 cm/s; the flow rate is 194 ml/min ± 25ml/min (n=8) or 6.5 ml/min/kg ± 1.9

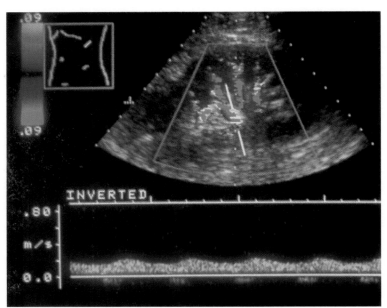

Fig. 8.**11** Image of a kidney along its longitudinal axis (oblique dorsolumbar section). The frequency spectrum from the intrarenal veins near the hilum shows flow that is hardly modulated by respiration, directed away from the transducer (blue color coding), and showing rhythmic fluctuation in its maximum velocity due to pulsation in the accompanying arteries

Fig. 8.**12** The same situation as in Figure 8.**8**. Here, the Doppler frequency spectrum from the distal left renal vein shows marked respiratory and cardiac modulation. The maximum velocity is 28 cm/s; the average velocity was calculated at 15 cm/s

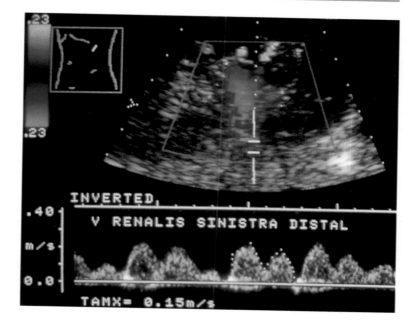

Fig. 8.**13** A transverse upper abdominal section somewhat distal to the origin of the renal arteries. Ventral to a vertebral body (WK) and dorsal to the aorta (AO, coded red), a retroaortic left renal vein with a narrow lumen can be seen (arrows). Due to the sector scanner that was used, the flow direction in the distal segment near the confluence is coded blue, while in the retroaortic section, the flow is coded red (1 = cross-section of the superior mesenteric artery)

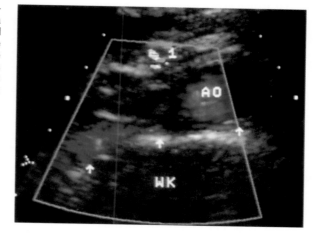

Fig. 8.**14** Color flow duplex sonogram of the superior mesenteric vein proximal to the venous confluence, in a paramedian section from the left. Flow directed away from the transducer is coded blue. The Doppler frequency spectrum shows respiration-dependent flow velocity changes

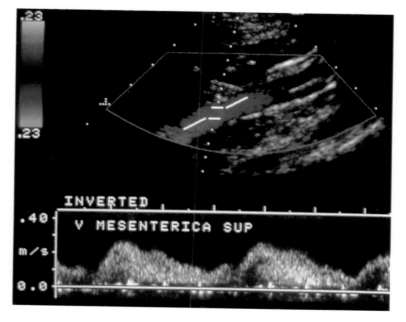

ml/min/kg body weight (n = 26) (Zoli et al. 1986, Ohnishi et al. 1987a).

In normal findings, examination of the superior (and inferior) mesenteric vein is only capable of detecting blood flow in the trunk region in order to exclude a mesenteric venous thrombosis or vascular dilatation associated with portal hypertension.

Evaluation

■ Conventional and Color Flow Duplex Sonography

B-mode. Particular attention needs to be given to the sonographic appearance of the *vascular caliber* of the abdominal veins. Simultaneous sonographic examination of the parenchymatous organs that are being drained is important in evaluating and interpreting the findings. The organs can be assessed with regard to their size (hepatomegaly, splenomegaly, renal swelling), form, contour, surface, echo pattern, and echo intensity (fatty liver, cirrhotic transformation, intraparenchymal space-occupying lesions).

Compressing the venous lumen with the transducer, as used in the area of the peripheral veins to exclude thrombosis, is not always successful in the abdominal cavity. Due to the anatomical situation, and mainly in slim patients, only a few venous segments (caval vein, splenic vein and left renal vein ventral to the aorta, and the superior mesenteric vein near the confluence), can be compressed through the abdominal wall. If complete compression is successful, then a thrombosis can be excluded in these venous segments.

When *measuring the vascular caliber,* which together with measurement of the flow velocity forms the basis for calculating the flow rate, care must be taken to ensure that measurements of the vascular cross-section are made at the same point at which the velocity is also measured, and that no pressure is exerted on the venous lumen with the transducer (see the section on sources of error below, p. 302).

Doppler mode. The following Doppler-sonographic criteria are assessed in the abdominal veins:

- Patency (detection of venous flow)
- Flow direction
- Flow velocity
- Cardiac and respiratory modulation
- Quantitatively determined flow rate (ml/min)

If it is possible to register physiological flow patterns for the individual venous segments, then a hemodynamically effective pathological finding can be excluded.

Color flow Doppler mode. When anechoic or hypoechoic structures are identified in the abdominal cavity that are difficult to classify, the color flow Doppler mode provides quick and simple visual differentiation between cystic space-occupying lesions, dilated bile ducts, tumors, lymphomas, and abscesses. This facility is even more important in the presence of atypical vascular anatomy, which may be congenital, or may occur as a collateral venous system, postoperatively, or in association with an (invasive) tumor (Ralls et al. 1989).

Distinguishing between venous and arterial flow is possible due to the visibly pulsatile representation of flow in the arteries in comparison with the more uniform, partly respiration-modulated color coding in the veins. However, it should be noted that the often curved or winding courses of the vascular axes in the abdominal veins produce a different color coding of the normal flow direction in neighboring venous segments, and also result in an angle-dependent, varying color saturation—although the velocity of venous flow in the displayed venous segments remains constant. It is often difficult to distinguish arterial and venous signals purely on the basis of their color coding. In these cases, additional assessment of the Doppler spectra is recommended.

Pathological Findings

Principle and Methodological Aspects

Due to a decrease in blood flow directed toward the heart, a hemodynamically active flow obstruction in the region of the abdominal veins causes venous congestion and a *pressure increase* in the preceding venous segments. Depending on the surrounding structures, the venous segments that are affected become dilated, and show clearly reduced or absent respiratory fluctuations in the vascular diameter. Table 8.**2** lists the various causes of venous flow obstruction in the region of the abdominal veins that can be recognized with (duplex) sonography.

Increased resistance in the region of the portal vein or subsequent veins is termed *portal hypertension*. Etiologically, the following *causes* for this can be distinguished: *prehepatic*—e.g., thrombosis of the portal or splenic veins, compression of the portal vein, arteriovenous fistulas (splenic vessels); *intrahepatic*—e.g., toxic or pharmacological liver damage, myeloproliferative and lymphoreticular diseases, arterioportal fistula, liver cirrhosis and fibrosis, primary carcinoma of the liver, liver metastases, hepatitis; and *posthepatic*—e.g., Budd–Chiari syndrome, thrombosis of the vena cava, tumor compression, cardiogenic (tricuspid insufficiency, right heart failure) (Seitz 1990, Dubois et al. 1993).

The most frequent cause of portal hypertension is intrahepatic flow obstruction, which in 80–90% of cases is due to *cirrhosis of the liver.* As in the case of intrahepatic metastases, cirrhosis causes an increase in the arterial perfusion via the hepatic artery, and a decrease in the liver's portal blood flow findings (Leen et al. 1993). Specialized findings in this complex functional disturbance, the effects of which can be seen in various abdominal veins, are described in the relevant sections below concerning the specific vascular regions.

In the presence of tumors in the upper abdomen, attention must be given to any *compression* or *thrombosis* of the vessels, or *intraluminal tumorous protrusions.* Particularly with pancreatic carcinomas, a targeted search needs to be made for possible involvement of the portal vein, splenic vein, superior mesenteric vein—possibly including its accompanying artery—and the celiac trunk (Seitz 1989). The use of intraluminal ultrasound probes may further improve the ability to assess vascular wall infiltrations by malignant tumors (Lehner et al. 1992).

Possible causes of *arteriovenous* or *venovenous fistulas* (between branches of the *portal vein* and the *hepatic vein),* which may also be responsible for venous flow obstruction, are:

- Trauma
- Surgery
- Organ punctures
- Perforated aneurysms
- Idiopathic causes

Because of the pressure gradients involved, obstruction causes the blood to travel from the artery to the vein, or from the portal vein to the hepatic vein.

In the *renal veins,* a flow obstruction in the region of the left renal vein can lead to an obstruction in the ipsilateral testicular or ovarian vein. This may then be interpreted incorrectly as an indirect sign of thrombosis in the renal vein, for example, although more often a different genesis for a varicocele can be assumed (see Chapter 9 on the vasculature of the male genitalia).

■ Continuous Wave Doppler Sonography

Continuous wave Doppler sonography with pen-shaped probes can be used on its own to detect dilated umbilical veins and paraumbilical veins that have opened due to portal hypertension in order to provide collateral circulation *(Cruveilhier–Baumgarten syndrome).* In their developed form, they represent the clinical picture of *Medusa's head.*

■ Conventional and Color Flow Duplex Sonography

B-mode. An increase in the vascular caliber of the large abdominal veins reflects a *pressure increase* in the venous system. If increased pressure persists for

Table 8.**2**   Causes of morphological or functional venous flow obstruction in the region of the abdominal veins

- Thrombi that are wall-adherent, or protruding into the blood vessel from confluent veins
- Occluding thrombi (postoperative, paraneoplastic)
- Processes that destroy the vascular wall (tumor invasion)
- Compression from without: tumors, inflammatory structures, fibroses
- Arteriovenous fistulas
- Right cardiac insufficiency (pericardial effusion, cor pulmonale)

some time, dilated venous collateral circulation may result. This can be detected with B-mode imaging, and represents a morphological indication of functionally effective venous congestion.

Thrombotic material in the abdominal veins can be sonographically recognized if it has an increased reflection pattern. In fresh, occluding thromboses, the vein is dilated, noncompressible, and shows no fluctuation in its caliber with respiration. In partial, nonoccluding thromboses, which are seen as mural or floating structures in the venous lumen, the Doppler or color flow Doppler mode allows the detection of residual flow even when it is still marginal.

In the following descriptions of sonographically detectable pathological findings in the individual abdominal veins, the role of B-mode imaging is discussed under *duplex sonography.* In all of the abdominal drainage pathways, a dilated caliber in individual abdominal veins or the imaging of varicose collateral veins indicates obstructed flow.

Doppler mode. The pulsed wave Doppler mode is mainly used in *quantitative* sonographic examinations. Pathological flow phenomena in the anatomically closely adjacent veins of the abdominal cavity, including the collateral systems, can be differentiated and delimited using pulsed wave Doppler sonography with regard to flow direction, respiratory modulation, and flow velocity (Ohnishi et al. 1985 b, Taylor 1988, Hosoki et al. 1990).

In pathological sets of findings (e.g., liver cirrhosis), as well as with normal findings as described above, care must be taken to note the effect of ingested food or medicines when measuring the flow velocity in the portal vein (Zoli et al. 1986, Pugliese et al. 1987, Meifort et al. 1990, Alvarez et al. 1991), in the mesenteric veins (Tsunoda et al. 1988), and in varicose collateral veins (Ohnishi and Sato 1990). Although the onset of action with nitrates is faster, both nitrates and á-blockers cause a reduction in flow and pressure in the portal vein, and this can be confirmed with duplex sonography when carrying out short-term and longer-term follow-up assessments of the effectiveness of treatment (Zoli et al. 1985, Gaitini et al. 1991, Yang et al. 1991; p. 274 and 283).

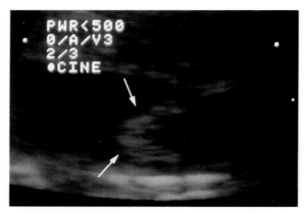

Fig. 8.**15** A paramedian sagittal section of the inferior vena cava, showing a thrombotic protrusion (arrows) extending from the right common iliac vein into the distal segment of the caval vein and clearly floating inside the venous lumen during the examination

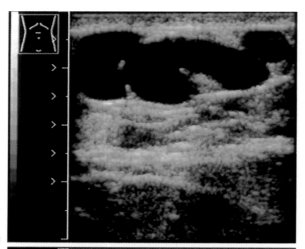

a

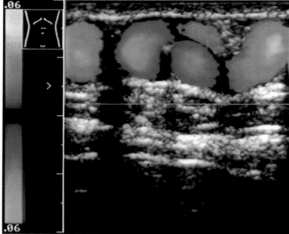

b

Fig. 8.**16** **a** Large epifascial collateral veins in a patient with post-traumatic total occlusion of the inferior vena cava of 10 years standing. In a transverse section above the navel, the vessel coils manifest themselves subcutaneously **b** Color Doppler sonography is used to visualize the slow flow-velocity associated with the low respiratory modulation in the convoluted veins

Color flow Doppler mode. In addition to the methods for distinguishing between patent veins through which blood is flowing and occluded veins, or recognizing a nonoccluding thrombus due to the flow phenomena presented by the color-coded venous flow, the color flow Doppler mode can also be used to identify the flow direction in the collateral veins in portal hypertension due to various causes (Wermke 1992).

Findings

As mentioned above, the use of continuous wave and pulsed wave Doppler sonography *without* including a B-mode procedure is no longer of any practical clinical relevance in examining the region of the abdominal veins. The present section on pathological findings therefore only describes examination findings obtained with *conventional and color flow Duplex sonography*. The classification of the findings here under the headings *B-mode, Doppler mode,* and *color flow Doppler mode* is intended to emphasize the particular strengths of each of these procedures in relation to specific findings, without implying that any artificial boundaries can be set when the various ultrasound modalities are used together.

Inferior Vena Cava

B-mode. In patients with *right cardiac insufficiency* and elevated central venous pressure, *venous congestion,* accompanied by a dilatation of the vascular lumina, occurs in the abdominal veins due to a reduction in the blood flow directed toward the heart. In the region of the inferior vena cava, the venous lumen appears bloated, and is larger than the caliber of the abdominal aorta. During respiratory modulation, a *decreased* or *absent inspiratory decrease in caliber* is observed.

Thrombi in the inferior vena cava can present as structures in the vascular lumen that range from being almost anechoic to medium echogenicity (Fig. 8.**15**). In longer-term occlusion of the inferior vena cava, dilated venous coils may be seen, representing the formation of a collateral circulatory system (Fig. 8.**16a, b**). The flow in these collateral vessels is largely uninfluenced by respiration and often displays a continuous flow toward the heart (Fig. 8.**17**). Caval vein thromboses that are associated with tumors—particularly with (right-sided) renal tumors spreading through the renal veins into the inferior vena cava, but also with other intra-abdominal neoplasms—often have the same echogenicity as the primary tumor (Didier et al. 1987, Suggs et al. 1991, Krakamp et al. 1993).

In favorable examination conditions, B-mode sonography can display the position and form of *caval vein filters* (Vorwerk et al. 1987, Urban et al. 1992), which can be sources for iatrogenically induced thromboses. However, it is not always possible to assess the patency using sonography alone. When ultrasound contrast media are also used, it is possible to

Fig. 8.**17** Venous collaterals in the abdominal wall, 6–9 mm below the skin, in a thrombotic occlusion of the inferior vena cava that had persisted for one year. The Doppler frequency spectrum shows continuous flow, with no respiratory modulation

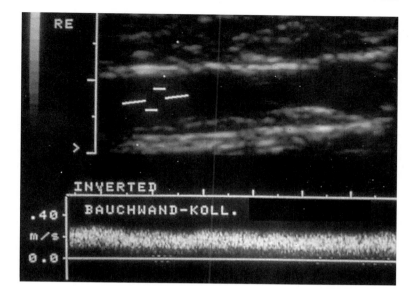

evaluate the patency even without Doppler imaging (Vorwerk et al. 1990).

Doppler mode. During systolic contraction, hemodynamically effective *tricuspid insufficiency* causes a reflux in the right auricle, which is propagated into the vena cava and can be detected due to an increased backflow component there. In the frequency spectrum, the retrograde flow direction is seen below the baseline (Smith et al. 1985).

A *completely occluding thrombosis* allows no flow signals to be registered from the vena cava, while *a partial thrombosis* can result in venous stenosis, with a velocity increase in the partially thrombosed segment.

Venous congestion may occur in abdominal veins that join the caval vein distal to the thrombosis. This can be recognized with Doppler sonography due to decreased flow velocity or respiratory modulation, or both, or even due to a reversal in the flow direction when collateral systems open up other drainage pathways, following the pressure gradient.

Color Doppler energy mode. The field of application of the power mode to display pathological phenomena in the abdominal veins is similar to that of the color Doppler mode. Its capacity to present slower flow velocities, in particular, means it is well suited to visualize parietal flow velocities around thrombi of the renal, hepatic, and mesenteric veins, the vena cava, and also for the recognition of small collaterals with slow flow velocities.

In a *caval vein filter* and preceding and subsequent venous segments, normal respiration-modulated flow can be recognized in a freely patent blood vessel, which may show whirlpool-like zones in the filter region. A thrombosis induced by foreign material here shows the same general diagnostic criteria as those applying to caval vein thrombosis.

Color flow Doppler mode. In *incomplete thrombi with blood flowing around them,* the advantage of color-coding the flow is the visible separation it provides between the thrombus material, with its weak to medium intensity, and the lumen of the caval vein, through which blood is still flowing. Particularly with thrombi that are virtually anechoic, this definitive differentiation is not always possible in B-mode imaging, nor in duplex sonography due to the numerous artifacts.

In *thrombotic occlusions of the caval vein,* which are often only slightly echogenic to begin with, it is possible to detect a vascular occlusion with color flow Doppler sonography due to the absent color coding of the flow. Thrombotic caval vein occlusions can result from:

– An ascending pelvic (leg) venous thrombosis
– Local wall damage (surgery, trauma)
– An appositional thrombus due to thrombi in the renal veins
– Infiltrating tumor growth

To exclude color artifacts, it is obligatory to image the appropriate vascular segment in at least two planes.

Although mostly asymptomatic and rarely diagnosed, *malformations* of the inferior vena cava can be associated with other venous variations and malformations and sometimes occur with deep venous thrombosis of the pelvic veins. They should be recognized before major abdominal surgery or implantation of cava filter to avoid intraoperative or peri-interventional complications. In a case study, Mori et al. (1996) report an unusual absence of the inferior vena cava in a patient with a repaired omphalocele. Two sets of bilateral paravertebral veins served as the channels of systemic venous return from the lower half of the body. These veins were narrower than typical azygos or hemiazygos continuations in the absence of the inferior vena cava.

Hepatic Veins

B-mode and Doppler mode. In *cirrhotic transformation processes* in the liver, the vascular caliber of the hepatic veins fluctuates along their course. This is accompanied by axial shifts and a rarefaction of the veins in the periphery of the liver. The Doppler frequency spectrum may fail to detect the typical cardiac modulation, and its amplitude may be generally reduced (Bolondi et al. 1991 a). In children, advanced forms of liver cirrhosis can lead to extensive shrinkage of the large intrahepatic veins, which can then no longer be displayed with duplex sonography (Keller et al. 1989).

By contrast, *dilated hepatic veins,* which can be followed far into the periphery of the liver, indicate hepatic congestion, which may occur in association with elevated central venous pressure (right heart insufficiency, pericardial effusion, cor pulmonale). The respiratory flow modulation is reduced or absent; the inferior vena cava is dilated, and also shows reduced respiratory modulation. During acute congestion, a waistlike constriction of the hepatic veins occurs before they pass through the liver capsule, which is expanded in chronic congestion (Weiss and Weiss 1990).

Budd–Chiari Syndrome

In *occlusion* of one or more hepatic veins (Budd–Chiari syndrome), sonography is not initially capable of imaging any of them, or will only give an incomplete image of the occluded veins in an enlarged organ. It is sometimes still possible to detect a slight, barely modulated residual flow, sometimes with a retrograde direction, in the affected hepatic veins (Hosoki et al. 1989, Bolondi et al. 1991 b). Over time, the thrombosed veins are seen as more echogenic cords, which present wall thickenings and irregularities during the recanalization stage (Sukigara et al. 1989). The extent to which the echogenicity definitely indicates the age of the thrombosis is not certain.

In an extensive thrombosis, the *inferior vena cava* may be partially or completely involved by the occlusion (Menu et al. 1985). Sonographically, a reduced caliber can be expected, with intravasal thrombi and possibly preocclusive dilatation. According to Bolondi et al. (1991 b), detecting a retrograde flow direction in the inferior vena cava makes the diagnosis of Budd–Chiari syndrome more probable.

Color flow Doppler mode. The cessation of flow in the *Budd–Chiari syndrome* can be recognized and precisely localized by the absence of any color coding of the venous flow. Retrograde flow may also be visually detected in the hepatic veins (Grant et al. 1989, Ralls et al. 1992).

Rare aneurysms of the hepatic veins show slower flow, with separation phenomena. If a fistular connection to the portal vein or hepatic artery develops, massively disturbed flow from these vessels into the hepatic veins or the aneurysm will be seen.

Intrahepatic portal-hepatic venous shunt can occur without chronic liver disease and usually consists of one or more focally dilated or multiple small vessels. Color Doppler sonography can detect blood flow in various directions in the shunt vessels (Kakitsubata et al. 1996).

Portal Vein

B-mode. *Cavernous transformations of the portal vein* can be recognized by the tangled, winding course of the venous segments in the area of the vein. In addition to a possible etiology due to angiomatous malformation, it has also been suggested that cavernous transformations may function as a portoportal collateral circulatory system, seen in approximately 30% of patients with portal vein thrombosis, according to observations made by Weltin et al. (1984).

In *portal hypertension,* frequent portosystemic anastomoses of the portal vein that can be displayed sonographically occur via:

- The *gastric veins* and *esophageal veins* (esophageal plexus), via the azygos veins and hemiazygos veins, into the inferior vena cava.
- The *middle and inferior rectal veins* (venous rectal plexus), via the internal iliac vein and the common iliac vein, into the inferior vena cava.
- The *paraumbilical veins* in the round ligament of the liver, via the epigastric veins and through the external iliac vein to the inferior vena cava; *and* through the brachiocephalic veins to the superior vena cava (Fig. 8.**18**).

Additional possible collateral systems are shown in Figure 8.**19**.

In patients with portal hypertension, the distal portal vein is dilated in its extrahepatic segment to over 15 mm, with marked differences in vascular caliber from the narrower, intrahepatic portal blood vessels. Additional diagnostic criteria are splenomegaly and the detection of free fluid in the abdominal cavity (ascites), as well as the sonographically recognizable findings listed in Table 8.**3**.

Thromboses of the portal vein, seen as reflective structures of weak to medium intensity in the vein or its branches, allow sonographic diagnosis of prehepatic sources of increased pressure in the portal system. Fresh thromboses are recognizable from an increase in the portal vein caliber amounting to approximately 15–25 mm (Miller and Berland 1985), and from rather hypoechoic intravasal thrombus material, which in reduced echogenicity conditions may make it difficult to distinguish internal structures in the lumen. In the portal vein region also, it is not yet clear to what extent the age of a thrombus can be determined solely by assessing its echogenicity.

Alpern et al. (1987) report on a series of very small-caliber thrombosed portal veins, and mention a reduction in caliber associated with increasing age of

Fig. 8.**18** A color flow duplex sonogram of a reopened umbilical vein in a
patient with a thrombosis of the portal
vein that had persisted for an extended
period (right paraumbilical transducer
position). The vascular diameter of the
collateral veins is approximately 10 mm.
The frequency spectrum shows almost
continuous flow, with a maximum flow
velocity of circa 20 cm/s

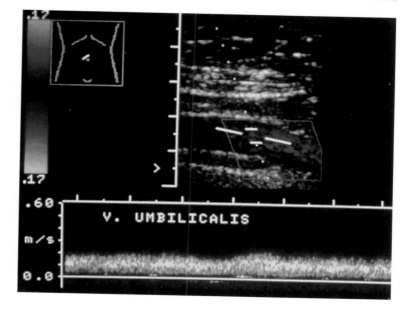

Fig. 8.**19** Various potential collateral
systems (1–5) occurring in obstructions
of the portal vein with accompanying
portal hypertension. The following
venous systems are typical circulatory
bypass systems: 1—umbilical and
paraumbilical veins, clinically appearing
as the Medusa's head;
2—gastric veins as varices of the
esophagus and the gastric fundus;
3—a venous splenorenal short-circuit,
with varices in the hilum of the spleen;
4—collateral veins through the inferior
mesenteric vein; 5—gastric varices
along the greater gastric curvature

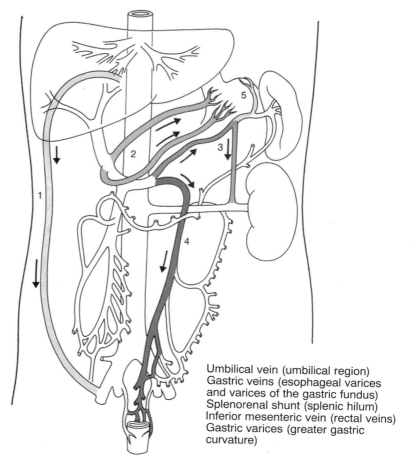

Umbilical vein (umbilical region)
Gastric veins (esophageal varices
and varices of the gastric fundus)
Splenorenal shunt (splenic hilum)
Inferior mesenteric vein (rectal veins)
Gastric varices (greater gastric
curvature)

the thrombosis as being a possible cause. Wang et al.
(1991) suggest that, in the presence of tumorous
thrombi from a hepatocellular carcinoma, it may
sometimes be possible to register the pulsatile signals
of tumorous blood vessels from the thrombus.

Doppler mode. *Portal hypertension* can be accompanied by a reduction in the average and maximum
flow velocity in the portal vein, which can be detected
with Doppler sonography. Ohnishi et al. (1985 a) report a maximal flow velocity of 17 cm/s ± 3.9 cm/s and
a flow rate of 8.5 ml/min/kg ± 2.9 ml/min/kg (n = 19).

Table 8.**3** Sonographic and Doppler-sonographic criteria and findings associated with portal hypertension

Portal vein
- Extrahepatic dilatation of the portal vein (> 13–15 mm)
- Round cross-section
- Decreased or absent respiratory modulation
- Decreased pendular or hepatofugal flow in the portal vein
- Obstruction of the portal vein (thrombi, tumor invasion)
- Pathological findings in the liver parenchyma (cirrhosis, tumor, Budd–Chiari syndrome)

Splenic vein
- Dilated splenic vein (> 10 mm)
- Decreased or absent respiratory modulation
- Hepatofugal flow in the splenic vein
- Splenomegaly

Additional findings
- Dilated mesenteric veins (> 15 mm)
- Dilated veins in the round ligament of the liver (> 2.5 mm)
- Hepatofugal flow in these veins (with or without dilation)
- Portocaval collateral veins–paraumbilical (Medusa's head), esophageal (esophageal varices), dilated gastric veins, hemorrhoidal, see Figs. 8.**17**–8.**19**)
- Ascites

When there is an increase in the intrahepatic pressure and a flow obstruction, stasis or reversal of the flow direction (hepatofugal flow) may take place in the individual vascular roots of the portal vein, or in the portal vein itself (Rector et al. 1988). In a group of 228 patients suffering from *cirrhosis of the liver,* Gaitini et al. (1991 b) confirmed the presence of retrograde flow in the portal vein, splenic vein, or superior mesenteric vein in 8.3% of cases.

A diagnosis of portal hypertension becomes increasingly more definite as more and more of the duplex-sonographic criteria listed in Table 8.3 are met. It should be borne in mind that decreased flow velocities in the portal vein can be due to various causes, and that extrahepatic factors such as right cardiac insufficiency can also influence these measurements.

In the presence of extensive esophageal varices or a gastrorenal shunt, flow in the left gastric vein may be directed hepatofugally, opposite to the physiological direction (Ohnishi et al. 1986).

As confirmed by Gaitini et al. (1991 a) when examining esophageal varices, the flow velocity in the collateral veins can also be modulated by respiration (Sukigara et al. 1988, Matre et al. 1990) and by pharmacological effects. In the left gastric vein, the flow velocity in 12 patients with liver cirrhosis decreased 120 minutes after oral administration of 40 mg *of propranolol,* on average from 15.4 cm/s ± 1.5 cm/s to 11.1 cm/s ± 0.9 cm/s. The percentage decrease in the

velocity (26.8% ± 3.5%) and flow rate (30.2% ± 4.4%) was significantly more marked in the left gastric vein in comparison to the portal vein (16.4% ± 3.6% and 20.9% ± 3.3%, respectively).

Winding varices near the hepatoduodenal ligament and around the wall of the gallbladder can be seen as rare collateral paths when there is a prehepatic occlusion of the portal vein (Ralls et al. 1988, Helbich et al. 1992).

Fistulas between the portal vein and the renal veins (see the color flow Doppler mode section), or between the hepatic artery and the portal vein, are rare findings. They mainly occur as complications of an aneurysmal vascular wall dilatation (Chagnon et al. 1986, Endress et al. 1989, McLoughlin et al. 1995).

Cirrhosis of the Liver

In 11 patients with cirrhosis of the liver, 60 minutes after *ingesting food* (1100 kcal), Gaiani et al. (1989) observed a smaller increase in the diameter of the portal vein (3% versus 14%), the flow velocity (3.2% versus 24%), and the flow rate in the portal vein (8.5% versus 59%) than in 12 normal individuals (Table 8.**4**). In individual cases, a decrease in the average flow velocity was even registered. This is explained by a redistribution of the blood flow from the mesenteric veins into portosystemic collaterals. Other research groups, however, have not been able to reproduce this reduction in the increased flow, and instead have described a postprandial increase in the flow rate of around 51% (n = 10, Alvarez et al. 1991). In addition to the reasons given by the authors for these discrepancies (composition of the food, time interval until measurement), another factor might be that individual collateral systems can decisively affect the hemodynamic parameters at specific measuring points (see the section on sources of error in flow rate measurements, p. 302).

In comparison with healthy individuals, the absolute flow rate in the portal vein does not show definite and consistent changes (Kawasaki et al. 1990). It is decisively influenced by the individual anatomical and functional collateralization capacity. The direction and velocity of flow in individual branches of the portal vein can therefore vary (Dao et al. 1993).

A continuous decrease in flow velocity in the portal vein follows the administration of β-blockers. Three hours after oral ingestion of 40 mg of *propranolol* or 100 mg *of atenolol,* Zoli et al. (1986 a) recorded a decrease of approximately 22–33% (n = 8). The flow rate decreased by approximately 230–400 ml/min. These findings are consistent with those obtained by other groups (Nakayama et al. 1983, Gaitini et al. 1991 a). With a shorter onset time, *isosorbide dinitrate* produces a similar decrease in the velocity and rate of portovenous flow by ca. 31% ± 3% and 260 ml/min ± 30 ml/min when measured 15 minutes after sublingual administration of 5 mg (Zoli et al. 1986 a).

Table 8.**4**   Measurements (mean ± SD) of the vascular caliber, average flow velocity, and flow rate in the portal vein in normal individuals (n = 12) and patients with cirrhosis of the liver (n = 11), before and 60 minutes after ingesting a solid standardized meal. In each instance, the postprandial percentage increase is also shown (adapted from Gaiani et al. 1989)

	Normal individuals Fasting	Postprandial	Increase	Patients with liver cirrhosis Fasting	Postprandial	Increase
Caliber (mm)	10.5 ± 1.5	11.9 ± 1.7	14%	14.0 ± 2.2	14.3 ± 2.1	3%
V_{mean} (cm/s)	16 ± 4.1	19.7 ± 4.8	24%	12.4 ± 2.3	12.7 ± 3.1	3.2%
Flow rate (ml/min)	832 ± 245	1312 ± 433	59%	1160 ± 426	1262 ± 480	8.5%

Transjugular Intrahepatic Portosystemic Shunt (TIPSS)

A nonsurgical approach for the management of portal hypertension can be the implantation of stents, creating an artificial shunt between the portal veins and the hepatic veins. The stents are placed in the liver parenchyma from a jugular vein via vena cava and a hepatic vein and are therefore called *transjugular intrahepatic portosystemic shunts* (TIPSS). Although this method is relatively new and has to be evaluated in larger series of patients, one can already establish that color Doppler sonography is useful for the assessment of intrahepatic hemodynamics in these patients, and shunt patency and shunt stenosis or occlusion can be detected by color Doppler or power mode sonography (Blum et al. 1995).

Banti's Syndrome

The term *Banti's syndrome* was formerly thought to be a complex of symptoms relating to a specific disease. Today, it is used to include a number of conditions that are not yet clearly classified (hepatoportal sclerosis, idiopathic portal hypertension, noncirrhotic portal hypertension, benign intrahepatic portal hypertension), and which are accompanied by splenomegaly, hypersplenism with pancytopenia, mild functional disturbances of the liver, and portal hypertension (Ohnishi et al. 1984, overview in Ohnishi et al. 1987b). Hemodynamically, an elevated flow rate is seen in the splenic artery and vein, and also an increase in intrahepatic portovenous resistance (Ohnishi et al. 1989a, 1989b). A series of measurements carried out by this group in 17 patients (Ohnishi et al. 1987b) detected a flow rate increase in the splenic vein of approximately two and a half to four times (312 ml/min ± 66 ml/min to 502 ml/ min ± 114 ml/min, compared to 120 ml/min ± 47 ml/ min in healthy subjects).

Color flow Doppler mode. Pathological flow patterns in the form of hepatofugal or pendular flow can be quickly recognized visually from the conflicting color coding, or the change in colors presented by pendular flow. A clear reduction in velocity correlates with darker colors, even when the pulse repetition rate is set at a low level.

Intra- and extrahepatic masses such as solid lesions, abscesses, large subcapsular hematomas, metastases, and hepatocellular carcinomas may induce portal flow reversal. A differentiation between these entities based on the portal flow direction only is not possible (Miller et al. 1996).

Thrombosis of the portal vein can be diagnosed using morphological criteria, and also from the absent or retrograde flow (Furuse et al. 1992).

Aneurysms of the portal vein can be diagnosed by a dilatation of the venous lumen, which shows a slow, whirlpool-like intraluminal flow. *Fistulas* between branches of the portal vein and the hepatic vein (or caval vein), which have as yet only been described rarely, show flow from the portal vein to the hepatic veins, with a double-peaked frequency spectrum in the draining veins (Tanaka et al. 1992). If these fistulas appear in association with liver disease accompanied by portal hypertension, they may be seen as indicating a collateral circulatory system. However, the congenital form, with simultaneous aneurysms of the portal vein, is more common. In four patients, the shunt volume was between 80 ml/min and 360 ml/min, corresponding to 9–29% of the portovenous flow rate (Kudo et al. 1993).

In surgically created *shunt connections,* the patency of the shunt can be evaluated with duplex sonography using the flow direction and the flow velocity (Lafortune et al. 1987, Johansen and Paun 1990, Longo et al. 1993; see the section on diagnostic effectiveness, p. 305).

Splenic Vein

B-mode. *Thromboses of the splenic vein* show a dilated, noncompressible vein, with absent respiratory modulation of the caliber. Due to the close anatomical relationships of the neighboring organs, thromboses may occur in the context of pancreatitis with an inflammatory infiltration of the venous wall, in pancreatic cancer, in cirrhosis of the liver, after liver transplantation, or idiopathically. In fresh thromboses, the diameter is usually between 9 mm and 12 mm.

Doppler and color flow Doppler mode. In obstructions of the splenic vein (thrombosis, tumorous infiltration), complete occlusion is detected by Doppler sonography due to the absent venous flow, while an incompletely occluded splenic vein still retains detectable residual flow, which may present with an elevated and stenotic flow velocity (Reuss 1990).

A dilated splenic vein with no detectable thrombus, and accompanied by reduced flow velocity or with retrograde hepatofugal flow, is one of the criteria for *portal hypertension*. In this case, the mesenteric venous blood is transported to the vena cava through portosystemic collaterals. In the splenic hilum, dilated veins sometimes show anastomoses to veins of the stomach or esophagus.

An *arteriovenous fistula* between the splenic artery and the splenic vein is a rare finding, which is accompanied by a dilated or aneurysmal splenic vein that can cause portal hypertension (Canterero et al. 1989). Often, the cause is an aneurysm of the splenic artery, with a fistular canal proceeding into the accompanying vein. This condition appears to be more prevalent in women. The changes in the afferent artery of the fistula, which can be detected with duplex sonography, are described in Chapter 7 (p. 259).

Renal Vein and Intrarenal Veins

B-mode. As with the other abdominal veins described above, detection of a *thrombosis* is particularly important in the region of the renal veins. Nonspecific criteria can be assessed with B-mode sonography, and then confirmed using the Doppler procedure (Table 8.**5**). In the acute phase, the kidney appears enlarged, and the parenchymal seam shows reduced echogenicity, and cannot be clearly distinguished from the renal sinus. Over time, the echogenicity of the edematous parenchyma increases after cellular infiltration, and the swelling of the organ reverses. Depending on the age of the thrombosis, weakly echogenic or more intensely reflective structures can be seen in the renal vein (Fig. 8.**20 a**).

Dilatation of the renal veins with no detectable signs of thrombosis is typical of venous congestion in the inferior vena cava. Possible causes that may be considered are right cardiac insufficiency, or a compression or thrombosis of the inferior vena cava.

In compression syndromes of the left renal vein, which can be accompanied prestenotically by a dilata-

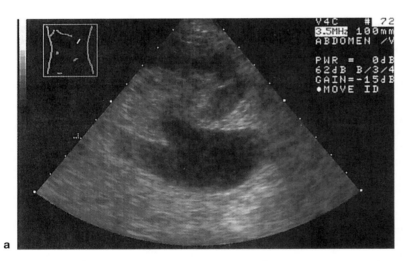

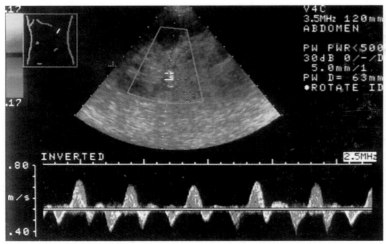

Fig. 8.**20**   **a** Intercostal view from the left flank showing the middle and caudal part of the kidney. The renal vein is massively distended in the hilar area to more than 20 mm in diameter. **b** The intraparenchymal artery alternates between red and blue in this doppler sonographic image. The Doppler frequency spectrum suggests absolute arrythmia with clearly recognizable pendular flow. The pendular flow can be regarded as typical for renal vein thrombosis (see Fig. 11.**58**)

Table 8.**5** Sonographic and Doppler-sonographic criteria indicating the presence of an acute thrombosis of the renal vein

Nonspecific criteria
- Increased kidney size
- Decreased echogenicity in the parenchyma
- An unclear border between the parenchymal seam and the renal sinus
- Dilatation of the intrarenal veins and renal vein

Specific criteria
- Internal echoes within the lumina of the renal veins
- Absence of flow in the intrarenal veins and renal vein
- A steep systolic increase and decrease in the flow velocity in the renal artery
- Continuous diastolic retrograde flow in the renal arteries

tion of the venous caliber, a search should be made for a dilated left testicular vein or left ovarian vein (Justich 1982). In retroaortic renal veins (Figs. 8.**13**, 8.**22**), and when these are combined with a renal venous ring, compression of the left renal vein between the abdominal aorta and the spinal column has been suggested as a possible cause of varicocele (Riedl 1980; see p. 317).

Doppler and color flow Doppler mode. If the arterial inflow is undisturbed, an absence of blood flow in the intrarenal veins and the main renal vein confirms the diagnosis of *thrombosis* (Krumme et al. 1993, Laplante et al. 1993). The change in the frequency spectra from the afferent arteries is also characteristic for the acute phase: a pendular flow typically arises in the intraparenchymal arteries, which can be recognized with color flow Doppler sonography and documented in the Doppler frequency spectrum (Fig. 8.**20 b**; Fig. 11.**58**, p. 447).

In *malformations* such as anomalous circumrenal veins, which may mimic perirenal fluid collection, pulsed wave Doppler or color Doppler sonography or power mode detects unexpected flow (Baker and Carroll 1995).

After *biopsy of native kidneys,* the systematic use of color Doppler sonography facilitates diagnosis of *arteriovenous fistula,* which occur as a complication in up to 5% of cases (Rollino et al. 1994, Özbek et al. 1995).

Mesenteric Veins

Thromboses in the region of the mesenteric veins are accompanied by a dilatation in the caliber, structures within the lumen, and absent flow signals.

Portal hypertension can lead to an increase in vascular caliber, slower flow, and reduced respiratory flow modulation in the superior mesenteric vein (Bolondi et al. 1984, Ohnishi et al. 1987 a). Some patients can be expected to show hepatofugal flow through a

splenorenal shunt, with retrograde (hepatopetal) flow in the splenic vein (Ohnishi et al. 1985).

Lateral or ventral displacements of the superior mesenteric vein (with or without thrombosis) may indicate *space-occupying lesions* in the pancreatic region. When there is invasive growth, infiltration of the vascular wall can be expected.

Other Abdominal Veins

Among the *epifascial abdominal veins* that can be classified as belonging to the superficial venous system of the abdomen, the *umbilical veins,* as mentioned above, are diagnostically relevant as a collateral system *(Cruveilhier–Baumgarten syndrome).* Their winding course can already be imaged sonographically in the area of the abdominal wall (Caturelli et al. 1989). In addition, in the area of the lower abdomen, dilated subcutaneous veins can also be detected, which function as collaterals due to thrombotic occlusions of the pelvic veins or inferior vena cava (Figs. 8.**17**, 8.**19**).

Duplex sonography can detect thrombotic occlusion of the *inferior epigastric vein,* a rare condition, due to the vessel's dilated caliber, lack of compressibility, and the absence of venous flow signals (Fig. 8.**21**).

Thromboses of the ovarian veins, which are seen more often during pregnancy and postpartum, can be recognized when the ovarian vein has no blood flow and is no longer dilated. Accompanying conditions that can be expected are perivascular inflammatory tissue swelling and thickened adnexae (Kubale 1992).

As mentioned above, a dilated left ovarian vein caused by venous congestion may be a sign of obstructed flow in the left renal vein (Justich 1982).

Organ Transplants

Primary vascular complications following organ transplantation can affect the arterial and venous blood vessels in the implanted organ (see Chapter 7, p. 266). In the venous component, *thromboses* and *anastomotic stenoses* can be expected. Thrombotic obstructions in the transplant veins show the same signs that are diagnostically relevant in the original abdominal veins (Keller et al. 1989). In pancreas transplants that develop transplant thromboses in the afferent artery, Snider et al. (1991) were able to detect typical changes, such as a steep systolic increase and decrease in the flow velocity associated with a holodiastolic retrograde flow, as described above in connection with thromboses of the renal veins (p. 298).

Low-grade stenoses due to suturing cause localized flow acceleration; higher-grade stenoses can present signs of venous pressure elevation.

Diagnostic criteria for *arteriovenous fistulas,* especially those following transplant punctures (Plainfosse et al. 1992, Deane et al. 1992, Gainza et al. 1995, Ahn and Cohen 1995), are described in Chapter 7 (pp. 259, 266). The greater the shunt volume flowing

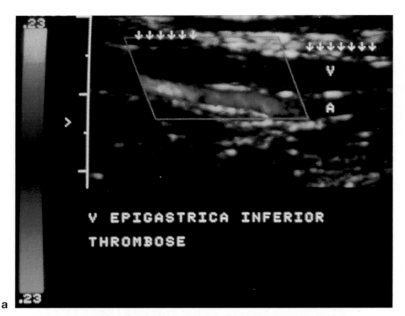

a

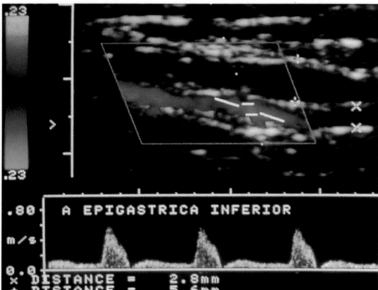

b

Fig. 8.**21**   A thrombotic occlusion of an inferior epigastric vein (V), with the accompanying artery (A, coded red) in the left lower abdominal region (**a**). The vein (arrows), appearing above the inferior epigastric artery, is dilated and not compressible. The Doppler frequency spectrum from the inferior epigastric artery shows a normal flow profile (**b**). The caliber of the thrombosed vein is ca. 5–6 mm; the caliber of the artery is about 3 mm

through the fistular canal, the stronger will be the recognizable pulsatile modulation of the flow in the efferent vein. In the arteriovenous fistula, there is an increased color saturation with high peak systolic velocity and a low resistive index in the supplying artery. Like the pathomorphological changes seen in the region of the original blood vessels, the draining veins in organ transplants also show dilatation, and possibly aneurysmal expansion.

As observed in the portal vein after *liver transplants,* gas bubbles can be observed as strongly echogenic, mobile structures without an ultrasound shadow, which are transported with the blood flow in the portal vein (Taylor et al. 1986, Chezmar et al. 1989).

Evaluation

■ Conventional and Color Flow Duplex Sonography

B-mode. Pathological findings in organs (e.g., the liver or spleen) that indicate venous congestion should prompt a search for collateral blood vessels, which provide targeted morphological clues to any functional change in the pressure relationships, particularly in the area of liver perfusion.

Thrombi can be recognized sonographically when they produce a dilatation of the veins and when intraluminal thrombotic structures of varying echogenicity are seen. When detecting thrombosis of an abdominal vein, or a tumor with invasive growth into a vessel (Suggs et al. 1992), a careful note should be made of the exact location and extent of the venous occlusion.

When interpreting the vascular findings in pathological cases, consideration should always be given to the overall collateral system associated with a particular drainage area, in order to detect pathologically dilated bypassing circulatory systems, such as portosystemic collaterals, which can only provide indications of a hemodynamic functional limitation in B-mode imaging. Ascites due to portal hypertension must not be overlooked.

If the *caval vein* is being examined to assess whether there is a *thrombosis*, then also the *pelvic veins* must be examined and documented. Vice versa, every *thrombosis of the pelvic veins* requires an ultrasound evaluation of the inferior vena cava in order to exclude or confirm its involvement (pp. 207–210). Duplex sonography appears to provide better results than those obtained with phlebography when measuring the length of thrombi in the caval vein.

Venous aneurysms are rare findings that can already be suspected with B-mode imaging, and which can be differentiated from cystic space-occupying lesions using Doppler sonography (see below) (Hagiwara et al. 1991).

Doppler mode. Absent, slower, or retrograde flow can be detected in the abdominal veins with duplex sonography, and the direction and velocity of flow in any collateral veins that may be present can be displayed. A necessary precondition for this is adequate B-mode imaging of the veins and a facility for angle-adapted placement of the sample volume in order to carry out flow velocity measurements.

Color flow Doppler mode. In addition to locating the abdominal veins and displaying their course with the color signal, the ability of the color flow Doppler mode to recognize slow venous flow makes visual assessment of the flow direction easier, and thus allows the detection of venous segments with partial or complete retrograde flow, as well as rare collateral systems such as varices of the gallbladder (Ralls et al. 1988).

Color flow Doppler sonography is also used for imaging blood flow through the marginal seam of a thrombus that also has blood flowing around it. This phenomenon can be seen in incomplete thrombotic occlusions or in reopened thrombotic sections detected during the recanalization stage (Seitz and Reuss 1992).

It is possible to differentiate between a venous aneurysm and a cyst due to the disturbed, and usually slower and whirlpool-like, flow in the vascular wall protrusion of an aneurysm. In the region of the liver, an attempt should be made to distinguish between aneurysms of the portal vein and the hepatic vein and to identify any additional portohepatic fistulas (Tanaka et al. 1992).

Sources of Error and Diagnostic Effectiveness

Sources of Error

External Examination Conditions

Unfavorable external examination conditions, such as:

- Obesity
- Superimposition of intestinal gas
- Restlessness of the patient
- Unfavorable planes for applying ultrasound and large incident angles of over 60° in (color) Doppler mode

can affect the quality and the informativeness of an ultrasound examination (Gill 1985). Basically, the potential errors when examining the abdominal veins are the same as those relating to the abdominal arteries (Chapter 7, pp. 269–272). In addition to *positional anomalies* and *positional variants* of the abdominal veins, particular mention should be made of *movement artifacts* and *superimposition* of venous Doppler signals by peristaltic background noise and arterial frequency spectra (see below).

Equipment Configuration

To obtain as much Doppler and color flow Doppler information as possible, particularly with the low flow velocities that occur in the abdominal veins, the equipment has to be optimally adjusted for the Doppler and color flow Doppler modes with regard to the following factors:

- Adequate amplification
- Low velocity scale (pulse repetition frequency)
- Correct angle adaptation
- Smallest (high-pass) filter selection

A slow recording speed is useful for documenting Doppler frequency spectra in order to demonstrate venous flow with respiratory and cardiac modulations over an extended period.

Due to the generally relatively slow speed of venous flow, and with suitably low adjustment of the pulse repetition frequency, it is not always possible to avoid *aliasing* in the color coding, which makes visual detection of the direction and velocity of blood flow more difficult.

With regard to B-mode imaging, it should be mentioned that detecting thrombi in the abdominal veins can be more difficult when the thrombotic material is of low echogenicity and the vein cannot be compressed due to the anatomical situation, or when the thrombus has a structure that is isodense to the surrounding tissue (e.g., splenic vein versus the pancreas parenchyma, hepatic vein and portal vein versus the liver parenchyma).

Isolated thrombotic occlusions of individual venous branches in the portal vein or hepatic vein can also be overlooked using a duplex system (Alpern et al. 1987). The following typical error patterns should be given special mention here.

Positional Anomalies and Positional Variants

Positional anomalies, multiple vessels, and variants in the vascular course sometimes complicate the anatomical classification of anechoic structures that can clearly be identified as veins using the Doppler frequency spectrum (Fig. 8.**22**).

When there are multiple veins that normally only appear singly, there is a danger that when blood flow is confirmed in one vein, a paired and possibly thrombosed vein may be overlooked.

Arterial Superimposition

Two different factors can influence the reception of the signal when registering venous flow speeds.

On the one hand, the venous flow can be modulated by arterial pulsations if the vein is anatomically located very close to an arterial blood vessel. This results in the masking of physiological flow changes in the vein (respiratory or cardiac modulation).

On the other hand, the spatial proximity of the intra-abdominal arteries and veins can also mean that, during registration, it is not possible to separate the arterial and venous flow signals. When there are opposite flow directions, the Doppler frequency spectra are registered both above and below the baseline; when the flow direction is the same, superimposition of the frequency spectra occurs (Fig. 8.**23**).

Collateral Veins

The formation of venous bypass circulation varies from individual to individual. According to examinations conducted by Sherlock et al. (1981), there is no correlation between the extent of esophageal varices and pressure in the portal vein. However, in appropriate situations, a targeted search for this evidence of venous bypass circulation should be carried out.

Color flow Doppler sonography makes it easier to recognize and classify vascular anomalies in *congenital vascular variants*. A clear distinction between atypical caval vein segments and weakly echogenic space-occupying lesions (tumors, lymphomas) is possible, provided that there is blood flow detectable by Doppler or color flow Doppler sonography in these venous segments.

Venous Compression

The danger of compressing the abdominal veins by exerting excessive pressure on the abdominal wall with the transducer during the examination, especially in obese patients, was mentioned above. Excess pressure causes:

– An artificial increase in the flow velocity
– An artificial decrease in the venous caliber

In addition, it can adversely affect flow rate measurements.

Flow Rate Measurement

Quantitative measurements of the flow rate are based on average flow velocity measurements and the venous caliber. In these calculations, all the technical limitations that are involved in detecting the Doppler shift (frequency) and in converting it into flow velocities (e.g., the incident angle, the placement and size of the sample volume, or gate) have to be properly taken into account. These limitations mean that only an approximate indication of the *absolute* flow velocities is possible (Gill 1979, Burns and Jaffe 1985, Robinson et al. 1986, Tessler et al. 1990).

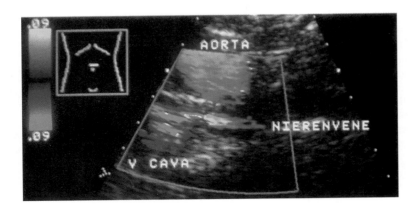

Fig. 8.**22**   Color flow duplex sonogram of the abdominal aorta (coded red) and inferior vena cava (coded blue), in a transverse upper abdominal section. A retroaortic positional variant is seen in the form of a left renal vein, with its confluence proceeding into the inferior vena cava dorsal to the aorta

Fig. 8.**23** A subxiphoid upper abdominal section, showing the close topographical relationship of the splenic vein, left renal vein, and right renal artery (each coded red), all showing almost parallel flow (directed away from the transducer; aliasing in the artery) in some areas (**a**). Arterial flow in the aorta (right edge of the picture, in cross-section) and the superior mesenteric artery, which lies in a ventral position, is color-coded blue (**b**). There is superimposition of the Doppler frequency spectra from the right renal artery and the left renal vein (arrows). Due to the identical flow direction, both of the frequency spectra with flow directed away from the transducer are displayed above the baseline. After the color coding is altered (see the color bar at the top left), the blood flow in the splenic vein and renal vein is shown in blue, and the abdominal aorta and superior mesenteric artery are shown in red

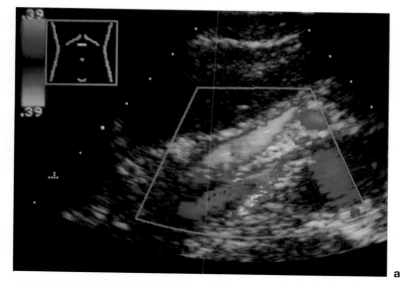

a

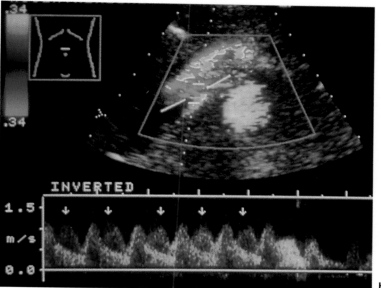

b

When the flow rate is being calculated, further possible sources of error are associated with measurements of the venous cross-section. Precise measurement of the diameter while simultaneously recording the velocity is not possible, and the venous diameter or cross-section is subject to respiratory fluctuations. In addition, simple geometric cross-sectional figures (circles, ellipses) are only rarely representative for volume measurements. The accuracy of measurements diminishes when the diameter is less than 4 mm (Barbara et al. 1990), multiplying the potential errors in measuring the venous diameter.

Color Doppler Energy Mode

Because the direction and velocity cannot be coded in power mode's amplitude-weighted flow display, it can be difficult to differentiate between those blood ves-

sels that differ precisely in these criteria (see Fig. 8.**23**). Figures 8.**24 a** and **b** present an example of similar representation of the flow in branches of the hepatic and portal veins, although the flow direction is opposite in these vessels. Color imaging that goes beyond the sonographically recognizable vessel boundaries in B-mode also makes it harder to distinguish between the single vessels in power mode.

Moreover, the imaging of the flow in Doppler energy mode depends very much on the selection of technical, equipment-specific measurements or settings such as the filter, persistence, amplification, etc., whereby false settings can give rise to movement artifacts.

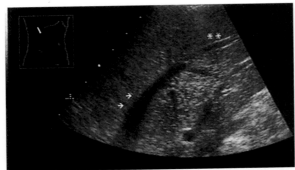

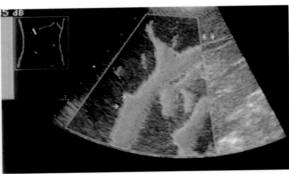

a

b

Fig. 8.**24**   **a** Intercostal view from the right flank showing a larger hepatic vein (→ →), identifiable by the missing wall. On the right edge of the image, cross sections of a portal vein branch can be seen (**). Although the portal vein is depicted as an extension of the hepatic vein, its B-mode sonographic identification is possible on the basis of the criteria described. Both vessels are sonographically sepa-rated by liver tissue and only seem to be in the same plane. **b** In the same section as in **a,** power mode shows apparent continuity of the vessels caused by the lack of coding for direction and speed. Moreover, this technique does not allow clear differentiation between blood-carrying vessels and liver parenchyma. The topographic resolution in power mode is inadequate in this representation

Diagnostic Effectiveness

Inferior Vena Cava

Using conventional duplex sonography to register frequency spectra for evaluation purposes. Smith et al. reported as early as 1985 that it was possible to image the inferior vena cava in 93% of cases (n = 85). Successful recording of the flow dynamics of the inferior vena cava using the color flow Doppler mode should be possible even more frequently.

Normal Doppler spectrum and intense color coding of the blood vessel excludes the presence of a hemodynamically effective *thrombosis* of the inferior vena cava. A thrombosis can already be identified in abdominal sonography using the B-mode criteria mentioned above (overview in Krakamp et al. 1993), and it is even possible in the intrauterine fetus (Rypens et al. 1993).

Caval vein filters can normally be imaged in follow-up examinations after implantation (n = 64) (Urban et al. 1992). However, our own experience shows that, especially in obese patients, sonographic imaging is not always satisfactory. With the assistance of color flow Doppler sonography, it is possible to depict whirlpool-like phenomena at filters implanted into the inferior vena cava, and to assess their patency. Initial experience with ultrasound contrast media shows that these allow imaging of the flow phenomena at filters (Vorwerk et al. 1990). It remains to be seen whether this procedure will be capable of enhancing the diagnostic facilities in combination with color flow duplex sonography.

If spectral changes in the inferior vena cava that are typical of *tricuspid insufficiency* are observed, it can be assumed that a hemodynamically effective valvular insufficiency is present. However, lower levels of insufficiency in combination with atrial fibril-lation can be overlooked by this indirect method of detection (Smith et al. 1985).

Color flow Doppler sonography makes it easier to recognize and classify vascular anomalies in *congenital vascular variants.* A clear distinction between atypical caval vein segments and weakly echogenic space-occupying lesions (tumors, lymphomas) is possible, provided that there is blood flow detectable by Doppler or color flow Doppler sonography in these venous segments.

Hepatic Veins

In physiological conditions, conclusive imaging of the hepatic veins is almost always possible. With duplex and color flow Doppler sonography, the flow direction from the liver periphery to the vena cava, as well as cardiac and respiratory flow modulations, can be documented.

The absolute flow velocities and the waveform of the Doppler frequency spectrum are variable, and subject to pronounced respiratory modulation (Abu-Yousef 1992).

In the presence of liver *cirrhosis,* 50–60% of patients show the above mentioned alterations in the hepatic veins.

In seven of eight patients, Bolondi et al. (1991 b) diagnosed the *Budd–Chiari syndrome* using Doppler-sonographic criteria such as absent flow, retrograde flow, or flow lacking respiratory modulation in the hepatic veins, along with retrograde flow in the inferior vena cava.

In a number of case studies, Grant et al. (1989) and Segal et al. (1996) were able to demonstrate that color flow duplex sonography was suitable both for primary identification of patients with Budd–Chiari syndrome and also for conducting postoperative follow-up studies after *shunt placement* (see the section on the portal vein, below). According to information provided by Ralls et al. (1992), the color-coded procedure also appears to be better than conventional duplex sonography in registering complex flow directions in the individual branches of the hepatic and portal veins, as well as in the collateral circulation.

Portal Vein

It is possible to image the vascular system of the portal vein with adequate certainty, including measurements of the vascular caliber and flow velocity, in 74–95% of cases (Kawamura et al. 1983, Nelson et al. 1987, Patriquin et al. 1987, Seitz and Kubale 1988, Sabb et al. 1990 a).

When interpreting the measurements, it should also be noted that food ingestion can lead to an increase in the flow velocity or cross-section, or both, in the portal vein (Table 8.**4**).

Morphological and *hemodynamic* indications of *portal hypertension* can be recognized using duplex sonography, and they appear in approximately half of patients suffering from cirrhosis of the liver (Patriquin et al. 1987, Taylor 1988, Gibson et al. 1992). In 20 patients evaluated with angiography, Alpern et al. (1987) estimated the specificity and sensitivity of conventional duplex-sonographic examinations for recording the flow direction and patency of the portal vein as being 93% and 83%, respectively.

With color flow duplex sonography, successful detection of a cavernous transformation or thrombosis of the portal vein, seen in pancreatitis, pancreatic carcinoma, liver tumors, or after liver transplantations (Gaitini et al. 1990, Pozniak and Baus 1991), is possible with a sensitivity of up to 100% (n = 40, Tanaka et al. 1993).

Color flow duplex sonography is superior to the sole use of B-mode imaging in detecting *portosystemic collaterals* in the presence of portal hypertension. In a group of 121 patients with 41 portosystemic shunts involving the left branches of the portal vein, Sugiura et al. (1992), using B-mode sonographic imaging, were able to depict only 45% of the collaterals that had previously been identified with the color flow Doppler mode. Research is in progress to identify a set of sonographic risk factors that might predict possible bleeding recurrences in esophageal varices (Schmassmann et al. 1993).

In a group of 75 patients in whom the findings obtained with color flow duplex sonography with regard to the *patency of the portal vein* were checked by angiography or intraoperatively, correspondence between the color flow Doppler findings and the control examinations was found in a total of nine thromboses of the portal vein in 69 patients (61 patient veins, eight thrombosed). In detecting thromboses, 89% of the findings obtained with the duplex procedure were correctly positive, and 92% were correctly negative (Tessler et al. 1991).

Using *endoscopic* Doppler sonography to measure the flow velocity in esophageal and gastric fundal varices in combination with sonographic criteria (Matre et al. 1990, Abdel-Wahab et al. 1993) may be able to contribute to further differentiation of the hemodynamic effects of portal hypertension in some patients (Jaspersen 1993). Transabdominally, the problem is that the clinically important esophageal varices cannot always be directly imaged in a paracardiac orientation because of a possible rupture.

Further indications for transabdominal examinations are in follow-up observations of spontaneous or surgical portocaval (side-to-side, side-to-end), mesocaval, splenorenal (proximal, distal), or intrahepatic portosystemic *shunt connections* (Terwey et al. 1981, Grunert et al. 1989, Ralls et al. 1990 b, Longo et al. 1993).

In 46 different *surgically placed portosystemic shunts,* of which 38 were patent when evaluated with duplex sonography, Lafortune et al. (1987) were able to examine the anastomotic region directly, and register Doppler frequency spectra, in 33 cases. In the remaining cases, in which the anastomotic regions could not be directly depicted, the authors interpreted a hepatofugal (retrograde) flow direction in the portal vein as being a sign that the portosystemic short-circuit connection was patent, since this criterion had been met in all of the shunts through which blood was flowing.

In a group of 42 patients, Johansen and Paun (1990) were even able to achieve a sensitivity of 100% when using duplex sonography to assess the patency of surgically placed shunts.

Case studies on the detection of *arterioportal short-circuit connections* with duplex sonography have been reported (Lafortune et al. 1986, Stiglbauer et al. 1988, Hausegger et al. 1991, Lin et al. 1992). In a group of 40 patients with hepatocellular carcinoma, Tanaka et al. (1993) were able to recognize three of five angiographically confirmed arterioportal shunts (60%) using color flow duplex sonography.

Splenic Vein

The splenic vein can usually be displayed, at least in the segment located ventral to the abdominal aorta and relatively close to the abdominal wall. In normal individuals who were not preselected, Sabbá et al. (1990 b) registered Doppler frequency spectra from the distal segment prior to the venous confluence in 69% of cases.

In unfavorable ultrasound conditions associated with acute pancreatitis, with areas of reduced echogenicity and edema in the pancreas, it can be difficult to distinguish the vascular wall of the splenic vein. Hyperechoic thrombi can have almost the same echogenicity as the pancreas, and can therefore be more difficult to differentiate.

When evaluating *splenomegaly,* various causes mentioned earlier (cardiac or hepatogenic portal hypertension, thrombosis of the portal vein, thrombosis of the splenic vein) can be detected with duplex sonography.

Renal Vein and Intrarenal Veins

Due to the relatively high flow rate in the arterial and venous vascular systems, which comprises roughly 20–25% of the cardiac output, the renal vein can be imaged in approximately 80–90% of examinations. If there is no thrombosis of the renal vein, Doppler frequency spectra can be registered from the intrarenal veins in almost all patients.

Examinations of the renal veins are mainly restricted to the detection or exclusion of *thrombosis*. With a patent arterial vascular system, the image of a dilated venous caliber, with internal structures and an absent venous flow signal, confirms the diagnosis when appropriate arterial signals are present. It is also obligatory to examine the confluence between a thrombosed renal vein and the inferior vena cava, in order not to overlook any involvement of caval vein or miss a thrombotic protrusion originating in the renal vein and extending into the caval vein (Didier et al. 1987). Alterations in the arterial vascular system occurring in connection with thromboses of the renal veins are discussed above (Fig. 8.**20 a, b**).

The overall sensitivity of color Doppler sonography in predicting combined renal vein and inferior vena cava tumor thrombi, isolated renal vein or vena cava tumor thrombi is reported to be 95%, 100%, and 89%, respectively. The specificity in this study of 23 patients with 19 renal cell carcinomas, 4 Wilms' tumors, and one rhabdoid tumor for exclusion of renal and vena cava thrombus was about 85% (McGahan et al. 1993).

Mesenteric Veins

Examining the *superior mesenteric vein* with duplex and color flow Doppler sonography is limited to the distal segment of the vein, since smaller mesenteric veins can only be depicted in exceptional cases.

It is not always possible to image the *inferior mesenteric vein* over an extended distance. Therefore duplex-sonographic examinations are not routinely used at present.

The clinical value of conventional and color flow duplex sonography lies in the ability to identify thromboses of the *mesenteric veins* as a possible cause of acute abdomen.

In 80% of 111 patients examined, Kubale et al. (1992) were able to image the venous confluence and the 6 cm of the superior mesenteric vein adjacent to the confluence using color flow duplex sonography. They were also able to recognize isolated thromboses of the mesenteric veins.

Organ Transplants

Following organ transplants, examinations using duplex and color flow duplex sonography have become increasingly important in the postoperative evalua-

tion of morphological criteria (e.g., anastomotic region, organ swelling, hematoma, seroma) and in monitoring functional parameters (arterial and venous inflow and outflow). Although initial experience suggests that transplant monitoring is a useful field of application, the validity of duplex and color flow duplex sonographic imaging with regard to obtaining a reliable and reproducible diagnostic evaluation, however, cannot yet be conclusively determined.

In the postoperative monitoring of *kidney transplants,* procedures using duplex sonography have acquired importance for functional evaluation of the renal veins (diagnosing thromboses) (Reuther et al. 1989, Krumme et al. 1993).

With regard to *liver transplants,* conventional and color flow duplex sonography provide a *postoperative* facility for imaging the anastomoses of the inferior vena cava and the portal vein and, if necessary, evaluating the hemodynamic effects of sudden changes in the vascular caliber in relation to possible stenosis, or detecting partial or occlusive thrombi in these vessels (Barton et al. 1989). However, the method should also be used *preoperatively* in order to identify existing vascular anomalies, such as thrombi, atresia, or agenesis of the vena cava (Taylor and Burns 1985). Detecting gas bubbles in the vena cava does not appear to have any prognostic significance (Chezmar et al. 1989).

Duplex sonography, and color flow duplex sonography in particular, are able to detect vascular complications following *pancreas transplants* (thromboses, vascular stenoses, arteriovenous fistulas), as well as any sonographic indications of potential transplant complications (parenchymal irregularities, liquid around the transplant). Snider et al. (1991) report their initial experiences with color flow duplex sonography in accurately confirming the presence of five transplant thromboses in eight examined patients and excluding three with certainty.

Possible complications after *bone marrow transplantation* include the occurrence of a rejection reaction or veno-occlusive disease. In addition to causing an elevated resistance index in the hepatic artery (p. 266), *veno-occlusive disease* can be accompanied by flow changes, including retrograde flow in the portal vein (Kriegshauser et al. 1988, Brown et al. 1990). However, Herbetko et al. (1992) were not able to identify any typical Doppler sonographic findings in 19 patients with veno-occlusive disease.

9 Vasculature of the Male Genitalia

Examination

Special Equipment and Documentation

■ Continuous Wave and Pulsed Wave Doppler Sonography

The *systolic blood pressure* in the arteries of the penis can be measured using a unidirectional or bidirectional Doppler system. Due to the relatively superficial location of the arteries, transmission frequencies of 8–10 MHz are used.

Continuous wave Doppler probes with a frequency of around 8 MHz are suitable for registering Doppler analogue curves in the genital blood vessels.

In duplex equipment, pulsed wave Doppler systems that allow good spatial discrimination of the blood flow in the various closely adjacent blood vessels are used.

■ Conventional and Color Flow Duplex Sonography

With the combination of pulsed wave Doppler mode and B-mode scanners in *duplex systems,* it is possible to register flow velocities in narrowly defined vascular segments and noninvasively obtain relevant information about the arterial and venous blood supply to the penis and scrotal contents (Lue et al. 1985, Müller et al. 1990, Rajfer et al. 1990, Dauzat 1991, Schwartz et al. 1991, Broderick and Lue 1991). Due to the required penetration depth of up to approximately 4 cm, linear transducers with a B-mode transmission frequency of 7.5–10 MHz are used for this purpose. To depict the *internal iliac artery* in the area of its origin, sector or vector transducers with a B-mode transmission frequency of around 3.5 MHz are preferable when examining obese patients, while slim patients can also be evaluated with linear transducers, using a B-mode transmission frequency of around 5 MHz (Fig. 5.**22**, p. 154).

When using duplex sonography to exclude vascular impotence, representative data should be registered in the Doppler frequency spectrum from all the arteries that can be imaged, and particularly from both of the deep arteries of the penis and the dorsal veins of the penis. The color coding of the various *flow directions* should be selected in such a way that arterial flow is shown in red and venous flow appears in blue.

Flow velocity measurements carried out after injections into the spongy body of the penis must be explicitly designated as such in the documentation.

Examination Conditions

Patient and Examiner

Examinations of the penile arteries and veins and of the scrotal arteries are usually carried out with the patient in a relaxed supine position. The pampiniform plexus can also be examined while the patient is standing in order to produce more noticeable filling of the veins due to the higher hydrostatic pressure. Since patients usually find the examination embarrassing, special attention needs to be given to respecting the patient's privacy as far as possible. A brief explanation of the examination procedure must be given to relieve any unfounded fears the patient may have in connection with the initial assessment, and to encourage him to relax.

Before any planned injection into the spongy body of the penis, the patient must be informed at an early stage—and not only for legal reasons—about potential risks and side effects associated with the drug injection (damage to the urethra, infection, prolonged erection, priapism, fibrosis of the spongy body). After hospital admission, a period of at least 24 hours should be allowed for this purpose, so that the patient has an opportunity for comprehensive discussion in a relaxed atmosphere concerning any questions and fears that may arise.

Conducting the Examination

Vasculature of the Penis

Systolic Blood Pressure Measurement

Using a small, narrow (blood pressure) cuff, which is loosely attached around the base of the penile shaft, and after the penile arteries have been located with Doppler sonography, arterial inflow is interrupted by suprasystolic vascular compression. As the compression is slowly released, the Doppler equipment is used to record the point at which the signal reappears, i.e., when the arterial blood pressure in the selected penile artery exceeds the occluding pressure, allowing arterial flow to be detected again. The value read from the blood pressure equipment is used to calculate a relation to the systemic systolic blood pressure, measured at a (nonstenotic) arm artery. The calculated value is termed the penobrachial index (PBI).

Continuous-Wave Doppler Sonography

The penile *arteries* are first localized at the base of the nonerect shaft of the penis. The superficial penile arteries here have to be distinguished from the deeper penile arteries. For the deep arteries of the penis, the probe is placed along both sides of the penile shaft. Contact can be made with the dorsal arteries of the penis on the dorsal side of the organ. They can be followed by placing the probe at an angle of approximately 45° to the suspected course, and then locating and optimizing the arterial signal. Differentiation is also possible by applying light pressure with the probe, which allows the flow velocity in the dorsal arteries of the penis to decrease without influencing the deep arteries of the penis, which are enclosed by the cavernous body.

Registering the flow in the deep and superficial dorsal *veins* of the penis using Doppler sonography is best carried out after stimulation of the arterial inflow just prior to maximum penile rigidity. Special care should be taken to ensure that the probe is guided without applying any firm contact pressure in order to avoid compressing the vein.

Conventional and Color Flow Duplex Sonography

B-mode. In the B-mode image, the individual *spongy bodies* should be depicted in longitudinal section and cross-section (Fig. 9.**1**). Morphological differentiation between the intracavernous structures *(penile septum, urethra, arterial walls, veins)* provides the orientation required to carry out flow measurements in a targeted fashion.

Doppler and color flow Doppler mode. When examining the *penile arteries* with duplex sonography, measurements of the various flow velocity parameters are made using *longitudinal* penile sections. Depending on the arterial segment being evaluated, the transducer can be positioned on the penis from a dorsal (Fig. 9.**2a**), ventral, or lateral (Fig. 9.**2b**) direction. Dorsal and ventral longitudinal sections in particular often allow the imaging of arterial sections that are located quite distantly in a proximal direction, and lateral sections allow simultaneous registration of both deep arteries of the penis. In a transverse sectional plane, simultaneous imaging of both deep arteries of the penis and the dorsal arteries of the penis is possible (Figs. 9.**3**, 9.**4**).

In pathological findings, a complete examination of the vasculature also includes imaging of the proximal segments of both internal iliac arteries, so as not to overlook stenoses or occlusions at this predilection point (the area of origin). The evaluation of these arteries is described on page 147 (Figs. 5.**15,** p. 147, 5.**22,** p. 154).

Color flow Doppler sonography and the *energy-mode* (Fig. 9.**2**) make it possible to locate the vessels quickly and determine the arterial course and flow direction, so that angle-corrected flow velocity measurements can be made. Color flow Doppler sonography can also exclude stenoses or occlusions within the intracavernous segments of the penile arteries.

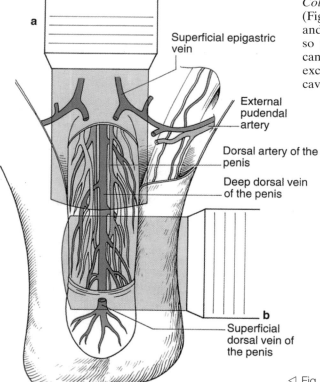

a

Superficial epigastric vein

External pudendal artery

Dorsal artery of the penis

Deep dorsal vein of the penis

b

Superficial dorsal vein of the penis

◁ Fig. 9.**1**   The penile vasculature. Examples of transducer positions: a = cross-section, b = longitudinal section

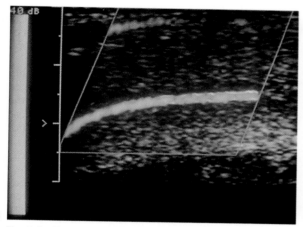

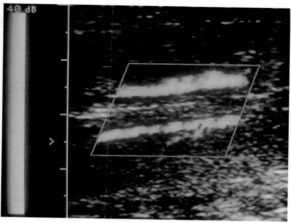

a

b

Fig. 9.**2** Energy mode image of the deep penile arteries from a dorsal direction (**a**) and in a lateral view. In the sagittal plane of the section, the septum of the penis can be differ- entiated from the cavernous bodies. The deep arteries of the penis lie very close together in this plane

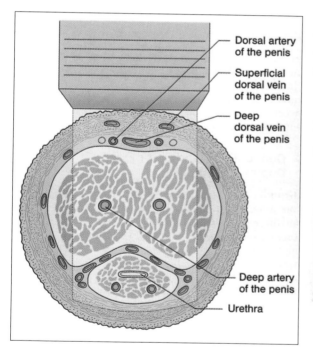

Fig. 9.**3** The penis in cross-section with the cavernous bodies, spongy body, urethra, deep and dorsal arteries of the penis, and superficial and deep dorsal veins of the penis

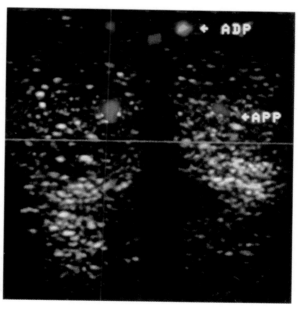

Fig. 9.**4** Color flow duplex sonogram from the middle third of the penis, in a dorsal (cross-sectional) orientation, show- ing the paired dorsal arteries of the penis near the trans- ducer (ADP) and the deep arteries of the penis in the cavernous bodies (APP)

Spongy Body Injections

Injecting vasoactive substances into the cavernous bodies causes a relaxation of the smooth muscle cells in the spongy bodies and their arteries. The resulting decrease in the peripheral resistance causes an in- crease in the arterial inflow and, after some 5–30 minutes, a consequent pharmacologically induced erection which lasts for approximately 30–120 minutes.

Vasoactive substances that are used include 12.5– 60.0 mg *papaverine* in 1–5 ml of 0.9% NaCl solution (Wespes et al. 1987, Shabsigh et al. 1990, Chiang et al. 1991, Meuleman et al. 1992, Levine and Coogan 1996) and 10–20 μg *alprostadil* (Broderick et al. 1992), possibly in combination with an α_1-antagonist (e.g. phentolamine) (Stief and Wetterauer 1988, Brandstet- ter et al. 1993, Mills and Sethia 1996).

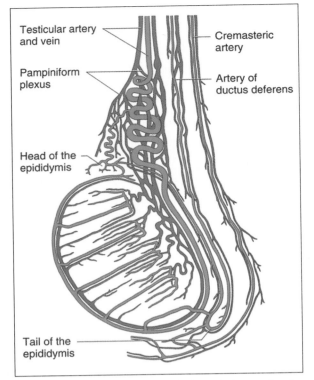

Fig. 9.**5** Schematic overview of the arteries and veins contributing to the blood supply of the testicles and epididymis

It should be explicitly stated here that *alprostadil is not currently approved for intracavernous application in many countries.* Due to this problematic legal situation, a necessary precondition for using this form of pharmacological stimulation is obtaining the patient's consent after providing detailed and comprehensive information about the procedure itself (Schroeder-Printzen et al. 1992). In addition, it goes without saying that the indications for using these preparations must be strictly followed, and that any known contraindications and side effects should be given due consideration during treatment.

After first placing a tourniquet at the base of the penile shaft for approximately two minutes, in order to prevent dilution of the drug, and after thorough disinfection of the intended injection point, the spongy body injection is administered with a thin needle inserted *laterally* into one or both spongy bodies. In addition to the need to avoid subcutaneous injection, with the accompanying danger of skin necrosis, damage to the urethra must be particularly avoided.

Providing additional visual, acoustic, or tactile stimuli can encourage erection (Lee et al. 1993, Shabsigh et al. 1990, Montorsi et al. 1996).

After the intracavernous injection, flow velocity measurements and vascular caliber measurements should be made with duplex sonography in the individual penile blood vessels during the various stages of tumescence. The largest increase in blood inflow is observed during the individually varying period until the start of the intumescent phase.

Physical Exercise

In patients with stenoses or occlusions in the region of the large pelvic arteries, a redistribution of the blood flow via the collateral circulation may cause a decrease in the penile circulation, due to a steal phenomenon. This redistribution to the disadvantage of the penile arteries can be provoked by exercising the leg muscles, which causes an accompanying decrease in the peripheral resistance in the vascular system of the extremities.

To verify this steal phenomenon, the patient can be asked to perform 10–20 knee bends after the pharmacologically induced maximum erection has been reached. A second assessment of the extent of erection is then made, using clinical and Doppler-sonographic criteria to estimate the effect of the leg muscle exercise on the penile blood flow.

Vasculature of the Scrotum

■ Continuous Wave Doppler Sonography

Although it is generally possible to identify the testicular artery with Doppler sonography, clear anatomical classification of arterial signals from the scrotum is uncertain because it is not possible to distinguish between intratesticular and extratesticular arteries without using the appropriate B-mode image.

■ Conventional and Color Flow Duplex Sonography

B-mode. The scrotal contents are readily accessible to sonographic examination. The testicular parenchyma and the spermatic cord should be completely imaged in longitudinal sections and cross-sections in order to depict the arteries and veins in various planes and measure the venous caliber. Attention should be given to nonhomogeneous echoes in the testicular parenchyma, changes in the epididymis, and dilatation of the serous sheath covering the testicles and epididymis.

Doppler and color flow Doppler mode. Examination of the scrotal contents should start on the healthy side in order to provide individual comparisons if there is any pathology and to allow optimal adjustment of the equipment (scales, filter, amplifications).

Along the course of the spermatic cord, cross-sections should be used to assess any ectasias of the pampiniform plexus and evaluate dilatation during the Valsalva maneuver.

When using color flow Doppler sonography to depict the intratesticular and extratesticular vasculature (Dauzat 1991, Coley et al. 1996) and evaluate the arterial and venous blood flow in the testicle and epididymis, precise anatomical classification of the flow signals is important to allow differentiation between inflammation and ischemia in cases of acute and painful scrotal swelling.

Sequence of the Examination

The sequence of the examination is determined by the indication for examining the flow dynamics in the vasculature of the male genitalia. In men with erectile dysfunction, duplex sonography can be used to clarify a vascular cause or element in the erectile dysfunction by examining the *penile vasculature*. In these patients, flow velocity measurements should always be carried out after physical rest initially (no climbing of stairs before the examination). Examinations after intra-cavernous drug injection, and also after exercise, can be carried out later.

In patients with an *acute scrotum* or swelling with no pain, the vascular examination can concentrate on the scrotal blood vessels. If a left-sided varicocele is detected, a search should be carried out for a flow obstruction in the ipsilateral renal vein, using duplex sonography.

Normal Findings

Principle

The indications for examining the *penile arteries* and *veins* include evaluating possible involvement of the vasculature after traumatic damage to the penis (fracture, hematoma) and clarifying a vascular cause of erectile dysfunction, which may be due to pathologically reduced inflow or increased outflow, during a provoked erection.

At the beginning of an *erection*, psychological and reflex stimuli from parasympathetic and sympathetic neural fibers result in a relaxation of the smooth muscles of the spongy body, and a *dilatation* of the arteries in the cavernous bodies and spongy body of the urethra. This causes increased arterial inflow into the penis and increased filling of the venous sinus with blood. Due to the increasing pressure within the completely filling cavernous structures, which can reach suprasystolic values at full rigidity, the veins are passively compressed at their passage through the fibrous tunic of the penile spongy body, reducing the venous outflow. In addition, closure of the extra-cavernous and particularly the intracavernous arterio-venous shunts, through which blood flows in the resting state, causes a further significant reduction in the outflow (Wagner 1984).

In the spongy body, *arteriovenous connections* remain open even during erection, so that the tumescence of the spongy body increases, although it does not reach the same rigidity as the cavernous bodies.

An ancillary function contributing to the intra-cavernous pressure elevation is attributed to an increase in tone in the bulbocavernous and ischio-cavernous muscles, which can probably also cause a blockage of the venous drainage (Porst 1987).

■ Continuous Wave and Pulsed Wave Doppler Sonography

With the help of the arterial signals that are registered and a small (blood pressure) cuff, it is possible to calculate the *penobrachial index* (PBI), which represents the quotient formed from the occluding pressure values in the penile arteries and a brachial artery. The hemodynamic principle is the same as that used in pressure measurements at the ankle arteries, described earlier (pp. 137–145).

In addition, arterial Doppler signals can be qualitatively evaluated in relation to their acoustic intensity, and their velocity waveform can be evaluated both by formal analysis and also quantitatively.

■ Conventional and Color Flow Duplex Sonography

B-mode. In the context of diagnosing *vascular impotence*, the B-mode image primarily serves to allow precise and separate placement of the pulsed wave Doppler mode in the individual blood vessels, so that Doppler frequency spectra can be registered and quantitatively evaluated in a targeted fashion.

In examinations of the *testicles* and *spermatic cord*, the narrow-caliber blood vessels are seen as hypoechoic tracts in B-mode sonographic imaging. Using the Doppler procedure here allows the detection of blood flow in these structures, so that they can be differentiated from cystic or paravascular fluid accumulations.

Doppler mode. Using Doppler frequency spectral analysis allows noninvasive flow velocity measurements in the genital vasculature. In the *penile arteries*, in addition to direct examination of the intra-cavernous arterial segments, indirect assessment of the preceding arterial vascular system is possible when measuring the velocity in vascular segments located further proximally (Porst 1987). During stimulated erection, the quantitative measurements that are possible are more or less limited to the following:

- Maximum systolic Doppler frequency (flow velocity)
- Maximum end-diastolic flow velocity
- Absolute vascular caliber
- Relative increase in caliber

The alteration in the intracavernous pressure relationships that appears during erection also influences the wave form in the Doppler frequency spectra, due to a change in the peripheral resistance and sampling location (Schwartz et al. 1989, Fitzgerald and Foley 1991, Kim et al. 1994) (see p. 314).

Color flow Doppler mode. Under physiological conditions, or at least after the injection of a vasoactive substance into the spongy body, the imaging of the *penile arteries* with color flow Doppler sonography shows pronounced flow in all four of the blood vessels that supply the spongy bodies (Fig. 9.**3**). This can be followed in color-coded form as far as the distal section of the penis. It is helpful to adjust the color coding of the flow velocities (pulse repetition rate) if the arterial color is saturated and intense without the presence of any aliasing (Fig. 9.**6**).

The uniform color signal provided by this adjustment of the equipment allows the exclusion of vascular stenoses and occlusions in the intracavernous segment of the penis.

Due to its very slow flow velocity, color flow Doppler signals from the *dorsal vein of the penis* cannot usually be registered in the nonerect state. The flow dynamics in this vein are of interest when differentiating an erectile dysfunction due to arterial causes from one that is venous in nature (Fig. 9.**12**).

In the *scrotal vasculature,* it is important to detect intratesticular arterial (and venous) flow in order to exclude torsion (absent flow) or inflammation (hyperemia). The Valsalva maneuver can be used to exclude a varicocele.

Anatomy and Findings

Vasculature of the Penis

Anatomy. The *internal pudendal artery,* the most caudally located visceral branch of the internal iliac artery, leaves the pelvis through the infrapiriform foramen, continues around the spine of the ischium outside the small pelvis, and proceeds through the lesser ischiadic foramen to reach the ischiorectal

fossa. Covered by the obturator fascia (Alcock's canal), the artery divides at the posterior margin of the urogenital diaphragm to form the perineal artery and the penile artery. The *penile artery* proceeds along the inferior ramus of the pubic bone in a ventral direction, and divides into two terminal branches. The *deep artery of the penis* moves from a medial to a ventral direction in the spongy body, while giving off side branches (Fig. 9.**6**), and forms anastomoses with the contralateral artery and the *dorsal artery of the penis.* The latter continues beneath the transverse perineal ligament to the dorsal side of the penis, and stretches into the balanus in the form of a helicine artery between the fibroelastic sheath and the deep fascia of the penis. Blood flow into the cavernous bodies is predominantly via the deep arteries of the penis (Fig. 9.**1**).

Anatomical variants of the superficial (singular or multiple dorsal arteries of the penis) and deep penile arteries (absent or multiple manifestations of the deep artery of the penis) are to be expected.

Venous drainage from the balanus, the spongy body of the urethra, and the cavernous bodies takes place through the *deep dorsal vein of the penis,* which takes up the circumflex veins and flows via the venous plexus and the internal iliac vein into the caval vein; and also through the *superficial dorsal vein of the penis,* from which blood flows into the great saphenous vein via the external pudendal veins. In addition, three or four *deep veins of the penis* drain each of the cavernous bodies via the vesicoprostatic plexus.

■ Continuous Wave Doppler Sonography

In healthy males with normal erectile function, the systolic penile blood pressure should at most be 40 mmHg lower than the systolic brachial pressure, and the *penobrachial index* (PBI) should be more than 0.66–0.8, and should not decrease during physical exercise (Jetvich 1984, Flanigan et al. 1985, Egger et al. 1988, Schoop 1988).

When the penis is flaccid, the Doppler signal obtained from the penile arteries corresponds to that registered from a vascular system with high peripheral resistance. A brief period of systolic forward flow with a steep increase and decrease in the flow velocity is followed by a short backflow component. No or only a very slow end-diastolic flow can be detected (Fig. 9.**7**).

In addition to vacuum constriction (Broderick et al. 1992), another possible *provocation maneuver* is intracavernous injection of vasoactive substances. In addition to their use in determining the degree of tumescence and the "erectile angle" (Wespes et al. 1987), the effects of vasoactive substances on the penile blood flow can be quantified from the elevation in the flow velocity within the deep and dorsal arteries of the penis, with an increase in the diastolic forward flow. The physiological reaction to this stimulation test is described in detail in connection with the duplex-sonographic diagnosis, below.

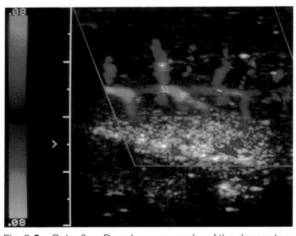

Fig. 9.**6** Color flow Doppler sonography of the deep artery of the penis in longitudinal section showing the origins of smaller branch arteries that lead into the cavernous body. The flow direction in the arteries flowing toward the transducer is coded blue; the flow away from the probe is coded red

Fig. 9.**7** An analogue curve recording penile artery flow in a flaccid penis. During systole, a steep rise and fall in the flow velocity is seen. In the diastole, there is only a slow, peripherally directed flow

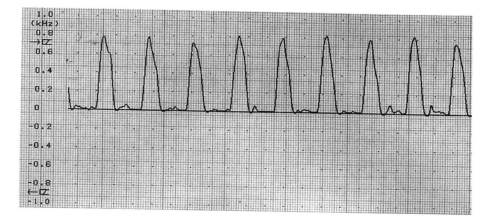

■ Conventional and Color Flow Duplex Sonography

B-mode. Sonographically, the paired cavernous bodies of the *penis,* which have a round to oval shape in a transverse section, present a homogeneous echo pattern of average density (Fig. 9.**8**). The more hyperechoic penile septum divides the two spongy bodies, which are covered by the hyperechoic tunica albuginea. The spongy body of the urethra, with the balanus, covers the urethra, and is imaged homogeneously in a slightly more hypoechoic form than the cavernous bodies.

Sonographic imaging of the penile arteries and veins in longitudinal section over an extended distance is often possible. The arteries can be recognized by their more hyperechoic, pulsatile vascular wall (Fig. 9.**9 a**), so that evidence of the mural structures in this region is also available.

Doppler and color flow Doppler mode. In the unstimulated state, the flow velocity in the deep artery of the penis is approximately 4–20 cm/s (up to 43 cm/s), with a maximal cross-sectional diameter of 0.3–0.9 mm. The velocity in the dorsal artery of the penis is approximately 10–20 cm/s (Shabsigh et al. 1989, Chiang et al. 1991, Meuleman et al. 1992). In physiological conditions, the paired arteries only show slight side-to-side differences in maximum systolic velocity and vascular caliber. In a series of measurements, Benson and Vickers (1989) found that the average deviation from the maximum systolic velocity amounted to 4 ± 3 cm/s in the deep arteries of the penis (Table 9.**1**).

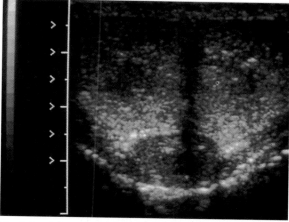

Fig. 9.**8** A B-mode image of the penis from a dorsal orientation. Both corpora cavernosa and the spongy body of the penis can be differentiated

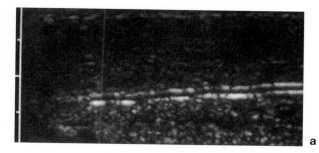

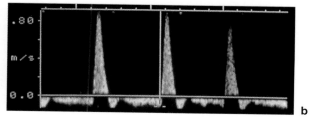

Fig. 9.**9** **a** A B-mode longitudinal section of the middle third of the right cavernous body, from a dorsal direction. The right deep artery of the penis is seen in the middle of the spongy body, recognizable by its hyperechoic mural structure

b A Doppler frequency spectrum from the deep artery of the penis after administering 10 µg of alprostadil, showing a steep systolic rise and fall in the flow velocity and retrograde diastolic flow. The spectrum was recorded at the time of maximum erection and indicates a very high peripheral resistance

Table 9.**1** Normal values for duplex-sonographic measurements in the deep artery of the penis and dorsal artery of the penis before and after intracavernous injection of 12.5–60.0 mg papaverine. The maximum systolic flow velocity (V_{max}) and maximum cross-sectional diameter are given according to values in the literature (mean ± SD)

Authors	Vessels (n)	Before papaverine		After papaverine	
		V_{max} (cm/s)	Diameter (mm)	V_{max} (cm/s)	Diameter (mm)
Deep artery of the penis					
Benson u. Vickers (1989)	30	–	–	47 ± 15	–
Shabsigh et al. (1989)	14–80	13 ± 4	0.5 ± 0.2	34 ± 14	0.9 ± 0.2
Chiang et al. (1991)	32	7.7 ± 3.9	0.4 ± 0.1	43.5 ± 13.1	0.7 ± 0.2
Pickard et al. (1991)	12	13.3 ± 7.2	–	32.2 ± 17.3	–
Meuleman et al. (1992)	31	21 ± 9	–	45 ± 23	–
Brandstetter et al. (1993)	22	–	–	35 ± 13	–
Dorsal artery of the penis					
Shabsigh et al. (1989)	55–76	20 ± 10	–	52 ± 25	–

Spongy Body Injection

During the first ten minutes after the injection of vasoactive substances into the spongy bodies, a reduction in the peripheral resistance of the penile arteries is observed, together with an elevation in the maximum systolic flow velocity in the *deep artery of the penis* to a value of 20–70 cm/s, and an increase in the average vascular diameter by around 20–186% to a value of 0.7–1.2 mm (Shabsigh et al. 1989, Schwartz et al. 1990). A maximum systolic flow velocity of at least 25–30 cm/s (Fig. 9.**10**), with an increase in the vascular diameter of around 75%, is considered to be physiological (Lee et al. 1993, Valji and Bookstein 1993) (Table 9.**1**). Cormio et al. (1996a) describe a peak systolic velocity of 25–35 cm/s as borderline and use an even higher cutoff volume of 35 cm/s to define arterial

sufficiency. The resistance index decreases from a value of 0.97 ± 0.08 before the injection into the spongy body to values of around 0.73 ± 0.06 in the first five minutes after the injection. On maximum erection, it increases again to values of around 0.99 ± 0.05 (Meuleman et al. 1992).

The *dorsal artery of the penis* reaches a maximum systolic flow velocity of 30–100 cm/s (Fig. 9.**11**), and the vascular diameter is approximately 1.8 ± 0.04 mm (Lue et al. 1987, Robinson et al. 1989, Schwartz et al. 1990, Chiang et al. 1991, Lee et al. 1992).

With increasing penile tumescence and rigidity, the Doppler frequency spectra from the deep artery of the penis alter (Schwartz et al. 1989). When erection begins, the systolic and end-diastolic peak velocity increases, with an accompanying intracavernous pres-

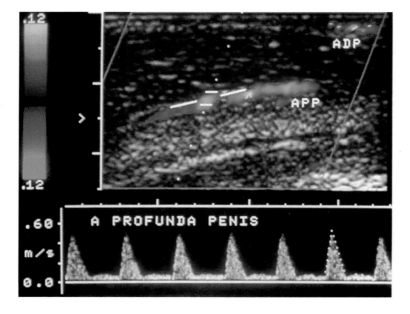

Fig. 9.**10** The (right) deep artery of the penis (APP) in longitudinal section, 18 min after an intracavernous injection of 20 µg alprostadil. The maximum systolic and end-diastolic flow velocities, based on the Doppler frequency spectrum, are 51 cm/s and 7 cm/s, respectively. A section of the right dorsal artery of the penis (ADP) can be seen at the top right

sure of around 11–25 mmHg. The waveform is monophasic. With increasing intracavernous pressure (40 mmHg), the end-diastolic velocity decreases, with constant systolic flow. After an additional pressure increase (63–83 mmHg), the end-diastolic velocity initially decreases to values that can no longer be measured, and then begins to move in a retrograde direction (Fig. 9.**9b**). Up to a pressure of approximately 106 mmHg in the cavernous bodies, the systolic flow velocity remains elevated, and decreases when a rigid erection is attained. During the erection, the waveform shows an increasingly steep ascending systolic velocity, with a simultaneous steep drop.

After *physical exercise* (ten knee bends), the rigidity and flow velocity in the penile arteries are largely unaffected (Shabsigh et al. 1989).

The *deep dorsal vein* and *superficial dorsal vein of the penis* have a slow flow velocity. The Valsalva maneuver causes a cessation in the venous flow (Fig. 9.**12**).

Vasculature of the Scrotum

Anatomy. The *testicular artery* (internal spermatic artery) originates from the abdominal aorta in paired form, and is surrounded by the venous pampiniform plexus in its winding course within the spermatic cord. After its entry into the white fibrous sheath (tunica albuginea), it divides into branches that initially course around the testicles (capsular arteries) and then split into subsidiary branches, which pass centripetally into the parenchyma. Within the spermatic cord, arteries of smaller caliber are found: the *artery of the ductus deferens* originates from the inferior vesical artery, and the *cremasteric artery* originates from the inferior epigastric artery. These arteries supply the ductus deferens, epididymis, peritesticular sheaths, and cremaster muscle.

The veins of the testicles gather at the hilum to form the venous network known as the *pampiniform plexus* along the spermatic cord, and flow into the inferior vena cava as the *right testicular vein,* and into the left renal vein as the *left testicular vein.*

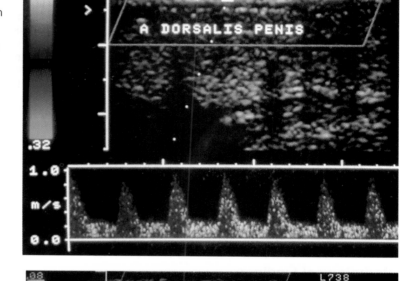

Fig. 9.**11** A color flow duplex sonogram showing the right dorsal artery of the penis in the middle third, from a dorsal direction in a longitudinal section (same patient as in Fig. 9.**10**), with elevated blood flow. Fifteen minutes after the injection of 20 μg of alprostadil into the spongy bodies, the maximum systolic and end-diastolic flow velocities were 93 cm/s and 26 cm/s, respectively

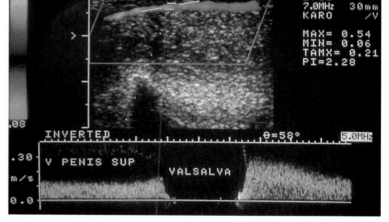

Fig. 9.**12** Deep dorsal vein of the penis from a dorsal orientation, together with the Doppler frequency spectrum. During the Valsalva maneuver, venous flow ceases. After the maneuver, with a subsequent decrease in the intra-abdominal pressure, a short and slight elevation in the flow velocity directed toward the heart is observed

■ Continuous Wave Doppler Sonography

When an erectile dysfunction has *arterial causes,* bidirectional Doppler sonography, particularly after intracavernous injection of vasoactive substances, facilitates the detection of a reduced maximum systolic flow

However, there is more often an obstruction in the preceding pelvic vascular system (common or internal iliac artery), so that using color flow Doppler sonography in the region of the penile arteries and veins is mainly useful for quick identification of the vessels. An analysis of the flow velocity waveform provides the basis for further diagnostic assessments (Krysiewicz and Mellinger 1989, Quam et al. 1989).

■ Continuous Wave Doppler Sonography

As described above (p. 308), continuous wave Doppler sonography is not used for the detection of intrascrotal and intratesticular flow, because of the lack of orientation in the scrotum without visual control.

■ Continuous Wave Doppler Sonography

Doppler sonography can successfully detect the blood flow in the vasculature of the penis. An initial assessment of the arterial blood flow in the arteries of the

---

Findings

Vasculature of the Penis

■ Continuous Wave Doppler Sonography

The effect of intracavernous drug injection and the corresponding reactions in the various parameters measured, as described below, is decisive in the diagnosis of the vascular causes of impotence and in differentiating between arterial and venous causes of erectile dysfunction.

In the presence of an *obstruction* in the *arterial vascular system* (stenosis, occlusion, hypoplasia), the arterial signals in the penile vasculature show a decreased amplitude, along with a delayed rise and fall in the Doppler curves. The end-diastolic component is clearly reduced, or is completely absent. Sometimes no flow whatsoever can be detected in the deep arteries of the penis. The penobrachial pressure index is below 0.5–0.6.

Venous insufficiency is consistent with a strong arterial signal from both deep arteries of the penis, an absent erection after an injection into the spongy body, and the detection of elevated continuous flow in the deep or superficial dorsal vein of the penis.

■ Conventional and Color Flow Duplex Sonography

Spongy Body Injection

In the presence of a functionally effective obstruction of the arterial vascular system or venous insufficiency, intracavernous injection of *papaverine* or *alprostadil* does not cause adequate erection of the spongy body.

An obstruction of the *arterial vascular system* can be assumed if, after pharmacological stimulation:

– There is no erection, or only inadequate erection
– Only weak (acoustic) signals from the penile vasculature can be registered
– The maximum systolic flow velocity in the deep arteries of the penis is less than 25 cm/s (Fig. 9.**13**)
– The arterial caliber only increases inadequately (<60%) in comparison to the resting state

After an injection into the spongy body, indications of *venous insufficiency* exist when there is:

– No erection, or inadequate erection
– High systolic flow velocities in the penile arteries
– A clear increase in the caliber of the penile arteries (>75%)
– End-diastolic flow velocity in the deep artery of the penis >5 cm/s
– Continuous high flow in the dorsal penile veins

Postoperative Monitoring

In order to evaluate the success of various revascularization operations, postoperative evaluation of the patency of *penile vascular anastomoses* is important. It is possible to distinguish between: side-to-side anastomosis of the dorsal artery of the penis and the deep dorsal vein, with a simultaneous end-to-side anastomosis of the inferior epigastric artery and the dorsal vasculature (Hauri's operation); end-to-side anastomosis of the inferior epigastric artery and the deep dorsal vein (Virag's operation); and an interposed venous graft be-

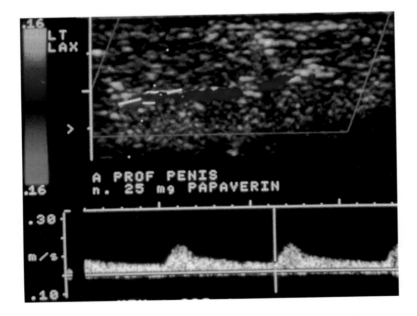

Fig. 9.**13** A color flow duplex sonogram of the middle segment of the left deep artery of the penis, from a dorsal direction (longitudinal section). After an intracavernous injection of 25 mg papaverine, there is an insufficient increase in the arterial inflow due to an occlusion of both internal iliac arteries. The maximum systolic and end-diastolic velocities were only 16 cm/s and 4 cm/s, respectively

Fig. 9.**14**   The testicle in cross-section with a color flow Doppler sonography of an artery in the parenchyma. The spectrum shows the low resistance in the artery with low systolic (12 cm/s) and end-diastolic (4 cm/s) flow velocities

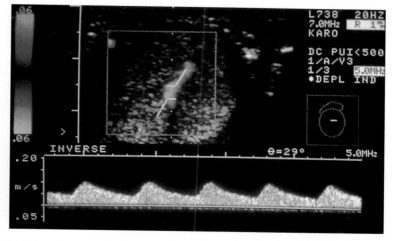

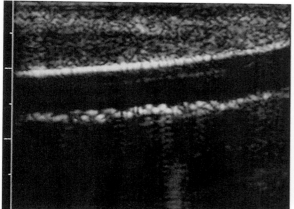

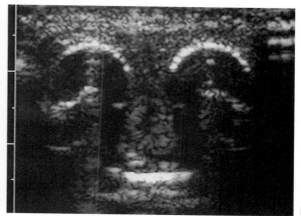

a   b

Fig. 9.**15**   B-mode sonographic longitudinal (**a**) and cross-section (**b**) of the penis after the implantation of a penile prosthesis in the cavernous bodies

tween the femoral artery and the deep or dorsal artery of the penis (Crespo's operation; overview in Porst 1987). Using color flow duplex sonography, the anastomotic region can be visualized without any additional manipulation (spongy body injection), and morphological assessments the success of the operation can be made (Gehl et al. 1990). B-mode imaging can be used to check the placement and postoperative status after implantation of penis prostheses (Fig. 9.**15 a, b**).

Vasculature of the Scrotum

■ Conventional and Color Flow Duplex Sonography

Varicocele

Vascular convolutions in the region of the spermatic cord are signs of a varicocele (Braedel et al. 1991). Due to the clearer filling of the anechoic, winding venous segments, better imaging is possible with the patient in a standing position (Fig. 9.**16**). When the intra-abdominal pressure is increased during the Valsalva maneuver, a sudden retrograde flow is seen in

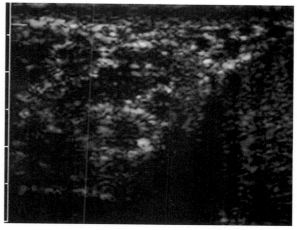

Fig. 9.**16**   Testicular varicocele on the left side: in a longitudinal section of the spermatic cord, sonography shows multiple oval hypoechoic structures indicating dilated veins

these ectatic veins, which can be registered with either color flow Doppler sonography (Fig. 9.**17**) or pulsed wave Doppler mode (Fobbe and Wolf 1988, Fitzgerald and Foley 1991).

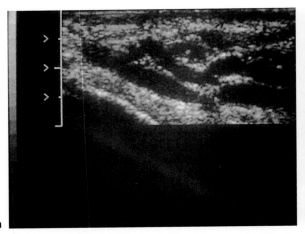

a

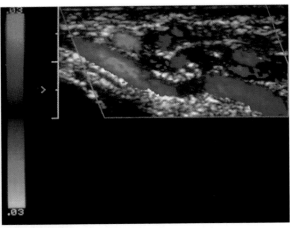

b

Fig. 9.**17**   B-mode (**a**) and color flow sonography (**b**) of a patient with a varicocele on the left side. In B-mode, the clearly dilated veins of the pampiniform plexus can be rec-ognized. During the Valsalva maneuver, it is possible to distinguish reflux in the dilated venous convolutions with color flow Doppler sonography (**b**)

Acute Scrotum

On the basis of the flow relationships in the testicular region, it is possible to distinguish between *testicular torsion* and *epididymitis* or *orchitis* in the acute scrotum. In the mechanical constriction of the spermatic cord vasculature that occurs in torsion, no arterial flow signals are detectable in the testicular parenchyma. By contrast, inflammatory hyperemia causes an increased flow velocity, which is well documented with color flow duplex sonography; when seen in conjunction with a possible reactive hydrocele, reduced echogenicity in the testicle, or an enlarged epididymis (epididymitis), this increased flow velocity suggests the correct diagnosis (Jensen et al. 1990, Lerner et al. 1990, Ralls et al. 1990, Horstman et al. 1991).

Evaluation

Continuous Wave Doppler Sonography

Since the unknown incident angle of the Doppler beam to the vasculature means that quantitative velocity measurements are not possible, the registered Doppler signals can only be evaluated qualitatively in terms of signal intensity (strong, weak, no signal), steepness of the systolic rise, backflow component, and end-diastolic flow.

A penobrachial pressure index below 0.5–0.6 indicates an increased reduction in the pressure due to vascular system obstructions in the pelvic or penile arteries, although it does not allow the level of the obstruction to be identified.

Conventional and Color Flow Duplex Sonography

If an intracavernous injection of vasoactive drugs does not result in adequate erection, and only a slight increase in the caliber of the penile arteries, accompanied by a low maximum systolic flow velocity, can be registered with duplex sonography, then there is an *inadequate arterial blood supply* to the penis.

The indications of a *steal phenomenon* are:

- Moderate to full erection during physical resting, plus:
- Loss of rigidity during or shortly after leg muscle exercise, plus:
- Initially sufficient increase in caliber, plus:
- Normal maximum systolic flow velocity in the penile arteries

Venous insufficiency in the spongy body is consistent with the following: an incomplete erection following an injection into the spongy body, inadequate rigidity of the penis after a maximum increase in vascular caliber and a high peak velocity (Müller et al. 1989, Shabsigh et al. 1989), and also an end-diastolic flow velocity higher than 5 cm/s in the deep arteries of the penis (Quam et al. 1989).

In the scrotal region, it is diagnostically important with regard to the testicular parenchyma to differentiate between ischemia, normal arterial blood flow, and hyperemia.

Sources of Error and Diagnostic Effectiveness

Sources of Error

Identification of the individual penile arteries and veins exclusively using continuous wave Doppler sonography involves a degree of uncertainty because it is not always possible to distinguish between the superficial and deep penile blood vessels.

Caliber measurements of the relatively thin penile arteries, which have a diameter of between 0.3 mm and 1.1 mm, cannot always be carried out without difficulty, due to the technically pre-set resolution of the equipment used. In addition, the diameters of the intracavernous arteries during erection are subject to physiological pressure fluctuations, which complicate the reproducibility of measurements and their correlation to the brief erectile phase (Meuleman et al. 1992).

When examining the dorsal veins of the penis, and especially the superficial dorsal vein of the penis, care must be taken not to *compress* the blood vessel *artificially* by applying too much pressure with the probe, as this can produce a flow velocity measurement that is too high.

In unfavorable examination conditions (poor delimitation of the arterial wall, absent erection), it can sometimes be difficult to determine the *incident angle* of the Doppler beam in order to calculate the absolute flow velocities.

When registering flow signals from the region of the testicular parenchyma, frequency-adapted adjustment, particularly of the color parameters, is very important. A high *scale adjustment* (high pulse repetition frequency) or the use of color coding that is too low (insufficient gain) may result in slow velocities being missed.

Diagnostic Effectiveness

Vasculature of the Penis

■ Continuous Wave Doppler Sonography

In comparison to selective angiograms, it has been claimed that the penile arteries can be detected in 95% of cases, and that the dorsal arteries of the penis can be detected in 100% of cases (Gall and Holzki 1991). However, it may be difficult to identify the paired arteries.

According to von Egger et al. (1988), a penobrachial pressure index higher than 0.66 indicates an arterial obstruction in the region of the afferent pelvic vascular system, so that an arterial erectile dysfunction is unlikely. However, these findings have not been reproduced by other authors, who have only described a correlation between the penobrachial index and results obtained in selective pelvic angiograms in 39% of cases (Robinson et al. 1989, von Wallenberg Pachaly

et al. 1989). Since penobrachial indices of between 0.6 and 0.8 have been measured in both potent and impotent men (Schwartz et al. 1990), sole use of this index is associated with diagnostic uncertainty. Values higher than 0.8 suggest that an insufficient arterial blood supply is unlikely (Schoop 1988).

When evaluating the reaction to an injection into the spongy body, a normal erection and physiological Doppler signals can exclude an obstruction of the arterial vascular system with a specificity of 95% (Porst 1987).

■ Conventional and Color Flow Duplex Sonography

In patients with an *erectile dysfunction,* examinations using duplex and color flow duplex sonography can determine whether the cause of the erectile disturbance is arterial or venous, and thus decide the further diagnostic course of action (selective intra-arterial angiography or cavernography) or required treatment (Shabsigh et al. 1989, Chiang et al. 1991, Pickard et al. 1991, DeWire 1996).

In comparison to selective angiography, it is possible to recognize pathological changes in the penile arteries using color flow duplex sonography with a *sensitivity* of 82.4–83.3% (Chiang et al. 1991, Brandstetter et al. 1993). Chiang arrived at a correlation of the results with intra-arterial vascular images as high as 87.5%. This corresponds to the results obtained by von Wallenberg Pachaly's group (1989), which showed correspondence in 91% of the cases in 23 patients when the results obtained with duplex sonography and angiography were compared.

An obstruction of the vascular system in the region of the afferent pelvic and penile arteries can be excluded with a *specificity* of 87.5–100% using color flow duplex sonography (Chiang et al. 1991, Rosen et al. 1991, Benson et al. 1993, Brandstetter et al. 1993).

When all of the relevant criteria are taken into account, *venous insufficiency* can be recognized with a *sensitivity* of approximately 80–100% using color flow duplex sonography. However, it can only be excluded using duplex sonography with a *specificity* of 55–70% (Quam et al. 1989, Fitzgerald and Foley 1991). In a study with 82 men for the diagnosis of venous dysfunction, color Doppler sonography compared to cavernosometry had a *sensitivity* and *specificity* of 100% and 66.6% respectively, and an accuracy rate of 91% with a positive predictive value of 0.89 (Karadeniz et al. 1995).

The relatively low specificity is because a pathological increase in the venous flow in the deep and/or superficial dorsal veins of the penis cannot always be registered, while the deep veins of the penis may also be responsible for an increased outflow. Moreover, it is not always possible to detect arteriovenous shunts, particularly in the region of the penile root.

Vasculature of the Scrotum

■ Conventional and Color Flow Duplex Sonography

It is generally possible to image the scrotal vasculature and intratesticular arteries, due to their superficial location (Middleton et al. 1989). At present, diagnoses involving an assessment of the blood supply in the testicles are predominantly still restricted to qualitative criteria relating to the presence or absence of flow. In an acute scrotum, acute ischemia can be excluded with a specificity of up to 100% (Fitzgerald and Foley 1991). In patients with scrotal trauma, ultrasound may demonstrate testicular fracture, hematoceles, and areas of hemorrhage or testicular infarction (Barloon et al. 1996).

When differentiating in smaller groups of patients between acute inflammation of the testicle or epididymis with subsequent hyperemia, on the one hand, and testicular torsion with acute ischemia and absent intratesticular blood flow on the other, color flow duplex examinations by a suitably experienced examiner have yielded a sensitivity of 86–100% and a specificity of 100% (Ralls et al. 1990, Lerner et al. 1990, Middleton et al. 1990, Fitzgerald and Foley 1991, Derwire et al. 1992). Initial results also indicate that (color flow) Doppler sonography may be useful when examining the vasculature of tumors in the testes (see Chapter 10 on tumor vascularization, p. 329). The clinical and diagnostic value of these methods has therefore not yet been conclusively assessed.

In a canine model Lee et al. (1996) found color Doppler energy sonography was not significantly more sensitive than color Doppler velocity sonography for the diagnosis of spermatic cord torsion. In this study with five dogs and different degrees of spermatic cord torsion, acutely complete occlusion of arterial inflow occurred at 450–540 degrees of torsion.

Brown et al. (1995) describe in a study with 31 patients that peak systolic velocities of more than 15 cm/s produced diagnostic accuracy of 90% for orchitis and 93% for epididymitis. Testicular peak systolic velocity ratios between the right and left side ≥ 1.9 and epididymal peak systolic velocity ratios ≥ 1.7 were diagnostic of acute inflammation.

It is possible to detect a *varicocele* using color flow duplex sonography in 93–100% of patients examined (Fobbe and Wolf 1988, Petros et al. 1991). In a study with 63 men presenting with infertility color Doppler ultrasound had a sensitivity and specificity of 97% and 94% for detection of a varicocele compared with spermatic venography as a reference strategy (Trum et al. 1996). However, it should be noted that even in a group of healthy, fertile men (n = 26), 42% will have a dilatation of the veins in the pampiniform plexus to a diameter of 2–3 mm, which would lead to a diagnosis of varicocele when using color flow duplex sonography (Cvitanic et al. 1993).

To sum up, the color flow Doppler mode is a procedure that expands the available range of sonographic diagnosis in scrotal diseases by allowing an evaluation of the blood flow. However, the encouraging indications of the validity of the procedure based on initial examinations using color flow duplex sonography still have to be confirmed in larger groups of patients.

10   Tumor Vascularization

Examination

Special Equipment and Documentation

The technical equipment required to detect the vasculature of tumors and arterial flow patterns in benign and malignant neoplasms is determined by the location of the tumor or the organ affected. For tumors of the *thyroid gland, breast,* and *testicles,* as well as for *superficial tumors of the soft tissues,* linear transducers with a B-mode transmission frequency of between 10 and 7 MHz, and with an integrated pulsed wave Doppler mode, are usually used. In tumors of the *abdominal cavity,* the *small pelvis,* and the *prostate,* suitable sector, vector, or convex probes with a lower transmission frequency are used to provide greater depth of penetration. In addition, transvaginal and transrectal probes can be used, which allow a significantly higher transmission frequency of around 10 MHz, due to the smaller depth of penetration required.

Although Doppler flow curves can be obtained with conventional duplex systems, without the color flow Doppler mode or power mode it is only possible to locate the vasculature of tumors by means of a time-consuming process of scanning the entire tumor with the sample volume of the pulsed wave Doppler mode. A facility for imaging the neovascularization using the color flow Doppler or amplitude mode should therefore always be easily available when evaluating tumor vascularization (Karasch and Schmidt-Decker 1996). In the future, three-dimensional color flow or energy mode maps of low velocity flow through small vessels detected with high resolution velocity estimation techniques may improve our knowledge of organ and tumor vasculature (Ferrara et al. 1996).

Examination Conditions

The external examination conditions correspond to the general guidelines for the ultrasound examination of specific regional vascular systems described in the individual chapters above.

Patient and Examiner

Since tumor vasculature usually consists of relatively narrow-caliber arteries and veins, and due to their winding course these can usually only be depicted over a short distance in a single plane, it is particularly important to choose examination positions for both the patient and the examiner that will not create movement artifacts. Usually, patients are placed in a stable, supine position, with the examiner seated next to them (abdomen, small pelvis) or behind them (neck region). When neovascularization is being imaged endosonographically using the transvaginal, transrectal, or esophageal methods, the positioning of the patient has to be appropriate to the endoscopic technique used.

Conducting the Examination

■ Continuous Wave and Pulsed Wave Doppler Sonography

It is rare in clinical practice today to use continuous wave or pulsed wave Doppler sonography with pen-shaped probes on its own, without the spatial guidance provided in B-mode imaging; occasionally, it is used to register flow signals in breast tumors. However, since even in this area of application the duplex procedure is more effective than using continuous wave sonography on its own, no further discussion of the method will be given here.

■ Conventional and Color Flow Duplex Sonography

B-mode. Sonographic findings that indicate a pathological organ or parenchymal region, or a noticeable space-occupying lesion in addition to enlarged or atypical lymph nodes along the lymphatic vessels, raise a suspicion of neoplasm. Using the following procedures, the neoplasm can then be evaluated in relation to its vasculature and perfusion.

Doppler mode. The Doppler mode is the basis for every quantification of intratumoral flow characteristics. Both the absolute velocities and the calculated indices (pulsatility index, resistance index) are evaluated. Using the two procedures described below, an initial attempt is usually made to depict the arterial and if possible also the venous perfusion in order to register Doppler frequency spectra from the tumor vasculature in a targeted fashion. Using conventional duplex sonography alone, without color-coded scanning for flow signals, does not allow assessment of the flow direction.

Color flow Doppler mode. After the tissue that is suspected of containing a neovascularization has been adjusted for, the color flow Doppler mode is used to locate the blood vessels, the location and course of which can rarely be displayed with B-mode sonography. The transducer has to be moved systematically, and a large number of sectional planes through the relevant tissue area have to be examined, in order to:

- Register *extended* segments of larger arteries in a single plane
- Select sections carefully to depict them in such a way that the vascular axis forms only a small angle with the ultrasound beam

This ensures optimal color coding and the ability to carry out velocity measurements.

During the examination, the *pulse repetition frequency* has to be adjusted for both high and low values, to allow visual assessment of fast and slow flow velocities. The color flow Doppler field from which ultrasound data are acquired (the color box) should be limited to the region of interest in order to enhance the sensitivity.

Color flow Doppler energy mode. Since the vasculature in tumors often follows a curving course, it is not always possible to ensure optimal examination condi-

tions with the color flow Doppler mode. In these cases, the advantages *of the power mode* can be used to depict flow in a *less angle-dependent* manner, so that blood vessels which the ultrasound waves meet at a large incident angle can also be depicted. The lack of directional coding is not a disadvantage here. In the power mode, care should also be taken to set specific limits in advance for the local region from which data are to be registered.

Sequence of the Examination

There is no established sequence for the individual examination steps, although an attempt should first be made to obtain a *B-mode* image of the tumor. Then color flow Doppler sonography or an amplitude-weighted image of the tumor vasculature can be used. *Doppler frequency spectra* can then be recorded from representative segments.

In more extensive neovascularization, flow spectra should also be registered from the afferent organ artery in order to detect any elevated flow in this vascular segment. In organ arteries that are paired, a comparison with the frequency spectra in the contralateral artery can be helpful. In the vasculature of the kidneys and the breasts, an attempt should also be made to carry out side-to-side comparisons of the intraparenchymal spectra.

Findings

The findings in specific organs using B-mode, Doppler sonography, and color flow Doppler sonography are discussed separately in this section where appropriate. The B-mode findings, which are placed first, can only give an extremely brief summary of the sonographic diagnostic criteria, and are no substitute for a detailed study of general textbooks of B-mode sonographic imaging.

Areas of examination for which there are, as yet, no confirmed findings based on large patient groups about the value of color flow Doppler and power mode sonography will not be discussed further at this point. Initial findings seem, however, to indicate the future usefulness of this combined approach with *rectal wall tumors* (Sudakoff et al. 1996), *colorectal cancer* (Leen et al. 1996), *bladder tumors* (Horstman et al. 1995), *gestational trophoblastic disease* (Kawano et al. 1996), or *acute cholecystitis* (Schiller et al. 1996), for example. The same also applies for the application of *sonographic contrast agents* in the diagnosis of tumors (Leen et al. 1994b, Cosgrove 1996, Ernst et al. 1996, Kedar et al. 1996).

Principle and Methodological Aspects

B-mode. As in the examination of the blood vessels supplying the brain and the peripheral and abdominal vasculature, the spatial orientation provided by B-mode sonography forms the basis for all further evaluations using Doppler sonography or color (color flow Doppler mode, power mode). In the case of tumors, the B-mode allows a diagnosis of "space-occupying lesion," which can justify the subsequent use of additional Doppler modes. A questionable neovascularization requires the exact assignment of flow signals to specific tissue structures.

The usual ultrasound procedures are used to examine the individual organ systems. Methodological aspects relevant to the Doppler-sonographic examination of the vasculature in specific abdominal organs are discussed in Chapters 7 and 8 on the abdominal arteries and veins (pp. 238 ff, 282 ff).

It may be worth repeating here that some sonographic findings, such as confirmation of an *abdominal venous thrombosis,* for example, should always suggest the involvement of a tumoral space-occupying lesion as a possible cause of occlusion (cf. p. 291). As possible etiologies, which are not always causally distinguishable from one other, the following factors can be considered:

- Compression of the venous drainage area by tumor
- Appositional thrombus growth along the venous vasculature
- Tumoral invasion in continuity into a vein
- Paraneoplastic syndrome

Doppler mode. Even at an early stage during their growth, neoplastic tissue structures require simultaneous formation of new blood vessels to meet the new tissue's nutritional requirements. Angiographic and histological studies have shown that tumor vasculature often:

- Has fewer smooth muscle cells
- Forms small, confluent sinusoids
- Shows arteriovenous shunt connections

All of these characteristics usually result in a relatively low peripheral resistance within the vascular bed of the tumor. In addition, arteriovenous shunts involving larger blood vessels lead to higher flow velocities in the afferent and efferent arteries and veins. However, since the peripheral resistance does not have to be the same in every area of a tumor, qualitatively different flow patterns can be registered (Fig. 10.**3 c, d,** p. 328).

Hemodynamic changes within the vasculature of the tumor can be visually depicted using the color flow Doppler or power mode, and can be quantified using Doppler sonography.

The value of individual measurements (maximum systolic and end-diastolic flow velocity, resistance index, pulsatility index), the ability of these measurements to distinguish between benign and malignant transformations, and their diagnostic validity, are discussed separately (pp. 332 ff). However, it should be mentioned here that when assessing the benign or malignant nature of *ovarian, cervical,* and *uterine* tumors using arterial flow patterns, fluctuations in the blood supply should be expected in connection with the normal menstrual cycle, and can also be related to age. Perfusion decreases with increasing age, as well as after menopause. This should always be taken into account when interpreting measurements.

Color flow Doppler mode. In color flow Doppler sonography, it is necessary to clarify the following aspects associated with space-occupying lesions visually:

- Relative hyperperfusion in comparison to the surrounding tissue
- Areas of the tumor showing more pronounced blood flow
- Typical vascular formations (halo)
- Visible afferent and efferent vessels

Color flow Doppler energy mode. As a complement to the color flow Doppler mode, amplitude-weighted flow imaging in the power mode may provide a better image of regional hyperperfusion. Since relatively slow flow can also be detected, the advantages of this procedure may be its more precise delimitation of the marginal zones in neovascularization.

At present, a conclusive assessment of the facilities provided by the power mode is not yet possible. It is therefore only mentioned separately here for areas in which sufficiently established experience is already available, or further information concerning tumor vascularizations can be expected with it.

Organ Findings

The criteria listed below for the B-mode, Doppler mode, color flow Doppler mode, and power mode can contribute to the classification of tumors according to their sonographic appearance, and to the assessment of the benign or malignant nature of tumors. However, it is not realistic to expect a clear histological classification of a space-occupying lesion purely on the basis of a single sonographic examination, and this would be beyond the capabilities of the imaging method. Every sonographic finding therefore needs to be integrated into a constantly critical assessment of all the information available from other examination procedures, and if necessary this should lead to the initiation of further examinations. For the patient, both false-positive and false-negative results may have far-reaching psychological and physical consequences.

Thyroid Gland Tumors

B-mode. A partial or extensive hypoechoic marginal seam (halo) of varying width around a tumor is typical of *benign adenomas of the thyroid gland* (Fig. 10.**1**). The tumor itself may be hypoechoic, isoechoic, or hyperechoic in relation to the surrounding tissue, and in these cases often corresponds to a *microfollicular, normofollicular,* or *macrofollicular adenoma,* respectively.

Almost all malignant tumors, such as *follicular* and *papillary carcinomas,* as well as rare *medullary carcinomas of the thyroid gland* and *malignant lymphomas,* appear as mostly hypoechoic, solitary, or more rarely multifocal, tumors, which have a generally ill-defined border. Intratumoral anechoic cystic structures or hyperechoic calcifications can cause the sonographic image to vary. Infiltration of the organ capsule, crossing of organ boundaries, and cervical lymph nodes are late criteria indicating malignancy.

Doppler and color flow Doppler mode. The recognizable halo in B-mode imaging corresponds to clearly identifiable blood vessels in Doppler and color flow Doppler sonography.

In *Graves' disease,* hyperthyroidism is accompanied by diffuse hypervascularization in the form of a typically enlarged, hypoechoic thyroid gland in which intraparenchymal velocities of up to 30 cm/s are registered. The flow in the thyroid artery can also rise to 60–100 ml/min compared to normal values of 8–13 ml/min (Castagnone et al. 1996).

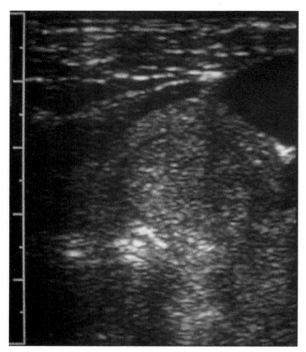

Fig. 10.**1** In the B-mode sonographic image, a round structure that is hypoechoic in comparison with the surrounding tissue is detected in the area of the right thyroid gland, which is surrounded by a hypoechoic seam (halo), corresponding histologically to an adenoma

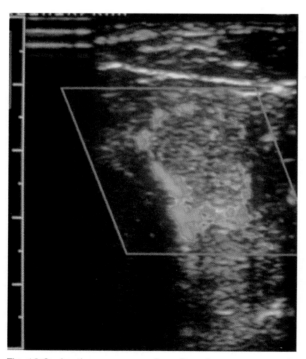

Fig. 10.**2** In the corresponding Doppler energy mode image, the perinodal vessels are well registered. Color coding of almost all sections of the halo is possible because of the low angle-dependency

Color flow Doppler energy mode. Using the power mode, it is possible to depict the vasculature of both benign and malignant space-occupying lesions. In some cases, the blood vessels can also be displayed over a longer distance in a single plane (Fig. 10.**2**).

Parathyroid Gland Tumors

B-mode. The size of normal parathyroid glands is $5 \times 4 \times 2$ mm, but in *hyperparathyroidism,* solitary adenomas, primary hyperplasias, multiple adenomas, and carcinomas can be detected as a pathological correlate. Benign *adenomas* are usually homogeneously more hypoechoic than normal thyroid gland tissue, and after increasing in size they can form regressive transformations and calcifications. True *cysts* and *pseudocysts* are relatively easy to recognize as expressions of regressive changes in an adenoma.

Larger *carcinomas* show nonhomogeneous internal structures, and relatively slow infiltration of surrounding tissue structures.

Doppler and color flow Doppler mode. Like adenomas of the thyroid gland, parathyroid adenomas appear to show a marginally-weighted hypervascularization, and it is not possible at present to distinguish this from carcinoma.

Breast Tumors

B-mode. When there are suspicious findings on palpation of the breast, sonography can distinguish cystic foci from solid foci, which must always be depicted in two planes perpendicular to one another. The marginal contours, internal echoes, and any enhancement and *lateral* or *central shadowing* associated with a space-occupying lesion, are evaluated.

Fibroadenoma, the most common benign, solid tumor, usually has a smooth border and is hypoechoic, with a displacing marginal seam. In approximately half the cases, it shows lateral ultrasound extinction, and in some 25% of cases distal ultrasound amplification. The internal echoes from *lipomas* are more likely to be hyperechoic.

Carcinomas can present a very varied sonographic pattern. The following can be considered suspicious:

- An ill-defined margin with striped protrusions extending into the breast tissue
- Nonhomogeneous internal echoes
- A hyperechoic marginal seam
- Microcalcifications

When surgical treatment is planned to preserve the breast, it is important *preoperatively* to determine the tumor size and also the distance between the tumor and the surface of the skin, the mamilla, and the pectoral fascia. *Postoperatively,* sonographic procedures can be used during the follow-up after the removal of malignant tumors to evaluate the mastectomy region and the axilla. Also, following operations preserving the breast and requiring radiotherapy, ultrasound can be used to monitor for any recurrence.

Doppler and color flow Doppler mode. When the maximum systolic Doppler shift in sonographically suspicious areas exceeds that measured in the contralateral breast by more than 2 kHz, the presence of tumor-associated hypervascularization can be assumed (Scoutt et al. 1990). In 45 *ductal carcinomas,* Dock et al. (1993) found an average flow velocity of 42 cm/s, while the same parameter was only 14 cm/s in 11 fibroadenomas evaluated (range: 0–92 cm/s and 0–41 cm/s, respectively).

Hepatic Tumors

B-mode. The most common benign space-occupying lesions are the usually strongly echogenic *hemangiomas,* which have to be distinguished from an equally strongly echogenic *focal fatty infiltration.* Approximately 90% of hepatocellular carcinomas, and many metastases, are also strongly echogenic. Several benign masses, including liver cell adenoma and sometimes even focal nodular hyperplasia, have a similar appearance. Usually, focal nodular hyperplasia appears as a neoformation in the liver parenchyma that has an isodense echo and occasionally a central scar structure.

Doppler and color flow Doppler mode. Hypervascularization in focal nodular hyperplasia often presents in the form of arteries that have a winding course and a high flow velocity in color flow Doppler sonography.

Color flow Doppler energy mode. Power Doppler sonography seems to be at least equal or even more sensitive in the depiction of the intratumoral vasculature of hemangiomas, hepatocellular carcinomas, and metastases of the liver compared to conventional color Doppler sonography (Choi et al. 1996, Lencioni et al. 1996).

Pancreatic Tumors

B-mode. In contrast to the relatively easily diagnosed pancreatic cysts, smaller malignant pancreatic tumors can be difficult to diagnose. Two-thirds of the tumors originate in the head of the pancreas with the remaining third almost equally distributed in the body and the tail of the pancreas. They usually have a round to oval form, irregular contours in the majority of cases, and the parenchyma, particularly of the smaller tumors, appears more hypoechoic. Dilated bile ducts and an extended pancreatic duct can provide indirect indications.

Doppler and color flow Doppler mode. Because the ultimately rather poor prognosis depends on the infiltration of the large intra-abdominal vessels, it is important during the examination to watch for infiltration of the celiac trunk (Haller's tripod), of the superior mesenteric arteries and veins, and of the vessels of the spleen and liver. Examinations of a total of 70

patients with pancreatic carcinomas have so far shown that a sonographically detectable increase in the envelopment or covering of the aforementioned vessels is associated with a higher probability of invasive growth into the vessel walls and with less chance of excision (Tomiyama et al. 1996, Wren et al. 1996).

Renal Tumors

B-mode. *Renal cell carcinoma,* the most common malignant tumor of the kidney, occurs with a frequency of approximately 0.3%, as determined in large studies using abdominal sonography. One sign of it is a bulge in the contour of an isoechoic or hypoechoic, and occasionally also a hyperechoic, space-occupying lesion (Fig. 10.3a). Cystic colliquations and calcifications can appear. Extension beyond the organ boundaries is an urgent indication of malignancy. An association with thromboses of the renal veins and the vena cava was already mentioned above (pp. 298 ff).

In addition to benign lipomas, leiomyomas, and hemangiomas, *angiomyolipoma* is the most common benign renal tumor. It is seen as a sharply-defined, smooth-edged, hyperechoic tumor that does not extend beyond the contours of the kidney. Intrarenal *metastases* usually have a more hypoechoic appearance than the surrounding renal parenchyma. *Carcinomas of the urothelium* can be recognized by a central echo reflection in structures of comparable echogenicity to the parenchyma, or in more hypoechoic structures. Their growth often follows the calices or the renal pelvis, not infrequently invading the proximal ureter.

Doppler and color flow Doppler mode. In comparison with benign space-occupying lesions, malignant tumors usually show higher intratumoral systolic and end-diastolic flow velocities (Fig. 10.3b, c, d). However, no reliable differences in the resistance index values have yet been identified between benign and malignant tumors (Taylor et al. 1988). It should be noted that increased vascularization often occurs in renal cell carcinomas, but not always (Kuijpers and Jaspers 1989, McGahan et al. 1993).

Uterine and Ovarian Tumors

In gynecological examinations, transabdominal and transvaginal sectional planes are now usually used. These provide complementary information, but neither is sufficient on its own. In addition to the improved resolution provided by its higher transmission frequencies, vaginal sonography, apart from the sonographic criteria involved, has the advantage of simultaneously being able to identify findings using targeted and image-guided palpation.

B-mode. The sonographic criteria indicating an ovarian tumor are increased or decreased echogenicity, irregular boundaries, and an increase in ovarian size to more than approximately 20 ml (Campbell et al. 1989).

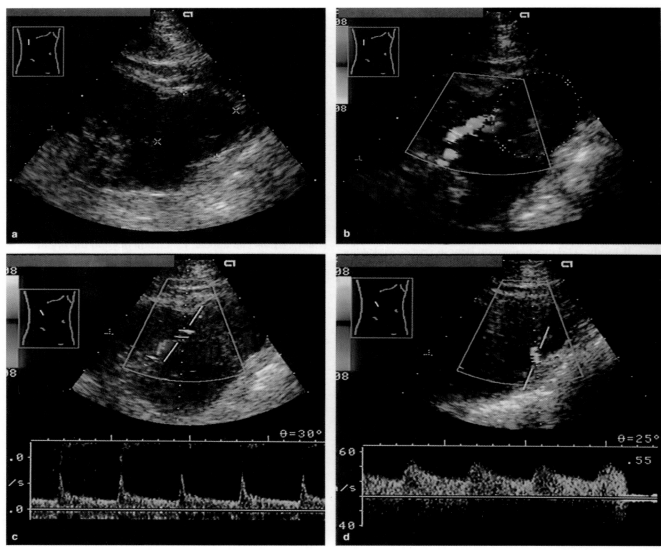

Fig. 10.**3   a** An almost round, space-occupying lesion detected at the lower right renal pole. It has almost the same echogenicity as the renal parenchyma, and extends beyond the renal contour. Histology identified it as a renal cell carcinoma. **b** Color flow Doppler sonography shows that the tumor is relatively avascular, and has an afferent tumor artery in the renal parenchyma. **c, d** The Doppler frequency spectra which can be registered from various segments of the intratumoral vasculature show differing characteristics. While the flow at the transition from the renal parenchyma to the tumoral tissue (**c**) presents a high peripheral resistance, at the edge of the tumor (**d**) there is a low-resistance flow, and relatively high flow velocities can be detected (maximum systolic and end-diastolic flow velocities 51 cm/s and 23 cm/s, respectively, resistance index 0.55, pulsatility index 0.84)

Doppler mode. When interpreting measurements obtained with Doppler sonography, it should be remembered that premenopausal women have lower peripheral resistances and resistance indices in the ovarian and cervical blood vessels and in the vasculature of the uterine body than postmenopausal women. For that reason there will be some overlap between malignant illnesses that are characterized by lowered resistance indices and resistance index values associated with benign findings in premenopausal women (Sohn et al. 1993 b).

The *pulsatility index* is higher in healthy women and women with benign space-occupying lesions than in those with ovarian carcinomas. However, the data reported in the literature show significant differences. For transvaginal flow measurements in healthy women (n = 30), women with benign tumors (n = 9) and women with ovarian carcinomas (n = 7), Bourne et al. (1989) report pulsatility index values for benign tumors ranging from 3.2 to 7.0 (Table 10.**1**). In 15 premenopausal and ten postmenopausal women without any age differentiation (25 benign and six malignant *adnexal masses*). Hamper et al. (1993) also calculated a pulsatility index range for benign tumors of 0.23–3.99, and Rehn et al. (1996) report a transvaginal study of 310 women with 259 benign and 51 malignant

Table 10.**1** Values for the average resistance index (RI) and pulsatility index (PI) in 30 healthy women and female patients with benign and malignant ovarian tumors. The calculated means are indicated in bold

| | Bourne 1989 | | Hamper 1993 | | | |
	PI	Range	RI	Range	PI	Range
Healthy	**5.1**	3.1 – 9.4	–		–	
Benign	–	3.2 ± 7.0	**0.77 ± 0.22**	0.2 – 1.0	**1.93 ± 1.02**	0.23–3.99
Malignant	–	0.3 ± 1.0	**0.5 ± 0.17**	0.27 ± 0.67	**0.77 ± 0.33**	0.31–1.09

tumors where a lower pulsatility index (0.94 ± 0.4) in malignant lesions than in benign lesions (1.06 ± 0.4, $p < 0.05$) was measured, although a remarkable overlap was found.

In 22 postmenopausal women with malignant ovarian tumors, Sohn et al. (1993 b) determined resistance index values of 0.46 ± 0.13, compared to 19 women with benign tumors with an average resistance index of 0.78 ± 0.20. In this study, the resistance index in premenopausal women with benign changes was markedly lower, at 0.49 ± 0.12, than in the research conducted by Hamper.

In uterine myomas, a very low peripheral resistance can be detected with Doppler sonography, which can give resistance indices of less than 0.35 (Sohn et al. 1993 b). This example shows clearly that Doppler criteria alone will never be able to substantiate the differentiation between benign and malignant tumors.

Color flow Doppler mode. As early as 1987, Shimamoto et al. reported the involution of hypoechoic areas in five invasive moles after chemotherapy. This regression was associated with the disappearance of high flow velocities in zones that had been detectable using color flow Doppler sonography.

In 10 cases of *uterine sarcoma* of a group of 2010 women examined by transvaginal color Doppler sonography 1 day before planned hysterectomy, Kurjak et al. (1995) noticed abnormal tumoral blood vessels with a mean resistance index 0.37 ± 0.03, ranging from 0.32 to 0.42, which was statistically significantly lower than that of the 150 normal and 1850 myomatous uteri.

Testicular Tumors

B-mode. Most tumoral space-occupying lesions in the parenchyma of the testicles are malignant processes. They are depicted as hypoechoic nodes in approximately 90% of cases, and can be differentiated from anechoic testicular or epididymal cysts and hypoechoic spermatoceles, focal orchitis. or inflammatory swellings of the epididymis (epididymitis), with or without extension into the testicle (epididymo-orchitis). In acute inflammations of the testicle, side-to-side comparisons show a usually clear swelling of the testicle, which is accompanied by a focal or generalized decrease in the echogenicity.

Although definitive differentiation is not possible, seminomas are more likely to have a homogeneous internal pattern, sharper marginal contours, and no calcifications or cystic portions. Corresponding to their histologically different tissue components, tumors other than seminomas, by contrast, often present varying internal reflections with cystic, and—in comparison with the testicular parenchyma—sometimes more hyperechoic and partially also calcified areas. A rare differential diagnosis in hypoechoic, heterogeneous texture disturbances is focal vasculitis.

Doppler and color flow Doppler mode. With increasing tumor size, the intratumoral vascularization increases (Horstmann et al. 1992). However, this finding does not at present noticeably improve the sonographic classification of testicular tumors. Systematic measurements of quantitative flow parameters, an area in which research has already started for tumors of the female genitalia, have yet to be undertaken.

When there is focally more pronounced vascularization, the differential diagnosis must always include the possibility of inflammatory processes (orchitis, epididymo-orchitis). These are also accompanied by hypervascularization, which cannot always be detected over the entire testicle.

Gallbladder Tumors

In a case report, Ueno et al. (1996) described a gallbladder carcinoma with **B-mode** findings of a mildly thickened gallbladder wall. The lumen was filled with debris-like components, and the **color flow Doppler mode** revealed color signals in both the lesion and the gallbladder wall. These signals showed a high-speed pulsatile wave.

Lymph Nodes

Benign Lymph-Node Enlargements

B-mode. Enlarged lymph nodes usually appear as minimally echogenic round or oval structures in the context of lymphadenitis due to bacterial or viral infections (Fig. 10.**4 a**). In localized infections, the nodes, which are sonographically sharply delimited from the surrounding tissue, show a typical location along the course of the large lymph-node tracts or stations of

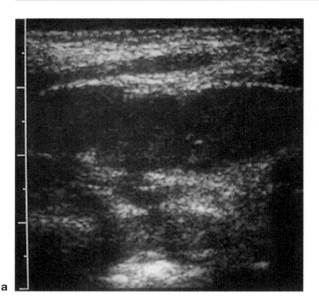

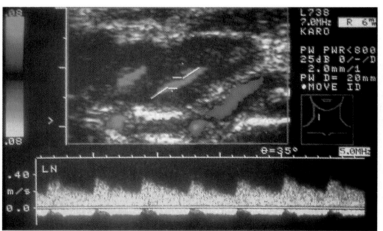

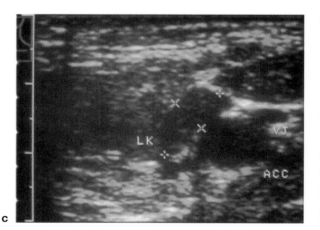

the drainage area. In systemic infections (acquired immunodeficiency syndrome, AIDS-related complex), they have a disseminated distribution pattern.

With color flow Doppler sonography, small blood vessels in the lymph-node parenchyma can be recognized. The corresponding frequency spectrum (Fig. 10.**4b**) shows low-resistance flow, with systolic flow velocities of up to 40 cm/s. In a series of measurements involving groups of 72 patients and 60 patients with lymphadenitis, the resistance and pulsatility indices were 0.68 ± 0.1 and 1.27 ± 0.26, respectively (Tschammler et al. 1991).

Fig. 10.**4** **a** Three acutely appearing lymph nodes that were painful on palpation in a 22-year-old woman with a peritonsillar abscess in the area of the right lateral side of the neck. **b** In color flow Doppler sonography, arteries are recognized within the lymph nodes (coded red). Their flow at maximum systolic and end-diastolic flow velocities of 38 cm/s and 19 cm/s, respectively, shows a low peripheral resistance (RI) of 0.5. **c** In a follow-up examination 14 days later, the lymph nodes, in which blood vessels were no longer detectable with color flow Doppler sonography, had significantly decreased in size

Malignant Lymph-Node Enlargements

B-mode. Sonographically, there are no definite criteria for differentiating between benign and malignant lymph-node enlargements. Malignancies can represent primary disease involving the lymphatic system, or may appear as lymph-node metastases from primary tumors elsewhere.

Doppler and color flow Doppler mode. As in the case of a reactively enlarged lymph node, the organ perfusion is elevated, so that a relatively high flow velocity can be detected in the vasculature. Due to the low peripheral resistance, the flow velocity can also show high diastolic values. In a series of measurements each involving more than 40 patients with malignant lymphomas and lymph-node metastases, Tschammler et al. (1991) calculated resistance indices of 0.72 ± 0.17 or 0.95 ± 0.29, and pulsatility indices of 1.72 ± 1.41 or 3.67 ± 4.52, respectively.

Furthermore, the following criteria seem to be suspicious of malignancy in superficial lymph nodes:

- Avascular areas
- Displacement of intranodal vessels
- Accessory peripheral vessels
- Aberrant course of central vessels

Using these patterns of intranodal color Doppler signals in a total of 130 superficial lymph nodes, 96% of the 73 neoplastic lymph nodes showed at least one pathological vascular pattern (Tschammler et al. 1996). Malignancy could be excluded in 95% of 57 reactive lymph nodes using these four criteria. Most reactive lymph nodes in contrast demonstrated a vascular hilus and/or vessels running at the long axis of the lymph node with branches to the cortex.

Metastases

B-mode. The sonographic results vary, depending on the type of the primary tumor and the location of the metastasis. In the thyroid gland and the kidney, metastases appear mainly more hypoechoic (Fig. 10.**5 a, b**) than the organ parenchyma, while metastases in the liver, for example, can have an internal pattern that is either more hyperechoic or more hypoechoic.

Doppler and color flow Doppler mode. The degree of perfusion is decisively affected, of course, by the histology and location of the secondary tumor. *Metastases of the liver* often show only minimal blood flow, so that they appear to be relatively avascular when examined with Doppler and color flow Doppler sonography.

The examples in Figure 10.**5 d** show clearly hypervascularized metastases from a renal cell carcinoma in the region of an already previously resected left lobe of the thyroid gland, with two qualitatively different frequency spectra.

Evaluation

The research currently available indicates that a tumoral space-occupying lesion may be *malignant* when:

- The number of blood vessels in a tumor increases
- Typical vasculature patterns or locations are present (centralized, decentralized, peripheral, interruption in the vascular course)
- The systolic (and diastolic) flow velocity in the tumor vasculature increases
- The resistance and pulsatility indices decrease

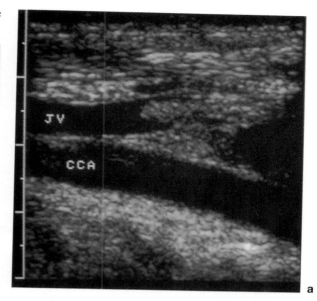

a

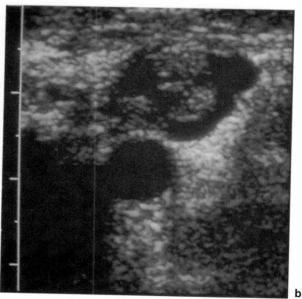

b

Fig. 10.**5** Proximal left jugular vein and common carotid artery (CCA) visualized with B-mode ultrasound in a longitudinal plane with insonation from an anterolateral position (**a**) and in a transverse plane (**b**).

Fig. 10.**5 c** u. **d** ▷

When classifying findings diagnostically, comparisons should always be made with the (healthy) contralateral vascular region if possible. Also, other local factors (tumor histology and size) and systemic factors influencing the blood flow in an organ system need to be taken into consideration because these can modulate the physiological flow level in a space-occupying lesion.

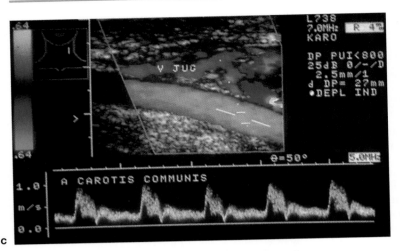

c

d

Fig. 10.**5c** u. **d** Normal morphology of the CCA. In the jugular vein, an inhomogeneous intraluminal structure almost fills up the lumen of the vessel. The surface of this tumor is irregular. Compared to the surrounding tissue structures, the echogenity of the intravenous masses are equal or less echogenic. The Doppler spectra of the common carotid artery show normal flow pattern with a maximal systolic and end-diastolic velocity of 85 and 24 cm/s, respectively (**c**). The Doppler spectra of a tumor artery of the intravenous masses represent a low resistance flow with a relatively high end-diastolic velocity (**d**). The red color in the distal jugular vein codes the reverse flow due to total tumor obstruction of the vein in the proximal part. The aliasing phenomenon in the tumor area, in spite of a high pulse repetition frequency, represents the high velocity in tumor arteries

Sources of Error and Diagnostic Effectiveness

Sources of Error

Potential sources of error that need to be considered when evaluating tumor vasculature using duplex sonography basically include all of the disruptive factors described in detail above in Chapters 7 and 8 (pp. 269 ff, 301 ff). Since the caliber of the tumor vasculature is usually very narrow, and the arteries can only be imaged over a short distance, the following phenomena produce particularly unfavorable effects on imaging in the abdominal cavity:

– Obesity
– Superimposed intestinal gas
– Peristaltic noise

In addition,

– Restlessness of the patient, and
– Artifacts due to movement

can also complicate the registration of Doppler frequency spectra, or make it impossible.

■ Doppler and Color Flow Doppler Parameters

Particularly with the often quite low flow velocities that are found in the small arteries of a tumor, the system parameters in Doppler and color flow Doppler mode have to be optimized to provide:

– Adequate amplification
– Low pulse repetition frequency
– Correct angle adaptation
– Smallest (high-pass) filter adjustment

in order to obtain as much Doppler and color flow Doppler information as possible.

With color flow Doppler sonography, differentiation between arterial and venous flow in a tumor cannot be made with certainty, since the blood vessels—which can usually only be imaged over a short distance—often present only a dot-like color image of the vascular cross-section. This differs from the situation in the region of the peripheral arteries and veins.

To distinguish between arteries and veins, the pulsed wave (or continuous wave) Doppler mode becomes necessary once again.

■ Color Flow Doppler Energy Mode

Due to the color coding used in the power mode, it is not possible to distinguish between arterial and venous flow, even in large blood vessels (compare Fig. 8.**24a, b,** p. 304), and differentiating between arteries and veins is therefore not possible in the parenchyma either. Frequency spectrum analysis is required for this purpose.

Diagnostic Effectiveness

The use of duplex and color flow duplex sonography to identify and differentiate between malignant and benign tumors, predominantly in the parenchymatous organs, is not yet standard practice. However, reported experiences indicate that Doppler procedures are helpful in detecting hypervascularized regions in the *thyroid gland, parathyroid glands, parotid glands, breasts, liver, ovaries, testicles, prostate, lymph nodes,* and also in tumors of the soft tissues (Fig. 10.**6**). In smaller series of measurements, high sensitivities and specificities have sometimes been reported (Tables 10.**2**–10.**4**).

At present, more extensive validation studies concerning the possible differentiation between benign and malignant tumors are not yet available for many organs (Taylor et al. 1993, Kurjak and Kupesic 1995). The use of varying criteria to characterize and classify flow (Tables 10.**2**–10.**4**) shows that there is uncertainty as to what constitutes a reliable and acceptable flow evaluation.

In all of the organ systems, duplex-sonographic measurements of tumor vascularization can only ever be a single link in the chain of information used to assess the benign or malignant nature of a space-occupying lesion. However, greater significance always attaches to a positive finding, i.e. a suspicion of malignancy, than to a negative result.

The extent to which color-coded depiction of tumor vasculature can provide information that is definitely useful in the *differential diagnosis,* above and beyond the information already available from B-mode imaging, remains to be determined in larger studies. It is also uncertain at present whether the degree of hypervascularization in individual tumors correlates with their degree of malignancy, or whether therapeutic success (Shimamoto et al. 1987) or tumor recurrence can be detected on the basis of characteristic flow patterns.

On the other hand, detecting intratumoral flow is already able to provide important information, even in relatively rare findings. Figure 10.**6** shows a finding of leiomyosarcoma that was noticed in B-mode as a hypoechoic tumor with very irregular internal echoes located in the musculature of the calf. Due to the pathologically elevated diastolic flow velocity, with resulting low resistance indices while resting, a malignant space-occupying lesion was suspected.

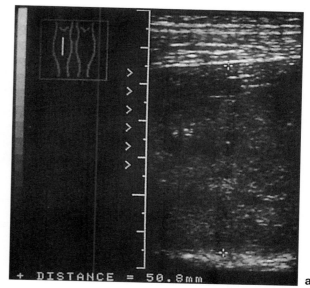

a

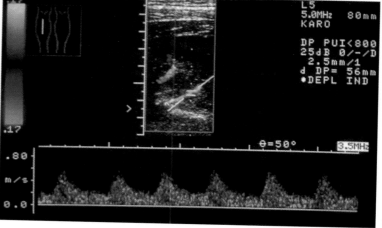

b

Fig. 10.**6**   **a** In a longitudinal section above the right calf, a hypoechoic, space-occupying tumor can be seen, surrounded by normal muscle tissue. **b** With color flow Doppler sonography, longer intratumoral arteries can be seen, with a flow pattern indicating a low peripheral resistance. Histologically, the tumor was classified as a grade II leiomyosarcoma

Table 10.**2** Examination series, some only descriptive, studying the differentiation between benign and malignant **breast tumors** using duplex and color flow duplex sonography. The sensitivity and specificity are indicated when available (V_{max} = maximum systolic velocity; F_{max} = maximum systolic frequency)

First author	Patients (n)	Tumors Benign	Malignant	Sensitivity (%)	Specificity (%)	Criterion
Wells 1977	9	6	3	–	–	Side-to-side comparison of signal
Burns 1982	350	349	52	90	84	F_{max}, $F_{end\text{-}diast}$, F_{mean}, PI, Comparison with the contralateral side
Dock 1991	44	13	31	65	85	$V_{max} > 40$ cm/s
Madjar 1991	37	–	21	81	–	Number of blood vessels, F_{max}
Schild 1991	59	24	35	83	67	$V_{end\text{-}diast}$*
Luska 1992	72	35	38	30	–	Number of blood vessels
Sohn 1992	151	59	92	–	–	RI
Dock 1993	75	20	55	82	64	$V_{max} > 20$ cm/s
Heilenkötter 1993	24	10	14	93	100	–
Huber 1994	57	32	25	92	78	Mean color value**

* Threshold value not indicated.
** Computer-assisted quantitation of color pixels.
V_{max}: In comparison to the contralateral side.

Table 10.**3** Examination series, some only descriptive, series of examinations concerning the differentiation between benign and malignant **hepatic tumors** using duplex and color flow duplex sonography. The sensitivity and the specificity are indicated when available (V_{max} = systolic maximal velocity; F_{max} = systolic maximal frequency)

First author	Patients (n)	Tumors Benign	Malignant	Sensitivity (%)	Specificity (%)	Criterion
Taylor 1988	68	18	50	100	?	$F_{max} > 3$ kHz
Börner 1990	50	30	48*	?	?	"Parenchyma coloring"
Dock 1991	43	3	40	?	?	V_{max}
Shimamoto 1992	15	0	15	?	?	F_{max}, RI
Nino-Murcia 1992	108	24	94	76	69	Intralocal vascularization
Tanaka 1992a	26	9	19	79	–	Flow in the tumor
Leen 1994a	90	0	23**	100	77.6	DPI > 0.3

* Several foci possible for each patient.   ** Hepatic metastases from colorectal tumors.
? Only a descriptive depiction is provided.   DPI: hepatic arterial to total liver blood flow ratio.

Thyroid gland tumors. In an investigation carried out by Stern et al. (1994), the criterion of perinodular *vascular halo* in color flow Doppler sonography, applied to *adenomatous nodules and adenomas,* showed a sensitivity and a specificity of 96% and 93%, respectively (n = 28). Clear central vascularization was detectable in all of the eight malignant tumors in this study. However, perifocal and central neovascularizations are to be expected in autonomous adenomas, as well as in carcinomas (Fobbe and Wolf 1988, Hübsch et al. 1992).

Parathyroid gland tumors. *B-mode.* Solitary cervical adenomas can be detected sonographically with a certainty of 85–90%. It is more difficult to confirm the presence of hyperplasia or ectopic (mediastinal, retroesophageal) or multiple adenomas, and to distinguish hypoechoic, dorsally located thyroid gland tumors, especially in adenomatous goiter.

Parathyroid gland carcinomas are difficult to differentiate from thyroid gland carcinomas and large, regressively transformed adenomas. Larger studies investigating the value of additional information obtained with *Doppler* and *color flow Doppler sonography* are not yet available.

Breast tumors. When palpable nodes are detected, *B-mode sonography* can be used on its own, independently of mammography, to differentiate between solid tumors and cysts, and can therefore help avoid unnecessary biopsies of clearly cystic structures. Sufficient certainty in the evaluation of the benign or malignant nature of *solid* tumors is not yet possible (Bassett et al. 1991). However, an awareness of the findings on inspection or palpation, such as a *flattening* or a *retraction* of the skin, can provide supplementary information when classifying sonographic findings.

Additional signs of malignancy are:

- Interruptions in the pectoral fascia
- Lymph nodes in the axilla
- Clavicular and nuchal lymph nodes
- Foci in the liver

Doppler and color flow Doppler mode. In 75 women with 55 malignant and 20 benign breast tumors, with a threshold value of 20 cm/s for maximum systolic flow velocity in malignancies and without angle correction, Dock et al. (1993) calculated a sensitivity and specificity of 82% and 64%, respectively. Computer-assisted analyses of the color flow Doppler signals from breast carcinomas may further improve the reproducibility and precision of the evaluation (Huber et al. 1994). Table 10.2 lists a selection of additional studies that have discussed the distinction between benign and malignant tumors. Here too, the wide variety of differentiation criteria used shows that there are no generally accepted parameters for distinguishing benign from malignant lesions.

The extent to which color flow Doppler information can be used on its own to evaluate the biological behavior of a breast tumor (lymph node status, S-phase of the tumor cells) has also not yet been determined (Sohn et al. 1993 a).

It should be noted that *fibroadenomas* in particular can present with a relatively high flow velocity, which cannot always be distinguished from flow profiles attributable to malignant breast tumors.

Hepatic tumors. Although several research groups have investigated the perfusion of malignant and benign hepatic tumors, few validating studies of the Doppler sonographic criteria are available yet to suggest that differentiation between malignant and benign hepatic tumors is possible. The currently available data are listed in Table 10.3. In a study of patients (n=51) with 88 histologically proven hepatocellular carcinoma, power mode was superior to color flow Doppler imaging in the depiction of tumor vascularity. The small size and the deep location of the lesion significantly reduced the detection rate of blood flow signals with the color Doppler mode compared to the power mode (Lencioni et al. 1996). In 19 consecutive patients with malignant hilar obstruction, color Dopp-

ler sonography was equal to angiography and CT portography or diagnosis of atrophy, level of bile duct obstruction, hepatic involvement, or venous invasion (Hann et al. 1996).

Renal tumors. *B-mode.* Due to the increased use of procedures providing sectional images, and the resulting improvement in image resolution, the rate of accidental detection of small and therefore potentially curable *renal cell carcinomas* has significantly increased. In a retrospective examination of their patients, Smith et al. (1989) found that for the periods 1974–1977 and 1982–1985, 5.3% (four of 74) and 25.4% (31 of 122), respectively, of all the detected renal cell carcinomas were smaller than 3 cm. Ten of the 31 cases in the later period (32.3%) were detected by sonography.

In selected patients, the administration of an ultrasound contrast medium that accumulates in hypervascularized tissue sections may assist in revealing malignant tumor areas (Takase et al. 1994).

Doppler and color flow Doppler mode. In a prospective study in 70 patients, Kier et al. (1990) showed that using a systolic Doppler shift of 2.5 kHz to distinguish between benign (n=33) and malignant (n=37) renal tumors provided a sensitivity and specificity of 70% and 92%, respectively. Even earlier, in a retrospective analysis by the same group involving 49 tumors, similar indications of the validity of the method were reported (sensitivity 77%, specificity 100%) (Ramos et al. 1988).

Uterine and ovarian tumors. Sohn et al. (1993 b) note that it is only possible to use a reduced resistance index to differentiate between benign and malignant tumors of the ovary and of the uterine body or cervix in postmenopausal women who are not undergoing hormone substitution treatment. In premenopausal women, lower resistance indices *dependent on the menstrual cycle* are always found. Postmenopausal hormone substitution also causes the resistance index values to fall, so that it is not possible to differentiate between benign and malignant tumors in these women with certainty on the basis of the resistance index. In addition, quite apart from the above factors, the resistance index and pulsatility index values in some malignant tumors overlap with those of benign space-occupying lesions (Hamper et al. 1993, Antonic and Raker 1996)—for example, myomas of the uterus and hemorrhagic corpus luteum cysts (Tables 10.1, 10.4).

With the above qualifications, the threshold value of the resistance index is stated to be 0.4–0.5 (Bourne et al. 1989, Kurjak et al. 1995; Tepper et al. 1995, Buy et al. 1996), and the cut off value for the pulsatility index is 1. In a study involving 24 malignant ovarian tumors, the pulsatility index, with a value of 0.74 ± 0.2, was significantly lower than in 40 benign space-occupying lesions that had a pulsatility index of 1.88 ± 0.74 (Eppel et al. 1992).

In transvaginal color Doppler sonography performed in 217 patients with adnexal masses prior to explorative laparotomy, flow was detectable in 82

Table 10.**4** Examination series, some only descriptive, concerning the differentiation between benign and malignant tumors in **various organ systems** using duplex and color flow duplex sonography. The sensitivity and the specificity are indicated when available (V_{max} = systolic maximal velocity; F_{max} = systolic maximal frequency)

First author	Patients (n)	Tumors Benign	Malignant	Sensitivity (%)	Specificity (%)	Criterion
Thyroid gland						
Thomas 1989	123	84	6	?	?	"Hypervascularization"
Hübsch 1992	65	42	23	43	67	f
Stern 1994	40	35	8	?	?	Central inhomogeneous vascularization, sonographically irregular peripheral structures
Salivary glands						
Martinoli 1994	111	98	13	?	?	"Hypervascularization"
Kidney						
Kier 1990	70	33	37	70	94	$F_{max} \geqslant 2.5$ kHz
Ovary						
Sohn 1993 b	41*	19	22	?	?	RI
Fleischer 1993	50	25	25	?	?	
Hamper 1993	31	25	6	?	?	RI, PI
Stein 1995	161	123	46	98	90	Combination of B-mode, Doppler and color flow Doppler modes
Tepper 1995	217*	179	38	88	85	RI < 0.47
Anandakumar 1996	146*	122	34	77	68	Color Doppler
Leeners 1996	101	80	23	74	74	RI, PI, V_{max}
Buy 1996	115	98	31***	88	97	RI < 0.4, PI ≤ 1, $V_{max} \geqslant 15$ cm/s
Uterus						
Skoder 1991	70	30	35	?	?	PI
Sohn 1993 b	64*	28	36	?	?	RI
Testicles						
Horstmann 1992	28	0	28	?	?	V_{max}, RI
Prostate						
Rifkin 1993	619	469	132**	93	86	RI, flow pattern

? Only a descriptive depiction is given.
** Several foci possible per patient.
f: Form of the frequency spectrum.

* Transvaginal examinations.
*** Plus 3 borderline tumors.
V_{max}: Blood vessel location.

(49.7%) of the 165 benign tumors, in 12 (85.7%) of 14 tumors of low malignant potential, and in 25 (65.8%) of women with malignant tumors. The mean resistive index was 0.39 ± 0.05 for malignant tumors, compared with 0.49 ± 0.06 and 0.55 ± 0.07 for the low malignant potential and benign tumors, respectively. These differences were statistically significant ($P < 0.01$). With a cutoff value for the resistive index of 0.47, sensitivity and specificity were calculated to be 88% and 85%, respectively (Tepper et al. 1995).

In 146 patients with 156 surgically removed and histologically confirmed ovarian tumors, transvaginal scanning with color Doppler imaging prior to surgery, the sensitivity and specificity of color Doppler imaging in identifying malignant ovarian tumors were 76.5% and 68%, respectively. Positive and negative predictive values were 40% and 91.2%, respectively (Anandakumar et al. 1996).

In a transvaginal color Doppler study of 2010 women (150 normal and 1850 myomatous uteri) prior to hysterectomy, Kurjak et al. (1995) diagnosed 10 cases of *uterine sarcoma* with a mean resistance index of 0.37 ± 0.03, ranging from 0.32 to 0.42. Using a cutoff point of 0.40 for resistance index it was possible to distinguish between benign and malignant myometrial tumors with a sensitivity and specificity of 90.91% and 99.82%, respectively, and a positive and negative predictive value of 71.43% and 99.96%, respectively. There was no significant difference between RI in the right and left uterine arteries in each separate group; however, there was a decline in these values from normal, through myomatous, to sarcomatous uteri.

Testicular tumors. The value of Doppler and color flow Doppler procedures here lies in differentiating between hypervascularized and hypovascularized acute swellings of the scrotal contents, in order to distinguish inflammatory swellings from torsion (p. 322).

Although B-mode sonographic imaging is the preferred method for detecting or excluding parenchymatous malignancies, and provides valuable assistance in the staging of testicular tumors (using the TNM classification), the differentiation between various germ cell and non-germ cell tumors remains uncertain. It remains to be seen whether the use of duplex-sonographic procedures will be able to provide new methods of analysis here.

Prostatic tumors. In a follow-up programm of 31 patients who had undergone radical prostatectomy for prostatic cancer, color Doppler imaging used during transrectal sonography improved the detection of early recurrent or residual prostatic cancer compared with transrectal sonography alone. The validity of color Doppler mode during transrectal sonography in this study had a sensitivity and specificity of 86% and 100%, respectively, with positive and negative predictive values of 100% and 82%, respectively, compared to transrectal sonography with a sensitivity and specificity of 71% and 89%, respectively, with positive and negative predictive values of 91% and 67%, respectively (Sudakoff et al. 1996).

Orbit and eye tumors. Compared with malignant tumors in other organs, malignant intraocular tumors demonstrate a high degree of neovascularization, which can be detected with Doppler sonography. Initial experience with conventional (Guthoff et al. 1989) and color flow duplex sonography in the diagnosis of choroidal melanomas (Guthoff et al. 1991, Wolff-Kormann et al. 1992) and the follow-up after bulb-preserving radiotherapy show that a systolic peak frequency of 0.3–2.7 kHz is likely in tumorous vessels of untreated melanomas, while no intratumorous vessels can be identified after successful radiation therapy.

Lymph nodes. Examinations involving 105 benign and 115 malignant lymph-node enlargements, using a threshold value of 1.8 for the pulsatility index and 0.9 for the resistance index, showed a sensitivity of 53% and a specificity of 97% (Tschammler et al. 1991).

Metastases. The extent of neovascularization in metastases is determined by the primary tumor, on the one hand, and by the location of the metastasis, on the other. At present, there are very few reports describing findings obtained with Doppler and color flow Doppler sonography in metastases. Quantitative assessments of the flow patterns in metastases are therefore not yet possible. It also remains to be seen whether Doppler-sonographic methods will in the future be able to help differentiate individual primary tumors on the basis of findings made in the metastases.

A combination of intraoperative B-mode sonography and the Doppler procedure was able to further improve the early detection of intrahepatic metastases from colorectal tumors in 23 patients (Leen et al. 1994).

11 Case Histories

Anatomical Overview

(Fig. 11.**1a–h**)

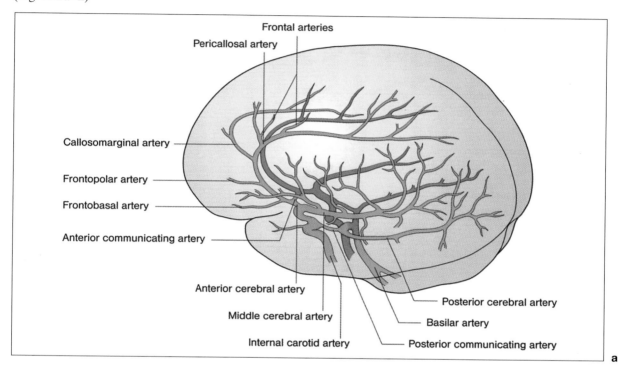

Frontal arteries
Pericallosal artery
Callosomarginal artery
Frontopolar artery
Frontobasal artery
Anterior communicating artery
Anterior cerebral artery
Middle cerebral artery
Internal carotid artery
Posterior cerebral artery
Basilar artery
Posterior communicating artery

a

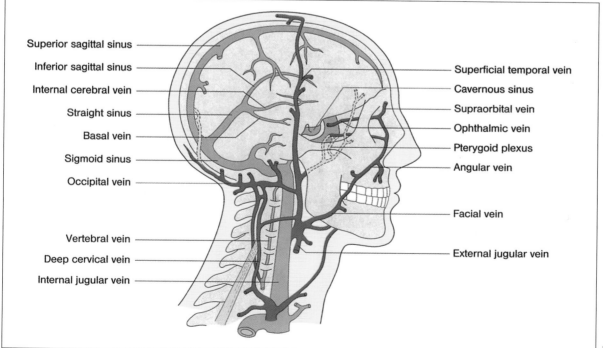

Superior sagittal sinus
Inferior sagittal sinus
Internal cerebral vein
Straight sinus
Basal vein
Sigmoid sinus
Occipital vein
Vertebral vein
Deep cervical vein
Internal jugular vein
Superficial temporal vein
Cavernous sinus
Supraorbital vein
Ophthalmic vein
Pterygoid plexus
Angular vein
Facial vein
External jugular vein

b

Fig. 11.**1c–h** ▷

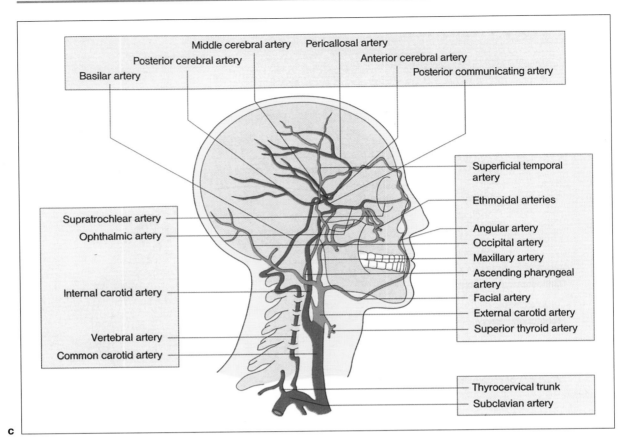

Basilar artery
Posterior cerebral artery
Middle cerebral artery
Pericallosal artery
Anterior cerebral artery
Posterior communicating artery

Superficial temporal artery
Ethmoidal arteries
Supratrochlear artery
Ophthalmic artery
Angular artery
Occipital artery
Maxillary artery
Ascending pharyngeal artery
Internal carotid artery
Facial artery
External carotid artery
Superior thyroid artery
Vertebral artery
Common carotid artery
Thyrocervical trunk
Subclavian artery

c

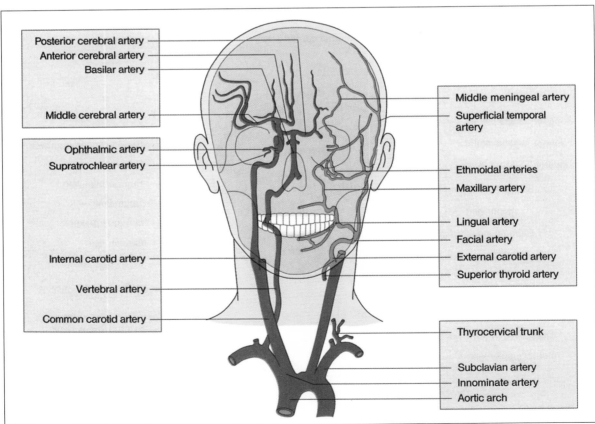

Posterior cerebral artery
Anterior cerebral artery
Basilar artery
Middle cerebral artery
Middle meningeal artery
Superficial temporal artery
Ophthalmic artery
Supratrochlear artery
Ethmoidal arteries
Maxillary artery
Lingual artery
Facial artery
Internal carotid artery
External carotid artery
Superior thyroid artery
Vertebral artery
Common carotid artery
Thyrocervical trunk
Subclavian artery
Innominate artery
Aortic arch

d

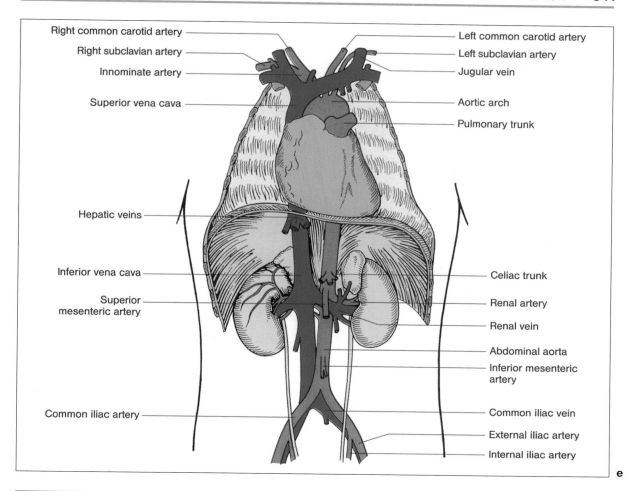

Right common carotid artery

Right subclavian artery

Innominate artery

Superior vena cava

Hepatic veins

Inferior vena cava

Superior
mesenteric artery

Common iliac artery

Left common carotid artery

Left subclavian artery

Jugular vein

Aortic arch

Pulmonary trunk

Celiac trunk

Renal artery

Renal vein

Abdominal aorta

Inferior mesenteric
artery

Common iliac vein

External iliac artery

Internal iliac artery

e

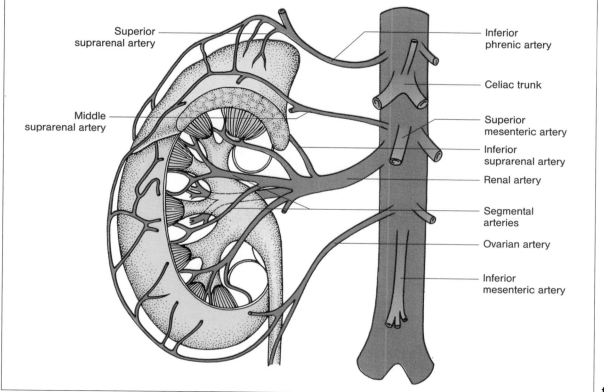

Superior
suprarenal artery

Middle
suprarenal artery

Inferior
phrenic artery

Celiac trunk

Superior
mesenteric artery

Inferior
suprarenal artery

Renal artery

Segmental
arteries

Ovarian artery

Inferior
mesenteric artery

f

Fig. 11.**1 g** and **h** ▷

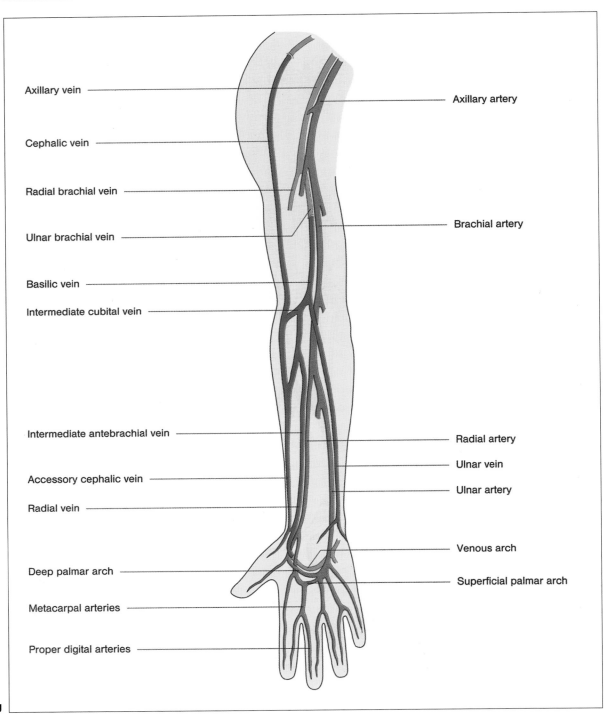

Axillary vein

Cephalic vein

Radial brachial vein

Ulnar brachial vein

Basilic vein

Intermediate cubital vein

Intermediate antebrachial vein

Accessory cephalic vein

Radial vein

Deep palmar arch

Metacarpal arteries

Proper digital arteries

Axillary artery

Brachial artery

Radial artery

Ulnar vein

Ulnar artery

Venous arch

Superficial palmar arch

g

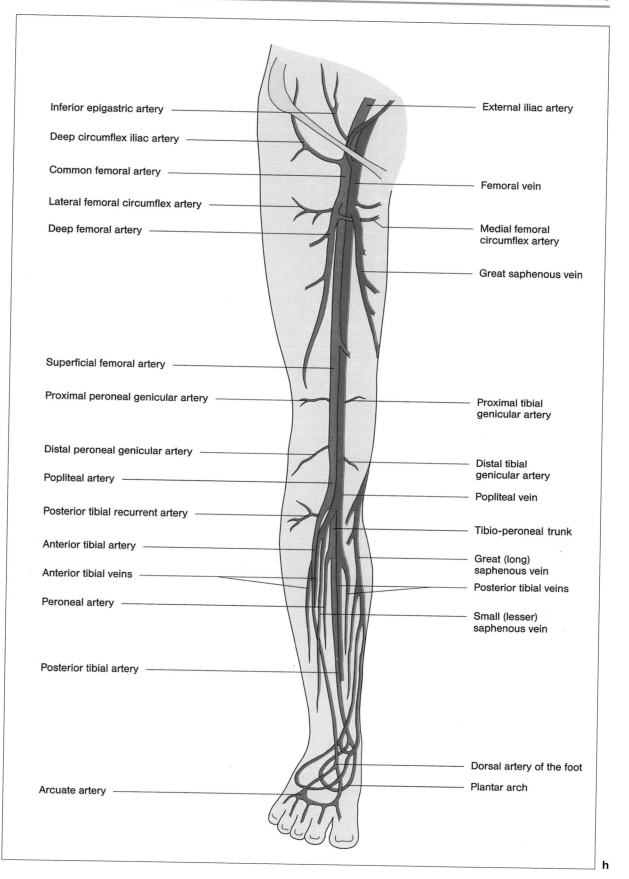

Inferior epigastric artery

Deep circumflex iliac artery

Common femoral artery

Lateral femoral circumflex artery

Deep femoral artery

Superficial femoral artery

Proximal peroneal genicular artery

Distal peroneal genicular artery

Popliteal artery

Posterior tibial recurrent artery

Anterior tibial artery

Anterior tibial veins

Peroneal artery

Posterior tibial artery

Arcuate artery

External iliac artery

Femoral vein

Medial femoral circumflex artery

Great saphenous vein

Proximal tibial genicular artery

Distal tibial genicular artery

Popliteal vein

Tibio-peroneal trunk

Great (long) saphenous vein

Posterior tibial veins

Small (lesser) saphenous vein

Dorsal artery of the foot

Plantar arch

h

Extracranial Arteries

Extracranial Carotid Artery Disease

Stenosis at the Origin of the Left Common Carotid Artery
(Fig. 11.2)

Patient: H.N., male, age 27.
Clinical diagnosis: amaurosis fugax attacks on the left eye.

CAD:	–	Vascular surgery:	–	Diabetes mellitus:	–
PAOD:	–	Heredity:	–	Hyperlipidemia:	–
TIA, stroke:	+	Hypertension:	+	Smoking:	+

Doppler sonogram *(analogue registration)* (**b**) and *(spectral analysis)* (**c**). The right supratrochlear artery has physiological blood flow. On the left, small diastolic residual flow with a physiological flow direction is seen.

A normal carotid artery system appears on the right. On the left, the CCA and ICA show a clear reduction, particularly in the diastolic flow velocity, but also in the systolic flow velocity. There are no flow acceleration, changes in waveform, or turbulence phenomena suggesting stenosis in the left neck region.

Vertebral and subclavian arteries are bilaterally normal.

With pulsed wave Doppler sonography (**c**), a pathological spectrum can be detected, which is a sign of stenosis at the origin of the CCA on the left. The maximum flow velocity is elevated (4.8 kHz on the left, compared to 2.5 kHz on the right), and there are bidirectional signal components around the baseline with slow flow velocities.

Angiogram (**a**). Confirmation of a high-grade stenosis at the origin of the left CCA, with a normal distal carotid system.

Comment. In contrast to the findings presented, turbulence phenomena (disturbed flow) are sometimes registered in the supraclavicular and neck regions, originating from stenoses near the aorta. This is particularly obvious if the proximal segment of the blood vessel is insonated (probe reversal, p. 61).

a

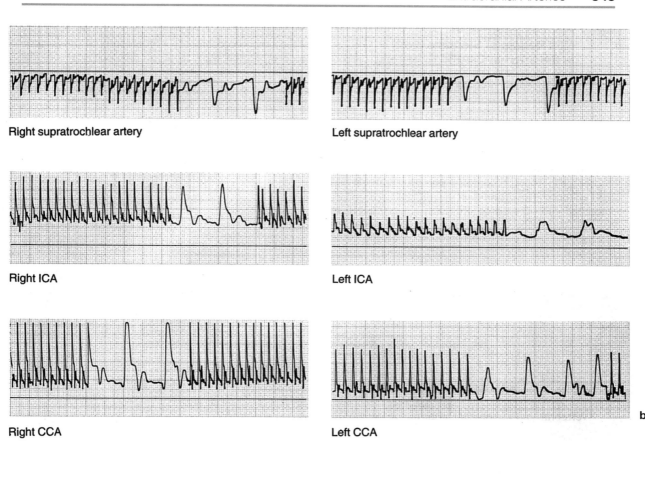

Right supratrochlear artery

Left supratrochlear artery

Right ICA

Left ICA

Right CCA

Left CCA

b

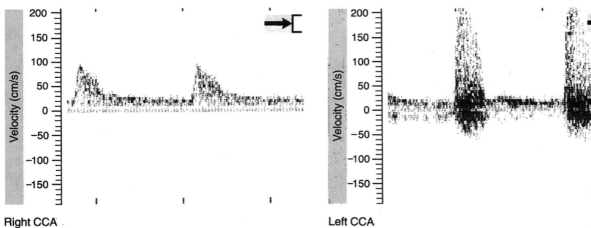

Right CCA

Left CCA

c

▬ Plaques (Right) and
Mild Degree of Stenosis (Left)
in the Internal Carotid Artery (Fig. 11.3)

Patient: R.M., female, age 63.
Clinical diagnosis: neurologically asymptomatic.

CAD:	–	Vascular surgery:	–	Diabetes mellitus:	–
PAOD:	+	Heredity:	–	Hyperlipidemia:	–
TIA, stroke:	–	Hypertension:	–	Smoking:	+

Doppler sonogram *(spectral analysis)* **(c)**. The supratrochlear artery bilaterally presents with physiological blood flow, and the waveforms in both extracranial carotid systems are normal. On the left immediately after the bifurcation, an increase in the systolic and diastolic flow velocity is seen, accompanied by acoustic turbulence phenomena without a significant change in the Doppler spectrum. On the right side at the same location, only a broadening of the systolic window can be seen. Note the different scales of the ordinate!

Color flow duplex sonogram (b). *Left:* From the distal segment of the CCA into the proximal ICA, there is inhomogeneous plaque formation, without any significant echo-shadow. The flow separation zone usually seen in the carotid sinus is absent. Instead, as far as the end of the plaque formation, a color reversal is seen as a sign of incipient flow turbulence. It is possible to interpret the color change in the distal segment of the ICA as a niche formation. Also noticeable with color bleaching is a flow acceleration, which serves as an initial sign of a stenosis in the ICA. Estimated degree of stenosis: mild, not hemodynamically significant ($\leqslant$ 70–80%).

Right: Proceeding from the terminal segment of the CCA and extending into the ICA, there is inhomogeneous plaque formation without any significant echo-shadow. Color bleaching is observed as a sign of flow acceleration. The often detectable flow separation boundary in the carotid sinus is absent.

Angiogram (a). *Left:* From a lateral viewpoint, extensive plaque formation is seen, with a relatively smooth surface, proceeding from the distal segment of the CCA to the upper proximal section of the ICA. In its distal segment, the plaque formation appears to have ill-defined boundaries, which may be a possible expression of a slight niche formation. The estimate of the degree of stenosis is affected by whether the local or distal degree of stenosis is used as the basis for determining this value. The local stenosis grade is approximately 70%, and the distal approximately 60%.

Right: From the distal end of the CCA up to and beyond the proximal ICA, there is a relatively smooth-surfaced and not significantly lumen-constricting plaque formation at the posterior wall of the blood vessel, without any detectable ulceration and without any significant degree of stenosis whether the grade of the stenosis is calculated locally or distally (< 50%).

Comment. Localized vascular wall changes that do not significantly obstruct the lumen can be imaged in several sections in the echotomogram. When compared to the angiogram, ultrasound produces a higher detection rate, and surface structures can be better imaged. With color flow sonography, it is also often possible to image ulcerative changes on the surface of the plaques, and the image is probably better than with angiography. Spectral analysis of Doppler signals on its own can at best provide indirect indications of a plaque formation that is not yet hemodynamically significant, e.g., abnormal audiosignals, a reduction in the width of the systolic window, or an increased appearance of low-frequency signal components. Carotid plaque formation at the origin of the ICA is often encountered with a loss of flow separation in the carotid sinus. However, this is not always the case, and separation phenomena may also be missed in the anatomical variant of an absent carotid sinus.

ECA   ICA   CCA   ICA

Left   Right

Moderate Degree of Stenosis of the Right Internal Carotid Artery
(Fig. 11.**4**)

Patient: S.F., male, age 58.
Clinical diagnosis: stroke in the left hemisphere.

CAD:	– Vascular surgery:	– Diabetes mellitus: –
PAOD:	– Heredity:	– Hyperlipidemia: +
TIA, stroke: +	Hypertension:	– Smoking: +

Doppler sonogram *(spectral analysis)* (**c**). The supratrochlear artery bilaterally has physiological flow. The carotid artery system on the left is normal. Carotid system on the right: CCA and ECA are normal. In the ICA, there is flow acceleration with systolic deceleration in the proximal ICA (stenosis region). There are poststenotic irregularities in the audiosignal, with preserved, but reduced flow waveforms.

At the site of stenosis, there is a broadening of the spectrum and an increase of the systolic peak frequency to 7 kHz.

Color flow duplex sonogram (b). At the transition of the distal CCA into the proximal ICA and extending beyond the carotid sinus, echo-intense plaque formation is seen, with a smooth surface clearly constricting the lumen in the longitudinal section. There is a clear bleaching of color at the origin of the ICA and a loss of the separation zone in the carotid sinus, with an irregular color mixture distal to the stenosis and superior to the bifurcation in the ICA.

Angiogram (a). Moderate degree of stenosis of the ICA after the bifurcation, with a smooth surface.

Comment. The stenosis can be well depicted using the echo-impulse imaging B-mode and can be graded as a moderate stenosis according to the flow parameters from either the Doppler sonogram or the color flow duplex analysis. In the angiogram as well, a moderate degree of stenosis is seen, particularly when the distal degree of stenosis is measured (70%). So far the patient has remained free of symptoms of ischemia in its territory.

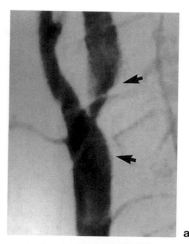

a

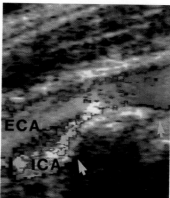

b

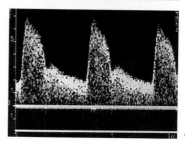

c

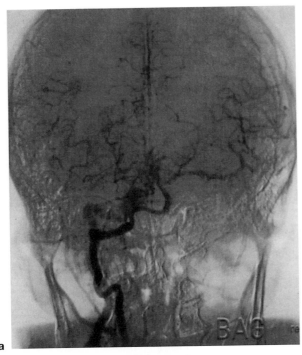

a

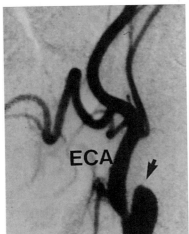

b

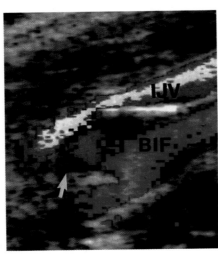

c

Left

Bilateral Carotid Disease: Occlusion of the Left Internal Carotid Artery, Mild Degree of Stenosis of the Right Internal Carotid Artery
(Fig. 11.**5**)

Patient: W.S., male, age 63.
Clinical diagnosis: asymptomatic aneurysm of the abdominal aorta.

CAD: –	Vascular surgery: –	Diabetes mellitus: –
PAOD: –	Heredity: –	Hyperlipidemia: –
TIA, stroke: –	Hypertension: –	Smoking: +

Doppler sonogram *(analogue registration)* (**d, h**) and *(spectral analysis)* (**f**). The left supratrochlear and supraorbital arteries have no spontaneously detectable flow, but show orthograde perfusion during compression of the ipsilateral superficial temporal artery. The right supratrochlear artery has flow direction toward the probe without any compression effect. In a side-to-side comparison, the left CCA presents a slightly reduced systolic but a clearly reduced diastolic flow velocity. The left ICA is not detectable, the ECA is smaller than on the right, and is almost identical to the ipsilateral CCA (not shown). On the right above the bifurcation, there is clear flow acceleration—both systolic and diastolic—in the proximal segment of the ICA. Spectral analysis shows a clear flow acceleration (peak frequency of 5500 Hz at a transmission frequency of 4 MHz) with a clear broadening of the systolic window. Proceeding distally, with pronounced irregularities in the audiosignal a systolic deceleration occurs, at moderately high diastolic flow velocities. An almost normal flow signal appears in the submandibular region. The vertebral and subclavian arteries are bilaterally normal.

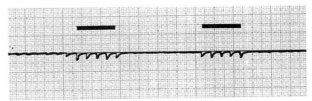

Left supratrochlear artery

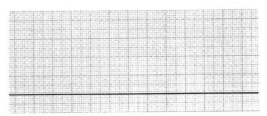

Left ICA

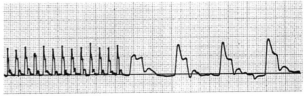

Left CCA

d

B-mode image (g). On the right, there is an extensive, flat heterogeneous plaque formation, with a smooth surface at the posterior wall of the CCA, extending into the proximal ICA without significant constriction of the lumen.

Color flow duplex sonogram (c). On the left, a typical forward-and-backward flow is seen, with systolic and diastolic homogeneous color coding, i.e., red or blue, respectively.

Angiogram (a, b, e). On the left, there is a cap-shaped occlusion of the ICA in the bifurcation region (**b**). On the right, no significant abnormality can be demonstrated (**e**). Intracranial collateralization of the territory of the left middle cerebral artery corresponds to the occlusion of the left ICA (**a**), predominantly through the right ICA.

Comment. In processes involving several blood vessels and affecting both carotid systems, the estimation of the degree of stenosis in the lower-grade flow obstruction is affected by the higher-grade one, since a compensatory collateral function has to be taken into account. Depending on the flow character of the Doppler sonogram in the present case, one would consider a moderate stenosis of the right ICA. Both B-mode image and angiogram suggest that there is only a low-grade plaque formation. Correspondingly the contralateral Doppler signals sometimes clearly change after recanalization of a high-grade stenosis, e.g., in normal conditions, the preoperative signs of a pseudo-low-grade stenosis may disappear, or parameters associated with a high-grade lesion may appear to be less pronounced.

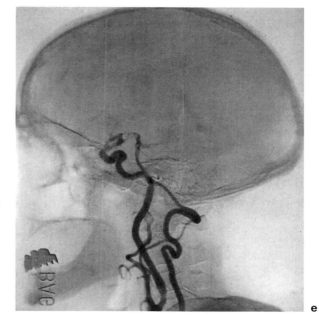

e

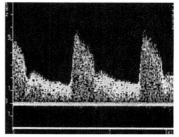

f

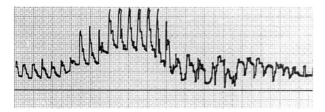

Right supratrochlear artery

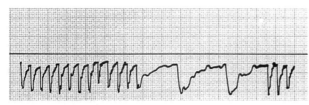

Bifurcation   Right ICA

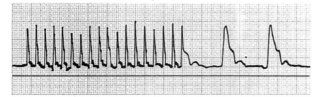

Right CCA   h

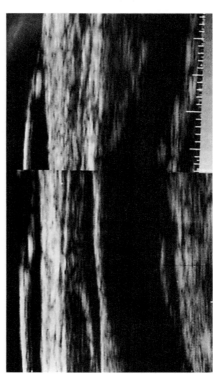

Right **g**

Carotid Bifurcation Disease: Combined Pseudo-Occlusion of the Left Internal Carotid Artery and Stenosis of the Left External Carotid Artery

(Fig. 11.**6**)

Patient: K.M., male, age 58.
Clinical diagnosis: recent cerebral infarction of the left hemisphere.

CAD:	–	Vascular surgery:	–	Diabetes mellitus:	–
PAOD:	+	Heredity:	+	Hyperlipidemia:	–
TIA, stroke:	+	Hypertension:	–	Smoking:	–

Doppler sonogram *(spectral analysis)* (**b**). The left supratrochlear artery shows retrograde flow, which decreases when the ipsilateral facial artery is compressed, but does not change flow direction. The right supratrochlear artery has orthograde flow. The entire extracranial carotid system on the right appears normal. The left CCA presents a clear reduction in diastolic flow and a broad systolic peak. Initially, there was no audiosignal detectable in the proximal ICA. Only after a lengthy search, and then only at certain points, was it possible to identify a high-frequency signal with few components of pronounced amplitude, predominantly up to 5 kHz (peak frequency > 10 kHz), without any compression effect (superficial temporal artery and facial artery). At the same location in the neck region, a loud signal from the ECA predominates with a clear flow acceleration, systolic deceleration, and high-frequency components up to peak frequencies above 9 kHz in the spectrum. There is a definite effect if the branches of the external carotid artery mentioned are compressed.

The vertebral and the subclavian arteries are bilaterally normal.

Color flow duplex sonogram (**c**). Above the carotid bifurcation, a tail-like course of the internal carotid artery with backflow components is detectable, although distally neither a flow signal nor a detectable high-frequency Doppler signal are recognizable in the spectrum. At the same time, the vortex zone from the stenosis of the external carotid artery, which is also located above the bifurcation, is depicted at certain points, without any significant flow acceleration in the color signal.

In the cross-section image more clearly than in the longitudinal section, the thread-like stenosis of the ICA can be seen in comparison with the larger lumen or less constricted ECA.

Angiogram (**a**). Angiography shows a typical pseudo-occlusion with a tail-like narrowing of the internal carotid artery above the carotid bifurcation, without complete closure of the distal neck section. Many branches of the external carotid artery are superimposed; the ascending pharyngeal artery should not be misinterpreted as a distal segment of the internal carotid artery. In the late angiogram, contrast medium uptake in the ICA extending all the way into the middle cerebral artery is demonstrable.

Comment. In processes involving the carotid bifurcation, it is sometimes difficult to differentiate between a flow obstruction in the ICA and one in the ECA. When the ECA functions as a collateral, the flow profile is often very similar to the physiological flow within the internal carotid artery. If a stenosis of the bifurcation is present, then only the presence of a compression effect can assist in the identification process. It must be determined whether the ICA is in fact also stenosed or occluded. If a signal from the ICA is not received, a high-grade or subtotal stenosis may nevertheless still be present. However, due to the significant flow changes and limitations of the audiosignal caused by the stenosis of the external carotid artery, this cannot always be verified. In such cases, B-mode imaging alone is not helpful, particularly when the examiner only uses a longitudinal section. In the present case, the cross-sectional color flow duplex study and the careful FFT spectral analysis do confirm the presence of a pseudo-occlusion in the ICA, in contrast to what might have been, suspected to be a complete occlusion. Also, in the angiogram, it is often only possible to determine whether a blood vessel is still patent or already totally blocked if delayed sequences using increased contrast medium dosages for intra-arterial injections are selected.

Retrograde flow in the fronto-orbital branches of the ophthalmic artery can be supplied through the contralateral ECA branches in disease of the carotid bifurcation. This is a characteristic, but not a necessary finding particularly in hemodynamically significant stenoses of the internal carotid artery.

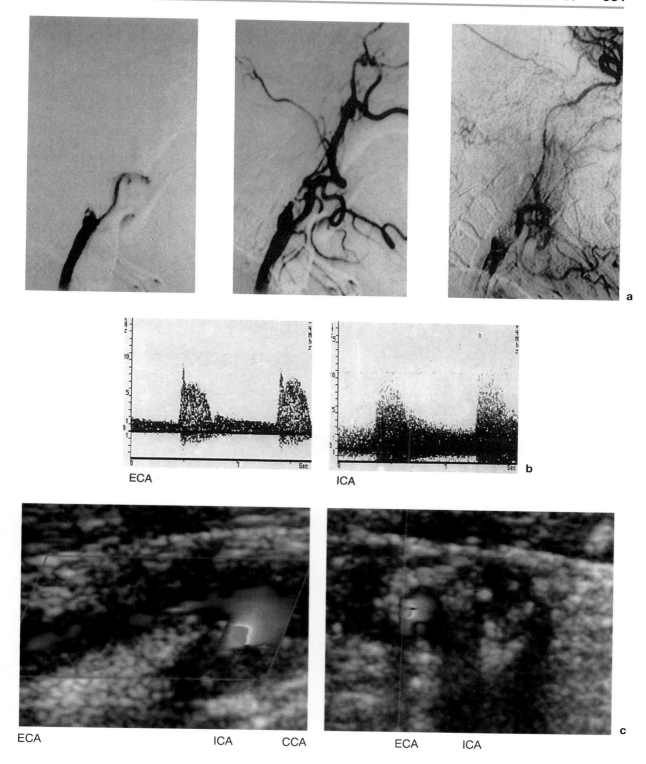

ECA   ICA   b

ECA   ICA   CCA   ECA   ICA   c

Aneurysm of the Right Carotid Bifurcation (Fig. 11.7)

Patient: H.H., male, age 43.
Clinical diagnosis: TIAs and right carotid endarterectomy three years ago, now recurrent TIAs with sensorimotor hemiparesis on the left.

CAD:	+	Vascular surgery:	+	Diabetes mellitus:	+
PAOD:	+	Heredity:	+	Hyperlipidemia:	+
TIA, stroke:	+	Hypertension:	+	Smoking:	+

Doppler sonogram *(spectral analysis)* (**a**). At the origin of the ICA, there is a turbulent flow signal with a low amplitude. Spectral analysis confirms a biphasic distribution with low systolic and diastolic peak velocities.

Color flow duplex sonogram (**b**). In the area of the aneurysm, there is a clear change in the flow direction during individual phases of the cardiac cycle, especially in the systole and early diastole. During the diastole, no flow is seen within the aneurysm itself.

Angiogram (**c**). Marked dilatation of the carotid artery with wall irregularities, especially in the distal common carotid artery.

Comment. Aneurysms of the extracranial carotid vascular system can be superbly depicted in B-mode imaging and color flow duplex sonography. With Doppler sonography, their presence may be missed or only assumed.

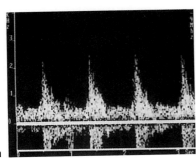

a

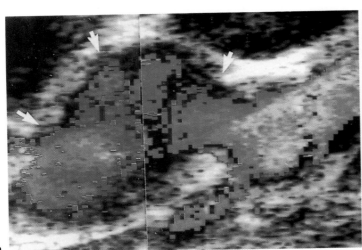

b

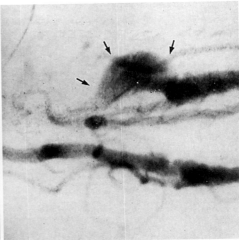

c

Dissection of the Left Internal Carotid Artery (Fig. 11.**8**)

Patient: A.T., male, age 39.
Clinical diagnosis: Horner's syndrome on the left, TIA with sensorimotor hemiparesis on the right.

CAD:	– Vascular surgery:	– Diabetes mellitus:	–
PAOD:	– Heredity:	– Hyperlipidemia:	–
TIA, stroke:	+ Hypertension:	– Smoking:	–

Doppler sonogram *(spectral analysis)* (**a, b**). The supratrochlear artery shows asymmetric, orthograde flow with a reduced flow amplitude on the left, but no change following compression of the branches of the ECA.

The entire extracranial carotid system on the right is normal. In a side-to-side comparison, the flow velocity in the left CCA is reduced. From the bifurcation up to the submandibular region in the ICA, there is a detectable slosh phenomenon with synchronous forward-and-backward components in the systole but an absent signal during diastole (upper spectrum).

The vertebral and the subclavian arteries are normal on both sides.

In a follow-up examination after three weeks, the left ICA has completely recanalized (lower spectrum).

Color flow duplex sonogram (**b**). In the longitudinal section, forward-and-backward signal components in blue–red color coding are seen next to one another in the proxima ICA. Distally, an area free of flow signals marks the proximal end of the dissection. Corresponding Doppler signals characterize partial recanalization with systolic forward-and-backward signal components, but with diastolic forward flow preservation.

Angiogram (**c**). Proximally, there is a thread-like occlusion/subtotal stenosis of the ICA without a connection to the intracranial vasculature (thick arrow). The carotid siphon is supplied into the preclinoid segment through the anastomosis of the maxillary artery from the ophthalmic artery, in which blood flows in a retrograde direction. By way of this anastomosis, good filling of the middle cerebral artery also results. After recanalization, the high-grade submandibular stenosis is still seen in the follow-up examination as the point of origin for the dissection (thin arrow). It has irregularities in the intraosseous canal, but the perfusion of the carotid sinus is once again normal and physiological.

Comment. Much more often than previously suspected, the highly characteristic Doppler-sonographic correlate of a dissection is seen in young patients with TIA/stroke. The dissection can appear spontaneously, or may follow traumatic events accompanied by the fully developed picture of focal ischemia with facial and neck pain and Horner's syndrome. It can also appear with very few symptoms, or may even be completely asymptomatic. In combination with the color flow duplex sonogram, it is almost always possible to make the diagnosis when the flow signal is carefully followed *over the entire neck region*. Monthly follow-up assessments are important, since in the majority of the cases spontaneous recanalization occurs.

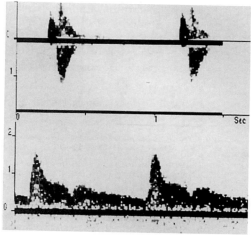

a

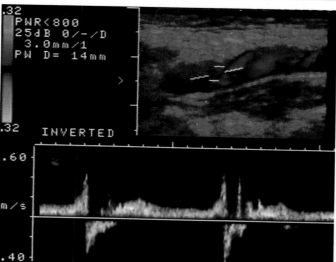

b

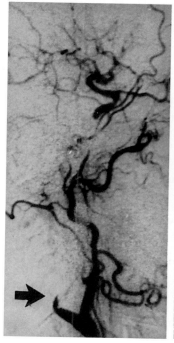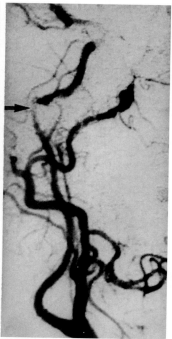

c

Ipsilateral Multilevel Carotid Disease: Stenosis at the Siphon and the Bifurcation of the Left Internal Carotid Artery in the Neck
(Fig. 11.**9**)

Patient: H.S., male, age 64.
Clinical diagnosis: stroke in the left hemisphere.

CAD:	–	Vascular surgery:	–	Diabetes mellitus:	–
PAOD:	–	Heredity:	–	Hyperlipidemia:	+
TIA, stroke:	+	Hypertension:	+	Smoking:	+

Doppler sonogram *(analogue registration)* (**c**) and *(spectral analysis)* (**d**). The systolic flow velocity in the left supratrochlear artery is lower than on the right. Flow is bilaterally orthograde. When compared to the right side, the left CCA, with a slight change in the waveform, clearly has a reduced diastolic flow velocity. Above the bifurcation, there is a localized diastolic flow acceleration with an increase in the systolic amplitude modulation. Further cranially, the flow signals are again reduced.

In the spectrum at the site of the extracranial stenosis (**d**), a clear broadening of the window can be seen, with an accumulation of low-frequency signals, especially during the diastole, and slightly elevated frequency values.

The right carotid system is normal. The vertebral and subclavian arteries are bilaterally normal.

Transorbital and extracranial pulsed wave Doppler sonograms (**e**). Using the transorbital access on the right, symmetrical spectra from the carotid segment can be registered at a depth of 70 mm, before (C_4 segment, curve deflection upward) and after (C_2 segment, curve deflection downward) the origin of the ophthalmic artery (average flow velocity 26 cm/s or 666 Hz). On the left, the proximal stenosis of the siphon can be recognized from the flow acceleration, with an increase in the average flow velocity (46 cm/s or 1177 Hz) in the presence of an extremely disturbed audio-signal, extending into the poststenotic segment. With a slight tilting of the probe at the same location, the distal carotid siphon can also be evaluated. A stenosis is not detectable here.

Angiogram (**a**). A moderate degree of stenosis distal to the bifurcation of the ICA and additional mild-grade stenoses at the origin of the ophthalmic artery in the intracranial carotid segment (carotid siphon) are demonstrated.

Color flow duplex sonogram (**b**). At the origin of the ICA, there is a small, smooth, homogeneous plaque formation, with moderate constriction of the lumen. Slight flow acceleration is demonstrated by the color bleaching in the color-coded image. At the carotid sinus, the usual flow separation zone, with blue-coded color components, is missing.

Comment. The mild flow acceleration at the origin of the ICA and the plaque formation in the B-mode image do not explain the clear asymmetry between the two diastolic flow velocities in the CCAs. From this finding, it has to be concluded that a second flow obstruction in-

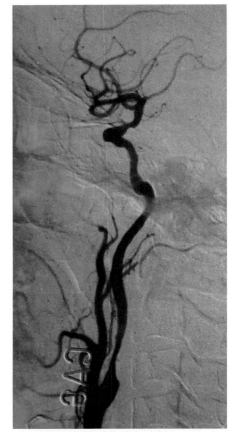

a

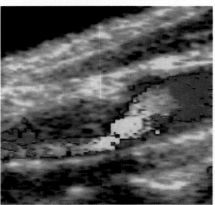

b

troducing a hemodynamically significant obstruction should, exist beyond the immediate area of direct ultrasound application. A comparison of the fronto-orbital terminal branches of the ophthalmic artery, which show greater flow on the right, suggests that the location of the stenosis should be sought on the left before the origin of this blood vessel. Alternatively, a stenosis at the origin of the left CCA might be possible. Pulsed wave Doppler sonography with an examination of both regions (immediately supra-aortal from the supraclavicular fossa, and in the carotid siphon from a transorbital approach) allows further differentiation. However, as in the present example, flow obstructions in the proximal siphon segment of the carotid artery are easier to detect than distal plaque formations.

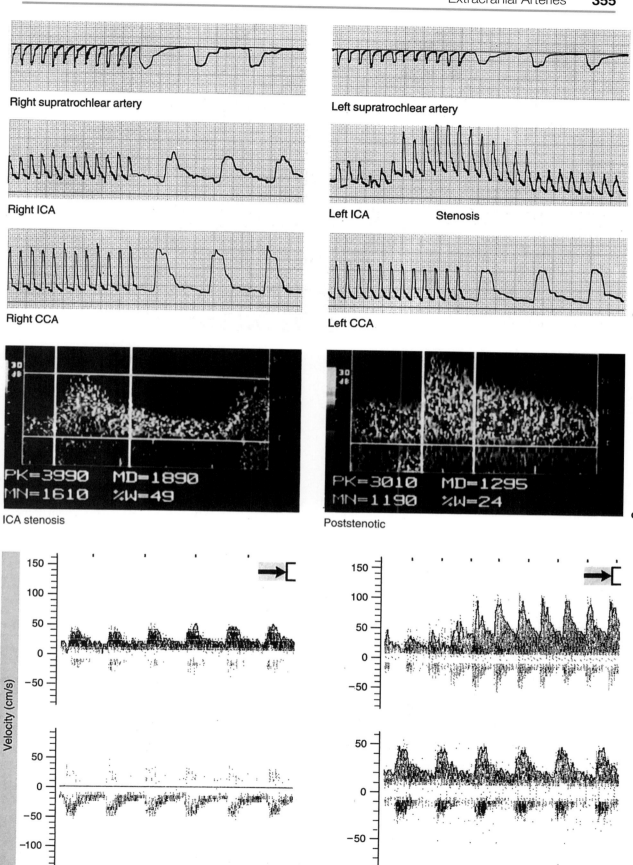

Right supratrochlear artery

Left supratrochlear artery

Right ICA

Left ICA   Stenosis

Right CCA

Left CCA   c

PK=3990   MD=1890
MN=1610   %W=49

ICA stenosis

PK=3010   MD=1295
MN=1190   %W=24

Poststenotic   d

Right ICA siphon

Left ICA siphon   e

Severe Stenosis of the Distal Internal Carotid Artery at the Carotid Siphon Proximal to the Origin of the Ophthalmic Artery
(Fig. 11.**10**)

Patient: M.Z., male, age 53.
Clinical diagnosis: stroke in the left hemisphere.

CAD:	–	Vascular surgery:	–	Diabetes mellitus:	–
PAOD:	–	Heredity:	–	Hyperlipidemia:	+
TIA, stroke:	+	Hypertension:	–	Smoking:	+

Doppler sonogram *(analogue registration)* (**b**) and *(spectral analysis)* (**c, d**). The left supratrochlear artery has retrograde blood flow, and shows only a slight flow reduction on compression of the external carotid artery branch. Flow is orthograde on the right side. The CCA and ICA on the left show clearly less pronounced diastolic flow when compared to the contralateral side. Amplitude modulation is not reduced. There are no direct signs of flow obstruction. The frequency analysis also shows asymmetries with higher peak frequencies on the right side (CCA 2310 Hz/1925 Hz and ICA 3150 Hz/2030 Hz).

The vertebral and the subclavian arteries are normal on both sides.

With extracranial application of the *pulsed wave Doppler,* the stenosis in the extracranial carotid segment can be identified in the submandibular region above the coiling, with a broadening of the spectrum. There is a flow increase in the peak and mean flow velocities, and bidirectional signals are observed around the baseline. From a transorbital (C_4 segment) access, extreme changes in the flow signal are also evident as a sign of an additional stenosis in the carotid siphon.

Angiogram (**a**). A high-grade stenosis of the left ICA is seen after loop formation (coiling) before the vessel enters the base of the skull. In addition, there is a stenosis in the carotid siphon (arrows).

Comment. A high-grade extracranial vascular obstruction at the base of the skull can be diagnosed using continuous wave Doppler sonography only by indirect criteria. If hemodynamic compromise is small, no changes may be evident in the extracranial vascular system. Color flow duplex system analysis may easily miss this condition if only images are displayed but additional recordings of the FFT-spectra—especially of the CCA and in a side-to-side comparison—are disregarded. When detecting a high-grade stenosis that lies extracranially at the base of the skull or intracranially in the siphon prior to the origin of the ophthalmic artery (C_4 or C_5 segment), or in the distal section of the internal carotid artery after its origin (C_1 or C_2 segment), asymmetry in the CCA's Doppler waveforms, together with the findings at the terminal branches of the ophthalmic artery, are particularly important. Positional changes in the ICA can result in individually differing flow signals, so that, in a side-to-side comparison, variations in the waveforms of these arteries are not so important as those in the CCAs. In the differential diagnosis, this finding has to be distinguished from a proximal stenosis at the origin of the common carotid artery. The situation can be clarified with extracranial pulsed wave Doppler sonography (Fig. 11.**2**).

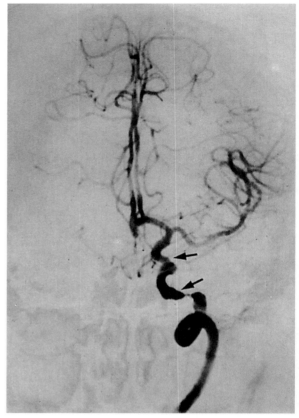

a

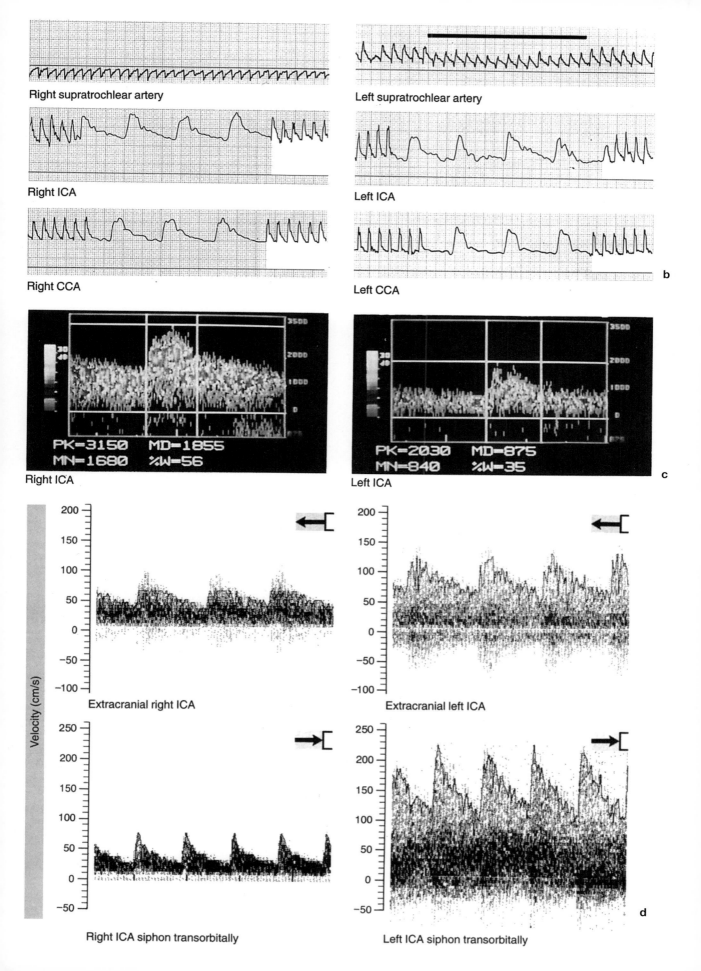

Right supratrochlear artery

Left supratrochlear artery

Right ICA

Left ICA

Right CCA

Left CCA

b

PK=3150 MD=1855
MN=1680 %W=56

Right ICA

PK=2030 MD=875
MN=840 %W=35

Left ICA

c

Extracranial right ICA

Extracranial left ICA

Right ICA siphon transorbitally

Left ICA siphon transorbitally

d

Velocity (cm/s)

Vertebral, Subclavian and Innominate Artery Diseases

Temporary Subclavian Steal Phenomenon on the Left
(Fig. 11.**11**)

Patient: H.S., female, age 65.
Clinical diagnosis: syncope of undetermined etiology.

CAD:	– Vascular surgery:	+ Diabetes mellitus:	–
PAOD:	– Heredity:	+ Hyperlipidemia:	–
TIA, stroke: –	Hypertension:	– Smoking:	+

Doppler sonogram *(analogue registration)* (**b**) and *(spectral analysis)* (**c**). The left vertebral artery shows an early systolic decrease in flow, with blood flow running cranially during the diastole. After the ipsilateral arteries of the upper arm open during the phase of reactive hyperemia, the diastolic flow directed toward the brain momentarily decreases. The left subclavian artery presents with a complete transformation of its usual waveform, and particularly with absent early diastolic reflux. The right vertebral artery is normal without impairment of the flow signal when the left upper arm is compressed with the blood pressure cuff. The right subclavian artery is normal.

The *pulsed wave Doppler (transnuchal insonation)* (**c**) depicts the confluence of both vertebral arteries at the origin of the basilar artery at a depth of 80 mm. The flow signal from the right vertebral artery (65 mm depth) and the left vertebral artery at the same level varies in a comparable manner to that seen with the extracranial continuous wave Doppler. At a depth of 90–100 mm, only signals from the basilar artery can be recorded.

Angiogram (**a**). The complex hemodynamics of this patient's intracranial circulation are clarified by selective angiography of the right vertebral artery. In the first image (**1**), there is, in addition to filling of the basilar artery, a retrograde flow into the left vertebral artery. During the subsequent phase (**2**), the right posterior cerebral artery is highlighted, while in (**3**) retrograde inflow into the left vertebral artery is once again seen. In the final phase (**4**), both posterior cerebral arteries are again displayed.

Comment. The pendular flow in the vertebral artery is an expression of a temporary subclavian steal phenomenon when the ipsilateral subclavian artery has a high-grade obstruction proximal to the origin of the vertebral artery; in addition, hypoplasia of the vertebral artery is often involved. The contralateral vertebral artery commonly serves as a collateral supplying the brain, and as in this case, exclusively supplies the basilar artery. There is no vertebrovertebral shunt.

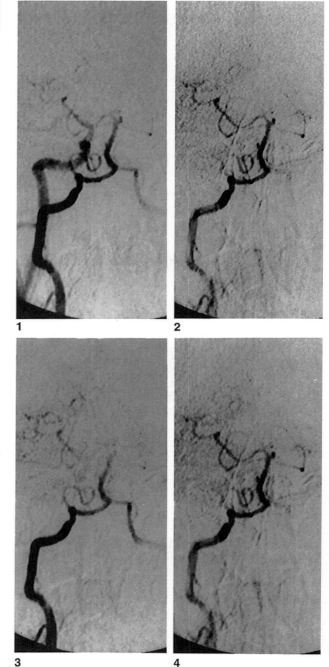

1   2

3   4   a

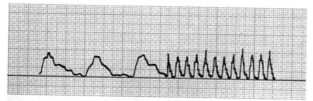

Right vertebral artery at the origin

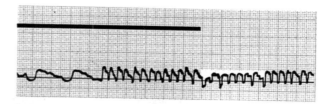

Left vertebral artery at the origin

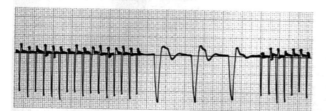

Proximal right subclavian artery

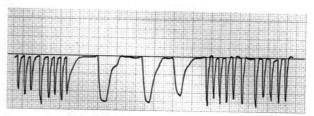

Proximal left subclavian artery

b

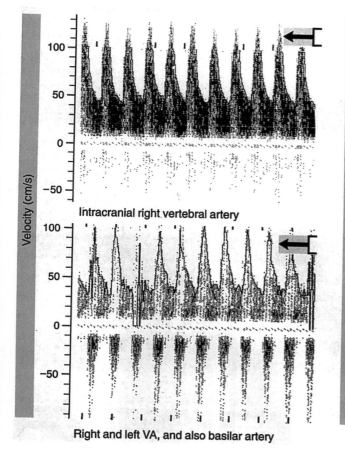

Intracranial right vertebral artery

Right and left VA, and also basilar artery

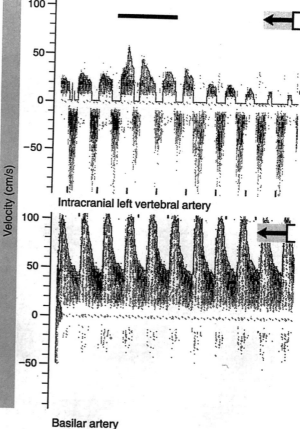

Intracranial left vertebral artery

Basilar artery

c

Permanent Subclavian Steal Phenomenon on the Left
(Fig. 11.**12**)

Patient: E.E., male, age 58.
Clinical diagnosis: asymptomatic.

CAD:	+	Vascular surgery:	–	Diabetes mellitus: +
PAOD:	–	Heredity:	–	Hyperlipidemia: –
TIA, stroke: –		Hypertension:	–	Smoking: –

Extracranial continuous wave Doppler sonogram *(analogue registration)* (**b**). The left vertebral artery has a pathological waveform and a positive compression effect: diastolic residual flow is absent during vascular compression of the left upper arm, and after reopening there is a transient diastolic flow acceleration for several cardiac cycles (reactive hyperemia). The left subclavian artery shows a similar waveform and compression effect (not shown). The right vertebral artery has a normal flow signal but also shows a positive compression effect during *left-sided* upper arm compression (vertebrovertebral shunt). The right subclavian artery is normal.

Extracranial pulsed wave Doppler sonogram *(spectral analysis)* (**c**). At the origin of the left subclavian artery close to the aorta, no signal can be registered. Distally, as in the vertebral artery, a pathological waveform is seen. In comparison, the right subclavian artery appears to be normally depicted at its origin from the innominate artery.

Angiogram (**a**). During retrograde angiography, the right vertebral artery, basilar artery, also the left vertebral artery are filled from the brachial artery. In later images, proximal occlusion of the subclavian artery is seen, with the blood supply of the distal subclavian artery proceeding through the ipsilateral retrograde vertebral artery. (The images shown are 1.0 s, 1.7 s, and 2.3 s after contrast medium injection).

Comment. A subclavian steal phenomenon that can only be demonstrated by either Doppler sonography, angiography, or both, but does not cause any neurological or peripheral deficits in the affected upper extremity should not be described as *subclavian steal syndrome and needs no further treatment.*

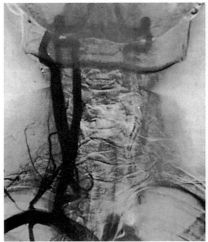
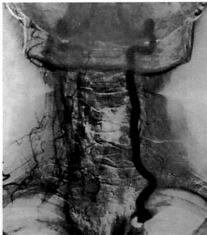
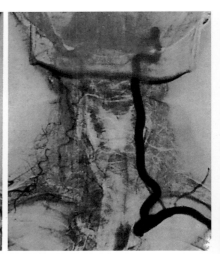

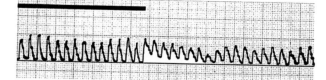

Right vertebral artery at the mastoid process

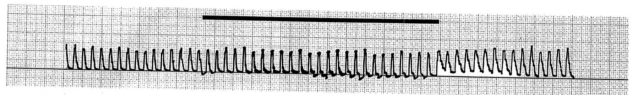

Left vertebral artery at the mastoid process

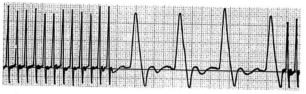

Distal right subclavian artery

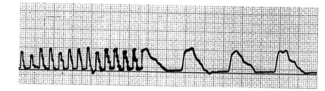

Distal left subclavian artery

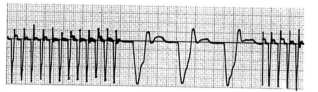

Proximal right subclavian artery

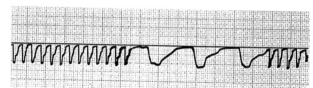

Proximal left subclavian artery

b

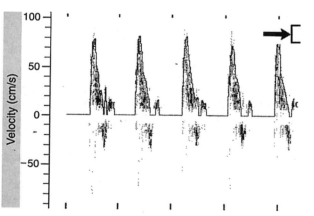

Right subclavian artery (normal, depth: 55 mm)

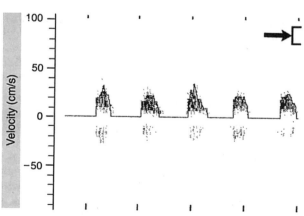

Distal left subclavian artery (occlusion, depth: 50 mm)

c

Severe Stenosis of the Innominate Artery and Bilateral Occlusion of the Subclavian Arteries
(Fig. 11.**13**)

Patient: L.K., male, age 42.
Clinical diagnosis: Takayasu's arteritis.
Blood pressure on the left is not measurable, on the right 80/?

CAD: –	Vascular surgery: –	Diabetes mellitus: –
PAOD: –	Heredity: –	Hyperlipidemia: –
TIA, stroke: –	Hypertension: +	Smoking: +

Doppler sonogram *(analogue registration)* (**b**). The supratrochlear artery presents with orthograde blood flow direction bilaterally, however, acoustically, a breath-like sound was already audible on the right side. The right CCA, like the ICA, shows a pathological flow pattern, with minimal amplitude modulation and a small systolic peak velocity. After vascular compression of the right upper arm with the blood pressure cuff, the Doppler signal changed only acoustically.

The left CCA and ICA show a high diastolic flow signal. The right vertebral artery has a poor signal with systolic deceleration but without any definite compression effect. The right subclavian artery is missing.

The left vertebral artery also suggests systolic deceleration without a compression effect. The left subclavian artery cannot be detected with certainty.

A *second registration made 14 days later* (not illustrated) shows pendular flow in the right ICA, with a temporary carotid steal phenomenon during the phase of reactive hyperemia. Similar reactions are present in the CCA and ECA, accompanied by spontaneous systolic deceleration. In addition, an increase in the pulsatile flow on vascular compression of the upper arm was observed. Spontaneous changes of this sort are often encountered when there is variable flow equilibrium in a large blood vessel.

Transcranial (d) and extracranial pulsed wave Doppler sonogram *(spectral analysis)* (**c**). In the left CCA, the frequency spectrum is normal at the origin of the blood vessel from the aortic arch. The high-grade stenosis of the innominate artery can be clearly recognized from the broad frequency band and high systolic peak velocity.

There is a suggestion of intermediate flow in both intracranial vertebral arteries and in the basilar artery, which can be registered at a sample volume depth of 105 mm and a low *mean flow velocity* (32 cm/s or 820 Hz). Systolic deceleration is also present in the right MCA, with a reduced mean flow velocity (512 Hz or 20 cm/s on the right; 1538 Hz or 60 cm/s on the left) and in the right ACA (358 Hz or 14 cm/s on the right). There is no change in the signals of the ACA, MCA, and BA on vascular compression of the right upper arm. The elevated flow in the left ACA is a sign of intracranial collateralization through the circle of Willis (2102 Hz or 82 cm/s).

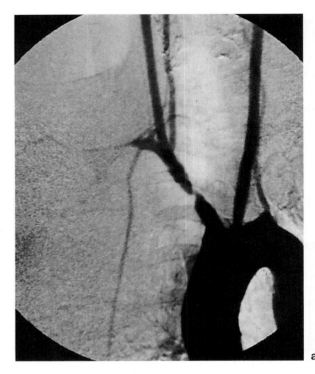

a

Angiogram (a). A high-grade stenosis of the innominate artery and a bilateral occlusion of the subclavian artery are seen, with the left vertebral artery originating directly from the aortic arch.

Comment. Obstructive processes near the aorta can be evaluated with continuous wave Doppler sonography only by indirect criteria. In this instance, *pulsed wave Doppler sonography* provides supplementary assistance. The low perfusion in the ipsilateral carotid and vertebral arterial systems that follow the severe obstruction of the innominate artery results in a poor detection of further wall changes based on hemodynamic criteria alone. *B-mode imaging* should therefore also be used to provide supplementary information, since surgical reconstruction of intrathoracic lesions improves perfusion and may generate secondary embolism from distal lesions. The *transcranial Doppler* can confirm the intracranial collateralization and also detect (an extremely rare occurrence) a retrograde perfusion of the arteries at the base of the brain (especially in the basilar artery).

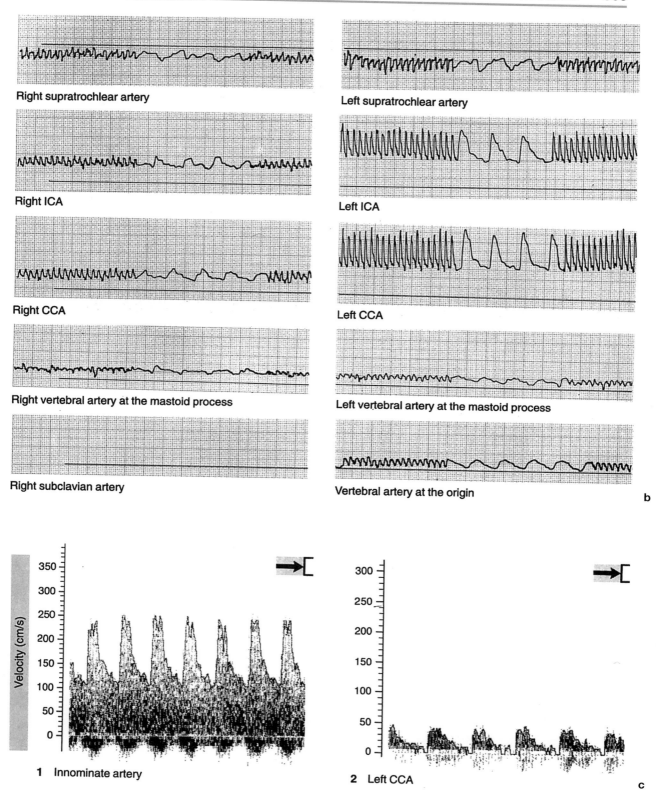

Right supratrochlear artery

Left supratrochlear artery

Right ICA

Left ICA

Right CCA

Left CCA

Right vertebral artery at the mastoid process

Left vertebral artery at the mastoid process

Right subclavian artery

Vertebral artery at the origin

b

1 Innominate artery

Velocity (cm/s)

2 Left CCA

c

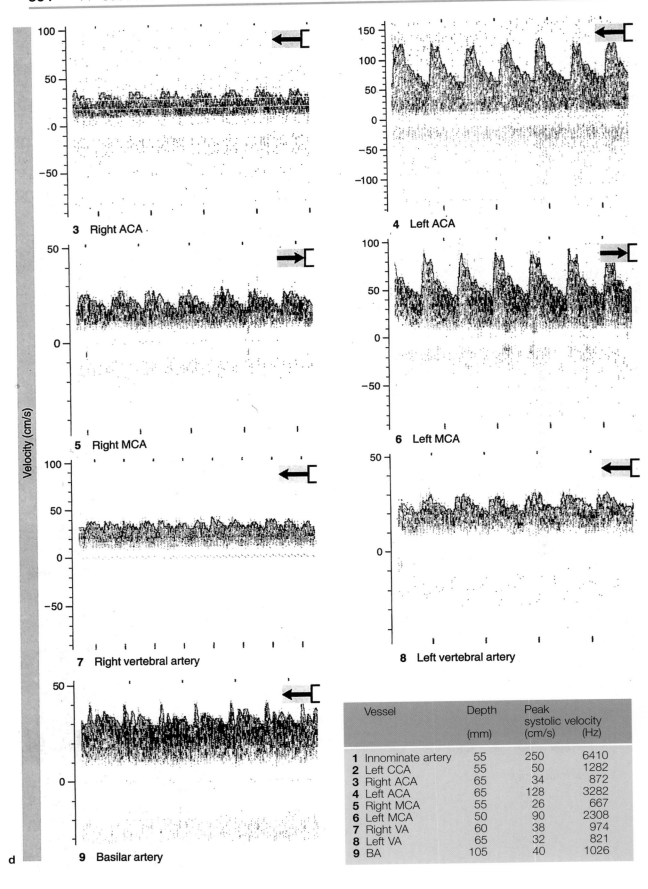

Velocity (cm/s)

3 Right ACA

4 Left ACA

5 Right MCA

6 Left MCA

7 Right vertebral artery

8 Left vertebral artery

9 Basilar artery

d

Vessel	Depth (mm)	Peak systolic velocity (cm/s)	(Hz)
1 Innominate artery	55	250	6410
2 Left CCA	55	50	1282
3 Right ACA	65	34	872
4 Left ACA	65	128	3282
5 Right MCA	55	26	667
6 Left MCA	50	90	2308
7 Right VA	60	38	974
8 Left VA	65	32	821
9 BA	105	40	1026

Stenosis of the Vertebrobasilar Arteries
(Fig. 11.**14**)

Patient: N.W., male, age 60.
Clinical diagnosis: recurrent TIA in the brain stem.

CAD: – Vascular surgery: – Diabetes mellitus: –
PAOD: – Heredity: – Hyperlipidemia: –
TIA, stroke: + Hypertension: + Smoking: –

Extracranial Doppler sonogram. The fronto-orbital branches of the ophthalmic arteries and both extracranial carotid systems are normal. The vertebral arteries present clear asymmetries, with a small systolic and low diastolic flow signal in the left vertebral artery, compared to a normal signal in the right vertebral artery.

Transnuchal Doppler sonograms *(spectral analysis)* (**a**). A clear flow acceleration is seen in the intracranial transition zone between both vertebral and the basilar arteries. Transnuchal application of ultrasound produces bidirectional high-amplitude signals in the low-frequency range, with high systolic flow velocities as a sign of stenosis (**2**). Distally, there is an almost completely normal Doppler signal (**1**).

Angiogram. In conventional X-ray angiography (**b**) and magnetic resonance angiography (**c**), a high-grade stenosis at the transition from the vertebral artery to the basilar artery is seen, with subsequent good filling of the basilar artery and its branches up to the top of the basilar artery. The left vertebral artery is hypoplastic.

Comment. It is often not possible to locate the stenosis between the confluence of the two vertebral arteries and the basilar artery using the Doppler-sonographic findings alone—even color flow duplex sonograms may fail to display the correct anatomy. In this case, magnetic resonance angiography (MRA) provides excellent noninvasive confirmation of the Doppler findings, which in turn compensate for potential MRA artifacts, in some cases making invasive angiography unnecessary.

Vessel	Depth (mm)	Flow velocity (cm/s) Peak	Mean
1 Distal BA	105	82	56
2 VA/BA confluence	85	142	88
3 Right VA	65	70	44

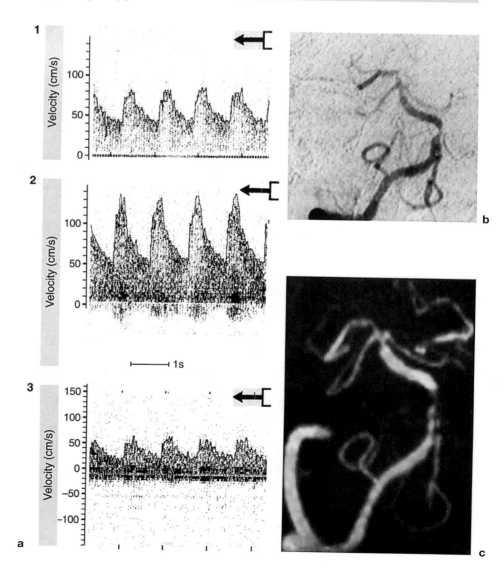

Stenosis of the Middle Cerebral Artery (Stenosis of the Left M₁ Segment)
(Fig. 11.15)

Patient: R.M., female, age 43.
Clinical diagnosis: stroke in the left middle cerebral artery territory.

CAD: – Vascular surgery: – Diabetes mellitus: –
PAOD: – Heredity: – Hyperlipidemia: –
TIA, stroke: + Hypertension: – Smoking: –

Extracranial Doppler sonogram (*analogue registration*) (**b**). The supratrochlear arteries bilaterally show orthograde blood flow direction, but there is a slight asymmetry of the diastolic signal which favors the left side. When compared side-to-side, flow curves in the ICA are almost the same with regard to their amplitude modulation, while the diastolic flow in both blood vessels is slightly reduced on the left. The left CCA has a clear reduction in its amplitude modulation and flow velocity.

The vertebral and the subclavian arteries are bilaterally normal.

Transcranial Doppler sonogram (*spectral analysis*) (**c**). Left: Between 55 mm and 65 mm, there is a flow reduction accompanied by extreme turbulences (**4**). At 45 mm, amplitude-intensive, slow flow velocities are noted, with a peak systolic flow acceleration (**2**). Poststenotically, the flow signal is attenuated.

Right: In the middle cerebral artery, the flow velocity relationships are normal (**1, 3**).

Angiogram (**a**). Moderate stenosis of the M₁ segment in the left middle cerebral artery.

Comment. Even high-grade, hemodynamically effective flow obstructions of the intracranial arteries at the base of the skull can be overlooked by extracranial Doppler sonography when a functioning collateral through the contralateral side or through the posterior cerebrovascular system is available, or alternatively, if other collaterals are supplied from the preceding vascular segment. In the present case, the intracranial and extracranial findings complement one another, to the extent that extracranial flow changes exist that can be classified more precisely with transcranial Doppler sonography.

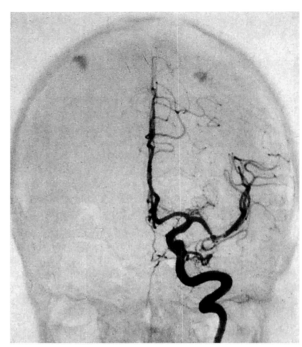

a

Artery	Depth (mm)	Flow velocity (cm/s)	
		Peak	Mean
1 Right distal MCA	45	64	42
2 Left distal MCA	45	216	132
3 Right proximal MCA	65	88	56
4 Left proximal MCA with bifurcation of the ICA	65	76	42

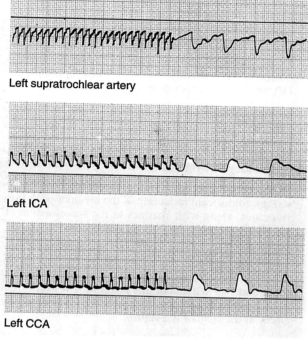

Right supratrochlear artery

Left supratrochlear artery

Right ICA

Left ICA

Right CCA

Left CCA

b

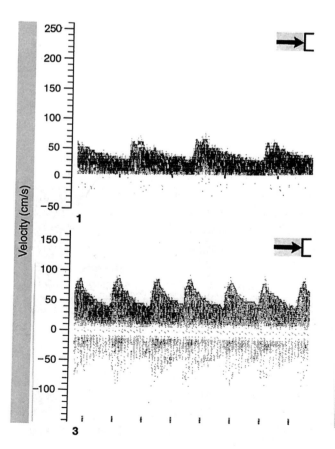

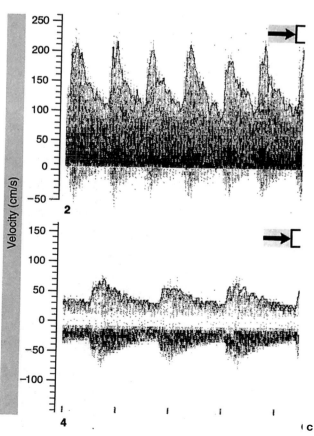

c

Dilatative Arteriopathy
(Fig. 11.**16**)

Patient: W.K., male, age 71.
Clinical diagnosis: vascular dementia.

CAD: +	Vascular surgery: –	Diabetes mellitus: +
PAOD: +	Heredity: –	Hyperlipidemia: –
TIA, stroke: +	Hypertension: +	Smoking: +

Extracranial Doppler sonogram (*analogue registration*) (**d**). Orthograde flow is bilaterally present in the supratrochlear arteries. CCA and ICA show a significant reduction of the flow waveform as a sign of dilatative arteriopathy (not shown).

Transcranial Doppler sonogram (**e**). Significant reduction of the mean or peak flow velocities, along with changes in the waveform in all the arteries near the base of the skull.

Computer tomogram (**c**). After application of contrast medium, point-shaped, hyperdense areas are observed in a parasellar location, in the suprasellar cisterns, and also in a prepontine position. These are signs of a dilatation in both carotid systems and in the basilar artery at the base of the skull.

B-mode image (**b**). Dilatation of the carotid segment.

Angiogram (**a**). Severe dilatative arteriopathy of both the extracranial and intracranial vascular segments, especially in the carotid, but also in the vertebral and basilar arteries.

Comment. While the normal aging process is often accompanied by a change in the flow velocity waveforms extracranially, which is frequently less pronounced intracranially or limited to only the basilar artery, the current findings indicate a severe pathological wall change in all intracranial and extracranial brain arteries.

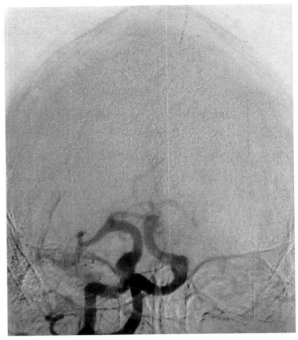

a

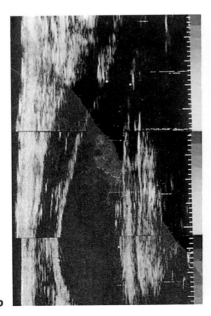

b

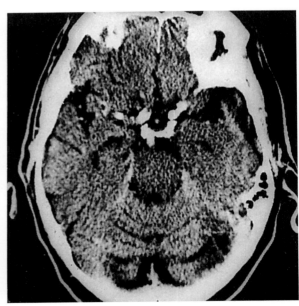

c

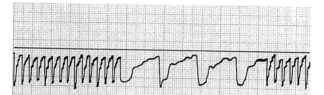

Right supratrochlear artery

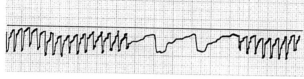

Left supratrochlear artery

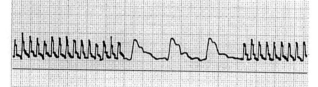

Right ICA

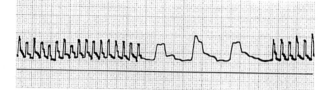

Left ICA

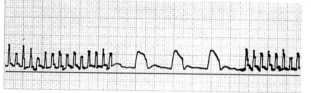

Right CCA

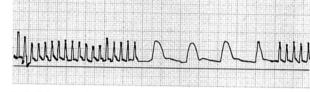

Left CCA

d

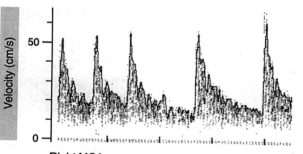

Right MCA
(max. systolic flow velocity: 56 cm/s,
average flow velocity: 24 cm/s)

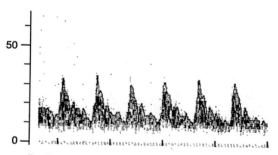

Basilar artery
(max. systolic flow velocity: 34 cm/s,
average flow velocity: 16 cm/s)

e

Intracerebral Angioma of the Left Hemisphere
(Fig. 11.**17**)

Patient: G.S., male, age 42.
Clinical diagnosis: cerebral seizures.

CAD:	−	Vascular surgery:	−	Diabetes mellitus:	+
PAOD:	−	Heredity:	−	Hyperlipidemia:	−
TIA, stroke:	−	Hypertension:	−	Smoking:	+

Doppler sonogram *(analogue registration)* (**c**). Bilateral orthograde flow in the supratrochlear arteries, with reduced diastolic flow on the left. The right extracranial carotid system is normal. In the left CCA and left ICA, elevated diastolic flow velocity and reduced amplitude modulation is noted throughout the entire length of the vascular segment.

Transorbital and transcranial Doppler sonograms *(spectral analysis)* (**d**). There is a clear flow acceleration within the intracranial carotid siphon on the left. With transorbital ultrasound, elevated peak (right 108 cm/s or 2769 Hz, left 148 cm/s or 3795 Hz) and mean flow velocities (right 76 cm/s or 1949 Hz, left 112 cm/s or 2872 Hz) are seen, and also bidirectionally high signal amplitudes of slow flow velocities near the baseline, which is a sign of a functional stenosis with an elevated flow volume. The ipsilateral anterior cerebral artery also shows an elevated peak (left 220 cm/s or 5641 Hz) and mean flow velocity (left 142 cm/s or 3641 Hz), with reversed flow direction.

Angiogram (a, b). Arteriovenous malformation in the left temporal region, which is mainly supplied through the middle cerebral artery from both carotid arteries.

Comment. In arteriovenous malformations that have a large shunt volume, the ipsilateral ophthalmic artery, along with its fronto-orbital branches, can present with reduced blood flow.

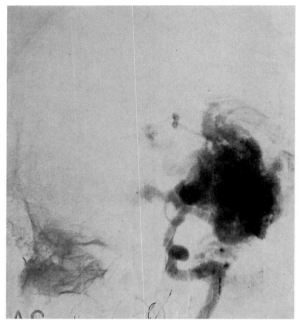

a

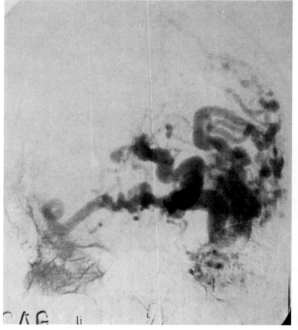

b

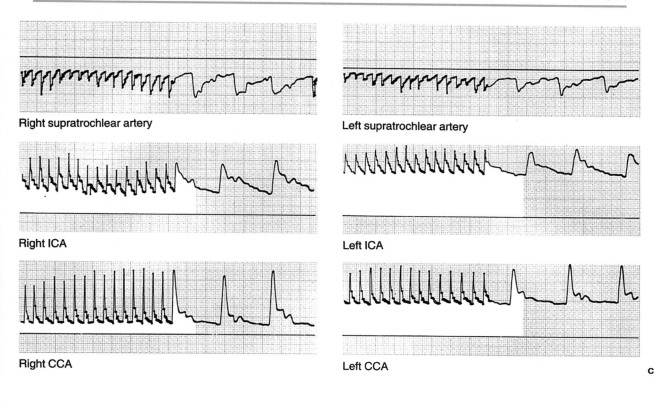

Right supratrochlear artery

Left supratrochlear artery

Right ICA

Left ICA

Right CCA

Left CCA

c

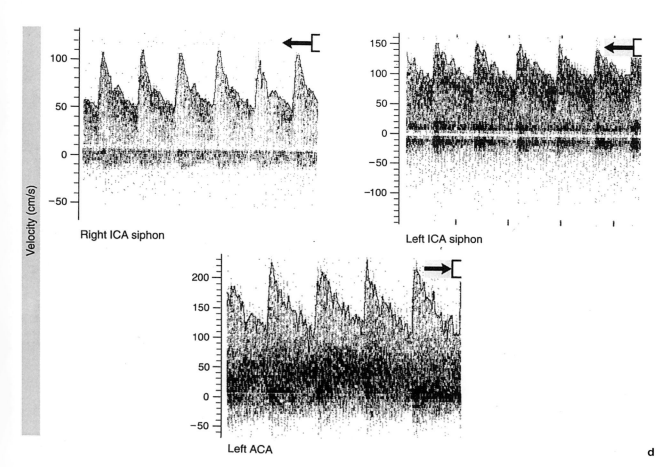

Right ICA siphon

Left ICA siphon

Left ACA

Velocity (cm/s)

d

Occlusion of the Basilar Artery
(Fig. 11.**18**)

Patient: A.E., female, age 29.
Clinical diagnosis: unclear dizzy spells, neurologically asymptomatic.

CAD: –	Vascular surgery: –	Diabetes mellitus: –
PAOD: –	Heredity: –	Hyperlipidemia: –
TIA, stroke: –	Hypertension: –	Smoking: –

Doppler sonogram. The supratrochlear arteries present with bilateral orthograde flow. Both extracranial carotid systems are normal. In relation to the patient's age, both vertebral arteries have a relatively moderate flow signal, without any clear abnormalities.

Color flow duplex sonograms (a). Bilaterally, both the vertebral artery and the vertebral vein have a narrow lumen, with a flow signal appropriate to the patient's age in its form and configuration, with the exception of the low diastolic flow component.

Transnuchal Doppler sonogram (*spectral analysis*) **(b).** In the proximal vertebral artery, an attenuated, pulsatile signal is detected, with low flow velocities (**3**) that is just still capable of being registered, and which disappears when the sample volume is placed further distally (**2**). When the sample volume is placed as far distally as possible, (**1**), an extremely attenuated signal appears, with a suggestion of pulsatility oriented toward the probe, which is a sign of retrograde perfusion of the basilar artery through the circle of Willis.

Angiogram/MRA (c–e). The MRA (**c, e**) is consistent with the findings obtained with Doppler sonography: there is an extinction phenomenon in the middle to upper region of the basilar artery, and its proximal segment as far as the origin of the right anterior inferior cerebellar artery and the terminal segment of the left vertebral artery are poorly imaged. The right vertebral artery is not displayed. The upper part of **c** shows the planning of a saturation pulse of the blood flow in the terminal segment of the carotid artery within the sagittal plane. This technique allows functional MR angiography to be carried out. The lower part of **c** shows the MR angiogram after saturation of both of the carotid artery's areas of distribution. There is an absence of arterial filling at the head of the basilar artery, serving as an additional indication of retrograde perfusion. Conventional angiography (**d**) confirms the occlusion of the basilar artery in its middle segment.

Comment. The noninvasive diagnostic techniques and complementary imaging of results provided by sonography in combination with MRA findings are increasingly allowing the diagnosis of vascular processes such as occlusion of the basilar artery without previous suspicion of the condition based on the clinical symptomatology, the set of risk factors, and the patient's age. This is important for the individual prognosis, and possibly also for planning the therapeutic approach, but it also plays a role in the further scientific analysis of such conditions, earlier considered to be rarities due to their unusual course.

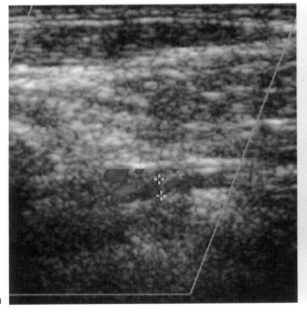

a

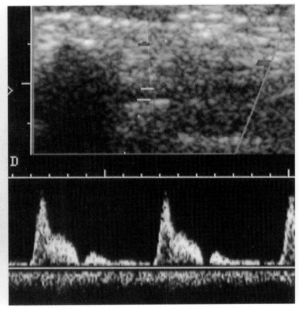

1

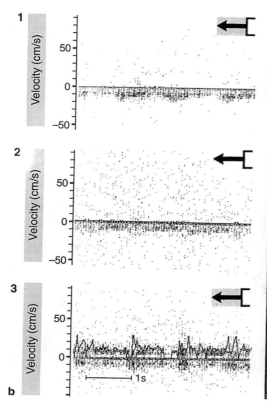

b

1s

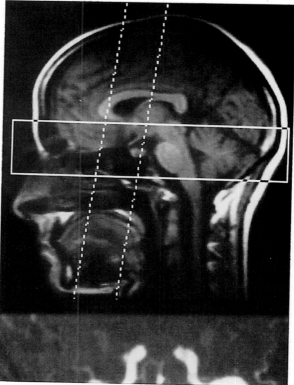

c

| Vessel | Depth | Flow direction |
| | | Peak systolic velocity |
	(mm)	(cm/s)
1 BA	110	Retrograde
2 BA	105	Retrograde
3 VA	75	24

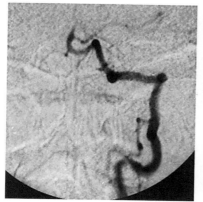

d

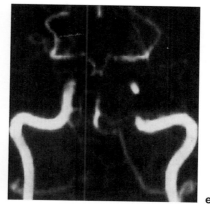

e

Peripheral Arteries

Pelvic Level

Stenosis of the Iliac Artery ≤ 50%
(Fig. 11.**19**)

Patient: E.R., male, age 44.
Clinical diagnosis: peripheral arterial occlusive disease **(PAOD), stenosis of the right iliac artery,** stenosis of the left iliac artery.

CAD:	–	Vascular surgery:	–	Diabetes mellitus:	–
PAOD:	+	Heredity:	–	Hyperlipidemia:	–
TIA, stroke:	–	Hypertension:	–	Smoking:	+

Case report. For the previous two years the patients had suffered hip and thigh claudication on the left side, and for the previous six months thigh claudication on the right. Systemic lysis with streptokinase showed a clear improvement. However, bilateral residual stenoses still remained, more pronounced on the left than on the right. Subsequently, treatment with a balloon catheter on the left was undertaken.

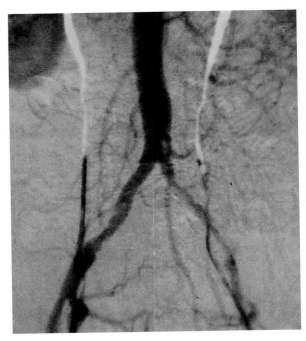

a

Examination results:

	Pulse status* Right	Left	Auscultation** Right	Left
Abdomen			+	+
Common femoral artery	(+)	+	+	(+)
Superficial femoral artery			–	–
Popliteal artery	(+)	(+)	–	–
Posterior tibial artery	(+)	((+))		
Dorsal artery of the foot	–	((+))		

* Pulse status: + = normal pulse, (+) = slightly attenuated pulse, ((+)) = significantly attenuated pulse, – = pulse not palpable.
** Auscultation: – = no flow sounds, (+) = short, soft flow sound (mainly systolic), + = clear flow sound (usually extending into the diastole).
This key also applies to the subsequent tables. (The omission of a bracket on one side of the plus sign implies an intermediate grade.)

Doppler sonogram *(analogue registration)* **(b)**. *Right:* The common femoral artery has a slightly reduced amplitude and a missing late-diastolic reverse flow component. The right popliteal artery also shows a reduced amplitude height, a delayed increase in its steepness, and a missing reverse flow component. The same applies to the right posterior tibial artery. In the dorsal artery of the foot, a moderate registration is seen, with an unfavorable ultrasound angle, but with a minimal reverse flow component.

Angiogram (a). Central venous digital subtraction angiography (DSA). *Right:* low-grade stenosis of the common iliac artery. *Left:* After completion of angioplasty, wall changes in the dilated region persist, along with a minimal residual pressure gradient.

Comment. It was possible to open up the high-grade obstructions in the pelvic region in this still relatively young patient to a large extent without surgery, using systemic lysis and angioplasty. The residual pressure gradient extending to the ankle amounted to only 10 mmHg on the right and 30 mmHg on the left. On the right, the pressure at the distal lower leg was higher than at the proximal thigh. This indicates that there is no additional intervening flow obstruction and that the collateral blood supply is very good. On the left, the pressures at the proximal thigh and the distal lower leg are nearly the same (slight fluctuations can be due to intraindividual changes in blood pressure). The pain-free walking distance increased from 200 m to 1000 m in this patient.

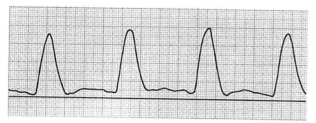

Right common femoral artery

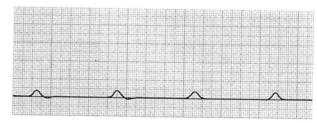

Right posterior tibial artery

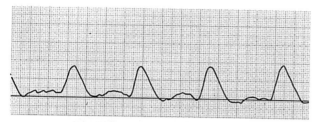

Right popliteal artery

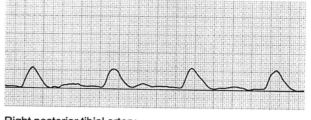

Right dorsal artery of the foot

b CW Doppler sonograms

Cuff position	Right registration point		Left registration point		
	Posterior tibial artery	Dorsal artery of the foot	Posterior tibial artery	Dorsal artery of the foot	
Proximal thigh	105	95	100	95	
Distal lower leg	115	120	95	90	

Systolic pressure measurement (Doppler ultrasound technique; mmHg)

Example using color flow duplex sonography. In a *different* patient (H.I., male, age 63), imaging showed a hemodynamically effective low-grade stenosis at the origin of the left external iliac artery obtained with color flow duplex sonography. In the color flow Doppler mode, the arterial lumen constriction can be recognized by a plaque protruding into the lumen from a dorsal direction (**c**). Aliasing here indicates an elevated average flow velocity. As a result of vascular wall vibrations, discrete color artifacts are formed in a projection on the perivascular tissue.

In a different sonographic sectional plane, the slightly broadened *Doppler frequency spectrum* documents the elevated maximal systolic flow velocity (270 cm/s), with longer diastolic forward flow and still retained triphasic wave form (**d**).

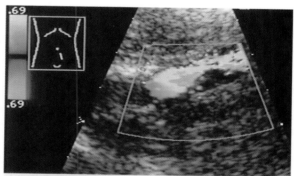

c

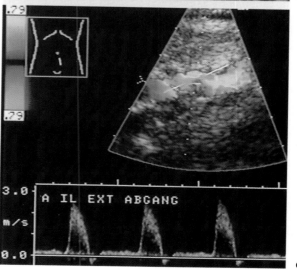

d

Stenosis of the Iliac Artery > 50%
(Fig. 11.**20**)

Patient: M.E., male, age 56.

Clinical diagnosis: **PAOD, high-grade stenosis of the right iliac artery;** differential diagnosis: occlusion of the left iliac artery.

CAD:	–	Vascular surgery:	– Diabetes mellitus: –
PAOD:	+	Heredity:	– Hyperlipidemia: –
TIA, stroke:	–	Hypertension:	– Smoking: +

Case report. For the previous nine months, there had been claudication in the region of the right buttock, mainly when climbing stairs. Level walking distance was 500–600 m (120 steps/min). The sole risk factor is a high consumption of nicotine (40 cigarettes per day, since teenage years).

Examination results:

	Pulse status Right	Left	Auscultation Right	Left
Abdomen			–	–
Common femoral artery	((+))	+	–	–
Superficial femoral artery			–	–
Popliteal artery	((+))	+	–	–
Posterior tibial artery	–	+		
Dorsal artery of the foot	–	+		

Doppler sonogram *(analogue registration)* (**c**). *Right:* Over the common femoral artery, a delayed systolic increase in steepness, decreased amplitude height, and absent late diastolic reverse flow are seen. The popliteal artery also presents a clearly reduced amplitude height and absent reverse flow component. Similar results are found over the posterior tibial artery and the dorsal artery of the foot. *Left:* Normal flow pulses are present.

Angiogram. *Right:* There is a high-grade stenosis at the origin of the common iliac artery and an additional occlusion at the origin of the internal iliac artery (**a**). Following angioplasty, one notes only a minimal constriction of the lumen (**b**). *Left:* A minimal stenosis is detected at the origin of the common iliac artery. The drainage relationships in both legs appear to be normal.

Comment. Usually, an isolated stenosis of the iliac artery is so well-compensated for through collaterals that the arm–ankle pressure gradient on the right is not very large (35 mmHg). In this case, it is seen that the pressure still shows a tendency to increase between the proximal thigh and the distal lower leg. This is a sign of good collateral filling without a subsequent flow obstruction being present.

The stenosis of the left iliac artery has hardly any hemodynamic effect. Since the patient insisted on re-

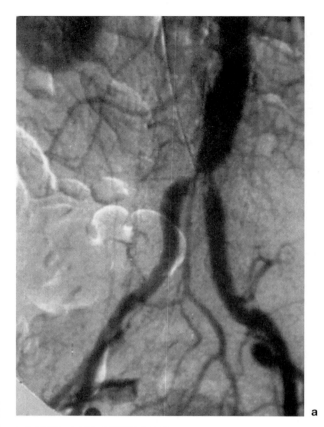

a

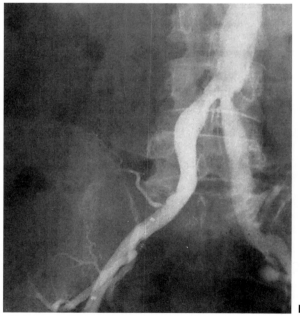

b

canalization therapy despite being able to walk a good distance, angioplasty was successfully conducted on the right side, and a complete absence of complaints was achieved.

Clinically, it was not possible to distinguish between a stenosis of the highest possible grade and an occlusion because no flow sound could be auscultated.

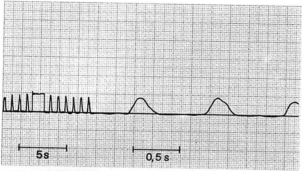

Right common femoral artery

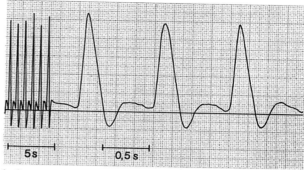

Left common femoral artery

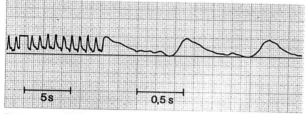

Right posterior tibial artery

a CW Doppler sonograms

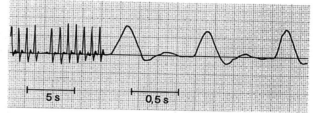

Left posterior tibial artery

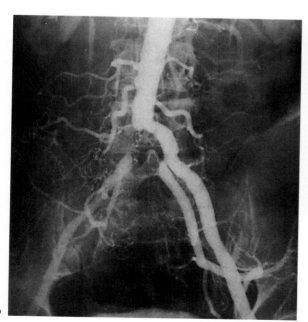

b

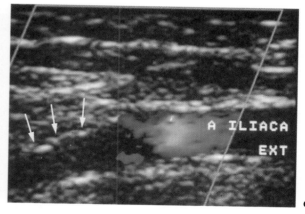

A ILIACA EXT

c

Thigh Level

Stenosis of the Femoral Artery
(Fig. 11.**22**)

Patient: K.K., male, age 66.
Clinical diagnosis: amputation of the right thigh following trauma, **PAOD, stenosis of the left superficial femoral artery.**

CAD:	+	Vascular surgery:	–
PAOD:	+	Heredity:	+
TIA, stroke:	–	Hypertension:	–

Diabetes mellitus: –
Hyperlipidemia: +
Smoking: +

Case report. On the right side, the thigh had been amputated following a war injury in 1945. In 1971, a bypass was completed using a Y prosthesis. During the previous year, there had been incidents of calf claudication on the left side, with the walking distance limited to approximately 100 m. Clinically, there was a suspicion of stenosis of the left femoral artery, to be reopened using angioplasty.

Examination results:

	Pulse status Right	Left	Auscultation Right	Left
Abdomen			–	–
Common femoral artery	+	+	+	+
Superficial femoral artery			–	+
Popliteal artery		+		–
Posterior tibial artery		(+)		
Dorsal artery of the foot		(+)		

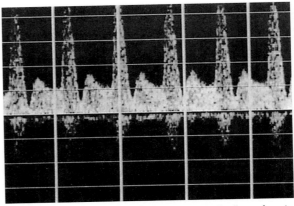

Left superficial femoral artery intrastenotically **prior** to catheter intervention (ordinates: 1 graduation = 2 kHz)

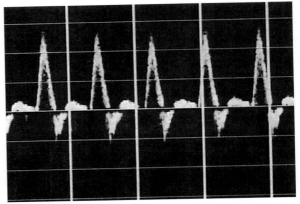

Left superficial femoral artery intrastenotically **after** catheter intervention (ordinates: 1 graduation = 1 kHz)

Doppler sonogram *(analogue registration)* (**c**). Both *before* and *after* angioplasty, the *common femoral artery* shows a more or less unchanged, normal wave form.

Intrastenotically, a systolic peak reversal in the *superficial femoral artery* can be registered.

In the *popliteal artery* and the *posterior tibial artery,* reduced amplitudes of the flow velocity are noted prior to the catheter intervention. The wave form in the popliteal artery is still triphasic, and in the posterior tibial artery it is monophasic.

After dilatation of the femoral artery stenosis, normalization of the flow waveforms in the following vascular system confirms the success of the procedure.

Doppler frequency spectrum. Registering the Doppler frequency shift from the stenotic region reveals a *massive flow disturbance,* with vortex formations and *high systolic maximal frequencies* up to 9 kHz (**a**). The wave form is no longer triphasic, and in the systole flow above and below the baseline is registered.

After angioplasty, the Doppler frequency spectrum from the dilated region again shows a narrow frequency band, with a triphasic waveform and a systolic maximal frequency of approximately 2.7 kHz (**b**).

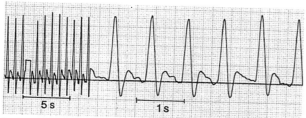

Left common femoral artery **prior** to catheter treatment

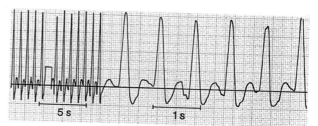

Left common femoral artery **after** catheter treatment

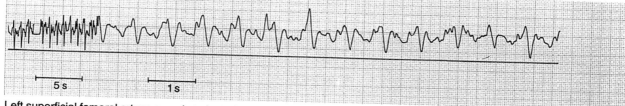

Left superficial femoral artery, sample volume in the stenosis with peak reversal, **prior** to catheter treatment

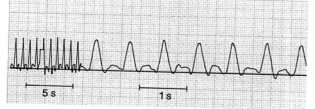

Left popliteal artery **prior** to catheter treatment

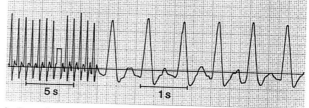

Left popliteal artery **after** catheter treatment

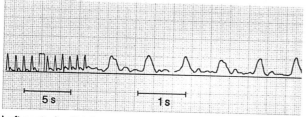

Left posterior tibial artery **prior** to catheter treatment

Left posterior tibial artery **after** catheter treatment

c CW Doppler sonograms

Cuff position	Right registration point		Left registration point		
	Posterior tibial artery	Dorsal artery of the foot	Posterior tibial artery	Dorsal artery of the foot	Systolic pressure measurement (Doppler ultrasound technique; mmHg) before catheter treatment
Proximal thigh			175	165	
Distal lower leg			90	100	

Cuff position	Right registration point		Left registration point		
	Posterior tibial artery	Dorsal artery of the foot	Posterior tibial artery	Dorsal artery of the foot	Systolic pressure measurement (Doppler ultrasound technique; mmHg) after catheter treatment
Proximal thigh			155	155	
Distal lower leg			110	110	

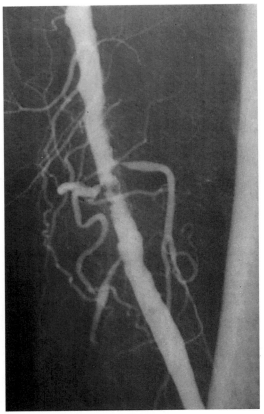

d

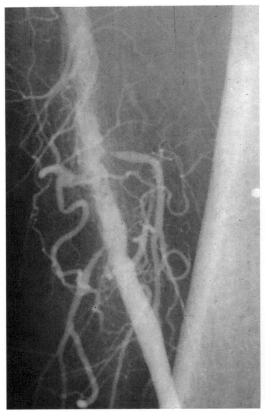

e

Angiogram. *Left:* (**d**) There is a significant stenosis of the superficial femoral artery in the distal third. An occlusion of the anterior tibial artery is seen in the distal segment. Stenoses of the posterior tibial artery are present. (**e**) After angioplasty, the stenosis of the femoral artery is extensively dilated.

Comment. All methods of measurement were able to locate the stenosis of the femoral artery in these patients. The stenosis was already clinically suspected due to a bruit in the adductor canal. The jump in pressure between the proximal thigh (175 mmHg) and the distal lower leg (90 mmHg) amounted to 85 mmHg prior to the lumen-opening procedure, and 45 mmHg after percutaneous transluminal angioplasty. The hemodynamic effect of the treatment was shown using functional measurements, and also in the angiogram. The walking distance increased to 300 m.

Example using color flow duplex sonography, in a different patient with a stenosis over a short distance in the left femoral artery in the middle third of the blood vessel (**f**). With the help of the *color flow Doppler mode,* the flow velocity can be measured in a more targeted manner, due to the lighter colors in the stenoses. The color change due to aliasing (light yellow to white to light blue) indicates the region of maximum flow acceleration.

The *Doppler frequency spectrum* from this vascular segment is monophasic, and high maximum systolic, end-diastolic, and average velocities of 401 cm/s, 105 cm/s, and 178 cm/s respectively can be measured. The pulsatility index lies in the pathologic range (1.66).

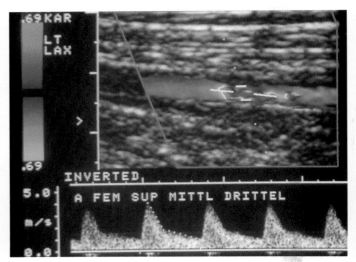

f

Occlusion of the Femoral Artery (Short)
(Fig. 11.**23**)

Patient: I.G., male, age 50.
Clinical diagnosis: **PAOD stage IIa on the left,** status after implantation of a **bifurcation prosthesis, occlusion of the left femoral artery.**

CAD: –	Vascular surgery: +	Diabetes mellitus: –
PAOD: +	Heredity: –	Hyperlipidemia: +
TIA, stroke: –	Hypertension: –	Smoking: +

Case report. For approximately the previous five years, the patient had suffered calf claudication, initially on the left side, and later also on the right. Ten months previously, the pain-free walking distance decreased to below 100 m, and a bifurcation prosthesis was implanted due to bilateral occlusions of the external iliac arteries.

Following an initially unlimited walking ability, four months later a calf claudication on the left appeared after the patient had walked a distance of approximately 150 m. Fibrinolytic therapy for an occlusion of the femoral artery with an ultrahigh dose of streptokinase followed over a period of two days. This treatment was initially successful. After a few days, a reocclusion of the superficial femoral artery occurred, with a retained increase in the pain-free walking distance up to 600 m.

On admission, there was renewed claudication in the left thigh after a distance of 200 m.

Examination results. Well-healed scars following the implantation of a Y prothesis were seen.

	Pulse status		Auscultation	
	Right	Left	Right	Left
Abdomen			–	–
Common femoral artery	+	+	–	+
Superficial femoral artery			–	–
Popliteal artery	+	–	–	–
Posterior tibial artery	+	–		
Dorsal artery of the foot	+	–		

Color flow duplex sonogram. Using a longitudinal section almost parallel to the axis of the left superficial femoral artery, the short arterial occlusion (18.8 mm) can be seen at the transition from the middle to the distal third of the blood vessel, due to the absent color coding of the lumen (**a**).

Proximal to the occlusion, only a narrow color band with a slow flow velocity is depicted. *Distal* to the occlusion, there is a typical finding of a confluent collateral blood vessel with a high flow velocity (aliasing), recognizable by the light colors, in the region of arterial *refilling* (*). In this blood vessel, the flow direction is oriented toward the transducer. Dorsolateral to the femoral artery, the femoral vein is seen as an accompanying structure.

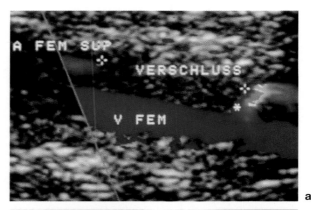
a

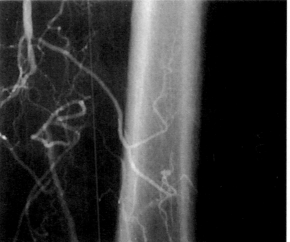
b

Angiogram. In the selective image during percutaneous transluminal angioplasty, the approximately 20-mm long occlusion of the superficial femoral artery is shown at the transition from the middle to distal third of the blood vessel (**b**). There are three patent arteries in the lower leg.

Comment. Recanalization of the approximately 2-cm long occlusion of the superficial femoral artery succeeded without leaving a residual stenosis. However, four months after the balloon dilatation, a renewed angioplasty was necessary to remove a filiform stenosis of the femoral artery due to a myointimal proliferation in the region of the previous balloon dilatation. This procedure was also successful.

With regard to the examination technique itself, it should be added that, particularly when dealing with longer vascular occlusions that involve the superficial femoral artery, using duplex sonography for orientation purposes through the accompanying femoral vein is very helpful in anatomically classifying the different structures in the thigh.

The color flow duplex sonogram (**a**) shows the close spatial relationship between these two vessels, so that in addition to B-mode imaging even of occluded vessels, based on their vascular walls and internal luminal structures, it is also possible to differentiate successfully between the original artery and collateral blood vessels.

Occlusion of the Femoral Artery (Longer)
(Fig. 11.**24**)

Patient: F.B., male, age 52.
Clinical diagnosis: **PAOD stage IIa on the right, reocclusion of the right superficial femoral artery.**

CAD:	+	Vascular surgery:	–	Diabetes mellitus: +
PAOD:	+	Heredity:	–	Hyperlipidemia: +
TIA, stroke:	–	Hypertension:	+	Smoking: +

Case report. Approximately 18 months previously, symptoms appeared for the first time in the form of a calf claudication on the right side. This was treated by an angioplasty of an occlusion of the right superficial femoral artery, with good results. For approximately three months, a clear decrease in the pain-free walking distance occurred, down to 50–80 m. The patient had occasional nocturnal pain in the right lower leg.

Examination results:

	Pulse status Right	Left	Auscultation Right	Left
Abdomen			–	–
Common femoral artery	+	+	–	+
Superficial femoral artery			–	–
Popliteal artery	–	+	–	–
Posterior tibial artery	–	+		
Dorsal artery of the foot	–	–		

Systolic pressure measurement (Doppler ultrasound technique; mmHg):

	Right	Left
Brachial artery	190	
Posterior tibial artery	–	170
Anterior tibial artery	80	150

Color flow duplex sonogram. A ventromedial sectional plane proceeding from the medial side of the thigh on the right shows the bifurcation of the femoral artery, with the proximal segment of the superficial femoral artery (coded red) (**c**).

The *beginning of the occlusion* can be localized at the gap in the color coding. Proximal to the occlusion, a stenosis is recognized in a collateral given off laterally, due to the visible lumen constriction and the aliasing, which indicates an elevated flow velocity.

Over a 16-cm stretch in the superficial femoral artery, no blood flow can be registered using color flow Doppler sonography, or in the pulsed wave Doppler mode. This is consistent with vascular occlusion. In the region of the distal third of the blood vessel, peripheral refilling is detected, with two collaterals joining (light blue) (**d**).

Angiogram. Catheter angiography during percutaneous transluminal angioplasty shows the proximal beginning of the occlusion of the right superficial femoral artery, several centimeters distal to the bifurcation of the femoral artery (**a**). Directly proximal to the occlusion, a collateral with a stenosis at its origin proceeds laterally. The trunk of the deep femoral artery is also occluded a few centimeters after its origin.

Angiography demonstrates the peripheral refilling of the superficial femoral artery via collateral blood vessels in its distal third (**b**).

Comment. For this patient, a noninvasive individual treatment concept was developed, consisting of intraarterial arteriography in preparation for a simultaneous catheter intervention.

After an initially successful catheter dilatation, there was still persistent occlusion of the deep femoral artery, and a further reocclusion of the superficial femoral artery occurred five months later. This too was successfully recanalized.

Following catheter dilatation of newly appearing residual stenoses of the ipsilateral femoral artery four and nine months later, only wall deposits without any hemodynamic effects were detectable with color flow duplex sonography in the right superficial femoral artery, 24 months after the final balloon dilatation. In retrospect, the concept of repeated angioplasty has proved an effective one.

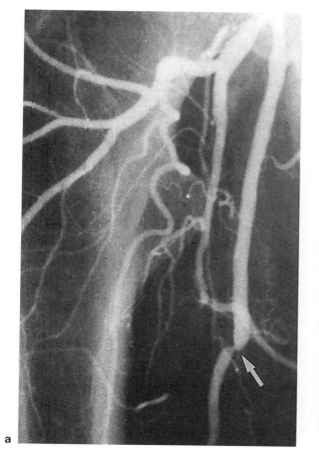

a

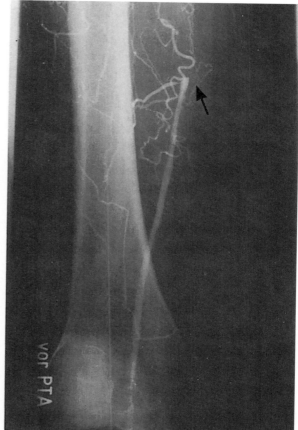

b

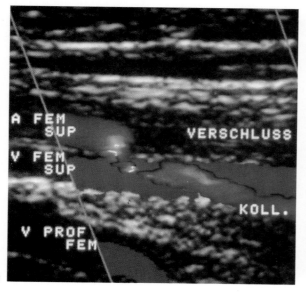

c

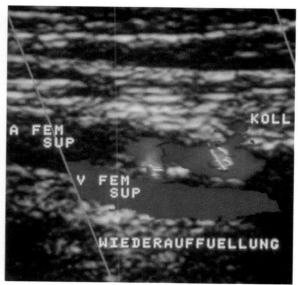

d

Occlusion of the Popliteal Artery
(Fig. 11.**25**)

Patient: G.K., male, age 59.
Clinical diagnosis: **PAOD stage IIa on the right,** suspected **stenosis of the right popliteal artery,** and **stenosis of the left femoral artery.**

CAD: –	Vascular surgery: +	Diabetes mellitus: +
PAOD: +	Heredity: –	Hyperlipidemia: +
TIA, stroke: –	Hypertension: +	Smoking: +

Case report. Bilateral calf claudication had persisted for about one year, on the right more than on the left. Angioplasty of a stenosis of the right popliteal artery had been carried out ten months previously. Suspicion of restenosis.

Examination results. A 59-year-old patient in a good general state of health and with good nutritional status, with the following angiological findings:

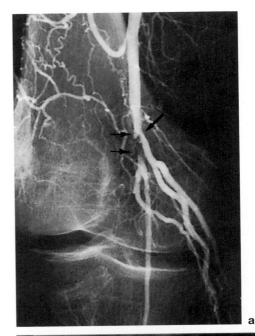

a

	Pulse status		Auscultation	
	Right	Left	Right	Left
Abdomen			–	–
Common femoral artery	+	+	–	–
Superficial femoral artery			–	+
Popliteal artery	+	(+	+	–
Posterior tibial artery	–	(+		
Dorsal artery of the foot	(+)	(+)		

Color flow duplex sonogram. In a longitudinal section of the popliteal artery proceeding from a dorsal direction (with the patient in prone position), a short *occlusion of the artery over a distance of some 13.5 mm* (thin arrows) (**b**) is seen above the interarticular space of the knee. Hyperechoic structures in the (dorsal) section of the vascular occlusion near the transducer cause an ultrasound shadow.

Proximal to the beginning of the occlusion, a collateral blood vessel of the popliteal artery is given off. The aliasing of the collateral at the origin is greater than expected, due to a change in the ultrasound application angle. Because of a preselected high range for the average flow velocity of the color flow Doppler mode (92 cm/s), stenosis of the collateral blood vessel (thick arrow) can be deduced, solely based on the color coding.

Angiogram. Selective catheter angiography during percutaneous transluminal angioplasty shows the *short occlusion in the proximal segment of the popliteal artery* (thin arrows) (**a**). The collateral, which proceeds dorsally, also shows a stenosis at the origin (thick arrow) on the angiogram.

Comment. Clinically, a *stenosis* of the popliteal artery was suspected on the basis of a systolic bruit in the hollow of the knee. Color flow duplex sonography, however, detected an *occlusion* of the popliteal artery. The bruit was most probably caused by a stenosis of the collateral blood vessel.

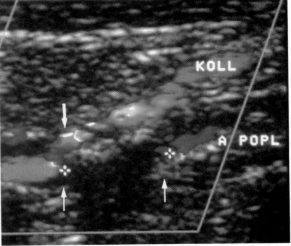

b

In the *differential diagnosis,* this set of findings obtained with color flow duplex sonography, with massive ultrasound extinction over a short distance due to a hyperechoic plaque near the transducer, should also suggest consideration of a hemodynamically effective high-grade stenosis. The differentiation process must then involve a careful examination of this vascular region using the pulsed wave or continuous wave Doppler mode, searching for a "stenosis signal." Indirect criteria of a distal or perhaps proximal decrease in the flow velocity, as well as morphologically and functionally detectable collateral circulation, is not always helpful in discriminating between stenosis and occlusion.

To provide orientation and assist in *localizing the level* of pathological vascular processes involving both the arteries and veins in the region of the popliteal artery, the *interarticular space of the knee* can be used as a reference point, since this bony structure usually presents a good image in sonographic examinations.

Aneurysm of the Femoral Artery
(Fig. 11.**26**)

Patient: H.T., female, age 76.
Clinical diagnosis: **PAOD;** distal **aneurysm of the left femoral artery,** and **stenosis of the left femoral artery.**

CAD:	+	Vascular surgery:	+	Diabetes mellitus:	–
PAOD:	+	Heredity:	–	Hyperlipidemia:	+
TIA, stroke:	+	Hypertension:	–	Smoking:	+

Case report. Seven years previously, an open recanalization in the region of the *left* superficial femoral artery had been carried out, and a Dacron patch dilatation was conducted to alleviate a stenosis of the femoral artery. At presentation, there was reported increasing swelling in the area of the medial thigh, without any intermittent claudication.

Examination results. There is a well-healed, 15-cm long scar located distally at the medial side of the thigh.

	Pulse status Right	Left	Auscultation Right	Left
Abdomen			–	–
Common femoral artery	+	+	–	–
Superficial femoral artery		*	–	+
Popliteal artery	+	+	–	–
Posterior tibial artery	+	(+)		
Dorsal artery of the foot	(+)	(+)		

Systolic pressure measurement (Doppler ultrasound technique; mmHg):

	Right	Left
Brachial artery	160	
Posterior tibial artery	160	125
Anterior tibial artery	155	115

Color flow duplex sonogram. A sonographic longitudinal section through the left distal superficial femoral artery shows two aneurysms (**a**), in the distal section and in the femoropopliteal transition zone. In this plane, the larger *distal* aneurysm is approximately 5.9 cm × 7.5 cm in size. The lumen, through which there is still blood flowing, is partially shown in the image, and its diameter measures approximately 2 cm. A sonographically nonhomogeneous massive partial thrombosis of the aneurysmal cavity can be seen. There is a hemodynamically effective stenosis distal to the aneurysm in the femoropopliteal vascular segment (not shown).

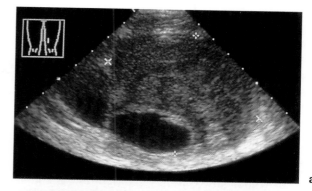

a

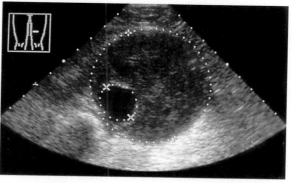

b

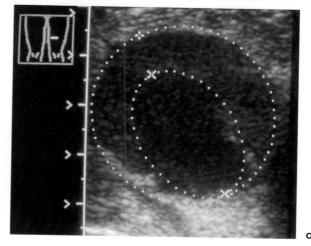

c

A corresponding cross-section through the distal aneurysm, with the sectional plane perpendicular to the vascular axis, also shows the extreme wall dilatation (6.8 cm × 6.7 cm) and a lumen (2.3 cm × 2 cm) through which blood is still flowing (**b**).

In a cross-section *of the proximal* aneurysm, also partially thrombosed, its maximum size can be measured as 3.6 cm × 3.4 cm (**c**). Here the lumen with blood flowing through it shows a distension of 2.8 cm × 1.3 cm.

Fig. 11.**26 d–h** ▷

Angiogram. The intra-arterial angiographic image shows two aneurysms of the left superficial femoral artery, also involving the popliteal artery. The artery is already dilated proximal to the aneurysms (**d**).

Doppler frequency spectrum. From the *right* proximal superficial femoral artery, an almost normal spectrum was registered, except for a missing forward flow component in the early diastole (**e**). The hemodynamic effects of an early *extrasystole* (second spectrum) can be seen, with a lowered systolic velocity (67 cm/s) and average flow velocity, corresponding to a lowered cardiac stroke volume and, assuming a constant vascular cross-section, also causing a lowered flow rate in the femoral artery. In the subsequent cardiac cycle (third spectrum), the systolic (111 cm/s) and average flow velocities are somewhat higher due to compensation, and then return to the steady prior values.

On the side of the aneurysms, the frequency spectrum in the *left* proximal femoral artery indicates pendular flow, with a physiological systolic flow direction and a clear holodiastolic backflow component, which corresponds to retrograde flow from the aneurysmal cavities (**f**).

Postoperatively, a triphasic Doppler frequency spectrum was again registered at the same location from the proximal superficial femoral artery (**g**).

A triphasic wave form was also registered in the flow passing through the Gore-Tex prosthesis in the distal thigh region (**h**).

Comment. In comparison with the angiogram (**d**), the B-mode findings (**a–c**) show the clear advantages of sonography in diagnosing aneurysms. While angiography is only able to depict the internal vascular lumen *through which blood is flowing,* sonography shows the full extent of the aneurysm, including its thrombosed segments. In this patient, the cross-sections of the thrombosed segments are several times the residual lumen (3605 mm² compared to 345 mm² in **b**).

Compared to the typical predilection points for aneurysms, the location here in the superficial femoral artery is quite rare (Fig. 11.**27**). The cause of the wall dilatation in this vascular region is therefore very likely to be the earlier recanalization and patch operation.

The indication for surgery was based in particular on risk of embolism and thrombosis, as well as an increased risk of aneurysmal rupture.

During the electively conducted aneurysmal resection, with the implantation of a ring-stabilized Gore-Tex prosthesis, the sonographic findings were confirmed intraoperatively.

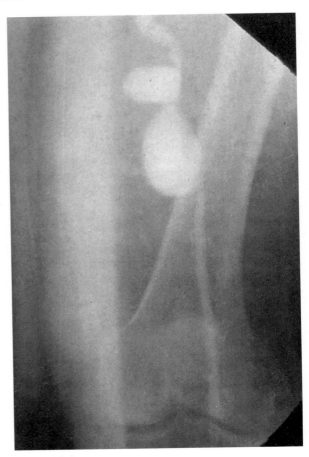

Fig. 11.**26 d**

Postoperatively, all pulses in the left lower extremity were strongly palpable. The systolic pressure at the ankle arteries, above the anterior and posterior tibial arteries, corresponded at 190 mmHg to the pressure measured above the brachial artery.

We are grateful to Dr. Borchers of the Department of Radiological Diagnostics in Gummersbach Hospital for his kind permission to use the angiogram.

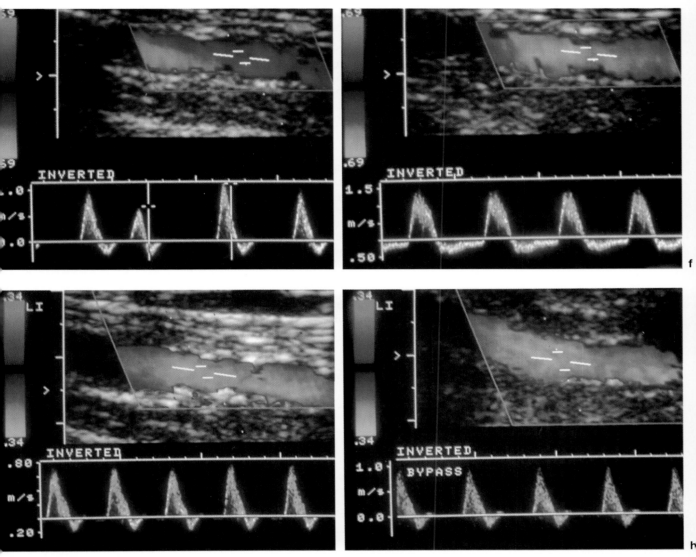

Fig. 11.**26 e–h**

Aneurysm of the Popliteal Artery
(Fig. 11.**27**)

Patient: M.B., male, age 70.

Clinical diagnosis: **dilatative/obliterative arteriopathy,** with previous bilateral **aneurysms of the popliteal arteries.** At presentation, there was an acutely appearing **occlusion of the right popliteal artery.**

CAD: +   Vascular surgery: +   Diabetes mellitus: –
PAOD: +   Heredity: –   Hyperlipidemia: +
TIA, stroke: –   Hypertension: +   Smoking: +

Case report. Ten years previously, the first incident of exertion-dependent pains in the right calf region occurred. Two years later, successful catheter fibrinolysis and catheter dilatation of an acute occlusion in the *right* popliteal artery was carried out. There were also signs of incipient dilatative arteriopathy at this time. Eight years later, an angioplasty of a stenosis of the *left* popliteal artery was carried out, and *bilateral* aneurysms of the popliteal arteries were diagnosed, more pronounced on the right than on the left.

Three months earlier, there had been a sudden increase in calf claudication on the right side, without any pain while at rest.

Examination results:

	Pulse status Right	Left	Auscultation Right	Left
Abdomen			–	–
Common femoral artery	+	+	–	+
Superficial femoral artery			–	–
Popliteal artery	–	+	–	–
Posterior tibial artery	–	(+)		
Dorsal artery of the foot	–	(+)		

Systolic pressure measurement (Doppler ultrasound technique; mmHg):

	Right	Left
Brachial artery	145	
Posterior tibial artery	110	150
Anterior tibial artery	85	150

Angiogram. An image obtained using central venous digital subtraction angiography (DSA) two years earlier shows an aneurysm of the right popliteal artery, with wall changes in the femoropopliteal vascular segment (**a**). The current DSA image shows an occlusion of the right popliteal artery, beginning in the femoropopliteal transition zone (**d**). On the contralateral side, there is an incipient aneurysmal dilatation of the popliteal artery.

Color flow duplex sonogram. The color flow duplex sonogram corresponding to the first angiographic examination (**a**) shows the large aneurysm of the popliteal artery in longitudinal (**b**) and transverse section (**c**). The thrombosed lumen, with a residual lumen (coded red) of 5.4 mm through which blood is still flowing, measures 30.5 mm in its sagittal diameter (**b**) and 33.5 mm × 30.3 mm in its cross-sectional diameter (**c**).

Two years later, this true aneurysm is occluded. In the longitudinal section of the right popliteal artery proceeding from a dorsal orientation, hypoechoic internal structures can be recognized in the expanded vascular wall dilatation (**e**). In transverse section, the cross-sectional diameter 40 mm × 29 mm (**f**). Next to the aneurysm, color-coded flow is seen, belonging to the popliteal vein, which is displaced medially but not compressed.

Comment. The region of the popliteal arteries is a predilection point for aneurysmal vascular dilatations. In an occlusion of the popliteal artery, the possibility of a thrombosed popliteal aneurysm should always be included in the differential diagnosis.

The preferred method for detecting or excluding an aneurysm is B-mode imaging. Color flow Doppler sonography provides fast information about segments through which there is still blood flowing, and the extent of the thrombosis.

Since an aneurysmal structural deficiency of the arterial wall rarely occurs in isolation, when *one* aneurysm is found, additional arterial segments should always be examined. The contralateral vascular system, as well as the abdominal aorta and the iliac arteries, in particular, should be evaluated so as not to overlook possible dilatations of the lumen in these vessels.

In the case presented here, the patient declined to have the aneurysm of the popliteal artery corrected by vascular surgery. Despite the administration of oral anticoagulants, a thrombotic occlusion of the aneurysm occurred within two years.

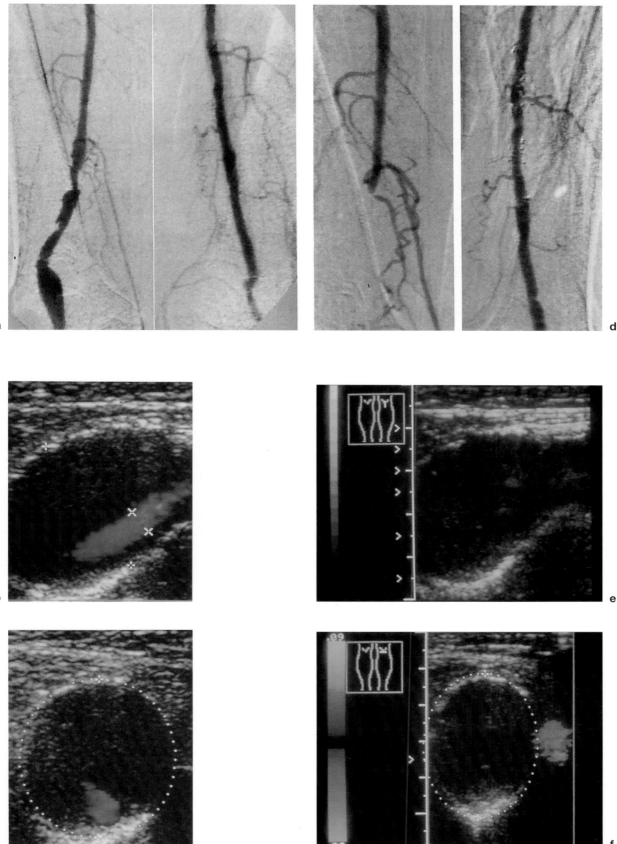

a

b

c

d

e

f

False Aneurysm of the Common Femoral Artery (Spontaneous Course)
(Fig. 11.**28**)

Patient: B.D., female, age 67.
Clinical diagnosis: **PAOD stage IV on the right, amputation at the left thigh.** Stenosis of the right femoral artery and occlusion of the right popliteal artery. Following percutaneous transluminal angioplasty of the vascular obstructions on the right side, a **false aneurysm** formed in the right groin region.

CAD:	– Vascular surgery:	+ Diabetes mellitus:	+
PAOD:	+ Heredity:	+ Hyperlipidemia:	+
TIA, stroke:	– Hypertension:	+ Smoking:	–

Case report. The patient had a ten-year history of incidents of bilateral calf claudication. Two years previously, a left femoropopliteal bypass was placed due to a perforating ulcer of the foot on the left side. Postoperatively, an occlusion of the bypass occurred. Four months later, an amputation at the left thigh was carried out.

Ten months previously, a thromboendarterectomy of the right superficial femoral artery was carried out. For three months, there had been traumatic lesions (injuries due to chafing) in the region of the right Achilles' tendon and the right heel. A lumbar sympathectomy was carried out three months earlier.

Examination results at admission. A well-healed thigh stump was present on the left.

On the right side, there was a well-healed scar at the medial side of the thigh, a contraction of the knee joint (160°), and a stiffening of the talocalcaneal joint. Dry, mummified necroses were observed above the Achilles' tendon and above the heel, with a reddened marginal seam, a perforating ulcer above the metatarsal head both dorsally and ventrally, and hyperkeratoses.

	Pulse status		Auscultation	
	Right	Left	Right	Left
Abdomen			–	–
Common femoral artery	+	+	–	–
Superficial femoral artery		Amputation	+	Amputation
Popliteal artery	–			
Posterior tibial artery	–			
Dorsal artery of the foot	–			

Systolic pressure measurement (Doppler ultrasound technique; mmHg):

	Right	Left
Brachial artery	145	
Posterior tibial artery	120	Amputation
Anterior tibial artery	115	Amputation

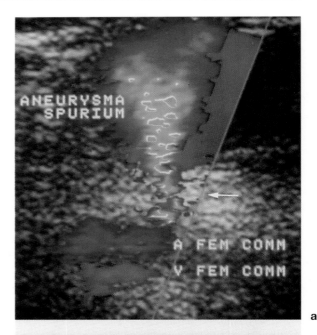

a

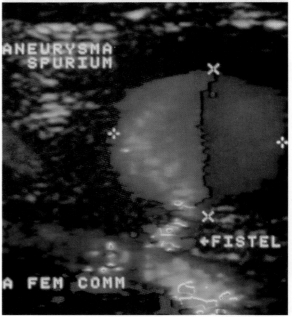

b

After angioplasty of the stenoses of the femoral artery and the occlusion of the popliteal artery, there was a patent hematoma in the right groin, a pulsating tumor next to the puncture point that was painful on pressure, and a systolic sound above the groin, with a clinical suspicion *of false aneurysm.*

Color flow duplex sonogram. The course of spontaneous thrombosis of a false aneurysm in the region of the common femoral artery was observed over six weeks. All the images (**a–e**) represent sagittal longitudinal sections above the right groin.

Four days after PTA, a patent false aneurysm (**a**) was observed (only the proximal segment is shown here), located ventral to the common femoral artery and vein. It was approximately 60 mm × 20 mm × 20 mm in size, and was connected to the artery by a fistular canal (arrow).

Eleven days after catheter intervention, the section of the false aneurysm through which blood was flowing had decreased in size, and a thrombosed seam was observed at the margins (**b**). In the frequency spectrum of flow in the fistular canal (**c**), there was a shorter systolic inflow into the aneurysm (shown below the baseline) and a longer diastolic outflow (above the baseline), to-and-fro sign.

After a further ten days, increasing thrombosis of the aneurysmal cavity was seen (**d**). At the follow-up examination *39 days after angioplasty,* this cavity was completely occluded (**e**). The ellipse marks the region in the aneurysmal cavity corresponding to the lumen through which blood had recently flowed, which is still hypoechoic in this examination.

Comment. Due to her extensive necroses, the patient still remained hospitalized even after the successful angioplasty. It was therefore possible for the spontaneous thrombosis of the false aneurysm to be awaited on an in-patient basis.

A form of treatment which has shown good success rates involves compression of the fistular canal with the transducer, under the guidance of color flow Doppler sonography (see the following case report). If it persists for a longer period, a false aneurysm will require vascular surgery.

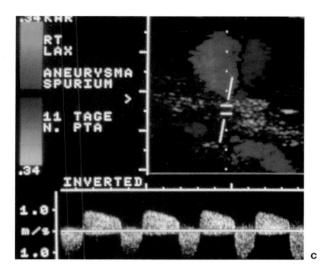

c

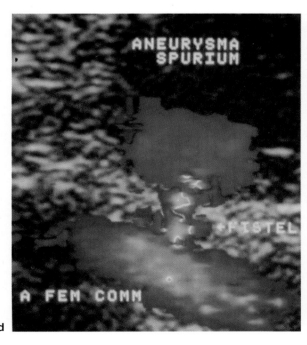

d

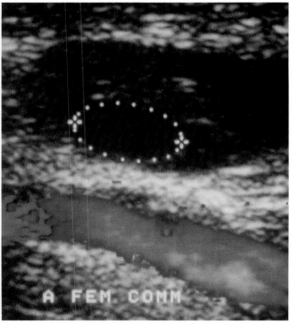

e

False Aneurysm (Compression Therapy)
(Fig. 11.**29**)

Patient: J.E., male, age 63.
Clinical diagnosis: **PAOD stage IIa,** with a **stenosis of the left femoral artery** and occlusions of the arteries in the lower leg.

After percutaneous transluminal angioplasty (PTA) of a high-grade stenosis of the right renal artery, there was a suspicion of false aneurysm formation in the region of the right groin.

CAD: –	PTA: +	Diabetes mellitus: –
PAOD: +	Heredity: –	Hyperlipidemia: +
TIA, stroke: –	Hypertension: +	Smoking: +

Case report. Four days previously, PTA of a stenosis in the right renal artery has been carried out in this patient suffering from arterial hypertension, renal failure (compensated retention), and bilateral stenoses of the renal arteries. After the patient had been on a long walk, a painful, pulsating tumor suddenly appeared, with a patent hematoma in the region of the right groin.

Examination results. There was a pulsating, full swelling as large as a walnut in the region of the right groin, which was painful on pressure, and a softer, fist-sized hematoma at the medial side of the thigh. Systolic flow sound above the right common femoral artery was observed, with the following additional findings:

	Pulse status		Auscultation	
	Right	Left	Right	Left
Abdomen			+	+
Common femoral artery	+	+	+	–
Superficial femoral artery			–	+
Popliteal artery	+	(+)	–	–
Posterior tibial artery	+	+		
Dorsal artery of the foot	+	–		

Systolic pressure measurement (Doppler ultrasound technique; mmHg):

	Right	Left
Brachial artery	160	
Posterior tibial artery	174	186
Anterior tibial artery	196	148

Color flow duplex sonogram. The illustrations show *compression treatment of a false aneurysm* in the region of the common femoral artery. All the images (**a–d**) are sagittal longitudinal sections above the right groin at the level of the puncture point.

With B-mode sonography, the longitudinal section in the groin region depicts a pulsating, hypoechoic, space-occupying lesion, approximately 45 mm × 30 mm × 14 mm in size (**a**). Dorsal to this, there is a connection, with a fluctuating caliber, to the common femoral artery (arrow), representing the reopened canal.

With color flow Doppler sonography, a fountain-like inflow of blood through the fistular canal (arrow) into the aneurysmal cavity during the systole is seen (**b**). Aliasing in the neck of the aneurysm reflects the relatively high flow velocity in the inflow jet. In this sectional plane, the jet can be readily followed into the cavity of the hematoma (light color mosaic), and has to be differentiated from the usual eddy-like flow, with slower velocities, that is found here.

After *compression of the aneurysm and the fistular canal* with the transducer lasting some 25 minutes, the cavity of the hematoma has clearly decreased in size, to 21.9 mm × 12.9 mm × 12 mm (**c**). The dorsally located fistular canal (arrow) still has a diameter of 2.6 mm.

With color flow Doppler sonography, occlusion of the false aneurysm and the fistular canal can be documented by the absent blood flow (**d**). It is important to use varying adjustments of the color parameters in this process, so as not to overlook fast as well as slow flow velocities.

After compression treatment, the common femoral artery (coded red) also shows a normal flow pattern at the puncture point.

Comment. With color flow duplex-sonographic guidance, targeted compression of the fistular canal and the aneurysmal cavity is possible. According to our own experience and communications we have received, compression therapy produces thrombosis of the false aneurysm in most patients. However, if an arteriovenous fistula is also present, this method is only rarely successful.

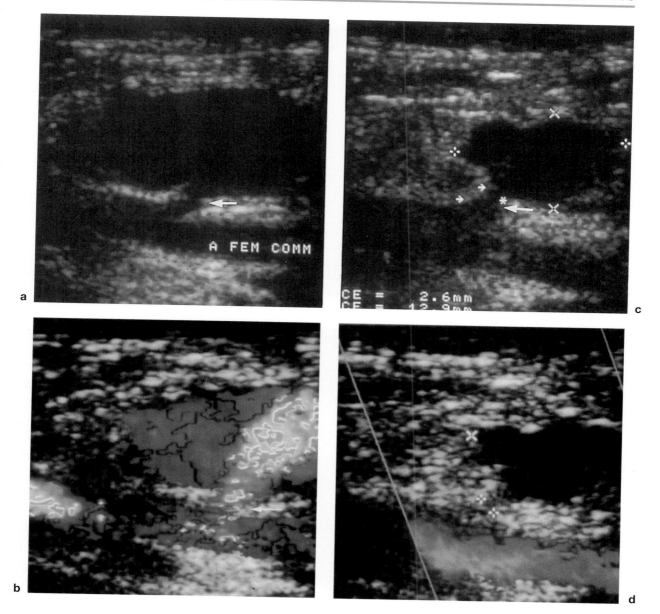

a

A FEM COMM

b

c

CE = 2.6mm
CE = 12.9mm

d

Stenosis at a Bypass Insertion Point
(Fig. 11.**30**)

Patient: W.U., male, age 45.

Clinical diagnosis: **PAOD stage IIb on the right, amputation at the left thigh,** suspicion of *right-sided stenoses in the pelvic vascular system and a stenosis at the bypass insertion point; on the left side,* occlusion at the origin of the iliac artery and occlusion at the origin of the femoral artery.

CAD:	+	Vascular surgery:	+	Diabetes mellitus:	–
PAOD:	+	Heredity:	+	Hyperlipidemia:	+
TIA, stroke:	–	Hypertension:	+	Smoking:	+

Case report. A laparotomy had been carried out 22 years earlier due to abdominal gunshot wounds, and a nephrectomy on the right followed somewhat later. The patient had been known to be suffering from PAOD for four years. Three years previously, an iliofemoral crossover bypass from left to right had been carried out, and four months later an amputation at the left thigh proved necessary. The year after that, a thromboendarterectomy of the left carotid vascular system and a femorocrural bypass of the right great saphenous vein followed. In the same year, a myocardial infarction occurred, and—with persistent pains at the stump remaining after amputation—a trial denudation of the arteries in the groin was conducted. Three months later, the stump was revised due to a suspicion of neurinoma, and a reamputation of the thigh was carried out. Two months previously, neurolysis had followed (femoral nerve and lateral cutaneous femoral nerve), and an arthroscopic operation on the lateral right meniscus was conducted. The patient was known to abuse alcohol.

In the region of the supra-aortal arteries, a high-grade stenosis of the left subclavian artery, with an ipsilateral occlusion of the vertebral artery, were detectable both clinically and with duplex sonography.

Examination results. A 45-year-old patient in a poor general state of health, with the following angiological findings:

	Pulse status Right	Pulse status Left	Auscultation Right	Auscultation Left
Abdomen			–	+
Common femoral artery	+	–	+	+
Superficial femoral artery		Amputation	+	Amputation
Popliteal artery	(+)		–	
Posterior tibial artery	+			
Dorsal artery of the foot	+			

Systolic pressure measurement (Doppler ultrasound technique; mmHg):

	Right	Left
Brachial artery	130	
Posterior tibial artery	130	Amputation
Anterior tibial artery	130	Amputation

Color flow duplex sonogram. In a sagittal longitudinal section of the right groin, the anastomotic region is seen, with a clear constriction of the lumen when in B-mode imaging (**a**).

Color flow Doppler sonography is able to image the high flow velocity in the stenosis and also displays (in shades of green) the different flow velocities that are simultaneously present intrastenotically, and here mainly poststenotically (**b**). The maximum systolic, end-diastolic, and time-standardized velocities were 306 cm/s, 93 cm/s, and 147 cm/s respectively. Poststenotically, in the dilated vascular segment, there is a flow jet (greenish–yellow), which is located mainly parallel to the vascular axis and secondary flow areas and has an eddy-like flow (dark blue–red).

Angiogram. Intra-arterial digital subtraction angiography (transaxillary access) shows the bypass stenosis on the right, with a poststenotic dilatation of the femorocrural bypass (**c**).

Comment. This patient, with similar systemic and ankle artery pressures, basically shows good compensation for the arterial obstruction. Planning appropriate treatment for the stenoses at the insertion point of a bypass has to take into account both the hemodynamic effect and also the possible progression of the stenosis, and the accompanying danger of a bypass occlusion, which would lead to a worse prognosis.

In the present case, the patient, who was not satisfactorily mobile due to significant fluctuations in the size of the stump, as well as an existing right-sided claudication, through compensatory additional exertion of the leg, declined a planned correction of the anastomotic stenosis.

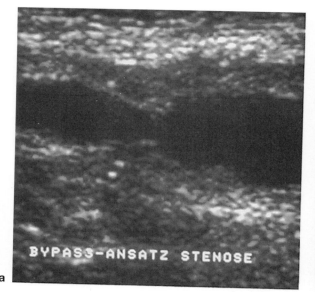

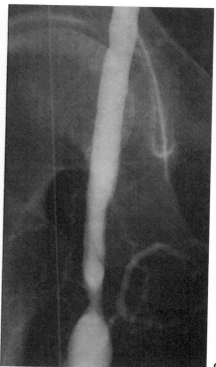

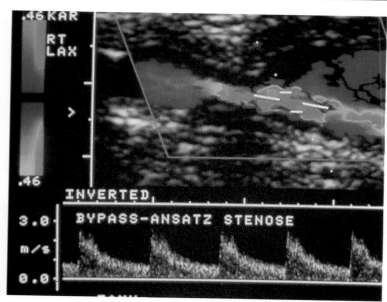

Stent Stenosis in the Superficial Femoral Artery (Percutaneous Transluminal Angioplasty)
(Fig. 11.**31**)

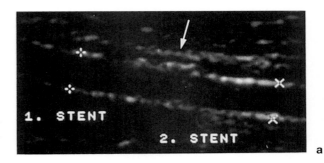

Patient: D.J., male, age 53.
Clinical diagnosis: **PAOD stage IIa.** *Right:* **Stenosis of the femoral artery** following angioplasty of an occlusion of the superficial femoral artery, with stent implantation due to an early relapse. *Left:* stenosis of the femoral artery.

CAD:	–	PTA:	+	Diabetes mellitus: +
PAOD:	+	Heredity:	–	Hyperlipidemia: +
TIA, stroke:	–	Hypertension:	–	Smoking: +

Case report. The patient had been suffering from claudication of the right calf for the previous four years. The symptoms had become worse approximately ten months earlier, and percutaneous transluminal angioplasty (PTA) of a distal occlusion in the right superficial femoral artery was carried out.

Reocclusion occurred the same day, and five days later, a renewed PTA was undertaken. Lysis was carried out with rt-PA (alteplase, 17.5 mg), and two overlapping stents were implanted, with a diameter of 5 mm and a length of 40 mm in each instance. The procedure led to good clinical success.

Four months later, right-sided calf claudication occurred again, and percutaneous transluminal angioplasty of a stenosis of the right superficial femoral artery in the stent region was carried out.

After a further six months, follow-up examinations with color flow duplex sonography detected a residual stenosis in the overlapping zone of the stents, which was again dilated using a balloon catheter. At the time of writing, the patient occasionally has pain and "paleness" in the region of the toes during longer walks.

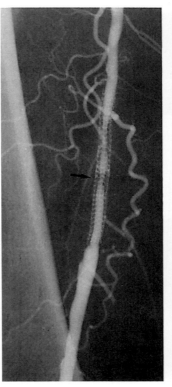

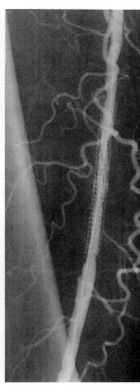

Examination results:

	Pulse status Right	Left	Auscultation Right	Left
Abdomen			–	–
Common femoral artery	+	+	–	–
Superficial femoral artery			+	+
Popliteal artery	(+)	+	–	–
Posterior tibial artery	–	(+)		
Dorsal artery of the foot	(+)	+		

Systolic pressure measurement (Doppler ultrasound technique; mmHg):

	Right	Left
Brachial artery	120	
Posterior tibial artery	110	125
Anterior tibial artery	120	115

Color flow duplex sonogram. The illustrations show longitudinal sections through the right superficial femoral artery in its distal third, in the region of the implanted stents (**a, d–g**).

In B-mode imaging, the stents are depicted as hyperechoic structures within the vascular wall (**a**). The beginning of the distal stent (second stent) can be identified from a clearly recognizable thickening of the mesh (arrow). Intraluminal structures cannot be depicted with certainty.

Aliasing (white and light blue colors) in color flow Doppler sonography raised a suspicion of a hemodynamically effective stenosis in, or shortly distal to, the overlapping zone of the stents (**d**). In the Doppler frequency spectrum, there was an angle-corrected, elevated flow velocity in this region, which in the systole was maximally 323 cm/s, with a time-averaged maximal velocity of 89 cm/s (pulsatility index 3.84).

After percutaneous transluminal angioplasty (PTA), the maximum systolic flow velocity at the same location decreased to 189 cm/s (**e**). In B-mode imaging, the overlapping zone of the stents was again visualized by its structural thickening (arrow). With the color coding, a discretely elevated local flow velocity (light yellow) was still visible.

The outflow region of the distal stent also showed a higher average flow velocity in comparison with the subsequent (larger-caliber) vascular segment (**f**). The color in the stent segment here is lighter than in the native artery, and shows transitions to shades of yellow. The color reflects an increase in the luminal caliber of the artery distal to the stents, which is also evident in the angiograms (**b, c**).

Angiograms. These show the selective intra-arterial catheter angiography of the distal superficial femoral artery in the context of percutaneous transluminal angioplasty (PTA) of the stent stenosis (**b**). Slightly distal to the overlapping zone of the two stents, which had been implanted ten months previously, there is a concentric lumen constriction of the superficial femoral artery (arrow). Medial to the femoral artery, there is a corkscrew-like, winding collateral.

After angioplasty, the superficial femoral artery was patent (**c**), although the dilated stent region still showed wall irregularities. The course of the collateral vessels confirms that the projection is identical in both angiograms.

Comment. Two indications influenced the decision to use angioplasty in the recurring stenosis in the stent region. First, the patient presented both subjective and objective symptoms of peripheral arterial occlusive disease. A second, and no less important indication, was the prophylactic aspect of preventing a potential occlusion of the superficial femoral artery due to progression of the lumen constriction in the stent region. With regard to possible recanalization with angioplasty, an occlusion has a significantly poorer prognosis.

When the hemodynamic compensation for the vascular obstruction was estimated, pressure measure-

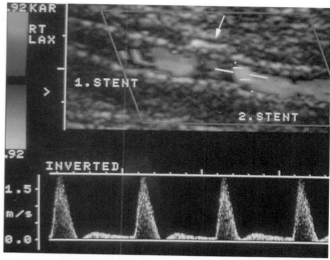

e

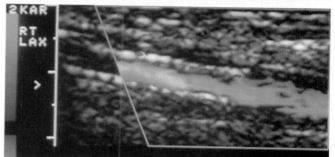

f

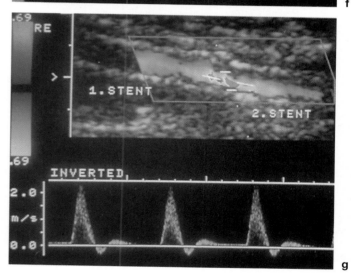

g

ments at the right ankle arteries of 110 mmHg and 120 mmHg represented misleadingly high values, due to the probable media sclerosis caused by many years of diabetes mellitus. They did not reflect the true compensation or circulatory reserve.

In the clinical follow-up, the arterial vascular system in the thigh was still patent when examined with color flow duplex sonography 22 months after the last angioplasty of the stent stenoses on the right side. The only finding was a persistent locally elevated flow velocity in the overlapping region of the vascular prostheses, after PTA (**g**).

Bypass Rupture
(Fig. 11.**32**)

Patient: E.-M.R., female, age 62.
Clinical diagnosis: **PAOD; restenosis of the right femoral artery** and a **bypass aneurysm on the left.**

CAD:	–	PTA:	+	Diabetes mellitus: –
PAOD:	+	Heredity:	–	Hyperlipidemia: +
TIA, stroke:	–	Hypertension:	+	Smoking: +

Case report. Bilateral calf claudication had first occurred ten years previously. Five years after that, a thromboendarterectomy on the left and implantation of a femoropopliteal polytetrafluoroethylene (PTFE) bypass on the left was conducted, with good clinical success. Seven months previously, angioplasty of several stenoses of the right femoral artery was carried out, and a bypass rupture (aneurysm) was diagnosed with duplex sonography. The patient's current hospital admission was in order to implant a stent on the left.

Examination results. A pulsating, space-occupying lesion was observed in the middle of the left thigh.

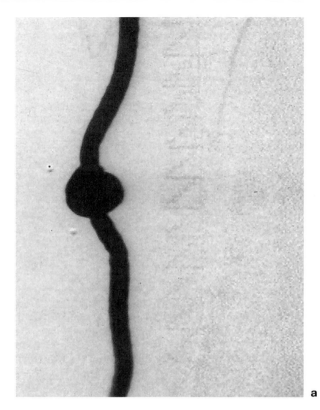
a

	Pulse status		Auscultation	
	Right	Left	Right	Left
Abdomen			–	–
Common femoral artery	+	+	–	–
Superficial femoral artery			+	–
Popliteal artery	((+))	+	–	–
Posterior tibial artery	(+)	+		
Dorsal artery of the foot	(+)	+		

Systolic pressure measurement (Doppler ultrasound technique; mmHg):

	Right	Left
Brachial artery	130	
Posterior tibial artery	105	150
Anterior tibial artery	110	130

Angiogram. Using intra-arterial DSA, a bypass rupture on the left side can be seen, with the formation of a false aneurysm in the middle third of the thigh (**a**). In the same session, a stent was implanted in order to bridge the ruptured bypass in the aneurysmal dilatation.

Color flow duplex sonogram *after stent implantation.* In a *longitudinal section,* B-mode sonographic imaging can readily distinguish the false aneurysm, due to its stronger reflections (**b**). In *cross-section,* the vascular prosthesis can be seen in the middle of the aneurysmal cavity (**c**). The false aneurysm measured approximately 21 mm × 23 mm × 19 mm. The stent's sonographic diameter was approximately 6 mm.

In longitudinal section (**d**) and cross-section (**e**), the *color flow Doppler mode* demonstrates that blood is still flowing through the aneurysmal cavity, in addition to the stent. In the stent region, the flow is strong, while in the aneurysmal cavity there are slower components with eddy-like flow visible in the darker colors, with color changes passing through black.

Comment. As an alternative to immediate surgery, a therapeutic attempt to provide internal support and possibly cause the aneurysmal cavity to thrombose proved unsuccessful. It was ultimately necessary to resect the aneurysm and bridge the bypass using an end-to-end anastomosis with a polytetrafluoroethylene (PTFE) prosthesis.

With regard to the stenosis of the right femoral artery, conservative therapy in the form of interval training succeeded in increasing the patient's pain-free walking distance to over 1000 m.

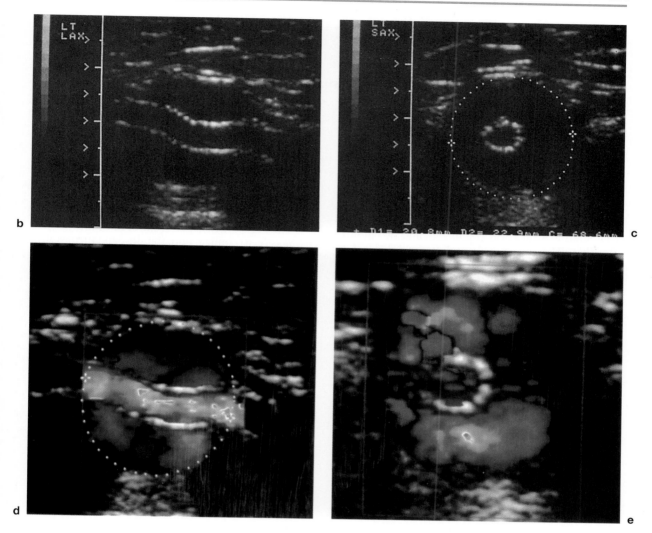

Dissection of the Common Femoral Artery
(Fig. 11.**33**)

Patient: R.H., male, age 45.

Clinical diagnosis: **PAOD bilateral stage IIa, stenosis of the right femoral artery,** and **stenosis of the left iliac artery.**

CAD:	–	PTA:	+	Diabetes mellitus:	–
PAOD:	+	Heredity:	+	Hyperlipidemia:	–
TIA, stroke:	–	Hypertension:	–	Smoking:	+

Case **report.** The patient had been suffering from calf claudication on the left over the previous three years. Two years previously, percutaneous transluminal angioplasty (PTA), initially of a stenosis in the left superficial femoral artery, was conducted, and two months after that, PTA of the ipsilateral external iliac artery. Initially, good clinical results were obtained in both instances. For the previous seven months, intermittent claudication had again been occurring in the region of the left calf, and three months after that also in the right calf. Currently, the chief complaint had been claudication of the right calf.

Examination results:

	Pulse status		Auscultation	
	Right	Left	Right	Left
Abdomen			–	+
Common femoral artery	+	(+)	–	+
Superficial femoral artery			+	–
Popliteal artery	(+)	(+)	–	–
Posterior tibial artery	(+)	(+)		
Dorsal artery of the foot	–	(+)		

Systolic pressure measurement (Doppler ultrasound technique; mmHg):

	Right	Left
Brachial artery	145	
Posterior tibial artery	120	115
Anterior tibial artery	–	105

Color flow duplex sonogram. *Left:* In a longitudinal section of the common femoral artery from a ventral orientation, with *poor delimitation in B-mode,* two hypoechoic pulsating structures are seen, which appear to be separated from one another by an oscillating membrane (arrows) (**a**).

Color flow Doppler sonography (**b**) shows two separate lumina with blood flowing through them. The flow velocity appears to be significantly elevated in the dorsal lumen; aliasing still occurs, although the color scale has been deliberately adjusted for high flow velocities (110 cm/s average velocity). Due to the color coding of the flow, inflow into the dissected lumen and reentry into the blood vessel can be recognized.

The registered *spectra* from both of the vascular channels (**c, d**) confirm both the slow and the fast flow velocity of around 100 cm/s systolic in the ventral lumen, and over 320 cm/s systolic in the dorsal lumen. This corresponds to a dissection with an overall hemodynamically effective constriction of the blood vessel.

Right: There is a high-grade effective stenosis of the superficial femoral artery at the transition zone from the proximal to the middle third of the blood vessel (not shown).

Comment. The clinical impression of the dissection was that of a moderate obstruction of the pelvic vascular system, most likely originating during the balloon dilatation of the left pelvic and leg vascular systems two years previously. There was currently no obligatory indication for surgical correction.

To alleviate the complaints, predominating on the right side, the ipsilateral stenosis of the femoral artery was dilated, allowing the patient's walking distance, at a rate of 120 steps/min, to increase to over 700 m.

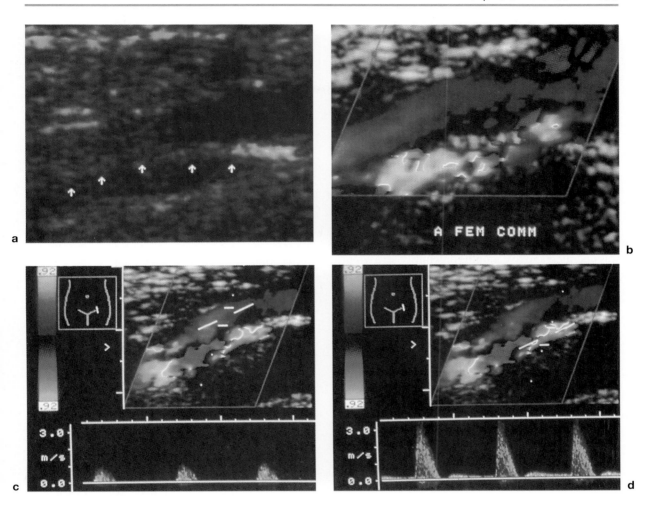

Arteries of the Lower Leg and Foot

▬▬ Stenosis of the Tibiofibular Trunk (Percutaneous Transluminal Angioplasty)
(Fig. 11.**34**)

Patient: H.R., male, age 66.
Clinical diagnosis: **PAOD bilateral stage IV,** suspicion of **low-grade bilateral stenoses of the superficial femoral artery and popliteal artery,** with **bilateral occlusions of the arteries in the lower leg.**

CAD:	+	Vascular surgery:	+	Diabetes mellitus:	–
PAOD:	+	Heredity:	–	Hyperlipidemia:	–
TIA, stroke:	–	Hypertension:	+	Smoking:	+

Case report. Ten years before, the right carotid artery was operated and five months previously, an asymptomatic occlusion of the ipsilateral common carotid artery occurred. An aortocoronary venous bypass (ACVB) was carried out three months previously, due to three-vessel disease of the coronary arteries. Occlusions of the arteries of the lower leg and toes had been occurring since approximately ten months before, and it was suspected that these were due to emboli associated with a dilatative and obliterative arteriopathy, in the presence a previously known saccular aneurysm of the abdominal aorta (70 mm × 41 mm × 39 mm).

Examination results. Edemas in the forefoot, which were more pronounced on the left than on the right, were seen. There were necroses of the tips of the toes, in the third digit on the right and on digits three and five on the left. There was inflammatory swelling and pain in the toes while at rest.

When the patient was admitted, the *pressure at the ankle arteries* could not be measured due to the edema in the feet. The brachial artery showed a pressure of 140 mmHg.

	Pulse status		Auscultation	
	Right	Left	Right	Left
Abdomen			–	–
Common femoral artery	+	+	+	+
Superficial femoral artery			+	+
Popliteal artery	+	+	+	+
Posterior tibial artery	–	–		
Dorsal artery of the foot	–	–		

Color flow duplex sonogram *before* (**a**) and *after* (**c**) percutaneous transluminal angioplasty. The illustrations show sagittal longitudinal sections from a dorsal orientation, which are adapted to the vascular axes in the region of the proximal lower leg, and the tibiofibular trunk.

Before angioplasty, color flow Doppler sonography showed aliasing in the trunk (**a**). The presence of a high-grade, hemodynamically effective stenosis, with maximum systolic and end-diastolic flow velocities of 338 cm/s and 43 cm/s, respectively, was confirmed.

After angioplasty, and using the *same* Doppler angle of 45° (!) with a still monophasic frequency spectrum, there was a clear decrease in the maximum systolic and end-diastolic flow velocities in the tibiofibular trunk to 83 cm/s and 11 cm/s, respectively (**c**). With a lower color scale, with a maximum of the average velocity at 34 cm/s compared to 69 cm/s, aliasing in the recanalized vascular segment still persisted.

Angiograms, before (**b**) and after (**d**) angioplasty. With selective application of an angiographic catheter, imaging (**b**) and simultaneous dilatation of the stenosis identified by duplex sonography in the tibiofibular trunk (arrow) was possible.

The second angiogram shows a satisfactory result, with clear lesions of the intima (**d**), and a persistent occlusion of the posterior and anterior tibial artery, both with peripheral refilling. There is also a stenosis of the fibular artery.

Comment. After angioplasty and conservative therapeutic measures (local treatment, antibiotics), the lesion in the third digit on the left healed. At the fifth toe, an amputation of the terminal joint was followed by satisfactory wound healing.

Five months after the above catheter intervention, the patency of the tibiofibular trunk was demonstrated during an angioplasty carried out due to fresh stenoses of the left superficial femoral artery.

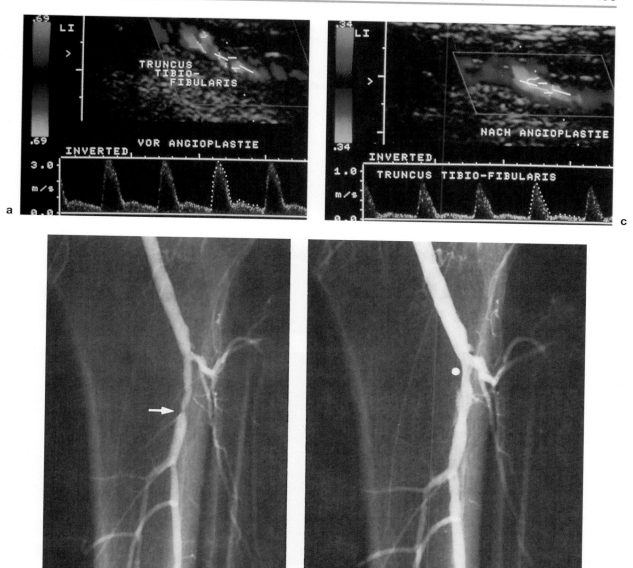

Occlusions of the Lower Leg Arteries
(Fig. 11.**35**)

Patient: H.M., male, age 51.
Clinical diagnosis: **PAOD** of the *acral* type, **occlusions of the lower leg arteries** (right anterior tibial artery and left anterior and posterior tibial artery).

CAD:	+	PTA:	– Diabetes mellitus: –
PAOD:	+	Heredity:	– Hyperlipidemia: +
TIA, stroke:	–	Hypertension:	– Smoking: +

Case report. For the previous five years, the patient had experienced exercise-dependent pain and paresthesia in the left foot. Prescribed foot supports had not led to any improvement in this condition.

Examination results. During Ratschow's test, the *left* forefoot and heel became pale, and when compared to the contralateral side showed delayed and excessive hyperemia.

	Pulse status		Auscultation	
	Right	Left	Right	Left
Abdomen			–	–
Common femoral artery	+	+	–	–
Superficial femoral artery			–	–
Popliteal artery	+	+	–	–
Posterior tibial artery	+	+		
Dorsal artery of the foot	–	–		

Systolic pressure measurement (Doppler ultrasound technique; mmHg):

	Right	Left
Brachial artery	135	
Posterior tibial artery	150	140
Anterior tibial artery	140	130

Angiogram. Intra-arterial needle angiography on the *left side* showed a normal image of the pelvic and thigh arteries. Stenoses of the posterior tibial artery are seen (upper arrow), and there is an occlusion of this vessel proximal to the talocrural articulation (**a**). There is strong refilling of the plantar artery of the foot and an occlusion of the anterior tibial artery at the level of the ankle.

Color flow duplex sonogram. The *left* posterior tibial artery (arrow in **a**) has a stenosis in its distal segment, showing local aliasing and a maximum systolic flow velocity of 103 cm/s; the pulsatility index was 2.8 (**b**). Distally, there was an occlusion of the blood vessel, with a hypoechoic image of the occluded lumen (arrows), and a collateral blood vessel given off preocclusively (**c**).

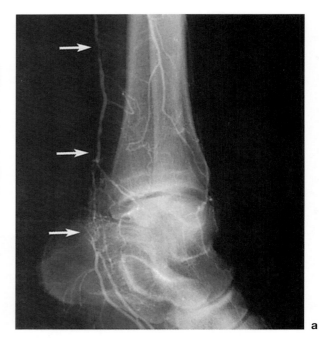

a

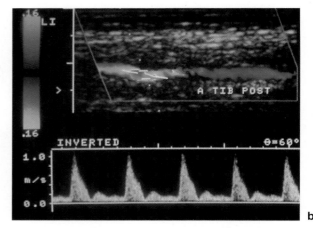

b

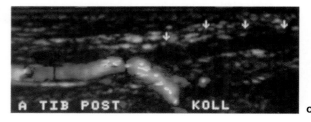

c

Comment. Particularly when atherogenic risk factors are present, the appearance of exercise-dependent pains in the sole of the foot should prompt consideration of a peripherally located obstruction of the arterial vascular system. A diagnosis indicating occlusions of the arteries in the distal lower leg or the foot rarely requires recanalization techniques. However, it is decisively important in relation to prognostic and prophylactic aspects (avoiding external noxa).

Shoulder and Arm Arteries

Stenosis of the Brachial Artery
(Fig. 11.**36**)

Patient: P.F., male, age 57.
Clinical diagnosis: **PAOD** with **stenosis of the right brachial artery** and **bilateral occlusions of the ulnar artery.** Status after amputation at the right thigh, with occlusion of the iliac and femoral arteries; stenosis of the popliteal artery and occlusions of the arteries in the left lower leg. Aneurysm of the abdominal aorta.

CAD:	–	PTA:	+ Diabetes mellitus: –
PAOD:	+	Heredity:	– Hyperlipidemia: +
TIA, stroke:	–	Hypertension:	– Smoking: +

Case report. Twenty-two years previously, calf claudication on the left had been accompanied by pallor of the toes when in a flat position. The diagnosis made at that time was occlusion of the left popliteal artery and bilateral occlusions of the lower leg arteries.

Since then, several percutaneous transluminal catheter treatments and recanalization attempts involving the leg arteries were conducted bilaterally. A lumbar sympathectomy on the left was completed 21 years ago. 4 years later, a right-sided lumbar and thoracic sympathectomy also followed. 1 year thereafter, an amputation at the level of the right thigh occurred.

Two years previously, a partially thrombosed infrarenal aneurysm of the abdominal aorta (70 mm × 35 mm × 30 mm), with a lumen of approximately 14 mm in its transverse diameter and with blood still flowing through it, had been detected sonographically.

There was no anamnestic arm claudication in the patient history.

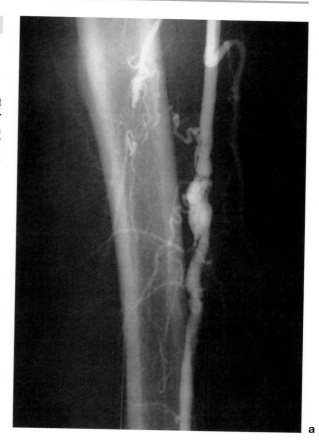

a

Angiogram. In selective angiography of the brachial artery with hand enlargement, clear fluctuations in the lumen of the brachial artery are seen in its middle third. Stenosed and aneurysmally dilated segments are seen (**a**). In the subsequent course, there was an occlusion of the ulnar artery and the radial artery, each in the distal third of the lower arm, with peripheral refilling of the radial artery and occlusions of the digital arteries in the fourth and fifth digits, ulnar and radial (not shown).

Fig. 11.**36 b, c** ▷

Examination results:

	Pulse status Right	Left	Auscultation Right	Left
Common carotid artery	+	+	–	–
Superior temporal artery	+	+		
Facial artery	+	+		
Subclavian artery	+	+	–	–
Distal brachial artery	((+))	+	+	–
Radial artery	(+)	(+)		
Ulnar artery	–	–		
Abdomen	Broad pulse		–	–
Common femoral artery	+	+	–	–
Superficial femoral artery	Amputation		Amputation	–
Popliteal artery		+		+
Posterior tibial artery		–		+
Dorsal artery of the foot		–		

Systolic pressure measurement (Doppler ultrasound technique; mmHg):

	Right	Left
Brachial artery	115	150
Ulnar artery	80	110
Radial artery	85	140
Posterior tibial artery	Amputation	–
Anterior tibial artery	Amputation	90

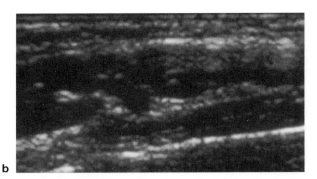

b

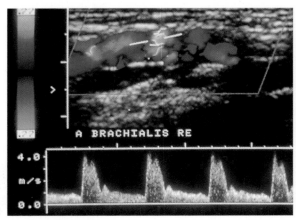

c

Fig. 11.**36 b, c**

Color flow duplex sonogram. In the *B-mode* image, a longitudinal section through the middle segment of the brachial artery shows the distension of the vascular lumen, with the vascular wall not continuously distinguishable (**b**).

The lighter colors and aliasing at a high velocity scale indicate a stenosis in the *color flow Doppler sonogram* (**c**). The blood flow in the aneurysmal sac shows disturbed flow.

With the monophasically changed *Doppler frequency spectrum,* the clearly elevated systolic (411 cm/s) and end-diastolic (94 cm/s) flow velocities can be calculated.

Comment. In this patient, dilatative/obliterative arteriopathy can be assumed in addition to a previously known occurrence of obliterating endangiitis, and which has also led to the stenotic and aneurysmal phenomena in the region of the right brachial artery. The obstruction in the vascular system of the right arm was discovered simply through a routinely administered side-to-side comparison of the measured blood pressure. This led to the further diagnostic procedures. The aneurysm of the abdominal aorta had already been known of for the previous two years.

In principle, the arteries of the upper extremity are at least as readily accessible to examination using (color flow) duplex sonography as the arterial vascular system in the pelvis and the leg. However, occlusive arterial disease in the region of the arm arteries is a very much less frequent occurrence.

Occlusions of the Brachial Artery and Arteries of the Lower Arm (Neurovascular Compression Syndrome)
(Fig. 11.**37**)

Patient: S.W., female, age 25.
Clinical diagnosis: **low-grade stenosis of the subclavian artery, proximal occlusions of the lower arm arteries on the right,** with suspicion of a **thoracic outlet syndrome,** suspicion of bilateral neck ribs.

CAD:	–	PTA:	– Diabetes mellitus: –
PAOD:	–	Heredity:	– Hyperlipidemia: –
TIA, stroke:	–	Hypertension:	– Smoking: +

Case report. The past medical history showed no significant prior illnesses. Three weeks previously, while lifting a heavy load (moving house), a tearing pain suddenly appeared in the right lower arm.

One week later, *pains in the right small finger and the hypothenar* appeared on slight exertion of the lower arm (rapid typing on a keyboard). The orthopedist providing treatment diagnosed a capsule irritation and prescribed *cryotherapy* and dressings with ointment, which did not produce lasting mitigation of the pain. Only ice application briefly alleviated the pain during arm exertion.

The symptoms were then classified as tendovaginitis, and *vibration massage treatment of the connective tissue* of the lower arm with ultrasound was initiated. Subsequently, there were also pains in the thumb region on exertion. The patient did not notice any white or blue discoloration of the fingers.

In addition to pain while using the arm, a feeling of numbness and coldness had appeared for the first time in the area of the entire hand three days before presentation in our clinic.

Examination results. A 25-year-old patient with good nutritional status and in a good general state of health. The heart and lungs were normal on percussion and auscultation, and there were no cardiac arrhythmias. The blood pressure on the right was 100/80, and 120/85 mmHg on the left. The following angiological findings were obtained:

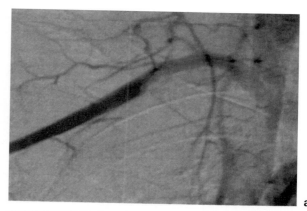

a

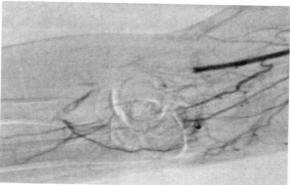

b

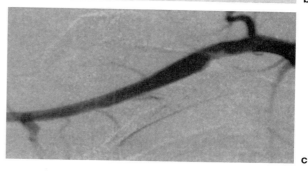

c

The *Allen test* on the right proved to be pathological.

Bilaterally in the region of the supraclavicular fossa, a bony structure with a connection to the spinal column was palpated in the slim patient.

The *mechanical oscillogram* at rest showed no oscillations above the right lower arm, with normal findings on the left.

Measuring the pressure at the finger (strain gauge) yielded the following values (in mmHg):

In the lower extremities, the pulse status and auscultation findings were normal.

	Pulse status		Auscultation	
	Right	Left	Right	Left
Common carotid artery	+	+	–	–
Superior temporal artery	+	+		
Facial artery	+	+		
Subclavian artery	(+	+	+	–
Distal brachial artery	(+	+	–	–
Radial artery	–	+		
Ulnar artery	((+))	+		

	Right	Left
Brachial artery	100	120
Digit II	33	120
Digit III	50	100
Digit IV	33	100

Fig. 11.**37 d–i** ▷

Fig. 11.**37 d**

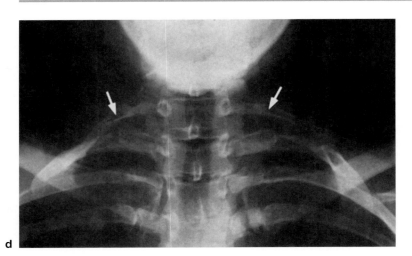

d

Angiograms. Selective catheterization with intra-arterial digital subtraction angiography (DSA) shows the subclavian artery, axillary artery, and brachial artery (**a**). In the region of the subclavian artery, there is a poorly delimited stenosis with irregular wall changes.

The brachial artery is occluded approximately at the height of the elbow joint. There are collaterals poorly refilling the proximal radial artery, ulnar artery, and interosseous arteries (**b**). There are subsequent occlusions of the radial artery and ulnar artery, with only the distal radial artery being refilled.

After lysis with rt-PA (alteplase, 13 mg) *using a catheter,* a clear decrease in the stenosis of the subclavian artery, with persistent wall deposits, is seen. The occlusion in the brachial artery has decreased in length, although complete patency of the vessel segment has not been achieved (**c**).

Radiograph of the upper thoracic aperture. Regular transparency and a typical bone structure of the skeletal components depicted, with normal soft tissues, are seen (**d**). The transverse processes of the cervical spine appear normal up to C6.

At C7, there are approximately 5 cm–long *neck ribs* (arrows). The uncinate processes are normal.

Color flow duplex sonogram. With regard to the clinical diagnosis of proximal occlusions of the arteries in the lower arm, color flow duplex sonography diagnosed an occlusion of the distal brachial artery (not shown), which was followed by selective lysis using an angiographic catheter. The results of this lysis treatment, monitored using color flow duplex sonography, were:

After selective lysis with an angiographic catheter, a hemodynamically patent subclavian artery (**e**) and brachial artery (**f**) were seen.

Compared to the result obtained immediately after lysis with the catheter (**c**), an examination with color flow duplex sonography one day after lysis, due to a delayed effect of the fibrinolytic process, shows the bifurcation at the lower arm with a patent radial artery and ulnar artery (**g**). The elbow joint is seen at the lower edge of the image.

The ulnar artery then becomes occluded approximately 6 cm after its origin (**h**). The vascular occlusion can be recognized from the interruption in the color; the vascular structures (wall and internal lumen components) are clearly visible. The preocclusive Doppler frequency spectrum shows a flow pattern corresponding to high peripheral resistance. Distally, peripheral refilling of the artery (not shown) was observed with duplex sonography.

The patent radial artery can be followed as far as the joints of the hand (**i**). The Doppler frequency spectrum indicates lower peripheral resistance in the artery's area of distribution.

Comment. The past medical history, suggesting a progressive set of symptoms developing in phases, leads one to suspect that recurrent emboli from the stenosis of the subclavian artery were passing into the arteries of the lower arm. Heavy lifting three weeks earlier might have caused trauma to the subclavian artery, with a parietal thrombus that scattered over the following period.

The successful selective lysis with the angiographic catheter supports the suspicion that the illness, at least the occlusive symptoms, began in the relatively recent past. It can be assumed that there had already been long-term (micro-)traumatization of the subclavian artery due to the neck ribs.

The patient was treated by exarticulation of the right first rib and the neck rib from a transaxillary access point, in order to remove the cause of the traumatic damage to the subclavian artery. Macroscopically, the neck rib and first rib showed a synostotic connection, with a club-like protrusion pointing in a ventral direction.

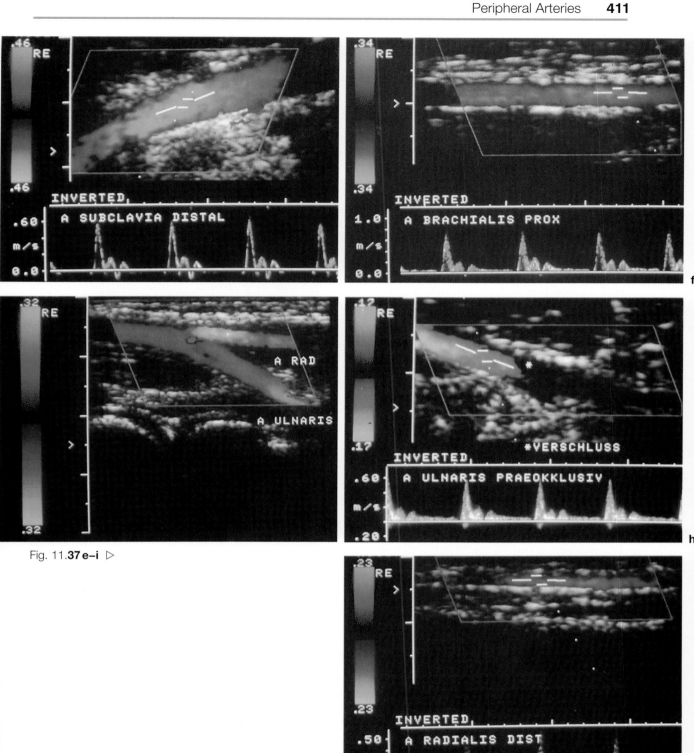

Fig. 11.**37 e–i** ▷

Occlusions of the Digital Arteries
(Fig. 11.**38**)

Patient: H.M., male, age 47. Clinical diagnosis: **occlusions of the digital arteries**
in the feet and the right hand, with an atypical rheumatoid arthritis as an underlying condition.

CAD: –	Vascular surgery: –	Diabetes mellitus: –
PAOD: –	Heredity: +	Hyperlipidemia: –
TIA, stroke: –	Hypertension: –	Smoking: +

Case report. Since 1971, the patient had had painful swelling in the region of the right talocalcaneal joint, later also involving the right joints of the hand. From 1975, the joints of the left hand were also affected. In 1982, a spontaneous ulcer of the lower leg appeared in the pretibial region on the right. In 1985, livid discolorations of the first and second right toes occurred. Gangrene developed. Since 1986, there had been a blue discoloration of the fourth right finger.

Laboratory results indicated the presence of a highly inflammatory process. Rheumatoid factors were negative. An atypical rheumatoid arthritis was diagnosed. Vascular occlusions were confirmed in 1987.

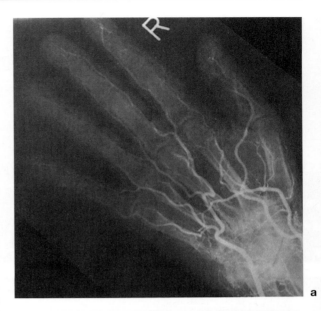

a

Examination results:

	Pulse status Right	Left	Auscultation Right	Left
Subclavian artery	+	+	–	–
Brachial artery	+	+		
Radial artery	+	+		
Ulnar artery	+	+		

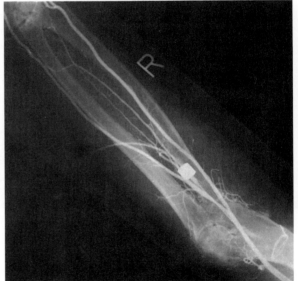

b

Doppler sonogram *(analogue registration)* (**c**). The Doppler sonograms of the radial artery and ulnar artery show a high diastolic resting flow rate and an absent reverse flow component. This is caused by the "warm hand" phenomenon, and partly by inflammation and the resultant peripheral hyperperfusion. The digital arteries could only be located on the radial side of the second, third, and fifth fingers.

Angiograms (**a, b**). *Angiography of the right brachial artery:* There are occlusions of the radial digital artery of the second digit, and also of the ulnar digital artery of the fifth digit. Stenoses of both the radial and the ulnar digital arteries of the fourth digit were noted. There was significant arthrosis in the metacarpal region.

Comment. Subsequent to an underlying case of atypical rheumatoid arthritis, the patient developed an acral occlusive disease, leading the ulcer formation in the feet and hand. A suspicion of crystalline emboli was excluded by examining a resected piece of skin from an ulcer on the tibia. Pressure measurement of the digital arteries also confirmed reduced pressures on the right. The blue, discolored finger (fourth digit), in particular, had a pressure of only 50 mmHg, indicating that both digital arteries were affected. The patient received sympatholytics and an antibiotic. With this therapy and careful treatment of the wound, remission of the necroses was achieved.

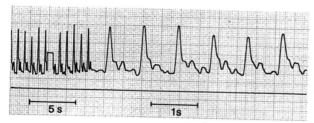

Right radial artery

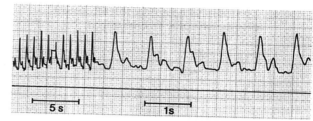

Right ulnar artery

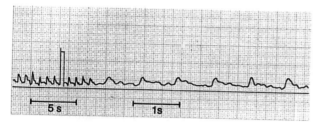

Digital artery, second right digit, radial side

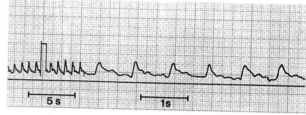

Digital artery, third right digit, radial side

c CW Doppler sonograms

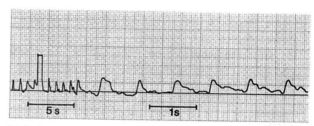

Digital artery, fifth right digit, radial side

Cuff position	Right radial artery	Left radial artery
Upper arm	115	120
Lower arm	130	135

Systolic pressure measurement (Doppler ultrasound and strain gauge technique; mmHg)

Cuff position	Right digital artery	Left digital artery
2nd basal joint of the finger	90	90
3rd basal joint of the finger	90	110
4th basal joint of the finger	50	110
5th basal joint of the finger	110	115

Stenosis of a Hemodialysis Shunt (Percutaneous Transluminal Angioplasty)
(Fig. 11.**39**)

Patient: R.S., male, age 63.
Clinical diagnosis: **stenosis of a dialysis shunt,** upper right arm.

CAD:	+	Vascular surgery:	+	Diabetes mellitus:	–
PAOD:	+	Heredity:	+	Hyperlipidemia:	–
TIA, stroke:	–	Hypertension:	+	Smoking:	–

Case report. The patient was suffering from dialysis-dependent renal failure with chronic glomerulonephritis, and had undergone dialysis treatment for the previous eight years. He presented with status after six shunt operations, involving both arms, that had been carried out due to recurrent shunt occlusions. The most recent of these, 14 months previously, had involved the placement of a Brescia–Cimino fistula in the crook of the right arm, with an anastomosis between the brachial artery and the cephalic vein. For the previous two weeks, there had been a continuous reduction in the arterial blood volume presented for hemodialysis, with a maximum flow rate of 150 ml/min at the dialyzer.

Examination results. The pulse on the right side above the brachial artery was strong on palpation. Above the radial artery, it appeared weakened. When *the shunt was palpated,* a strong thrill above the cephalic vein was observed approximately 5–10 cm proximal to the anastomosis. The proximal segment of the cephalic vein was only moderately filled, and was easily compressible. Smaller cutaneous collateral veins at the distal upper arm were noted. On *auscultation,* a pronounced, pulse-synchronous sound was audible above the venous shunt component, beginning shortly above the anastomosis and extending approximately 10 cm in a proximal direction.

Color flow duplex sonogram. There is an arteriovenous shunt in the crook of the right arm. The afferent brachial artery and anastomosis appear normal. Proximal to the anastomosis, increasing stenosis is seen (**a**). The flow direction is indicated by the arrow. The venous wall appears to be broadened (asterisk), due to massive fibrosis/hyperplasia of the intima (***). The residual lumen, with blood flowing through it, shows good visual delimitation due to the color coding.

The flow rate measurement in the nonstenotic venous shunt segment showed an integral flow velocity averaged over time of 13.2 cm/s. With a vascular diameter of 5.2 mm, one can therefore calculate a flow rate ($\pi \cdot r^2 \cdot$ flow velocity [cm/s]) of 168 ml/min.

Two weeks after *angioplasty* of the stenosis (**c**), the shunt vein shows a normal diameter. Without no obstacle to flow, the flow volume was calculated at 420 ml.

Shunt angiogram. Due to the normal-looking anastomosis in color flow duplex sonography, only the venous component was imaged using digital subtraction angiography (DSA) after a puncture near the anastomosis (**b**). The shunt stenosis (asterisks) is seen, as well as the branching of a venous collateral directly before the stenosis (bent arrow; main direction of flow: straight arrows). In the same session, angioplasty of the shunt using a 6-mm balloon was carried out.

The angiographic follow-up study after percutaneous transluminal angioplasty (PTA) (**d**) shows normalization of the diameter of the lumen and a functional exclusion of the collateral, as a result of the blood flow through the shunt vein, which is now unobstructed.

Comment. The stenotic area of the shunt corresponded to the region used for punctures during dialysis. Adequate dialysis aims to achieve a usable minimum flow rate of 250–300 ml/min. In the present patient, due to the precise color flow duplex-sonographic localization and hemodynamic effect of the stenosis, and the objective clinical observation of a reduced shunt flow, it was possible to identify the indication for PTA and plan the strategy of the interventional procedure. The shunt angiography therefore did not have a diagnostic role, but instead was simply used for overall guidance during the angioplasty itself. After successful PTA of the stenosis, problem-free dialysis using a two-needle system once more became possible.

We are grateful to Dr. P. Landwehr of the Institute and Polyclinic of Radiological Diagnostics at the University of Cologne for kindly giving permission to use this case and providing the illustrations.

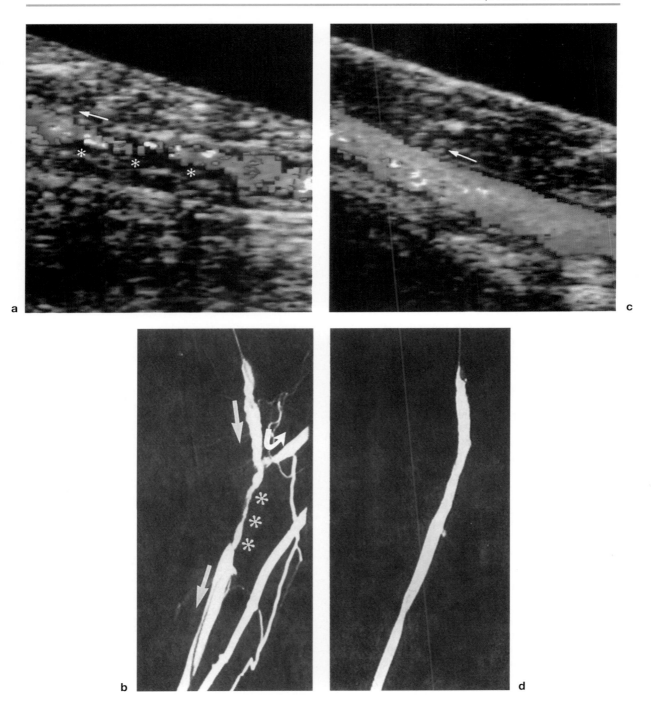

Peripheral Veins

Thrombotic Processes

Iliac Vein Thrombosis
(Fig. 11.**40**)

Patient: P.S., male, age 50.
Clinical diagnosis: **postthrombotic syndrome of the left extremity.**

Case report. In 1978, the patient had suffered a deep venous thrombosis on the left following a meniscus operation. An ulcer of the lower leg developed in 1979, which opened repeatedly. A venous exeresis at the left leg was carried out. Currently, there was hyperpigmentation of the left lower leg and an induration of the calf. The patient presented with status after a healed ulcer of the lower leg. He was substantially obese (107 kg with a height of 169 cm). Treatment was given with intermittent compression and fast compression bandages.

Examination results. The left lower leg showed hyperpigmentation, induration, and ulcer scars. There are also scars present due to venous stripping.

Doppler sonogram *(analogue registration)* (**c**): *Right:* There is spontaneous modulation of the respiration-dependent venous wave in the groin, and reflux during forced inspiration and during the Valsalva maneuver. There are pronounced flow peaks (A sounds) on compression in the thigh. *Left:* Continuous flow is seen during normal respiration with slight modulation during deepened respiration and a cessation of flow during the Valsalva maneuver. When the thigh is compressed, there are no flow peaks in the direction of the heart (central), due to the proximal flow obstruction. On decompression, there is suction in a peripheral direction due to the pressure gradient associated with valvular insufficiency.

Phlebogram (a, b). Ascending leg phlebography was carried out bilaterally.

Right: The deep venous system is intact, both in the pelvis and the leg, and there is no indication of a varicosis. *Left:* The image shows the status after a thrombosis of the proximal and middle superficial femoral vein, with poor recanalization, as well as a thrombosis of the external iliac vein, which is also poorly recanalized. The common femoral vein is patent.

Comment. As a consequence of the venous thrombosis, the patient has developed a marked postthrombotic syndrome on the left side. The Doppler-sonographic findings, showing continuous flow above the left groin, reflect the elevated pressure in the venous system of the left leg.

The phlebogram confirms the poor recanalization of the thrombosis in the thigh and pelvic region. A well-developed suprapubic collateral has formed to compensate (spontaneous Palma-like shunt. Fig. 11.**41**).

For the imaging of iliac vein thromboses using color flow duplex sonography, see Figure 6.**20a–c**, p. 211.

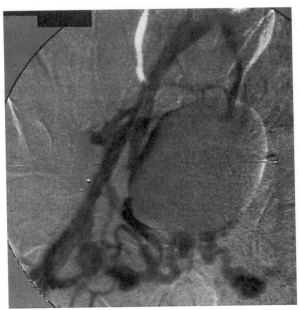

a

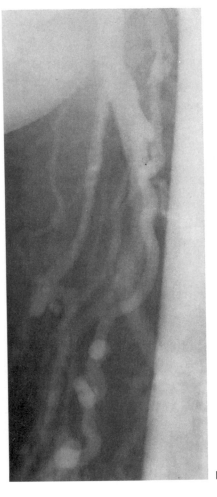

b

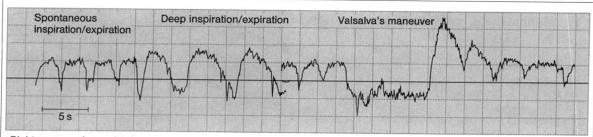

Spontaneous
inspiration/expiration

Deep inspiration/expiration

Valsalva's maneuver

5 s

Right common femoral vein

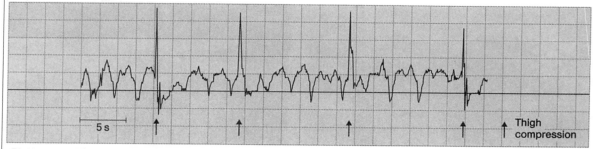

5 s

↑ Thigh
compression

Right common femoral vein, A-sounds

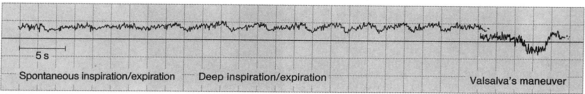

5 s

Spontaneous inspiration/expiration

Deep inspiration/expiration

Valsalva's maneuver

Left common femoral vein

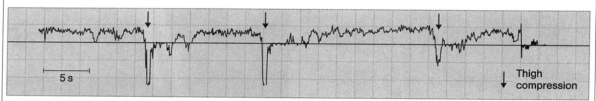

5 s

↓ Thigh
compression

Left common femoral vein, A-sounds

c

Thrombosis of the Iliac and Leg Veins (Spontaneous Palma-like Shunt)
(Fig. 11.**41**)

Patient: R.F., female, age 38.
Clinical diagnosis: status after postpartal **thrombosis of the left pelvic, thigh, and lower leg veins** 15 years previously, with a postthrombotic syndrome. There was a primary varicosis of the right great saphenous vein.

Case report. Fifteen years previously, a thrombosis of the *left* leg and iliac veins developed postpartally, and conservative treatment was given. Compression therapy was not consistently followed subsequently. Two years previously, a herniotomy was carried out on the left side, followed by a surgical neurolysis six months later. The patient now presented with pains in the left groin area, radiating into the thigh, and with the lower leg showing a tendency toward swelling.

Examination results. There was a slight increase in the circumference of the *left* leg and a suprapubic bypass circulation system had developed. In addition, varices of the great saphenous vein and pain when exerting pressure on the scar in the left groin were noted. Varicosis of the great saphenous vein was present in the *right* thigh and lower leg.

B-mode imaging. A nearly sagittal suprapubic section shows a dilated collateral vein, with blood flowing through it very slowly from left to right (**a**). On the left, color flow duplex sonography (not shown) demonstrated a recanalized external iliac vein, common femoral vein, and superficial femoral vein, as well as recanalized veins of the lower leg, with an occluded common iliac vein accompanied by iliac vein collaterals.

Phlebogram. On the left, there are partially recanalized lower leg veins, superficial femoral vein and iliac veins, and an occlusion of the common iliac vein. Contrast medium is draining via a dilated great saphenous vein into a dilated complex of suprapubic collateral veins, and passing to the contralateral side (**b**).

Comment. Due to the complete obstruction of the left pelvic vascular system, the patient has developed dilated suprapubic collateral veins (spontaneous Palma-like collateralization) toward the contralateral side. These provide substantial venous drainage for the left leg, and should therefore not be removed.

The patient was treated by an initial use of intensive methods to counteract the venous congestion (tight wrapping of the legs, intermittent compression treatment), and these were subsequently supported by a custom-made knee-length compression stocking.

We are grateful to Mr. G.G. Anding, Dr. H.-J. Walter, Mr. M. Schubert and Dr. H. Fallenski of Lüdenscheid for their kind permission to use the phlebogram.

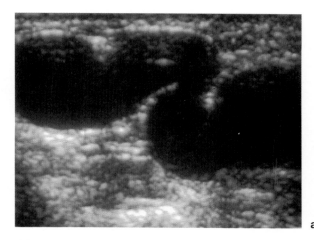

a

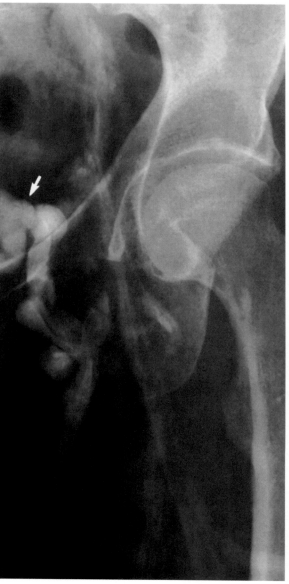

b

Thrombosis of the Femoral Vein
(Fig. 11.**42**)

Patient: H.K., male, age 63.
Clinical diagnosis: **acute thrombosis of the right deep leg veins.** Suspected rethrombosis after previous thrombosis of the right leg and pelvic veins.

Case report. Three years previously, a thrombosis of the deep leg and iliac veins occurred on the *right,* involving the common femoral vein, the superficial femoral vein, the popliteal vein, and also all three pairs of veins in the lower leg. At that time, successful fibrinolytic therapy was given, followed by oral anticoagulation treatment for 12 months, without any compression therapy. At presentation, an acutely appearing swelling of the *right* thigh and lower leg had been noticed four days before.

Examination results. On the *right,* there was disseminated hyperpigmentation in the lower leg region, with pretibial edema. In comparison with the left side, the circumference of the thigh and the lower leg was enlarged by approximately 1 cm. On palpation, the thigh was soft to the touch; the lower leg was somewhat firmer. There were no visible varices, and no suprapubic collateral vascular system.

Color flow duplex sonogram. Initially, *B-mode* imaging shows the right common femoral artery and vein in transverse section in the groin region (**c**). Compared to the artery shown laterally, the vein has a larger caliber. Even when pressure was applied with the transducer (right section of the image), hardly any compression of the venous lumen was possible.

In the same sectional plane, *color flow Doppler sonography* shows blood flow in the artery in red (**b**). Within the medially located vein, there is only sparse, marginal flow in the dorsal vascular segment (coded blue).

The corresponding longitudinal section through the right groin shows dorsal marginal flow in the segment of the common femoral vein furthest removed from the transducer (**d**).

In a slightly medial section, the confluence of the great saphenous vein (VSM) and common femoral vein (**e**) is seen. In this segment, the partially thrombosed common femoral vein has respiration-modulated physiological blood flow. The superficial femoral vein (VFS) is completely thrombosed. Further examination showed absent compressibility of the deep veins in the thigh and the lower leg (not shown).

Phlebogram. Imaging of the deep veins of the lower leg or thigh on the right side was not possible. Contrast drainage was exclusively through the epifascial venous system. Contrast is seen in the great saphenous vein, with drainage through the common femoral vein, presenting a clear, cone-shaped gap (**a**). Drainage continues freely through the iliac veins into the vena cava. There is a *venous thrombosis* of the right lower leg and thigh, with a cone-shaped thrombus in the common femoral vein.

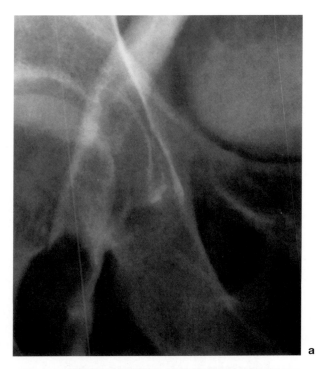

a

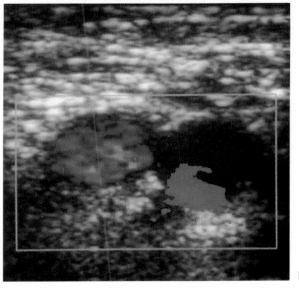

b

Comment. Using B-mode imaging alone, it is often difficult to detect a rethrombosis in the context of a postthrombotically altered venous system. In this set of findings, decreased or absent compressibility of the vein is not a definite indication of acute thrombosis. If detectable, the clearly dilated venous lumen is a more reliable sign. Color flow duplex sonography provides an additional margin of diagnostic certainty when evaluating absent or only sparse marginal flow near thrombi that have blood flowing round them. It is helpful to carry out comparisons with findings precisely documented earlier using phlebography or duplex sonography, to provide additional criteria for classifying the current findings on the basis of the location and extent of prior postthrombotic changes.

Fig. 11.**42 c–e** ▷

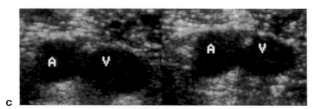

c

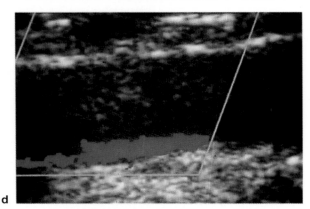

d

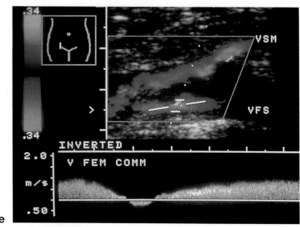

e

Fig. 11.**42c–e**

███ **Partially Occluded Thrombosis
of the Popliteal Vein**
(Fig. 11.**43**)

Patient: M.P., male, age 54.
Clinical diagnosis: **bilateral primary varicosis, insufficient perforating veins** (blow-outs), and **lymphedema** of the left lower leg, significant obesity.

Case report. The patient had been suffering from bilateral primary varicosis for the previous ten years. Four weeks previously, erysipelas developed in the left lower leg, with lymphangitis extending to the groin. Due to the difficult examination conditions when using continuous wave Doppler sonography (obesity), color flow duplex sonography was used, particularly to evaluate the valvular function of the superficial and deep venous system.

Examination results. The patient, with a height of 179 cm and weight of 110 kg, was obese.
The *left lower leg* appeared reddened, scaly, and swollen; the swelling was more pronounced distally (back of the foot) than proximally. Congestive dermatosis was observed, as a symptom of chronic venous insufficiency. There was a varicose, crown-like formation of the veins (cockpit varices) at the margins of the foot. All three Cockett groups in the left lower leg showed blow-outs.

Color flow duplex sonogram. The sagittal longitudinal section (**b**), proceeding from a dorsal direction (hollow of the knee) with color-coded flow visualization, depicts a partially occluded thrombus in the proximal popliteal vein. Using B-mode imaging in this sectional plane only allows moderate delimitation of the thrombotic material.
In the corresponding transverse section (**c**), the base of the thrombus, which is adherent to the wall, can be seen. In this plane, too, the round thrombus is recognizable through the marginal blood flow; the flow is modulated by respiration, and there is *no* hemodynamically effective venous flow obstruction at this location.
In a longitudinal section (**e**) and cross-section (**f**) from a dorsal direction in the same segment of the popliteal vein *16 days later,* color flow duplex sonography shows complete (spontaneous) lysis of the thrombotic material.

Phlebogram. Ascending leg phlebography on the left shows the floating thrombus (**a**) protruding into the lumen of the popliteal vein, with a paired distal superficial femoral vein.
In addition, varicosis of the trunk of the great saphenous vein was seen, with insufficient perforating veins of the lower, middle, and upper Cockett groups (not shown).
The *follow-up phlebography* after heparin-supported fibrinolysis (**d**) confirms the findings obtained with color flow duplex sonography (**e, f**), i.e., complete recanalization of the popliteal vein.

Comment. The underlying condition in this patient was chronic venous insufficiency, superimposed on in-

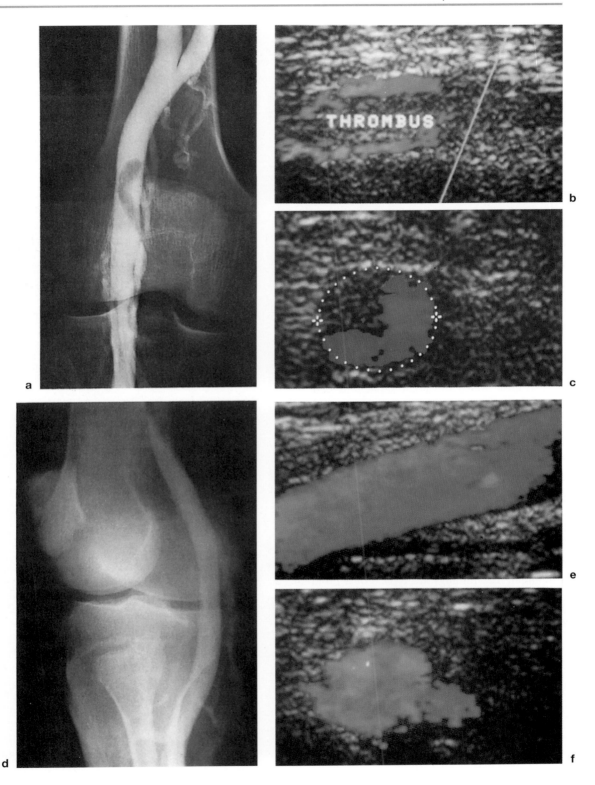

sufficiencies of the perforating veins. A previously undiscovered lymphedema was complicated by erysipelas, leading to a clear increase in swelling of the left lower leg, so that hospital admission was required.

The fresh thrombus confirmed in the left popliteal vein is an *accidental finding* with color flow duplex sonography.

The patient was initially treated with intravenous full heparinization, with a subsequent, overlapping period of transfer to an oral anticoagulant, and the course was typical for heparin-supported spontaneous lysis of the thrombus.

Valvular Insufficiency

Varicosis of the Great Saphenous Vein
(Fig. 11.**44**)

Patient: A.Q., male, age 49.
Clinical diagnosis: **varicosis of the main trunk and side branch of the right great saphenous vein.** Chronic venous insufficiency.

Case report. For the previous ten years, the patient had been suffering increasing varicosis, which was more pronounced on the right than on the left. One year previously, a swelling of the right leg occurred, and there was a suspicion of thrombosis or thrombophlebitis, as there was a sense of tightness in both calves. Clinically, the right leg presented with an increased circumference and signs of chronic venous congestion, induration, and brown pigmentation. Varicosis of the right great saphenous vein was observed.

Examination results. There was an increase in the circumference of the right leg, with chronic signs of venous congestion, and a slight increase in consistency was detected.

Doppler sonogram *(analogue registration)* (**b**). *Right:* Registration above the great saphenous vein, 7 cm below the confluence: there is respiratory modulation during both normal and deep inspiration, with pendular flow present as a sign of valvular insufficiency. The Valsalva maneuver was positive, and showed a reverse flow phenomenon.

Phlebogram (**a**). *Right:* Ascending leg phlebography depicts an intact deep venous system in the pelvis and leg, and varicosis of the trunk of the great saphenous vein, with insufficient lower and middle Cockett perforating veins. During the Valsalva maneuver, clearly insufficient venous valves were detected at the origin and in the course of the great saphenous vein.

Comment. The findings are typical of a pronounced trunk varicosis of the great saphenous vein. Surgery was proposed, but was delayed due to the patient's significant obesity.

The clinical suspicion of a deep vein thrombosis was not confirmed. The congestion is a symptom of chronic venous insufficiency, caused by pronounced primary varicosis, with insufficient venous valves and insufficiency of the perforating veins.

Example using color flow duplex sonography. The findings in a *different* patient (H.C., male, age 51) with varicosis of the great saphenous vein on the right, and with an insufficient valve at the vessel's origin were documented with color flow duplex sonography.

During spontaneous, quiet respiration, the normal blood flow in the proximal great saphenous vein directed toward the heart is coded blue in a longitudinal section (**c**). The frequency spectrum shows good respiratory modulation.

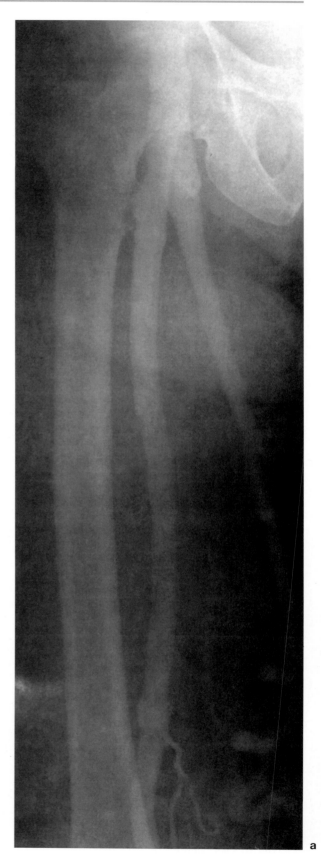

a

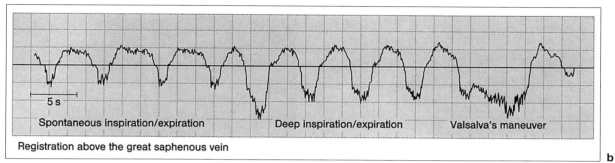

Spontaneous inspiration/expiration   Deep inspiration/expiration   Valsalva's maneuver

5 s

Registration above the great saphenous vein

b

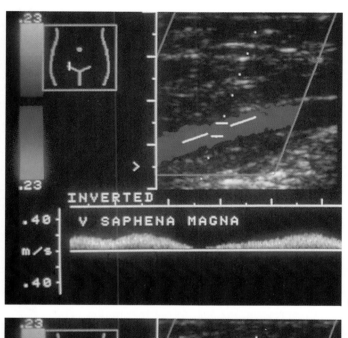

c

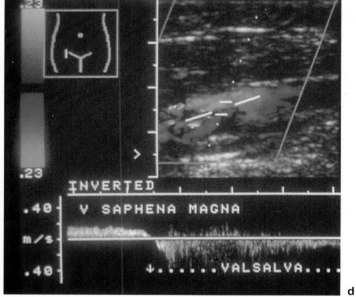

d

During the Valsalva maneuver, the red color coding in the vein shows a reversal in the flow direction, with retrograde flow (**d**). Also typical is the local appearance of red and blue pixels, representing eddies, e.g., at the valvular structures. The retrograde blood flow during the Valsalva maneuver is also shown below the baseline in the frequency spectrum.

Insufficient Perforating Veins
(Fig 11.**45**)

Patient: K.G., male, age 51.
Clinical diagnosis: **chronic venous insufficiency with an insufficiency of perforating veins of the left leg.**

Case report. The patient had been suffering varicosis of the left lower leg since the age of 40. Initially, there were minimal symptoms. There was a discrete supra-malleolar tendency toward venous congestion. Three years previously, brown spots had developed on the left distal medial lower leg, with a small lesion exuding pus. As yet, neither phlebography, sclerotherapy, nor surgery had been carried out. There was a familial predisposition to varicosis.

Examination results. In the left ankle region, there was a clear increase in the circumference of about 3 cm in comparison with the right side. The patient presented with status following a healed ulcer of the lower leg, with pigmented skin discoloration and moderate varicosis in the thigh region and in the course of the left great saphenous vein.

Doppler sonogram (*analogue registration*) (**b**). Above the left groin, there is a slight reflux during the Valsalva maneuver: normal, steep A sounds with a narrow peak were noted indicating an intact deep venous system.

The probe was placed above the blow-out of the perforating vein. Quick compression of the *distal* calf caused forward flow toward the probe, and retrograde suction when the compression was released. During this maneuver, afferent flow from the superficial varices was ligated using tourniquets.

Compression *proximal* to the registration point only resulted in minimal flow toward the probe, and on decompression there was reflux moving away from the probe. The slight reaction to proximal compression indicates that the valves of the deep venous system are intact.

Phlebogram. Ascending phlebography of the left leg (**a**) shows a freely patent venous system in the deep lower leg veins, popliteal vein, superficial and deep femoral veins, and iliac vein. Under the Valsalva maneuver, retrograde low-contrast imaging of the great saphenous vein was seen, but not of the small saphenous vein. There are insufficient perforating veins of the upper Cockett type, Sherman type, and Boyd type.

Comment. The patient was suffering from chronic venous insufficiency, superimposed on a slight insufficiency of the great saphenous vein and a pronounced insufficiency of the perforating veins.

During standing exercise, blood flowed out of the deep venous system through the insufficient valves of the perforating veins into the varicose superficial venous region. The result was chronic venous insufficiency, and clearly damaged skin with brown discoloration had formed due to the chronic venous congestion. This represents an indication for ligation the perforating veins.

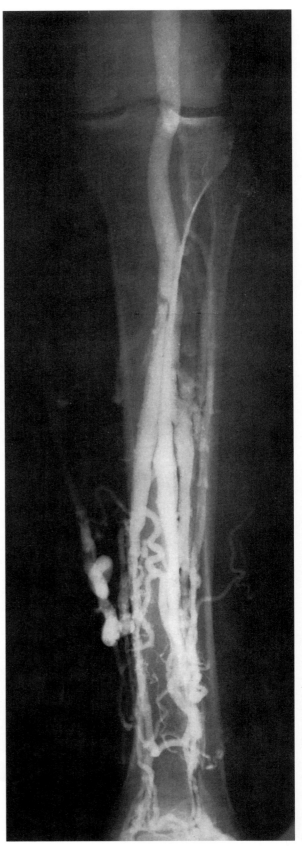

a

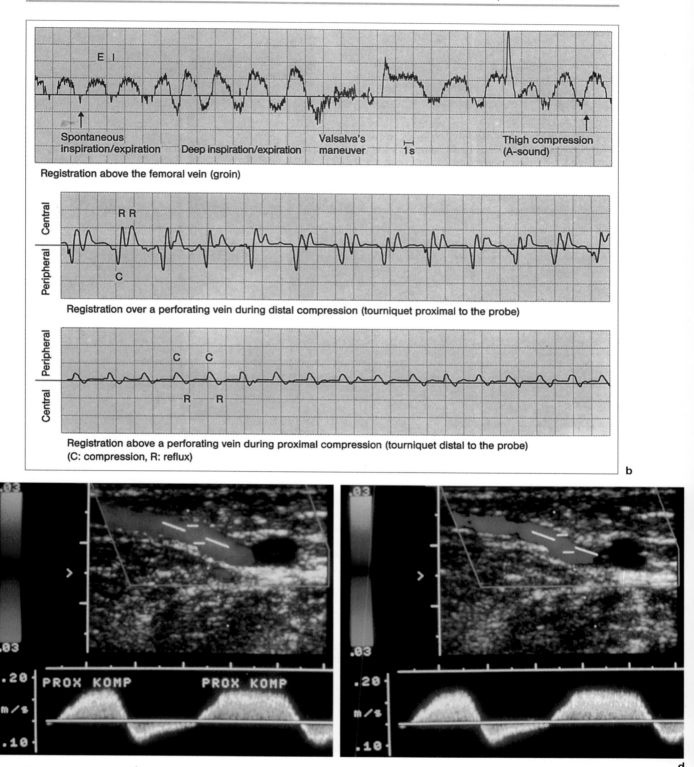

Registration above the femoral vein (groin)

E I · Spontaneous inspiration/expiration · Deep inspiration/expiration · Valsalva's maneuver · 1s · Thigh compression (A-sound)

Registration over a perforating vein during distal compression (tourniquet proximal to the probe)

R R · C · Central · Peripheral

Registration above a perforating vein during proximal compression (tourniquet distal to the probe)
(C: compression, R: reflux)

C C · R R · Peripheral · Central

b

PROX KOMP · PROX KOMP

d

Example using color flow duplex sonography. In a *different* patient (R.F., female, age 38) with an insufficiency of the perforating veins in the right lower leg region, pendular flow was demonstrated visually by the varying color coding of the opposite flow directions.

In adapted sectional planes along the anatomical course of a perforating vein above a *blow-out,* flow from the depths proceeding to the surface was recorded (coded red) during *proximal compression* of the calf above the registration point (**c**). After the pressure on the calf had been *released* (**d**), inflow into the deep vein proceeding toward the interior followed (coded blue), suggesting decompression. In both spectra flow above the baseline is coded red, and the reverse flow after release of compression is coded blue.

(The proximal compression is only effective if there are insufficient valves in the deep peripheral veins or when there is no valve in the region between the compression and perforating veins.)

Aneurysm

Aneurysm of a Great Saphenous Vein Bypass
(Fig. 11.**46**)

Patient: I.R., male, age 51.
Clinical diagnosis: **PAOD on the right,** status after **multiple bypass implantations** and **revisions.** Currently, **patent great saphenous vein bypass on the right.**

CAD:	–	Vascular surgery:	+	Diabetes mellitus:	–
PAOD:	+	Heredity:	–	Hyperlipidemia:	–
TIA, stroke:	–	Hypertension:	+	Smoking:	+

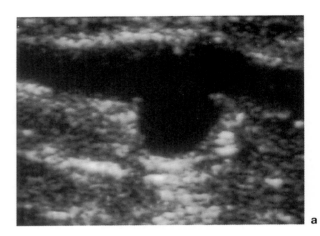

a

Case **report.** Three years previously, claudication in the right calf had occurred for the first time. Nine months earlier, a femoropopliteal Gore-Tex bypass was implanted. An occlusion of the bypass occurred two months previously, and fibrinolysis proved unsuccessful. A renewed implantation of a femoropopliteal bypass was carried out, requiring a triple revision (open recanalization, thrombectomies, venous patch surgery of the anastomoses). Finally, a femoropopliteal venous bypass (autologous left great saphenous vein) with an infraglenoid anastomosis was implanted. Currently, the patient was experiencing paresthesias and dysesthesias in the right lower leg region. There was no claudication.

Examination results. There was a pulsating resistance in the region of the right proximal thigh, located medially, that was painful on pressure. The scars at the right and left thigh and lower leg were mainly free of any irritation. The right lower leg was swollen due to postoperative lymphedema.

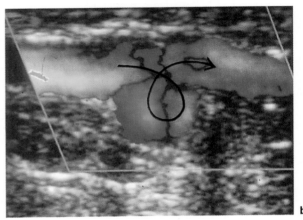

b

Color flow duplex sonogram. With *B-mode imaging,* and using a longitudinal section adjusted to the course of the bypass, an aneurysmal dilatation of the venous bypass with a diameter of approximately 8×11 mm, without any thrombotic covering of the aneurysmal wall, is seen in the proximal right thigh (**a**). At the bypass wall located closest to the transducer, there is a small, hyperechoic structure, representing a remaining venous valve.

Color flow Doppler sonography (**b**) shows that there is blood flowing freely through the bypass. In the *aneurysmal cavity,* the color change from blue through black to red, using a lower preselected scale, indicates slow, eddy-like flow. In this sectional plane, the main flow direction is marked by a curved arrow.

Angiogram. Central venous digital subtraction angiography (DSA) shows the bypass aneurysm in the proximal right thigh in two planes (**c**). There is an occlusion at the origin of the superficial femoral artery.

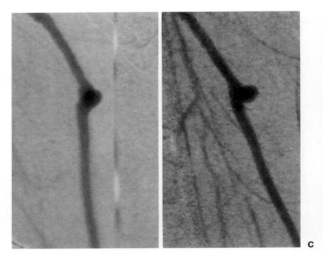

c

Comment. Due to the small size of the bypass aneurysm and the multiple operations the patient had previously undergone, it initially seemed justified to wait cautiously while providing oral anticoagulation, closely monitoring the patient's status with color flow duplex sonography. If the size of the aneurysm increases, vascular surgical intervention may eventually become necessary.

Upper Extremities

Subclavian Vein Thrombosis
(Fig. 11.**47**)

Patient: W.L., male, age 36.
Clinical diagnosis: **thrombosis of the right subclavian vein.**

Case report. For the previous five days, there had been swelling of the right arm, which had begun in the hand. The patient was unable to recall any possible cause of the condition, such as a vigorous or unusual exertion of the arm, or trauma.

Examination results. A bluish, livid discoloration of the *right* hand and forearm was seen, which became pronounced when the arms were allowed to hang at the sides. Emptying of the veins was delayed on arm elevation. The right upper arm and forearm were approximately 2–3 cm larger in circumference in comparison with the contralateral side. Palpation detected somewhat firmer subfascial tissue, with an increase in consistency in comparison with the left side. Collateral veins were not visible in the thoracic region.

Phlebogram of the arm. There is normal venous drainage in the axillary vein, and an occlusion over a short distance in the proximal right subclavian vein (**a**).

The follow-up phlebogram after fibrinolysis documents continuing unobstructed venous drainage through the axillary vein, and a recanalization of the subclavian vein with a residual thrombotic fragment in the proximal vascular segment (**b**).

Color flow duplex sonogram *during the first fibrinolytic cycle.* The deep and superficial veins of the forearm and upper arm were completely compressible (not shown). In an almost frontal sectional plane proceeding from a supraclavicular direction, the proximal

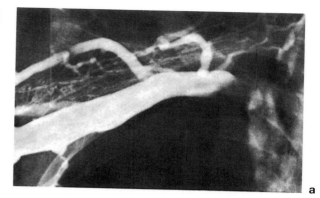

a

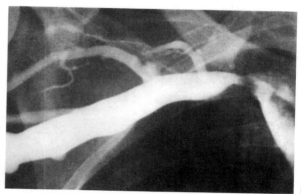

b

right subclavian vein and the jugular vein can be seen. Blood is already flowing round the thrombus in the proximal subclavian vein (asterisk). The high flow velocity in the partially recanalized vein is reflected by the light shades of blue (**c**).

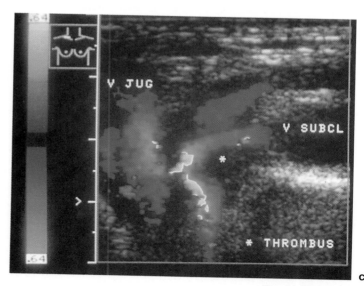

c

Fig. 11.**47 d–h** ▷

The *Doppler frequency spectrum* from the re-canalized venous segment (**d**) documents a high flow velocity of maximally 100–130 cm/s, which is not modulated by respiration.

The right *jugular vein* is not thrombosed, and shows flow affected by both respiratory and cardiac modulation (**e**).

The *M-mode* confirms the absence of a collapse of the right axillary vein during forced panting in the *sniff test* (**f**). The fine indentations in the venous wall reflections above and below the anechoic lumen correspond to the high respiration frequency, and are caused by concomitant transducer movements due to respiration-dependent skin and tissue movement above the supraclavicular fossa.

In the contralateral *left* axillary vein, the M-mode image shows a decrease in the venous caliber during rapid, panting respiration (**g**).

After fibrinolytic therapy, the right subclavian vein is functionally patent again. In the same sectional plane as in Figures **c** and **d,** the patent proximal segment of the subclavian vein is seen with color flow duplex sonography (**h**). The *Doppler frequency spectrum* once again shows a normal cardiac modulation of the flow velocity, whereas a continuous flow was registered previously (**d**). Also, the flow velocity has decreased in comparison to the earlier findings due to the wider venous lumen through which the blood is flowing.

Comment. With two fibrinolytic cycles administered over four hours using six million units of streptokinase in each instance (ultrahigh short-time lysis), successful thrombolysis of the subclavian vein occlusion was achieved. Oral anticoagulation treatment followed.

The cause of the subclavian vein thrombosis remained unclear. The patient history showed no evidence of an effort thrombosis, nor were there any indications of neurovascular compression or a paraneoplastic syndrome.

Fig. 11.**47 d** and **e**

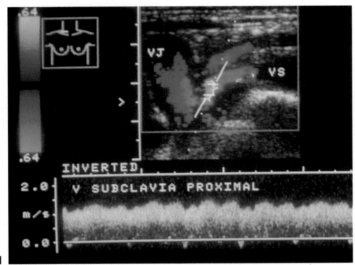

d

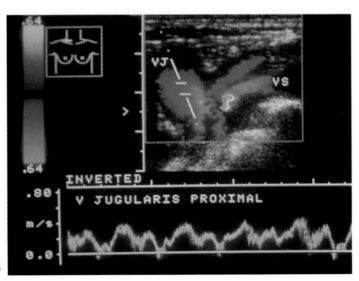

e

Fig. 11.**47 f–h**

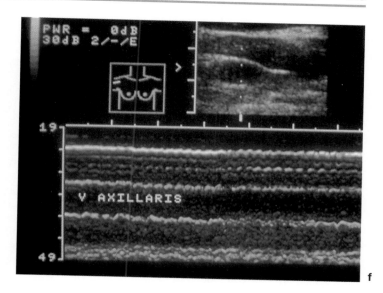

f

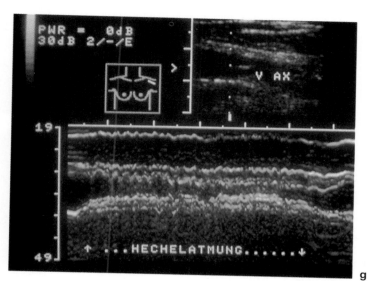

g

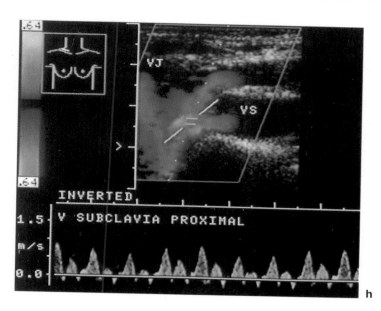

h

Abdominal Arteries

Abdominal Aorta

Aortic Aneurysm
(Fig. 11.**48**; two cases)

First patient: C.T., male, age 71.
Clinical diagnosis: PAOD, with a suspicion of **abdominal aortic aneurysm.** In addition, occlusion of the *right* iliac artery and superficial femoral artery, and occlusions of the lower leg arteries. On the *left*, occlusion of the superficial femoral artery and occlusions of the lower leg arteries.

CAD:	–	Vascular surgery:	–	Diabetes mellitus:	–
PAOD:	+	Heredity:	–	Hyperlipidemia:	–
TIA, stroke:	–	Hypertension:	+	Smoking:	+

Case report. The patient appeared prematurely aged, and gave an impression of reduced general health. This was his first ever angiological examination. With the exception of medically untreated hypertension, no prior illnesses were known of. Due to the patient's immobilization, intermittent claudication was not observed.

Examination results. There was a pulsating tumor in the epigastric region, which was painful on pressure. The feet were cold and had a livid discoloration, with scaly, dry skin.

	Pulse status Right | Left		Auscultation Right | Left	
Abdomen	Broad Pulse		–	+
Common femoral artery	((+))	+	–	–
Superficial femoral artery			–	–
Popliteal artery	–	((+))	–	–
Posterior tibial artery	–	–		
Dorsal artery of the foot	–	–		

Systolic pressure measurement (Doppler ultrasound technique; mmHg):

	Right	Left
Brachial artery	210	
Posterior tibial artery	50	95
Anterior tibial artery	60	120

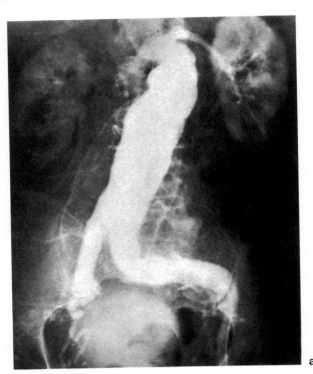

a

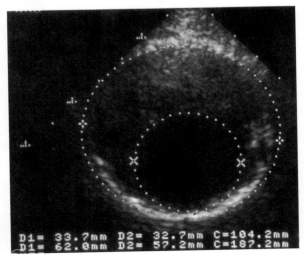

D1= 33.7mm D2= 32.7mm C=104.2mm
D1= 62.0mm D2= 57.2mm C=187.2mm   b

Angiogram. In conventional angiography of the abdominal aorta (**a**), an infrarenal aneurysm of the abdominal aorta is seen over a considerable distance, extending into both common iliac arteries. This confirms the clinical diagnosis.

B-mode image. A transverse section distal to the renal arteries shows a large aneurysm of the abdominal aorta (62 mm × 57 mm), with a clear partial thrombosis (**b**), some hyperechoic material, and a lumen (34 mm × 33 mm) with blood flowing through it dorsally.

Second patient: H.K., male, age 64.

Clinical diagnosis: PAOD. Suspicion of an **aneurysm of the abdominal aorta** accompanied by an occlusion on the *right* of the iliac artery, common femoral artery, and superficial femoral artery. On the *left*, an aneurysm of the common femoral artery and popliteal artery. **Dilatative/obliterative arteriopathy.**

CAD: +   Vascular surgery: –   Diabetes mellitus: –
PAOD: +   Heredity: –   Hyperlipidemia: –
TIA, stroke: +   Hypertension: –   Smoking: +

Case report. The patient presented with a brachiofacial-weighted spastic hemiparesis following an ischemic stroke three years earlier. The patient was mobile predominantly in a wheelchair, there were therefore no symptoms indicating intermittent claudication.

Examination results. There was a pulsating tumor in the epigastric region that was painful on pressure. Spastic hemiparesis was seen on the right side. The right lower leg was cool, and there were no lesions.

Color flow duplex sonogram. The transverse section through the abdominal aorta distal to the renal arteries (**c**) shows an infrarenal aneurysm of the abdominal aorta, with a cross-sectional diameter of 52 mm × 51 mm. There is a massive partial thrombosis, with a left lateral marginal lumen through which blood is flowing (21 mm × 21 mm), which is coded red.

In the sagittal longitudinal section, color coding shows blood flow through the lumen (**d**). Flow direction from the left (cranial), toward the transducer, is coded red; flow direction toward the right (caudal), away from the transducer, is coded blue.

	Pulse status Right	Left	Auscultation Right	Left
Abdomen	Broad Pulse		–	+
Common femoral artery	–	+	–	+
Superficial femoral artery			–	–
Popliteal artery	–	Broad	–	+
Posterior tibial artery	–	+		
Dorsal artery of the foot	–	+		

Systolic pressure measurement (Doppler ultrasound technique; mmHg):

	Right	Left
Brachial artery	165	
Posterior tibial artery	65	170
Anterior tibial artery	75	160

Comment. The size of the aneurysms in both patients represented an indication to proceed with surgical elimination of the aneurysm. In the second patient, the procedure was carried out without any complications. Prior to the planned operation, the first patient suffered an ischemic stroke in the left brain, and surgical implantation of a prosthesis was therefore postponed.

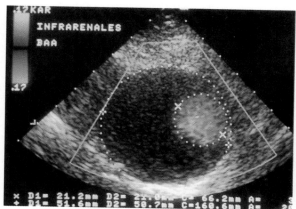

c

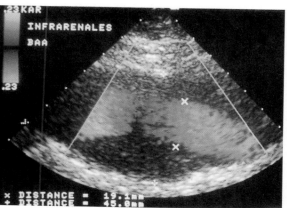

d

Supplementary images (e, f) from a *different* patient (S.B., male, age 62), in a longitudinal section proximal to the navel, show both branches of an aortobifemoral bifurcation prosthesis, which are still surrounded by the aneurysmal sac (46 mm × 40 mm) left in situ. Using various adjustments to the color-coded B-mode image, both branches of the prosthesis can be distinguished as relatively hyperechoic rings, located dorsolaterally on the left in the earlier aneurysm.

Postoperatively, depending on the surgical technique used, an aneurysmal sac left in situ may continue to give the impression of being a dilatation of the vascular wall. In such cases, special attention needs to be given to the differential diagnosis of liquid (perivascular) structures (hematoma, seroma, infection). Differentiation between prosthetic material and the original wall structures is also important. In some cases, the sonographic findings can only be properly evaluated if the surgical procedure used is known.

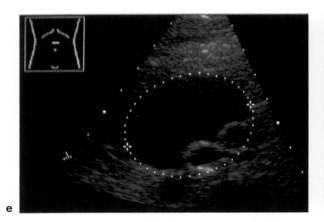

e

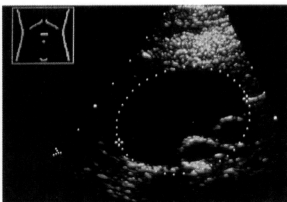

f

Aortic Stenosis
(Fig. 11.**49**)

Patient: M.P., female, age 80.
Clinical diagnosis: **bilateral arterial occlusive disease, stage IV** on the left. **Bilateral stenoses of the iliac arteries.** The differential diagnosis suggests an **aortic stenosis; bilateral occlusions of the lower leg arteries.**

CAD: +	PTA: –	Diabetes mellitus: +	
PAOD: +	Heredity: –	Hyperlipidemia: +	
TIA, stroke: –	Hypertension: +	Smoking: –	

Case report. Six months previously, the patient had received a traumatic lesion to the left first toe, which was accompanied by delayed healing. For the previous six weeks and currently, there was a moist lesion on the fourth digit on the left, showing no tendency to heal. The patient was experiencing pain in the left foot when resting it a horizontal position.

Examination results:

	Pulse status Right	Pulse status Left	Auscultation Right	Auscultation Left
Abdomen			+	–
Common femoral artery	((+))	((+))	+	+
Superficial femoral artery			–	–
Popliteal artery	–	–	–	–
Posterior tibial artery	–	–		
Dorsal artery of the foot	–	–		

Systolic pressure measurement (Doppler ultrasound technique; mmHg):

	Right	Left
Brachial artery	160	
Posterior tibial artery	50	40
Anterior tibial artery	55	40

Angiogram. There is a high-grade concentric stenosis of the infrarenal abdominal aorta, proximal to the bifurcation (**a**). Stenoses of the right superficial femoral artery, an occlusion at the origin of the left superficial femoral artery, and bilateral occlusions of the lower leg arteries were also noted (not shown).

Color flow duplex sonogram. A sagittal longitudinal section of the infrarenal abdominal aorta using color flow Doppler sonography displays the spindle-shaped aortic stenosis (**b**).

The high flow velocity in the stenosis jet during the systole is represented by light colors (yellow/white/blue), contrasting with the uniform prestenotic color coding of the aortic lumen.

The *Doppler frequency spectrum* shows maximum systolic and end-diastolic flow velocities in the stenosis of 438 cm/s and 33 cm/s, respectively. The pulsatility index is 3.02 (**b**).

Comment. The Doppler pressure of 40 mmHg in the left foot indicated that the resting circulation was at a dangerously low level. Further evidence of insufficient blood supply to the toes was provided by the le-

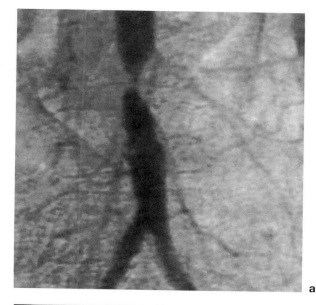

a

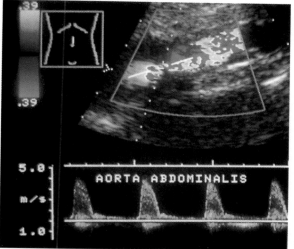

b

sion in the fourth toe. Recanalization procedures were indicated.

Treatment consisted of successful catheter angioplasty of the aortic stenosis, which eliminated the systolic and average pressure gradients of 68 mmHg and 13 mmHg, respectively. Despite the subsequent occlusions of the femoral artery and lower leg arteries, blood flow to the left foot improved to such an extent that a demarcation zone formed between the necrotic distal and middle joints in the fourth digit. The toe, which now had sufficient perfusion to promote wound healing, was successfully amputated.

As there was arterial hypertension, accompanied by many years of diabetes mellitus, and there was concern due to the patient's age, systemic fibrinolytic therapy was not used to remove the vascular obstruction, although in view of the location of the stenosis this might have been considered as a possible treatment.

Aortic Stenosis in Combination with Iliac Artery Stenosis
(Fig. 11.**50**)

Patient: H.-U.N., male, age 55.
Clinical diagnosis: **bilateral stage IIb arterial occlusive disease, bilateral occlusion of the iliac artery.** Differential diagnosis includes a possible **aortic occlusion.**

CAD:	–	PTA:	– Diabetes mellitus: –
PAOD:	+	Heredity:	– Hyperlipidemia: +
TIA, stroke:	–	Hypertension:	– Smoking: +

Case report. For the previous 18 months, the patient had suffered increasing bilateral calf claudication. Exercise-dependent pains, radiating into the thighs, had now also developed bilaterally in the region of the buttocks.

After *subsequent balloon dilatation,* all pulses in the region of the lower extremities could be palpated. Flow sounds could still not be auscultated. At a constant systemic blood pressure, the pressure at the ankle arteries could be elevated to 120 mmHg.

Angiogram *before lysis and angioplasty.* Central venous digital subtraction angiography (DSA) identified a high-grade distal aortic stenosis (arrow) and an extensive stenosis of the left common iliac artery (**a**).

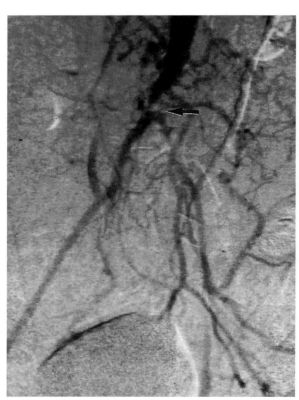

a

Examination results before lysis:

	Pulse status Right	Left	Auscultation Right	Left
Abdomen			–	–
Common femoral artery	–	–	–	–
Superficial femoral artery			–	–
Popliteal artery	–	–	–	–
Posterior tibial artery	–	–		
Dorsal artery of the foot	–	–		

Systolic pressure measurement (Doppler ultrasound technique; mmHg):

	Right	Left
Brachial artery	150	
Posterior tibial artery	75	70
Anterior tibial artery	65	65

Examination results after lysis:

	Pulse status Right	Left	Auscultation Right	Left
Abdomen			–	–
Common femoral artery	+	+	–	–
Superficial femoral artery			–	–
Popliteal artery	+	+	–	–
Posterior tibial artery	–	(+)		
Dorsal artery of the foot	+	+)		

Systolic pressure measurement (Doppler ultrasound technique; mmHg):

	Right	Left
Brachial artery	120	
Posterior tibial artery	100	95
Anterior tibial artery	105	90

Color flow duplex sonogram *before lysis and angioplasty.* The illustration shows the *distal abdominal aorta* in a sagittal ultrasound plane (**b**). The vascular lumen (above the arrows) is almost completely filled with arteriosclerotic material. The color coding indicates the ventrally located lumen through which blood is still flowing.

The Doppler spectrum corresponds to a high-grade stenosis, with high maximum systolic, end-diastolic, average flow velocities of 519 cm/s, 103 cm/s, and 252 cm/s (pulsatility index = 1.6), respectively, and it shows pronounced spectral broadening. In addition, there are retrograde flow components below the baseline that represent parts of the turbulent flow. With somewhat lower velocities, the frequency spectrum from the right common iliac artery had a similar profile (not shown), corresponding to a high-grade, hemodynamically effective stenosis.

In longitudinal sections through the arteries of the groin region, the *Doppler frequency spectra* from the right (**e**) and left (**f**) common femoral arteries show a monophasic wave form, with reduced maximum systolic flow velocities of 36 cm/s and 32 cm/s due to the preceding obstructions of the vascular system.

In the right common femoral artery, hyperechoic plaques are seen at the dorsal wall, from which ultrasound shadows continue into the tissue. The ventral wall, which is located close to the transducer, and also the contralateral wall (**f**), appear to be thickened and to have stronger reflections, suggesting vascular sclerosis.

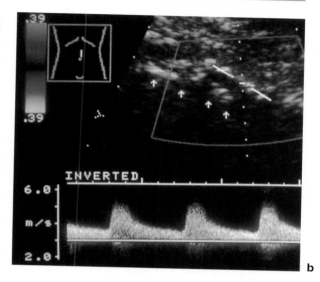

b

Angiogram. *After systemic lysis with streptokinase,* the stenosis of the left iliac artery has been eliminated (**c**). In the distal abdominal aorta, there is still an eccentric stenosis (arrow).

After subsequent percutaneous transluminal angioplasty using the kissing-balloon technique, a freely patent aortic bifurcation and left common iliac artery are seen (**d**).

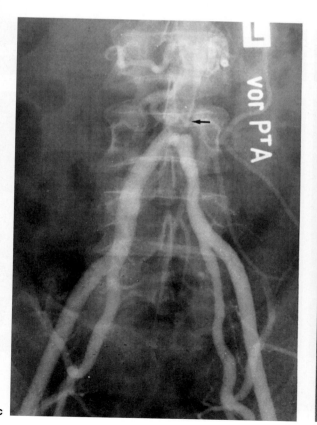

c

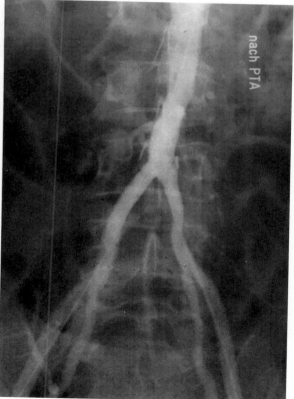

d

Fig. 11.**50 e–h** ▷

Color flow duplex sonogram. *After fibrinolysis and angioplasty,* the right (**g**) and left (**h**) common femoral arteries are shown again in the same sectional planes as in Figures **e** and **f** with now triphasic flow profiles. However, there is still clear flow on the left. The maximum systolic flow velocities have increased to 119 cm/s and 89 cm/s.

Comment. Even when they have already existed for a longer period, stenoses of the aorta and iliac arteries are particularly amenable to fibrinolytic therapy. In the present case, clearly increased peripheral pressure in the ankle arteries was observed, as well as an increased pain-free walking distance clinically after two fibrinolytic cycles had been given. Subsequent angio-plasty of the distal abdominal aorta provided further improvement in the lysis, to the extent that hemodynamically effective obstructions of the vascular system were no longer present (systemic pressure and pressure at the ankle arteries of equal magnitude).

With color flow duplex sonography, it was *not* possible to recognize the stenosis of the left common iliac artery accompanied by a distal aortic stenosis solely using *indirect* ultrasound at the arteries of the groin, as is usual in continuous wave Doppler sonography, since both femoral arteries presented a monophasic wave profile with a reduced systolic flow velocity. Only the direct and complete examination of both pelvic vascular systems was capable of identifying the additional stenosis of the left iliac artery.

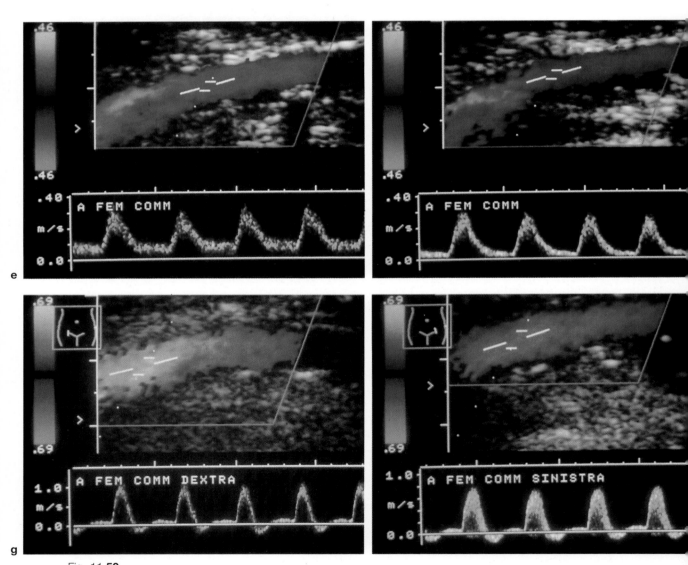

Fig. 11.**50**
e Right common femoral artery *prior* to angioplasty of the aorta
f Left common femoral artery *prior* to angioplasty of the aorta
g Right common femoral artery *after* angioplasty of the aorta
h Left common femoral artery *after* angioplasty of the aorta

Aortic Occlusion
(Fig. 11.**51**)

Patient: M.E., female, age 39.
Clinical diagnosis: **bilateral stage IIb arterial occlusive disease,** claudication of the buttocks; suspicion of a **bilateral occlusion of the iliac arteries;** the differential diagnosis suggests **aortic occlusion.**

CAD: –	Vascular surgery: –	Diabetes mellitus: –
PAOD: +	Heredity: –	Hyperlipidemia: –
TIA, stroke: –	Hypertension: –	Smoking: +

The patient had used hormonal birth control for the previous four years.

Case report. For the previous 15 months, the patient had been experiencing pain during exercise in the pelvic and thigh regions (especially when climbing stairs), initially more on the left side than on the right. A *computed tomogram of the lumbar spine* had detected a *"sclerosis of the iliac artery."* However, no further angiological examinations had been undertaken, and the indication for digital subtraction angiography examination had not been acted on.

Three months earlier, an orthopedic diagnosis of "pseudoradicular sciatic pain in the region of the lumbar spine" with an "arthrosis of the articular facets" had been made. The recommended treatment had consisted of injections specifically into the area of the articular facets, and physiotherapy to encourage stabilization of the lumbar spine and eliminate its lordosis.

On admission, the female patient complained of pains on exercise, which radiated from the lumbar spine into both buttocks and thighs. The pains were associated with a feeling of numbness in the buttock region, and also with calf pains after more strenuous exercise. There was no pain during rest. A standardized walking test (pace of 120 steps/min) showed thigh claudication after 100–150 m.

Examination results:

	Pulse status Right	Pulse status Left	Auscultation Right	Auscultation Left
Abdomen			–	–
Common femoral artery	–	–	–	–
Superficial femoral artery			–	–
Popliteal artery	((+))	((+))	–	–
Posterior tibial artery	–	–		
Dorsal artery of the foot	–	–		

Systolic pressure measurement (Doppler ultrasound technique; mmHg):

	Right	Left
Brachial artery	160	
Posterior tibial artery	90	90
Anterior tibial artery	80	75

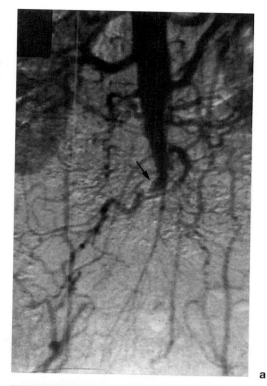

a

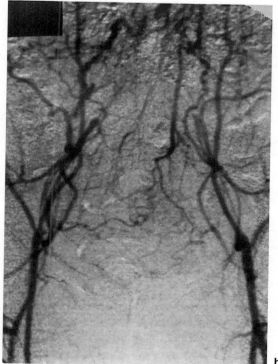

b

Angiogram. Central venous digital subtraction angiography (DSA) demonstrates an infrarenal occlusion of the abdominal aorta and the common iliac arteries bilaterally (**a**). There is bilateral refilling of the external iliac arteries (**b**), and the thigh and lower leg vascular system was patent (not shown).

Fig. 11.**51 c–f** ▷

Color flow duplex sonogram. In a sagittal, right paramedian sectional plane with good delimitation of the wall structures in the region of the infrarenal abdominal aorta, an almost completely occluded lumen can be recognized from the absent color coding and weakly echoic internal structures. In the dorsally located lumen through which blood is still flowing, the preocclusive *flow velocity spectrum* is monophasically altered, and shows low maximum systolic and end-diastolic flow velocities of 31 cm/s and 11 cm/s, respectively (**c**).

The distal segment of the aorta, in which blood is still flowing, continues into an artery serving as a collateral vessel, which has high maximum systolic and end-diastolic flow velocities of 414 cm/s and 132 cm/s, respectively (**d**). The registration point corresponds to the vascular region indicated with an arrow in **a**.

In the right (**e**) and left (**f**) common femoral arteries, the *Doppler frequency spectrum* is postocclusively altered to a monophasic form (maximum systolic and end-diastolic flow velocities on the *right*, 95 cm/s and 29 cm/s, pulsatility index 1.25; on the *left*, 61 cm/s and 11 cm/s, pulsatility index 1.99).

Comment. The patient's symptoms predominantly consisted of claudication in the buttocks and the thighs; the time at which the suspected aortic occlusion had originally occurred could not be determined. Even higher-grade (proximal) stenoses of the iliac arteries or distal aortic stenoses can be well compensated for, only becoming symptomatic from time to time during more vigorous exercise.

Due to the patient's relatively young age, the initial treatment consisted of systematic fibrinolysis, but this had to be discontinued without success, due to allergic reactions.

Since the patient initially refused local recanalization, and the compensation of the aortic occlusion still proved to be adequate, conservative therapy in the form of steadily increasing interval training exercise was given.

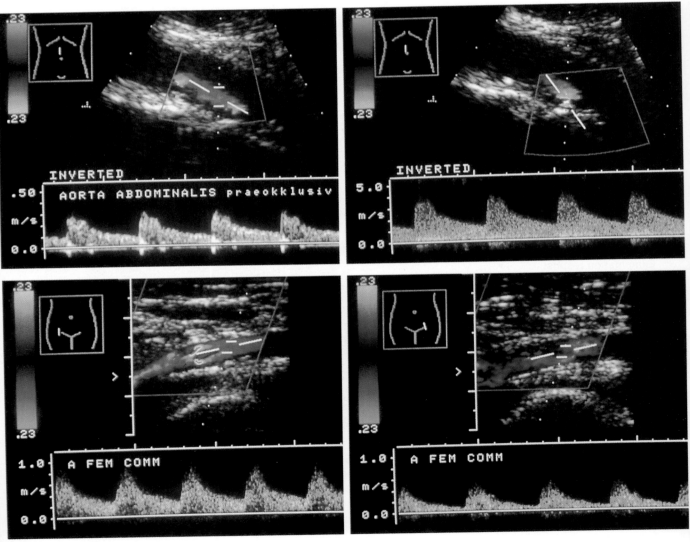

Fig. 11.**51 c–f**

Upper Abdominal Arteries

Stenosis of the Celiac Trunk
(Fig. 11.**52**)

Patient: S.B., female, age 51.

Clinical diagnosis: **PAOD** with a **stenosis of the left iliac artery; abdominal angina,** and suspicion of *vascular system obstruction in the upper abdominal arteries.*

CAD: +	Vascular surgery: –	Diabetes mellitus: –
PAOD: +	Heredity: –	Hyperlipidemia: +
TIA, stroke: –	Hypertension: +	Smoking: +

Case report. The patient had been suffering from arterial hypertension for the previous ten years. A myocardial infarction had occurred seven years previously. For approximately the past 12 months, she had had recurrent pain in the upper abdomen, which had been diagnosed gastroscopically as due to gastritis attacks. For approximately the past six months, there had also been increasing *postprandial pain* in the area of the middle abdomen. There had been an involuntary weight loss of 10 kg over the previous few months.

For nine months, she had been suffering *claudication of the calves and buttocks* on the left, with increasing intensity particularly when she was climbing stairs.

Examination results. A 51-year-old woman in a good general state of health, with an underweight nutritional status (height 162 cm, weight 46 kg). Paraumbilical vascular bruits were audible, and the following angiological findings were present in the region of the peripheral arteries:

	Pulse status Right	Pulse status Left	Auscultation Right	Auscultation Left
Abdomen			+	+
Common femoral artery	+	(+)	–	+
Superficial femoral artery			–	–
Popliteal artery	+	(+)	–	–
Posterior tibial artery	+	(+)		
Dorsal artery of the foot	+	(+)		

Color flow duplex sonogram. The examination of the upper abdominal arteries with color flow duplex sonography detected stenoses of the celiac trunk, superior mesenteric artery, and both renal arteries. The stenoses of the renal arteries are not shown here.

In a paramedian sagittal section, pronounced aliasing near the origin of the celiac trunk, and still just recognizable in the superior mesenteric artery, indicates the presence of stenoses (**b**).

In the *Doppler frequency spectrum* obtained from the celiac trunk, extremely high frequencies are seen, corresponding to maximum systolic, end-diastolic, and time-standardized maximum velocities of 610 cm/s, 348 cm/s, and 443 cm/s, respectively (resistance index = 0.59).

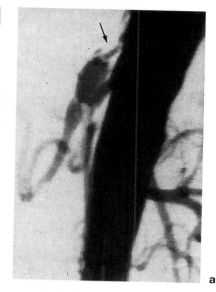

a

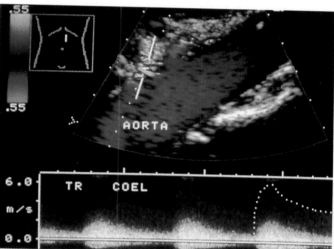

b

Angiogram. Using intra-arterial digital subtraction angiography with the beam approaching from a lateral direction, high-grade stenoses of the celiac trunk (arrow) and superior mesenteric artery are seen (**a**).

Comment. The etiology of this condition indicates that this is an early manifestation of obliterative arteriosclerosis, with symptomatic involvement of the coronary arteries, the upper abdominal arteries including both arterial renal vascular systems, and the peripheral arteries. In addition to the risk factors of hypertension and hyperlipidemia, there had also been a significant abuse of nicotine over a period of more than 30 years, with consumption of up to 35 cigarettes daily.

With this type of multilocular stenosis of the infradiaphragmatic aortic branches, vascular surgery treatment to eliminate the vascular constrictions is appropriate, although percutaneous transluminal angioplasty procedures could be used by a suitably experienced physician.

▬ Arcuate Ligament Syndrome
(Fig. 11.**53**)

Patient: S.E., female, age 45.
Clinical diagnosis: suspicion of a flow obstruction in the region of the upper abdominal arteries.

CAD: –	Vascular surgery: –	Diabetes mellitus: –
PAOD: –	Heredity: –	Hyperlipidemia: –
TIA, stroke: –	Hypertension: –	Smoking: +

Case report. For the previous two years, the patient had been suffering recurring duodenal ulcers, with postprandial, cramp-like pains in the upper abdomen radiating into the back and the left flank. The pains had been worsening for approximately the previous 15 months. The patient had lost 10 kg in weight during the previous 12 months. Gastroduodenoscopy had shown normal results.

Examination results. The patient was 162 cm in height and weighed 52 kg. The heart and lungs were clear upon percussion and auscultation. Blood pressure was bilaterally 100/70 mmHg.

There was a systolic flow sound in the upper abdomen, which became softer on inspiration and was noticeably louder during expiration.

The upper and lower extremities showed a normal pulse status and normal findings on auscultation.

Color flow duplex sonogram. In a sagittal sectional plane (fasting), the celiac trunk is seen at its origin from the abdominal aorta. With color flow Doppler sonography, an end-expiratory, curve-shaped stretching of the proximal celiac trunk is seen (**a**), with aliasing present in the color flow Doppler mode.

During *inspiration,* the Doppler frequency spectrum (**b**) is normal, with maximum systolic and end-diastolic flow velocities of 188 cm/s and 39 cm/s and a resistance index of 0.79.

After deep expiration, there is a clear increase in the systolic and particularly the end-diastolic flow velocities, to values of 330 cm/s and 123 cm/s, respectively (**c**). The resistance index has decreased to 0.67. At the region of the origin, the superior and inferior mesenteric arteries had normal blood flow (not shown).

Comment. The respiratory fluctuations of the flow velocity in the celiac trunk led to the suspected diagnosis of an *arcuate ligament syndrome.*

Due to the weight loss and pain the patient was experiencing, decompression was carried out using vascular surgery (splitting the ligament), which confirmed this rare finding, a ligamentous constriction of the celiac trunk.

Six weeks after surgery, a hemodynamically only slight stenosis of the celiac trunk was seen, which was not influenced by respiration and which had a maximum systolic flow velocity of 267 cm/s (not shown).

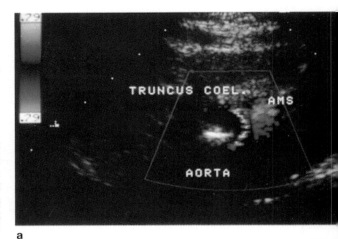

a

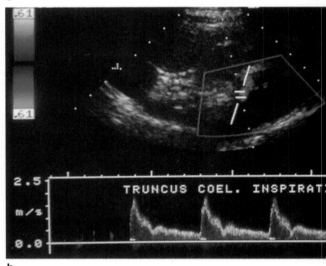

b

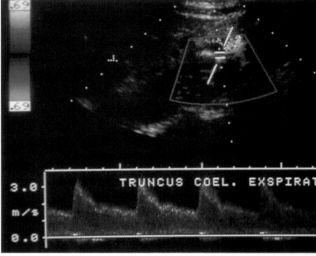

c

Renal Arteries

Stenosis of the Renal Artery
(Fig. 11.**54**)

Patient: H.K., male, age 51.
Clinical diagnosis: **arterial hypertension; PAOD** with
suspected stenosis of the left femoral artery.

CAD:	– Vascular surgery:	+ Diabetes mellitus:	–
PAOD:	+ Heredity:	– Hyperlipidemia:	+
TIA, stroke:	– Hypertension:	+ Smoking:	+

Case report. The patient had been suffering bilateral
calf claudication for the previous seven years. Six
years previously, transposition of the subclavian and
carotid arteries on the left had been carried out due to
an occlusion of the subclavian artery. One year earlier,
a recanalization of the right carotid artery with an
ICA stenosis had been carried out. Arterial hypertension had been present for the previous four years.

Examination results. Under medication (calcium antagonist), the blood pressure was 200/110 mmHg. A
systolic flow sound was audible in the paraumbilical
region on the right.

Angiogram. The intra-arterial DSA image shows the
stenosis in the proximal segment of the right renal
artery (**a**), with a normal-appearing left renal artery.

Color flow duplex sonogram. A subxiphoid transverse
section images the abdominal aorta (AO), with the
origin of the right renal artery (**b**). Ventral to the
aorta, there is a section through the superior mesenteric artery (AMS) and the distal splenic vein (VL,
coded blue). Color flow duplex sonography demonstrates the lumen constriction in the renal artery near
its origin, with a poststenotic turbulent flow accompanied by aliasing.

The *Doppler frequency spectrum* documents high
systolic and end-diastolic flow velocities of 411 cm/s
and 216 cm/s in the jet of the high-grade stenosis of
the renal artery (**c**). The frequency spectrum, which
for technical reasons is registering physiological blood
flow below the baseline, is broadened, and shows
frequency bands that are rich in intensity due to the
regions of turbulent flow both around the baseline
and above it.

Comment. Due to the high-grade lumen constriction
of the right renal artery and the arterial hypertension
first identified four years previously, angioplasty of
the stenosis was indicated. However, the patient did
not agree to undergo the procedure.

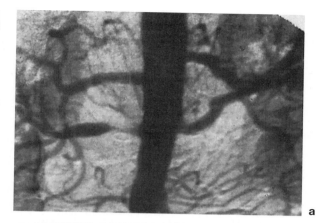
a

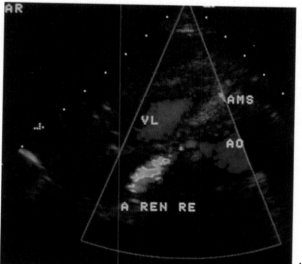
b

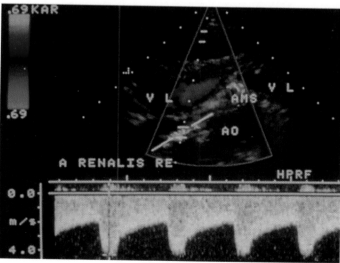
c

Stenosis of the Renal Artery (Percutaneous Transluminal Angioplasty)
(Fig. 11.**55**)

Patient: W.N., male, age 54.
Clinical diagnosis: **arterial hypertension; PAOD,** with suspicion of a **stenosis of the iliac artery, right more than left.**

CAD:	+	PTA:	+	Diabetes mellitus: –
PAOD:	+	Heredity:	–	Hyperlipidemia: +
TIA, stroke:	–	Hypertension:	+	Smoking: +

Case report. Three years previously, the patient had undergone an angioplasty of a stenosis of the right iliac artery. The patient was now being admitted to hospital to clarify possible causes of the arterial hypertension that had developed during the previous few months.

Examination results. The blood pressure under medication (á-blocker and diuretic) was 130/60 mmHg.

	Pulse status		Auscultation	
	Right	Left	Right	Left
Abdomen			+	+
Common femoral artery	(+)	+	+	+
Superficial femoral artery			–	–
Popliteal artery	(+)	+	–	–
Posterior tibial artery	+)	+		
Dorsal artery of the foot	+)	+		

Systolic pressure measurement (Doppler ultrasound technique; mmHg):

	Right	Left
Brachial artery	130	
Posterior tibial artery	100	130
Anterior tibial artery	110	120

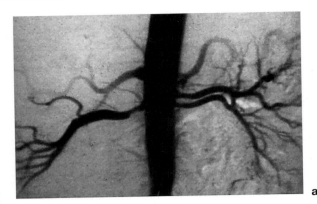

a

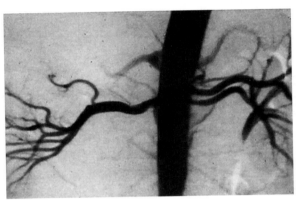

b

Angiogram. The intra-arterial DSA demonstrates the stenosis of the *right* renal artery before (**a**) and after (**b**) percutaneous transluminal angioplasty. On the *left,* there are two renal arteries, without no stenosis.

Color flow duplex sonogram. Subxiphoid transverse sections show the abdominal aorta with the origin of the right renal artery (**c–f**). The *Doppler spectrum* documents high systolic and end-diastolic flow velocities of 449 cm/s and 150 cm/s in the stenosis of the renal artery near the origin (**c**).

After balloon dilatation, strong blood flow is seen with color flow Doppler sonography (variance image) in the right renal artery, without any local acceleration (**d**).

In the *B-mode image,* a plaque (arrow) can still be detected in the branching segment of the right renal artery after PTA (**e**) (from Karasch et al. 1993 a, Chapter 6).

Using the *color flow Doppler mode* with high spatial resolution, it is very clear that the plaque is surrounded by the flowing blood (**f**). The frequency spectrum indicated a disturbed, but not accelerated, flow velocity of a maximum of 135 cm/s (not shown).

Comment. The values for the blood pressure, which had previously been controlled with a β-blocker and a diuretic, returned to normal, without any further need for medication, after balloon dilatation of the renal artery stenosis.

In the same session, successful dilatation of a stenosis in the ipsilateral common iliac artery was also carried out, with access via a puncture of the right common femoral artery.

A comparison of the intra-arterial angiogram after balloon dilatation (**b**) with the B-mode image of the right renal artery (**e**) shows that sonography may also be superior to angiographic imaging with regard to morphological considerations. The plaque, which can be precisely delimited at the origin of the right renal artery with sonography, cannot be imaged angiographically in this patient.

Evaluating the hemodynamic relevance of this type of lumen constriction following angioplasty remains a special task for pulsed wave or continuous wave Doppler procedures, although in this case it was also possible to recognize the absent hemodynamic effectiveness with color flow Doppler sonography.

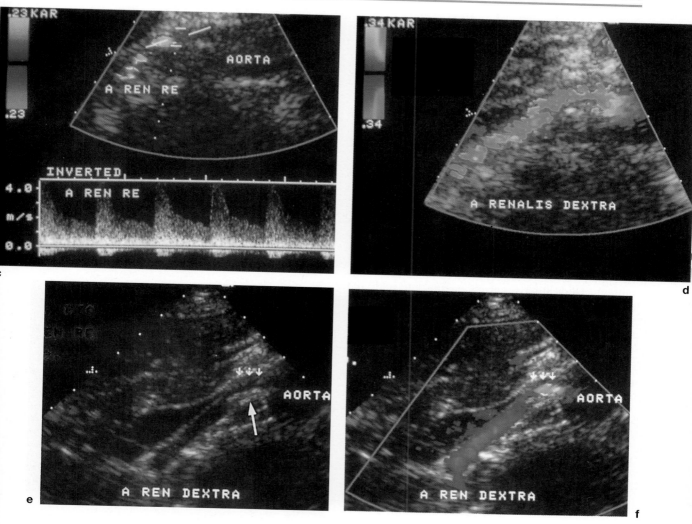

Occlusion of the Renal Artery (Percutaneous Transluminal Angioplasty with Stent Implantation)
(Fig. 11.56)

Patient: B.H., male, age 40.
Clinical diagnosis: newly appearing **arterial hypertension; bilateral stage II PAOD; confirmation/exclusion of renovascular hypertension.** Suspicion of **stenosis of the right iliac artery** and **occlusion of the left iliac artery.**

CAD: –	Vascular surgery: +	Diabetes mellitus: –
PAOD: +	Heredity: –	Hyperlipidemia: –
TIA, stroke: –	Hypertension: +	Smoking: +

Case report. The patient had first suffered claudication in the left calf at the age of 34. At that time, successful percutaneous transluminal angioplasty of an occlusion of the left popliteal artery had been carried out. Five years later, renewed complaints of intermittent claudication occurred on the left side, and arterial hypertension appeared, with blood pressure values up to 230/135 mmHg. Reducing the blood pressure with a triple combination of antihypertensive medications had not been adequately successful.

Pathological laboratory values: aldosterone 397 pg/ml (normal value: up to 125 pg/ml while lying down), renin (direct) 137 pg/ml (normal value: up to 30 pg/ml while lying down).

Examination results: BP 200/100 mmHg

	Pulse status		Auscultation	
	Right	Left	Right	Left
Abdomen			+	+
Common femoral artery	(+)	((+))	+	–
Superficial femoral artery			–	–
Popliteal artery	(+)	–	–	–
Posterior tibial artery	(+)	–		
Dorsal artery of the foot	(+)	–		

Systolic pressure measurement (Doppler ultrasound technique; mmHg):

	Right	Left
Brachial artery	165	
Posterior tibial artery	125	90
Anterior tibial artery	135	85

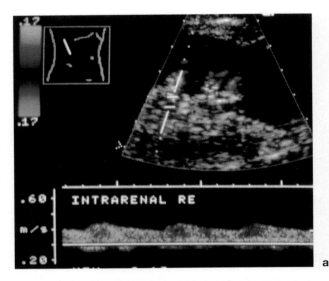

a

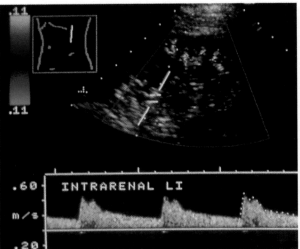

b

Color flow duplex sonogram. Blood flow was not detectable in the proximal right renal artery. Distally, there was refilling, with normal blood flow in the left renal artery (not shown). Registering *intrarenal* Doppler spectra on the right side (**a**) shows reduced flow velocity, particularly during systole (maximum systolic and end-diastolic flow velocities 31 cm/s and 12 cm/s, resistance index = 0.45). *On the left side* (**b**), a normal wave form is seen (maximum systolic and end-diastolic flow velocities 53 cm/s and 17 cm/s, resistance index = 0.68). The *size* of the right kidney was 121 mm × 47 mm, and the left kidney was 136 mm × 68 mm.

Angiography confirmed the occlusion of the right renal artery (not shown).

Six days after *percutaneous transluminal angioplasty* of the occlusion of the renal artery and implantation of a stent (6 mm/35 mm), the examination showed the following findings:

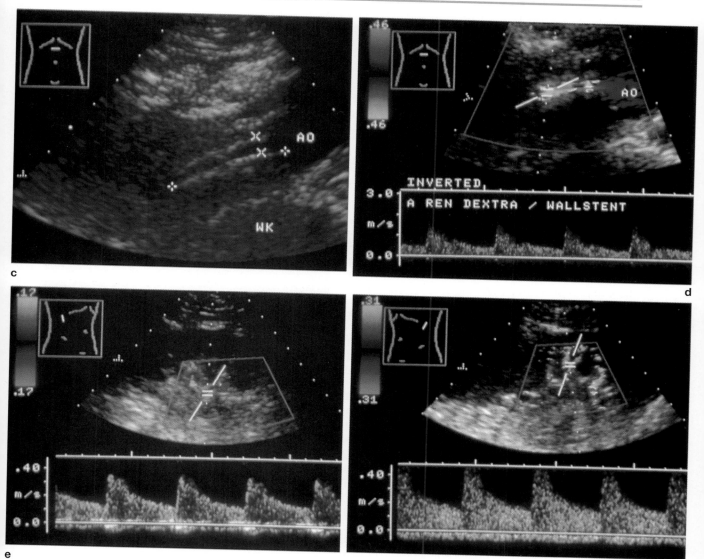

In a transverse section through the upper abdomen using B-mode imaging, the abdominal aorta and the proximal right renal artery can be seen (**c**). The implanted stent can be identified, with a diameter of 5.3 mm and a length of 35.6 mm (AO = aorta; WK = vertebral body).

The maximum systolic and end-diastolic flow velocities in the stent region on the right, with a normal vascular system in the contralateral kidney, are 187 cm/s and 49 cm/s (**d**).

After angioplasty, normal systolic and end-diastolic flow patterns are seen *intrarenally* on the right (**e**) and on the left (**f**) (maximum systolic and end-diastolic flow velocities on the *right* 36 cm/s and 12 cm/s, resistance index = 0.67; on the *left,* 46 cm/s and 18 cm/s, resistance index = 0.61).

Comment. The catheter intervention successfully recanalized the occlusion of the renal artery. A stent was implanted to provide internal support for the arterial wall. The patient was thus restored to a condition of normotonic blood pressure, and no longer required antihypertensive medication.

This case shows that, in some circumstances, revascularizing an occlusion of a renal artery can still turn out to be successful several months later. Follow-up observations with color flow duplex sonography indicated that blood was still freely flowing through the stent eight months after implantation.

In view of the peripheral arterial occlusive disease, percutaneous transluminal angioplasty of the occlusion of the left common iliac artery was carried out in order to increase the patient's overall pain-free walking distance.

Renal Infarction (After Percutaneous Transluminal Angioplasty)
(Fig. 11.**57**)

Patient: A.S., female, age 53.
Clinical diagnosis: **PAOD, arterial hypertension,** status after placement of an **aortobifemoral bifurcation prosthesis,** status after PTA of a stenosis of the left renal artery; recurrent stenosis had been excluded.

CAD:	–	PTA:	+	Diabetes mellitus:	–
PAOD:	+	Heredity:	–	Hyperlipidemia:	–
TIA, stroke:	–	Hypertension:	+	Smoking:	+

Case report. The patient had been suffering from arterial hypertension for the previous 12 years. Six years previously, a thromboendarterectomy of the left pelvic vascular system and lumbar sympathectomy on the ipsilateral side had been carried out. Twenty months after that, there had been a recurrence of stenosis in the external iliac artery, and again 19 months after that, a further implantation of a bifemoral bifurcation prosthesis had been carried out in the presence of an occlusion of the left external iliac artery.

Three years before, percutaneous transluminal angioplasty of a high-grade stenosis of the left renal artery was still followed by a residual constriction. However, the blood pressure situation was normalized. Approximately one year previously, there had been a renewed increase in blood pressure, which could ultimately only be moderately controlled using a combination of three antihypertensive medications (calcium antagonist, diuretic, β-blocker).

Examination results. Multiple scars were seen following several surgical interventions in the abdominal wall. There are no vascular bruits in the abdomen. There was a patent vascular system in the pelvis and legs bilaterally.

The preliminary examination of the renal vascular system with color flow duplex sonography showed a filiform recurrent stenosis near the origin of the left renal artery, with maximum systolic and end-diastolic flow velocities of 519 cm/s and 241 cm/s, respectively, and an intrarenal resistance index of 0.53 (not shown).

Angiogram. During percutaneous transluminal angioplasty with transaxillary access, an occlusion of a branch of a segment artery supplying the inferior pole of the kidney occurred. Selective angiography shows the perfusion defect in the region of the caudal renal pole (**a**). Near the capsule, interlobar arteries with blood still flowing through them can still be seen (arrow).

Color flow duplex sonogram. In a transverse section through the flank proceeding in a dorsolateral direction and *after angioplasty,* there is a cone-shaped re-

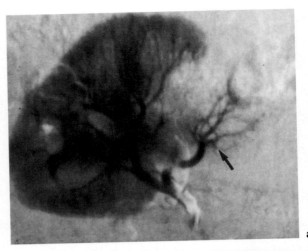

a

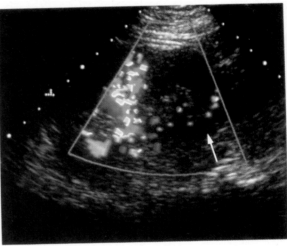

b

gion lacking detectable blood flow in the area of the caudal pole of the left kidney (**b**). Near the renal capsule, a small blood vessel can still be identified, running from the hilum toward the periphery (arrow). Its blood flow can only be partly reproduced in the freeze frame.

Several days after dilatation with the catheter, the maximum systolic and end-diastolic flow velocities in the proximal segment of the left renal artery amounted to 222 cm/s and 74 cm/s, respectively. Due to the relatively high systolic flow velocity following PTA, the intrarenal resistance index was 0.70 (not shown).

Comment. Renal infarctions are rare complications in the context of an angioplasty of the renal arteries, resulting from *embolic occlusions* or *vascular wall dissections* in the area of the peripheral vascular system. Renal infarction is generally accompanied by acutely appearing severe pain in the area of the renal bed.

Renal Veins

Renal Vein Thrombosis
(Fig. 11.**58**)

Patient: M.M., male, age 45.
Clinical diagnosis: kidney transplantation two days previously, with failure of normal renal functioning to start.

Case report. The patient had been suffering from terminal renal failure with chronic glomerulonephritis, and has been undergoing hemodialysis treatment for the previous ten years. Kidney transplantation had been carried out two days previously, with subsequent failure of the new kidney to start proper renal functioning.

Examination results. The patient was anuric, and palpation presented a clearly enlarged transplant with a firm consistency. Significant pain occurred on applying pressure during palpation.

Laboratory results. The serum showed a creatinine of 7.9 mg/dl; urea 138 mg/dl, potassium 5.4 mmol/l, LDH = 205 U/l.

Color flow duplex sonogram. With the patient in supine position, the renal transplant (NTX) is seen, showing an intrarenal segmental artery (coded red). Color flow Doppler sonography shows a rarefaction of the arterial vasculature, and an absence of venous flow signals.

With elevated peripheral resistance, the *Doppler frequency spectrum* from the segmental artery shows a systolic antegrade flow profile, but an *early diastolic retrograde flow profile,* which does not return to the baseline until the beginning of the next systole.

Histology. A renal transplant with fresh, lumen-obstructing thrombi in the intrarenal renal veins and the renal veins near the hilum. The branches of the renal artery were freely patent, and there was no indication of an acute rejection reaction.

Comment. On the basis of the color flow duplex-sonographic findings, immediate surgical revision was carried out, with no need for any further angiographic diagnosis beforehand.

Intraoperatively, there was an already livid, discolored organ, with a hard thrombosis of the renal vein. Therefore, revascularization was no longer possible, and a nephrectomy of the transplant had to be carried out.

We are grateful to Dr. B. Krumme of Medical Department IV (Nephrology) at the Medical University Clinic, Freiburg, for kind permission to use this case and the accompanying illustration.

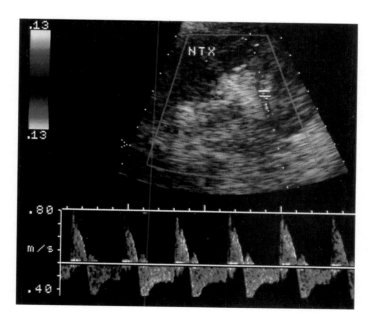

12 Glossary

The following terminological overview attempts to summarize brief definitions of selected important specialist terms that are relevant to angiological ultrasound procedures. The selection is mainly restricted to equipment-related technical concepts, to meet users' actual needs. Cross-references are given in italics.

Acoustic energy: mechanical energy that is transported in the form of an acoustic wave.
Unit: joule = watt × second.

Acoustic impedance: the product formed from the tissue density and the ultrasound velocity. Changing the acoustic resistance leads to a reflection of acoustic energy at an interface.

Acoustic power: acoustic energy that is transported per unit of time in the form of a wave.
Unit: joule/second = watt.

Acoustic shadow: the region of limited echo reflection located behind a strongly reflecting interface (e.g., calcified plaque).

Acoustic wave: the transport of energy by the propagation of mechanical oscillations. Both spatially and with respect to time, this is a periodic process, in which neighboring locations in the same phase define the acoustic wavelength. In an acoustic wave front, all the points are in the same phase.

Aliasing: the situation that arises when the *Nyquist* sampling limit is exceeded by the frequency of the input signal.

A-mode (amplitude mode): depicting the amplitudes of the echo signals as a function of the transit time between echogenic structures and the transducer: abscissa = time axis, ordinate = axis showing the ampli-

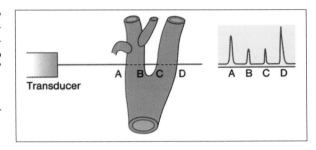

tude of the echo signal. Corresponding to the transit time of the echo, the distance on the time axis indicates the location of the reflection.

Amplitude: the magnitude of the wave variable, such as velocity, displacement, or acceleration.

Analogue: the transfer methodology of a system in which the initial value is a continuous function of the entered value. This is in contrast to digital.

Analysis, three-dimensional: the depiction of ultrasound sectional images (B-mode or Doppler mode) in three spatial dimensions.

Analysis, four-dimensional: the depiction of ultrasound sectional images (B-mode or Doppler mode) in spatial (three-dimensional) and time dimensions in order to show wall and plaque movements.

Angle: The important angle is the one formed by the ultrasound beam and the vascular axis. According to the *Doppler formula,* the frequency of the reflected ultrasound signal is reduced proportionately to the cosine of the angle. For various technical and physical reasons, it can be difficult to measure this angle, e.g., in a *duplex system.* In addition, inaccurate estimates are also made at vascular bends.

to **Angle** ▷

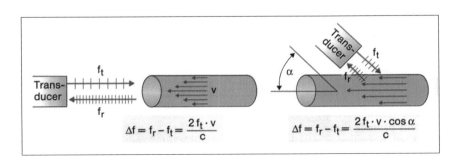

Angle of incidence: angle between the ultrasound axis and the structure to be examined, e.g., axis of a blood vessel.

Array: a spatial arrangement of several transducer components:

– Linear
– Annular
– Circular

Artifact: an interfering echo that does not belong to a true reflector, and is usually caused by multiple reflection or scattering.

Audio analysis: the simplest, purely acoustic reproduction of the amplified Doppler signal, using a loudspeaker or headphones.

Autocorrelation: multiplication of a wave by a time-shifted section of the same wave.

Axial resolution: the minimum distance between two reflectors, in the direction of ultrasound propagation, at which they are still shown as separate entities. This is also called depth resolution or range resolution, and among other factors depends on the transmission frequency (e.g., at 8 MHz: 0.3 mm in muscle).

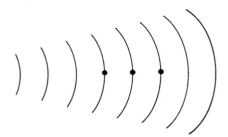

Backscatter: the energy reradiated by a scatterer in a direction opposite to that of the incident wave.

Beam: an acoustic field produced by the transducer with the axis located at the point of maximal acoustic pressure (far-field). The transducer form also determines the form of the beam.

Beam angle: the angle between the direction of propagation of ultrasound and the direction of anatomical structures reflecting ultrasound. For B-mode imaging, perpendicular insonation is optimal; for Doppler mode studies, 30–60° are optimal.

Bernouilli, Daniel (1700–1782): Swiss mathematician, sometimes referred to as the founder of mathematical physics, who unified the study of hydrodynamics under what became known later as the principle of the conservation of energy. Daniel Bernouilli was one of an extraordinary dynasty of eight mathematicians spanning three generations of the same family. Among his contributions to hemodynamics are the *Bernouilli effect* and the *Bernouilli equation*.

Bernouilli effect: the reduction in pressure that accompanies an increase in velocity of fluid flow.

Bernouilli equation: the equation that states that the total fluid energy along a streamline of fluid flow is constant.

Bidirectional: This refers to the ability of Doppler equipment to distinguish between positive and negative Doppler shifts (flow toward the probe or away from the probe).

B-mode (brightness mode): a two-dimensional ultrasound image produced by depicting echo signals on the screen. The amplitude is indicated by the brightness modulation (gray scales). The point of origin of the image is located on the abscissa, and, depending on the echo's transit time (= depth), indicates the reflection's location. Placing single scans one next to the other creates a two-dimensional image (echotomogram = B-scan).

Boundary layer: the thin layer of stationary fluid adjacent to the walls of the containing vessel.

Bruit: the name given to sounds sometimes associated with disturbed and turbulent flow. They arise from the periodic variation in shear stress on the vessel wall, which causes it to vibrate.

Cathode-ray oscilloscope: electronic registration equipment used to record a rapidly occurring fluctuation in voltage. It mainly consists of a cathode-ray tube with x–y deflection and an amplification unit. The electrode beam is made visible on a fluorescent screen, with the abscissa corresponding to the time and the ordinate representing the amplified applied voltage. The beam's movement without any inertia also makes it possible to register brief or high-frequency fluctuations in voltage (see *M-mode*).

Color flow duplex sonography: a new procedure for imaging structural and hemodynamic relationships

◁   to **B-mode**

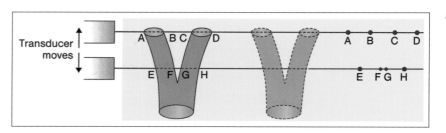

within tissue in a two-dimensional, almost real-time image.

Compliance: the rate of change of cross-sectional area with pressure ("area" compliance); or the rate of change of diameter with pressure ("diameter" compliance).

Continuous wave (CW) Doppler: a Doppler method in which ultrasound waves are continuously transmitted. It functions simultaneously with two piezoelectric crystals (transmitter/receiver). Its advantage is that there is no physical upper limit (Nyquist frequency) for clearly measuring the Doppler shift. High flow velocities can therefore be measured. Its disadvantage is that no information about the location of a measurement is possible.

Critical Reynolds number: the Reynolds number around which the transition from laminar to turbulent flow takes place.

Critical stenosis: a stenosis of sufficient diameter to reduce flow rate and pressure significantly. Sometimes called "hemodynamically significant" stenosis.

Curved-array transducer: a transducer consisting of a curved arrangement of piezoelectric crystals lying next to one another. The image field that results is trapezoid in form (see pp. 17–19).

DC: direct current.

Decibel (dB): the relative measurement for the signal intensity. For the intensity (I), or intensity ratio, and amplitude (A), or the amplitude ratio, the following formulas apply:

$$x \text{ (dB)} = 10 \times \log I_1/I_2$$
$$x \text{ (dB)} = 20 \times \log A_1/A_2$$

Density: specific weight. Tissue that is sonographically "dense" is strongly echogenic, due to the presence of structures that have large differences in impedance.

Digital: numerical, having to do with numbers. In contrast to analogue methods, a continuously modulated input signal is transformed into a series of discrete symbols that lie next to one another or one after another, e.g., the conversion of spoken language into Morse code.

Directionality:

– Unidirectional equipment functions without detecting the flow direction.
– Using a phase detector in bidirectional systems, the flow direction is indicated in relation to the transducer as a positive or negative Doppler shift value.

Disturbed flow: deviations from laminar flow consisting of oscillating variations in the direction of the formation of vortices. Disturbance of blood flow may be caused by various reasons, e.g., high peak velocities, by curving, branching, and divergence of vessels, or by projections into the vessel lumen.

Doppler, Christian Andreas (1803–1853): a mathematician and physicist, who was born in Salzburg and died in Venice. In 1842, Doppler published "Concerning the Colored Light of the Double Stars and Several Other Celestial Bodies," in which he described alterations in stellar light that were dependent on the movement of a star either toward the Earth or away from it. The frequency shift that forms the basis of this phenomenon is called the Doppler effect, and it applies to all types of waves. He enunciated his principle in 1842, but unfortunately confused its interpretation. The acoustic Doppler effect was demonstrated in 1845 by Buys Ballot using a trumpeter riding on a steam locomotive.

Doppler angle: the angle between the direction of propagation of the ultrasound and the direction of vessel flow. As an approximation, the angle between the axis of the ultrasound beam and the axis of the vessel lumen is generally used (see *angle*).

Doppler effect: a frequency change in ultrasound waves due to the relative movement of the receiver or the transmitter, when one component of this movement lies in the ultrasound axis.

Doppler formula:

$$\Delta f = \frac{2 f_t \times v \times \cos\alpha}{c}$$

where:
c = ultrasound velocity in tissue (1540 m/s)
v = velocity of red blood cells in blood vessels
f_t = transmission frequency
f_r = reception frequency
Δf = Doppler shift = $f_r - f_t$
α = the angle between the ultrasound and vascular axes

Example:
c = 1540 m/s
v = 1 m/s
f_t = 2.5 MHz
α = 0°; cos = 1
Δf = 3247 Hz
(see *angle*)

Doppler shift: the difference between the transmission frequency f_t and the reception frequency f_r:

$$\Delta f = f_r - f_t$$

Doppler shift frequency, peak: the highest Doppler shift frequency at a moment in time or in an individual Doppler spectrum, corresponding to the fastest moving target in the Doppler sample volume.

Doppler shift frequency, mean: the average Doppler shift frequency above and below which half of the total power in the spectrum resides.

Doppler shift frequency, mode: the Doppler shift frequency with the greatest power in a given spectrum.

Doppler spectrum: In the Doppler frequency spectrum, the frequencies are indicated that appear within a measurement interval, along with how often they occur, i.e., their amplitudes. The Doppler spectrum thus reflects the number of corpuscular elements in a measured vascular segment that can be classified as having certain flow velocities. The imaging is usually accomplished using a *filter bank analysis* or *fast Fourier transform analysis*. Three forms of graphic representation are commonly used:

– Abscissa = time
– Ordinate = frequency (flow velocity)
– Point density = (brightness or color level) = amplitude

In the second form of representation, the frequency is indicated on the abscissa, the amplitude on the ordinate, and the time on the Z axis. In the *power spectrum*, the square of the amplitude is plotted.

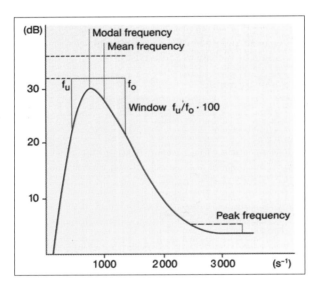

Duplex system: the combination of echo impulse and Doppler procedures to locate the Doppler *sample volume* in the B-mode image in real time, using a single *transducer*. In equipment that has a separately mounted Doppler probe on the side of the transducer, it should be noted that the refractive relationships differ in B-mode (perpendicular) and in Doppler mode (oblique). This leads to errors in the location of the sample volume shown on the monitor. The true position of the sample volume is in reality usually displaced to the side or ahead of the location indicated on the monitor.

Dynamic focusing: This is used to improve the lateral resolution in linear and phased-array systems. While the echo pulse is being received, the effective focal zone is displaced (receiving focus following a single transmitted pulse).

Dynamic range:
– *For amplifiers:* maximal transmittable signal range above the noise level (in dB).
– *For displays:* the range of the input signal (in dB) which lies between the saturation level and the noise level.

Echo: an acoustic signal resulting from scattering or reflection of acoustic waves at an interface.

Envelope: the curve that joins the peaks of oscillations following one another in a wave.

Far field (Fraunhofer zone): the region of the acoustic field beyond the focus.

Fast Fourier transform (FFT): a quick version of the *Fourier transformation,* adapted for computers.

Filter: an electronic circuit designed to allow signals of certain frequencies to pass and to stop signals of other frequencies.

Filter, high-pass: a device that allows high-frequency but not low-frequency variations to pass through. An example is the electrical filter used in Doppler devices to eliminate low-frequency Doppler shifts caused by clutter.

Filter, low-pass: a device that allows low-frequency but not high-frequency variations to pass through. An example is a stenosis, which has the effect of damping rapid variations in the pressure and flow waveforms.

Filter bank analysis: This is often used in the spectral analysis of a Doppler signal. Through a certain number of high-pass and low-pass filters that are switched in parallel and have the same bandwidth, the range of Doppler frequencies appearing (0–20 kHz) is covered. Since the filters more or less represent large frequency ranges overlapping at their limits, artificial changes in the spectrum result.

Flow separation: If a body of fluid within a vessel has a particularly high momentum (because it has entered the vessel as a jet through a tight stenosis, for example), its boundary will separate from the laminae of surrounding fluid. The region of flow separation is marked by a high velocity gradient and hence shear stress.

Focal zone: the area of an ultrasound beam in which the lateral resolution is the best. Continuous wave Doppler systems are usually not very focused. In systems that use a mechanical transducer, the focal zone is fixed between the near field and the far field, while it is variable (usually selectable in steps) in systems with an electronic transducer. Depending on the focusing, different segments of a blood vessel with a constant diameter are faded out: e.g., flow regions near the wall may be in the center of focus, but not those in the far field.

Fourier, Jean Baptiste (1768–1830): son of a French tailor who was persuaded to leave his Benedictine monastery to become a professor of mathematics at the age of 21. Among his many original and outstanding contributions to mathematics and physics was heat analysis. Some say that his death was speeded by his insistence that heat was good for the health and his later habit of living in rooms "hotter than the Sahara desert."

Fourier transformation: a mathematical algorithm that is used to represent a periodic function as a sum of sinusoidal functions of different frequencies, named after Jean Baptiste Fourier (1768–1830); French physicist and mathematician. The calculation allows the presentation of the Doppler signal in terms of the relative power of the various Doppler shift frequencies.

Freeze mode: a method of displaying images that allows functions to be depicted in real time as individual images.

Frequency: in periodic signals, the number of oscillations per unit of time (unit: Hz, Hertz; kHz kilohertz, MHz megahertz; 1 cycle/s = 1 Hz, 1000 Hz = 1 kHz, 1,000,000 Hz = 1 MHz).

Frequency modulation: the principle of information transfer by modulating the signal. Usually, groups of impulses are used that are successive over time, in the form of impulse frequency modulation or pulse frequency modulation.

Gray scale: the brightness-modulated depiction of echo signal amplitudes in several gray scales ranging between black and white (16–128). Alternatively, color scales can also be used. The adjustment of brightness differences in an gray scale can be carried out in either a linear or a logarithmic fashion.

Harmonic imaging: the possibility of improving the registration of weak Doppler signals from reflections occurring at interfaces by using a separate analysis of signals, which, after the application of an *ultrasound contrast medium* (oscillating MAB), reflect harmonic frequencies *(2f,* i.e., at a transmission frequency of 4 MHz, harmonic frequencies of 8 MHz are received and recorded in so-called sonoangiograms) (see p. 12). This method is also used to analyze movement phenomena and flow structures (unresolved flow), which would otherwise not be detectable in small blood vessels. Practical experience with this methodology is still scarce. In comparison with the standard procedure, disadvantageous effects may occur due to examination times that can take several times longer.

Heat effect: A portion of the acoustic energy that impinges on the tissue is converted into heat by absorption, the extent of the temperature elevation depending on the ultrasound intensity, the duration of the ultrasound application, its frequency, and also the charac-

teristics of the tissue affected. This is an important parameter, to which special attention needs to be given during long monitoring and follow-up examinations.

High intensity transient signal (HITS): frequency bands of short duration with a high signal amplitude in the Doppler spectrum. A controversial interpretation has recently suggested that such signals are little more than artifacts, i.e. transmitted perturbations or microcavitation phenomena, or the equivalent of microemboli. For transcranial Doppler monitoring, the following criteria can be used to differentiate a true signal from an artifact:

- Duration < 0.1 s
- Amplitude > 3 dB above background
- Unidirectional
- Variable distribution in subsequent cardiac cycles
- A discrete, metallic audio signal

Impedance: density × acoustic velocity (also termed wave resistance).

Intensity: the acoustic power in the propagation direction. Units:

I_{SATA} [mW/cm²] = spatial average and temporal average intensity
I_{SPTP} [mW/cm²] = spatial peak and temporal peak intensity
I_{SPTA} [mW/cm²] = spatial peak and temporal average intensity

According to the American Institute for Ultrasound in Medicine (AIUM), $I_{SPTA} < 100$ mW/cm² is not harmful. According to studies conducted by the AIUM, there is no proof of any biological changes associated with the MHz spectrum commonly used in medical diagnostics.

Intima media thickness (IMT): a statistical measurement parameter used to define the extent of atherogenesis in its early phase (discrimination threshold in the resolution area of ultrasound equipment: ≥ 100 μm).

Intravascular sonography: the introduction of suitable transducers into blood vessels for B-mode and Doppler mode imaging.

Laminar flow: ideal flow, with its highest velocity in the center and its lowest velocity at the wall of the blood vessel (see *velocity profile*).

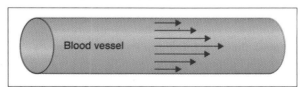

Lateral resolution: minimal distance between two reflecting structures perpendicular to the ultrasound beam at which they can still be depicted as separate entities. The lateral resolution improves at a higher transmission frequency, with a shorter focal zone and with a transducer that covers a greater surface area (important when applying the echo impulse technique to the carotid artery).

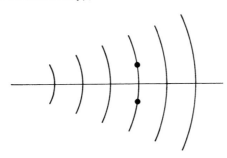

Linear-array transducer: a transducer that has a straight arrangement of the piezoelectric elements, which results in a rectangular-shaped image field (see p. 18).

Mean frequency: the average value of all the frequency components in a spectrum, calculated according to varying formulas (see the illustration under *Doppler spectrum*, above). (This value can be determined for every time interval analyzed).

M-mode (motion mode): continuous depiction of changes in the *A-mode* or *B-mode* (ordinate) over time (abscissa) in order to document moving (pulsatile) echo structures (vascular wall, heart valves). The signal intensity can be documented on paper using a gray scale, or directly with a *cathode-ray oscilloscope*. It is important for measurements of the systolic and diastolic arterial diameter.

Mode frequency: indicates the frequency with the highest amplitude, and is the value that can be measured most reliably (see the illustration under *Doppler spectrum*, above). (This value can be determined for every interval analyzed).

Multi-gate pulsed wave (PW) Doppler system: Several sample volumes are placed along the length of the ultrasound beam simultaneously, and are analyzed independently of each other. This is necessary to register a *velocity profile*.

Near field (Fresnel zone): the region of the acoustic field from the transducer to the focus.

Newtonian fluid: a fluid in which the viscous force opposing flow is proportional to the velocity gradient.

Nyquist theorem (aliasing): In order to classify one wave unambiguously with the assistance of another, the applied frequency must be at least twice as large, so that stroboscopic effects are avoided (F_{max} = PRF/2). This can be important, for example, when measuring high Doppler frequencies (F_{max}) with pulsed wave Doppler procedures in which the *pulse*

repetition frequency (PRF) determines the upper limit of the velocity measurement. According to the *Doppler formula*, the maximum flow velocity (V_{max}) is calculated as follows:

$$V_{max} = \frac{PRF}{2} \cdot \frac{c}{2 \cdot f_t \cdot \cos \alpha}$$

This means that higher flow velocities can still be measured by increasing the PRF, or, if this is no longer feasible, by optimizing the angle at which ultrasound is applied (without *aliasing).*

Peak frequency: maximum frequency 3–6 dB above the noise level, or the frequency which borders 95–98% of the spectrum integral in a positive (upward) direction (see the illustration under *Doppler spectrum*, above). It is methodologically restricted by artifacts resulting from inaccurate parameters that have measurement errors of a magnitude of 1 kHz (can be determined for every analyzed time-interval).

Phased-array transducer: transducer with delayed regulation of the crystal elements in order to form a sector-shaped image (see p. 18).

Piezoelectric elements: crystals (e.g., quartz, barium titanate) that change their form in the presence of an electric voltage (transmission mode) or, vice versa, convert changes in their form into an electric voltage (receiving mode).

Pixel: area elements, or also volume elements, that result from the spatial analysis of object and image during imaging with digital systems.

Poiseuille, Jean Leonard M. (1797–1869): French physician and physicist who performed the first experiments demonstrating viscosity in a fluid and its relationship to the pressure gradient in a tube. Poiseuille also refined the techniques of Hales for measuring pressure in the arterial circulation.

Power: the energy delivered by a wave in unit time, measured in watts.

Power mode (intensity imaging, amplitude-mode imaging): application of various imaging techniques to enhance signal-noise reduction in the color image.

Power spectrum: a graph using the square of the amplitude in order to show the intensity component of the flowing corpuscular elements in the blood.

Pulsatility index (PI): Several pulsatility indices (PI), or resistance indices (RI) are suggested in the literature as methods of describing peripheral vascular resistance:

- (a) According to Pourcelot, the RI = (A – D)/A
- (b) According to Gosling, the ratio A/B
- (c) According to Gosling, PI = A/B
- (d) Or, for monophasic curves, PI = (A – D)/B

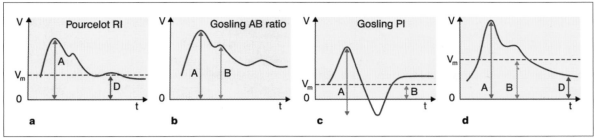

a b c d

to **Pulsatility index**

Pulse duration: interval (usually 0.5–2.0 μs) after which the absolute value of the acoustic pressure exceeds 32% (–10 dB) of the maximum value until the point in time at which the acoustic pressure again falls below this value.

Pulsed wave (PW) Doppler (range-gated Doppler): Ultrasound pulses are discontinuously transmitted at a *pulse repetition frequency* (PRF). A *sample volume* from which the Doppler shift is to be measured can be defined. This method has the advantage of topographically specifying the region from which a Doppler signal will be registered. It has a disadvantage in that the upper boundary for clear measurement of the Doppler shift is restricted by an upper velocity range that can be only partially evaluated (see *Nyquist theorem*).

Pulse repetition frequency (PRF): an important parameter in pulsed wave Doppler procedures. The sample volume depth determines the maximal PRF due to the specific transit time (T) of sound through tissue.

$$PRF = 1/T.$$

The maximum penetration depth Z is calculated by the formula:

$$Z = \frac{c}{2} \times T$$

(c = velocity of sound in tissue). The PRF, in turn, determines the maximum measurable Doppler frequency, amounting to half of the PRF, due to the outward and return path. If the PRF is not twice as large as the Doppler shift that is to be measured, a stroboscopic effect occurs (see *Nyquist theorem*). This means that the true frequency can no longer be ascertained. Modern systems automatically calculate the PRF for every adjusted depth.

Quadrature detector: equipment designed to determine the *Doppler shift,* including a directional analysis. In order to separate the two components (toward the probe, away from the probe) the received signal is split. One part is multiplied by the sine component of the transmission frequency, and the other by the cosine component. Both signals are filtered. At the exit point of the quadrature detector, both channels contain all the information from the received signal, but differ from one another in the phase. With a 90° phase shift, both voltages can be examined in such a manner that during further evaluation in spectral analysis the forward flow and return flow components of the blood circulation can be separated in relation to the probe.

Real-time ultrasound: displaying the ultrasound image at an appropriate rate above the flicker-fusion threshold (> 15–20 images/s).

Red blood cell density imaging: Instead of depicting the frequency information in the color-coded Doppler sonogram, echo amplitudes obtained from erythro-

to **Quadrature detector**

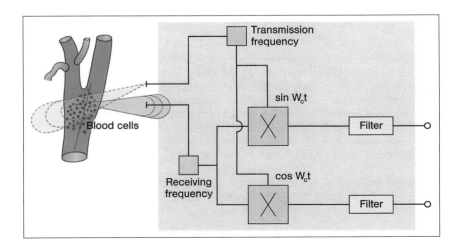

cyte densities are shown in the sample volume. This means that signals can also be recognized with perpendicular application of the ultrasound.

Reflection (echo): an acoustic signal caused by a reflected beam. The incident angle is equal to the angle of reflection.

Reverberation (multiple reflections): In multiple interface reflections, repeating echoes can cause an untrue image, or poor *axial resolution*. Due to a longer transit time, these disturbing echoes appear displaced in relation to the true echo reflections in the B-mode image, and can complicate the interpretation of the image.

Reynolds, Sir Osborne (1842–1912): English mechanical civil engineer who pioneered the study of vortical and turbulent flow in liquids and laid the theoretical basis for subsequent study of the behavior of viscous fluids. Reynolds also built a steam engine to determine the mechanical equivalent of heat and held patents for the design of marine turbines.

Reynolds number: a number expressing the balance of inertial and viscous forces acting on a flowing fluid. Reynolds numbers higher than a critical value result in disturbed or turbulent flow.

Sample volume: the region in which a Doppler shift is measured. The axial and lateral dimensions (the size and form of the sample volume) are dependent on the duration of the pulse and on the geometry of the ultrasound beam. A variable sample volume can be selected by changing the time of measurement or the number of oscillations per transmitted impulse, or both. The greater the number of oscillations per transmitted impulse, the better the sensitivity, but the poorer the axial resolution.

Scattering: In contrast to interface reflections, which depict a directed transmission of ultrasound, scattering describes a diffuse directional distribution of the ultrasound energy subsequent to its interaction with the tissue structure:

$$f = \frac{30}{\text{depth/cm}} \cdot \text{MHz}$$

where f represents the transmission frequency. When it reaches the corpuscular elements of the blood, ultrasound energy is completely scattered in all directions. The energy from these reflected signals that returns to the transducer is therefore much less than in the case of interface reflections (e.g., a vascular wall). The signals need to be more intensely amplified, which causes an unfavorable signal-to-noise ratio.

Sectional image: a two-dimensional depiction of tissue structures using the echo impulse technique (e.g., cross-sections and longitudinal sections).

Sector scan: a transducer with electronic or mechanical regulation used to generate a sector-shaped sectional image (see p. 18).

Signal movement artifacts (clutter, degradation): superimpositions of low-amplitude Doppler signals by signal-intense interface reflections of the vascular walls (e.g., poststenotic) or capillary/tissue movements. Methods available to compensate for this are ultrasound contrast media and *harmonic imaging*.

Single-gate pulsed wave (PW) Doppler: Along the length of the ultrasound beam, a sample volume can be placed at varying depths, and can be adjusted in its width to match the vascular segments to be analyzed. It is not adequate for obtaining a *velocity profile*.

Sound velocity:

c = f × λ
f = sound frequency (Hz)
λ = wavelength (m)

Spectral broadening (window): as an indication of turbulence, this is defined differently by various authors. It is usually calculated over a certain amplitude distance from the modal frequency (see the illustration under *Doppler spectrum,* above).

Spectrum, see *Doppler spectrum*.

Stenosis evaluation (carotid artery): Various angiographic classification schemes are used for stenoses of the carotid artery (local, distal), as well as the hemodynamic parameters used in duplex analysis.

Local stenosis: % stenosis = [1− (c/b)] × 100%
Distal stenosis: % stenosis = [1− (c/a)] × 100%

a = poststenotic lumen width, b = suspected original width of the lumen in the stenotic region, c = lumen width in the stenosis.
The mathematically calculated and experimentally confirmed relationship between these parameters—from conventional angiograms—is as follows:

Distal stenosis (%) = 1.67 × local stenosis (%)

Time gain compensation (TGC): electronic compensation for weakened reflections from deeper tissue layers.

Tone: sound with a pure sinus oscillation of a single frequency in the audible frequency range.

Transducer: using the piezoelectric crystals (= transducer elements) in the transducer of an ultrasound device (= transmitting–receiving unit), electrical energy is transformed into mechanical energy, and vice versa. Various arrangements are available for the B-mode procedure (see pp. 17–19).

- Annular array: elements arranged in a ring shape.
- Linear array: a series of crystal elements which are simultaneously controlled, apart from the accom-

panying focusing, and which construct wave fronts consisting of acoustic pulses moving in a forward direction.
- Phased array: a series of crystal elements that are individually regulated. Using a time delay in the impulses, acoustic wave fronts are emitted that allow the construction of a sector-shaped image (swiveling ultrasound beam).

Transmission frequency: the exact frequency of the acoustic waves emitted from a transducer, the so-called *mean frequency,* lies in the range of the manufacturer's indicated transmission frequency, allowing for small deviations upward and downward. For example, a designated transmission frequency of 3.0 MHz may apply when the actual middle frequency is 2.8 MHz. (The term is synonymous with nominal frequency).

Turbulence: disorganized flow chaotically oriented in many directions simultaneously.

Turbulent flow: According to Reynolds, flow that deviates from Poiseuille's law can be defined as a dimensionless number (Reynolds number; RN) that depends on the liquid's density and its viscosity. When the RN > 2000, turbulent flow in the moving liquid's components appears.

Ultrasound: Frequencies above the range of human hearing (16,000 Hz to 1 GHz) are called ultrasound. In examining biological tissues, frequencies of approximately 0.5–20.0 MHz are used. The acoustic properties of tissue are exploited diagnostically in this process, e.g.:

- Acoustic quantities (conduction of sound waves, attenuation of sound waves, reflection, scattering, Doppler shift)
- Intrinsic quantities (density, compressibility, viscosity)
- Quantities related to the state (temperature and flow velocity)

Ultrasound beam: In addition to the main ultrasound beam within the acoustic field of a probe, there are many subsidiary beams of varying magnitudes that are dependent on the focusing *(focal zone)*. The width of the ultrasound beam determines the diameter of a sample volume, and therefore varies in the *near field* and the *far field*. Usually, the manufacturers of the equipment only provide insufficient information on this aspect.

Ultrasound contrast media: These are used to improve the Doppler signal analysis (see also *harmonic imaging, signal movement artifacts*), using:

- Microscopic air bubbles of varying sizes (12–250 μm)
- Air-bubble microspheres

- Microparticle suspensions (e.g., Echovist, Levovist) consisting of water-soluble microparticles as carriers for the microbubbles—depending on their size, these are stable either when passing through the lung (2–8 μm) or not (8–20 μm)

Possible areas of application lie in the differential diagnosis of subtotal stenoses versus occlusions of large blood vessels, analyzing flow structures in smaller blood vessels and tissue capillaries, diagnosing tumors, and in regions that strongly absorb ultrasound (transcranial sonography).

Ultrasound frequency: frequencies used in the field of vascular diagnostics:

- 1–2 MHz: for deep blood vessels (e.g., transcranial Doppler)
- 3–5 MHz: extracranial blood vessels, large arteries
- 8–10 MHz: for superficially located arteries (orbital arteries, arteries of the hands and feet)

Ultrasound probe: with one piezoelectric crystal *(pulsed wave Doppler)* or two *(continuous wave Doppler)* in the ultrasound probe, Doppler systems transform electrical energy into mechanical energy, and vice versa.

Vector: quantity defined by both magnitude and direction. Velocity is a vector quantity; the Doppler shift frequency is determined by the magnitude of the component of the velocity vector along a line between the source and the receiver of sound.

Velocity profile: With the varying flow velocities that occur in the cross-section of a blood vessel, measurements have to be carried out simultaneously at several measuring points in order to obtain the flow velocity profile. Physiologically, the flow velocity is at its lowest near the vessel wall, and at it is highest in the center (see *laminar flow*).

Viscosity: the tendency of a fluid to resist deformation, such as that required to maintain laminar flow. Viscous forces have their origin in the internal cohesion of the fluid.

Vortices: elements of rotational flow often seen with flow separation and disturbance. Rotating flow comprises a wide range of velocities aligned in both directions along a line passing through its center. Doppler shifts from vortical flow are thus characterized by spectral broadening and simultaneous forward and reverse flow.

Wall motion filter: a high-pass filter that distinguishes Doppler shifts in the low-frequency range, caused by wall movements, from the signals caused by the corpuscular flow elements that are actually of interest, and are often more than 20 dB weaker (see *scattering*). This allows the lower measurable flow velocity to be identified. However, it also leads to errors in the calculated parameters of the spectrum, e.g., overestimation of the *mean frequency.*

Wave, wavelength: a periodic process involving oscillation or vibration (λ).

Zero-crossing counter: transformation of the Doppler input signal into a sequence of rectangular pulses, according to the frequency (number) of times that zero is crossed (0 volt comparator).

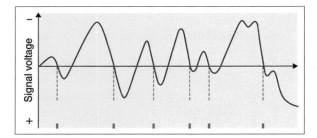

References

Chapter 1

Angelsen, B. A. J., K. Kristofferson: Combination of ultrasonic 2D-echo amplitude imaging and Doppler measurements. In Spencer, M. P., ed: Cardiac Doppler Diagnosis. Nijhoff, Dordrecht 1984

Arbeille, Ph., F. Lapierre, L. Pourcelot: L'échotomographie et l'analyse spectrale du signal Doppler dans le bilan des lesions carotidiennes. J. Malad. vasc. 9 (1984) 178

Baker, D. W., F. K. Forster, R. E. Daigle: Doppler Principles and Techniques in Ultrasound: Its Application in Medicine and Biology. Elsevier, Amsterdam 1978

Barber, F. E., D. W. Baker, A. W. C. Nation, D. E. Strandness, J. M. Reid: Ultrasonic duplex echo-Doppler-scanner. IEEE Trans. biomed. Engng 21 (1974) 109–113

Barnes, R. W., H. E. Russell, K. K. Wu, J. C. Hoak: Accuracy of Doppler ultrasound in clinically suspected venous thrombosis of the calf. Surg. Gynecol. Obstet. 143 (1976) 425–428

Barnes, R. W., L. Nix, S. E. Rittgers: Audible interpretation of carotid Doppler signals. Arch. Surg. 116 (1981) 1185–1189

Baumgartner, R. W., J. Mathis, M. Sturzenegger, H. P. Mattle: A validation study on the intraobserver reproducibility of transcranial color-coded duplex sonography velocity measurements. Ultrasound Med. Biol. 20 (1994) 233–237

Bäzner, H., M. Daffertshofer, M. Koniezko, M. Hennerici: Modification of low frequency spontaneous oscillations in blood flow velocity in large- and small-artery disease. J. Neuroimag. 5 (1995) 212–218

Blackshear, W. M., D. J. Phillips, B. L. Thiele, J. H. Hirsch, P. M. Chikos, M. R. Marinelli, K. J. Ward, D. E. Strandness: Detection of carotid occlusive disease by ultrasonic imaging and pulsed Doppler spectrum analysis. Surgery 86 (1979) 698–706

Bollinger, A., U. Hoffmann, U. K. Franzeck: Evaluation of flux motion in man by the laser Doppler technique. Blood Vessels 28 (1991) 21–26

Bönhof, J. A.: Schriftliche Dokumentation sonographischer Befunde. Ultraschall Med. 13 (1992) 283–288

Bosward, K. L., S. B. Barnett, A. K. Wood, M. J. Edwards, G. Kossoff: Heating of guinea-pig fetal brain during exposure to pulsed ultrasound. Ultrasound Med. Biol. 19 (1993) 415–424

Burns, P. N.: Principles of Doppler and color flow. Radiol. Med. Torino 85 (Suppl. 1) (1993) 3–16

Burns, P. N.: Ultrasound contrast agents in radiological diagnosis. Radiol. Med. Torino 87 (1994) 71–82

Burns, P. N.: Harmonic imaging with ultrasound contrast agents. Clin. Radiol. 51 (1996) 50–55

Bushong, St. C., B. R. Archer: Diagnostic Ultrasound. Physics, Biology, and Instrumentation. Mosby, St. Louis 1991

Cape, E. G., H. W. Sung, A. P. Yoganathan: Basics of color Doppler imaging. In Lanzer, P., A. P. Yoganathan: Vascular Imaging by Color Doppler and Magnetic Resonance. Springer, Berlin 1991

Casty, M., M. Anliker: Technik und praktische Anwendung mehrkanaliger, gepulster Doppler-Ultraschallgeräte zur quantitativen Flußmessung in großen Gefäßen. In Kriessmann, A., A. Bollinger: Ultraschall-Doppler-Diagnostik in der Angiologie. Thieme, Stuttgart 1978

Casty, M.: Technik der Ultraschall-Doppler-Geräte. In Kriessmann, A., A. Bollinger, H. Keller: Praxis der Doppler-Sonographie. Thieme, Stuttgart 1982; 2. Aufl. 1990

Cavaye, D. M., R. A. White: Intravascular Ultrasound Imaging. Raven Press, New York 1993

Delcker, A., B. Turowski: Diagnostic value of 3D-transcranial contrast duplex sonography. J. Neuroimag. 7 (1997) 139–144

Dirnagl, U., U. Lindauer, A. Villringer: Nitric oxide blockade enhances vasomotion in the cerebral microcirculation of anaesthetized rats. Microvasc. Res. 45 (1993) 318–323

Eden, A.: The beginnings of Doppler. In Aaslid, R. ed: Transcranial Doppler Sonography. Springer, Wien 1986

Fehske, W.: Praxis der konventionellen und farbcodierten Doppler-Echokardiographie. Huber, Bern 1988

Felix, W. R., B. Siegel, R. J. Gibson, J. Williams, G. L. Popky, A. L. Edelstein, J. R. Justin: Pulsed Doppler ultrasound detection of flow disturbances in arteriosclerosis. J. clin. Ultrasound 4 (1976) 275–282

Goldberg, B. B., J.-B. Liu, F. Forsberg: Ultrasound contrast agents: A review. Ultrasound Med. Biol. 20 (1994) 319–333

Görtler, M., R. Niethammer, Y. Katib, B. Widder, K. S. Piper, M. Herrmann: Ultrasound contrast agents: Iatrogenic emboli which can cause focal dilatations or microneurysms in cerebral arterioles? Stroke 26 (1995) 732 (abstract)

Hatle, L., B. Angelsen: Doppler Ultrasound in Cardiology, 2nd ed. Lea & Febiger, Philadelphia 1985

Hennerici, M.: Nicht-invasive Diagnostik des Frühstadiums arteriosklerotischer Karotisprozesse mit dem Duplex-System. Vasa 12 (1983) 228–232

Hennerici, M., H.-J. Freund: Efficacy of CW-Doppler and duplex system examinations for the evaluation of extracranial carotid disease. J. clin. Ultrasound 12 (1984) 155–161

Hennerici, M. G., S. Meairs, M. Daffertshofer, W. Neff, U. Schminke, A. Schwartz: Investigation of wall motion, local hemodynamics, and atherosclerosis plaque geometry in carotid arteries using 4-D ultrasound and high resolution magnetic resonance angiography. In Moskowitz, M. A., L. R. Caplan, eds: Cerebrovascular Diseases. 19. Princeton Stroke Conference. Butterworth-Heinemann Verlag 1995

Hennerici, M., W. Rautenberg, G. Sitzer, A. Schwartz: Transcranial Doppler ultrasound for the assessment of intracranial arterial flow velocity. – Part I. Examination technique and normal values. Surg. Neurol. 27 (1987) 439–448

Herment, A., P. Dumee: Comparison of blood flow imaging methods. Eur. J. Ultrasound 1 (1994) 345–353

Herment, A., J. P. Guglielmi: Principles of color imaging of blood flow. Eur. J. Ultrasound 1 (1994) 197–204

Hundley, W. G., G. J. Renaldo, J. E. Levasseur, H. E. Kontos: Vasomotion in cerebral microcirculation of awake rabbits. Amer. J. Physiol. 254 (1988) H67–H71

Keller, H. M., W. E. Meier, M. Anliker, D. A. Kumpe: Non-invasive measurement of velocity profiles and flow in the common carotid artery by pulsed Doppler ultrasound. Stroke 7 (1976) 370–377

Ku, D. N., D. P. Giddenns, D. J. Phillips, D. E. Jr. Strandness: Hemodynamics of the normal human carotid bifurcation: In vitro and in vivo studies. Ultrasound Med. Biol. 11 (1985) 13–26

Lewis, R. R., M. G. Beasley, D. E. Hyams, R. G. Gosling: Imaging the carotid bifurcation using continuous wave Dopplershift ultrasound and spectral analysis. Stroke 9 (1978) 465–471

Meairs, S., J. Röther, W. Neff, M. Hennerici: New and future developments in cerebrovascular ultrasound, magnetic resonance angiography, and related techniques. J. clin. Ultrasound 23 (1995) 139–149

Mitchell, D. G.: Color Doppler imaging: Principles, limitations, and artifacts. Radiology 177 (1990) 1–10

Otis, S., M. Rush, R. Boyajian: Contrast-enhanced transcranial imaging. Results of an American phase-two study. Stroke 26 (1995) 203–209

Peronneau, P., A. J. Hinglais, M. Pellet, F. Legere: Velocimetre sanguin par effet Doppler à émission ultrasonore pulsée. L'onde electr. 50 (1970) 369–384

Picot, P. A., D. W. Rickey, R. Mitchell, R. N. Rankin, A. Fenster: Three-dimensional colour Doppler imaging. Ultrasound Med. Biol. 19 (1993) 95–104

Porter, T. R., F. Xie, D. Kricsfeld, R. W. Armbruster: Improved myocardial contrast with second harmonic transient ultrasound response imaging in humans using intravenous perfluorocarbon-exposed sonicated dextrose albumin. J. Amer. Coll. Cardiol. 27 (1996) 1497–1501

Ranke, C., A. Creutzig, K. Alexander: Duplex scanning of the peripheral arteries: correlation of the peak velocity ratio with angiographic diameter reduction. Ultrasound Med. Biol. 18 (1992) 433–440

Reneman, R. S., H. F. Clarke, N. Simmons, M. P. Spencer: In vivo comparison of electromagnetic and Doppler flowmeters: with special attention to the processing of the analogue Doppler flow signal. Cardiovasc. Res. 7 (1973) 557–566

Reneman, R. S., T. van Merode, P. Hick, A. P. G. Hoeks: Flow velocity patterns in and distensibility of the carotid artery bulb in subjects of various ages. Circulation 71 (1985) 500–509

Reneman, R. S., T. van Merode, P. Hick, A. P. G. Hoeks: Cardiovascular applications of multi-gate pulsed Doppler systems. Ultrasound Med. Biol. 12 (1986) 357–370

Ries, S., W. Steinke, K. W. Neff, M. Hennerici: Echocontrast-enhanced transcranial color-coded sonography for the diagnosis of transverse sinus venous thrombosis. Stroke 28 (1997) 696–700

Sandmann, W., P. Peronneau, B. Ulrich, C. P. Bournat, M. Xhaard, K. H. Gisbertz: Die Messung von Turbulenz mit Ultraschall-Doppler-Verfahren am Strömungsmodell, am Hund und am Menschen. In Zeitler, E.: Hypertonie – Risikofaktor in der Angiologie. Witzstrock, Baden-Baden 1975

Schechner, J. S., I. M. Braverman: Synchronous vasomotion in the human cutaneous microvasculature provides evidence for central modulation. Microvasc. Res. 42 (1992) 27–32

Schwarz, K. Q., G. P. Bezante, X. Chen, R. Schlief: Quantitative echo contrast concentration measurement by Doppler sonography. Ultrasound Med. Biol. 19 (1993) 289–297

Seitz, K., R. Kubale: Duplexsonographie der abdominellen und retroperitonealen Gefäße. Verlag Chemie, Weinheim 1988

Sitzer, M., G. Fürst, M. Siebler, H. Steinmetz: Usefulness of an intravenous contrast medium in the characterization of high-grade internal carotid stenosis with color Doppler-assisted duplex imaging. Stroke 25 (1994) 385–389

Sold, G.: Zweidimensionale Echokardiographie. Urban & Schwarzenberg, München 1986

Spencer, M. P., J. M. Reid: Quantitation of carotid stenoses with continuous wave (CW) Doppler ultrasound. Stroke 10 (1979) 326–330

Steinke, W., C. Kloetzsch, M. Hennerici: Variability of flow patterns in the normal carotid bifurcation. Atherosclerosis 84 (1990 a) 121–128

Steinke, W., C. Kloetzsch, M. Hennerici: Carotid artery disease assessed by color Doppler flow imaging. Amer. J. Neuroradiol. 11 (1990 b) 259–266

Wells, P. N. T.: Physical and technical aspects of colour flow ultrasound. In Labs, K. H., K. A. Jäger, D. E. Fitzgerald, J. P. Woodcock, D. Neuerburg-Heusler, eds: Diagnostic Vascular Ultrasound. Arnold, London 1992

WFUMB Symposium on Safety and Standardization of Ultrasound in Medicine. Ultrasound Med. Biol. 18 (1992) 731–814

Widder, B.: Doppler- und Duplexsonographie der hirnversorgenden Arterien. Springer, Berlin-Heidelberg 1995

Widder, B., B. Arnolds, S. Drews, M. Fischer, W. Heiß, M. Marshall, D. Neuerburg-Heusler, P. Nissen, F. Reimer, G. M. von Reutern, H. Straub, R. Winter: Terminologie der Ultraschall-Gefäßdiagnostik. Ultraschall Med. 11 (1990) 214–218

Zagzebski, J. A., E. L. Madsen: Physics and instrumentation in Doppler and B-mode ultrasonography. In Zwiebel, W. J.: Introduction to Vascular Ultrasonography. Grune & Stratton, New York 1982

Zarins, C. K., D. P. Giddens, B. K. Bharadvaj, V. S. Sottiurai, R. F. Mabon, S. Glagov: Carotid bifurcation atherosclerosis: Quantitative correlation of plaque localization with flow velocity profiles and wall shear stress. Circulat. Res. 53 (1983) 502–514

Chapter 2

Abu Rahma, A. F., C. Stuart: The value of a combined study of carotid duplex ultrasound and oculopneumoplethysmography (OPG-Gee) in the diagnosis of carotid artery stenosis. Amer. Surg. 53 (1987) 80–83

Ackerman, R. H.: Perspective on noninvasive diagnosis of carotid disease. Neurology 29 (1979) 615–622

Ackerstaff, R. G. A., H. Hoeneveld, J. M. Slowikowski, F. L. Moll, B. C. Eikelboom, J. W. Ludwig: Ultrasonic duplex scanning in atherosclerotic disease of the innominate, subclavian and vertebral arteries. A comparative study with angiography. Ultrasound Med. Biol. 10 (1984) 409–418

Alexandrov, A. V., C. F. Bladin, R. Maggisano, J. W. Norris: Measuring carotid stenosis: Time for a reappraisal. Stroke 24 (1993) 1292–1296

Amarenco, P., A. Cohen, M. Baudrimont, M.-G. Bousser: Transesophageal echocardiographic detection of aortic arch disease in patients with cerebral infarction. Stroke 23 (1992) 1005–1009

Amarenco, P., C. Duyckaerts, C. Tzourio, D. Hénin, M. G. Bousser, J. J. Hauw: The prevalence of ulcerated plaques in the aortic arch in patients with stroke. New. Eng. J. Med. 326 (1992) 221–225

Amarenco, P., A. Cohen, C. Tzouric et al.: Atherosclerotic disease of the aortic arch and the risk of ischemic stroke. New. Engl. J. Med. 331 (1994) 1474–1479

Amarenco, P., O. Heinzlef, C. Lucas et al.: The risk of cerebral infarct in patients with aortic arch plaques: A prospective study. Stroke 26 (1995) 184 (abstract)

Anderson, C. M., D. Saloner, R. E. Lee, et al.: Assessment of carotid artery stenosis by MR angiography: Comparison with X-ray angiography and color-coded Doppler ultrasound. Amer. J. Neuroradiol. 13 (1992) 989–1003

Arbeille, Ph., F. Lapierre, L. Pourcelot: Evaluation des sténoses carotidiennes par les ultrasons. Encyclop. méd.-chir. 322 (1985) 1–8

Baker, D. W., F. K. Forster, R. E. Daigle: Doppler Principles and Techniques in Ultrasound: Its Application in Medicine and Biology. Elsevier, Amsterdam 1978

Bandyk, D. F., A. W. Levine, L. Pohl, J. B. Towne: Classification of carotid bifurcation disease using quantitative Doppler spectrum analysis. Arch. Surg. 120 (1985) 306–314

Barnes, R. W., L. Nix, S. E. Rittgers: Audible interpretation of carotid Doppler signals. Arch. Surg. 116 (1981) 1185–1189

Barnett, H. J. M.: Delayed cerebral ischemic episodes distal to occlusion of major cerebral arteries. Neurology 28 (1978) 769–774

Barnett, H. J. M., C. P. Warlow: Carotid endarterectomy and the measurement of stenosis. Stroke 24 (1993) 1281–1284

Bartels, E., K. A. Flügel: Advantages of color Doppler imaging for the evaluation of vertebral arteries. J. Neuroimag. 3 (1993) 229–233

Bartels, E., K. A. Flügel: Evaluation of extracranial vertebral artery dissection with duplex color-flow imaging. Stroke 27 (1996) 290–295

Bartels, E., H. H. Fuchs, K. A. Flügel: Duplex ultrasonography of vertebral arteries: Examination, technique, normal values, and clinical applications. Angiography 43 (1992) 169–180

Baud, J. M., J. M. de Bray, P. Delanoy: La reproductibilité ultrasonore des plaques carotidiennes dans la caractérisation. JEMU 6 (1996)

Beach, K., R. Lawrence, D. Phillips, J. Primozich, E. Strandness: The systolic velocity criterion for diagnosing significant internal carotid stenoses. J. vasc. Technol. 13 (1989) 65–68

Biller, J., W. L. Hinktgen, H. P. Adams, W. R. K. Smoker, J. C. Godersky, G. J. Toffol: Cervico cephalic arterial dissections – a 10 year experience. Arch. Neurol. 43 (1986) 1234–1238

Blackshear, W. M., D. J. Phillips, D. L. Thiele, J. H. Hirsch, P. M. Chikos, M. R. Marinelli, K. J. Ward, D. E. Strandness: Detection of carotid occlusive disease by ultrasonic imaging and pulsed Doppler spectrum analysis. Surgery 86 (1979) 698–706

Blackshear, W. M., D. J. Phillips, P. M. Chikos, J. D. Harley, B. L. Thiele, D. E. Strandness: Carotid artery velocity patterns in normal and stenotic vessels. Stroke 11 (1980) 67–71

Blackshear, W. M., K. B. Seifert, S. Lamb et al.: Pulsed Doppler frequency and carotid stenosis. J. surg. Res. 42 (1987) 179–184

Bladin, C. F., A. V. Alexandrov, J. W. Norris: How should we measure carotid stenosis? Lancet 344 (1994) 69

Bluth, E. I., D. Kay, C. R. B. Merritt, M. A. Sullivan, G. Farr, N. G. Mills, M. Foreman, K. Sloan, M. Schlater, J. Steward: Sonographic characterization of carotid plaque: Detection of hemorrhage. Amer. J. Neuroradiol. 7 (1986) 311–315

Bock, R. W., A. C. Gray-Weale, P. A. Mock, D. A. Robinson, M. B. Irwig, R. J. Lusby: The natural history of asymptomatic carotid artery disease. J. vasc. Surg. 17 (1993) 160–171

Bonithon-Kopp, C., P. J. Touboul, C. Berr, C. Magne, P. Ducimetiere: Factors of carotid arterial enlargement in a population aged 59 to 71 years: The EVA study. Stroke 27 (1996) 654–660

Bornstein, N. M., Z. G. Beloev, J. W. Norris: The limitation of diagnosis of carotid occlusion by Doppler ultrasound. Ann. Surg. 207 (1988) 315–317

Brown, P. M., K. W. Johnston, Y. Douville: Detection of occlusive disease of the carotid artery with continuous-wave Doppler spectral analysis. Surg. Gynecol. Obstet. 155 (1982 a) 183–186

Brown, P. M., K. W. Johnston, M. Kassam, R. S. Cobbold: A critical study of ultrasound Doppler spectral analysis for detecting carotid disease. Ultrasound Med. Biol. 8 (1982 b) 515–523

Büdingen, H. J., M. Hennerici, K. Voigt, K. Kendel, H.-J. Freund: Die Diagnostik von Stenosen oder Verschlüssen der A. carotis interna mit der direktionellen Ultraschall-Doppler-Sonographie der A. supratrochlearis. Dtsch. med. Wschr. 101 (1976) 269–275

Büdingen, H. J., G.-M. von Reutern: Ultraschalldiagnostik der hirnversorgenden Arterien. Thieme, Stuttgart 1989; 2. Aufl. 1993

Burns, R. P., J. B. Cofer, W. L. Russell et al.: Extracranial carotid vascular disease: Detection by real-time B-mode ultrasonography. Sth. med. J. 78 (1985) 518–522

Caes, F., T. Vierendeels, E. Janssens-Willem, B. Cham et al.: Comparison of auscultation, continuous-wave Doppler imaging, intravenous digital subtraction angiography and conventional angiography in diagnosis of carotid artery disease. Angiology 38 (1987) 799–806

Cape, C. A., R. L. DeSaussure, J. Nixon: Carotid ultrasonography in carotid artery disease. Sth. med. J. 77 (1984) 183–186

Carpenter, J. P., F. J. Lexa, J. T. Davis: Determination of duplex Doppler ultrasound criteria appropriate to the North American Symptomatic Carotid Endarterectomy Trial. Stroke 27 (1996) 695–699

Chang, Y. J., S. K. Lin, S. J. Ryu, Y. Y. Wai: Common carotid artery occlusion: Evaluation with duplex sonography. Amer. J. Neuroradiol. 16 (1995) 1099–1105

Comerota, A. J., J. J. Cranley, S. E. Cook: Real-time B-mode carotid imaging in diagnosis of cerebrovascular disease. Surgery 89 (1981) 718–729

Comerota, A. J., J. J. Cranley, M. L. Katz et al.: Real-time B-mode carotid imaging: A three-year multicenter experience. J. vasc. Surg. 1 (1984) 85–95

Croft, R. J., L. D. Ellam, M. J. G. Harrison: Accuracy of carotid angiography in the assessment of atheroma of the internal carotid artery. Lancet 1980/I, 997–999

Daffertshofer, M., M. Hennerici: Spektrumanalyse von Doppler-Signalen. Ultraschall 11 (1990) 219–226

D'Agostino, R. B. Jr., G. Burke, D. O'Leary, M. Rewers, J. Selby, P. J. Savage, M. F. Saad, R. N. Bergman, G. Howard, L. Wagenknecht, S. M. Haffner: Ethnic differences in carotid wall thickness: The insulin resistance atherosclerosis study. Stroke 27 (1996) 1744–1749

Daigle, R. J., A. T. Stavros, R. Platon, D. B. Anderst, P. D. Nurre: Velocity criteria for differentiation of 60%–79% carotid stenoses from 80% or greater stenoses. J. vasc. Technol. 7 (1988) 176–183

Daiss, W., H. C. Diener, A. Thron et al.: Diagnosis of stenoses and occlusion of extracranial arteries. Comparison of Doppler sonography, duplex scan, and cerebral angiography. Dtsch. med. Wschr. 109 (1984) 1595–1599

Dauzat, M., J. P. Laroche, J. M. de Bray, G. Deklunder, A. Couture, J. B. Cesari, F. Barral: Ultrasonographie vasculaire diagnostique. Vigot, Paris 1991

De Bray, J. M., J. M. Baud, M. Dauzat on behalf of the Consensus Conference: Concerning the morphology and the risk of carotid plaques. Cerebrovascular Diseases (1997) (in print)

De Bray, J. M., F. Galland, P. Lhoste, et al.: Colour Doppler imaging duplex sonography and angiography of carotid bifurcations: Prospective and double-blind study. Neuroradiology 37 (1995 b) 219–224

De Bray, J. M., B. Glatt for the International Consensus Conference: Quantification of atheromatous stenosis in the extracranial internal carotid artery. Cerebrovascular Diseases 5 (1995) 414–426

Decker, K., H. J. Schlegel: Normbilder und Normvarianten der A. ophthalmica im Röntgenbild. Albrecht v. Graefes Arch. Ophthalmol. 159 (1957) 302–310

Delcker, A., H. C. Diener, H. Wilhelm: Influence of vascular risk factors for atherosclerotic carotid artery plaque progression. Stroke 26 (1995) 2016–2022

Di Chiro, G.: Ophthalmic arteriography. Radiology 77 (1961) 948–957

Doorly, T. P. G., P. I. Atkinson, V. Kingston et al.: Carotid ultrasonic arteriography combined with real-time spectral analysis: A comparison with angiography. J. cardiovasc. Surg. 23 (1982) 243–246

Dreisbach, J. N., C. E. Seibert, S. F. Smazal et al.: Duplex sonography in the evaluation of carotid artery disease. Amer. J. Neuroradiol. 4 (1983) 678–680

Easton, J. D.: Accuracy of high resolution ultrasound imaging for quantitative assessment for early carotid atherosclerosis. Cerebrovascular Diseases 4 (1994) 109–113

Eicke, B. M., C. H. Tegeler, G. Dalley, L. G. Myers: Angle correction in transcranial Doppler sonography. J. Neuroimag. 4 (1994) 29–33

Eikelboom, B. C., T. R. Riles, R. Mintzer, F. G. Baumann, G. DeFillip, J. Lin, A. M. Imparato: Inaccuracy of angiography in diagnosis of carotid ulceration. Stroke 14 (1983) 882–885

Eliasziw, M., J. Y. Streifler, A. J. Fox, V. C. Hachinski, G. G. Ferguson, H. J. M. Barnett: Significance of plaque ulceration in symptomatic patients with high-grade carotid stenosis. Stroke 25 (1994) 304–308

Erdoes, L. S., J. M. Marek, J. L. Mills, S. S. Berman, T. Whitehill, G. C. Hunter, W. Feinberg, W. Krupski: The relative contributions of carotid duplex scanning, magnetic resonance angiography, and cerebral arteriography to clinical decisionmaking: A prospective study in patients with carotid occlusive disease. J. vasc. Surg. 23 (1996) 950–956

Erickson, S. J., M. W. Mewissen, W. D. Foley, et al.: Stenosis of the internal carotid artery: Assessment using color Doppler imaging compared with angiography. Amer. J. Roentgenol. 152 (1988) 1299–1305

European Carotid Surgery Trialists' Collaborative Group: MRC European Carotid Surgery Trial: Interim results for symptomatic patients with severe (70–90%) and with mild (0–29%) carotid stenosis. Lancet 337 (1991) 1235–1243

Fischer, M., T. Stegmann, H. Becker et al.: Doppler frequency spectrum analysis and digital subtraction angiography in carotid artery disease. Thorac. cardiovasc. Surg. 33 (1985) 304–307

Fisher, C. M., R. G. Ojemann, G. H. Roberson: Spontaneous dissection of cervico-cerebral arteries. Canad. J. Neurol. Sci. 5 (1978) 9–19

Floriani, M., S. Bonardelli, N. Portolani et al.: Direct in indirect evaluation of lesions obstructing the carotid bifurcation. A comparison of Doppler spectrum analysis with angiography. Int. Angiol. 6 (1987) 375–382

Gabrielsen, T. O., T. Greitz: Normal size of the internal carotid, middle cerebral and anterior cerebral arteries. Acta radiol. 10 (1970) 1

Geroulakos, G., J. Domjan, A. Nicolaides: Ultrasonic carotid artery plaque structure and the risk of cerebral infarction on computed tomography. Vasc. Surg. 20 (1994) 263–266

Glagov, S., H. S. Bassiouny, D. P. Giddens, C. K. Zarins: Intimal thickening: Morphogenesis, functional significance and detection. J. vasc. Investig. 1 (1995) 2–14

Glagov, S., C. K. Zarins: Quantitating atherosclerosis – problems of definition. In Bond, M. G., W. Insull, S. Glagov, A. B. Chandler, J. F. Cornhill: Clinical Diagnosis of Atherosclerosis. Springer, Berlin 1983

Görtler, M., R. Niethammer, B. Widder: Differentiating subtotal carotid artery stenoses from occlusions by colour coded duplex sonography. J. Neurol. 241 (1994) 301–305

Griewing, B., C. Morgenstern, F. Driesner, G. Kallwellis, M. L. Walker, C. Kessler: Cerebrovascular disease assessed by color-flow and power Doppler ultrasonography. Comparison with digital subtraction angiography in internal carotid artery stenosis. Stroke 27 (1996) 95–100

Grosveld, W. J. H. M., J. A. Lawson, B. C. Eikelboom, J. M. v. d. Windt, R. G. A. Ackerstaff: Clinical and hemodynamic significance of innominate artery lesions evaluated by ultrasonography and digital angiography. Stroke 19 (1988) 958–962

Guo, Z., A. Fenster: Three-dimensional power Doppler imaging: A phantom study to quantify vessel stenosis. Ultrasound Med. Biol. 22 (1996) 1059–1069

Hames, T. K., K. N. Humphries, V. M. Gazzard et al.: The role of continuous-wave Doppler imaging in a vascular unit. Cardiovasc. Res. 19 (1985 a) 631–635

Hames, T. K., D. A. Ratliff, K. N. Humphries et al.: The accuracy of duplex scanning in the evaluation of early carotid disease. Ultrasound Med. Biol. 11 (1985 b) 819–825

Harward, T. R., E. F. Bernstein, A. Fronek: Continuous-wave versus range-gated pulsed Doppler power frequency spectrum analysis in the detection of carotid arterial occlusive disease. Ann. Surg. 204 (1986) 32–37

Hass, W. K., W. S. Fields, R. R. North, I. I. Kricheff, N. E. Chase, R. B. Bauer: Joint study of extracranial arterial occlusion. II. Arteriography, techniques, sites, and complications. J. Amer. med. Ass. 203 (1968) 961–968

Hayreh, S. S., R. Dass: The ophthalmic artery. I. Origin and intracranial and intra-canalicular course. Brit. J. Ophthalmol. 46 (1962 a) 46

Hayreh, S. S., R. Dass: The ophthalmic artery II. Intraorbital course. Brit. J. Ophthalmol. 46 (1962 b) 165

Hedges, T. R: Ocular Ischemia. In Caplan, L. R. ed: Brain Ischemia. Basic Concepts and Clinical Relevance. Springer, Berlin 1995 (61–73)

Hedges, T. R., E. Reichel, J. S. Duker, C. A. Puliafito, P. A. Heggerick: Color Doppler imaging identifies different mechanisms of central retinal artery occlusion. Invest. Ophthalmol. Visual Sci. 34 (1993) 842

Hennerici, M., A. Aulich: Ultraschall-Doppler-Sonographie. Mediuz 3 (1979) 168–179

Hennerici, M., A. Aulich, W. Sandmann, H.-J. Freund: Incidence of asymptomatic extracranial arterial disease. Stroke 12 (1981 a) 750–758

Hennerici, M., A. Aulich, W. Sandmann, J. Lerut: Stenosen und Verschlüsse des Truncus brachiocephalicus. Dtsch. med. Wschr. 106 (1981 b) 1697–1703

Hennerici, M.: Nicht-invasive Diagnostik des Frühstadiums arteriosklerotischer Karotisprozesse mit dem Duplex-System. Vasa 12 (1983) 228–232

Hennerici, M., H.-J. Freund: Efficacy of CW-Doppler and duplex system examinations for the evaluation of extracranial carotid disease. J. clin. Ultrasound 12 (1984 a) 155–161

Hennerici, M., G. Reifschneider, U. Trockel, A. Aulich: Detection of early atherosclerotic lesions by duplex scanning of the carotid artery. J. clin. Ultrasound 12 (1984 b) 455–464

Hennerici, M., M. Daffertshofer, G. Esser, A. Aulich: Spektralanalyse kontinuierlicher Dopplersignale der extrakraniellen Karotis. Vasa 14 (1985 a) 131–138

Hennerici, M., W. Rautenberg, U. Trockel, R. G. Kladetzky: Spontaneous progression and regression of small carotid atheroma. Lancet 1985 b/I, 1415–1419

Hennerici, M., W. Rautenberg, U. Trockel, R. G. Kladetzky: Entwicklung nichtstenosierender extrakranieller Karotis-Plaques – Eine prospektive Verlaufsuntersuchung mit dem Duplexsystem. Ultraschall 6 (1985 c) 68–73

Hennerici, M.: Clinical applications of high resolution B-scan imaging with pulsed Doppler profiles (10 MHz). In Spencer, M. P.: Ultrasonic Diagnosis of Cerebrovascular Disease. Nijhoff Den Haag 1987 a

Hennerici, M.: Hochauflösende Ultraschall-Duplexsystemanalyse der extrakraniellen Karotisstrombahn. In Hartmann, A., H. Wassmann: Hirninfarkt. Urban & Schwarzenberg, München 1987 b

Hennerici, M., A. Aulich, H.-J. Freund: Carotid system syndromes. In Vinken, P. J., G. W. Bruyn, H. L. Klawans: Handbook of Clinical Neurology, Vol. 53. Elsevier, Amsterdam 1988

Hennerici, M., C. Klemm, W. Rautenberg: The subclavian steal phenomenon: A common vascular disorder with rare neurologic deficits. Neurology 38 (1988) 669–673

Hennerici, M., W. Steinke, W. Rautenberg: High-resistance Doppler flow pattern in extracranial carotid dissection. Arch. Neurol. 46 (1989) 670–672

Hennerici, M., W. Steinke: Carotid plaque developments—aspects of hemodynamic and vessel wall platelet interaction. Cerebrovascular Diseases 1 (1991) 142–148

Hennerici, M. G.: High intensity transcranial signals (HITS): a questionable ‚jackpot‘ for the prediction of stroke risk. J. Heart Valve Dis. 3 (1994) 124–125

Hennerici, M. G., W. Steinke: Accuracy of high-resolution ultrasound imaging for quantitative assessment of early carotid atherosclerosis. Cerebrovascular Diseases 4 (1994) 109–113

Hennerici, M. G., S. Meairs, M. Daffertshofer, W. Neff, U. Schminke, A. Schwartz: Investigation of wall motion, local hemodynamics, and atherosclerotic plaque geometry in carotid arteries using 4-D ultrasound and high resolution magnetic resonance angiography. In Moskowitz, M. A., L. R. Caplan, eds: Cerebrovascular Diseases. 19. Princeton Stroke Conference. Butterworth-Heinemann Verlag, London 1995

Horowitz, D. R., S. Tuhrim, J. Budd, M. E. Goldman: Aortic plaques in patients with brain ischemia: Diagnosis by transesophageal echocardiography. Neurology 42 (1992) 1602–1604

Howard, G., W. H. Baker, L. E. Chambless, V. J. Howard, A. M. Jones, J. F. Toole, for the Asymptomatic Carotid Atherosclerosis Study Investigators: An approach for the use of Doppler ultrasound as a screening tool for the hemodynamic significant stenosis (despite heterogeneity of Doppler performance). A multicenter experience. Stroke 27 (1996) 1951–1957

Humphrey, P., P. Bradbury: Continuous wave Doppler ultrasonography in the detection of carotid stenosis and occlusions. J. Neurol. Neurosurg. Psychiat. 47 (1984) 1128–1130

Humphrey, P., P. Sandercock, J. Slattery: A simple method to improve the accuracy of non-invasive ultrasound in selecting TIA patients for cerebral angiography. J. Neurol. Neurosurg. Psychiat. 53 (1990) 966–971

Huston III, J., B. D. Lewis, D. O. Wiebers, F. B. Meyer, S. J. Riederer, A. L. Weaver: Carotid artery: Prospective blinded comparison of two-dimensional time-of-flight MR angiography with conventional angiography and duplex US. Radiology 186 (1993) 339–344

Jacobs, N., E. G. Grant, D. Schellinger, M. C. Byrd, J. D. Richardson, S. L. Cohan: Duplex carotid sonography: Criteria for stenosis, accuracy, and pitfalls. Radiology 154 (1985) 305–391

Johnson, K. W., W. H. Baker, S. J. Burnham, A. C. Hayes, C. A. Kupper, M. A. Poole: Quantitative analysis of continuous-wave Doppler spectral broadening for the diagnosis of carotid disease: Results of a multicenter study. J. vasc. Surg. 4 (1986) 493–504

Jones, E. F., J. M. Kalman, P. Calafiore, A. M. Tonkin, G. A. Donnan: Proximal aortic atheroma: An independent risk factor for cerebral ischemia. Stroke 26 (1995) 218–224

Jones, M., J. Biller, A. R. Cowley, G. Howard, W. M. McKinney, J. F. Toole: Extracranial carotid arteriosclerosis. Diagnosis with continuous-wave Doppler and real-time ultrasound studies. Arch. Neurol. 39 (1982) 393–394

Karino, T., H. L. Goldsmith: Particle flow behaviour in models of bending vessels. II. Effects of branching angle and diameter ratio on flow pattern. Biorheology 22 (1985) 87–105

Keagy, B. A., W. F. Pharr, D. Thomas et al.: A quantitative method for the evaluation of spectral analysis patterns in carotid artery stenosis. Ultrasound Med. Biol. 8 (1982) 625–630

Keller, H., W. Meier, Y. Yonekawa, D. Kumpe: Noninvasive angiography for the diagnosis of carotid artery disease using Doppler ultrasound (carotid artery Doppler). Stroke 7 (1976 a) 354–363

Keller, H. M., W. E. Meier, D. A. Kumpe: Noninvasive angiography for the diagnosis of vertebral artery disease using Doppler ultrasound (vertebral artery Doppler). Stroke 7 (1976 b) 364–369

Keller, H. L.: Varianten der Arteria carotis interna, der A. meningea media und der A. ophthalmica im Carotisangiogramm. Fortschr. Röntgenschr. 95 (1961) 472–482

Kessler, C., C. V. Maravic, M. V. Maravic, D. Kömpf: Color Doppler flow imaging of the carotid arteries. Neuroradiology 3 (1991) 114–117

Krayenbühl, H., M. G. Yasargil, P. Huber: Zerebrale Angiographie für Klinik und Praxis, 3. Aufl. Thieme, Stuttgart 1979

Lally, K., K. W. Johnston, R. S. C. Cobbold: Limitations in the accuracy of peak frequency measurement in the diagnosis of carotid disease. J. clin. Ultrasound 12 (1984) 403–409

Lane, R. L., L. K. Dart, M. Appleberg: The limitations of B-mode imaging of the carotid bifurcation: A comparison with three flow-dependent noninvasive tests. Brit. J. Radiol. 55 (1982) 817–820

Langlois, Y., G. O. Roederer, A. Chan et al.: Evaluating carotid artery disease: The concordance between pulsed Doppler-spectrum analysis and angiography. Ultrasound Med. Biol. 9 (1983) 51–63

Laster, R. E., J. D. Acker, H. H. III Halford, T. C. Nauer: Assessment for evaluation of cervical carotid bifurcation disease. Amer. J. Neuroradiol. 14 (1993) 681–688

Lieb, W. E., P. M. Flaherty, R. C. Sergott, R. D. Medlock, G. C. Brown, T. Bosley, P. J. Savino: Color Doppler imaging provides accurate assessment of orbital blood flow in occlusive carotid artery disease. Ophthalmology 98 (1991 a) 548–552

Lieb, W. E., S. M. Cohen, D. A. Merton, J. A. Shields, D. G. Mitchell, B. B. Goldberg: Color Doppler imaging of the eye and orbit: Technique and normal vascular anatomy. Arch. Ophthalmol. 109 (1991 b) 527–531

Lindegaard, K. F., S. J. Bakke, A. Grip et al.: Pulsed Doppler techniques for measuring instantaneous maximum and mean flow velocities in carotid arteries. Ultrasound Med. Biol. 10 (1984) 419–426

Londrey, G., D. Spadone, K. Hodgson et al.: Does color-flow imaging improve the accuracy of duplex carotid evaluation? J. vasc. Surg. 13 (1991) 659

Markus, H. S., M. M. Brown: Asymptomatic cerebral embolic signals in carotid artery stenosis. Cerebrovascular Diseases 4 (1994) 235 (abstract)

Marosi, L., H. Ehringer: Die extrakranielle Arteria carotis im hochauflösenden Ultraschallechtzeit-Darstellungssystem: Morphologische Befunde bei gesunden jungen Erwachsenen. Ultraschall 5 (1984) 174–181

Mattle, H. P., C. Kent, R. R. Edelman, D. J. Atkinson, J. J. Skilman: Evaluation of the extracranial carotid arteries: Correlation of magnetic resonance angiography, duplex ultrasonography, and conventional angiography. J. vasc. Surg. 6 (1991) 838–844

Meairs, S., J. Röther, W. Neff, M. Hennerici: New and future developments in cerebrovascular ultrasound, magnetic resonance angiography and related techniques. J. clin. Ultrasound 23 (1995) 139–149

Melis-Kisman, E., J. M. F. Mol: L'application de l'effect Doppler à l'exploration cerebrovasculaire (rapport preliminaire). Rev. Neurol. 122 (1970) 470–477

Merland, J. J.: Artériographie Supersélective de la Carotide Externe. Thesis, Paris 1973

Middleton, W. D., W. D. Foley, T. L. Lawson: Color-flow Doppler imaging of carotid artery abnormalities. Amer. J. Roentgenol. 150 (1988 a) 419–425

Middleton, W. D., W. D. Foley, T. L. Lawson: Flow reversal in the normal carotid bifurcation: Color Doppler flow imaging analysis. Radiology 167 (1988 b) 207–209

Miyazaki, M., K. Kato: Measurement of cerebral blood flow by ultrasonic Doppler technique. Jap. Circulat. J. 29 (1965 a) 375–382

Miyazaki, M., K. Kato: Measurement of cerebral blood flow by ultrasonic Doppler technique: Hemodynamic comparison of right and left carotid artery in patients with hemiplegia. Jap. Circulat. J. 29 (1965 b) 383–386

Mokri, B., T. M. Sundt, O. W. Houser, D. G. Piepgras: Spontaneous dissection of the cervical internal carotid artery. Ann. Neurol. 19 (1986) 126–138

Moneta, G., J. M. Edwards, R. W. Chitwood, L. M. Taylor, R. W. Lee, C. A. Cummings, J. M. Porter: Correlation of North American Symptomatic Carotid Endarterectomy Trial (NASCET) angiography definition of 70% to 99% internal carotid artery stenosis with duplex scanning. J. vasc. Surg. 17 (1993) 152–159

Moore, W. S., S. Ziomek, W. J. Quinones-Baldrich et al.: Can clinical evaluation and noninvasive testing substitute for arteriography in the evaluation of carotid artery disease. Ann. Surg. 208 (1988) 91–94

Müller, H. R.: Direktionelle Doppler-Sonographie der Arteria frontalis medialis. Z. EEG EMG 2 (1971) 24–32

Müller, H. R., E. W. Radue, A. Saia, C. Pallotti, M. Buser: Carotid blood flow measurement by means of ultrasonic techniques: limitations and clinical use. In Hartmann, A., S. Hoyer, eds: Cerebral Blood Flow and Metabolism Measurement. Springer, Berlin 1985

Neuburg-Heusler, D.: Dopplersonographische Diagnostik der extrakraniellen Verschlußkrankheit. Vasa Suppl. 12 (1984) 59–70

Neuburg-Heusler, D., G. M. von Reutern: Qualitätssicherung dopplersonographischer Verfahren. Ultraschall 6 (1985) 270–278

Neuburg-Heusler, D., M. Todt, F. J. Roth: Quantitative computergestützte Frequenzanalyse bei Karotisstenosen. In Maurer, H. J., ed: Berichtsband Deutsch-Japanischer Kongreß für Angiologie. Demeter, Gräfelfing 1985

Neuburg-Heusler, D.: Doppler-Sonographie der Hirnarterien. In Simon, H., W. Schoop, eds: Diagnostik in der Kardiologie und Angiologie. Thieme, Stuttgart 1986

Norrving, H., S. Cronquist: Doppler examinations of the carotid arteries. Acta Neurol. Scand. 64 (1981) 241–252

North American Symptomatic Carotid Endarterectomy Trial Collaborators: Beneficial effect of carotid endarterectomy in symptomatic patients with high-grade carotid stenosis. New Engl. J. Med. 325 (1991) 445–453

O'Leary, D. H., J. F. Polak, R. A. Kronmal, P. J. Savage, N. O. Borhani, S. J. Kittner, R. Tracy, J. M. Gardin, T. R. Price, C. D. Furberg, for the Cardiovascular Health Study Collaborative Research Group: Thickening of the carotid wall: A marker for atherosclerosis in the elderly? Stroke 27 (1996) 224–231

Padget, D. H.: The Circle of Willis, its Embryology and Anatomy. Comestock, Ithaca, N.Y. 1944

Patel, M. R., K. M. Kuntz, A. K. Roman, K. Ducksoo, J. Kramer, J. F. Polak, J. J. Skillman, A. D. Whittemore, R. R. Edelman, K. C. Kent: Preoperative assessment of the carotid bifurcation. Can magnetic resonance angiography and duplex ultrasonography replace contrast arteriography? Stroke 26 (1995) 1753–1758

Perry, M. O., M. F. Silane, D. Calcagno et al.: Quantitative Doppler spectrum analysis of extracranial carotid artery stenosis. N. Y. St. J. Med. 85 (1985) 577–580

Pfadenhauer, K., H. Müller: Color-coded duplex ultrasound of the vertebral artery: Normal findings and pathologic findings in obstruction of the vertebral artery and remaining cerebral arteries. Ultraschall Med. 16 (1995) 228–233

Picot, P. A., D. W. Rickey, R. Mitchell, R. N. Rankin, A. Fenster: Three-dimensional colour Doppler imaging. Ultrasound Med. Biol. 19 (1993) 95–104

Pignoli, P., E. Tremoli, A. Poli, P. Oreste, R. Paoletti: Intimal plus medial thickness of the arterial wall: A direct measurement with ultrasound imaging. Circulation 74 (1986) 1399–1406

Planiol, Th., L. Pourcelot, J. M. Pottiers, E. Degiovanni: Étude de la circulation carotidiénne par les methodes ultrasoniques et la thermographie. Rev. neurol. 126 (1972) 127–141

Polak, J. F., R. L. Bajakian, D. H. O'Leary, M. R. Anderson, M. D. Donaldson, F. A. Jolesz: Detection of internal carotid artery stenosis: Comparison of MR angiography, color Doppler sonography, and arteriography. Radiology 182 (1992 a) 35–40

Polak, J. F., G. R. Dobkin, D. H. O'Leary et al.: Internal carotid artery stenosis: Accuracy and reproducibility of color-Doppler assisted duplex imaging. Radiology 173 (1989) 793–800

Polak, J. F., P. Kalina, M. C. Donaldson, D. H. O'Leary, A. D. Whittemore, J. A. Mannick: Carotid endarterectomy: Preoperative evaluation of candidates with combined Doppler sonography and MR angiography. Radiology 186 (1993) 333–338

Poli, A., E. Tremoli, A. Colombo, M. Sirtori, P. Pignoli, R. Paoletti: Ultrasonographic measurement of the common carotid arterial wall thickness in hypercholesterolemic patients. Atherosclerosis. 70 (1988) 253–261

Pourcelot, L., J. L. Ribadeau-Dumas, D. Fagret, T. Planiol: Apport de l'examen Doppler dans le diagnostic du vol sous-clavier. Rev. neurol. 133 (1977) 309–323

Ratliff, D. A., P. J. Gallagher, T. K. Hames et al.: Characterization of carotid artery disease: Comparison of duplex scanning with histology. Ultrasound Med. Biol. 11 (1985) 835–840

Rauh, G., M. Fischereder, F. A. Spengel: Transesophageal echocardiography in patients with focal cerebral ischemia of unknown cause. Stroke 27 (1996) 691–694

Rautenberg, W., M. Hennerici: Pulsed Doppler assessment of innominate artery obstructive diseases. Stroke 19 (1988) 1514–1520

Reilly, L. M., R. J. Lusby, L. Hughes, L. D. Ferrel, R. J. Stoney, W. K. Ehrenfeld: Carotid plaques histology using real-time ultrasonography. Amer. J. Surg. 146 (1983) 188–193

Reneman, R. S., M. P. Spencer: Local Doppler audio spectra in normal and stenosed carotid arteries in man. Ultrasound Med. Biol. 5 (1979) 1–11

Reneman, R. S., T. van Merode, P. Hick, A. P. G. Hoeks: Flow velocity patterns in and distensibility of the carotid artery bulb in subjects of various ages. Circulation 71 (1985) 500–509

von Reutern, G.-M., H. J. v. Büdingen: Ultrasound Diagnosis of Cerebrovascular Disease: Doppler Sonography of the Extra- and Intracranial Arteries, Duplex Scanning. Thieme, Stuttgart 1993

von Reutern, G. M., H. J. Büdingen, H. J. Freund: Dopplersonographische Diagnostik von Stenosen und Verschlüssen der Vertebralarterien und des Subclavian Steal-Syndroms. Arch. Psychiat. Nervenkr. 222 (1976 a) 209–222

von Reutern, G. M., H. J. Büdingen, M. Hennerici, H.-J. Freund: Diagnose und Differenzierung von Stenosen und Verschlüssen der Arteria carotis mit der Doppler-Sonographie. Arch. Psychiat. Nervenkr. 222 (1976 b) 191–207

von Reutern, G. M., K. Voigt, E. Ortega-Suhrkamp, H. J. Büdingen: Dopplersonographische Befunde bei intrakraniellen vaskulären Störungen. Differentialdiagnose zur Obliteration der extrakraniellen Hirnarterien. Arch. Psychiat. Nervenkr. 223 (1977) 181–196

von Reutern, G. M., L. Pourcelot: Cardiac cycle dependent alternating flow in vertebral arteries with subclavian artery stenoses. Stroke 9 (1978) 229–236

von Reutern, G. M., P. Clarenbach: Valeur de l'exploration Doppler des collaterales cervicales et de l'ostium vertebral dans le diagnostic des stenoses et occlusions de l'artere vertebrale. Ultrasonics I (1980) 153–162

Ricotta, J. J., F. A. Bryan, M. G. Bond, A. Kurtz, D. H. O'Leary, J. K. Raines, A. S. Berson, M. E. Clouse, M. Calderon-Ortiz, J. F. Toole, J. A. DeWeese, S. N. Smullens, N. F. Gustafson: Multicenter validation study of real-time (B-Mode) ultrasound, arteriography, and pathologic examination. J. vasc. Surg. 6 (1987) 512–520

Ries, S., M. Daffertshofer, W. Steinke, M. G. Hennerici: Power amplitude duplex ultrasound imaging of vertebral artery dolichoectasia. Cerebrovascular Diseases 6 (1996) 374 (Stroke Vignette)

Ries, S., U. Schminke, M. Daffertshofer, M. Hennerici: High intensity transient signals (HITS) in patients with carotid artery disease. Europ. J. Med. Res. 1 (1995/96) 328–330

Ries, S., W. Steinke, G. Devuyst, N. Artemis, A. Valikovics, M. Hennerici: Amplitude modulated and frequency modulated color duplex flow imaging for the evaluation of normal and pathological vertebral arteries. Cerebrovascular Diseases 7 (Suppl. P4) (1977 b) 63

Riles, T. S., E. M. Eidelman, A. W. Litt, R. S. Pinto, F. Oldford, G. W. S. Thoe Schwartzenberg: Comparison of magnetic resonance angiography, conventional angiography, and duplex scanning. Stroke 23 (1992) 341–346

Riley, W. A., R. W. Barnes, W. B. Applegate, R. Dempsey, T. Hartwell, V. G. Davis, M. G. Bond, C. D. Furberg: Reproducibility of noninvasive ultrasonic measurement of carotid atherosclerosis. The asymptomatic carotid artery plaque study. Stroke, 23 (1992) 1062–1068

Ringelstein, E. B: Skepticism toward carotid ultrasonography: A virtue, an attitude, or fanaticism? Stroke 26 (1995) 1743–1746

Rittgers, St. E., B. M. Thornhill, R. W. Barnes: Quantitative analysis of carotid Doppler spectral waveforms: Diagnostic value of parameters. Ultrasound Med. Biol. 9 (1983) 255–264

Robinson, M. L., D. Sacks, G. S. Perlmutter, D. L. Marinelli: Diagnostic criteria for carotid duplex sonography. Amer. J. Roentgenol. 151 (1988) 1045–1049

Roederer, G. O., Y. E. Langlois, A. W. Chan, J. Primozich, R. A. Lawrence, P. M. Chikos, D. E. Strandness: Ultrasonic duplex scanning of extracranial carotid arteries: Improved accuracy using new features from the common carotid artery. J. cardiovasc. Ultrasound 1 (1982) 373–378

Rubin, D. C., G. D. Plotnick, M. W. Hawke: Intraaortic debris as a potential source of embolic stroke. Amer. J. Cardiol. 69 (1992) 819–820

Rubin, J. M., R. O. Bude, P. L. Carson, R. L. Bree, R. L. Adler: Power Doppler US: A potentially useful alternative to mean frequency-based color Doppler US. Radiology 199 (1994) 853–856

Rush, M., M. Thomas, J. Zyroff et al.: Duplex scanning with continuous-wave Doppler for carotid disease. J. clin. Ultrasound 13 (1985) 325–328

Salamon, G., G. Guerinel, F. Demard: Étude radioanatomique de l'artère carotide externe. Ann. Radiol. 11 (1968) 199

Sauve, J. S., K. E. Thorpe, D. L. Sackett, W. Taylor, H. J. Barnett, R. B. Haynes, A. J. Fox: Can bruits distinguish high-grade from moderate symptomatic carotid stenosis? The North American Symptomatic Carotid Endarterectomy Trial. Ann. Intern. Med. 120 (1994) 633–637

Schmid-Schönbein, H., K. Perktold: Physical factors in the pathogenesis of atheroma formation. In Caplan, L. R. ed: Brain Ischemia. Basic Concepts and Clinical Relevance. Springer, Berlin 1995

Sergott, R. C., P. M. Flaharty, W. E. Lieb, A. C. Ho, M. D. Kay, R. A. Mitra et al.: Color Doppler imaging identifies four syndromes of the retrobulbar calcification in patients with amaurosis fugax and central retinal artery occlusions. Trans. Am. Ophthalmol. Soc. (in press)

Seward, J. B., B. K. Khandheria, W. D. Edwards, J. K. Oh, W. K. Freeman, A. J. Tajik: Biplanar transesophageal echocardiography: Anatomic correlations, image orientation, and clinical applications. Mayo. Clin. Proc. 65 (1990) 1193–1213

Sheldon, C. D., J. A. Murie, R. O. Quin: Ultrasonic Doppler spectral broadening in the diagnosis of internal carotid artery stenosis. Ultrasound Med. Biol. 9 (1983) 575–580

Siebler, M., M. Sitzer, G. Rose, D. Bendfeldt, H. Steinmetz: Silent cerebral embolism caused by neurologically symptomatic high-grade carotid stenosis: Event rates before and after carotid endarterectomy. Brain 116 (1993) 1005–1015

Sillesen, H.: Carotid artery plaque composition – Relation to clinical presentation and ultrasound B-mode imaging. Europ. J. Endovasc. Surg. 10 (1995) 23–30

Sillesen, H., K. R. Bitsch, T. Schroeder et al.: Pulsed multigated Doppler ultrasonography in the diagnosis of carotid artery disease. Stroke 19 (1988) 846–851

Sitzer, M., G. Fürst, H. Fisher, M. Siebler, T. Fehlings, A. Kleinschmidt, T. Kahn, H. Steinmetz: Between-method correlation in quantifying internal carotid stenosis. Stroke 24 (1993) 1513–1518

Sitzer, M., G. Fürst, M. Siebler, H. Steinmetz: Usefulness of an intravenous contrast medium in the characterization of high-grade internal carotid stenosis with color Doppler-assisted duplex imaging. Stroke 25 (1994) 385–389

Sitzer, M., W. Müller, M. Siebler, W. Hort, H. W. Kniemeyer, L. Janke, H. Steinmetz: Plaque ulceration and lumen thrombus are the main sources of cerebral microemboli in high-grade internal carotid stenosis. Stroke 26 (1995) 1231–1233

Sliwka, U., W. Rautenberg, A. Schwartz, M. Hennerici: Multimodal ultrasound imaging of the vertebral circulation compared with intra-arterial angiography. J. Neurol. 239S (1992) 38

Spencer, M. P., J. M. Reid: Quantification of carotid stenoses with continuous wave (CW) Doppler ultrasound. Stroke 10 (1979) 326–330

Spencer, M., D. Whisler: Transorbital Doppler diagnosis of intracranial arterial stenosis. Stroke 17 (1986) 916–921

Steinke, W., A. Aulich, M. Hennerici: Diagnose und Verlauf von Carotisdissektionen. Dtsch. med. Wschr. 114 (1989) 1869–1875

Steinke, W., C. Kloetzsch, M. Hennerici: Carotid artery disease assessed by color Doppler flow imaging: Correlation with standard Doppler sonography and angiography. Amer. J. Neuroradiol. 11 (1990 a) 259–266

Steinke, W., C. Kloetzsch, M. Hennerici: Variability of flow patterns in the normal carotid bifurcation. Atherosclerosis 84 (1990 b) 121–127

Steinke, W., C. Kloetzsch, M. Hennerici: Carotid artery disease assessed by color Doppler flow imaging: Correlation with standard Doppler sonography and angiography. Amer. J. Neuroradiol. 154 (1990 c) 1061–1068

Steinke, W., C. Kloetzsch, M. Hennerici: Doppler color flow imaging after carotid endarterectomy. Europ. J. vasc. Surg. 5 (1991) 527–534

Steinke, W., T. Els, M. Hennerici: Comparison of flow disturbances in small carotid atheroma using a multi-gate pulsed Doppler system and Doppler color flow imaging. Ultrasound Med. Biol. 18 (1992 a) 11–18

Steinke, W., M. Hennerici, W. Rautenberg, J. P. Mohr: Symptomatic and asymptomatic high-grade carotid stenoses in Doppler color-flow imaging. Neurology 42 (1992 b) 131–138

Steinke, W., M. Hennerici: Three-dimensional ultrasound imaging of carotid artery plaques. J. cardiovasc. Technol. 8 (1989) 15–22

Steinke, W., S. Meairs, S. Ries, M. Hennerici: Sonographic assessment of carotid artery stenosis: Comparison of power Doppler imaging and color Doppler flow imaging. Stroke 27 (1996) 91–94

Steinke, W., W. Rautenberg, A. Schwartz, M. Hennerici: Noninvasive monitoring of internal carotid artery dissection. Stroke 25 (1994) 998–1005

Sterpetti, A. V., R. D. Schultz, R. J. Feldhaus: Ultrasonographic features of carotid plaque and the risk of subsequent neurological deficits. Surgery 104 (1988) 652–660

Terwey, B., H. Gabauer H.: Examination of the extracranial carotid artery with a high resolution B scanner. A comparison with carotid angiography. Fortschr. Röntgenstr. 135 (1981) 524–532

Terwey, B., H. Gahbauer, M. Montemayor, A. Proussalis, G. Zöllner: Die B-Bild Sonographie der Karotisbifurkation. Ultraschall 5 (1984) 190–201

Thomas, A. C., M. J. Davies, N. Dilly, P. Dilly, F. Franc: Potential errors in the estimation of coronary arterial stenoses from clinical arteriography with reference to the shape of the coronary arterial lumen. Brit. Heart J. 55 (1986) 129–139

Tismer, R., J. Böhlke: Die Anatomie der Carotisgabel – Ein Beitrag zur Real-Time-Sonographie der extracraniellen A. carotis. Ultraschall Klin. Prax., (Suppl. 1) (1986) 86 (abstract)

Touboul, P.-J., M.-G. Bousser, D. LaPlane, P. Castaigne: Duplex scanning of normal vertebral arteries. Stroke 17 (1986) 921–923

Touboul, P.-J., J. L. Mas, M. G. Bousser, D. Laplane: Duplex scanning in extracranial vertebral artery dissection. Stroke 18 (1987) 116–121

Trattnig, S., P. Hübsch, H. Schuster, D. Pölzleitner: Color-coded Doppler imaging of normal vertebral arteries. Stroke 21 (1990) 1222–1225

Trattnig, S., C. Matula, F. Karnel, K. Daha, M. Tschabitscher, B. Schwaighofer: Difficulties in examination of the origin of the vertebral artery by duplex and colour-coded Doppler sonography: Anatomical considerations. Neuroradiology 35 (1993) 296–299

Trockel, U., M. Hennerici, A. Aulich, W. Sandmann: The superiority of combined continuous wave Doppler examination over periorbital Doppler for the detection of extracranial carotid disease. J. Neurol. Neurosurg. Psychiat. 47 (1984 b) 43–50

Tunick, P. A., J. L. Perez, I. Kronzon: Protruding atheromas in the thoracic aorta and systemic embolization. Ann. Intern. Med. 115 (1991) 423–427

Valentine, R. J., S. I. Myers, R. T. Hagino, G. P. Clagett: Late outcome of patients with premature carotid atherosclerosis after carotid endarterectomy. Stroke 27 (1996) 1502–1506

Valton, L., V. Larrue, P. Arrue, G. Geraud, A. Bes: Asymptomatic cerebral embolic signals in patients with carotid stenosis: Correlation with appearance of plaque ulceration on angiography. Stroke 26 (1995) 813–815

Vanninen, R., H. Manninen, K. Koivisto, H. Tulla, K. Partanen, M. Puranen: Carotid stenosis by digital subtraction angiography: Reproducibility of the European Carotid Surgery Trial and the North American Symptomatic Carotid Endarterectomy Trial measurements, methods, and usual interpretation. Amer. J. Neuroradiol. 15 (1994) 1635–1641

Vanninen, R., H. Manninen, S. Soimakallio: Imaging of carotid artery stenosis: Clinical efficacy and cost-effectiveness. Amer. J. Neuroradiol. 16 (1995) 1875–1883

Vogelsang, H.: Über eine angiographisch selten nachzuweisende Anastomose zwischen dem A. carotis interna und dem A. carotis externa-Kreislauf. Nervenarzt 32 (1961) 518–520

Vollmar, J.: Rekonstruktive Chirurgie der Arterien, 2. Aufl. Thieme, Stuttgart 1975

Wernz, M.-G., A. Hetzel, B. Eckenweber, F. Beyersdorf: Pseudo-occlusion (PO) of the internal carotid artery (ICA): Carotid endarterectomy (CEA) without angiography? Cerebrovascular Diseases 7 (Suppl. 4) (1994) 1–88

Whiters, C. E., B. B. Gosink, A. M. Keightley, G. Casola, A. A. Lee, E. van Sonnenberg, J. F. Rotrock, P. D. Lyden: Duplex carotid sonography: Peak systolic velocity in quantifying internal carotid artery stenosis. J. Ultrasound Med. 9 (1990) 345–349

Widder, B.: Doppler-Sonographie der hirnversorgenden Arterien. Springer, Berlin 1985

Widder, B., G. Berger, J. Hackspacher: Reproduzierbarkeit sonographischer Kriterien zur Charakterisierung von Karotisstenosen. Ultraschall in Med. 11 (1990) 56–61

Widder, B., K. Paulat, J. Hachspacher: Morphological characterisation of carotid artery stenoses by ultrasound duplex scanning. Ultrasound Med. Biol. 16 (1990) 349–354

Widder, B., G. M. von Reutern, D. Neuerburg-Heusler: Morphologische und dopplersonographische Kriterien zur Bestimmung von Stenosierungsgraden an der A. carotis interna. Ultraschall 7 (1986b) 70–75

Widder, B., J. M. Friedrich, K. Paulat et al.: Bestimmung von Stenosierungsgraden bei Karotisstenosen: Ultraschall und iv-DSA im Vergleich zum Operationsbefund. Ultraschall Med. 8 (1987) 82–86

Winter, R., S. Bieden, T. Staudacher, H. Betz, R. Reuther: Vertebral artery Doppler sonography. Europ. Arch. Psychiat. Neurol. Sci. 237 (1987) 21–28

Withers, C. E., B. B. Gosink, A. M. Keightley et al.: Duplex carotid sonography. Peak systolic velocity in quantifying internal carotid artery stenosis. J. Ultrasound Med. 9 (1990) 345–349

Wolverson, M. K., E. Heiberg, M. Sundaram et al.: Carotid atherosclerosis: High-resolution real-time sonography correlated with angiography. Amer. J. Roentgenol. 140 (1983) 355–361

Young, G. R., P. R. D. Humphrey: Skepticism and carotid ultrasonography. Stroke 27 (1996) 768–769

Zbornikova, V., C. Lassvik, I. Johannson I.: Prospective evaluation of the accuracy of duplex scanning with spectral analysis in carotid artery disease. Clin. Physiol. 5 (1985a) 257–269

Zbornikova, V., C. Lassvik, I. Johansson: Duplex scanning and periorbital pulsed Doppler in the diagnosis of external carotid artery disease: Analysis of causes of error. Clin. Physiol. 5 (1985b) 271–279

Zbornikova, V., C. Lassvik: Duplex scanning in presumably normal persons of different ages. Ultrasound Med. Biol. 12 (1986) 371–378

Zeitler, E., H. W. Greiling, H. J. Roth, G. Friedmann: Computertomographie, B-Scan-Sonographie und cerebrale Angiographie bei Carotis-Obliterationen. Dtsch. med. Wschr. 105 (1980) 715–719

Zierler, R. E., D. J. Phillips, E. W. Beach, J. F. Primozich, D. E. Strandness: Noninvasive assessment of normal carotid bifurcation hemodynamics with color-flow ultrasound imaging. Ultrasound Med. Biol. 13 (1987) 471–476

Zwiebel, W. J., C. W. Austin, J. F. Sackett et al.: Correlation of high-resolution B-mode and continous-wave Doppler with arteriography in the diagnosis of carotid stenosis. Radiology 149 (1983) 523–532

Zwiebel, W. J.: Cerebrovascular Ultrasound Doppler and B-mode Techniques. Yearbook, Chicago 1986

Chapter 3

Aaslid, R., T.-M. Markwalder, H. Nornes: Non-invasive transcranial Doppler ultrasound recording of flow velocity in basal cerebral arteries. J. Neurosurg. 57 (1982) 769–774

Aaslid, R., P. Huber, H. Nornes: Evaluation of Cerebrovascular spasm with transcranial Doppler ultrasound. J. Neurosurg. 60 (1984) 37–41

Aaslid, R.: Visually evoked dynamic blood flow response of the human cerebral circulation. Stroke 17 (1987) 771–775

Aaslid, R., K.-F. Lindegaard, W. Sorteberg, H. Nornes: Cerebral autoregulation dynamics in humans. Stroke 20 (1989) 45–52

Arnolds, B. J., G.-M. von Reutern: Transcranial Doppler sonography. Examination technique and normal reference values. Ultrasound Med. Biol. 12 (1986a) 115–123

Arnolds, B. J. A., M. Ochme, G.-M. von Reutern: Detection of intracranial stenosis and occlusion with transcranial Doppler sonography. J. cardiovasc. Ultrasonogr. 5 (1986b) 4 (abstract)

Babikian, V. L., C. Hyde, V. Pochay, M. R. Winter: Clinical correlates of HITS detected on transcranial Doppler sonography in patients with cerebrovascular disease. Stroke 25 (1994) 1570–1573

Baumgartner, R. W., I. B. Baumgartner: Transcranial Doppler and color-coded duplex ultrasound. Ultraschall Med. 17 (1996) 50–54

Baumgartner, R. W., G. Schroth, K. Kothbauer, M. Sturzenegger, H. P. Mattle: Transcranial color-coded duplex sonography in cerebral aneurysms. Stroke 25 (1994) 749 (abstract)

Baumgartner, R. W., J. Mathis, M. Sturzenegger, H. P. Mattle: A validation study on the intraobserver reproducibility of transcranial color-coded duplex sonography velocity measurements. Ultrasound Med. Biol. 20 (1994) 233–237

Baumgartner, R. W., H. P. Mattle, G. Schroth: Transcranial colour-coded duplex sonography of cerebral arteriovenous malformations. Neuroradiology 38 (1996) 734–737

Baumgartner, R. W., C. Schmid, I. Baumgartner: Comparative study of power-based vs. mean frequency-based transcranial color-coded duplex sonography in normal adults. Stroke 27 (1996) 101–104

Baumgartner, R. W., H. P. Mattle, R. Aaslid, M. Kaps: Transcranial color-coded duplex sonography in arterial cerebrovascular disease. Cerebrovascular Diseases 7 (1997) 57–63

Bäzner, H., S. Ries, M. Daffertshofer, M. Hennerici: Localizing the emboligenic focus – documentation of microemboli generated across an MCA stenosis by a new multi-gate technology. Cerebrovascular Diseases 6 (Suppl. 3) (1996) 62 (abstract)

Becker, V. H., B. Eckert, A. Thie: Isolated symptomatic stenosis of the middle cerebral artery in younger adults. Europ. Neurol. 36 (1996) 65–70

Becker, V.-U., H. C. Hansen, U. Brewitt, A. Thie: Visually evoked cerebral blood flow velocity changes in different states of brain dysfunction. Stroke 27 (1996) 446–449

Bishop, C. C. R., S. Powell, M. Insall, D. Rutt, N. L. Brownse: Effect of internal carotid artery occlusion on middle cerebral artery blood flow at rest and in response to hypercapnia. Lancet 1986/I, 710–712

Boeri, R., A. Passerini: The megadolichobasilar anomaly. J. neurol. Sci. 1 (1964) 475–484

Bogdahn, W., G. Becker, J. Winkler, D. Greiner, J. Perez, B. Meurers: Transcranial color-coded real-time sonography in adults. Stroke 21 (1990) 1680–1688

Bogdahn, U., G. Becker, R. Schlief, J. Reddig, W. Hassel: Contrast-enhanced transcranial color-coded real-time sonography. Stroke 24 (1993) 676–684

Büdingen, H. J., T. Staudacher: Die Identifizierung der A. basilaris mit der transkraniellen Doppler-Sonographie. Ultraschall 8 (1987) 95–101

Büdingen, H. J., G.-M. von Reutern: Ultraschalldiagnostik der hirnversorgenden Arterien. Thieme, Stuttgart 1989; 2. Aufl. 1993

Busch, W.: Beitrag zur Morphologie und Pathologie der Arteria basilaris (Untersuchungsergebnisse bei 1000 Gehirnen). Arch. Psychiat. Nervenkr. 208 (1966) 326–344

Conrad, B., J. Klingelhöfer: Dynamics of regional cerebral blood flow for various visual stimuli. Exp. Brain. Res. 77 (1989) 437–441

Daffertshofer, M., R. R. Diehl, G.-U. Ziems, M. Hennerici: Orthostatic changes of cerebral blood flow velocity in patients with autonomic dysfunction. J. neurol. Sci. 104 (1991) 32–38

Daffertshofer, M., S. Ries, U. Schminke, M. Hennerici: High-intensity transient signals in patients with cerebral ischemia. Stroke 27 (1996) 1844–1849

De Bray, J. M., B. Glatt for the International Consensus Conference: Quantification of atheromatous stenosis in the extracranial internal carotid artery. Cerebrovascular Diseases 5 (1995) 414–426

Démolis, P., Y. R. Tran Dinh, J.-F. Giudicelli: Relationships between cerebral regional blood flow velocities and volumetric blood flows and their respective reactivities to acetazolamide. Stroke 27 (1996) 1835–1839

Di Tullio, M., R. L. Sacco, N. Venketasubramanian, D. Sherman, J. P. Mohr, S. Homma: Comparison of diagnostic techniques for the detection of a patent foramen ovale in stroke patients. Stroke 24 (1993) 1020–1024

Diehl, R. R., B. Diehl, M. Sitzer, M. Hennerici: Spontaneous oscillations in cerebral blood flow velocity in normal humans and in patients with carotid artery disease. Neurosci. Lett. 127 (1991) 5–8

Droste, D. W., A. G. Harders, E. Rastogi: A transcranial Doppler study of blood flow velocity in the middle cerebral arteries performed at rest and during mental activities. Stroke 20 (1989) 1005–1011

Droste, D. W., T. Hansberg, V. Kemény, D. Hammel, G. Schulte-Altedorneburg, D. G. Nabavi, M. Kaps, H. H. Scheld, E. B. Ringelstein: Oxygen inhalation can differentiate gaseous from non-gaseous microemboli detected by transcranial Doppler ultrasound Cerebrovascular Diseases 7 (Suppl. 4) (1997) 13 (abstract)

Eicke, B. M., C. H. Tegeler, G. Dalley, L. G. Myers: Angle correction in transcranial Doppler sonography. J. Neuroimag. 4 (1994) 29–33

Eng, C. C., A. Lam, S. Byrd et al.: The diagnosis and management of a perianesthetic cerebral aneurysmal rupture aided with transcranial Doppler ultrasonography. Anesthesiology 78 (1993) 191–194

Fischer, E.: Die Lageabweichung der vorderen Hirnarterie im Gefäßbild. Zbl. Neurochir. 3 (1938) 300–313

Fisher, C. M.: The circle of Willis. Anatomical variations. Vasc. Dis. 2 (1965) 99–105

Fujii, K., D. D. Heistad, F. F. Faraci: Flow-mediated dilatation of the basilar artery in vivo. Circulat. Res. 69 (1991) 697–705

Gerraty, R. P., D. N. Bowser, B. Infeld, P. J. Mitchell, S. M. Davis: Microemboli during carotid angiography. Association with stroke risk factors or subsequent magnetic resonance imaging changes? Stroke 27 (1996) 1543–1547

Giller, C. A.: Is angle correction correct? J. Neuroimag. 4 (1994) 51–52

Harders, A.: Neurosurgical Applications of Transcranial Doppler Sonography. Springer, Wien 1986

Harders, A. G., J. M. Gilsbach: Time course of blood velocity changes related to vasospasm in the circle of Willis measured by transcranial Doppler ultrasound. J. Neurosurg. 66 (1987) 745–751

Hassler, W.: Hemodynamic Aspects of Cerebral Angiomas. Springer, Wien 1986

Hedges, T. R.: Occular Ischemia. In Caplan, L. R., ed: Brain Ischemia. Basic Concepts and Clinical Relevance. Springer, Berlin 1995

Hennerici, M., W. Rautenberg, A. Schwartz: Transcranial Doppler ultrasound for the assessment of intracranial arterial flow velocity. Part I. Evaluation of intracranial arterial disease. Surg. Neurol. 27 (1987a) 523–532

Hennerici, M., W. Rautenberg, G. Sitzer, A. Schwartz: Transcranial Doppler ultrasound for the assessment of intracranial arterial flow velocity. Part I. Examination technique and normal values. Surg. Neurol. 27 (1987b) 439–448

Hennerici, M., C. Klemm, W. Rautenberg: The subclavian steal phenomenon: A common vascular disorder with rare neurologic deficits. Neurology 38 (1988) 669–673

Hennerici, M., J. P. Mohr, W. Rautenberg, W. Steinke: Ultrasound imaging and Doppler sonography in the diagnosis of cerebrovascular diseases. In Barnett, H. J. M., J. P. Mohr, F. M. Yatsu, eds: Stroke. Churchill Livingstone, Edinburgh 1992

Hennerici, M.: Can carotid endarterectomy be improved by neurovascular monitoring? Stroke 24 (1993) 637–638

Hennerici, M. G., M. Daffertshofer: Noninvasive vascular testing. In Fisher, M., J. Bogousslavsky, eds: Current Review of Cerebrovascular Disease. 1993 (pp. 121–137)

Hennerici, M. G.: High intensity transcranial signals (HITS): a questionable 'jackpot' for the prediction of stroke risk. J. Heart Valve Dis. 3 (1994) 124–125

Hennerici, M.: High intensity transient signals (HITS): Evolution or revolution in understanding cerebral embolism? Europ. Neurol. 35 (1995) 249–253

Herman, L. H., A. Z. Ostrowski, E. S. Guardian: Perforating branches of the middle cerebral artery: An anatomical study. Arch. Neurol. 8 (1963) 32–34

Huber, P., J. Handa: Effect of contrast material, hypercapnia, hyperventilation, hypertonic glucose and papaverine on the diameter of the cerebral arteries: Angiographic determination in man. Invest. Radiol. 2 (1967) 17–32

Jain, K. K.: Some observations on the anatomy of the middle cerebral artery. Canad. J. Surg. 7 (1964) 134–139

Job, F. P., A. Grafen, F. A. Flachskampf, E. B. Ringelstein, P. Hanrath: Stellenwert der transkraniellen Kontrastdopplersonographie in der Diagnostik des klinisch relevanten offenen Foramen ovale (PFO) bei der kardialen Emboliequellensuche. Z. Kardiol. 82, S1 (1993) 123

Kenton, A. R., P. J. Martin, D. H. Evans: Power Doppler: An advance over colour Doppler for transcranial imaging? Ultrasound Med. Biol. 22 (1996) 313–317

Klötzsch, C., O. Popescu, P. Berlit: Assessement of the posterior communicating artery by transcranial color-coded duplex sonography. Stroke 27 (1996) 486–489

Kontos, H. A.: Validity of cerebral arterial blood flow calculations from velocity measurements. Stroke 20 (1989) 1–3

Krayenbühl, H., M. G. Yasargil: Die vaskulären Erkrankungen im Gebiet der Arteria vertebralis und Arteria basilaris. Thieme, Stuttgart 1957

Krayenbühl, H., M. G. Yasargil, P. Huber: Zerebrale Angiographie für Klinik und Praxis, 3. Aufl. Thieme, Stuttgart 1979

Kushner, M. J., A. Rosenquist, A. Alavi, M. Rosen, R. Dann, F. Fazekas, T. Bosley, J. Greenberg, M. Reivich: Cerebral metabolism and patterned visual stimulation: A positron emission tomographic study of the human visual cortex. Neurology 38 (1988) 89–95

Lang, J.: Clinical Anatomy of the Posterior Cranial Fossa and its Foramina. Thieme, Stuttgart, New York 1991

Laumer, R., R. Steinmeier, F. Gönner, T. Vogtmann, R. Priem, R. Fahlbusch: Cerebral hemodynamics in subarachnoid hemorrhage evaluated by transcranial Doppler sonography. Part 1. Reliability of flow velocities in clinical management. Neurosurgery 33 (1993) 1–9

Lechat, P., J. L. Mas, G. Lascault, P. Loron, M. Theard, M. Klimczak, G. Drobinski, D. Thomas, Y. Grosgoeat: Prevalence of patent foramen ovale in patients with stroke. New Engl. J. Med. 318 (1988) 1148–1152

Lennihan, L., G. W. Petty, M. E. Fink, R. A. Solomon, J. P. Mohr: Transcranial Doppler detection of anterior cerebral artery vasospasm. J. Neurol. Neurosurg. Psychiat. 56 (1993) 906–909

Ley-Pozo, J., E. B. Ringelstein: Noninvasive detection of occlusive disease of the carotid siphon and middle cerebral artery. Ann. Neurol. 28 (1990) 640–647

Lindegaard, K.-F., S. J. Bakke, P. Grolimund, R. Aaslid, P. Huber, H. Nornes: Assessment of intracranial hemodynamics in carotid artery disease by transcranial Doppler ultrasound. J. Neurosurg. 63 (1985) 890–898

Lindegaard, K.-F., S. J. Bakke, R. Aaslid, H. Nornes: Doppler diagnosis of intracranial artery occlusive disorders. J. Neurol. Neurosurg. Psychiat. 49 (1986) 510–518

Markus, H. S., M. J. G. Harrison: Estimation of cerebrovascular reactivity using transcranial Doppler, including the use of breathholding as the vasodilatory stimulus. Stroke 23 (1992) 668–673

Markus, H. S., M. M. Brown: Differentiation between different pathological cerebral embolic materials using transcranial Doppler in an in vitro model. Stroke, 24 (1993) 1–5

Markus, H., J. M. Bland, G. Rose, M. Sitzer, M. Siebler: How good is intercenter agreement in the identification of embolic signals in carotid artery disease? Stroke 27 (1996) 1249–1252

Martin, P. J., M. E. Gaunt, A. R. Naylor, D. T. Hope, V. Orpe, D. H. Evans: Intracranial aneurysms and arteriovenous malformations: Transcranial colour-coded sonography as a diagnostic aid. Ultrasound in Med. Biol. 20 (1994) 689–698

Mess, W. H., B. M. Titulaer, R. G. A. Ackerstaff: An in vivo model to detect microemboli with multidepth technique, preliminary results. Cerebrovascular Diseases 6 (Suppl. 3) (1996) 60 (abstract)

von Mitterwallner, F.: Variationsstatistische Untersuchungen an den basalen Hirngefäßen. Acta anat. 24 (1955) 51–88

Mull, M., A. Aulich, M. Hennerici: Transcranial Doppler ultrasound for the assessment of intracranial arterial flow velocity. Transcranial Doppler ultrasonography versus arteriography for assessment of the vertebrobasilar circulation. J. clin. Ultrasound 18 (1990) 539–549

Müller, H. R., M. Casty, R. Moll, R. Zehnder: Response of middle cerebral artery volume flow to orthostasis. Cerebrovascular Diseases 1 (1991) 82–89

Nabavi, D. G., D. Georgiadis, T. Mumme, C. Schmid, T. G. Mackay, H. H. Scheld, E. B. Ringelstein: Clinical relevance of intracranial microembolic signals in patients with left ventricular assist devices. Stroke 27 (1996) 891–896

Neff, K. W., K. J. Lehmann, S. Ries, A. Sommer, W. Steinke, A. Schwartz, M. Georgi, M. Hennerici: CTA of middle cerebral artery stenosis and occlusion: Is CTA as valid as MRA and TCD? Cerebrovascular Diseases 7 (Suppl. 4) (1997) 14 (abstract)

Nikutta, P., M. Schneider, G. Claus, S. M. Schellong, H. Kühn, D. Hausmann, A. Mügge, W. G. Daniel: Wie zuverlässig ist der transkranielle Dopplerultraschall in der Diagnostik des offenen Foramen ovale? Z. Kardiol. 82, S1 (1993) 123

Otis, S., M. Rush, R. Boyajian: Contrast-enhanced transcranial imaging. Results of an American phase-two study. Stroke 26 (1995) 203–209

Padget, D. H.: The Circle of Willis, Its Embryology and Anatomy. Comestock, Ithaca (N. Y.) 1944

Piepgras, A., P. Schmiedek, G. Leinsinger, R. L. Haberl, C. M. Kirsch, K. M. Einhäupl: A simple test to assess cerebrovascular reserve capacity using transcranial Doppler sonography and acetazolamide. Stroke 21 (1990) 1306–1311

Postert, T., T. Büttner, S. Meves, C. Börnke, H. Przuntek: Power Doppler compared to color-coded duplex sonography in the assessment of the basal cerebral circulation. J. Neuroimag. (1997) (In print)

Poulin, M. J., P. A. Robbins: Indexes of flow and cross-sectional area of the middle cerebral artery using Doppler ultrasound during hypoxia and hypercapnia in humans. Stroke 27 (1996) 2244–2250

President's Commission: Guidelines for the determination of brain death. J. Amer. med. Assoc. 246 (1981) 2184–2187

Rautenberg, W., A. Schwartz, M. Hennerici: Transkranielle Doppler-Sonographie während der zerebralen Angiographie. In Widder, B.: Transkranielle Doppler-Sonographie bei zerebrovaskulären Erkrankungen. Springer, Berlin 1987

Rautenberg, W., M. Hennerici: Pulsed Doppler assessment of innominate artery obstructive diseases. Stroke 19 (1988) 1514–1520

Rautenberg, W., A. Schwartz, M. Mull, A. Aulich, M. Hennerici: Noninvasive detection of intracranial stenoses and occlusions. Stroke 21 (1990) I 49

von Reutern, M.: Zerebraler Zirkulationsstillstand: Diagnostik mit der Dopplersonographie. Dt. Ärztebl. 88 (1991) B-2842–B-2848

Ries, S., W. Steinke, K. W. Neff, M. Hennerici: Echocontrast-enhanced transcranial color-coded sonography for the diagnosis of transverse sinus venous thrombosis. Stroke 28 (1997) 696–700

Ringelstein, E. B., W. Grosse, M. Matentzoglu, W. M. Glöckner: Noninvasive assessment of the cerebral vasomotor reactivity by means of transcranial Doppler sonography during hyper- and hypocapnia. Klin. Wschr. 64 (1986) 194–195

Rosenkranz, K., W. Zendel, R. Langer, T. Heim, P. Schubeus, A. Scholz, R. Schlief, R. Schurmann, R. Felix: Contrast-enhanced transcranial Doppler US with a new transpulmonary echo contrast agent based on saccharide microparticles. Radiology 187 (1993) 439–443

Röther, J., K.-U. Wentz, W. Rautenberg, A. Schwartz, M. Hennerici: Magnetic resonance angiography in vertebrobasilar ischemia. Stroke 24 (1993) 1310–15

Röther, J., A. Schwartz, K. U. Wentz, W. Rautenberg, M. Hennerici: Middle cerebral artery stenoses: Assessment by MRA and TDU. Cerebrovascular Diseases 4 (1994) 273–279

Santosh, C. G., J. E. Rimmington, J. J. K. Best. Functional magnetic resonance imaging at 1 T: Motor cortex, supplementary motor area and visual cortex activation. Brit. J. Radiol. 68 (1995) 369

Schmid-Schönbein, H., K. Perktold: Physical factors in the pathogenesis of atheroma formation. In Caplan, L. R., ed: Brain Ischemia. Basic Concepts and Clinical Relevance. Springer, Berlin 1995

Schminke, U., S. Ries, M. Daffertshofer, U. Staedt, M. Hennerici: Patent foramen ovale: A potential source of cerebral embolism? Cerebrovascular Diseases 5 (1995) 133–138

Schwartz, A., M. Hennerici: Transkranielle Dopplersonographie bei hochsitzenden extrakraniellen Karotisprozessen. In Otto, R. Ch., P. Schnaars: Ultraschalldiagnostik 85. Thieme, Stuttgart 1986 a

Schwartz, A., M. Hennerici: Non-invasive transcranial Doppler ultrasound in intracranial angiomas. Neurology 36 (1986 b) 626–635

Schwartz, A., W. Rautenberg, M. Hennerici: Dolichoectatic intracranial arteries: Review of selected aspects. Cerebrovascular Diseases 3 (1993) 273–279

Sedzimir, C. B.: An angiographic test of collateral circulation through the anterior segment of the circle of Willis. J. Neurol. Neurosurg. Psychiat. 22 (1959) 64–68

Seidel, G., M. Kaps, T. Gerriets: Potential and limitations of transcranial color-coded sonography in stroke patients. Stroke 26 (1995) 2061–2066

Seiler, R. W., P. Grolimund, R. Aaslid, P. Huber, H. Nornes: Cerebral vasospasm evaluated by transcranial ultrasound correlated with clinical grade and CT-visualized subarachnoid hemorrhage. J. Neurosurg. 64 (1986) 594–600

Sitzer, M., R. R. Diehl, M. Hennerici: Visually evoked cerebral blood flow responses: Normal and pathological conditions. J. Neuroimag. 2 (1992) 65–70

Spencer, M. P., G. I. Thomas, S. C. Nicholls, L. R. Sauvage: Detection of middle cerebral artery emboli during carotid endarterectomy using transcranial Doppler ultrasonography. Stroke 21 (1990) 415–423

Sunderland, S.: Neurovascular relations and anomalies at the base of the brain. J. Neurol. Neurosurg. Psychiat. 11 (1948) 243–254

Teague, S. M., J. K. Sharma: Detection of paradoxical cerebral echo contrast embolization by transcranial Doppler ultrasound. Stroke 22 (1991) 740–745

Tönnis, W., W. Schiefer: Zirkulationsstörungen des Gehirns im Serienangiogramm. Springer, Berlin 1959

Wardlaw, J. M., J. C. Cannon: Color transcranial power Doppler ultrasound of intracranial aneurysms. J. Neurosurg. 84 (1996) 459–461

Widder, B., K. Paulat, J. Hackspacher, E. Mayr: Trancranial Doppler CO_2 test for the detection of hemodynamically critical carotid artery stenoses and occlusions. Europ. Arch. Psychiat. Neurol. Sci. 236 (1986 a) 162–168

Wollschläger, P. B., B. Wollschläger: The anterior cerebral/internal carotid and middle cerebral/internal carotid arteries. Acta radiol., Diagn. 5 (1966) 615–620

Wollschläger, B., P. B. Wollschläger: The circle of Willis. In Newton, T. H., D. Potts, eds: Radiology of the Skull and Brain, Angiography, Vol. II. Mosby, St. Louis 1974

Zanette, E. M., G. Mancini, S. De Castro, M. Solaro, D. Cartoni, F. Chiarotti: Patent foramen ovale and transcranial Doppler. Comparison of different procedures. Stroke 27 (1996) 2251–2255

Zuilen van, E. V., W. H. Mess, C. Jansen, I. van der Tweel, J. van Gijn, R. G. A. Ackerstaff: Automatic embolus detection compared with human experts. A Doppler ultrasound study. Stroke 27 (1996) 1840–1843

Zuilen van, E. V., F. L. Moll, F. E. E. Vermeulen, H. W. Mauser, J. van Gijn, R. G. A. Ackerstaff: Detection of cerebral microemboli by means of transcranial Doppler monitoring before and after carotid endarterectomy. Stroke 26 (1995) 210–213

Chapter 4

Aaslid, R., D. W. Newel, R. Stoos, W. Sorteberg, K.-F. Lindegaard: Assessment of cerebral autoregulation dynamics from simultaneous arterial and venous transcranial Doppler recordings in humans. Stroke 22 (1991) 1148–1154

Albertyn, L. E., M. K. Alcock: Diagnosis of internal jugular vein thrombosis. Radiology 162 (1987) 505–508

Andeweg, J.: The anatomy of collateral venous flow from the brain and its value in aetiological interpretation of intracranial pathology. Neuroradiology 38 (1996) 621–628

Baumgartner, I., A. Bollinger: Zur diagnostischen Bedeutung der Jugularvenen. Vasa 20 (1991) 3–9

Becker, G., U. Bogdahn, C. Gehlberg, T. Fröhlich, E. Hofmann, R. Schlief: Transcranial color-coded real-time sonography on intracranial veins. Normal values of blood flow velocities and findings in superior sagittal sinus thrombosis. J. Neuroimag. 5 (1994) 87–94

Bloching, H., J. A. Reuss, K. Seitz, G. Rettenmaier: Thromboses of the subclavian vein and jugular vein and superior vena cava sonographic diagnosis and control of treatment using tissue-type plasminogen activator therapy. Ultraschall. Med. 10 (1989) 314–317

Bogdahn, U., G. Becker, R. Schlief, J. Reddig, W. Hassel: Contrast-enhanced transcranial color-coded real-time sonography: Results of a phase-two study. Stroke 24 (1993) 676–684

Bogdahn, U., G. Becker, A. Bauer, P. Jachimczak, A. Krone, R. Schlief: Functional imaging of the cerebral venous system by contrast-enhanced transcranial color-coded real-time sonography (TCCS). Neurology 45 (Suppl. 4 A) (1995) 225 (abstract)

Brownlow jr., R. L., W. M. McKinney: Ultrasonic evaluation of jugular venous valve competence. J. Ultrasound Med. 4 (1985) 169–172

Hennerici, M.: Ultrasound diagnosis of cerebrovenous flow disturbances. In Einhäupl, K., O. Kempski, A. Baethmann, eds: Cerebral Sinus Thrombosis: Experimental and Clinical Aspects. Plenum Press, New York, London 1990

Krünes, U., H. P. Dübel, P. Romaniuk, H. Warnke, U. Engelmann: Duplexsonographische Untersuchungen an häufig punktierten Venen bei herztransplantierten Patienten. Ultraschall Klin. Prax. 7 (1992) 147

Mallory, D. L., W. T. McGee, T. H. Shawker, M. Brenner, K. R. Bailey, R. G. Evans, M. M. Parker, J. C. Farmer, J. E. Parillo: Ultrasound guidance improves the success rate of internal jugular vein cannulation. A prospective, randomized trial. Chest 98 (1990) 157–160

Mattle, H. P., K. U. Wentz, R. R. Edelmann, B. Wallner, J. P. Finn, P. Barnes, D. J. Atkinson, J. Kleefield, H. M. Hoogewoud: Cerebral venography with MR. Radiology 178 (1991) 453–458

Müller, H. R., M. Casty, M. Buser, M. Haefele: Ultrasonic jugular venous flow measurement. J. cardiovasc. Ultrasonogr. 7 (1988) 25

Ries, S., W. Steinke, K. W. Neff, M. Hennerici: Echocontrast-enhanced transcranial color-coded sonography for the diagnosis of transverse sinus venous thrombosis. Stroke 28 (1997) 696–700

Steinke, W., M. Hennerici: Duplexsonographie der Vena jugularis interna. Ultraschall 10 (1989) 72–76

Terwey, B., C. Krier, P. Gerhardt: Die Darstellung der Jugularvenenthrombose mit Hilfe des hochauflösenden Ultraschallverfahrens. Fortschr. Röntgenstr. 134 (1981) 557–559

Valdueza, J. M., K. Schmierer, S. Mehraein, K. M. Einhäupl: Assessment of normal flow velocity in basal cerebral veins. A transcranial Doppler ultrasound study. Stroke 27 (1996) 1221–1225

Valdueza, J. M., M. Schultz, L. Harms, K. M. Einhäupl: Venous transcranial Doppler ultrasound monitoring in acute dural sinus thrombosis: Report of two cases. Stroke 26 (1995) 1196–1199

Wardlaw, J. M., G. T. Vaughan, A. J. W. Steers, R. J. Sellar: Transcranial Doppler ultrasound findings in venous sinus thrombosis. J. Neurosurg. 80 (1994) 332–335

Chapter 5

Abu-Yousef, M. M., J. A. Wiese, A. R. Shamma: The 'to-and-fro' sign: Duplex Doppler evidence of femoral artery pseudoaneurysm. Amer. J. Roentgenol. 150 (1988) 632–634

Allard, L., Y. E. Langlois, L.-G. Durand, G. O. Roederer, M. Beaudoin, G. Cloutier, P. Roy, P. Robillard: Computer analysis and pattern recognition of Doppler blood flow spectra for disease classification in the lower limb arteries. Ultrasound Med. Biol. 17 (1991) 211–223

Allard, L., G. Cloutier, L.-G. Durand, G. O. Roederer, Y. E. Langlois: Limitations of ultrasonic duplex scanning for diagnosing lower limb arterial stenoses in the presence of adjacent segment disease. J. vasc. Surg. 19 (1994) 650–657

Baker, A. R., D. R. Prytherch, D. H. Evans, P. R. F. Bell: Doppler ultrasound assessment of the femoropopliteal segment: Comparison of different methods using roc curve analysis. Ultrasound Med. Biol. 12 (1986) 473–482

Bandyk, D. F.: Monitoring during and after distal aterial reconstruction. In Bernstein, E. F., ed: Vascular Diagnosis. Mosby, St. Louis

Bandyk, D. F., R. F. Cato, J. B. Towne: A low flow velocity predicts failure of femoropopliteal and femorotibial bypass grafts. Surgery 98 (1985) 799–809

Bandyk, D. F., G. R. Seabrook, P. Moldenhauer, J. Lavin, J. Edwards, R. Cato, J. B. Towne: Hemodynamics of vein graft stenosis. J. vasc. Surg. 8 (1988) 688–695

Baumgartner, I., S. E. Maier, M. Koch, E. Schneider, G. K. von Schulthess, A. Bollinger: Magnetresonanzarteriographie, Duplexsonographie und konventionelle Arteriographie zur Beurteilung der peripheren Verschlußkrankheit. Fortschr. Röntgenstr. 159 (1993) 167–173

Baumgartner, I., I. Zwahlen, D.-D. Do, F. Redha, F. Mahler: Color-coded duplex sonography for evaluation of femoro-popliteal restenosis after percutaneous catheter atherectomy and subsequent transluminal balloon angioplasty. J. vasc. Investig. 2 (1996) 125–130

Bernstein, E. F.: Vascular Diagnosis. Mosby, St. Louis 1993

Bollinger, A., F. Mahler, O. Zehender: Kombinierte Druck- und Durchflußmessungen in der Beurteilung arterieller Durchblutungsstörungen. Dtsch. med. Wschr. 95 (1970) 1039–1043

Bollinger, A., M. Schlumpf, P. Butti, A. Grüntzig: Measurement of systolic ankle blood pressure with Doppler Ultrasound at rest and after exercise in patients with leg artery occlusions. Scand. J. clin. Lab. Invest. 31 (Suppl. 128) (1973) 123–127

Bollinger, A., J. P. Barras, F. Mahler: Measurement of foot artery blood pressure by micromanometry in normal subjects and in patients with arterial occlusive disease. Circulation 53 (1976) 506–512

Bowers, B. L., J. Valentine, S. I. Myers, A. Chervu, G. P. Clagett: The natural history of patients with claudication with toe pressures of 40 mmHg or less. J. vasc. Surg. 18 (1993) 506–511

Buchholz, J., F. Scherf, H. J. Bücker-Nott, J. Hilgenberg: Das iatrogene femorale Pseudoaneurysma – eine dringliche Operationsindikation. Vasa 20 (1991) 261–266

Busse, R., E. Wetterer, R. D. Bauer, Th. Pasch, Y. Summa: The genesis of pulse contours of the distal leg arteries in man. Pflügers Arch. 360 (1975) 63–79

Busse, R.: Kreislaufphysiologie. Thieme, Stuttgart 1982

Buth, J., B. Disselhoff, C. Sommeling, L. Stam: Color-flow duplex criteria for grading stenosis in infrainguinal vein grafts. J. vasc. Surg. 14 (1991) 716–728

Cachovan, M., U. Maass: Vergleichende Untersuchungen zur Beurteilung der Gehstrecke bei Claudicatio intermittens. In Mahler, F., B. Nachbur: Zerebrale Ischämie. Huber, Bern 1984

Carter, S. A.: Clinical measurement of systolic pressures in limbs with arterial occlusive disease. J. Amer. med. Ass. 207 (1969) 1869–1874

Carter, S. A.: Response of ankle systolic pressure to leg exercise in mild or questionable arterial disease. New Engl. J. Med. 287 (1972) 578–582

Carter, S. A.: Role of pressure measurements. In Bernstein, E. F., ed: Vascular Diagnosis. Mosby, St. Louis 1993

Cluley, S. R., B. J. Brener, L. Hollier, R. Schoenfeld, A. Novick, D. Vilkomerson, M. Ferrara-Ryan, V. Parsonnet: Transcutaneous ultrasonography can be used to guide and monitor balloon angioplasty. J. vasc. Surg. 17 (1993) 23–31

Cobet, U., R. Scharf, A. Klemenz, R. Millner, G. Blumenstein, E. Wiegand: Rechnergestütztes, richtungssensitives Ultraschall-Dopplersystem zur quantitativen Blutflußcharakterisierung in Arterien. Z. klin. Med. 41 (1986) 523–527

Coleman, S. S., B. J. Anson: Arterial patterns in the hand based upon a study of 650 specimens. Surg. Gynecol. Obstet. 113 (1961) 409–424

Cossman, D. V., J. E. Ellison, W. H. Wagner, R. M. Carroll, R. L. Treiman, R. F. Foran, P. M. Levin, J. L. Cohen: Comparison of contrast arteriography to arterial mapping with color-flow duplex imaging in the lower extremities. J. vasc. Surg. 10 (1989) 522–529

Coughlin, B. F., D. M. Paushter: Peripheral pseudoaneurysms: Evaluation with Duplex US. Radiology 168 (1988) 339–342

Dean, S. M., J. W. Olin, M. Piedmonte, M. Grubb, J. R. Young: Ultrasound-guided compression closure of postcatheterization pseudoaneurysms during concurrent anticoagulation: A review of seventy-seven patients. J. vasc. Surg. 23 (1996) 28–35

Do, D. D., T. Zehnder, F. Mahler: Farbkodierte Duplexsonographie bei iatrogenen Aneursymata spuria in der Leiste. Dtsch. med. Wschr. 118 (1993) 656–660

Dousset, V., N. Grenier, C. Douws, P. Senuita, G. Sassouste, L. Ada, L. Potaux: Hemodialysis grafts: Color Doppler flow imaging correlated with digital subtraction angiography and functional status. Radiology 181 (1991) 89–94

Edwards, J. M., J. M. Porter: Evaluation of upper extremity ischemia. In Bernstein, E. F., ed: Vascular diagnosis. Mosby, St. Louis 1993

Eichlisberger, R., B. Frauchiger, H. Schmitt, K. Jäger: Aneurysma spurium nach arterieller Katheterisierung: Diagnose und Verlaufskontrolle. Ultraschall Med. 13 (1992) 54–58

Fellmeth, B. D., A. C. Roberts, J. J. Bookstein, J. A. Freischlag, J. R. Forsythe, N. K. Buckner, R. J. Hye: Postangiographic femoral artery injuries: Nonsurgical repair with US-guided compression. Radiology 178 (1991) 671–675

Fields, W. S., V. Maslenikov, J. S. Meyer, W. K. Hass, R. D. Remington, M. Macdonald: Joint study of extracranial arterial occlusion. J. Amer. med. Ass. 211 (1970) 1993–2003

FitzGerald, D. E., J. Carr: Peripheral arterial disease: Assessment by arteriography and alternative noninvasive measurements. Amer. J. Roentgenol. 128 (1977) (385–388)

Flanigan, D. P., B. Gray, J. J. Schuler, J. A. Schwartz, K. W. Post: Correlation of Doppler-derived high thigh pressure and intra-arterial pressure in the assessment of aorto-iliac occlusive disease. Brit. J. Surg. 68 (1981) 423–425

Franzeck, U. K., E. F. Bernstein, A. Fronek: The effect of sensing site on the limb segmental blood pressure determination. Arch. Surg. 116 (1981) 912–916

Fronek, A.: Noninvasive diagnostics in vascular disease. McGraw-Hill Book Company, New York 1989

Fronek, A., K. Johansen, R. B. Dilley, E. F. Bernstein: Ultrasonographically monitoring postocclusive reactive hyperemia in the diagnosis of peripheral arterial occlusive disease. Circulation 48 (1973) 149–152

Fronek, A., M. Coel, E. F. Bernstein: Quantitative ultrasonographic studies of lower extremity flow velocities in health and disease. Circulation 53 (1976) 957–960

Fronek, A., M. Coel, E. F. Bernstein: The importance of combined multisegmental pressure and Doppler flow velocity studies in the diagnosis of peripheral arterial occlusive disease. Surgery 84 (1978) 840–847

Frühwald, F., D. E. Blackwell: Atlas of Color-coded Doppler sonography. Vascular and Soft Tissue Structures of the Upper Extremity, Thoracic Outlet and Neck. Springer, Wien 1992

Gooding, G. A. W., S. Perez, J. H. Rapp, W. C. Krupski: Lower-extremity vascular grafts placed for peripheral vascular disease: Prospective evaluation with duplex Doppler sonography. Radiology 180 (1991) 379–386

Gosling, R. G., D. H. King: Continuous wave ultrasound as an alternative and complement to X-rays in vascular examinations. In Reneman, R.: Cardiovascular Applications of Ultrasound. North-Holland, Amsterdam 1974

Gosling, R. G., D. H. King: Processing arterial Doppler signals for clinical data. In Vlieger et al.: Handbook of Clinical Ultrasound. Wiley, New York 1978

Grigg, M. J., A. N. Nicolaides, J. H. N. Wolfe: Detection and grading of femorodistal vein graft stenoses: Duplex velocity measurements compared with angiography. J. vasc. Surg. 8 (1988) 661–666

Gross-Fengels, W., D. Beyer, R. Lorenz, R. Kristen: Darstellung iatrogener Aneurysmen und AV-Fisteln der unteren Extemität mit IV-DSA und Sonographie. Röntgen-Bl. 40 (1987) 131–136

Grüntzig, A., M. Schlumpf: The validity and reliability of post-stenotic blood pressure measurement by Doppler ultrasonic sphygmomanometry. Vasa 3 (1974) 65–71

Gundersen, J.: Segmental measurements of systolic blood pressure in the extremities including the tumb and the great toe. Acta chir. scand. 426 (Suppl.) (1972) 1

Habscheid, W., P. Landwehr: Das Aneurysma spurium der Arteria femoralis nach Herzkatheteruntersuchung: eine prospektive Sonographiestudie. Z. Kardiol. 78 (1989) 573–577

Hafferl, A.: Lehrbuch der topographischen Anatomie. Springer, Berlin 1969

Hajarizadeh, H., C. R. LaRosa, P. Cardullo, M. J. Rohrer, B. S. Cutler: Ultrasound-guided compression of iatrogenic femoral pseudo-aneurysm failure, recurrence, and long-term results. J. vasc. Surg. 22 (1995) 425–433

Hass, W. K., W. S. Fields, R. R. North, I. I. Kricheff, N. E. Chase, R. B. Bauer: Joint study of extracranial arterial occlusion. II. Arteriography, techniques, sites and complications. J. Amer. med. Ass. 203 (1968) 961–968

Hatsukami, T. S., J. Primozich, R. E. Zierler, D. E. Strandness jr.: Color Doppler characteristics in normal lower extremity arteries. Ultrasound Med. Biol. 18 (1992) 167–172

Heberer, G., G. Rau, W. Schoop: Angiologie. Grundlagen, Klinik und Praxis. Thieme, Stuttgart 1974

Helvie, M. A., J. M. Rubin, T. M. Silver, T. F. Kresowik: The distinction between femoral artery pseudoaneurysms and other causes of groin masses: Value of duplex Dopplersonography. Amer. J. Roentgenol. 150 (1988) 1177–1180

Henderson, J., J. Chambers, T. A. Jeddy, J. Chamberlain, T. A. Whittingham: Serial investigation of balloon angioplasty induced changes in the superficial femoral artery using colour duplex ultrasonography. Brit. J. Radiol. 67 (1994) 546–551

Hepp, W., N. Pallua: Das Anastomosenaneurysma der Leiste: Chirurgische Therapie und Ergebnisse. In Sandmann, W., H. W. Kniemeyer, R. Kolvenbach: Aneurysmen der großen Arterien. Diagnostik und Therapie. Huber, Bern 1991

Hirai, M.: Arterial insufficiency of the hand evaluated by digital blood pressure and arteriography findings. Circulation 58 (1978) 902–908

Humphries, K. N., T. K. Hames, S. W. J. Smith, V. A. Cannon, A. D. B. Chant: Quantitative assessment of the common femoral to popliteal arterial segment using continuous wave Doppler ultrasound. Ultrasound Med. Biol. 6 (1980) 99–105

Idu, M. M., J. D. Blankensteyn, P. de Gier, E. Truyen, J. Buth: Impact of a color-flow duplex surveillance program on infrainguinal vein graft patency: A five-year experience. J. vasc. Surg. 17 (1993) 42–53

Jäger, K. A., D. J. Phillips, R. L. Martin, C. Hanson, G. O. Roederer, Y. E. Langlois, H. J. Ricketts, D. E. Strandness: Noninvasive mapping of lower limb arterial lesions. Ultrasound Med. Biol. 11 (1985) 515–521

Jäger, K. A., J. Landmann (Hrsg.): Praxis der angiologischen Diagnostik. Stufendiagnostik und rationelles Vorgehen bei arterieller und venöser Durchblutungsstörung. Springer, Berlin 1994

Johns, J. P., L. E. Pupa, St. R. Bailey: Spontaneous thrombosis of iatrogenic femoral artery pseudoaneurysms: Documentation with color Doppler and two-dimensional ultrasonography. J. vasc. Surg. 14 (1991) 24–29

Johnson, E. L., P. G. Yock, V. K. Hargrave, J. P. Srebro, S. M. Manubens, W. Seitz, T. A. Ports: Assessment of severity of coronary stenoses using a Doppler catheter. Validation of a method based on the continuity equation. Circulation 80 (1969) 625–635

Johnston, K. W., M. Kassam, R. S. C. Cobbold: Relationship between Doppler pulsatility index and direct femoral pressure measurements in the diagnosis of aortoiliac occlusive disease. Ultrasound Med. Biol. 9 (1983) 271–281

Johnston, K. W., M. Kassam, J. Koers, R. S. C. Cobbold, D. MacHattie: Comparative study of four methods for quantifying Doppler ultrasound waveforms from the femoral artery. Ultrasound Med. Biol. 10 (1984) 1–12

Karasch, Th., J. Veit, F. Herrmann, D. Neuerburg-Heusler: Duplexsonographische Messungen der maximalen und mittleren Strömungsgeschwindigkeit im interapparativen und interindividuellen Vergleich am Modell der Arteria poplitea. Vasa (Suppl. 32) (1991 a) 167–171

Karasch, Th., R. Rieser, D. Neuerburg-Heusler, F.-J. Roth: Vena-saphena-magna-Varikose als Leitsymptom einer iatrogenen arteriovenösen Fistel. Dtsch. med. Wschr. 116 (1991 b) 1871–1874

Karasch, Th., R. Rieser, B. Grün, A. L. Strauss, D. Neuerburg-Heusler, F.-J. Roth, H. Rieger: Bestimmung der Verschlußlänge in Extremitätenarterien. Farbduplexsonographie versus Angiographie. Ultraschall Med. 14 (1993) 247–254

Kathrein, H., P. König, S. Weimann, G. Judmaier, P. Dittrich: Nichtinvasive morphologische und funktionelle Beurteilung arteriovenöser Fisteln von Dialysepatienten mit der Duplexsonographie. Ultraschall 10 (1989) 33–40

Katzenschlager, R., A. Ahmadi, E. Minar, R. Koppensteiner, Th. Maca, K. Pikesch, A. Stümpflen, A. Ugurluoglu, H. Ehringer: Femoropopliteal artery: Initial and 6-month results of color duplex US-guided percutaneous transluminal angioplasty. Radiology 199 (1996) 331–334

Knoblich, S., W. Krings, M. Schulte, D. Neuerburg-Heusler: Messung des systolischen Blutdruckes an den Digital-Arterien, mit Hilfe der Strain-Gauge-Technik. Vasa (Suppl. 15) (1986) 13

Köhler, M., H. U. Hinger, W. Zahnow: Die Beurteilung der Kompensation bei chronischen Arterienverschlüssen mit Hilfe der Ultraschall-Doppler-Methode und der Venenverschlußplethysmographie. Z. Kreisl.-Forsch. 61 (1972) 401–412

Köhler, M., B. Lösse: Simultane Messungen des systolischen Blutdruckes mit der Ultraschall-Doppler-Technik und der blutigen Methode an der Arteria radialis des Menschen. Z. Kardiol. 68 (1979) 551–556

Köhler, M., M. Krüpe: Untersuchungen über Spezifität und Normalität der peripheren systolischen Druckmessung mit der Ultraschall-Doppler-Technik an gesunden angiographierten Extremitäten. Z. Kardiol. 74 (1985) 39–45

Koennecke, H.-C., G. Fobbe, M. M. Hamed, K.-J. Wolf: Diagnostik arterieller Gefäßerkrankungen der unteren Extremitäten mit der farbkodierten Duplexsonographie. Fortschr. Röntgenstr. 151 (1989) 42–46

Kohler, T. R., D. R. Nance, M. M. Cramer, N. Vandenburghe, D. E. Strandness jr.: Duplex scanning for diagnosis of aortoiliac and femoropopliteal disease: A prospective study. Circulation 76 (1987 a) 1074–1080

Kohler, T. R., St. C. Nicholls, R. E. Zierler, K. W. Beach, P. J. Schubart, D. E. Strandness jr.: Assessment of pressure gradient by Doppler ultrasound: Experimental and clinical observations. J. vasc. Surg. 6 (1987 b) 460–469

Kresowik, T. F., M. D. Khoury, B. V. Miller, M. D. Winniford, A. R. Shamma, W. J. Sharp, M. B. Blecha, J. D. Corson: A prospective study of the incidence and natural history of femoral vascular complications after percutaneous transluminal coronary angioplasty. J. vasc. Surg. 13 (1991) 328–336

Landwehr, P., K. Lackner: Farbkodierte Duplexsonographie vor und nach PTA der Arterien der unteren Extremität. Fortschr. Röntgenstr. 152 (1990 a) 35–41

Landwehr, P., A. Tschammler, R. M. Schaefer, K. Lackner: Wertigkeit der farbkodierten Duplexsonographie des Dialyseshunts. Fortschr. Röntgenstr. 153 (1990 b) 185–191

Langholz, J., O. Stolke, H. Heidrich, Ch. Behrendt, B. Blank, B. Feßler: Farbkodierte Duplexsonographie von Unterschenkelarterien. Darstellbarkeit in Zuordnung zu Fontainestadien. Vasa (Suppl. 33) (1991) 209

Langsfeld, M., J. Nepute, F. B. Hershey, L. Thorpe, A. I. Auer, B. Binnington, J. J. Hurley, G. J. Peterson, R. Schwartz, J. J. Woods: The use of deep duplex scanning to predict hemodynamically significant aortoiliac stenoses. J. vasc. Surg. 7 (1988) 395–399

Larch, E., R. Ahmadi, R. Koppensteiner, E. Minar, G. Schnürer, A. Stümpflen, H. Ehringer: Farbkodierte Duplexsonographie (FD) zur Beurteilung der Unterschenkelarterien (US) bei peripherer arterieller Verschlußkrankheit. Vasa (Suppl. 41) (1993) 14

Lawrence, P. F., S. Lorenzo-Rivero, J. L. Lyon: The incidence of iliac, femoral, and popliteal artery aneurysms in hospitalized patients. J. vasc. Surg. 22 (1995) 409–416

Legemate, D. A., C. Teeuwen, H. Hoeneveld, R. G. A. Ackerstaff, B. C. Eikelboom: Spectral analysis criteria in duplex scanning of aortoiliac and femoropopliteal arterial disease. Ultrasound Med. Biol. 17 (1991) 769–776

Legemate, D. A., C. Teeuwen, H. Hoeneveld, B. C. Eikelboom: How can the assessment of the hemodynamic significance of aortoiliac arterial stenosis by duplex scanning be improved? A comparative study with intraarterial pressure measurement. J. vasc. Surg. 17 (1993) 676–684

Leng, G. C., M. R. Whyman, P. T. Donnan, C. V. Ruckley, I. Gillespie, G. R. Fowkes, P. L. Allan: Accuracy and reproducibility of duplex ultrasonography in grading femoropopliteal stenoses. J. vasc. Surg. 17 (1993) 510–517

Loeprecht, H., H. Bruijnen: Femoro-popliteale Aneurysmen. In Sandmann, W., H. W. Kniemeyer, R. Kolvenbach: Aneurysmen der großen Arterien. Diagnostik und Therapie. Huber, Bern 1991

Ludwig, M., G. Trübestein, N. Leipner: Die Bedeutung der Dopplerdruckmessung unter Belastung bei vorliegender arterieller Verschlußkrankheit im Bereich der unteren Extremität. Herz Kreisl. 4 (1985) 189–191

Luska, G., U. Risch, M. Pellengahr, H. v. Boetticher: Farbcodierte dopplersonographische Untersuchungen zur Morphologie und Hämodynamik der Arterien des Beckens und der Beine bei gesunden Probanden. Fortschr. Röntgenstr. 153 (1990) 246–251

Luzsa, G.: Röntgenanatomie des Gefäßsystems. Akadémiai Kiadó, Budapest 1972

Lynch, T. G., R. W. Hobson, C. B. Wright, G. Garcia, R. Lind, S. Heintz, L. Hart: Interpretation of Doppler segmental pressures in peripheral vascular occlusive disease. Arch. Surg. 119 (1984) 465–467

Mahler, F., H. H. Brunner, A. Fronek, A. Bollinger: Der Knöchelarteriendruck während reaktiver Hyperämie bei Gefäßgesunden und Patienten mit arterieller Verschlußkrankheit. Schweiz. med. Wschr. 105 (1975) 1786–1788

Mahler, F., L. Koen, K. H. Johansen, E. F. Bernstein, A. Fronek: Postocclusion and postexercise flow velocity and ankle pressures in normals and marathon runners. Angiology 27 (1976) 721–729

Mahler, F.: Systolische Druckmessung nach Belastung. In Kriessmann, A., A. Bollinger, H. Keller: Praxis der Doppler-Sonographie. Thieme, Stuttgart 1990

Matsubara, J., D. Neuerburg-Heusler, W. Schoop: Über das Verhalten des poststenotischen systolischen Blutdruckes bei Änderungen des arteriellen Systemdruckes. In Müller-Wiefel, H.: Gefäßersatz. Witzstrock, Baden-Baden 1980

Mattos, M., P. S. van Bemmelen, K. J. Hodgson, D. E. Ramsey, L. D. Barkmeier, D. S. Sumner: Does correction of stenoses identified with color duplex scanning improve infrainguinal graft patency? J. vasc. Surg. 17 (1993) 54–66

Mewissen, M. W., E. V. Kinney, D. F. Bandyk, T. Reifsnyder, G. R. Seabrook, E. O. Lipchik, J. B. Towne: The role of duplex scanning versus angiography in predicting outcome after balloon angioplasty in the femoropopliteal artery. J. vasc. Surg. 15 (1992) 860–866

Middelton, W. D., S. Erickson, G. L. Melson: Perivascular color artifact: Pathologic significance and appearance on color Doppler US images. Radiology 171 (1989a) 647–652

Middelton, W. D., D. D. Picus, M. V. Marx, G. L. Melson: Color Doppler sonography of hemodialysis vascular access: Comparison with angiography. Amer. J. Radiol. 152 (1989b) 633–639

Mills, J. L., E. J. Harris, L. M. Taylor, W. C. Beckett, J. M. Porter: The importance of routine surveillance of distal bypass grafts with duplex scanning: A study of 379 reversed vein grafts. J. vasc. Surg. 12 (1990) 379–389

Mitchell, D. G., L. Needleman, M. Bezzi, B. B. Goldberg, A. B. Kurtz, R. G. Pennell, M. D. Rifkin, M. Vilaro, O. H. Baltarowich: Femoral artery pseudoaneurysm: Diagnosis with conventional duplex and color Doppler US. Radiology 165 (1987) 687–690

Moll, R., W. Habscheid, P. Landwehr: Häufigkeit des Aneurysma spurium der Arteria femoralis nach Herzkatheteruntersuchung und PTA. Fortschr. Röntgenstr. 154 (1991) 23–27

Moneta, G. L., R. A. Yeager, R. Antonovic, L. D. Hall, J. D. Caster, C. A. Cummings, J. M. Porter: Accuracy of lower extremity arterial duplex mapping. J. vasc. Surg. 15 (1992) 275–284

Moneta, G. L., R. A. Yeager, R. W. Lee, J. M. Porter: Noninvasive localization of arterial occlusive disease: A comparison of segmental Doppler pressures and arterial duplex mapping. J. vasc. Surg. 17 (1993) 578–582

Mönig, S. P., M. Walter, S. Sorgatz, H. Erasmi: True infrapopliteal artery aneurysms: Report of two cases and literature review. J. vasc. Surg. 24 (1996) 276–278

Mulligan, S. A., T. Matsuda, P. Lanzer, G. M. Gross, W. D. Routh, F. S. Keller, D. B. Koslin, L. L. Berland, M. D. Fields, M. Doyle, G. B. Cranney, J. Y. Lee, G. M. Pohost: Peripheral arterial occlusive disease: Prospective comparison of MR angiography and color duplex US with conventional angiography. Radiology 178 (1991) 695–700

Murphy, T. P., G. S. Dorfman, M. Segall, W. I. Carney jr.: Iatrogenic arterial dissection: Treatment by percutaneous transluminal angioplasty. Cardiovasc. intervent. Radiol. 14 (1991) 302–306

Neuerburg-Heusler, D., W. Schoop: Ultraschall-Doppler-Diagnostik in der Angiologie. Hippokrates 49 (1978) 231–246

Neuerburg-Heusler, D., Th. Voigt, F. J. Roth: Doppler-Sonographie der Leistenarterie. In Breddin, K.: Thrombose und Atherogenese. Pathophysiologie und Therapie der arteriellen Verschlußkrankheit. Bein-Beckenvenen-Thrombose. Witzstrock, Baden-Baden 1981

Neuerburg-Heusler, D., A. Schlieszus, M. Schulte: Topographische Diagnose der peripheren arteriellen Verschlußkrankheit (PAVK). Vergleich von segmentaler systolischer Druckmessung (SSP) und Oszillographie (OSZ) in Ruhe und nach zwei differenzierten Belastungen. Vasa (Suppl. 27) (1989) 369

Neuerburg-Heusler, D., Th. Karasch: Farbkodierte Duplexsonographie der Extremitätenarterien – Möglichkeiten in Diagnostik und Therapiekontrolle. Vasa (Suppl. 32) (1991) 113–121

Neuerburg-Heusler, D., Th. Karasch: Stenosegradbestimmung an peripheren Arterien. Hämodynamische und sonographische Grundlagen. Vasa 25 (1996) 109–113

Nicolaides, A. N., I. C. Gordon-Smith, J. Dayandas, H. H. G. Eastcott: The value of Doppler blood velocity tracings in the detection of aortoiliac disease in patients with intermittent claudication. Surgery 80 (1976) 774–778

Nielsen, P. E., G. Bell, N. A. Lassen: Strain gauge studies of distal blood pressure in normal subjects and in patients with peripheral arterial disease. Analysis of normal variation and reproducibility and comparison to intraarterial measurements. Scand. J. clin. Lab. Invest. 31 (Suppl. 128) (1973) 103–109

Ouriel, K., C. K. Zarins: Doppler ankle pressure: An evaluation of three methods of expression. Arch. Surg. 117 (1982) 1297–1300

Paulson, E. K., B. S. Hertzberg, S. S. Paine, B. A. Carroll: Femoral artery pseudoaneurysms: Value of color Doppler-sonography in predicting which ones will thrombose without treatment. Amer. J. Radiol. 159 (1992) 1077–1081

Phillips, D. J., W. Beach, J. Primozich, D. E. Strandness jr.: Should results of ultrasound Doppler studies be reported in units of frequency or velocity? Ultrasound Med. Biol. 15 (1989) 205–212

Polak, J. F., M. C. Donaldson, G. R. Dobkin, J. A. Mannick, D. H. O'Leary: Early detection of saphenous vein arterial bypass graft stenosis by color-assisted duplex sonography: A prospective study. Amer. J. Radiol. 154 (1990a) 857–861

Polak, J. F., M. I. Karmel, J. A. Mannick, D. H. O'Leary, M. C. Donaldson, A. D. Whittemore: Determination of the extent of lower-extremity peripheral arterial disease with color-assisted duplex sonography: Comparison with angiography. Amer. J. Roentgenol. 155 (1990b) 1085–1089

Ranke, C., A. Creutzig, K. Alexander: Duplex scanning of the peripheral arteries: Correlation of the peak velocity ratio with angiographic diameter reduction. Ultrasound Med. Biol. 18 (1992) 433–440

Rieger, H.: Pathophysiologie des akuten und chronischen Arterienverschlusses. In Alexander, K.: Gefäßkrankheiten. Urban & Schwarzenberg, München 1993

Rode, V., R. M. Schütz: Doppler-Ultraschall-Messungen: Relevanz ihrer Maßzahlen für die klinische und die Funktionsdiagnostik. Herz Kreisl. 9 (1977) 422–426

Ruland, O., N. Borkenhagen, Th. Prien: Der Doppler-Hohlhand-Test. Ultraschall 9 (1988) 63–66

Rutherford, R. B., D. H. Lowenstein, M. F. Klein: Combining segmental systolic pressures and plethysmography to diagnose arterial occlusive disease of the legs. Amer. J. Surg. 138 (1979) 211–218

Sacks, D., M. L. Robinson, D. L. Marinelli, G. S. Perlmutter: Peripheral arterial Doppler ultrasonography: Diagnostic criteria. J. Ultrasound Med. 11 (1992) 95–103

Sacks, D., M. L. Robinson, T. A. Summers, D. L. Marinelli: The value of duplex sonography after peripheral artery angioplasty in predicting subacute restenosis. Amer. J. Radiol. 162 (1994) 179–183

Scharf, R., U. Cobet, R. Millner: Die Wertigkeit von Ultraschall-Doppler-Pulskurven-Parametern bei der Beurteilung stenotischer Arterienerkrankungen. Ultraschall 9 (1988) 67–71

Schmitz-Rixen, Th., S. Horsch, H. Erasmi, R. Schmidt, H. Pichlmaier: Die arteriovenöse Fistel als Hilfsmittel zur Revaskularisation der Unterschenkelarterien. Akt. Chir. 21 (1986) 46–51

Schoop, W., H. Levy: Messung des systolischen Blutdrucks distal eines Extremitätenarterienverschlusses mit Hilfe der Ultraschall-Doppler-Technik. Verh. dtsch. Ges. Kreisl.-Forsch. 35 (1969) 455–461

Schoop, W.: Pathophysiologie der Arterien und der arteriellen Durchblutung. In Heberer, G., G. Rau, W. Schoop: Angiologie. Thieme, Stuttgart 1974

Seifert, H., K. Jäger, D. Mona, A. Segantini, A. Bollinger: Arteriovenöse Fistel und Aneurysma spurium nach arthroskopischer Meniskektomie. Vasa 16 (1987) 389–392

Skidmore, R., J. P. Woodcock: Physiological interpretation of Doppler-shift waveforms. II. Validation of the Laplace transform method for characterization of the common femoral blood-velocity/time waveform. Ultrasound Med. Biol. 6 (1980) 219–225

Sladen, J. G., J. D. S. Reid, P. L. Cooperberg, P. B. Harrison, T. M. Maxwell, M. O. Riggs, L. D. Sanders: Color flow duplex screening of infrainguinal grafts combining low- and high-velocity criteria. Amer. J. Surg. 158 (1989) 107–112

Smet, de A. A. E. A., E. J. M. Ermers, P. J. E. H. M. Kitslaar: Duplex velocity characteristics of aortoiliac stenoses. J. vasc. Surg. 23 (1996) 628–636

Sorrell, K. A., R. L. Freinberg, J. R. Wheeler, R. T. Gregory, S. O. Snyder, R. G. Gayle, N. F. Parent: Color-flow duplex-directed manual occlusion of femoral false aneurysms. J. vasc. Surg. 17 (1993) 571–577

Spijkerboer, A. M., P. C. Nass, J. C. de Valois, B. C. Eikelboom, T. Th. C. Overtoom, F. J. A. Beck, F. L. Moll: Iliac artery stenoses after percutaneous transluminal angioplasty: Follow-up with duplex ultrasonography. J. vasc. Surg. 23 (1996) 691–697

Steinkamp, H. J., R. Jochens, W. Zendel, C. Zwicker, W. Hepp, R. Felix: Katheterbedingte Femoralgefäßläsionen: Diagnose mittels B-mode-Sonographie, Dopplersonographie und Farbdopplersonographie. Ultraschall Med. 13 (1992) 221–227

Strandness, D. E., E. P. McCutcheon, R. F. Rushmer: Application of a transcutaneous Doppler flowmeter in evaluation of occlusive arterial disease. Surg. Gyn. Obst. 122 (1966) 1039–1045

Strano, A., A. Pinto, G. Alletto: Congenital arteriovenous fistulas of the limbs: Pathophysiological and diagnostic aspects. In Blov, St., D. A. Loose, J. Weber, eds: Vascular malformations. Einhorn-Presse Verlag, Reinbek 1989

Strauss, A. L., W. Schäberle, H. Rieger, D. Neuerburg-Heusler, F.-J. Roth, W. Schoop: Duplexsonographische Untersuchungen der A. profunda femoris. Z. Kardiol. 78 (1989) 567–572

Strauss, A. L., A. Scheffler, H. Rieger: Doppler-sonographische Bestimmung des Druckabfalls über peripheren Modell-Arterienstenosen. Vasa 19 (1990) 207–211

Strauss, A. L., F.-J. Roth, H. Rieger: Noninvasive assessment of pressure gradients across iliac artery stenoses: Duplex and catheter correlative study. J. Ultrasound Med. 12 (1993 a) 17–22

Strauss, A. L., D. Sandor, Th. Karasch, F.-J. Roth, D. R. C. Brocai, D. Neuerburg-Heusler, H. Rieger: Wertigkeit der Farbduplexsonographie in der arteriellen Gefäßdiagnostik. Vasa (Suppl. 41) (1993 b) 15

Sumner, D. S.: Evaluation of noninvasive testing procedures: data analysis and interpretation. In Bernstein, E. F., ed: Vascular Diagnosis. Mosby, St. Louis 1993

Sumner, D. S., D. E. Strandness jr.: The relationship between calf blood flow and ankle blood pressure in patients with intermittent claudication. Surgery 65 (1969) 763–771

Taylor, P. R., M. R. Tyrrell, M. Crofton, B. Bassan, M. Grigg, J. H. N. Wolfe, A. O. Mansfield, A. N. Nicolaides: Colour flow imaging in the detection of femoro-distal graft and native arters stenosis: Improved criteria. Europ. J. vasc. Surg. 6 (1992) 232–236

Thiele, B. L., D. F. Bandyk, R. E. Zierler, D. E. Strandness jr.: A systematic approach to the assessment of aortoiliac disease. Arch. Surg. 118 (1983) 477–481

Thulesius, O.: Beurteilung des Schweregrades arterieller Durchblutungsstörungen mit dem Doppler-Ultraschall-Gerät. In Bollinger, A., U.

Brunner: Meßmethoden bei arteriellen Durchblutungsstörungen. Huber, Bern 1971

Thulesius, O.: Systemic and ankle blood pressure before and after exercise in patients with arterial insufficiency. Angiology 29 (1978) 374–378

Tordoir, J. H. M., H. G. de Bruin, H. Hoeneveld, B. C. Eikelboom, P. J. E. H. M. Kitslaar: Duplex ultrasound scanning in the assessment of arteriovenous fistulas created for hemodialysis access: Comparison with digital subtraction angiography. J. vasc. Surg. 10 (1989) 122–128

Trattnig, S., A. Maier, Th. Sautner, B. Schwaighofer, M. Breitenseher, F. Karnel: Vena-saphena-Bypass-Stenosen. Frühdiagnose mittels farbkodierter Dopplersonographie. Ultraschall Med. 13 (1992) 67–70

Varga, Z. A., J. C. Locke-Edmunds, R. N. Baird and the Joint Vascular Research Group: A multicenter study of popliteal aneurysms. J. vasc. Surg. 20 (1994) 171–177

Vollmar, J.: Rekonstruktive Chirurgie der Arterie, 3. Aufl. Thieme, Stuttgart 1982

Walton, L., T. R. P. Martin, M. Collins: Prospective assessment of the aorto-iliac segment by visual interpretation of frequency analysed Doppler waveforms – A comparison with arteriography. Ultrasound Med. Biol. 10 (1984) 27–32

Wetterer, E., R. D. Bauer, Th. Pasch: Arteriensystem. In Schütz, E.: Physiologie des Kreislaufs, Bd. 1. Springer, Berlin 1971

Whyman, M. R., P. R. Hoskins, G. C. Leng, P. L. Allan, P. T. Donnan, V. Ruckley, F. G. R. Fowkes: Accuracy and reproducibility of duplex ultrasound imaging in a phantom model of femoral artery stenosis. J. vasc. Surg. 17 (1993) 524–530

Widder, B., G.-M. von Reutern, D. Neuerburg-Heusler: Morphologische und dopplersonographische Kriterien zur Bestimmung von Stenosierungsgraden an der A. carotis interna. Ultraschall 7 (1986) 70–75

Widder, B.: Bedeutung technischer Kenngrößen der farbkodierten Duplexsonographie für Gefäßuntersuchungen. Ultraschall Med. 14 (1993) 231–239

Windeck, P., K.-H. Labs, K. A. Jäger: How useful are acceleration- and deceleration-based Doppler indices? A trial on patients with percutaneous transluminal angioplasty. Ultrasound Med. Biol. 18 (1992) 525–534

Wittenberg, G., P. Landwehr, R. Moll, A. Tschammler, B. Buschmann, Th. Krahe: Interobserver-Variabilität von Dialyseshuntflußmessungen mit der farbkodierten Duplexsonographie. Fortschr. Röntgenstr. 159 (1993) 375–378

Wolf, K.-J., F. Fobbe: Farbkodierte Duplexsonographie. Thieme, Stuttgart 1993

Yao, S. T., J. T. Hobbs, W. T. Irvine: Ankle systolic pressure measurements in arterial disease affecting the lower extremities. Brit. J. Surg. 56 (1969) 676–679

Yao, S. T.: Haemodynamic studies in peripheral arterial disease. Brit. J. Surg. 57 (1970) 761–766

Chapter 6

Aitken, A. G. F., D. J. Godden: Real-time ultrasound diagnosis of deep vein thrombosis: A comparison with venography. Clin. Radiol. 38 (1987) 309–313

Aldridge, S. C., A. J. Comerota, M. L. Katz, J. H. Wolk, B. I. Goldman, J. V. White: Popliteal venous aneurysm: Report of two cases and review of the world literature. J. vasc. Surg. 18 (1993) 708–715

Appelman, P. T., T. E. DeJong, L. E. Lampmann: Deep venous thrombosis of the leg: US findings. Radiology 163 (1987) 743–746

Barnes, R. W., H. E. Russell, K. K. Wu, J. C. Hoak: Accuracy of Doppler ultrasound in clinically suspected venous thrombosis of the calf. Surg. Gynecol. Obstet. 143 (1976) 425–428

Barnes, R. W., M. L. Nix, C. L. Barnes, R. C. Lavender, W. E. Golden, B. H. Harmon, E. J. Ferris, C. L. Nelson: Perioperative asymptomatic venous thrombosis: Role of duplex scanning versus venography. J. vasc. Surg. 9 (1989) 251–260

Becker, D. M., J. T. Philbrick, P. L. Abbitt: Real-time ultrasonography for the diagnosis of lower extremity deep venous thrombosis. Arch. intern. Med. 149 (1989) 1731–1734

Belov, St., D. A. Loose, J. Weber, ed: Vascular Malinformations. Einhorn-Presse Verlag, Reinbek 1989

van Bemmelen, P. S., G. Bedford, K. Beach, D. E. Strandness jr.: Quantitative segmental evaluation of venous valvular reflux with duplex ultrasound scanning. J. vasc. Surg. 10 (1989) 425–431

van Bemmelen, P. S., G. Bedford, K. Beach, D. E. Strandness jr.: Status of the valves in the superficial and deep venous system in chronic venous disease. Surgery 109 (1990) 730–734

Bendick, P. J., J. L. Glover, R. W. Holden, R. S. Dilley: Pitfalls of the Doppler examination for venous thrombosis. Amer. Surg. 6 (1983) 320–323

Bendick, P. J., J. L. Glover, O. W. Brown, T. J. Ranval: Serial duplex ultrasound examinations for deep vein thrombosis in patients with suspected pulmonary embolism. J. vasc. Surg. 24 (1996) 732–737

Bjordal, R. J.: Die Zirkulation in insuffizienten Vv. perforantes der Wade bei venösen Störungen. In May, R., H. Partsch, J. Staubesand: Venae perforantes. Urban & Schwarzenberg, München 1981

Bollinger, A., F. Mahler, G. de Sepibus: Diagnostik peripherer Venenerkrankungen mit Doppler-Strömungsdetektoren. Dtsch. med. Wschr. 46 (1968) 2197–2201

Bollinger, A., W. Rutishauser, F. Mahler, A. Grüntzig: Zur Dynamik des Rückstroms aus der Vena femoralis. Kreisl.-Forsch. 59 (1970) 963–971

Bollinger, A., W. Wirth, U. Brunner: Klappenagenesie und -dysplasie der Beinvenen. Schweiz. med. Wschr. 101 (1971) 1348–1353

Bollinger, A.: Funktionelle Angiologie. Thieme, Stuttgart 1979

Borgstede, J. P., G. E. Clagett: Types, frequency, and significance of alternative diagnoses found during duplex Doppler venous examinations of the lower extremities. J. Ultrasound Med. 11 (1992) 85–89

Bork-Wölwer, L., Th. Wuppermann: Verbesserung der nichtinvasiven Diagnostik der V. saphena magna- und der V. saphena parva-Insuffizienz durch die Duplex-Sonographie. Vasa 20 (1991) 343–347

Chengelis, D. L., P. J. Bendick, J. L. Glover, O. W. Brown, T. J. Ranval: Progression of superficial venous thrombosis to deep vein thrombosis. J. vasc. Surg. 24 (1996) 745–749

Comerota, A. J., M. L. Katz, L. L. Greenwald, E. Loefmans, M. Czeredarczuk, J. V. White: Venous duplex imaging: Should it replace hemodynamic tests for deep venous thrombosis? J. vasc. Surg. 11 (1990) 53–61

Cronan, J. J., G. S. Dorfman, F. H. Scola, B. Schepps, J. Alexander: Deep venous thrombosis: US assessment using vein compression. Radiology 162 (1987) 191–194

Cronan, J. J., G. S. Dorfman, J. Grusmark: Lower-extremity deep venous thrombosis: Further experience with and refinements of US assessment. Radiology 168 (1988) 101–107

Cronan, J. J., V. Leen: Recurrent deep venous thrombosis: Limitations of US. Radiology 170 (1989) 739–742

Dauzat, M. M., J. P. Laroche, Ch. Charras, B. Blin, M.-M. Domingo-Faye, P. Saint-Luce, A. Domergue, F.-M. Lopez, Ch. Janbon: Real-time B-mode ultrasonography for better specificity in the noninvasive diagnosis of deep venous thrombosis. J. Ultrasound Med. 5 (1986) 625–631

Diepgen, T. L., I. D. Bassukas, O. P. Hornstein: Atypisches Klippel-Trenaunay-Weber Syndrom mit Osteohypotrophie. Akt. Dermatol. 14 (1988) 299–301

Dosick, St. M., W. S. Blakemore: The Role of Doppler ultrasound in acute deep vein thrombosis. Amer. J. Surg. 136 (1978) 265–268

Elias, A., G. Le Corff, J. L. Bouvier, M. Benichou, A. Serradimigni: Value of real-time B-mode ultrasound imaging in the diagnosis of deep vein thrombosis of the lower limbs. Int. Angiol. 6 (1987) 175–182

Falk, R. L., D. F. Smith: Thrombosis of upper extremity thoracic inlet veins: Diagnosis with duplex Doppler sonography. Amer. J. Radiol. 149 (1987) 677–682

Flanigan, P. D., J. J. Goodreau, J. S. Burnham, J. J. Bergan, S. T. J. Yao: Vascular-laboratory diagnosis of clinically suspected acute deep-vein thrombosis. Lancet 1978/II, 331

Fobbe, F., H.-C. Koennecke, M. El Bedewi, P. Heidt, J. Boese-Landgraf, K.-J. Wolf: Diagnostik der tiefen Beinvenenthrombose mit der farbkodierten Duplexsonographie. Fortschr. Röntgenstr. 151 (1989) 569–573

Fobbe, F., M. Ruhnke-Trautmann, D. van Gemmeren, C.-A. Hartmann, U. Kania, K.-J. Wolf: Altersbestimmung venöser Thromben im Ultraschall. Fortschr. Röntgenstr. 155 (1991) 344–348

Foley, W. D., W. D. Middleton, T. L. Lawson, F. A. Erickson, S. Macrander: Color Doppler ultrasound imaging of lower-extremity venous disease. Amer. J. Radiol. 152 (1989) 371–376

Fowl, R. J., G. B. Strothman, J. Blebea, G. J. Rosenthal, R. F. Kempczinski: Inappropriate use of venous duplex scans: An analysis of indications and results. J. vasc. Surg. 23 (1996) 881–886

Franzeck, U. K., M. Billeter, R. Schultheiss, A. Bollinger: Vergleich von klinischer Untersuchung, Doppler-Ultraschall und Farb-Duplex-Sonographie bei Patienten mit insuffizienten Perforansvenen. Vasa (Suppl. 41) (1993) 46

Friedman, S. G., K. V. Krishnasastry, W. Doscher, S. L. Deckoff: Primary venous aneurysms. Surgery 108 (1990) 92–95

Fronek, A.: Noninvasive Diagnostics in Vascular Disease. McGraw-Hill, New York 1989

Fürst, G., F.-P. Kuhn, R. P. Trappe, U. Mödder: Diagnostik der tiefen Beinvenenthrombose. Farb-Doppler-Sonographie versus Phlebographie. Fortschr. Röntgenstr. 152 (1990) 151–158

Gaitini, D., J. K. Kraftori, M. Pery, Y. L. Weich, A. Markel: High-resolution real-time ultrasonography in the diagnosis of deep vein thrombosis. Fortschr. Röntgenstr. 149 (1988) 26–30

Gaitini, D., J. K. Kraftori, M. Pery, A. Markel: Late changes in veins after deep venous thrombosis. Ultrasonic findings. Fortschr. Röntgenstr. 153 (1991) 68–72

van Gemmeren, D., F. Fobbe, M. Ruhnke-Trautmann, C. A. Hartmann, R. Gotzen, K. J. Wolf, A. Distler, K.-L. Schulte: Diagnostik tiefer Beinvenenthrombosen mit der farbkodierten Duplexsonographie und sonographische Altersbestimmung der Thrombose. Z. Kardiol. 80 (1991) 523–528

Gooding, G. A. W., D. R. Hightower, E. H. Moore, W. P. Dillon, M. J. Lipton: Obstruction of the superior vena cava of subclavian veins: Sonographic diagnosis. Radiology 159 (1986) 663–665

Grosser, S., G. Kreymann, A. Guthoff, C. Taube, A. Raedler, V. Tilsner, H. Greten: Farbkodierte Duplexsonographie bei Phlebothrombosen. Dtsch. med. Wschr. 115 (1990) 1939–1944

Grosser, S., G. Kreymann, J. Kühns: Stellenwert der farbkodierten Duplexsonographie bei der Diagnostik von akuten und chronischen venösen Erkrankungen der unteren Extremität. Ultraschall Med. 12 (1991) 222–227

Grüntzig, A., A. Bollinger, O. Zehender: Möglichkeiten und Grenzen der qualitativen Venendiagnostik mit Doppler-Ultraschall (Ergebnisse einer Blindstudie). Klin. Wschr. 49 (1971) 245–251

Habscheid, W., Th. Wilhelm: Diagnostik der tiefen Beinvenenthrombose durch Real-time-Sonographie. Dtsch. med. Wschr. 113 (1988) 586–591

Habscheid, W., P. Landwehr: Diagnostik der akuten tiefen Beinvenenthrombose mit der Kompressionssonographie. Ultraschall Med. 11 (1990) 268–273

Habscheid, W.: Die bildgebende Sonographie in der Diagnostik der tiefen Beinvenenthrombose. Ergebnisse der Angiologie und Phlebologie, Bd. 40. Schattauer, Stuttgart 1991

Habscheid, W.: Die Venenkatheterthrombose der oberen Thoraxapertur. Eine duplexsonographische Untersuchung bei Intensivpatienten. Intensivmed. u. Notfallmed. 29 (1992) 137–143

Habscheid, W., P. Landwehr: Duplexsonographie als Diagnoseverfahren bei Venenthrombosen der oberen Thoraxapertur. Med. Welt 43 (1992) 137–141

Hach, W., E. Girth, W. Lechner: Einteilung der Stammvarikose der V. saphena magna in 4 Stadien. Phlebol. u. Proktol. 6 (1977) 116–123

Hach, W.: Phlebographie der Bein- und Beckenvenen. Schnetztor, Konstanz 1985

Handl-Zeller, L., P. Hübsch, G. Hohenberg, A. Schratter, J. Wickenhauser: B-Bild und farbcodierte Dopplersonographie zur Bestrahlungsplanung bei Klippel-Trenaunay-Syndrom mit Kasabach-Merritt-Symptomatik. Ultraschall 10 (1989) 41–43

Hanel, K. C., W. M. Abbott, N. C. Reidy, D. Fulchino, A. Miller, D. C. Brewster, C. A. Athanasoulis: The role of two noninvasive tests in deep venous thrombosis. Ann. Surg. 194 (1981) 725–730

Heijboer, H., H. R. Büller, A. W. A. Lensing, A. G. G. Turpie, L. P. Colly, J. Wouter ten Cate: A comparison of real-time compression ultrasonography with impedance plethysmography for the diagnosis of deep-vein thrombosis in symptomatic outpatients. New Engl. J. Med. 329 (1993) 1365–1369

Herzog, P., M. Anastasiu, W. Wollbrink., W. Hermann, K.-H. Holtermüller: Real-time-Sonographie bei tiefer Becken- und Beinvenenthrombose. Ein prospektiver Vergleich zur Phlebographie. Med. Klin. 86 (1991) 132–137

Hill, St. I., R. E. Berry: Subclavian vein thrombosis: A continuing challenge. Surgery 108 (1990) 1–9

Hirschl, M., R. Bernt: Normalwerte, Reproduzierbarkeit und Aussagekraft duplexsonographischer Kriterien in der Venenfunktionsdiagnostik. Ultraschall Klin. Prax. 5 (1990) 81–84

Holmes, M. C. G.: Deep venous thrombosis of the lower limbs diagnosed by ultrasound. Med. J. Aust. 1 (1973) 427–430

Horattas, M. C., D. J. Wright, A. H. Fenton, D. M. Evans, M. A. Oddi, R. W. Kamienski, E. F. Shields: Changing concepts of deep venous thrombosis of the upper extremity – Report of a series and review of the literature. Surgery 104 (1988) 561–567

Huber, P., W. Häuptli, H. E. Schmitt, L. K. Widmer: Die Axillar-Subclavianvenenthrombose und ihre Folgen. Internist 28 (1987) 336–343

Hübsch, P. J. S., R. L. Stiglbauer, B. W. A. M. Schwaighofer, F. M. Kainberger, P. P. A. Barton: Internal jugular and subclavian vein thrombosis caused by central venous catheters. Evaluation using Doppler blood flow imaging. J. Ultrasound Med. 7 (1988) 629–636

Jorgensen, O., K. C. Hanel, A. M. Morgan, J. M. Hunt: The incidence of deep venous thrombosis in patients with superficial thrombophlebitis of the lower limbs. J. vasc. Surg. 18 (1993) 70–73

Karasch, Th., R. Rieser, D. Neuerburg-Heusler, F.-J. Roth: Vena-saphena-magna-Varikose als Leitsymptom einer iatrogenen arterio-venösen Fistel. Dtsch. med. Wschr. 116 (1991) 1871–1874

Karkow, W. S., J. J. Cranley, R. D. Cranley, C. D. Hafner, B. A. Ruoff: Extended study of aneurysm formation in umbilical vein grafts. J. vasc. Surg. 4 (1986) 486–492

Kerr, Th. M., J. J. Cranley, J. R. Johnson, K. S. Lutter, G. C. Riechmann, R. D. Cranley, M. A. True, M. Sampson: Analysis of 1084 consecutive lower extremities involved with acute venous thrombosis diagnosed by duplex scanning. Surgery 108 (1990a) 520–527

Kerr, T. M., K. S. Lutter, D. M. Moeller, K. A. Hasselfeld, L. R. Roedersheimer, P. J. McKenna, J. L. Winkler, K. Spirtoff, M. G. Sampson, J. J. Cranley: Upper extremity venous thrombosis diagnosed by duplex scanning. Amer. J. Surg. 160 (1990b) 202–206

Killewich, L. A., R. Martin, M. Cramer, K. W. Beach, D. E. Strandness jr.: An objective assessment of the physiologic changes in the post-thrombotic syndrome. Arch. Surg. 120 (1985) 424–426

Killewich, L. A., G. R. Bedford, K. W. Beach, D. E. Strandness jr.: Spontaneous lysis of deep venous thrombi: Rate and outcome. J. vasc. Surg. 9 (1989a) 89–97

Killewich, L. A., G. R. Bedford, K. W. Beach, D. E. Strandness jr.: Diagnosis of deep venous thrombosis. A prospective study comparing duplex sanning to contrast venography. Circulation 79 (1989b) 810–814

Killewich, L. A., J. D. Nunnelee, A. I. Auer: Value of lower extremity venous duplex examination in the diagnosis of pulmonary embolism. J. vasc. Surg. 17 (1993) 934–939

Kempczinski, R. F.: Presidential address: Lord Kelvin and the golden goose—Challenging times for the vascular laboratory. J. vasc. Surg. 19 (1994) 773–777

Knudson, G. J., D. A. Wiedmeyer, S. J. Erickson, W. D. Foley, T. L. Lawson, M. W. Mewissen, E. O. Lipchik: Color Doppler sonographic imaging in the assessment of upper-extremity deep venous thrombosis. Amer. J. Radiol. 154 (1990) 399–403

Köhler, M., D. Neuerburg: Recherche de la fonction valvulaire à l'aide de la technique ultrasonique et de l'auscultation pedant la manoeuvre de valsalva. Phlebologie 31 (1978) 269–272

Kriessmann, A., A. Bollinger, H. M. Keller: Praxis der Doppler-Sonographie. Thieme, Stuttgart 1990

Krings, W., J. Adolph, S. Diederich, S. Urhahne, P. Vassallo, P. E. Peters: Diagnostik der tiefen Becken- und Beinvenenthrombose mit hochauflösender Real-time- und CW-Doppler-Sonographie. Radiologe 30 (1990) 525–531

Krupski, W. C., A. Bass, R. B. Dilley, E. F. Bernstein, S. M. Otis: Propagation of deep venous thrombosis identified by duplex ultrasonography. J. vasc. Surg. 12 (1990) 467–475

Labropoulos, N., M. Leon, A. N. Nicolaides, O. Sowade, N. Volteas, F. Ortega, P. Chan: Venous reflux in patients with previous deep venous thrombosis: Correlation with ulceration and other symptoms. J. vasc. Surg. 20 (1994) 20–26

Langer, M., R. Langer: Radiologisch erfaßbare Veränderungen der Angiodysplasien Typ Klippel-Trenaunay und Typ Servelle-Martorell. Fortschr. Röntgenstr. 136 (1982) 577–582

Langholz, J., H. Heidrich: Sonographische Diagnose der tiefen Becken-/Beinvenenthrombose: Ist die farbkodierte Duplexsonographie „überflüssig"? Ultraschall Med. 12 (1991) 176–181

Langsfeld, M., F. B. Hershey, L. Thorpe, A. I. Auer, H. B. Binnington, J. J. Hurley, J. J. Woods: Duplex B-mode imaging for the diagnosis of deep venous thrombosis. Arch. Surg. 122 (1987) 587–591

Leipner, N., R. Janson, J. Kühr: Angiomatöse Dysplasie (Typ F. P. Weber). Fortschr. Röntgenstr. 137 (1982) 73–77

Lensing, A. W. A., P. Prandoni, D. Brandjes, P. M. Huisman, M. Vigo, G. Tomasella, J. Krekt, J. W. Ten Cate, M. V. Huisman, H. R. Büller: Detection of deep-vein thrombosis by real-time B-mode ultrasonography. New Engl. J. Med. 320 (1989) 342–345

van Limborgh, J.: La nomenclature des veines communicantes de l'extremité inferier. Rapport du Comité des Nomenclature de la Société Beneluxienne des Phlebologie. Anat. Inst., Amsterdam 1963

Lindner, D. J., J. M. Edwards, E. S. Phinney, L. M. Taylor, J. M. Porter: Long-term hemodynamic and clinical sequelae of lower extremity deep vein thrombosis. J. vasc. Surg. 4 (1986) 436–442

Lohr, J. M., T. M. Kerr, K. S. Lutter, R. D. Cranley, K. Spirtoff, J. J. Cranley: Lower extremity calf thrombosis: To treat or not to treat? J. vasc. Surg. 14 (1991) 18–23

Longley, D. G., D. E. Finaly, J. G. Letourneau: Sonography of the upper extremity and jugular veins. Amer. J. Radiol. 160 (1993) 957–962

Loose, D. A., J. Drewes: Venous aneurysm. Union Internationale de Phlebologie. 8éme Congress mondial. Bruxelles, 1983. Book of Abstracts, Nr. 149

Lutter, K. S., Th. M. Kerr, L. R. Roedersheimer, J. M. Lohr, M. G. Sampson, J. J. Cranley: Superficial thrombophlebitis diagnosed by duplex scanning. Surgery 110 (1991) 42–46

Malan, E.: Vascular Malformations (Angiodysplasias) Carlo Erba Foundation, Milan 1974

Mantoni, M.: Deep venous thrombosis: Longitudinal study with duplex US. Radiology 179 (1991) 271–273

Markel, A., R. A. Manzo, R. O. Bergelin, D. E. Strandness jr.: Valvular reflux after deep vein thrombosis: Incidence and time of occurrence. J. vasc. Surg. 15 (1992) 377–384

Marshall, M.: Die Duplex-Sonographie bei phlebologischen Fragestellungen in Praxis und Klinik. Ultraschall Klin. Prax. 5 (1990) 51–56

Marshall, M.: Differenzierte Beurteilung der „proximalen Beinveneninsuffizienz" durch die Duplex-Sonographie. Perfusion 6 (1991) 191–197

Martorell, F., J. Monserrat: Atresic iliac vein and Klippel-Trenaunay syndrome. Angiology 13 (1962) 265–267

Matteson, B., M. Langsfeld, C. Schermer, W. Johnson, E. Weinstein: Role of venous duplex scanning in patients with suspected pulmonary embolism. J. vasc. Surg. 24 (1996) 768–773

Mattos, M. A., G. L. Londrey, D. W. Leutz, K. J. Hodgson, D. E. Ramsey, L. D. Barkmeier, E. S. Stauffer, D. P. Spadone, D. S. Sumner: Color-flow duplex scanning for the surveillance and diagnosis of acute deep venous thrombosis. J. vasc. Surg. 15 (1992) 366–376

Mattos, M. A., G. Melendres, D. S. Sumner, D. B. Hood, L. D. Barkmeier, K. J. Hodgson, D. E. Ramsey: Prevalence and distribution of calf vein thrombosis in patients with symptomatic deep venous thrombosis: A color-flow duplex study. J. vasc. Surg. 24 (1996) 738–744

May, R., R. Nißl: Phlebographische Studien zur Anatomie der Beinvenen. Fortschr. Röntgenstr. 104 (1966) 171

May, R.: Chirurgie der Bein- und Beckenvenen. Thieme, Stuttgart 1974

Meissner, M. H., M. T. Caps, R. O. Bergelin, R. A. Manzo, D. E. Strandness jr.: Propagation, rethrombosis and new thrombus formation after acute deep venous thrombosis. J. vasc. Surg. 22 (1995) 558–567

Möllmann, M., W. Wagner, H. D. Böttcher, P. Lawin: Lagekontrolle zentralvenöser Katheter mittels Ultraschall. Ultraschall 8 (1987) 215–217

Murphy, T. P., J. J. Cronan: Evolution of deep venous thrombosis: A prospective evaluation with US. Radiology 177 (1990) 543–548

Nicholls, St. C., J. K. O'Brien, M. G. Sutton: Venous thromboembolism: Detection by duplex scanning. J. vasc. Surg. 23 (1996) 511–516

Nonnast-Daniel, B., R. P. Martin, O. Lindert, A. Mügge, J. Schaeffer, H. v. d. Lieth, E. Söchtig, M. Galanski, K.-M. Koch, W. G. Daniel: Colour doppler ultrasound assessment of arteriovenous haemodialysis fistulas. Lancet 339 (1992) 143–145

O'Leary, D. H., R. A. Kane, B. M. Chase: A prospective study of the efficacy of B-scan sonography in the detection of deep venous thrombosis in the lower extremities. J. clin. Ultrasound 16 (1988) 1–8

O'Shaughnessy, A. M., D. E. FitzGerald: Organization patterns of venous thrombus over time as demonstrated by duplex ultrasound. J. vasc. Investig. 2 (1996) 75–81

Partsch, H., O. Lofferer: Untersuchungen des venösen Rückstroms aus der unteren Extremität mit einem direktionalen Ultraschalldopplerdetektor. Wien. klin. Wschr. 83 (1971) 781–789

Partsch, H., O. Lofferer, A. Mostbeck: Zur Diagnostik von arteriovenösen Fisteln bei Angiodysplasien der Extremitäten. Vasa 4 (1975) 288–295

Partsch, H.: "A-sounds" or "S-sounds" for Doppler ultrasonic evaluation of pelvic vein thrombosis. Vasa 5 (1976) 16–19

Persson, A. V., C. Jones, R. Zide, E. R. Jewell: Use of the triplex scanner in diagnosis of deep venous thrombosis. Arch. Surg. 124 (1989) 593–596

Poppiti, R., G. Papanicolaou, S. Perese, F. A. Weaver: Limited B-mode venous imaging versus complete color-flow duplex venous scanning for detection of proximal deep venous thrombosis. J. vasc. Surg. 22 (1995) 553–557

Pulliam, C. W., S. L. Barr, A. B. Ewing: Venous duplex scanning in the diagnosis and treatment of progressive superficial thrombophlebitis. Ann. vasc. Surg. 5 (1991) 190–195

Quinn, K. L., F. N. Vandeman: Thrombosis of a duplicated superficial femoral vein. J. Ultrasound Med. 9 (1990) 235–238

Raghavendra, B. N., R. J. Rosen, St. Lam, Th. Riles, St. C. Horii: Deep venous thrombosis: Detection by high-resolution real-time ultrasonography. Radiology 152 (1984) 789–793

Raghavendra, B. N., St. C. Horii, S. Hilton, B. R. Subramanyam, R. J. Rosen, St. Lam: Deep venous thrombosis: Detection by probe compression of veins. J. Ultrasound Med. 5 (1986) 89–95

van Ramshorst, B., D. A. Legemate, J. F. Verzijlbergen, H. Hoeneveld, B. C. Eikelboom, J. C. de Valois, O. J. A. Th. Meuwissen: Duplex scanning in the diagnosis of acute deep vein thrombosis of lower extremity. Europ. J. vasc. Surg. 5 (1991) 255–260

van Ramshorst, B., P. S. van Bemmelen, H. Hoeneveld, B. C. Eikelboom: The development of valvular incompetence after deep vein thrombosis: A follow-up study with duplex scanning. J. vasc. Surg. 20 (1994) 1059–1066

Richter, G., S. Böhm, Ch. Görg, W. B. Schwerk: Verlaufsuntersuchungen zur Echogenität venöser Gerinnungsthromben. Ultraschall Klin. Prax. 7 (1992) 69–73

Ritter, H., J. Weber, D. A. Loose: Venöse Aneurysmen. Vasa 22 (1993) 105–112

Rodrigues, A. A., C. M. Witehead, R. L. McLaughlin, S. E. Umphrey, H. J. Welch, T. F. O'Donnell: Duplex-derived valve closure times fail to correlate with reflux flow volumes in patients with chronic venous insufficiency. J. vasc. Surg. 23 (1996) 606–610

Rollins, D. L., C. M. Semrow, M.L. Friedell, K. D. Calligaro, D. Buchbinder: Progress in the diagnosis of deep venous thrombosis: The efficacy of real-time B-mode ultrasonic imaging. J. vasc. Surg. 7 (1988) 638–641

Rose, St. C., W. J. Zwiebel, B. D. Nelson, D. L. Priest, R. A. Knighton, J. W. Brown, P. F. Lawrence, B. M. Stults, J. C. Reading, F. J. Miller: Symptomatic lower extremity deep venous thrombosis: Accuracy, limitations, and role of color duplex flow imaging in diagnosis. Radiology 175 (1990) 639–644

Ruoff, B. A., J. J. Cranley, L. A. Hannan, N. Aseffa, W. S. Karkow, K. G. Stedje, R. D. Cranley: Real-time duplex ultrasound mapping of the greater saphenous vein before in situ infrainguinal revascularization. J. vasc. Surg. 6 (1987) 107–113

Rutherford, R. B.: New approaches to the diagnosis of congenital vascular malformations. In: Belov, St., D. A. Loose, J. Weber, ed: Vascular Malinformations. Einhorn-Presse Verlag, Reinbek 1989

Schindler, J. M., M. Kaiser, A. Gerber, A. Vuilliomenet, A. Popovic, O. Bertel: Colour coded duplex sonography in suspected deep vein thrombosis of the leg. Br. Med. J. 301 (1990) 1369–1370

Schmitt, H. E.: Aszendierende Phlebographie bei tiefer Venenthrombose. Huber, Stuttgart 1977

Schönhofer, B., H. Bechtold, R. Renner, H.-D. Bundschu: Sonographische Befunde bei Varikophlebitis der Vena saphena magna. Dtsch. med. Wschr. 117 (1992) 51–55

Schoop, W.: Praktische Angiologie, 4. Aufl. Thieme, Stuttgart 1988

Schweizer, J., F. Oehmichen, H. G. Brandl, E. Altman: Farbkodierte Duplexsonographie und kontrastmittelverstärkte Duplexsonographie bei tiefer Beinvenenthrombose. Vasa 22 (1993) 22–25

Servelle, M.: Les malformations congénitales des veins. Revue de Chirurgie. 68 (1949) 88

Shami, S. K., S. Sarin, T. R. Cheatle, J. H. Scurr, P. D. Coleridge Smith: Venous ulcers and the superficial venous system. J. vasc. Surg. 17 (1993) 487–490

Sigel, B., W. R. Felix jr., G. L. Popky, J. Ipsen: Diagnosis of lower limb venous thrombosis by Doppler ultrasound techniques. Arch. Surg. 104 (1972) 174–179

Sigel, B., G. L. Popky, D. K. Wagner, J. P. Boland, D. E. McMapp, P. Feigl: Comparison of clinical and Doppler ultrasound evaluation of confirmed lower extremity venous disease. Surgery 69 (1986) 332

Skillman, J. J., K. C. Kent, D. H. Porter, D. Kim: Simultaneous occurrence of superficial and deep thrombophlebitis in the lower extremity. J. vasc. Surg. 11 (1990) 818

Söldner, J., W. Lösch, E.-I. Richter, E. Zeitler: Radiologische Diagnostik der Venenerkrankungen bei geriatrischen Patienten. Phlebol. u. Proktol. 22 (1993) 253–258

Stapff, M., G. Betzl, G. V. Küffer, D. Hahn, F. A. Spengel: Stellenwert der Duplex-Sonographie in der Diagnostik der tiefen Bein- und Beckenvenenthrombose. Bildgebung/Imaging 56 (1989) 52–56

Stiegler, H., G. Rotter, R. Standl, S. Mosavi, H. J. v. Kooten, B. Weichenhain, G. Baumann: Wertigkeit der Farb-Duplexsonographie in der Diagnose insuffizienter Vv. perforantes. Eine prospektive Untersuchung an 94 Patienten. Vasa 23 (1994) 109–113

Strandness jr., D. E., D. S. Summer: Ultrasonic velocity detector in the diagnosis of thrombophlebitis. Arch. Surg. 104 (1972) 180–183

Strandness jr., D. E., B. L. Thiele: Selected Topics in Venous Disorders. Futura, Mount Kisco (N. Y.) 1981

Strandness jr., D. E., Y. Langlois, M. Cramer, A. Randlett, B. L. Thiele: Long-term sequelae of acute venous thrombosis. J. Amer. med. Ass. 250 (1983) 1289–1292

Strandness jr., D. E.: Duplex Scanning in Vascular Disorders. Raven Press, New York 1990

Straub, H., M. Ludwig: Der Doppler-Kurs. Doppler-Sonographie der peripheren Arterien und Venen. Zuckschwerdt, München 1990

Strothman, G., J. Blebea, R. J. Fowl, G. Rosenthal: Contralateral duplex scanning for deep venous thrombosis is unnecessary in patients with symptoms. J. vasc. Surg. 22 (1995) 543–547

Sullivan, E. D., D. J. Peter, J. J. Cranley: Real-time B-mode venous ultrasound. J. vasc. Surg. 1 (1984) 465–471

Sumner, D. S., A. Lambeth: Reliability of Doppler ultrasound in the diagnosis of acute venous thrombosis both above and below the knee. Amer. J. Surg. 138 (1979) 205–209

Sumner, D. S.: Venous anatomy and pathophysiology. In Hershey, F. B., R. W. Barnes, D. S. Sumner: Noninvasive Diagnosis of Vascular Disease. Appleton Davies, Pasadena 1984

Talbot, S. R.: Use of real-time imaging in identifying deep venous obstruction. A primary report. Bruit 65 (1982) 41

Talbot, S. R.: B-mode evaluation of peripheral arteries and veins. In Zwiebel, W. J.: Introduction to Vascular Ultrasonography. Grune & Stratton, New York 1986 (p. 351)

Taylor, B. L., I. Yellowlees: Central venous cannulation using the infraclavicular axillary vein. Anesthesiology 72 (1990) 55–58

Thulesius, O.: Pathophysiologische Gesichtspunkte über den venösen Rückstrom. In Kriessmann, A., A. Bollinger: Ultraschall-Doppler-Diagnostik in der Angiologie. Thieme, Stuttgart 1978

Towner, K. M., A. E. McDonnell, J. K. Turcotte, Ch. K. Zarins: Noninvasive assessment of upper extremity deep venous obstructions. Bruit 5 (1981) 21–22

de Valois, J. C., C. C. van Schaik, F. Verzijlbergen, B. van Ramshorst, B. C. Eikelboom, O. J. A. Th. Meuwissen: Contrast venography: from gold standard to 'golden backup' in clinically suspected deep vein thrombosis. Europ. J. Radiol. 11 (1990) 131–137

Vasdekis, S. N., H. Clarke, A. N. Nicolaides: Quantification of venous reflux by means of duplex scanning. J. vasc. Surg. 10 (1989) 670–677

Voet, D., M. Afschrift: Floating thrombi: Diagnosis and follow-up by duplex ultrasound. Brit. J. Radiol. 64 (1991) 1010–1014

Vogel, P., F. C. Laing, R. B. Jeffrey, V. W. Wing: Deep venous thrombosis of the lower extremity: US evaluation. Radiology 163 (1987) 747–751

Vollmar, J. F., E. Paes, B. Irion, J. M. Friedrich, B. Heymer: Aneurysmatische Transformation des Venensystems bei venösen Angiodysplasien der Gliedmaßen. Vasa 18 (1989) 96–111

Voss, E. U.: Angiodysplasia of extremities. In Nobbe, F., H. Hammann: Gefäßchirurgie – Möglichkeiten und Grenzen. TM Verlag, Bad Oeynhausen 1984

Weber, J., R. May: Funktionelle Phlebologie. Thieme, Stuttgart 1990

Weingarten, M. S., C. C. Branas, M. Czeredarczuk, J. D. Schmidt, C. C. Wolferth: Distribution and quantification of venous reflux in lower extremity chronic venous stasis disease with duplex scanning. J. vasc. Surg. 18 (1993) 753–759

Weingarten, M. S., M. Czeredarczuk, S. Scovell, C. C. Branas, G. M. Mignogna, C. C. Wolferth: A correlation of air plethysmography and color-flow-assisted duplex scanning in the quantification of chronic venous insufficiency. J. vasc. Surg. 24 (1996) 750–754

Weissleder, R., G. Elizondo, D. D. Stark: Sonographic diagnosis of subclavian and internal jugular vein thrombosis. J. Ultrasound Med. 6 (1987) 577–587

Welch, H. J., J. L. Villavicencio: Primary varicose veins of the upper extremity: A report of three cases. J. vasc. Surg. 20 (1994) 839–843

Welch, H. J., C. M. Young, A. B. Semegran, M. D. Iafrati, W. C. Mackey, T. F. O'Donnell: Duplex assessment of venous reflux and chronic venous insufficiency: The significance of deep venous reflux. J. vasc. Surg. 24 (1996) 755–762

Wright, A. T., T. M. Kerr, D. M. Moeller, J. L. Winkler, J. J. Cranley: Analysis of 68 consecutive acute deep venous thrombi of the upper extremities diagnosed by duplex scanning. J. vasc. Techn. 13 (1989) 221–223

Wright, D. J., A. D. Shepard, M. McPharlin, C. B. Ernst: Pitfalls in lower extremity venous duplex scanning. J. vasc. Surg. 11 (1990) 675–679

Wuppermann, Th., U. Exler, J. Mellmann, M. Kestilä: Non-invasive quantitative measurement of regurgitation in insufficiency of the superior saphenous vein by Doppler-ultrasound: A comparison with clinical examination and phlebography. Vasa 10 (1981) 24–27

Yao, S. T., C. Gourmos, J. T. Hobbs: Detection of proximal-vein thrombosis by Doppler ultrasound flow-detection method. Lancet 1972/I, 1–4

Yucel, E. K., J. S. Fisher, Th. K. Egglin, St. C. Geller, A. C. Waltman: Isolated calf venous thrombosis: Diagnosis with compression US. Radiology 179 (1991) 443–446

Yucel, E. K., T. K. Egglin, A. C. Waltman: Extension of saphenous thrombophlebitis into the femoral vein: Demonstration by color flow compression sonography. J. Ultrasound Med. 11 (1992) 285–287

Zwiebel, W. J.: Introduction to Vascular Ultrasonography. Saunders, Philadelphia 1992

Chapter 7

Abad, J., E. G. Hidalgo, J. M. Cantarero, G. Parga, R. Fernandez, M. Gomes, F. Colina, E. Moreno: Hepatic artery anastomotic stenosis after transplantation: Treatment with percutaneous transluminal angioplasty. Radiology 171 (1989) 661–662

Aldoori, M. I., M. I. Qamar, A. E. Read, R. C. N. Williamson: Increased flow in the superior mesenteric artery in dumping syndrome. Brit. J. Surg. 72 (1985) 389–390

Alvarez, G., M. Gonzalez-Molina, M. Cabello, A. Gomes: Pulsed and continuous Doppler evaluation of renal dysfunction after kidney transplantation. Europ. J. Radiol. 12 (1991) 108–112

Arienti, V., C. Califano, G. Brusco, L. Boriani, F. Biagi, M. Giulia-Sama, S. Sottili, A. Domanico, G. R. Corazza, G. Gasbarrini: Doppler ultrasonographic evaluation of splanchnic blood flow in coeliac disease. Gut 39 (1996) 369–373

Arima, M., T. Ogino, S. Hosokawa, H. Ihara, F. Ikoma: Functional image diagnosis of kidney transplants using ultrasonic Doppler flow-metry and magnetic resonance imaging. Transplant. Proc. 21 (1989) 1907–1911

Avasthi, P. S., W. F. Voyles, E. R. Greene: Noninvasive diagnosis of renal artery stenosis by echo-Doppler velocimetry. Kidney int. 25 (1984) 824–829

Barth jr., W. H., D. R. Genest, L. E. Riley, F. D. Frigoletto jr., B. R. Benacerraf, M. F. Greene: Uterine arcuate artery Doppler and decidual microvascular pathology in pregnancies complicated by type I diabetes mellitus. Ultrasound Obstet. Gynecol. 8 (1996) 98–103

Barton, P., H. Jantsch, W. Pichler, H. Schurawitzki, R. Stiglbauer, G. Lechner: Ultraschalldiagnostik nach Lebertransplantationen. Fortschr. Röntgenschr. 151 (1989) 145–153

Baxter, G. M., H. Ireland, J. G. Moss, P. N. Harden, B. J. Junor, R. S. Rodger, J. D. Briggs: Colour Doppler ultrasound in renal transplant artery stenosis: Which Doppler index? Clin. Radiol. 50 (1995) 618–622

Baxter, G. M., F. Aitchison, D. Sheppard, J. G. Moss, M. J. McLeod, P. N. Harden, J. G. Love, M. Robertson, G. Taylor: Colour Doppler ultrasound in renal artery stenosis: Intrarenal waveform analysis. Br. J. Radiol. 69 (1996) 810–815

Ben-Shoshan, M., N. P. Rossi, M. E. Korns: Coarctation of the abdominal aorta. Arch Pathol. 95 (1973) 221–225

Berland, L. L., D. B. Koslin, W. D. Routh, F. S. Keller: Renal artery stenosis: Prospective evaluation of diagnosis with color duplex US compared with angiography. Radiology 174 (1990) 421–423

Bernaschek, G., J. Deutinger, M. Endler: Derzeitiger Stand der Vaginalsonographie – eine weltweite Umfrage. Geburtsh. u. Frauenheilk. 51 (1991) 729–733

Blum, U., B. Krumme, P. Flügel, A. Gabelmann, Th. Lennert, C. Buitrago-Tellez, P. Schollmeyer, M. Langer: Treatment of ostial renal-artery stenoses with vascular endoprotheses after unsuccessful balloon angioplasty. New Engl. J. Med. 336 (1997) 459–465

Bluth, E. I., S. M. Murphey, L. H. Hollier, M. A. Sullivan: Color flow Doppler in the evaluation of aortic aneurysms. Int. Angiol. 9 (1990) 8–10

Bowersox, J. C., R. M. Zwolak, D. B. Walsh, J. R. Schneider, A. Musson, E. LaBombard, J. L. Cronenwett: Duplex ultrasonography in the diagnosis of celiac and mesenteric artery occlusive disease. J. vasc. Surg. 14 (1991) 780–788

Branger, B., M. Dauzat, S. Ovtchinnikoff, F. Vecina, B. Zabadani, G. Mourad: Renal arterial fistula after transplant biopsy: An unusual complication detected by colour Doppler imaging. Transplant. Proc. 27 (1995) 2440

Breitenseher, M., F. Kainberger, P. Hübsch, S. Trattnig, M. Baldt, P. Barton, F. Karnel: Screening von Nierenarterienstenosen. Fortschr. Röntgenstr. 156 (1992) 228–231

Bunchman, T. E., H. S. J. Walker III, P. F. Joyce, M. E. Danter, M. J. Silberstein: Sonographic evaluation of renal artery aneurysm in childhood. Pediat. Radiol. 21 (1991) 312–313

Burns, P. N.: Doppler ultrasound in obstetrics. In Taylor, K. J. W., P. N. Burns, N. T. Wells: Clinical Applications of Doppler Ultrasound. Raven Press, New York 1988

Burns, P. N.: Harmonic imaging with ultrasound contrast agents. Clin. Radiol. 51 (Suppl. 1) (1996) 50–55

Cantarero, J. M., J. G. Llorente, E. G. Hidalgo, A. Hualde, R. Ferreiro: Splenic arteriovenous fistula: Diagnosis by duplex Doppler sonography. Amer. J. Roentgenol. 153 (1989) 1313–1314

Chow, W.-H., Y.-T. Tai, K.-L. Cheung, A. S. B. Yip: Spontaneous dynamic echoes in aortic dissection. J. clin. Ultrasound 18 (1990) 442–445

Coppens, M., P. Loquet, M. Kollen, F. De-Neubourg, P. Buytaert: Longitudinal evaluation of uteroplacental and umbilical blood flow changes in normal early pregnancy. Ultrasound Obstet. Gynecol. 7 (1996) 114–121

Crawford, E. S., K. H. Hess: Abdominal aortic aneurysms. New Engl. J. Med. 321 (1989) 1040–1042

Cronenwett, J. L., T. F. Murphy, G. B. Zelenock et al. Actuarial analysis of variables associated with rupture of small abdominal aortic aneurysms. Surgery 98 (1985) 472–483

Cronenwett, J. L., S. K. Sargent, M. H. Wall, M. L. Hawkes, D. H. Freeman, B. J. Dain, J. K. Curé, D. B. Walsh, R. M. Zwolak, M. D. McDaniel, J. R. Schneider: Variables that affect the expansion rate and outcome of small abdominal aortic aneurysms. J. Vasc. Surg. 11 (1990) 260–269

Curé, D., B. Walsh, R. M. Zwolak, M. D. McDaniel, J. R. Schneider: Variables that affect the expansion rate and outcome of small abdominal aortic aneurysms. J. vasc. Surg. 11 (1990) 260–269

Danse, E. M., B. E. Van-Beers, P. Goffette, A. N. Dardenne, P. F. Laterre, J. Pringot: Acute intestinal ischemia due to occlusion of the superior mesenteric artery: Detection with Doppler sonography. J. Ultrasound Med. 15 (1996) 323–326

Dawson, D. L.: Noninvasive assessment of renal artery stenosis. Semin. vasc. Surg. 9 (1996) 172–181

Deane, C., T. Cairns, H. Walters, A. Palmer, V. Parsons, V. Roberts, D. Taube: Diagnosis of renal transplant artery stenosis by color Doppler ultrasonography. Transplant. Proc. 22 (1990) 1395

Deane, C.: Doppler and color Doppler ultrasonography in renal transplants: Chronic rejection. J. clin. Ultrasound 20 (1992) 539–544

De Bakey, M. E., W. S. Henly, D. A. Cooley, G. C. Morris, E. S. Crawford, A. C. Beall: Surgical management of dissecting aneurysms of the aorta. J. thorac. cardiovasc. Surg. 49 (1965) 130–149

de Jong, N., F. J. Ten-Cate: New ultrasound contrast agents and technological innovations. Ultrasonics 34 (1996) 587–590

Desberg, A. L., D. M. Paushter, G. K. Lammert, J. C. Hale, R. B. Troy, A. C. Novick, J. V. Nally jr., A. M. Weltevreden: Renal artery stenosis: Evaluation with color Doppler flow imaging. Radiology 177 (1990) 749–753

de Toledo, L. S., T. Martinez-Berganza-Asensio, R. Cozcolluela-Cabrejas, M. A. de Gregorio-Ariza, P. Pardina-Cortina, L. Ripa-Saldias: Doppler-duplex ultrasound in renal colic. Europ. J. Radiol. 23 (1996) 143–148

Deutinger, J., R. Rudelstorfer, G. Bernaschek: Transvaginale gepulste Dopplermessungen von Strömungsgeschwindigkeiten in Beckengefäßen nach Zyklusstimulation. Geburtsh. u. Frauenheilk. 49 (1989 a) 33–36

Deutinger, J., R. Rudelstorfer, G. Bernaschek: Vergleich von transvaginalen und transabdominalen Doppler-Strömungsmessungen in Uterusgefäßen bei unauffälligem Schwangerschaftsverlauf. Ultraschall 10 (1989 b) 15–18

Distler, A., K.-P. Spies, F. Fobbe: Diagnostik der renovaskulären Hypertonie. Dtsch. Ärztebl. 89 (1992) B 596–603

Don, S., K. K. Kopecky, R. S. Filo, S. B. Leapman, J. V. Thomalla, J. A. Jones, E. C. Klatte: Duplex Doppler US of renal allografts: Causes of elevated resistive index. Radiology 171 (1989) 709–712

Drake, D. G., D. L. Day, J. G. Letourneau, B. A. Alford, R. K. Sibley, S. M. Mauer, T. E. Bunchman: Doppler evaluation of renal transplant in children: A prospective analysis with histopathological correlation. Amer. J. Roentgenol. 154 (1990) 785–787

Dudiak, C. M., C. G. Salomon, H. V. Posniak, M. C. Olson, M. E. Flisak: Sonography of the umbilical cord. Radiographics 15 (1995) 1035–1050

Edwards, J. M., M. J. Zaccardi, D. E. Strandness jr.: A preliminary study of the role of duplex scanning in defining the adequacy of treatment of patients with renal artery fibromuscular dysplasia. J. vasc. Surg. 15 (1992) 604–611

Elliott jr., J. P., T. L. Ashley: Aneurysmen der Viszeralarterien. In Sandmann, W., H. W. Kniemeyer: Aneurysmen der großen Arterien: Diagnostik und Therapie. Huber Bern 1991

Emamian, S. A., M. B. Nielsen, J. F. Pedersen, L. Ytte: Kidney dimensions at sonography: Correlation with age, sex, and habitus in 665 adult volunteers. Amer. J. Roentgenol. 160 (1993) 83–86

Endress, C., G. A. Kling, B. L. Medrazo: Diagnosis of hepatic artery aneurysm with portal vein fistula using image-directed doppler ultrasound. J. clin. Ultrasound 17 (1990) 206–208

Evans, D., J. Cochlin, C. Ferguson, P. J. Griffin, J. R. Salaman: Duplex Doppler studies in acute renal transplant rejection. Transplant. Proc. 21 (1989) 1897–1898

Feichtinger, W., M. Putz, P. Kemeter: Transvaginale Doppler-Sonographie zur Blutflußmessung im kleinen Becken. Ultraschall 9 (1988) 30–36

Feindt, P., P. Walter, G. Omior: Das Kompressionssyndrom des Truncus coeliacus: Ein seltenes Krankheitsbild. Vasa 21 (1992) 307–309

Ferretti, G., A. Salomone, P. L. Castagno: Renovascular hypertension: a non-invasive duplex scanning screening. Int. Angiol. 7 (1988) 219–223

Flinn, W. R., R. J. Rizzo, J. S. Park, G. P. Sandager: Duplex scanning for assessment of mesenteric ischemia. Surg. Clin. N. Amer. 70 (1990) 99–107

Flint, E. W., J. H. Sumkin, A. B. Zajko, A. Bowen: Duplex sonography of hepatic artery thrombosis after liver transplantation. Amer. J. Roentgenol. 151 (1988) 481–483

Flückiger, F., H. Steiner, S. Horn, M. Ratschek, E. Deu: Farbkodierte Duplexsonographie und Widerstandsindex bei Nierentransplantation mit Dysfunktion. Fortschr. Röntgenstr. 153 (1990) 692–697

Flynn, M. K., D. Levine: The noninvasive diagnosis and management of a uterine arteriovenous malformation. Obstet. Gynecol. 88 (1996) 650–652

Fobbe, F., L. Schudrowitsch, M. Diezel, K.-J. Wolf: Verlaufskontrollen von Nierentransplantaten mit der farbkodierten Duplexsonographie. Fortschr. Röntgenstr. 150 (1989) 76–79

Fransen, H., R. Kubale, K.-D. Wurche, A. Kalähne: Nicht-invasive Diagnostik von Milzarterienaneurysmata. Fortschr. Röntgenstr. 151 (1989) 532–535

Frauchiger, B., R. Zierler, R. O. Bergelin, J. A. Isaacson, D. E. Strandness jr.: Prognostic significance of intrarenal resistance indices in patients with renal artery interventions: A preliminary duplex sonographic study. Cardiovasc. Surg. 4 (1996) 324–330

Frazin, L. J., M. Siddiqui, K. Venugopalan, P. Pop, M. J. Vonesh, D. D. McPherson: Feasibility of transcolonic and transgastric abdominal vascular ultrasound. Amer. J. Card. Imaging. 8 (1994) 95–99

Fröhlich, E., P. Frühmorgen, E. Stahl, J. Treichel: Aneurysma spurium der Arteria pancreatioduodenalis. Ultraschall 9 (1988) 138–140

Funk, A., H. Jörn, H. Fendel: Abdominelle versus transvaginale Doppler-Sonographie der uterinen Gefäße. Ultraschall Klin. Prax. 7 (1992) 264–268

Gage, T. S., S. K. Sussman, F. U. Conard III, D. Hull, S. A. Bartus: Pseudoaneurysm of the inferior epigastric artery: Diagnosis and percutaneous treatment. Amer. J. Roentgenol. 155 (1990) 529–530

Gainza, F. J., I. Minguela, I. Lopez-Vidaur, L. M. Ruiz, I. Lampreabe: Evaluation of complications due to percutaneous renal biopsy in allografts and native kidneys with color-coded Doppler sonography. Clin. Nephrol. 43 (1995) 303–308

Geelkerken, R. H., T. A. Delahunt, L. J. Schultze-Kool, J. M. van-Baalen, J. Hermans, J. H. van-Bockel: Pitfalls in the diagnosis of origin stenosis of the coeliac and superior mesenteric arteries with transabdominal color duplex examination. Ultrasound Med. Biol. 22 (1996) 695–700

Giyanani, V. L., C. A. Krebs, C. A. Nall, R. L. Eisenberg, H. R. Parvey: Diagnosis of abdominal aortic dissection by imagedirected Doppler sonography. J. clin. Ultrasound 17 (1989) 445–448

Gleeson, F. V., M. M. Fitzpatrick, J. Somers, C. Kennedy, R. De Bruyn, T. M. Barratt: Duplex Doppler ultrasound in the investigation of occult nephropathy following haemolytic uraemic syndrome. Brit. J. Radiol. 65 (1992) 137–139

Goldberg, B. B., J. B. Liu, F. Forsberg: Ultrasound contrast agents: A review. Ultrasound Med. Biol. 20 (1994) 319–333

Grab, D., K. Sterzik, R. Terinde: Ultraschalluntersuchungen in der Schwangerschaft: Technische Grundlagen und klinische Sicherheit. Geburtsh. u. Frauenheilk. 52 (1992) 721–729

Grant, E. G., F. N. Tessler, R. Perrella: Clinical Doppler imaging. Amer. J. Roentgenol. 152 (1989) 707–717

Grech, P., Rowlands, M. Crofton: Aneurysm of the inferior pancreaticoduodenal artery diagnosed by real-time ultrasound and pulsed Doppler. Brit. J. Radiol. 62 (1989) 753–755

Grenier, N., C. Douws, D. Morel, J.-M. Ferriére, M. Le Guillou, L. Potaux, J. Broussin: Detection of vascular complications in renal allografts with color Doppler flow imaging. Radiology 178 (1991) 217–223

Gritzmann, N., I. Huk, H. Schurawitzki, B. Teleky: Duplexsonographie der Aorten- und Bifurkationsprothesen. Vasa (Suppl. 26) (1988) 33–37

Gross-Fengels, W., R. Lorenz, R. Schmidt, K. F. R. Neufang: Abdominelle Real-time-Sonographie bei Patienten mit Gefäßprothesen. Fortschr. Röntgenstr. 148 (1988) 131–136

Grün, B., H. Tschakert, J. Schaffeldt, M. Steinhoff: Asymptomatisches verkalktes Aneurysma der Arteria pancreaticoduodenalis inferior – differentialdiagnostische Überlegungen. Radiologe 29 (1989) 572–575

Grunert, D., M. Schöning, W. Rosendahl: Renal blood flow and flow velocity in children and adolescents: Duplex Doppler evaluation. Europ. J. Pediat. 149 (1990) 287–292

Hall, T. R., S. V. Mc Diarmid, E. G. Grant, M. I. Boechat, R. W. Busuttil: False-negative duplex Doppler studies in children with hepatic artery thrombosis after liver transplantation. Amer. J. Roentgenol. 154 (1990) 573–575

Handa, N., R. Fukunaga, A. Uehara, H. Etani, S. Yoneda, K. Kimura, T. Kamada: Echo-Doppler velocimeter in the diagnosis of hypertensive patients: The renal artery Doppler technique. Ultrasound Med. Biol. 12 (1986) 945–952

Handa, N., R. Fukunaga, H. Etani, S. Yoneda, K. Kimura, T. Kamada: Efficacy of echo-doppler examination for the evaluation of renovascular disease. Ultrasound Med. Biol. 14 (1988) 1–5

Harris, D. C. H., V. Antico, R. Allen, S. Gruenewald, S. Lawrence, J. H. Stewart, J. R. Chapman: Doppler assessment in renal transplantation. Transplant. Proc. 21 (1989) 1895–1896

Harward, T. R. S., S. Smith, J. M. Seeger: Detection of celic axis and superior mesenteric artery occlusive disease with use of abdominal duplex scanning. J. vasc. Surg. 17 (1993) 738–745

Hastie, S. J., C. A. Howie, M. J. Whittle, P. C. Rubin: Daily variability of umbilical and lateral uterine wall artery blood velocity waveform measurements. Brit. J. Obstet. Gynaecol. 95 (1988) 571–574

Hausegger, K. A., R. Fotter, E. Sorantin, F. Flückiger: Kongenitale intrahepatische arterio-portale Fisteln als Ursache einer nekrotisierenden Enteritis – dopplersonographische und angiographische Erfassung. Ultraschall Med. 12 (1991) 193–196

Hendrickx, Ph., U. Roth, H. von der Lieth: Wertigkeit der Angiodynographie zur Verlaufskontrolle chirurgischer Gefäßprothesen. Ultraschall Med. 12 (1991) 188–192

Hennerici, M., W. Steinke: Präoperative nicht-invasive zerebrovaskuläre Untersuchung bei Aneurysmapatienten. In Sandmann, W., H. W. Kniemeyer: Aneurysmen der großen Arterien: Diagnostik und Therapie. Huber Bern 1991

Herbetko, J., A. P. Grigg, A. R. Buckley, G. L. Phillips: Venoocclusive liver disease after bone marrow transplantation: Findings at duplex sonography. Amer. J. Roentgenol. 158 (1992) 1001–1005

Hess, C. F., G. Kölbel, M. C. Majer: Zur Reproduzierbarkeit der sonographischen Größenbestimmung intraabdomineller Raumforderungen. Fortschr. Röntgenstr. 149 (1988) 184–188

Hill, M. C., I. M. Lande, J. H. Grossman III: Duplex evaluation of fetoplacental and uteroplacental circulation. In Grant, E. G., E. M. White: Duplex Sonography. Springer, Berlin 1988

Hillman, B. J.: Imaging advances in the diagnosis of renovascular hypertension. Amer. J. Roentgenol. 153 (1989) 5–14

Hoffmann, U., J. M. Edwards, D. E. Strandness jr.: Stellenwert der Duplexsonographie in der Diagnostik von Nierenarterienstenosen. Vasa (Suppl. 32) (1990) 164–166

Hoffmann, U., J. M. Edwards, S. Carter, M. L. Goldman, J. D. Harley, M. J. Zaccardi, D. E. Strandness jr.: Role of duplex scanning for the detection of atherosclerotic renal artery disease. Kidney int. 39 (1991) 1232–1239

Hollenbeck, M., B. Grabensee: Duplexsonographische Befunde bei primärem Transplantatversagen. In Keller, E., B. Krumme: Farbkodierte Duplexsonographie in der Nephrologie. Springer, Berlin 1994

Hübsch, P. J. S., G. Mostbeck, P. P. Barton, N. Gritzmann, F. X. J. Fruehwald, H. Schurawitzki, J. Kovarik: Evaluation of arteriovenous fistulas and pseudoaneurysms in renal allografts following percutaneous needle biopsy. Color-coded Doppler sonography versus duplex Doppler sonography. J. Ultrasound Med. 9 (1990) 95–100

Hüneke, B.: Sitzungsbericht der Arbeitsgemeinschaft Dopplersonographie in der Geburtshilfe vom 10.9.90. Ultraschall Med. 12 (1991) 45–48

Imig, H., G. von Klinggräff, R. Horstmann: Sonographische Routinekontrollen nach Bauchaortenaneurysmaoperation. In Sandmann, W., H. W. Kniemeyer: Aneurysmen der großen Arterien: Diagnostik und Therapie. Huber Bern, 1991

Iwao, T., A. Toyonaga, K. Oho, T. Sakai, C. Tayama, H. Masumoto, M. Sato, K. Nakahara. K. Tanikawa: Postprandial splanchnic hemodynamic response in patients with cirrhosis of the liver: Evaluation with "triple-vessel" duplex US. Radiology. 201 (1996) 711–715

Jacobs, A., Th. Karasch, F.-J. Roth: Angiomorphologisch ungewöhnliche, klinisch asymptomatische Coarctatio der Aorta abdominalis. Fortschr. Röntgenstr. 161 (1994) 375

Jaffe, R., J. R. Woods: Doppler velocimetry of intraplacental fetal vessels in the second trimester: Improving the prediction of pregnancy complications in high-risk patients. Ultrasound Obstet. Gynecol. 8 (1996) 262–266

Jäger, K. A., G. S. Fortner, B. L. Thiele, D. E. Strandness: Noninvasive diagnosis of intestinal angina. J. clin. Ultrasound 12 (1984) 588–591

Jäger, K., A. Bollinger, C. Valli, R. Ammann: Measurements of mesenteric blood flow by duplex scanning. J. vasc. Surg. 3 (1986) 462–469

Jäger, K.: Neuere diagnostische Methoden zur nichtinvasiven Lokalisation und hämodynamischen Beurteilung arterieller Obstruktionen. Internist 30 (1989) 397–405

Jäger, K.: Moderne Möglichkeiten bei der Abklärung renovaskulärer Stenosen. Internist 32 (1991) 127–134

Jäger, K., B. Frauchiger, R. Eichlisberger, C. Beglinger: Evaluation of the gastrointestinal vascular system by duplex sonography. In Labs, K. H., K. A. Jäger, D. E. Fitzgerald, J. P. Woodcock, D. Neuerburg-Heusler: Diagnostic Vascular Ultrasound. Arnold, London 1992

Jain, K. A., R. B. Jeffrey jr., F. G. Sommer: Gynecologic vascular abnormalities: Diagnosis with Doppler US. Radiology 178 (1991) 549–551

Jansen, O., C. Strassburger, N. Marienhoff: Duplexsonographische Verlaufskontrolle eines Nierentransplantates beim hämolytischurämischen Syndrom (HUS). Fortschr. Röntgenstr. 153 (1990) 484–486

Järvinen, O., J. Laurikka, T. Sisto, J.-P. Salenius, M. R. Tarkka: Atherosclerosis of the visceral arteries. Vasa 24 (1995) 9–14

Joynt, L. K., J. F. Platt, J. M. Rubin, J. H. Ellis, R. O. Bude: Hepatic artery resistance before and after standard meal in subjects with diseases and healthy livers. Radiology 196 (1995) 489–492

Jurriaans, E., P. A. Dubbins: Renal Transplantation: The normal morphological and Doppler ultrasound examination. J. clin. Ultrasound 20 (1992) 495–506

Kamps, J., B. Kleuren, Th. Karasch, A. L. Strauss: Einfluß von Respiration und Valsalva-Test auf den Resistance-Index RI in der A. carotis interna, A. renalis und A. hepatica. Vasa 21 (1992) 444

Karasch, Th., M. Worringer, A. L. Strauss, D. Neuerburg-Heusler, F.-J. Roth, H. Rieger: Diagnostische Aussage und Wertigkeit der farbkodierten Duplexsonographie zum Nachweis von Nierenarterienstenosen. Vasa (Suppl. 35) (1992) 58–59

Karasch, Th., A. L. Strauss, B. Grün, M. Worringer, D. Neuerburg-Heusler, F.-J. Roth, H. Rieger: Farbkodierte Duplexsonographie in der Diagnostik von Nierenarterienstenosen. Dtsch. med. Wschr. 118 (1993 a) 1429–1436

Karasch, Th., A. L. Strauss, M. Worringer, D. Neuerburg-Heusler, F.-J. Roth, H. Rieger: Normwerte farbduplexsonographisch ermittelter Strömungsgeschwindigkeiten in der A. renalis. Vasa (Suppl. 41) (1993 b) 67

Karasch, Th., A. L. Strauss, M. Worringer, D. Neuerburg-Heusler, F.-J. Roth, H. Rieger: Vergleich der Farbduplexsonographie mit verschiedenen angiographischen Verfahren in der Diagnostik arteriosklerotischer Nierenarterienstenosen und -verschlüsse. Ultraschall Klin. Prax. 8 (1993 c) 180

Karasch, Th., D. Neuerburg-Heusler, A. Strauss, H. Rieger: Farbduplexsonographische Kriterien arteriosklerotischer Nierenarterienstenosen und -verschlüsse. In Keller, E., B. Krumme: Farbkodierte Duplexsonographie in der Nephrologie. Springer, Berlin 1994

Karasch, Th., A. Jakobs, A. L. Strauss, F.-J. Roth, H. Rieger: Coarctation of the abdominal aorta—First experience with colour-coded duplex sonography. J. vasc. Invest. 1 (1995) 150–153

Karasch, Th.: Koarktation der Aorta abdominalis. Dtsch. med. Wschr. 121 (1996) 159–164

Kathrein, H., R. Schuhmayer, G. Judmaier: Änderungen des Blutflusses in der Arteria mesenterica inferior bei entzündlichen Darmerkrankungen. Ultraschall Klin. Prax. 5 (1990) 187

Kaufmann, W., G. Bönner, K. A. Meurer, A. Helber: Vaskuläre Nephropathien. In Hornbostel, H., W. Kaufmann, W. Siegenthaler: Innere Medizin in Praxis und Klinik, 4. Aufl., Bd. II Thieme, Stuttgart 1992

Keen, R. R., J. S. Yao, P. Astleford, D. Blackburn, L. J. Frazin: Feasibility of transgastric ultrasonography of the abdominal aorta. J. vasc. Surg. 24 (1996) 834–842

Keener, T. S., D. R. Cyr, L. A. Mack, D. Barr, S. J. Althaus: Sonographic diagnosis of arteriovenous fistula in pancreas transplant. J. Ultrasound Med. 14 (1995) 149–152

Kelcz, F., M. A. Pozniak, J. D. Pirsch, T. D. Oberly: Pyramidal appearance and resistive index: Insensitive and nonspecific sonographic indicators of renal transplant rejection. Amer. J. Roentgenol. 155 (1990) 531–535

Kernohan, R. M., A. A. Barros, B. Cranley, H. L. M. Johnston: Further evidence supporting the existence of the celiac artery compression syndrome. Arch. Surg. 120 (1985) 1072–1076

Klassen, D. K., E. W. Hoen-Saric, M. R. Weir, J. C. Papadimitriou, C. B. Drachenberg, L. Johnson, E. J. Schweitzer, S. T. Bartlett: Isolated pancreas rejection in combined kidney pancreas transplantation. Transplantation. 61 (1996) 974–977

Kniemeyer, H. W., W. Sandmann, R. Jaeschock: Das inflammatorische Aortenaneurysma. In Sandmann, W., H. W. Kniemeyer: Aneurysmen der großen Arterien: Diagnostik und Therapie. Huber Bern 1991

Kohler, T. R., R. E. Zierler, R. C. Martin, S. C. Nicholls, R. O. Bergelin, A. Kazmers, K. W. Beach, D. E. Strandness jr.: Noninvasive diagnosis of renal artery stenosis by ultrasonic duplex scanning. J. vasc. Surg. 4 (1986) 450–456

Koito, K., T. Namieno, T. Nagakawa, K. Morita: Splenic artery prior to rupture in the pancreatic pseudocyst: Detection by endoscopic color Doppler ultrasonography. J. Ultrasound Med. 15 (1996) 721–724

Kooner, J. S., W. S. Peart, C. J. Mathias: The peptide release inhibitor, octreotide (SMS 201–995), prevents the haemodynamic changes following food ingestion in normal human subjects. Quart. J. exp. Physiol. 74 (1989) 569–572

Kremer, H., C. Haschka, B. Weigold, W. Zoller, F. Spengel, N. Zöllner: Wachstumskurven von Bauchaortenaneurysmata. Bildgebung 56 (1989) 64–67

Krumme, B.: Farbkodierte Duplexsonographie in der Diagnostik von Nierenarterienstenosen nach allogener Nierentransplantation. In Keller, E., B. Krumme: Farbkodierte Duplexsonographie in der Nephrologie. Springer, Berlin 1994

Kubale, R., B. Güttner, R. Kretschmer, K. Scherer: Klinische Wertigkeit der Duplexsonographie zur Diagnostik viszeraler Gefäßerkrankungen. Ultraschall Klin. Prax. 5 (1990) 186

Kubota, K., H. Billing, B.-G. Ericzon, U. Kelter, C. G. Groth: Duplex Doppler ultrasonography for monitoring liver transplants. Acta radiol. 31 (1990) 279–283

Kudielka, I., H. Raimann, C. Schatten, W. Eppel, B. Schurz, E. Reinold: Umbilikale Strömungsverhältnisse bei Gestosepatientinnen – gestosecharakteristisches Dopplerphänomen. Geburtsh. u. Frauenheilk. 52 (1992) 589–591

Kurjak, A., D. Jurkovic, Z. Alfirevic, I. Zalud: Transvaginal color Doppler imaging. J. clin. Ultrasound 18 (1990 a) 227–234

Kurjak, A., I. Zalud: Transvaginaler Farbdoppler für die Beurteilung von gynäkologischen Pathologien im kleinen Becken. Ultraschall Med. 11 (1990 b) 164–168

Kurjak, A., S. Kupesic-Urek, H. Schulmann, I. Zalud: Transvaginal color flow Doppler in the assessment of ovarian and uterine blood flow in infertile women. Fertil. Steril. 56 (1991) 870–873

Labs, K.-H.: Die Aussagekraft der Dopplerspektralanalyse für die praktische Diagnostik. Vasa (Suppl. 32) (1990) 85–101

Lederle, F. A., J. M. Walker, D. B. Reinke: Selective screening for abdominal aortic aneurysms with physical examinations and ultrasound. Arch. intern. Med. 148 (1988) 1753–1756

Leen, E., C. S. McArdle: Ultrasound contrast agents in liver imaging. Clin. Radiol. 51 (Suppl. 1) (1996) 35–39

van Leeuwen M. S., R. J. Hené, A. F. Muller, F. J. A. Beek, R. Meyer, R. Ganpat: Clinical value of Doppler ultrasound in the first two months after kidney transplant. Europ. J. Radiol. 14 (1992) 26–30

Leichtman, A. B., K. S. Sorrell, D. G. Wombolt, R. L. Hurwitz, M. H. Glickman: Duplex imaging of the renal transplant. Transplant. Proc. 21 (1989) 3607–3610

Leidig, E.: Pulsed Doppler ultrasound blood flow measurement in the superior mesenteric artery of the newborn. Pediat. Radiol. 19 (1989) 169–172

Lilly, M. P., T. R. S. Harward, W. R. Flinn, D. R. Blackburn, P. M. Astleford, J. S. T. Yao: Duplex ultrasound measurement of changes in mesenteric flow velocity with pharmacological and physiological alteration of intestinal blood flow in man. J. vasc. Surg. 9 (1989) 18–25

Lin, Z.-Y., W. Y. Chang, L.-Y. Wang, W.-P. Su, S.-N. Lu, S.-C. Chen, W. L. Chuang, M.-Y. Hsieh, J.-F. Tsai: Clinical utility of pulsed Doppler in the detection of arterioportal shunting in patients with hepatocellular carcinoma. J. Ultrasound Med. 11 (1992) 269–273

Lippert, H., R. Pabst: Arterial variations in man. Classification and frequency. Bergmann, München 1985

Long, M. G., J. E. Bowlbee, M. E. Hanson, R. M. J. Begent: Doppler time velocity wave from studies of the uterine artery and uterus. Brit. J. Obstet. Gynaecol. 96 (1989) 588–593

Longley, D. G., M. L. Skolnick, D. G. Sheahan: Acute allograft rejection in liver transplant recipients: Lack of correlation with loss of hepatic artery diastolic flow. Radiology 169 (1988) 417–420

Loyer, E., K. D. Eggli: Sonographic evaluation of superior mesenteric vascular relationship in malrotation. Pediat. Radiol. 19 (1989) 173–175

Ludwig, M., K. Kraft, W. Rücker, A. M. Hüther: Die Diagnose sehr früher arteriosklerotischer Gefäßwandveränderungen mit Hilfe der Duplexsonographie. Klin. Wschr. 67 (1989) 442–446

Luzsa, G.: Röntgenanatomie des Gefäßsystems. Akadémiai Kiadó, Budapest 1972

Mc Cormack, L. J., E. F. Poutasse, Th. F. Meany, T. J. Noto, H. P. Dustan: A pathologic-arteriographic correlation of renal artery disease. Amer. Heart J. 72 (1966) 188–198

McGrath, F. P., S. H. Lee, R. G. Gibney: Color Doppler imaging of the cystic artery. J. clin. Ultrasound 20 (1992) 433–438

Mallek, R., G. Mostbeck, R. Kain, P. Pokieser, A. Gebauer, Ch. Herold, F. Stockenhuber, D. Tscholakoff: Vaskuläre Nierentransplantatabstoßung – Ist eine duplexsonographische Diagnose möglich? Fortschr. Röntgenstr. 152 (1990 a) 283–286

Mallek, R., G. Mostbeck, R. Walter, D. Tscholakoff: Duplexsonographie der Visceralarterien: Angiographische Korrelation. Ultraschall Klin. Prax. 5 (1990 b) 186

Mann, S. J., T. G. Pickering: Detection of renovaskular hypertension. State of the art – 1992. Ann. intern. Med. 117 (1992) 845–853

Marchal, G., H. Rigauts, B. Van Damme, Y. Vanrentergham, M. Waer, H. Verbrugge, A. Baert, P. Michielsen: Duplex Doppler evaluation of renal allograft dysfunction. Transplant. Proc. 21 (1989) 1893–1894

Martin, R. S., P. W. Meacham, J. A. Ditesheim, J. L. Mulherin, W. H. Edwards: Renal artery aneurysm: Selective treatment for hypertension and prevention of rupture. J. vasc. Surg. 9 (1989) 26–34

Meyer, M., D. Paushter, D. R. Steinmuller: The use of duplex Doppler ultrasonography to evaluate renal allograft dysfunction. Transplantation 50 (1990) 974–978

Mirk, P., A. R. Cotroneo, G. Palazzoni, E. Bock: Doppler ultrasonography assessment of the inferior mesenteric artery. Feasibility study and definition of morphologic and flowmetric characteristics. Radiol. Med. Torino. 87 (1994) 275–282

Mitchell, D. G.: Color Doppler imaging: Principles, limitations, and artifacts. Radiology 177 (1990) 1–10

Moneta, G. L., D. C. Taylor, W. S. Helton, M. W. Mulholland, D. E. Strandness jr.: Duplex ultrasound measurements of postprandial intestinal blood flow: Effect of meal composition. Gastroenterology 95 (1988) 1294–1301

Müller-Schwefe, C., G. von Klinggräff, G. Riepe, A. Schröder: Das inflammatorische Bauchaortenaneurysma. Ultraschall Med. 12 (1991) 158–163

Münch, R., K. Jäger: Duplexsonographie – eine Erweiterung der Diagnostik mesenterialer Durchblutungsstörungen. Schweiz. Rdsch. Prax. 77 (1988) 51–54

Muller, A. F.: Role of duplex Doppler ultrasound in the assessment of patients with postprandial abdominal pain. Gut 33 (1992) 460–465

Musa, A. A., T. Hata, K. Hata, M. Kitao: Pelvic arteriovenous malformation diagnosed by color flow imaging. Amer. J. Roentgenol. 152 (1989) 1311–1312

Nakajima, M., H. Hoshino, E. Hayashi, K. Nagano, D. Nishimura, N. Katada, H. Sano, K. Okamoto, K. Kato: Pseudoaneurysm of the cystic artery associated with upper gastrointestinal bleeding. J. Gastroenterol. 31 (1996) 750–754

Nakamura, T., F. Moriyasu, N. Ban, O. Nishida, T. Tamada, T. Kawasaki, M. Sakai, H. Uchino: Quantitative measurement of abdominal aterial blood flow using image-directed Doppler ultrasonography: Superior mesenteric, splenic, and common hepatic arterial blood flow in normal adults. J. clin. Ultrasound 17 (1989) 261–268

Namieno, T., Y. Hata, J. Uchino, H. Kondoh, T. Shibata, T. Satoh: Spontaneous rupture of intrahepatic artery aneurysm with complicated vascular anomalies. Gastrointest. Radiol. 16 (1991) 172–174

Nghiem, D. D.: Pancreatic allograft thrombosis: Diagnostic and therapeutic importance of splenic venous flow velocity. Clin. Transplant. 9 (1995) 390–395

Nicholls, S. C., T. R. Kohler, R. L. Martin, D. E. Strandness jr.: Use of hemodynamic parameters in the diagnosis of mesenteric insufficiency. J. vasc. Surg. 3 (1986) 507–510

Nishioka, T., M. Ikegami, M. Imanishi, T. Ishii, T. Uemura, S. Kunikata, H. Kanda, T. Matsuura, T. Akiyama, T. Kurita: Renal transplant blood flow evaluation of color Doppler echography. Transplant. Proc. 21 (1989) 1919–1922

Norris, C. S., J. S. Pfeiffer, S. E. Rittgers, R. W. Barnes: Noninvasive evaluation of renal artery stenosis and renovascular resistence. Experimental and clinical studies. J. vasc. Surg. 1 (1984) 192–201

Özbek, S. S., S. K. Aytac, M. I. Erden, N. U. Sanlidilek: Intrarenal Doppler findings of upstream renal artery stenosis: a preliminary report. Ultrasound Med. Biol. 19 (1993) 3–12

Oh, H. K., W. Kupin, B. Madrazo, N. Turza, K. K. Venkat, A. Langnas, D. Parker, D. Visscher: Evaluation of renal allograft by quantitative duplex sonography and radioisotope renogramm. Transplant. Proc. 21 (1989) 1917–1918

Otto, R., J. Meier, T. Lüscher, W. Vetter: Sonographische Befunde bei Nierenerkrankungen mit Hypertonie. Dtsch. med. Wschr. 106 (1981) 539–543

Paes, E. H. J., J. F. Vollmar: Fehldiagnose des abdominellen Aortenaneurysma bei atypischer Erstmanifestation. In Sandmann, W., H. W. Kniemeyer: Aneurysmen der großen Arterien: Diagnostik und Therapie. Huber Bern 1991

Pandian, N. G.: Intravascular and intracardiac ultrasound imaging. An old concept, now on the road to reality. Circulation 80 (1989) 1091–1094

Patel, B., M. K. Wolverson, B. Mahanta: Pancreatic transplant rejection: Assessment with duplex US. Radiology 173 (1989) 131–135

Patriquin, H. B., S. O'Regan, P. Robitaille, H. Paltiel: Hemolytic-uremic syndrome: Intrarenal arterial Doppler patterns as a usefull guide to therapy. Radiology 172 (1989) 625–628

Patriquin, H., M. Lafortune, J.-C. Jéquier, S. O'Regan, L. Garel, J. Landriault, A. Fontaine, D. Filiatrault: Stenosis of renal artery: Assessment of slowed systole in the downstream circulation with Doppler sonography. Radiology 184 (1992) 479–485

Pelling, M., P. A. Dubbins: Doppler and color Doppler imaging in acute transplant failure. J. clin. Ultrasound 20 (1992) 507–516

Perchik, J. E., B. R. Baumgartner, M. E. Bernardino: Renal transplant rejection. Limited value of duplex Doppler sonography. Invest. Radiol. 26 (1991) 422–426

Pierce, M. E., R. Sewell: Identification of hepatic cirrhosis of duplex Doppler ultrasound. Value of the hepatic artery resistive index. Aus. Radiol. 34 (1990) 331–333

Pignoli, P., E. Tremoli, A. Poli, P. L. Oreste, R. Paoletti: Intimal plus medial thickness of arterial wall: A direct measurement with ultrasound imaging. Circulation 74 (1986) 1399–1406

Pinna, A. D., C. V. Smith, H. Furukawa, T. E. Starzl, J. J. Fung: Urgent revascularization of liver allografts after early hepatic artery thrombosis. Transplantation. 15 (1996) (62) 1584–1587

Plainfoss, M. C., V. M. Calonge, C. Beyloune-Mainardi, D. Glotz, A. Dubost: Vascular complications in the adult kidney transplant recipient. J. clin. Ultrasound 20 (1992) 517–527

Platt, J. F., J. M. Rubin, J. H. Ellis, M. A. DiPietro: Duplex Doppler US of the kidney: Differentiation of obstructive from nonobstructive dilatation. Radiology 171 (1989) 515–520

Platt, J. F.: Duplex Doppler evaluation of native kidney dysfunction: Obstructive and nonobstructive disease. Amer. J. Roentgenol. 158 (1992) 1035–1042

Pöllmann, H., B. Wallner, D. Wanjura: Die Duplexsonographie in der postoperativen Überwachung von Pankreastransplantaten – erste Erfahrungen. Ultraschall 10 (1989) 303–306

Pozniak, M. A., F. Kelcz, R. J. Stratta, T. D. Oberley: Extraneous factors affecting resistive index. Invest. Radiol. 23 (1988) 899–904

Pozniak, M. A., K. M. Baus: Hepatofugal arterial signal in the main portal vein: An indicator of intravascular tumor spread. Radiology 180 (1991) 663–666

Propeck, P. A., K. Scanlan: Reversed or absent hepatic arterial diastolic flow in liver transplants shown by duplex sonography: A poor predictor of subsequent hepatic artery thrombosis. Amer. J. Roentgenol. 159 (1992) 1199–1201

Qamar, M. I., A. E. Read, R. Skidmore, J. M. Evans, R. C. N. Williamson: Transcutaneous Doppler ultrasound measurement of coeliac axis blood flow in man. Brit. J. Surg. 72 (1985) 391–393

Qamar, M. I., A. E. Read, R. Skidmore, J. M. Evans, P. N. T. Wells: Transcutaneous Doppler ultrasound measurements of superior mesenteric artery blood flow in man. Gut 27 (1986a) 100–105

Qamar, M. I., A. E. Read, R. Skidmore, J. M. Evans, P. N. T. Wells: Pulsatility index of superior mesenteric artery blood velocity waveforms. Ultrasound Med. Biol. 12 (1986b) 773–776

Qamar, M. I., A. E. Read: Effects of exercise on mesenteric blood flow in man. Gut 28 (1987) 583–587

Qamar, M. I., A. E. Read: Effects of ingestion of carbohydrate, fat, protein, and water on the mesenteric blood flow in man. Scand. J. Gastroenterol. 23 (1988) 26 530

Reilly, L. M., A. D. Ammar, R. J. Stoney, W. K. Ehrenfeld: Late results following operative repair for celiac artery compression syndrome. J. vasc. Surg. 2 (1985) 79–91

Reuther, G., D. Wanjura, H. Bauer: Acute renal vein thrombosis in renal allografts: Detection with duplex Doppler US. Radiology 170 (1989) 557–558

Rifkin, M. D., L. Needleman, M. E. Pasto, A. B. Kurtz, P. M. Foy, E. McGlynn, C. Canino, O. H. Baltarowich, R. G. Pennell, B. B. Goldberg: Evaluation of renal transplant rejection by duplex Doppler examination. Value of the resistive index. Amer. J. Roentgenol. 148 (1987) 759–762

Rigauts, H., G. Marchal, M. Grieten, B. van Damme, Y. Vanrenterghem, M. Waer, H. Verbrugge: Duplex Doppler evaluation of renal allograft dysfunction: Doppler signal quantitation and pathologic correlation. J. belge Radiol. 73 (1990) 475–483

Rigsby, C. M., P. N. Burns, G. G. Weltin, B. Chen, M. Bia, K. J. W. Taylor: Doppler signal quantitation in renal allografts: Comparison in normal and rejecting transplants with pathologic correlation. Radiology 162 (1987) 39–42

Rittgers, S. E., C. S. Norris, R. W. Barnes: Detection of renal artery stenosis: experimental and clinical analysis of velocity waveforms. Ultrasound Med. Biol. 11 (1985) 523–531

Sabbá, C., G. Ferraioli, P. Genecin, L. Colombato, P. Buonamico, E. Lerner, K. J. W. Taylor, R. J. Groszmann: Evaluation of postprandial hyperemia in superior mesenteric artery and portal vein in healthy and cirrhotic humans: An operator-blind echo-Doppler study. Hepatology 13 (1991) 714–718

Sandager, G., W. R. Flinn, W. J. McCarthy, J. S. T. Yao, J. J. Bergan: Assessment of visceral arterial reconstruction using duplex scan. J. vasc. Technol. 9 (1987) 13–16

Santolaya-Forgas, J.: Physiology of the menstrual cycle by ultrasonography. J. Ultrasound Med. 11 (1992) 139–142

Sato, S., K. Ohnishi, S. Sugita, K. Okuda: Splenic artery and superior mesenteric artery blood flow: Nonsurgical Doppler US measurement in healthy subjects and patients with chronic liver disease. Radiology 164 (1987) 347–352

Schäberle, W., K. Seitz: Duplexsonographische Blutflußmessung in der Arteria mesenterica superior. Ultraschall Med. 12 (1991) 277–282

Schäberle, W., A. Strauss, D. Neuerburg-Heusler, F. J. Roth: Wertigkeit der Duplexsonographie in der Diagnostik der Nierenarterienstenose und ihre Eignung in der Verlaufskontrolle nach Angioplastie (PTA). Ultraschall Med. 13 (1992) 271–276

Schörner, W., H. Kempter, D. Banzer, C. Aviles, Th. Weiss, R. Felix: Transvenöse digitale Subtraktionsangiographie (DSA) zur Abklärung der Nierenarterienstenose bei arterieller Hypertonie. Radiologe 24 (1984) 171–176

Schoop, W.: Pathophysiologie der Arterien und der arteriellen Durchblutung. In Heberer, G., G. Rau, W. Schoop: Angiologie. Grundlagen, Klinik, Praxis. Thieme 1974

Schulte, K.-L., D. van Gemmeren, K. Liederwald, K.-P. Spies, F. Fobbe, K.-J. Wolf, A. Distler, R. Gotzen: Stellenwert der 24-Stunden-Blutdruckmessung in der Diagnostik von Nierenarterienstenosen im Vergleich zur farbkodierten Duplexsonographie und direkten Nierenangiographie. Nieren- und Hochdruckkr. 21 (1992) 460–462

Schurz, B., W. Eppel, R. Wenzl, J. Huber, G. Söregi, E. Reinhold: Gynäkologische Dopplersonographie. Ultraschall Med. 11 (1990) 176–179

Schwaighofer, B., F. Kainberger, R. Stiglbauer, P. Hübsch, O. Traindl, P. Barton: Duplex-Sonographie zur Beurteilung der normalen Strömungsverhältnisse am Nierentransplantat. Ultraschall 8 (1987) 178–179

Schwerk, W. B., I. K. Restrepo, H. Prinz: Semiquantitative Analysen intrarenaler arterieller Dopplerflußspektren bei gesunden Erwachsenen. Ultraschall Med. 14 (1993) 117–122

Schwerk, W. B., I. K. Restrepo, M. Stellwaag, K. J. Klose, C. Schade-Brittinger: Renal artery stenosis: Grading with image-directed Doppler US evaluation of renal resistive index. Radiology. 190 (1994) 785–790

Seitz, K.: Klinische Anwendung der Duplexsonographie im Abdomen. Ultraschall 10 (1989) 182–189

Sieber, C. C., C. Beglinger, K. Jäger, P. Hildebrand, G. A. Stalder: Regulation of postprandial mesenteric blood flow in humans: evidence for a cholinergic nervous reflex. Gut 32 (1991) 361–366

Sieber, C. C., K. Jäger: Duplex scanning – a useful tool for noninvasive assessment of visceral blood flow in man. Vasc. Med. Rev. 3 (1992a) 95–114

Sieber, C. C., C. Beglinger, K. Jäger, G. A. Stalder: Intestinal phase of superior mesenteric artery blood flow in man. Gut 33 (1992b) 497–501

Sievers, K. W., E. Löhr, K. H. Albrecht, T. Player: Diagnose einer Nierentransplantatstenose mit arteriovenöser Fistel durch die Duplex-Doppler-Sonographie. Ultraschall Klin. Prax. 4 (1989a) 216–218

Sievers, K. W., E. Löhr, W. R. Werner: Duplex doppler ultrasound in determination of renal artery stenosis. Urol. Radiol. 11 (1989b) 142–147

Sievers, K. W., K. H. Albrecht, J. V. Kaude, A. Wegener, H. Annweiler: Der Pulsationsflußindex (PFI) in der Diagnostik dysfunktioneller Nierentransplantate. Fortschr. Röntgenstr. 153 (1990) 698–701

Skodler, W. D., N. Vavra, H. Enzelsberger, H. Kucera, E. Reinhold: Farbdoppler-Flußmessungen bei Kollumkarzinomen. Ultraschall Med. 12 (1991) 146–148

Snider, J. F., D. W. Hunter, C. C. Kuni, W. R. Castaneda-Zuniga, J. G. Letourneau: Pancreatic transplantation: Radiologic evaluation of vascular complications. Radiology 178 (1991) 749–753

Snook, J. A., A. D. Wells, D. R. Prytherch, D. H. Evans, S. R. Bloom, D. A. Colin-Jones: Studies on the pathogenesis of the early dumping syndrome induced by intraduodenal instillation of hypertonic glucose. Gut 30 (1989) 1716–1720

Sohn, C., H. Fendel: Arterielle renale und uterine Durchblutung in normalen und gestotischen Schwangerschaften. Z. Geburtsh. Perinat. 192 (1988) 43–48

Sohn, C., W. Stolz: Die unterschiedliche Durchblutung mütterlicher Gefäße in normalen und gestotischen Schwangerschaften. Ultraschall Med. 12 (1991) 6–10

Sohn, C., W. Stolz, G. Bastert: Dopplersonographie in der Gynäkologie und Geburtshilfe. Thieme, Stuttgart 1993c

Song, H.-Y., K.-H. Choi, J.-H. Park, B.-I. Choi, Y.-S. Chung: Radiological evaluation of hepatic artery aneurysms. Gastrointest. Radiol. 14 (1989) 329–333

Soper, W. D., T. Bergman, T. Harward, C. Huang, L. Peterson, J. S. Wolf: Use of duplex ultrasound scanning in renal transplantation. Transplant. Proc. 21 (1989) 1903–1904

Spies, K.-P., K.-L. Schulte, D. van Gemmeren, M. El-Bedewi, F. Fobbe, E. Wudel, R. Gotzen, A. Distler: Colour coded duplex sonography, 99Tc-DTPA scintigraphy and plasma renin activity (PRA) before and after captopril in the diagnosis of renal artery stenosis. Clin. exp. Hypertens. Part. A 12 (1990) 490

Spies, K. P., F. Fobbe, M. El-Bedewi, K. J. Wolf, A. Distler, K. L. Schulte: Color-coded duplex sonography for noninvasive diagnosis and grading of renal artery stenosis. Amer. J. Hypertens. 8 (1995) 1222–1231

Stiglbauer, R., P. Hübsch, N. Gritzmann, B. Schwaighofer: Zur Diagnose arterioportaler Kurzschlußverbindungen der Leber mittels pepulster und farbkodierter Doppler-Sonographie. Fortschr. Röntgenstr. 149 (1988) 214–215

Strandness, D. E.: Duplex scanning in diagnosis of renovascular hypertension. Surg. Clin. N. Amer. 70 (1990) 109–117

Sturgiss, S. N., K. Martin, A. Whittingham, J. M. Davison: Assessment of the renal circulation during pregnancy with color Doppler ultrasonography. Amer. J. Obstet. Gynecol. 167 (1992) 1250–1254

Swartbol, P., B. O. T. Thorvinger, H. Pärsson, L. Norgren: Renal artery stenosis in patients with peripheral vascular disease and its correlation to hypertension. A retrospective study. Int. Angiol. 11 (1992) 195–199

Takebayashi, S., N. Aida, K. Matsui: Arteriovenous malformations of the kidneys: Diagnosis and follow-up with color Doppler sonography in six patients. Amer. J. Roentgenol. 157 (1991) 991–995

Tanaka, K., S. Inoue, K. Numata, Y. Takamura, S. Takebayashi, Y. Ohaki, K. Misugi: Color Doppler sonography of hepatocellular carcinoma before and after treatment by transcatheter arterial embolization. Amer. J. Roentgenol. 158 (1992 b) 541–546

Tanaka, K., K. Mitsui, M. Morimoto, K. Numata, S. Inoue, Y. Takamura, M. Masumura: Increased hepatic arterial blood flow in acute viral hepatitis: Assessment by color Doppler sonography. Hepatology. 18 (1993) 21–27

Taylor, D. C., M. D. Kettler, G. L. Moweta: Duplex ultrasound scanning in the diagnosis of renal artery stenosis: A prospective evaluation. J. vasc. Surg. 7 (1988) 363–369

Taylor, K. J. W., P. N. Burns, P. N. T. Wells, D. I. Conway, M. G. R. Hull: Ultrasound Doppler flow studies of the ovarian and uterine arteries. Brit. J. Obstet. Gynaecol. 92 (1985) 240–246

Taylor, K. J. W., S. S. Morse, C. M. Rigsby: Vascular complications in renal allografts. Detection with duplex Doppler US. Radiology 162 (1987 a) 31–38

Taylor, K. J. W., I. Ramos, S. S. Morse, K. L. Fortune, L. Hammers, C. R. Taylor: Focal liver masses: Differential diagnosis with pulsed Doppler US. Radiology 164 (1987 b) 643–647

Tekay, A., P. Jouppila: A longitudinal Doppler ultrasonographic assessment of the alterations in peripheral vascular resistance of uterine arteries and ultrasonographic findings of the involuting uterus during the puerperium. Amer. J. Obstet. Gynecol. 168 (1993) 190–198

Tessler, F. N., V. L. Schiller, R. R. Perrella, M. L. Sutherland, E. G. Grand: Transabdominal versus endovaginal pelvic sonography: Prospective Study. Radiology 170 (1989) 553–556

Thaler, I., D. Manor, S. Rottem, I. E. Timor-Trisch, J. M. Brandes, J. Itskovitz: Hemodynamic evaluation of the female pelvic vessels using a high-frequency transvaginal image-directed Doppler system. J. clin. Ultrasound 18 (1990) 364–369

Thomas, E. A., P. A. Dubbins: Duplex ultrasound of the abdominal aorta – A neglected tool in aortic dissection. Clin. Radiol. 42 (1990) 330–334

Torsello, G., H. Kniemeyer, R. Jaeschock, W. Sandmann: Simultane transaortale Nierenarteriendesobliteration bei Operation von Bauchaortenaneurysmen. In Sandmann, W., H. W. Kniemeyer: Aneurysmen der großen Arterien: Diagnostik und Therapie. Huber Bern 1991

Trattnig, S., K. Frenzel, M. Eilenberger, A. Khoss, B. Schwaighofer: Akute Nierenvenenthrombose bei Kindern: Früher Nachweis mit Duplex- und farbkodierter Dopplersonographie. Ultraschall Med. 14 (1993) 40–43

Tullis, M. J., R. E. Zierler, D. J. Glickerman, R. O. Bergelin, K. Cantwell-Gab, D. E. Strandness jr.: Results of percutaneous transluminal angioplasty for atherosclerotic renal artery stenosis: A follow-up study with duplex ultrasonography. J. Vasc. Surg. (25) 1997, 46–54

Ugolotti, U., A. Miselli, R. Mandrioli, P. Larini, A. Ross: Ultrasound diagnosis of superior mesenteric artery aneurysm: Two case reports. J. clin. Ultrasound 12 (1984) 581–584

Uzawa, M., E. Karasawa, N. Sugiura, N. Saotome, K. Kita, H. Fukuda, M. Miki, Y. Togawa, F. Kondou, S. Matsutani, M. Ohto: Doppler color flow imaging in the detection and quantitative measurement of the gastroduodenal artery blood flow. J. clin. Ultrasound 21 (1993) 9–17

Vassiliades, V. G., M. E. Bernardino: Percutaneous renal and adrenal biopsies. Cardiovasc. intervent. Radiol. 14 (1991) 50–54

Verma, B. S., A. K. Bose, H. C. Bhatia, R. Katoch: Superior mesenteric artery branch aneurysm diagnosed by ultrasound. Brit. J. Radiol. 64 (1991) 169–172

Vollmar, J.: Rekonstruktive Chirurgie der Arterien, 3. Aufl. Thieme, Stuttgart 1982

Wan, S. K. H., C. F. Ferguson, D. L. Cochlin, C. Evans, D. F. R. Griffiths: Duplex Doppler ultrasound in the diagnosis of acute renal allograft rejection. Clin. Radiol. 40 (1989) 573–576

Ward, R. E., S. T. Bartlett, J. O'Green Koenig, J. Ballenger, J. Neylan, M. Friend: The use of duplex scanning in evaluation of the posttransplant kidney. Transplant. Proc. 21 (1989) 1912–1916

Warshauer, D. M., B. Keefe, M. A. Mauro: Intrahepatic hepatic artery aneurysm: Computed tomography and color-flow Doppler ultrasound findings. Gastrointest. Radiol. 16 (1991) 175–177

Weigold, B., F. A. Spengel, N. Zöllner: Sonographische Untersuchungen zum Spontanverlauf von Bauchaortenaneurysmen. In Sandmann, W., H. W. Kniemeyer: Aneurysmen der großen Arterien: Diagnostik und Therapie. Huber Bern 1991

Wenz, W.: Abdominelle Angiographie. Springer, Berlin 1972

Williams, S., P. Gillespie, J. M. Little: Celiac axis compression syndrome: factors predicting a favorable outcome. Surgery 98 (1985) 879–887

Worthy, S. A., J. F. Olliff, S. P. Olliff, J. A. Buckels: Color flow Doppler ultrasound diagnosis of a pseudoaneurysm of the hepatic artery following liver transplantation. J. Clin. Ultrasound. 22 (1994) 461–465

Zaidi, J., W. Collins, S. Campbell, R. Pittrof, S. L. Tan: Blood flow changes in the intraovarian arteries during the periovulatory period: Relationship to the time of day. Ultrasound Obstet. Gynecol. 7 (1996) 135–140

Zierler, R. E., R. O. Bergelin, R. C. Davidson, K. Cantwell-Gab, N. L. Polissar, D. E. Strandness jr.: A prospective study of disease progression in patients with atherosclerotic renal artery stenosis. Amer. J. Hypertens. 9 (1996) 1055–1061

Zoller, W. G., M. Stapff: Diagnostische Möglichkeiten der abdominellen Duplexsonographie. Bildgebung/Imaging 56 (1989) 82–88

Zwicker, C., M. Langer, R. Langer: Sonographische Beurteilung von Gefäßinterponaten „in vitro". Ultraschall 9 (1988) 82–75

Chapter 8

Abdel-Wahab, M. F., G. Esmat, A. Farrag, Y. El-Boraey, G. Th. Strickland: Ultrasonographic prediction of esophageal varices in schistosomiasis mansoni. Amer. J. Gastroenterol. 88 (1993) 560–563

Abu-Yousef, M. M., S. G. Milam, R. M. Farner: Pulsatile portal vein flow: a sign of tricuspid regurgitation on duplex Doppler sonography. Amer. J. Roentgenol. 155 (1990) 785–788

Abu-Yousef, M. M.: Normal and respiratory variations of the hepatic and portal venous duplex Doppler waveforms with simultaneous electrocardiographic correlation. J. Ultrasound Med. 11 (1992) 263–268

Ahn, J., H. L. Cohen: Case of the day. Post-renal biopsy complication: Perinephric hematoma and arteriovenous fistula. J. Ultrasound Med. 14 (1995) 327–328

Alpern, M. B., J. M. Rubin, D. M. Williams, P. Capek: Porta hepatis: Duplex Doppler US with angiographic correlation. Radiology 162 (1987) 53–56

Alvarez, D., R. Mastal, A. Lennie, G. Soifer, D. Levi, R. Terg: Noninvasive measurement of portal venous blood flow in patients with cirrhosis: Effects of physiological and pharmacological stimuli. Dig. Dis. Sci. 36 (1991) 82–86

Baker, J. A., B. A. Carroll: The sonographic appearance of anomalous circumrenal vein mimicking perirenal fluid collection. J. Ultrasound Med. 14 (1995) 244–246

Barbara, L., L. Bolondi, J. Bosch et al. (Consensus conference, Bologna, Italy, 1989): The value of Doppler US in the study of hepatic hemodynamics. J. Hepatol. 10 (1990) 353–355

Barton, P., H. Jantsch, W. Pichler, H. Schurawitzki, R. Stiglbauer, G. Lechner: Ultraschalldiagnostik nach Lebertransplantationen. Fortschr. Röntgenstr. 151 (1989) 145–153

Blum, U., M. Rossle, K. Haag, A. Ochs, H. E. Blum, K. H. Hauenstein, F. Astinet. M. Langer: Budd-Chiari syndrome: Technical, hemodynamic, and clinical results of treatment with transjugular intrahepatic portosystemic shunt. Radiology 197 (1995) 805–811

Bolondi, L., A. Mazziotti, V. Arienti, et al.: Ultrasonographic study of portal venous system in portal hypertension and after portosystemic shunt operation. Surgery 95 (1984) 261–269

Bolondi, L., S. Gaiani, S. Li Bassi, G. Zironi, P. Casanova, L. Barbara: Effect of secretin on portal venous flow. Gut 31 (1990) 1306–1310

Bolondi, L., S. Li Bassi, S. Gaiani, G. Zironi, G. Benzi, V. Santi, L. Barbara: Liver cirrhosis: Changes of Doppler waveform of hepatic veins. Radiology 178 (1991 a) 513–516

Bolondi, L., S. Gaiani, S. Li Bassi, G. Zironi, F. Bonono, M. Brunetto, L. Barbara: Diagnosis of Budd-Chiari syndrome by pulsed Doppler ultrasound. Gastroenterology 100 (1991 b) 1324–1331

Brown, B. P., M. Abu-Yousef, R. Farner, D. LaBrecque, R. Gingich: Doppler sonography: a noninvasive method for evaluation of hepatic venoocclusive disease. Amer. J. Roentgenol. 154 (1990) 721–724

Brown, H. S., M. Halliwell, M. Qamar, A. E. Read, J. M. Evans, P. N. T. Wells: Measurement of normal portal venous blood flow by Doppler ultrasound. Gut 30 (1989) 503–509

Brüggemann, A., A. Schmid, M. Wüstner, B. Klinge, G. Lepsien: Sonographische Darstellung einer Dopplung der V. cava inferior. Ultraschall Klin. Prax. 8 (1993) 37–40

Burns, P. N., C. C. Jaffe: Quantitative flow measurements with Doppler ultrasound: techniques, accuracy, and limitations. Radiol. Clin. N. Amer. 23 (1985) 641–657

Burns, P., K. Taylor, A. T. Blei: Doppler flowmetry and portal hypertension. Gastroenterology 92 (1987) 824–826

Caturelli, E., G. Sperandeo, M. M. Squillante, A. Mangia, L. Gabbrielli: Cruveilhier-Baumgarten syndrome without esophageal varices: ultrasonographic diagnosis and echo-Doppler study. J. clin. Gastroenterol. 11 (1989) 357–361

Chagnon, S. F., C. A. Vallee, J. Barge, L. J. Chevalier, J. Le Gal, M. V. Blery: Aneurysmal portahepatic venous fistula: report of two cases. Radiology 159 (1986) 693–695

Chezmar, J. L., R. C. Nelson, M. E. Bernardino: Portal venous gas after hepatic transplantation: Sonographic detection and clinical significance. Amer. J. Roentgenol. 153 (1989) 1203–1205

Dao, T., S. Elfadel, G. Bouvard, N. Bouvard, I. Lecointe, I. Jardin-Grimaux, J.-C. Verwaerde, A. Valla: Assessment of portal contribution to liver perfusion by quantitative sequential scintigraphy and Doppler ultrasound in alcoholic cirrhosis. Diagnostic value in the detection of portal hypertension. J. clin. Gastroenterol. 16 (1993) 160–167

Deane, C., N. Cowan, J. Giles, H. Walters, I. Rifkin, A. Severn, V. Parsons: Arterio-venous fistulas in renal transplants: Color Doppler ultrasound observations. Urol. Radiol. 13 (1992) 211–217

Didier, D., A. Racle, J. P. Etievent, F. Weill: Tumor thrombus of the inferior vena cava secondary to malignant abdominal neoplasms: US and CT evaluation. Radiology 162 (1987) 83–89

Dubois, A., M. Dauzat, C. Pignodel, G. Pomier-Layrargues, C. Marty-Double, F.-M. Lopez, C. Janbon: Portal hypertension in lymphoproliferative and myeloproliferative disorders: Hemodynamic and histological correlations. Hepatology 17 (1993) 246–250

Endress, C., G. A. Kling, B. L. Medrazo: Diagnosis of hepatic artery aneurysm with portal vein fistula using image-directed doppler ultrasound. J. clin. Ultrasound 17 (1989) 206–208

Furuse, J., S. Matsutani, M. Yoshikawa, M. Ebara, H. Saisho, Y. Tsuchiya, M. Ohto: Diagnosis of portal vein tumor thrombus by pulsed Doppler ultrasonography. J. clin. Ultrasound 20 (1992) 439–446

Gaiani, S., L. Bolondi, S. Li Bassi, V. Santi, G. Zironi, L. Barbara: Effect of meal on portal hemodynamics in healthy humans and in patients with chronic liver disease. Hepatology 9 (1989) 815–819

Gaiani, S., L. Bolondi, D. Fenyves, G. Zironi, A. Rigamonti, L. Barbara: Effect of propranolol on portosystemic collateral circulation in patients with cirrhosis. Hepatology 14 (1991 a) 824–829

Gaiani, S., L. Bolondi, S. Li Bassi, G. Zironi, S. Siringo, L. Barbara: Prevalence of spontaneous hepatofugal portal flow in liver cirrhosis. Gastroenterology 100 (1991 b) 160–167

Gainza, F. J., I. Minguela, I. Lopez-Vidaur, L. M. Ruiz, I. Lampreabe: Evaluation of complications due to percutaneous renal biopsy in allografts and native kidneys with color-coded Doppler sonography. Clin. Nephrol. 43 (1995) 303–308

Gaitini, D., I. Thaler, J. K. Kaftori: Duplex sonography in the diagnosis of portal vein thrombosis. Fortschr. Röntgenstr. 153 (1990) 645–649

Gibson, P. R., R. N. Gibson, M. R. Ditchfield, J. D. Donlan: A comparison of duplex Doppler sonography of the ligamentum teres and portal vein with endoscopic demonstration of gastroesophageal varices in patients with chronic liver disease or portal hypertension, or both. J. Ultrasound Med. 11 (1991) 327–331

Gill, R. W.: Pulsed Doppler with B-mode imaging for quantitative blood flow measurement. Ultrasound Med. Biol. 5 (1979) 223–235

Gill, R. W.: Measurement of blood flow by ultrasound: Accuracy and sources of error. Ultrasound Med. Biol. 11 (1985) 625–641

Grant, E. G., R. Perrella, F. N. Tessler, J. Lois, R. Busutti: Budd-Chiari syndrome: The results of duplex and color Doppler imaging. Amer. J. Roentgenol. 152 (1989) 377–381

Grunert, D., B. Stier, M. Schöning: Kontrolle operativ angelegter portosystemischer Shunts im Kindesalter mit Hilfe der Computer-Duplexsonographie. Ultraschall 10 (1989) 295–302

Hagiwara, H., A. Kasahara, M. Kono, S. Kashio, A. Kaneko, A. Okuno, N. Hayashi, H. Fusamoto, T. Kamada: Extrahepatic portal vein aneurysm associated with a tortuous portal vein. Gastroenterology 100 (1991) 818–821

Hausegger, K. A., R. Fotter, E. Sorantin, F. Flückiger: Kongenitale intrahepatische arterioportale Fisteln als Ursache einer nekrotisierenden Enteritis – dopplersonographische und angiographische Erfassung. Ultraschall Med. 12 (1991) 193–196

Helbich, Th., F. Kainberger, K. A. Vergesslich, G. Granditsch, P. Hübsch, G. H. Mostbeck: Varikositas in der Gallenblasenwand bei Kindern mit portaler Hypertension. Ultraschall Klin. Prax. 7 (1992) 246

Herbetko, J., A. P. Grigg, A. R. Buckley, G. L. Phillips: Venoocclusive liver disease after bone marrow transplantation: Findings at duplex sonography. Amer. J. Roentgenol. 158 (1992) 1001–1005

Hosoki, T., C. Kuroda, K. Tokunaga, T. Marukawa, M. Masuike, T. Kozuka: Hepatic venous outflow obstruction: Evaluation with pulsed duplex sonography. Radiology 170 (1989) 733–737

Hosoki, T., J. Arisawa, T. Marukawa, K. Tokunaga, C. Kuroda, T. Kozuka, S. Nakano: Portal blood flow in congestive heart failure: Pulsed duplex sonographic findings. Radiology 174 (1990) 733–736

Jaspersen, D.: Die transendoskopische Dopplersonographie. Ultraschall med. 14 (1993) 123–125

Johansen, K., M. Paun: Duplex ultrasonography of the portal vein. Surg. Clin. N. Amer. 70 (1990) 181–190

Justig, E.: Das Kompressionssyndrom der linken Nierenvene. Fortschr. Röntgenstr. 136 (1982) 404–412

Kakitsubata, Y., S. Kakitsubata, H. Kiyomizu, T. Ogawa, K. Kato, K. Watanabe: Intrahepatic portal-hepatic venous shunts demonstrated by US, CT, and MR imaging. Acta radiol. 37 (1996) 680–684

Kawamura, S., K. Miyatake, K. Okamoto, S. Beppu, N. Kinoshita, H. Sakakibara, Y. Nimura: Analysis of the portal vein flow with twodimensional echo-Doppler-method. In Lerski, R. A., P. Morley: Ultrasound '82. Pergamon, Oxford 1983

Kawasaki, T., F. Moriyasu, T. Kimura, H. Someda, T. Tamada, Y. Yamashita, S. Ono, K. Kajimura, N. Hamato, M. Okuma: Hepatic function and portal hemodynamics in patients with liver cirrhosis. Amer. J. Gastroenterol. 85 (1990) 1160–1164

Kawasaki, T., F. Moriyasu, T. Kimura, H. Someda, Y. Fukuda, K. Ozawa: Changes in portal blood flow consequent to partial hepatectomy: Doppler estimation. Radiology 180 (1991) 373–377

Keller, M. S., K. J. W. Taylor, C. A. Riely: Pseudoportal Doppler signal in the partially obstructed inferior vena cava. Radiology 170 (1989) 475–477

Kudo, M., S. Tomita, H. Tochio, K. Minowa, A. Todo: Intrahepatic portosystemic venous shunt: Diagnosis by color Doppler imaging. Amer. J. Gastroenterol. 88 (1993) 723–729

Krakamp, B., P. Leidig, H. A. Dickmanns: Verschluß der Vene cava inferior – ein seltenes, jedoch schwerwiegendes Krankheitsbild, leicht mit Ultraschall zu diagnostizieren. Ultraschall Med. 14 (1993) 106–111

Kriegshauser, J. S., J. W. Charboneau, L. Letendre: Hepatic venoocclusive disease after bone-marrow transplantation: Diagnosis with duplex sonography. Amer. J. Roentgenol. 150 (1988) 289–290

Krumme, B., K. Gondolf, G. Kirste, P. Schollmeyer, E. Keller: Farbkodierte Duplexsonographie zur Diagnostik von Nierenvenenthrombosen in der Frühphase nach Nierentransplantation. Dtsch. med. Wschr. 118 (1993) 1629–1635

Kubale, R.: Abdominelle und retroperitoneale Gefäße. In Rettenmaier, G., Seitz, K.: Sonographische Differentialdiagnostik, Bd. II. Verlag Chemie, Weinheim 1992

Kubale, R., M. Grimbach, F. Walter, M. Girmann, B. Güttner, H. Boos, N. Heger: Nichtinvasive Diagnostik isolierter Thrombosen der V. mesenterica superior. Ultraschall Klin. Prax. 7 (1992) 238

Kubota, K., H. Billing, B.-G. Ericzon, U. Kelter, G. Groth: Duplex Doppler ultrasonography for monitoring liver transplants. Acta radiol. 31 (1990) 279–283

Lafortune, M., G. Breton, S. Charlebois: Arterioportal fistula demonstrated by pulsed doppler ultrasonography. J. Ultrasound Med. 5 (1986) 105–106

Lafortune, M., H. Patriquin, G. Pomier, P. M. Huet, A. Weber, P. Lavoie, H. Blanchard, G. Breton: Hemodynamic changes in portal circulation after portosystemic shunts: Use of duplex sonography in 43 patients. Amer. J. Roentgenol. 149 (1987) 701–706

Laplante, S., H. B. Patriquin, P. Robitaille, D. Filiatrault, A. Grignon, J. C. Decarie: Renal vein thrombosis in children: Evidence of early flow recovery with Doppler US. Radiology 189 (1993) 37–42

Leen, E., J. A. Goldberg, J. R. Anderson, J. Robertson, B. Moule, T. G. Cooke, C. S. McArdle: Hepatic perfusion changes in patients with liver metastases: Comparison with those patients with cirrhosis. Gut 34 (1993) 554–557

Lehner, K.. P. Gerhardt, R. Blasini: Intravasaler Ultraschall (IVUS): Beurteilung der Infiltration von Tumoren in die Gefäßwand. Fortschr. Röntgenstr. 156 (1992) 146–150

Lin, Z.-Y., W. Y. Chang, L.-Y. Wang, W.-P. Su, S.-N. Lu, S.-C. Chen, W.-L. Chuang, M-.Y. Hsieh, J.-F. Tsai: Clinical utility of pulsed Doppler in the detection of arterioportal shunting in patients with hepatocellular carcinoma. J. Ultrasound Med. 11 (1992) 269–273

Longo, J. M., J. I. Bilbao, H. P. Rosseau, L. Garcia-Villareal, J. P. Vinel, J. M. Zozaya, F. G. Joffre, J. Prieto: Transjugular intrahepatic portosystemic shunt: Evaluation with Doppler sonography. Radiology 186 (1993) 529–534

Matre, K., S. Ødegaard, T. Hausken: Endoscopic ultrasound Doppler probes for velocity measurements in vessels in the upper gastrointestinal tract using a multifrequency pulsed Doppler meter. Endoscopy 22 (1990) 268–270

McGahan, J. P., L. C. Blake, R. deVere-White, E. O. Gerscovich, W. E. Brant: Color flow sonographic mapping for intravascular extension of malignant renal tumors. J. Ultrasound Med. 12 (1993) 403–409

McLoughlin, R. F., S. M. Dashefsky, P. L. Cooperberg, J. R. Mathieson: Spontaneous portal-right renal vein shunt in portal hypertension. J. Ultrasound Med. 14 (1995) 959–961

Meckler, U., J. Bönhof, W. Caspary, N. Gritzmann, K.-H. Hennermann, P. Herzog, W. Stelzel, R. Strand, J. Tuma: Ultraschall des Abdomens. Diagnostischer Leitfaden. Deutscher Ärzte-Verlag, Köln 1992

Meifort, R., H.-M. Vogel, H. Henning: Duplexsonographische Pfortadermessungen bei Lebergesunden und Patienten mit chronischer Hepatitis nach Verabreichung einer vollresorbierbaren Testmahlzeit. Z. Gastroenterol. 28 (1990) 291–294

Menu, Y., D. Alison, J.-M. Lorphelin, D. Valla, J. Belghiti, H. Nahum: Budd-Chiari syndrome: US evaluation. Radiology 157 (1985) 761–764

Miller, M. A., D. M. Balfe, W. D. Middleton: Peripheral portal venous blood flow alterations induced by hepatic masses: Evaluation with color and pulsed Doppler sonography. J. Ultrasound Med. 15 (1996) 707–713

Miller, V. E., L. L. Berland: Pulsed Doppler duplex sonography and CT of portal vein thrombosis. Amer. J. Roentgenol. 145 (1985) 73–76

Mori, K., S. Matsuoka, Y. Hayabuchi, Y. Kuroda, T. Kitagawa: Absence of the inferior vena cava in a patient with omphalocele: Two-dimensional echocardiographic and cineangiographic findings. Heart Vessels 11 (1996) 104–109

Moriyasu, F., O. Nishida, N. Ban, T. Nakamura, K. Miura, M. Sakai, T. Miyake, H. Uchino: Measurement of portal vascular resistance in patients with portal hypertension. Gastroenterology 90 (1986a) 710–717

Moriyasu, F., N. Ban, O. Nishida, T. Nakamura, T. Miyake, H. Uchino, Y. Kanematsu, S. Koizumi: Clinical application of an ultrasonic duplex system in the quantitative measurement of portal blood flow. J. clin. Ultrasound 14 (1986b) 579–588

Mostbeck, G. H., G. R. Wittich, C. Herold, K. A. Vergesslich, R. M. Walter, S. Frotz, G. Sommer: Hemodynamic significance of the paraumbilical vein in portal hypertension: Assessment with duplex US. Radiology 170 (1989) 339–342

Nakayama, T., K. Ohnishi, M. Saito, H. Hatano, F. Nomura, K. Kono, K. Okuda: Effects of propranolol on portal vein pressure, portal blood flow, hepatic blood flow, and cardiac output in patients with chronic liver disease. Hepatology 3 (1983) 812

Nelson, R., K. E. Lovett, J. L. Chezmar, J. H. Moyers, W. E. Torres, F. B. Murphy, M. E. Bernardino: Comparison of pulsed Doppler sonography and angiography in patients with portal hypertension. Amer. J. Roentgenol. 149 (1987) 77–81

Ohnishi, K., M. Saito, H. Koen, T. Nakayama, F. Nomura, K. Okuda: Pulsed Doppler flow as an criterion of portal venous velocity: Comparison with cineangiographic measurements. Radiology 154 (1985a) 495–498

Ohnishi, K., M. Saito, S. Sato, T. Nakayama, M. Takashi, S. Iida, F. Nomura, H. Koen, K. Okuda: Direction of splenic venous flow assessed by pulsed Doppler flowmetry in patients with a large splenorenal shunt. Gastroenterology 89 (1985b) 180–185

Ohnishi, K., M. Saito, T. Nakayama, S. Iida, F. Nomura, H. Koen, K. Okuda: Portal venous hemodynamics in chronic liver disease: Effects of posture change and exercise. Radiology 155 (1985c) 757–761

Ohnishi, K., M. Saito, S. Sato, S. Sugita, H. Tanaka, K. Okuda: Clinical utility of pulsed Doppler flowmetry in patients with portal hypertension. Amer. J. Gastroenterol. 81 (1986) 1–8

Ohnishi, K., S. Sato, D. Pugliese, T. Tsunoda, M. Saito, K. Okuda: Changes of splanchnic circulation with progression of chronic liver dis-

ease studied by echo-Doppler flowmetry. Amer. J. Gastroenterol. 82 (1987a) 507–511

Ohnishi, K., M. Saito, S. Sato, H. Terabayashi, S. Iida, F. Nomura, M. Nakano, K. Okuda: Portal hemodynamics in ideopathic portal hypertension (Banti's syndrome). Comparison with chronic persistent hepatitis and normal subjects. Gastroenterology 92 (1987b) 751–758

Ohnishi, K., S. Sato, F. Nomura, S. Iida: Sphlanchnic hemodynamics in ideopathic portal hypertension: comparison with chronic persistent hepatitis. Amer. J. Gastroenterol. 84 (1989a) 403–408

Ohnishi, K., N. Chin, H. Tanaka, S. Iida, S. Sato, H. Terabayashi, F. Nomura: Differences in portal hemodynamics in cirrhosis and idiopathic portal hypertension. Amer. J. Gastroenterol. 84 (1989b) 409–412

Ohnishi, K., S. Sato: Effects of vasopressin on left gastric venous flow in cirrhotic patients with esophageal varices. Amer. J. Gastroenterol. 85 (1990) 293–295

Okuda, K., K. Kono, K. Ohnishi, K. Kimura, M. Omata, H. Koen, Y. Nakajima, H. Musha, T. Hirashima, M. Takashi, K. Takayasu: Clinical study of eighty-six cases of ideopathic portal hypertension and comparison with cirrhosis with splenomegaly. Gastroenterology 86 (1984) 600–610

Özbek, S. S., A. Memis, R. Killi, E. Karaca, C. Kabasakal, S. Mir: Image-directed and color Doppler ultrasonography in the diagnosis of postbiopsy arteriovenous fistulas of native kidneys. J. Clin. Ultrasound 23 (1995) 239–242

Patriquin, H., M. Lafortune, P. N. Burns, M. Dauzat: Duplex Doppler examination in portal hypertension: Technique and anatomy. Amer. J. Roentgenol. 149 (1987) 71–76

Pierro, J. A., M. Soleimanpour, J. L. Bory: Left retrocaval ureter associated with left inferior vena cava. Amer. J. Roentgenol. 155 (1990) 545–546

Plainfoss, M. C., V. M. Calonge, C. Beyloune-Mainardi, D. Glotz, A. Duboust: Vascular complications in the adult kidney transplant recipient. J. clin. Ultrasound 20 (1992) 517–527

Pozniak, M. A., K. M. Baus: Hepatofugal arterial signal in the main portal vein: An indicator of intravascular tumor spread. Radiology 180 (1991) 663–666

Pugliese, D., K. Ohnishi, T. Tsunoda, C. Sabba, O. Albanao: Portal hemodynamics after meal in normal subjects and patients with chronic liver disease studied by echo-Doppler flowmeter. Amer. J. Gastroenterol. 82 (1987) 1052–1056

Rabinovici, N., N. Navot: The relationship between respiration, pressure and flow distribution in the vena cava and portal and hepatic veins. Surg. Gynecol. Obstet. 151 (1980) 753–763

Ralls, P. W., D. S. Mayekawa, K. P. Lee, P. M. Colletti, M. B. Johnson, J. M. Halls: Gallbladder wall varices: Diagnosis with color flow Doppler sonography. J. clin. Ultrasound 16 (1988) 595–598

Ralls, P. W., D. S. Mayekawa, K. P. Lee, M. B. Johnson, J. Halls: The use of color Doppler sonography to distinguish dilated intrahepatic ducts from vascular structures. Amer. J. Roentgenol. 152 (1989) 291–292

Ralls, P. W.: Color Doppler sonography of the hepatic artery and portal venous system. Amer. J. Roentgenol. 155 (1990a) 517–525

Ralls, P. W., K. P. Lee, D. S. Mayekawa, W. D. Boswell jr., D. R. Radin, P. M. Colletti, J. M. Halls: Color Doppler sonography of portocaval shunts. J. clin. Ultrasound 18 (1990b) 379–381

Ralls, P. W., M. B. Lohnson, D. R. Radin, W. D. Boswell jr., K. P. Lee, J. M. Halls: Budd-Chiari syndrome: detection with color Doppler sonography. Amer. J. Roentgenol. 159 (1992) 113–116

Rector jr., W. G., J. C. Hoefs, K. F. Hossack, G. T. Everson: Hepatofugal portal flow in cirrhosis: observations on hepatic hemodynamics and the nature of the arterioportal communications. Hepatology 8 (1988) 16–20

Rettenmaier, G., K. Seitz: Sonographische Differentialdiagnostik, Bd. I. Verlag Chemie, Weinheim 1990

Rettenmaier, G., K. Seitz: Sonographische Differentialdiagnostik, Bd. II. Verlag Chemie, Weinheim 1992

Reuß, J.: Pankreastumoren. In Rettenmaier, G., Seitz, K.: Sonographische Differentialdiagnostik, Bd. I. Verlag Chemie, Weinheim 1990

Reuther, G., D. Wanjura, H. Bauer: Acute renal vein thrombosis in renal allografts: Detection with duplex Doppler US. Radiology 170 (1989) 557–558

Riedl, P.: Retrograde Nierenvenen – eine mögliche Ursache der linksseitigen Varikozele. Fortschr. Röntgenstr. 133 (1980) 477–479

Robinson, D. E., R. W. Gill, G. Kossoff: Quantitative sonography. Ultrasound Med. Biol. 12 (1986) 555–565

Rollino, C., G. Garofalo, D. Roccatello, T. Sorrentino, M. Sandrone, B. Basolo, G. Quattrocchio, C. Massara, M. Ferro, G. Picciotto, S. Rendine, G. Piccoli: Colour-coded Doppler sonography in monitoring native kidney biopsies. Nephrol. Dial. Transplant. 9 (1994) 1260–1263

Rypens, F., F. Avni, P. Braude, C. Matos, F. Rodesch, A. Pardou, J. Struyven: Calcified inferior vena cava thrombus in a fetus: Perinatal imaging. J. Ultrasound Med. 12 (1993) 55–58

Sabbá, C., G. Ferraioli, S. K. Sarin, E. Lerner, R. J. Groszmann, K. J. W. Taylor: Feasibility spectrum for Doppler flowmetry of splanchnic vessels in normal and cirrhotic populations. J. Ultrasound Med. 9 (1990 a) 705–710

Sabbá, C., G. G. Weltin, D. V. Cicchetti, G. Ferraioli, K. J. W. Taylor, T. Nakamura, F. Moriyasu, R. J. Groszmann: Observer variability in echo-Doppler measurements of portal flow in cirrhotic patients and normal volunteers. Gastroenterology 98 (1990 b) 1603–1611

Sabbá, C., G. Ferraioli, P. Genecin, L. Colombato, P. Buonamico, E. Lerner, K. J. W. Taylor, R. J. Groszmann: Evaluation of postprandial hyperemia in superior mesenteric artery and portal vein in healthy and cirrhotic humans: An operator-blind echo-Doppler study. Hepatology 13 (1991) 714–718

Saito, M., K. Ohnishi, T. Nakayama, F. Nomura, H. Kono, H. Koen, K. Okuda: Ultrasonic measurements of portal and splenic vein blood flows and their velocities in normal subjects and patients with chronic liver disease. Hepatology 3 (1983) 812

Schmassmann, A., M. Zuber, M. Livers, K. Jäger, H. R. Jenzer, H. F. Fehr: Recurrent bleeding after variceal hemorrhage: Predictive value of portal venous duplex sonography. Amer. J. Roentgenol. 160 (1993) 41–47

Segal, S., S. Shenhav, O. Segal, E. Zohav, O. Gemer: Budd-Chiari syndrome complicating severe preeclampsia in a parturient with primary antiphospholipid syndrome. Europ. J. Obstet. Gynecol. 68 (1996) 227–229

Seitz, K., R. Kubale: Duplexsonographie der abdominellen und retroperitonealen Gefäße. Verlag Chemie, Weinheim 1988

Seitz, K.: Klinische Anwendung der Duplexsonographie im Abdomen. Ultraschall 10 (1989) 182–189

Seitz, K.: Portale Hypertension. In Rettenmaier, G., K. Seitz: Sonographische Differentialdiagnostik, Bd. I. Verlag Chemie, Weinheim 1990

Seitz, K., J. Reuß: Farbdopplersonographie im Abdomen: Diagnostischer Wert und Problematik langsamer Flüsse. Ultraschall Klin. Prax. 7 (1992) 238

Shapiro, R. S., F. Winsberg, C. Maldjian, A. Stancato-Pasik: Variability of hepatic vein Doppler tracings in normal subjects. J. Ultrasound Med. 12 (1993) 701–703

Sherlock, S.: The portal venous system and portal hypertension. In Sherlock, S.: Disease of the Liver and Biliary System. Blackwell, Oxford 1981

Sieber, C. C., K. Jäger: Duplex scanning – a useful tool for noninvasive assessment of visceral blood flow in man. Vasc. Med. Rev. 3 (1992) 95–114

Smith, H.-J., P. Grøttum, S. Simonsen: Ultrasonic assessment of abdominal venous return. Acta radiol., Diagn. 26 (1985) 581–588

Snider, J. F., D. W. Hunter, C. C. Kuni, W. R. Castaneda-Zuniga, J. G. Letourneau: Pancreatic transplantation: Radiologic evaluation of vascular complications. Radiology 178 (1991) 749–753

Stavros, A. T., K. J. Sickler, R. R. Menter: Color duplex sonography of the nutcracker syndrome (aortomesenteric left renal vein compression). J. Ultrasound Med. 13 (1994) 569–574

Stiglbauer, R., P. Hübsch, M. Gritzmann, B. Schwaighofer: Zur Diagnose arterioportaler Kurzschlußverbindungen der Leber mittels pepulster und farbkodierter Doppler-Sonographie. Fortschr. Röntgenstr. 149 (1988) 214–215

Sugiura, N., E. Karasawa, N. Saotome, M. Miki, S. Matsutani, M. Ohto: Portosystemic collateral shunts originating from the left portal veins in portal hypertension: Demonstration by color Doppler flow imaging. J. clin. Ultrasound 20 (1992) 427–432

Sukigara, M., M. Ohata, T. Komazaki, R. Omoto: Assessment of the effect of respiration on the esophageal variceal blood flow using transesophageal real-time two-dimensional Doppler echography. Hepatology 8 (1988) 663–667

Sukigara, M., S. Kimura, H. Adachi, H. Asano, I. Koyama, T. Tsuji, R. Omoto: Primary Budd-Chiari syndrome: Demonstration by real-time two-dimensional Doppler echography. J. clin. Ultrasound 17 (1989) 615–621

Suggs, W. D., R. B. Smith, Th. F. Dodson, A. A. Salam, S. D. Graham jr.: Renal cell carcinoma with inferior vena caval involvement. J. vasc. Surg. 14 (1991) 413–418

Tanaka, S., T. Kitamura, M. Fujita, H. Iishi, H. Kasugai, K. Nakanishi, S. Okuda: Interahepatic venous and portal venous aneurysms examined by color Doppler flow imaging. J. clin. Ultrasound 20 (1992) 89–98

Tanaka, K., K. Numata, H. Okazaki, S. Nakamura, S. Inoue, Y. Takamura: Diagnosis of portal vein thrombosis in patients with hepatocellular carcinoma: Efficacy of color Doppler sonography compared with angiography. Amer. J. Roentgenol. 160 (1993) 1279–1283

Taylor, K. J. W., P. N. Burns: Duplex Doppler scanning in the pelvis and abdomen. Ultrasound Med. 11 (1985) 643–658

Taylor, K. J. W., S. S. Morse, G. G. Weltin, C. A. Riely, M. W. Flye: Liver transplant recipients: portable duplex US with correlative angiography. Radiology 159 (1986) 357–363

Taylor, K. J. W.: Gastrointestinal Doppler ultrasound. In Taylor, K. J. W., P. N. Burns, P. N. T. Wells: Clinical Applications of Doppler Ultrasound. Raven Press, New York 1988

Terwey, B., M. Bolkenius, P. Gerhardt, R. Daun: Prä- und postoperative Ultraschalluntersuchung des Pfortaderkreislaufs bei portaler Hypertension im Kindesalter. Z. Kinderchir. 32 (1981) 337–340

Tessler, F. N., C. Kimme-Smith, M. L. Sutherland, V. L. Schiller, R. R. Perrella, E. G. Grant: Inter- und intra-observer variability of Doppler peak velocity measurements: an in-vitro study. Ultrasound Med. Biol. 16 (1990) 653–657

Tessler, F. N., B. J. Gehring; A. S. Gomes, R. R. Perrella, N. Ragavendra, R. W. Busuttil, E. G. Grant: Diagnosis of portal vein thrombosis: value of color Doppler imaging. Amer. J. Roentgenol. 157 (1991) 293–296

Tsunoda, T., K. Ohnishi, H. Tanaka: Portal hemodynamic responses after oral intake of glucose in patients with cirrhosis. Amer. J. Gastroenterol. 83 (1988) 398–403

Tsushima, Y., M. Matsumoto, N. Sato, H. Ishizaka, K. Endo: Renal vein to portal vein collaterals in three cases of renal cell carcinoma extending into the inferior vena cava: consequences for chemoembolization. Cardiovasc. intervent. Radiol. 16 (1993) 189–192

Urban, M., W. Hruby, W. B. Winkler, H. Mosser, M. Baldt: Strebenbrüche bei der Langzeitkontrolle von Günther-Kavafiltern – Ergebnisse nach 64 Filterimplantationen. Fortschr. Röntgenstr. 156 (1992) 342–345

Vorwerk, D., R. W. Günther, K. Bohndorf: Erste Erfahrungen mit der Angiodynographie (farbcodierten Duplexsonographie) in der Kontrolle von Kavafiltern. Ultraschall 8 (1987) 259–262

Vorwerk, D., H.-B. Gehl, R. Schlief, A. Nelles, R. W. Günther: Dynamische kontrastmittelgestützte Ultraschallkavographie bei Kavafilterpatienten. Ultraschall Med. 11 (1990) 146–149

Wang, L.-Y., Z.-Y. Lin, W.-Y. Chang, S.-C. Chen, W.-L. Chuang, M.-Y. Hsieh, J.-F. Tsai, K. Okuda: Duplex pulsed Doppler sonography of portal vein thrombosis in hepatocellular carcinoma. J. Ultrasound Med. 10 (1991) 265–269

Weiss, H., A. Weiss: Ultraschallatlas 2. Internistische Ultraschalldiagnostik. Edition medizin. Weinheim 1990

Weltin, G., Taylor, K. J. W., Carter, A. R.: Duplex Doppler: An aid in the diagnosis of cavernous transformation of the portal vein. Amer. J. Roentgenol. 44 (1984) 999–1001

Wenz, W.: Abdominelle Angiographie. Springer, Berlin 1972

Wermke, W.: Portalverschlüsse bei chronischer Pankreatitis – Diagnostik mit der Farbdopplersonographie (FDS). Ultraschall Klin. Prax. 7 (1992) 239

Yang, S.-S., P. W. Ralls, J. Korula: The effect of oral nitroglycerin on portal blood velocity as measured by ultrasonic Doppler. A double blind, placebo controlled study. J. clin. Gastroenterol. 13 (1991) 173–177

Zoli, M., G. Marchesini, A. Marzocchi, C. Marrozzini, C. Dondi, E. Pisi: Portal pressure changes induced by medical treatment: US detection. Radiology 155 (1985) 763–766

Zoli, M., G. Marchesini, A. Brunori, M. R. Cordiani, E. Pisi: Portal venous flow in response to acute b-blocker and vasodilatatory treatment in patients with liver cirrhosis. Hepatology 6 (1986 a) 1248–1251

Zoli, M., G. Marchesini, M. R. Cordiani, P. Pisi, A. Brunori, A. Trono, E. Pisi: Echo-flow measurement of splanchnic blood flow in control and cirrhotic subjects. J. clin. Ultrasound 14 (1986 b) 429–435

Chapter 9

Bähren, W., C. Biehl, B. Danz: Frustrane Sklerotherapieversuche der V. spermatica interna. Retrospektive Analyse bei 1141 Patienten mit ideopathischer Varikozele. Fortschr. Röntgenstr. 157 (1992) 355–360

Barloon, T. J., A. M. Weissman, D. Kahn: Diagnostic imaging of patients with acute scrotal pain. Amer. Fam. Phycn. 53 (1996) 1734–1750

Benson, C. B., M. A. Vickers: Sexual impotence caused by vascular disease: Diagnosis with duplex sonography. Amer. J. Roentgenol. 153 (1989) 1149–1153

Benson, C. B., J. E. Aruny, M. A. Vickers jr.: Correlation of duplex sonography with arteriography in patients with erectile dysfunktion. Amer. J. Roentgenol. 160 (1993) 71–73

Bigot, J. M., A. Chatel, M. Dectot, C. Helenon: La phlebographie spermatique retrograde. Anatomie radiologique (a propos de 152 explorations). Ann. Radiol. 21 (1978) 515–523

Braedel, H. U., J. Steffens, M. Ziegler, M. S. Polsky: Betrachtungen zur Ausbildung des sekundären Venensystems des Bauchraumes unter besonderer Berücksichtigung der idiopathischen linksseitigen Varikozele. Fortschr. Röntgenstr. 155 (1991) 11–19

Brandstetter, K., J. U. Schwarzer, W. Bautz, U. Pickl, M. Lenz: Vergleich der Farbduplexsonographie mit der selektiven penilen DSA bei der Abklärung der erektilen Dysfunktion. Fortschr. Röntgenstr. 158 (1993) 405–409

Broderick, G. A., T. F. Lue: The penile blood flow study: Evaluation of vasculogenic impotence by duplex ultrasonography. In Jonas, U., W. F. Thon, C. G. Stief: Erectile Dysfunction. Springer, Berlin 1991

Broderick, G. A., J. P. McGahan, A. R. Stone, R. deVere White: The hemodynamics of vacuum constriction erections: Assessment by color Doppler ultrasound. J. Urol. 147 (1992) 57–61

Brown, J. M., L. W. Hammers, J. W. Barton, C. K. Holland, L. M. Scoutt, J. S. Pellerito, K. J. Taylor: Quantitative Doppler assessment of acute scrotal inflammation. Radiology 197 (1995) 427–431

Chiang, P. H., C. P. Chiang, C. C. Wu, C. J. Wang, M. T. Chen, C. H. Huang, D. K. Wu: Color duplex sonography in the assessment of impotence. Brit. J. Urol. 68 (1991) 181–186

Coley, B. D., D. P. Frush, D. S. Babcock, S. M. O'Hara, A. G. Lewis, M. J. Gelfand, K. E. Bove, C. A. Sheldon: Acute testicular torsion: Comparison of unenhanced and contrast-enhanced power Doppler US, color Doppler US, and radionuclide imaging. Radiology 199 (1996) 441–446

Cormio, L., H. Nisen, F. P. Selvaggi, M. Ruutu: A positive pharmacological erection test does not rule out arteriogenic erectile dysfunction. J. Urol. 156 (1996a) 1628–1630

Cormio, L., J. Edgren, M. Lepantalo, O. Lindfors, H. Nisen, O. Saarinen, M. Ruutu: Aortofemoral surgery and sexual function. Europ. J. Vasc. Endovasc. Surg. 11 (1996b) 453–457

Cvitanic, O. A., J. J. Cronan, M. Sigman, S. T. Landau: Varicoceles: postoperative prevalence – A prospective study with color Doppler US. Radiology 187 (1993) 711–714

Dauzat, M.: Examen ultrasonographique des vaisseaux génitaux masculins. In Dauzat, M.: Ultrasonographie vasculaire diagnostique. Théorie et pratique. Vigot, Paris 1991

DeWire, D. M.: Evaluation and treatment of erectile dysfunction. Amer. Fam. Phycn. 53 (1996) 2101–2108

DeWire, D. M., F. P. Begun, R. K. Lawson, S. Fitzgerald, W. D. Foley: Color Doppler ultrasonography in the evaluation of the acute scrotum. J. Urol. 147 (1992) 89–91

Egger, B., P. Stirnemann, P. Vock: Die Bedeutung des Penis-Arm-Druck-Indexes in der Beurteilung der Impotenz beim Gefäßpatienten. Vasa 17 (1988) 102–106

Fakhry, J., A. Khoury, K. Barakat: The hypoechoic band: A normal finding on testicular sonography. Amer. J. Roentgenol. 153 (1989) 321–323

Fitzgerald, S. W., W. D. Foley: Genitourinary system. In Lanzer, P., A. P. Yoganathan: Vascular Imaging by Color Doppler and Magnetic Resonance. Springer, Berlin 1991

Fobbe, F., K.-J. Wolf: Erste klinische Erfahrungen mit der Angiodynographie. Fortschr. Röntgenstr. 148 (1988) 259–264

Gall, H., G. Holzki: Doppler ultrasound investigations of penile vessels. In Jonas, U., W. F. Thon, C. G. Stief: Erectile Dysfunction. Springer, Berlin 1991

Gehl, H.-B., R. Sikora, K. Bohndorf, M. Sohn, D. Vorwerk: Darstellung von Penisgefäßanastomosen mittels farbkodierter Doppler-Sonographie: Vergleich mit CW-Doppler-Sonographie und Klinik. Ultraschall Med. 11 (1990) 155–160

Hasse, H. M.: Chronische arterielle Verschlußkrankheiten der Extremitäten. In Heberer, G., G. Rau, W. Schoop: Angiologie. Grundlagen, Klinik, Praxis, Thieme, Stuttgart 1974

Horstman, W. G., W. D. Middleton, G. L. Melson: Scrotal inflammatory disease: Color Doppler US findings. Radiology 179 (1991) 55–59

Horstman, W. G., G. L. Melson, W. D. Middleton, G. L. Andriole: Testicular tumors: Findings with color Doppler US. Radiology 185 (1992) 733–737

Jensen, M. C., K. P. Lee, J. M. Halls, P. W. Ralls: Color Doppler sonography in testicular torsion. J. clin. Ultrasound 18 (1990) 446–448

Jevtich, M. J.: Non-invasive vascular and neurologic tests in use for evaluation of angiogenic impotence. Int. Angiol. 3 (1984) 225–232

Justich, E.: Das Kompressionssyndrom der linken Nierenvene. Fortschr. Röntgenstr. 136 (1982) 404–412

Karadeniz, T., A. Ariman, M. Topsakal, A. Eksioglu, T. Engin, D. Basak: Value of color Doppler sonography in the diagnosis of venous impotence. Urol. int. 55 (1995) 143–146

Kim, S. H., J. S. Paick, S. E. Lee, B. I. Choi, K. M. Yeon, M. C. Han: Doppler sonography of deep cavernosal artery of the penis: Variation of peak systolic velocity according to sampling location. J. Ultrasound Med. 13 (1994) 591–594

Krysiewicz, S., B. C. Mellinger: The role of imaging in the diagnostic evaluation of impotence. Amer. J. Roentgenol. 153 (1989) 1133–1139

Lee, B., S. S. Sikka, E. R. Randrup, P. Villemarette, N. Baum, J. F. Hower, W. J. Hellstrom: Standardization of penile blood flow parameters in normal men using intracavernous prostaglandin E1 and visual sexual stimulation. J. Urol. 149 (1993) 49–52

Lee, F. T. jr., D. B. Winter, F. A. Madsen, J. A. Zagzebski, M. A. Pozniak, S. G. Chosy, K. A. Scanlan: Conventional color Doppler velocity sonography versus color Doppler energy sonography for the diagnosis of acute experimental torsion of the spermatic cord. Amer. J. Roentgenol. 167 (1996) 785–790

Lerner, R. M., R. A. Medorach, W. C. Hulbert, R. Rabinowitz: Color Doppler US in the evaluation of acute scrotal disease. Radiology 176 (1990) 355–358

Levine, L. A., C. L. Coogan: Penile vascular assessment using color duplex sonography in men with Peyronie's disease. J. Urol. 155 (1996) 1270–1273

Lue, T. F., H. Hricak, K. W. Marich, E. A. Tanagho: Vasculogenic impotence evaluated by high-resolution ultrasonography and pulsed Doppler spectrum analysis. Radiology 155 (1985) 777–781

Lue, T. F., S. C. Müller, K. P. Jünemann, G. R. Fournier jr., E. A. Tanagho: Hämodynamische Veränderungen während der Erektion und funktionelle, klinische Diagnostik der penilen Gefäße mittels Ultraschall und gepulstem Doppler. Akt. Urol. 18 (1987) 115–123

McClure, R. D., H. Hricak: Scrotal ultrasound in the infertile man: detection of subclinical unilateral and bilateral varicoceles. J. Urol. 135 (1986) 711–715

Meuleman, E. J. H., B. L. H. Bemelmans, W. N. J. C. van Asten, W. H. Doesburg, H. Skotnicki, F. M. J. Debruyne: Assessment of penile blood flow by duplex ultrasonography in 44 men with normal erectile potency in different phases of erection. J. Urol. 147 (1992) 51–56

Middleton, D., A. Thorne, G. L. Melson: Color Doppler ultrasound of the normal testis. Amer. J. Roentgenol. 152 (1989) 293–297

Middleton, W. D., B. A. Siegel, G. L. Melson, C. K. Yates, G. L. Andriole: Acute scrotal disorders: prospective comparison of color Doppler US and testicular scintigraphy. Radiology 177 (1990) 177–181

Mills, R. D., K. K. Sethia: Reproducibility of penile arterial colour duplex ultrasonography. Brit. J. Urol. 78 (1996) 109–112

Montorsi, F., G. Guazzoni, L. Barbieri, L. Galli, P. Rigatti, G. Pizzini, A. Miani: The effect of intracorporeal injection plus genital and audiovisual sexual stimulation versus second injection on penile color Doppler sonography parameters. J. Urol. 155 (1996) 536–540

Müller, S. C., G. E. Voges, H. v. Wallenberg Pachaly: Vaskuläre Diagnostik der erektilen Dysfunktion: Wertung verschiedener Methoden. Akt. Urol. 20 (1989) 123–131

Müller, S. C., H. v. Wallenberg Pachaly, G. E. Voges, H. H. Schild: Comparison of selective internal iliac pharmacoangiography, penile brachial index and duplex sonography with pulsed doppler analysis for the evaluation of vasculogenic (arteriogenic) impotence. J. Urol. 143 (1990) 928–932

Petros, J. A., G. L. Andriole, W. D. Middleton, D. A. Picus: Correlation of testicular color Doppler ultrasonography, physical examination and venography in the detection of left varicoceles in men with infertility. J. Urol. 145 (1991) 785–788

Pickard, R. S., C. P. Oates, K. K. Sethia, P. H. Powell: The role of color duplex ultrasonography in the diagnosis of vasculogenic impotence. Brit. J. Urol. 68 (1991) 537–540

Porst, H., M. Lenz, W. Bähren, J. E. Altwein: Gefäßveränderungen bei primärer und sekundärer Impotenz. Akt. Urol. 14 (1983) 281–285

Porst, H.: Erektile Impotenz. Ätiologie, Diagnostik, Therapie. Enke, Stuttgart 1987

Quam, J., B. King, E. M. James, R. W. Lewis, D. M. Brakke, D. M. Ilstrup, B. G. Parulkar, R. R. Hattery: Duplex and color Doppler sonographic evaluation of vasculogenic impotence. Amer. J. Roentgenol. 153 (1989) 1141–1147

Rajfer, J., V. Canan, F. J. Dorey, M. Mehringer: Correlation between penile angiography and duplex scanning of the cavernous arteries in impotent men. J. Urol. 143 (1990) 1128–1130

Ralls, P. W., M. C. Jensen, K. P. Lee, D. S. Mayekawa, M. B. Johnson, J. M. Halls: Color Doppler sonography in acute epididymitis and orchitis. J. clin. Ultrasound 18 (1990) 383–386

Riedl, P.: Retroaortale Nierenvenen – eine mögliche Ursache der linksseitigen Varikozele. Fortschr. Röntgenstr. 133 (1980) 477–479

Rieger, H.: Physiologie und Pathophysiologie des Gefäßsystems im Genitalbereich. In Breddin, K., D. Gross, H. Rieger: Angiologie und Hämostaseologie. Fischer, Stuttgart 1988

Rifkin, M. D., P. M. Foy, A. B. Kurtz, M. E. Pasto, B. B. Goldberg: The role of diagnostic ultrasonography in varicocele evaluation. J. Ultrasound Med. 2 (1983) 271–275

Robinson, L. Q., J. P. Woodcock, T. P. Stephenson: Duplex scanning in suspected vasculogenic impotence: A worthwhile exercise? Brit. J. Urol. 63 (1989) 432–436

Rosen, M. P., A. N. Schwartz, F. J. Levine, A. J. Greenfield: Radiologic assessment of impotence: Angiography, sonography, cavernosography, and scintigraphy. Amer. J. Roentgenol. 157 (1991) 923–931

Schild, H. H., S. C. Müller, M. Hermann, H. Kaltenborn, M. Kern: Penile Abflußvenenokklusion: Vergleich von erektiler Funktion und Kavernosometrie vor und nach perkutanen Eingriffen. Fortschr. Röntgenstr. 158 (1993) 59–61

Schoop, W.: Angiologie – Arterielle Verschlußkrankheit und Sexualfunktion. In Breddin, K., D. Gross, H. Rieger: Angiologie und Hämostaseologie. Fischer, Stuttgart 1988

Schroeder-Printzen, I., J. Göben, W. Weidner, R.-H. Ringert: Die Verwendung vasoaktiver Substanzen in der Diagnostik und Therapie der erektilen Dysfunktion – Rechtliche Aspekte. Akt. Urol. 23 (1992) 248–251

Schwartz, A. N., K. Y. Wang, L. A. Mack, M. Lowe, R. E. Berger, D. R. Cyr, M. Feldman: Evaluation or normal erectile function with color flow Doppler sonography. Amer. J. Roentgenol. 153 (1989) 1155–1160

Schwartz, A., M. Lowe, R. Ireton, R. Berger, M. L. Richardson, D. Graney: A comparison of penile brachial index and angiography: evaluation of corpora cavernosa arterial inflow. J. Urol. 143 (1990) 510–513

Schwartz, A. N., M. Lowe, R. E. Berger, K. Y. Wang, L. A. Mack, M. L. Richardson: Assessment of normal and abnormal erectile function: Color Doppler flow sonography versus conventional techniques. Radiology 180 (1991) 105–109

Shabsigh, R., I. J. Fishman, E. T. Quesada, C. K. Seale-Hawkins, J. K. Dunn: Evaluation of vasculogenic erectile impotence using penile duplex ultrasonography. J. Urol. 142 (1989) 1469–1474

Shabsigh, R., I. J. Fishman, Y. Shotland, I. Karacan, J. K. Dunn: Comparison of penile duplex ultrasonography with nocturnal penile tumescence monitoring for the evaluation of erectile impotence. J. Urol. 143 (1990) 924–927

Stief, C. G., U. Wetterauer: Erectile responses to intracavernous papaverine and phentolamine: comparison of single and combined delivery. J. Urol. 140 (1988) 1415–1416

Thetter, O.: Potenzprobleme nach Bauchaortenaneurysmaoperationen. In Sandmann, W., H. W. Kniemeyer: Aneurysmen der großen Arterien: Diagnostik und Therapie. Huber, Bern 1991

Trum, J. W., F. M. Gubler, R. Laan, F. van der Veen: The value of palpation, varicoscreen contact thermography and colour Doppler ultrasound in the diagnosis of varicocele. Hum. Reprod. 11 (1996) 1232–1235

Valji, K., J. J. Bookstein: Diagnosis of arteriogenic impotence: Efficacy of duplex sonography as a screening tool. Amer. J. Roentgenol. 160 (1993) 65–69

Voss, E. U., W. Langer: Störungen der Sexualfunktion nach aorto(-ilio)-femoralen Gefäßrekonstruktionen. In Breddin, K., D. Gross, H. Rieger: Angiologie und Hämostaseologie. Fischer, Stuttgart 1988

Wagner, G.: Vascular mechanisms involved in human erection. Int. Angiol. 3 (1984) 221–224

v. Wallenberg Pachaly, H., G. Voges, H. Schild, S. C. Müller: Vergleich von Beckenangiographie, gepulster Dopplersonographie und PBI in der Abklärung der erektilen Impotenz: Akt. Urol. 20 (1989) 24–28

Wespes, E., C. Delcour, C. Rondeux, J. Struyven, C. C. Schulman: The erectile angle: Objective criterion to evaluate the papaverine test in impotence. J. Urol. 138 (1987) 1171–1173

Zeitler, E., W. Seyferth: Gefäßpathologische Befunde im Angiogramm und ihr Einfluß auf die Sexualfunktion des Mannes. In Breddin, K., D. Gross, H. Rieger: Angiologie und Hämostaseologie. Fischer, Stuttgart 1988

Zorgniotti, A. W., W. W. Shaw, G. Padula, G. Rossi: Impotence associated with pudendal arteriovenous malformation. Int. Angiol. 3 (1984) 267–269

Chapter 10

Anandakumar, C., S. Chew, Y. C. Wong, D. Chia, S. S. Ratnam: Role of transvaginal ultrasound color flow imaging and Doppler waveform analysis in differentiating between benign and malignant ovarian tumors. Ultrasound Obstet. Gynecol. 7 (1996) 280–284

Antonic, J., S. Rakar: Validity of colour and pulsed Doppler US and tumour marker CA 125 in differentiation between benign and malignant ovarian masses. Eur. J. Gynaecol. Oncol. 17 (1996) 29–35

Bassett, L. W., M. Ysrael, R. H. Gold, C. Ysrael: Usefulness of mammography and sonography in women less than 35 years of age. Radiology 180 (1991) 831–835

Börner, N., T. Clement, P. Herzog, K. F. Kreitner, H. Miltenberger, H. Schild, J. Meyer: Colour-coded Doppler sonography of primary and secondary liver tumors. Ultraschall in Med. 11 (1990) 274–280

Bourne, T., S. Campbell, C. Steer, M. I. Whitehead, W. P. Collins: Transvaginal colour flow imaging: A possible new screening technique for ovarian cancer. Brit. Med. J. 299 (1989) 1367–1370

Bourne, T. H., T. Hillard, M. Whitehead, S. Campbell, W. P. Collins: Transvaginal ultrasonography with color flow imaging to monitor hormone replacement therapy in postmenopausal women. Brit. J. Radiol. 64 (1991 a) 657

Bourne, T. H., K. Reynolds, S. Campbell: Ovarian cancer screening. Europ. J. Cancer 27 (1991 b) 655–659

Buy, J. N., M. A. Ghossain, D. Hugol, K. Hassen, C. Sciot, J. B. Truc, P. Poitout, D. Vadrot: Characterization of adnexal masses: Combination of color Doppler and conventional sonography compared with spectral Doppler analysis alone and conventional sonography alone. Amer. J. Roentgenol. 166 (1996) 385–393

Campbell, S., V. Bhan, P. Royston, M. I. Whitehead, W. P. Collins: Transabdominal ultrasound screening for early ovarian cancer. Brit. Med. J. 299 (1989) 1363–1367

Castagnone, D., R. Rivolta, S. Rescalli, M. I. Baldini, R. Tozzi, L. Cantalamessa: Color Doppler sonography in Graves' disease: Value in assessing activity of disease and predicting outcome. Amer. J. Roentgenol. 166 (1996) 203–207

Choi, B. I., T. K. Kim, J. K. Han, J. W. Chung, J. H. Park, M. C. Han: Power versus conventional color Doppler sonography: Comparison in the depiction of vasculature in liver tumors. Radiology 200 (1996) 55–58

Cosgrove, D.: Ultrasound contrast enhancement of tumours. Clin. Radiol. 51 (Suppl. 1) (1996) 44–49

Dock, W., F. Grabenwöger, V. Metz, K. Eibenberger, M. T. Farrés: Tumor vascularization: Assessment with duplex sonography. Radiology 181 (1991) 241–244

Dock, W.: Duplex sonography of mammary tumors: A prospective study of 75 patients. J. Ultrasound Med. 2 (1993) 79–82

Eppel, W., R. Wenzl, E. Asseryanis, E. Reinold: Sonographie bei benignen und malignen Ovarialtumoren unter besonderer Berücksichtigung der Tumorvaskularität. Ultraschall Klin. Prax. 7 (1992) 247

Ernst, H., E. G. Hahn, T. Balzer, R. Schlief, N. Heyder: Color Doppler ultrasound of liver lesions: Signal enhancement after intravenous injection of the ultrasound contrast agent Levovist. J. Clin. Ultrasound 24 (1996) 31–35

Fleischer, A. C., W. H. Rodgers, D. M. Kepple, L. L. Williams, H. W. Jones: Color Doppler sonography of ovarian masses: A multiparameter analysis. J. Ultrasound Med. 12 (1993) 41–48

Hamper, U. M., S. Sheth, F. M. Abbas, N. B. Rosenshein, D. Aronson, R. J. Kurman: Transvaginal color Doppler sonography of adnexal masses: Differences in blood flow impedance in benign and malignant lesions. Amer. J. Roentgenol. 160 (1993) 1225–1228

Hann, L. E., Y. Fong, C. D. Shriver, J. F. Botet, K. T. Brown, D. S. Klimstra, L. H. Blumgart: Malignant hepatic hilar tumors: Can ultrasonography be used as an alternative to angiography with CT arterial portography for determination of resectability? J. Ultrasound Med. 15 (1996) 37–45

Heilenkötter, U., P. Jagella: Colour Doppler sonography of breast tumours requiring surgical removal—Description of an examination method. Geburtsh. u. Frauenheilk. 53 (1993) 247–252

Horstman, W. G., G. L. Melson, W. D. Middleton, G. L. Andriole: Testicular tumors: Findings with color Doppler US. Radiology 185 (1992) 733–737

Horstman, W. G., R. M. McFarland, J. D. Gorman: Color Doppler sonographic findings in patients with transitional cell carcinoma of the bladder and renal pelvis. J. Ultrasound Med. 14 (1995) 129–133

Huber, S., S. Delorme, M. V. Knoop, H. Junkermann, I. Zuna, D. von Fournier, G. van Kaick: Breast tumors: Computer-assisted quantitative assessment with Color Doppler US. Radiology 192 (1994) 797–801

Hübsch, P., B. Niederle, P. Barton, B. Pesau, M. Knittel, M. Schratter, M. Hermann, F. Längle: Colour–coded Doppler sonography of the thyroid gland: A progress in the diagnosis of carcinoma? Fortschr. Röntgenstr. 156 (1992) 125–129

Karasch, Th., S. Schmidt-Decker: Duplex sonography in the diagnosis of tumor vascularity. Dtsch. med. Wschr. 121 (1996) 876–884

Kawano, M., H. Masuzaki, T. Ishimaru: Transvaginal color Doppler studies in gestational trophoblastic disease. Ultrasound Obstet. Gynecol. 7 (1996) 197–200

Kedar, R. P., D. Cosgrove, V. R. McCready, J. C. Bamber, E. R. Carter: Microbubble contrast agent for color Doppler US: Effect on breast masses: Work in progress. Radiology 198 (1996) 679–686

Kier, R., K. J. W. Taylor, A. L. Feyock, I. M. Ramos: Renal masses: Characterization with Doppler US. Radiology 176 (1990) 703–707

Kuijpers, D., R. Jaspers: Renal masses: Differential diagnosis with pulsed Doppler US. Radiology 170 (1989) 59–60

Kurjak, A., S. Kupesic: Transvaginal color Doppler and pelvic tumor vascularity: Lessons learned and future challenges. Ultrasound. Obstet. Gynecol. 6 (1995) 145–159

Kurjak, A., S. Kupesic, H. Shalan, S. Jukic, D. Kosuta, M. Ilijas: Uterine sarcoma: A report of 10 cases studied by transvaginal color and pulsed Doppler sonography. Gynecol. Oncol. 59 (1995) 342–346

Leen, E., W. J. Angerson, H. Wotherspoon, B. Moule, T. G. Cooke, C. S. McArdle: Comparison of the Doppler perfusion index and intraoperative ultrasonography in diagnosing colorectal liver metastases. Evaluation with postoperative follow-up results. Ann. Surg. 220 (1994 a) 663–667

Leen, E., W. J. Angerson, H. W. Warren, P. O. Gorman, B. Moule, E. C. Carter: Improved sensitivity of colour Doppler flow imaging of colorectal hepatic metastases using galactose microparticles: A preliminary report. Brit. J. Surg. 81 (1994 b) 252–254

Leen, E., W. J. Angerson, T. G. Cooke, C. S. McArdle: Prognostic power of Doppler perfusion index in colorectal cancer. Correlation with survival. Ann. Surg. 223 (1996) 199–203

Leeners, B., R. L. Schild, A. Funk, S. Hauptmann, B. Kemp, W. Schroder, W. Rath: Colour Doppler sonography improves the preoperative diagnosis of ovarian tumours made using conventional transvaginal sonography. Europ. J. Obstet. Gynecol. Reprod. Biol. 64 (1996) 79–85

Lencioni, R., F. Pinto, N. Armillotta, C. Bartolozzi: Assessment of tumor vascularity in hepatocellular carcinoma. Comparison of power Doppler US and color Doppler US. Radiology 201 (1996) 353–358

Luska, G., D. Lott, U. Risch, H. v. Boetticher: Colour Doppler sonography for tumours of the breast. Fortschr. Röntgenstr. 156 (1992) 142–145

Madjar, H., W. Sauerbrei, S. Münch, H. Schillinger: Continuous-wave and pulsed Doppler studies of the breast. Clinical results and effect of transducer frequency. Ultrasound Med. Biol. 17 (1991) 31–39

Martinoli, C., L. E. Derchi, L. Solbiati, G. Rizzatto, E. Silvestri, M. Giannoni: Color Doppler sonography of salivary glands. Amer. J. Roentgenol. 163 (1994) 933–941

McGahan, J. P., L. C. Blake, R. deVere-White, E. O. Gerscovich, W. E. Brant: Color flow sonographic mapping of intravascular extension of malignant renal tumors. J. Ultrasound Med. 12 (1993) 403–409

Nino-Murcia, M., P. W. Ralls, R. B. Jeffrey jr., M. Johnson: Color flow Doppler characterization of focal hepatic lesions. Amer. J. Roentgenol. 159 (1992) 1195–1197

Ramos, I. M., K. J. W. Taylor, R. Kier, P. N. Burns, D. P. Snower, D. Carter: Tumor vascular signals in renal masses: Detection with Doppler US. Radiology 168 (1988) 633–637

Rehn, M., K. Lohmann, A. Rempen: Transvaginal ultrasonography of pelvic masses: Evaluation of B-mode technique and Doppler ultrasonography. Amer. J. Obstet. Gynecol. 175 (1996) 97–104

Rifkin, M. D., G. S. Sudakoff, A. A. Alexander: Prostate: Techniques, results, and potential applications of color Doppler US scanning. Radiology 186 (1993) 509–513

Schild, R., H. Fendel: Doppler sonographic differentiation of benign and malignant tumors of the breast. Geburtsh. u. Frauenheilk. 51 (1991) 969–972

Schiller, V. L., R. R. Turner, D. A. Sarti: Color doppler imaging of the gallbladder wall in acute cholecystitis: Sonographic-pathologic correlation. Abdom. Imaging 21 (1996) 233–237

Schurz, B., W. Eppel, R. Wenzl, J. Huber, G. Söregi, E. Reinhold: Gynäkologische Dopplersonographie. Ultraschall in Med. 11 (1990) 176–179

Scoutt, L., D. Luthringer, D. Carter, S. D. Flynn, K. H. Katz, P. Kornguth, C. Lee, J. Richter, K. J. W. Taylor: Correlation of Doppler US-detected neovascularity with histologic predictors of breast cancer behavior. Radiology 177 (1990) 287

Shimamoto, K., S. Sakuma, T. Ishigaki, N. Makino: Intratumoral blood flow: Evaluation with color Doppler echography. Radiology 165 (1987) 683–685

Shimamoto, K., S. Sakuma, T. Ishigaki, T. Ishiguchi, S. Itoh, H. Fukatsu: Hepatocellular carcinoma: Evaluation with color Doppler US and MR imaging. Radiology 182 (1992) 149–153

Skoder, W. D., N. Vavra, H. Enzelsberger, H. Kucera, E. Reinhold: Farbdoppler-Flußmessungen bei Kollumkarzinomen. Ultraschall in Med. 12 (1991) 146–148

Smith, S. J., M. A. Bosniak, A. J. Megibow, D. H. Hulnick, S. C. Horii, B. N. Raghavendra: Renal cell carcinoma: Earlier discovery and increased detection. Radiology 170 (1989) 699–703

Sohn, Ch., E. M. Grischke, D. Wallwiener, M. Kaufmann, D. v. Fournier, G. Bastert: Sonographic diagnosis of the blood flow of malignant and benign breast tumors. Geburtsh. u. Frauenheilk. 52 (1992) 397–403

Sohn, Ch., E. M. Grischke, W. Stolz, G. Bastert: Investigations on the correlation between blood-flow velocity and the biological behavior of mammary tumors. Ultraschall Klin. Prax. 8 (1993 a) 11–14

Sohn, Ch., G. Meyberg, D. v. Fournier, G. Bastert: Blood supply of malignant and benign tumors of the inner genital tract. Geburtsh. u. Frauenheilk. 53 (1993 b) 395–399

Stern, W. D., M. Laniado, W. Vogl, G. Weisser, A. Tolksdorf, W. Kaiser, G. Köveker, C. D. Claussen: Colour coded duplex sonography and contrast enhanced MRT of scintigraphically "cold" thyroid nodules. Fortschr. Röntgenstr. 160 (1994) 3–10

Sudakoff, G. S., A. Gasparaitis, E. Michelassi, R. Hurst, K. Hoffmann, C. Hackworth: Endorectal color Doppler imaging of primary and recurrent: Preliminary experience. Amer. J. Roentgenol. 166 (1996) 55–61

Sudakoff, G. S., R. Smith, N. J. Vogelzang, G. Steinberg, C. B. Brendler: Color Doppler imaging and transrectal sonography of the prostatic fossa after radical prostatectomy: Early experience. Amer. J. Roentgenol. 167 (1996) 883–888

Takase, K., S. Takahashi, S. Tazawa, Y. Terasawa, K. Sakamoto: Renal cell carcinoma associated with chronic renal failure: Evaluation with sonographic angiography. Radiology 192 (1994) 787–792

Tanaka, S., T. Kitamra, M. Fujita, H. Kasugai, A. Inoue, S. Ishiguro: Small hepatocellular carcinoma: Differentiation from adenomatous hyperplastic nodule with color Doppler flow imaging. Radiology 182 (1992 a) 161–165

Tanaka, K., S. Inoue, K. Numata, Y. Takamura, S. Takebayashi, Y. Ohaki, K. Misugi: Color Doppler sonography of hepatocellular carcinoma before and after treatment by transcatheter arterial embolization. Amer. J. Roentgenol. 158 (1992 b) 541–546

Taylor, K. J. W., I. Ramos, S. S. Morse, K. L. Fortune, L. Hammers, C. R. Taylor: Focal liver masses: Differential diagnosis with pulsed Doppler US. Radiology 164 (1987) 643–647

Taylor, K. J. W., I. Ramos, D. Carter, S. S. Morse, D. Snower, K. Fortune: Correlation of Doppler US tumor signals with neovascular morphological features. Radiology 166 (1988) 57–62

Taylor, K. J. W.: Doppler studies of tumor neovascularity. In Bernstein, E. F., ed: Vascular diagnosis. Mosby, St. Louis 1993

Tepper, R., L. Lerner-Geva, M. M. Altaras, S. Goldberger, G. Ben-Baruch, S. Markov, I. Cohen, Y. Beyth: Transvaginal color flow imaging in the diagnosis of ovarian tumors. J. Ultrasound Med. 14 (1995) 731–734

Thomas, Chr., W. Bautz, W. Müller-Schauenburg, U. Feine: Angio-dynography for focal thyroid disease. Fortschr. Röntgenstr. 150 (1989) 72–75

Tomiyama, T., N. Ueno, S. Tano, S. Wada, K. Kimura: Assessment of arterial invasion in pancreatic cancer using color Doppler ultra-sonography. Am. J. Gastroenterol. 91 (1996) 1410–1416

Tschammler, A., G. Gunzer, E. Reinhart, D. Höhmann, A. C. Feller, W. Müller, K. Lackner: Type diagnosis of enlarged lymph nodes by qualitative and semiquantitative evaluation of lymph node perfusion with colour-coded duplex sonography. Fortschr. Röntgenstr. 154 (1991) 414–418

Tschammler, A., H. Wirkner, G. Ott, D. Hahn: Vascular patterns in reactive and malignant lymphadenopathy. Europ. Radiol. 6 (1996) 473–480

Ueno, N., T. Tomiyama, S. Tano, S. Wada, K. Kimura: Diagnosis of gallbladder carcinoma with color Doppler ultrasonography. Amer. J. Gastroenterol. 91 (1996a) 1647–1649

Wren, S. M., P. W. Ralls, S. C. Stain, A. Kasiraman, C. L. Carpenter, D. Parekh: Assessment of resectability of pancreatic head and periampullary tumors by color flow Doppler sonography. Arch. Surg. 131 (1996) 812–817

Index

Note: page numbers in *italics* refer to figures and tables